Aquaponics Food Production Systems

Simon Goddek • Alyssa Joyce • Benz Kotzen
Gavin M. Burnell
Editors

Aquaponics Food Production Systems

Combined Aquaculture and Hydroponic Production Technologies for the Future

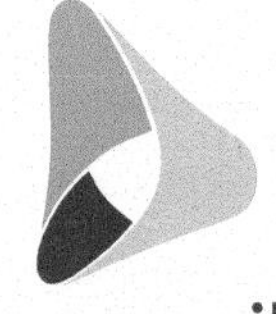

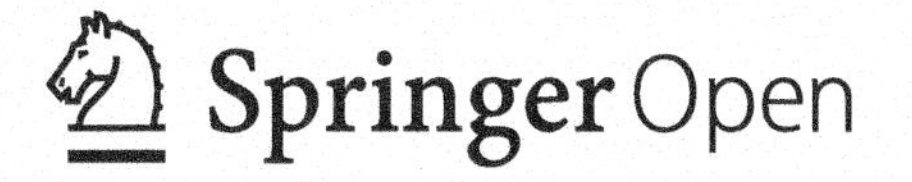

Funded by the Horizon 2020 Framework Programme
of the European Union

Editors
Simon Goddek
Mathematical and Statistical Methods (Biometris)
Wageningen University
Wageningen, The Netherlands

Benz Kotzen
School of Design
University of Greenwich
London, UK

Alyssa Joyce
Department of Marine Science
University of Gothenburg
Gothenburg, Sweden

Gavin M. Burnell
School of Biological, Earth and Environmental Sciences
University College Cork
Cork, Ireland

ISBN 978-3-030-15945-0 ISBN 978-3-030-15943-6 (eBook)
https://doi.org/10.1007/978-3-030-15943-6

This Springer imprint is published by the registered company Springer Nature Switzerland AG.
The registered company address is: Gewerbestrasse 11, 6330 Cham, Switzerland

Preface

It has been more than 45 years since the science fiction film Soylent Green (1973) first appeared in cinemas. The movie was prescient for its time and predicted many of our current environmental problems, including dying oceans, the greenhouse effect, overpopulation, and loss of biodiversity. Even though we hope that humans will not serve as a future nutrient source, the scenarios laid out in the movie are not that far from being realised. As researchers and citizens, we realise our duty of care to the environment and the rest of our world's ever-growing population. We are concerned that if we stand back and ignore the current trends in exploitation of resources and methods of production that our paradise of a planet will be doomed or at least far diminished, such that living on the sterile surfaces of the Moon or Mars will seem like a pleasant alternative. Generations to come will and should hold us individually and collectively responsible for the mess that we leave. The numerous authors of this book are in a lucky as well as in an unfortunate position, in that we can either help to solve problems or be held responsible by future generations for being part of the problem. When we started the COST Action FA1305 'The EU Aquaponics Hub – Realising Sustainable Integrated Fish and Vegetable Production for the EU', aquaponics was a niche technology that, at an industrial scale, could not compete with stand-alone hydroponics and aquaculture technologies. However, aquaponics technology in the past decade has taken great leaps forward in efficiency and hence economic viability through a wide range of technological advances. As our ability to understand the environmental costs of industrial farming increases, we are more capable of developing technologies to ensure that farming is more productive and less damaging to the environment. This positive outcome should be bolstered by the very encouraging signs that although young people are statistically not interested in being the farmers of the future, they do want to be future farmers if technology is involved and they can adapt these technologies to live closer to urban environments and have a better quality of life than in the rural past. Kids of all ages are fascinated by technology, and it is no wonder as technology solves many problems. At the same time though, kids (perhaps less so with teenagers) are also environmentally conscious and understand that the future of our planet lies in the melding of nature and technology. Technology allows us to be more productive, and although we have no certainty that we can and will effectively solve climate change, we still have hope that there will be a future where people will be healthy and fed with nutritious food. We, the authors of this book, realise that we are but small fry in a world of much bigger fish (sometimes sharks), but we are more than hopeful, indeed confident, that aquaponics has a role to play in the world's future food production.

Within the timeline of COST Action FA1305, our objective was to bring aquaponics closer to the public and to raise awareness of alternative growing methods. The Action's Management Committee had 90 experts from 28 EU countries, 2 near neighbour countries, and 2 international partner countries. We organised 7 training schools in different parts of Europe, involving 92 trainees from 21 countries, and 20 STSMs were awarded to 18 early career researchers from 12 countries. Most importantly, we published 59 videos based on the training schools, all of which are freely available on YouTube (https://www.youtube.com/EUAquaponicsHub). Action members collaborated in writing 24 papers (19 of which are open access), book chapters, monographs, and a white paper. The white paper identifies eight key recommendations based on the experience of the working group members, trends within current research and entrepreneurship, and the directions being investigated by ECIs. The recommendations are:

1. The promotion of continued research in aquaponics.
2. The development of financial incentives to enable the commercialisation of aquaponics.
3. The promotion of aquaponics as social enterprise in urban areas.
4. The promotion of aquaponics in the developing world and in refugee camps.
5. The development of EU-wide aquaponics legislation and planning guidance.
6. The development of aquaponics training courses in order to provide the necessary skilled workforce to enable aquaponics to expand in the EU.
7. The development of stricter health and safety protocols, including fish welfare.
8. The establishment of an EU Aquaponics Association, in order to promote aquaponics and aquaponics technology in the EU and to assist with knowledge transfer, and the promotion of high production and produce standards in EU aquaponics (Fig. 1).

The assembled knowledge and experience of the group is considerable, and it is therefore appropriate to take the opportunity at the end of the 4-year COST project to gather this into a book, which was originally proposed by Benz Kotzen and Gavin M. Burnell at the start and then with Simon Goddek and Alyssa Joyce. We are fortunate that Springer Nature particularly Alexandrine Cheronet has been enthusiastic about this publication and that the COST organisation has funded the book as open access so that it is available for anyone to download. We see it as part of our duty to ensure that as many people as possible can benefit from the knowledge and expertise. The book is the product of 68 researchers and practitioners from 29 countries (Australia, Austria, Belgium, Brazil, Croatia, Czech Republic, Denmark, Finland, France, Germany, Greece, Iceland, Ireland, Israel, Italy, Malta, the

Fig. 1 Group picture of the COST group in Murcia, Spain, 2017

Netherlands, North Macedonia, Norway, Portugal, Serbia, Slovenia, South Africa, Spain, Sweden, Switzerland, Turkey, the United Kingdom, and the United States). When asking the members of our COST Action as well as external experts whether they were willing to contribute to this book, the response was overwhelming. Putting a book together with 24 chapters within 1 year would not have been possible without the cooperative spirit of every single lead author and coauthor. The book is testament to their knowledge and enthusiasm. We offer our warmest appreciation to our scientific review committee including Ranka Junge (aquaponics and education), Lidia Robaina (fish feed), Ragnheidur Thorarinsdottir (commercial aquaponics), Harry Palm (aquaponics and aquaculture systems), Morris Villarroel (fish welfare), Haissam Jijakli (plant pathology), Amit Gross (aquaculture and recycling), Dieter Anseeuw (hydroponics), and Charlie Shultz (aquaponics). We would also like to thank all peer reviewers of the 24 chapters who improved the content of the chapters. Finally, yet importantly, the editors would also like to thank their families and partners who have been patient in the editing a large book such as this.

Wageningen, The Netherlands — Simon Goddek
Gothenburg, Sweden — Alyssa Joyce
London, UK — Benz Kotzen
Cork, Ireland — Gavin M. Burnell
February 2019

Acknowledgements

The editors, authors, and publishers would like to acknowledge the COST (European Cooperation in Science and Technology) organisation (https://www.cost.eu) initially for funding and supporting the 4-year COST Action 1305, 'The EU Aquaponics Hub – Realising Sustainable Integrated Fish and Vegetable Production for the EU', which was conceived and chaired by Benz Kotzen, University of Greenwich, and then finally for contributing funds to this publication, making it open-source and available to all to read. Without COST, who brought almost all of the authors together, in an amazing project, this book would not have been written, and without their final dissemination contribution, this book would not be available to everyone. We also acknowledge and greatly appreciate the support of Desertfoods International GmbH (www.desertfoods-international.com) and Developonics asbl (www.developonics.com) for the additional financial support required to enable the publication to be open-source. Additionally we applaud the efforts and great skill of Aquaponik Manufaktur GmbH (www.aquaponik-manufaktur.de) for producing a cohesive and attractive set of illustrations for the book, the Netherlands Organisation for Scientific Research (NWO; project number 438-17-402) for supporting Simon Goddek in his editorial work and writing, and the Swedish Research Council FORMAS grant 2017-00242 for similarly supporting Alyssa Joyce whilst she undertook editorial work and writing on this book. Finally, the editors are indebted to the enthusiasm and diligence of its authors, especially of the 22 lead authors of the 24 chapters in their sterling efforts to get this remarkable book delivered on time. A heartfelt well-done one and all!

Wageningen University, Wageningen, The Netherlands — Simon Goddek
University of Gothenburg, Gothenburg, Sweden — Alyssa Joyce
University of Greenwich, London, UK — Benz Kotzen
University College Cork, Cork, Ireland — Gavin M. Burnell

Contents

About the Editors

Simon Goddek Simon is an expert in the field of multi-loop aquaponics systems and an ecopreneur. In 2014, Simon started his PhD in the faculty of environmental engineering at the University of Iceland, completing it in the group Biobased Chemistry and Technology at Wageningen University & Research (the Netherlands). At the time of publication, he is a postdoc in the Mathematical and Statistical Methods group (Biometris), where he is involved in several projects in Europe (i.e. CITYFOOD) and Africa (e.g. desertfoods Namibia). His research focus in aquaponics includes numerical system simulation and modelling, decoupled multi-loop aquaponics systems, and anaerobic mineralization solutions.

Alyssa Joyce Alyssa is an assistant professor in the Department of Marine Sciences (aquaculture) at the University of Gothenburg, Sweden. In her group, several researchers are focused on the role of bacterial relationships in nutrient bioavailability and pathogen control in aquaponics systems. She was one of the Swedish representatives to the EU COST Network on aquaponics and is a partner in the CITYFOOD project developing aquaponics technology in urban environments.

Benz Kotzen Benz is an associate professor and head of Research and Enterprise in the School of Design, University of Greenwich, London, and a consultant landscape architect. He runs the rooftop Aquaponics Lab at the University. He developed and was chair of the EU Aquaponics Hub, whose remit was to raise the state of the art of aquaponics in the EU and facilitate collaborative aquaponics research. Urban agriculture including vertical aquaponic systems and growing exotic vegetables aquaponically and drylands restoration are key fields of research.

Gavin M. Burnell Gavin is an emeritus professor at the Aquaculture and Fisheries Development Centre, University College Cork, Ireland, and president of the European Aquaculture Society (2018–2020). He has been researching and promoting the concept of marine aquaponics as a contribution to the circular economy and sees an important role for this technology in outreach to urban communities. As a co-founder of AquaTT and editor of Aquaculture International, he is excited at the possibilities that aquaponics has in research, education, and training across disciplines.

The original version of this book was revised: The book was inadvertently published with few typesetting errors in Chapters 3 and 7 which have now been corrected. The correction to this book is available at https://doi.org/10.1007/978-3-030-15943-6_25.

Part I
Framework Conditions in a Resource Limited World

Chapter 1
Aquaponics and Global Food Challenges

Simon Goddek, Alyssa Joyce, Benz Kotzen, and Maria Dos-Santos

Abstract As the world's population grows, the demands for increased food production expand, and as the stresses on resources such as land, water and nutrients become ever greater, there is an urgent need to find alternative, sustainable and reliable methods to provide this food. The current strategies for supplying more produce are neither ecologically sound nor address the issues of the circular economy of reducing waste whilst meeting the WHO's Millennium Development Goals of eradicating hunger and poverty by 2015. Aquaponics, a technology that integrates aquaculture and hydroponics, provides part of the solution. Although aquaponics has developed considerably over recent decades, there are a number of key issues that still need to be fully addressed, including the development of energy-efficient systems with optimized nutrient recycling and suitable pathogen controls. There is also a key issue of achieving profitability, which includes effective value chains and efficient supply chain management. Legislation, licensing and policy are also keys to the success of future aquaponics, as are the issues of education and research, which are discussed across this book.

Keywords Aquaponics · Agriculture · Planetary boundaries · Food supply chain · Phosphorus

---

S. Goddek (✉)
Mathematical and Statistical Methods (Biometris), Wageningen University, Wageningen, The Netherlands
e-mail: simon.goddek@wur.nl; simon@goddek.nl

A. Joyce
Department of Marine Science, University of Gothenburg, Gothenburg, Sweden
e-mail: alyssa.joyce@gu.se

B. Kotzen
School of Design, University of Greenwich, London, UK
e-mail: b.kotzen@greenwich.ac.uk

M. Dos-Santos
ESCS-IPL, DINÂMIA'CET, ISCTE-Institute University of Lisbon, Lisbon, Portugal
e-mail: mjpls@iscte-iul.pt

S. Goddek et al. (eds.), *Aquaponics Food Production Systems*,
https://doi.org/10.1007/978-3-030-15943-6_1

1.1 Introduction

Food production relies on the availability of resources, such as land, freshwater, fossil energy and nutrients (Conijn et al. 2018), and current consumption or degradation of these resources exceeds their global regeneration rate (Van Vuuren et al. 2010). The concept of planetary boundaries (Fig. 1.1) aims to define the environmental limits within which humanity can safely operate with regard to scarce resources (Rockström et al. 2009). Biochemical flow boundaries that limit food supply are more stringent than climate change (Steffen et al. 2015). In addition to nutrient recycling, dietary changes and waste prevention are integrally necessary to transform current production (Conijn et al. 2018; Kahiluoto et al. 2014). Thus, a major global challenge is to shift the growth-based economic model towards a

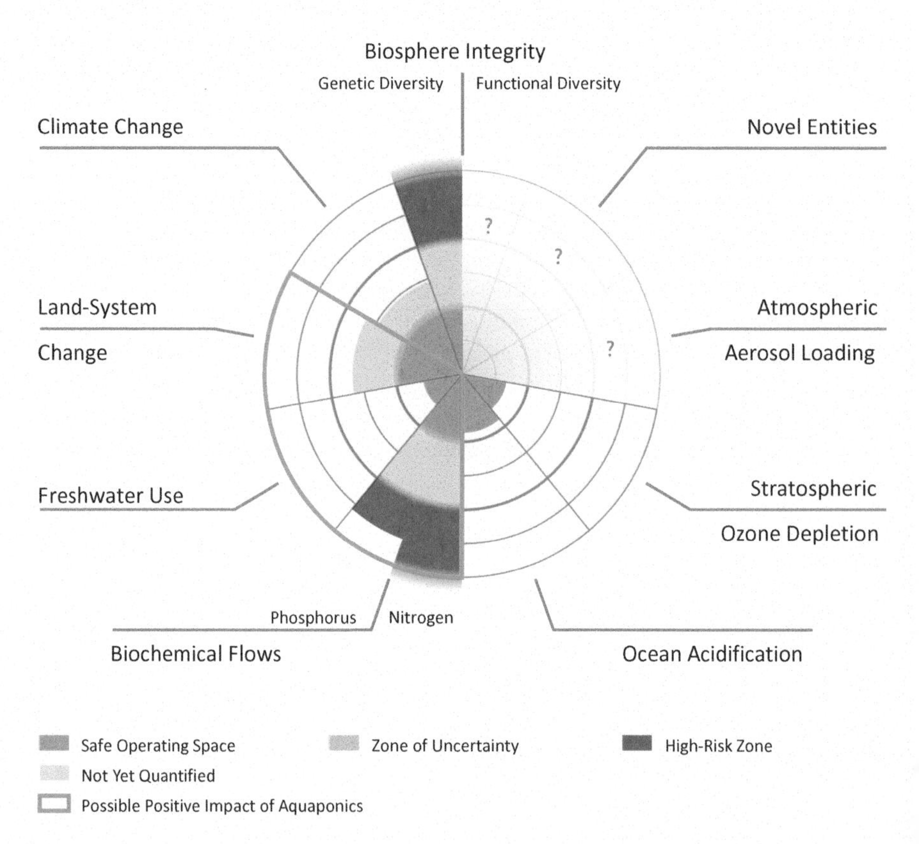

Fig. 1.1 Current status of the control variables for seven of the planetary boundaries as described by Steffen et al. (2015). The green zone is the safe operating space, the yellow represents the zone of uncertainty (increasing risk), the red is a high-risk zone, and the grey zone boundaries are those that have not yet been quantified. The variables outlined in blue (i.e. land-system change, freshwater use and biochemical flows) indicate the planetary boundaries that aquaponics can have a positive impact on

balanced eco-economic paradigm that replaces infinite growth with sustainable development (Manelli 2016). In order to maintain a balanced paradigm, innovative and more ecologically sound cropping systems are required, such that trade-offs between immediate human needs can be balanced whilst maintaining the capacity of the biosphere to provide the required goods and services (Ehrlich and Harte 2015).

In this context, aquaponics has been identified as a farming approach that, through nutrient and waste recycling, can aid in addressing both planetary boundaries (Fig. 1.1) and sustainable development goals, particularly for arid regions or areas with nonarable soils (Goddek and Körner 2019; Appelbaum and Kotzen 2016; Kotzen and Appelbaum 2010). Aquaponics is also proposed as a solution for using marginal lands in urban areas for food production closer to markets. At one time largely a backyard technology (Bernstein 2011), aquaponics is now growing rapidly into industrial-scale production as technical improvements in design and practice allow for significantly increased output capacities and production efficiencies. One such area of evolution is in the field of coupled vs. decoupled aquaponics systems. Traditional designs for one-loop aquaponics systems comprise both aquaculture and hydroponics units between which water recirculates. In such traditional systems, it is necessary to make compromises to the conditions of both subsystems in terms of pH, temperature and nutrient concentrations (Goddek et al. 2015; Kloas et al. 2015) (see Chap. 7). A decoupled aquaponics system, however, can reduce the need for trade-offs by separating the components, thus allowing the conditions in each subsystem to be optimized. Utilization of sludge digesters is another key way of maximizing efficiency through the reuse of solid wastes (Emerenciano et al. 2017; Goddek et al. 2018; Monsees et al. 2015). Although many of the largest facilities worldwide are still in arid regions (i.e. Arabian Peninsula, Australia and sub-Saharan Africa), this technology is also being adopted elsewhere as design advances have increasingly made aquaponics not just a water-saving enterprise but also an efficient energy and nutrient recycling system.

1.2 Supply and Demand

The 2030 Agenda for Sustainable Development emphasizes the need to tackle global challenges, ranging from climate change to poverty, with sustainable food production a high priority (Brandi 2017; UN 2017). As reflected in the UN's Sustainable Development Goal 2 (UN 2017), one of the greatest challenges facing the world is how to ensure that a growing global population, projected to rise to around 10 billion by 2050, will be able to meet its nutritional needs. To feed an additional two billion people by 2050, food production will need to increase by 50% globally (FAO 2017). Whilst more food will need to be produced, there is a shrinking rural labour force because of increasing urbanization (dos Santos 2016). The global rural population has diminished from 66.4% to 46.1% in the period from 1960 to 2015 (FAO 2017). Whilst, in 2017, urban populations represented more than 54% of the total world population, nearly all future growth of the world's population will occur in urban areas, such that by 2050, 66% of the global population will live in cities (UN 2014). This increasing urbanization of cities is accompanied by a simultaneously growing network of infrastructure systems, including transportation networks.

To ensure global food security, total food production will need to increase by more than 70% in the coming decades to meet the Millennium Development Goals (FAO 2009), which include the 'eradication of extreme poverty and hunger' and also 'ensuring environmental sustainability'. At the same time, food production will inevitably face other challenges, such as climate change, pollution, loss of biodiversity, loss of pollinators and degradation of arable lands. These conditions require the adoption of rapid technological advances, more efficient and sustainable production methods and also more efficient and sustainable food supply chains, given that approximately a billion people are already chronically malnourished, whilst agricultural systems continue to degrade land, water and biodiversity at a global scale (Foley et al. 2011; Godfray et al. 2010).

Recent studies show that current trends in agricultural yield improvements will not be sufficient to meet projected global food demand by 2050, and these further suggest that an expansion of agricultural areas will be necessary (Bajželj et al. 2014). However, the widespread degradation of land in conjunction with other environmental problems appears to make this impossible. Agricultural land currently covers more than one-third of the world's land area, yet less than a third of it is arable (approximately 10%) (World Bank 2018). Over the last three decades, the availability of agricultural land has been slowly decreasing, as evidenced by more than 50% decrease from 1970 to 2013. The effects of the loss of arable land cannot be remedied by converting natural areas into farmland as this very often results in erosion as well as habitat loss. Ploughing results in the loss of topsoil through wind and water erosion, resulting in reduced soil fertility, increased fertilizer use and then eventually to land degradation. Soil losses from land can then end up in ponds, dams, lakes and rivers, causing damage to these habitats.

In short, the global population is rapidly growing, urbanizing and becoming wealthier. Consequently, dietary patterns are also changing, thus creating greater demands for greenhouse gas (GHG) intensive foods, such as meat and dairy products, with correspondingly greater land and resource requirements (Garnett 2011). But whilst global consumption is growing, the world's available resources, i.e. land, water and minerals, remain finite (Garnett 2011). When looking at the full life-cycle analysis of different food products, however, both Weber and Matthews (2008) and Engelhaupt (2008) suggest that dietary shifts can be a more effective means of lowering an average household's food-related climate footprint than 'buying local'. Therefore, instead of looking at the reduction of supply chains, it has been argued that a dietary shift away from meat and dairy products towards nutrition-oriented agriculture can be more effective in reducing energy and footprints (Engelhaupt 2008; Garnett 2011).

The complexity of demand-supply imbalances is compounded by deteriorating environmental conditions, which makes food production increasingly difficult and/or unpredictable in many regions of the world. Agricultural practices cannot only undermine planetary boundaries (Fig. 1.1) but also aggravate the persistence and propagation of zoonotic diseases and other health risks (Garnett 2011). All these factors result in the global food system losing its resilience and becoming increasingly unstable (Suweis et al. 2015).

The ambitious 2015 deadline of the WHO's Millennium Development Goals (MDGs) to eradicate hunger and poverty, to improve health and to ensure environmental sustainability has now passed, and it has become clear that providing nutritious food for the undernourished as well as for affluent populations is not a simple task. In summary, changes in climate, loss of land and diminution in land quality, increasingly complex food chains, urban growth, pollution and other adverse environmental conditions dictate that there is an urgent need to not only find new ways of growing nutritious food economically but also locate food production facilities closer to consumers. Delivering on the MDGs will require changes in practice, such as reducing waste, carbon and ecological footprints, and aquaponics is one of the solutions that has the potential to deliver on these goals.

1.3 Scientific and Technological Challenges in Aquaponics

Whilst aquaponics is seen to be one of the key food production technologies which 'could change our lives' (van Woensel et al. 2015), in terms of sustainable and efficient food production, aquaponics can be streamlined and become even more efficient. One of the key problems in conventional aquaponics systems is that the nutrients in the effluent produced by fish are different than the optimal nutrient solution for plants. Decoupled aquaponics systems (DAPS), which use water from the fish but do not return the water to the fish after the plants, can improve on traditional designs by introducing mineralization components and sludge bioreactors containing microbes that convert organic matter into bioavailable forms of key minerals, especially phosphorus, magnesium, iron, manganese and sulphur that are deficient in typical fish effluent. Contrary to mineralization components in one-loop systems, the bioreactor effluent in DAPS is only fed to the plant component instead of being diluted in the whole system. Thus, decoupled systems that utilize sludge digesters make it possible to optimize the recycling of organic wastes from fish as nutrients for plant growth (Goddek 2017; Goddek et al. 2018). The wastes in such systems mainly comprise fish sludge (i.e. faeces and uneaten feed that is not in solution) and thus cannot be delivered directly in a hydroponics system. Bioreactors (see Chap. 10) are therefore an important component that can turn otherwise unusable sludge into hydroponic fertilizers or reuse organic wastes such as stems and roots from the plant production component into biogas for heat and electricity generation or DAPS designs that also provide independently controlled water cycling for each unit, thus allowing separation of the systems (RAS, hydroponic and digesters) as required for the control of nutrient flows. Water moves between components in an energy and nutrient conserving loop, so that nutrient loads and flows in each subsystem can be monitored and regulated to better match downstream requirements. For instance, phosphorous (P) is an essential but exhaustible fossil resource that is mined for fertilizer, but world supplies are currently being depleted at an alarming rate. Using digesters in decoupled aquaponics systems allows microbes to convert the phosphorus in fish waste into orthophosphates that can be utilized by plants, with high recovery rates (Goddek et al. 2016, 2018).

Although decoupled systems are very effective at reclaiming nutrients, with near-zero nutrient loss, the scale of production in each of the units is important given that nutrient flows from one part of the system need to be matched with the downstream production potential of other components. Modelling software and Supervisory Control and Data Acquisition (SCADAS) data acquisition systems therefore become important to analyse and report the flow, dimensions, mass balances and tolerances of each unit, making it possible to predict physical and economic parameters (e.g. nutrient loads, optimal fish-plant pairings, flow rates and costs to maintain specific environmental parameters). In Chap. 11, we will look in more detail at systems theory as applied to aquaponics systems and demonstrate how modelling can resolve some of the issues of scale, whilst innovative technological solutions can increase efficiency and hence profitability of such systems. Scaling is important not only to predict the economic viability but also to predict production outputs based on available nutrient ratios.

Another important issue, which requires further development, is the use and reuse of energy. Aquaponics systems are energy and infrastructure intensive. Depending on received solar radiation, the use of solar PV, solar thermal heat sources and (solar) desalination may still not be economically feasible but could all be potentially integrated into aquaponics systems. In Chap. 12, we present information about innovative technical and operational possibilities that have the capacity to overcome the inherent limitations of such systems, including exciting new opportunities for implementing aquaponics systems in arid areas.

In Chap. 2, we also discuss in more detail the range of environmental challenges that aquaponics can help address. Pathogen control, for instance, is very important, and contained RAS systems have a number of environmental advantages for fish production, and one of the advantages of decoupled aquaponics systems is the ability to circulate water between the components and to utilize independent controls wherein it is easier to detect, isolate and decontaminate individual units when there are pathogen threats. Probiotics that are beneficial in fish culture also appear beneficial for plant production and can increase production efficiency when circulated within a closed system (Sirakov et al. 2016). Such challenges are further explored in Chap. 5, where we discuss in more detail how innovation in aquaponics can result in (a) increased space utilization efficiency (less cost and materials, maximizing land use); (b) reduced input resources, e.g. fishmeal, and reduced negative outputs, e.g. waste discharge; and (c) reduced use of antibiotics and pesticides in self-contained systems.

There are still several aquaponic topic areas that require more research in order to exploit the full potential of these systems. From a scientific perspective, topics such as nitrogen cycling (Chap. 9), aerobic and anaerobic remineralization (Chap. 10), water and nutrient efficiency (Chap. 8), optimized aquaponic fish diets (Chap. 13) and plant pathogens and control strategies (Chap. 14) are all high priorities.

In summary, the following scientific and technological challenges need to be addressed:

1. *Nutrients*: As we have discussed, systems utilizing sludge digesters make it possible to optimize the recycling of organic waste from fish into nutrients for plant growth, such designs allow for optimized reclamation and recycling of nutrients to create a near-zero nutrient loss from the system.
2. *Water*: The reuse of nutrient-depleted water from greenhouses can also be optimized for reuse back in the fish component utilizing condensers.
3. *Energy*: Solar-powered designs also improve energy savings, particularly if preheated water from solar heaters in the greenhouses can be recirculated back to fish tanks for reuse.

The ability to recycle water, nutrients and energy makes aquaponics a potentially unique solution to a number of environmental issues facing conventional agriculture. This is discussed in Chap. 2.

1.4 Economic and Social Challenges

From an economic perspective, there are a number of limitations inherent in aquaponics systems that make specific commercial designs more or less viable (Goddek et al. 2015; Vermeulen and Kamstra 2013). One of the key issues is that stand-alone, independent hydroponics and aquaculture systems are more productive than traditional one-loop aquaponics systems (Graber and Junge 2009), as they do not require trade-offs between the fish and plant components. Traditional, classic single-loop aquaponics requires a compromise between the fish and plant components when attempting to optimize water quality and nutrient levels that inherently differ for the two parts (e.g. desired pH ranges and nutrient requirements and concentrations). In traditional aquaponics systems, savings in fertilizer requirements for plants do not make up for the harvest shortfalls caused by suboptimal conditions in the respective subsystems (Delaide et al. 2016).

Optimizing growth conditions for both plants (Delaide et al. 2016; Goddek and Vermeulen 2018) and fish is the biggest challenge to profitability, and current results indicate that this can be better achieved in multi-loop decoupled aquaponics systems because they are based on independent recirculating loops that involve (1) fish, (2) plants and (3) bioreactors (anaerobic or aerobic) for sludge digestion and a unidirectional water (nutrient) flow, which can improve macro- and micro-nutrient recovery and bioavailability, as well as optimization of water consumption (Goddek and Keesman 2018). Current studies show that this type of system allows for the maintenance of specific microorganism populations within each compartment for better disease management, and they are more economically efficient in so much as the systems not only reduce waste outflow but also reutilize otherwise unusable sludge, converting it to valuable outputs (e.g. biogas and fertilizer).

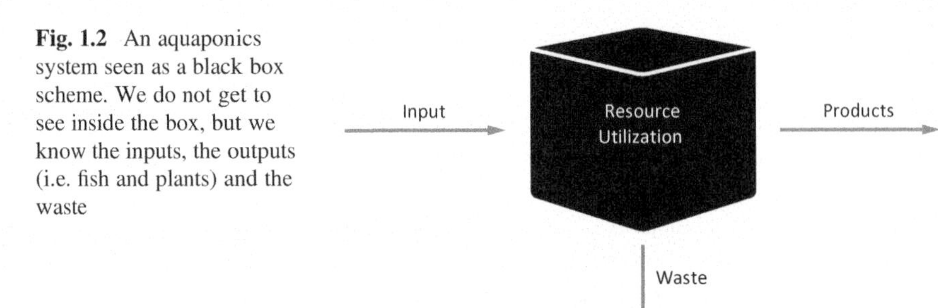

Fig. 1.2 An aquaponics system seen as a black box scheme. We do not get to see inside the box, but we know the inputs, the outputs (i.e. fish and plants) and the waste

Independent, RAS systems and hydroponics units also have a wide range of operational challenges that are discussed in detail in Chaps. 3 and 4. Increasingly, technological advances have allowed for higher productivity ratios (Fig. 1.2), which can be defined as a fraction of the system's outputs (i.e. fish and plants) over the system's input (i.e. fish feed and/or additional fertilization, energy input for lighting, heating and pumping CO_2 dosing and biocontrols).

When considering the many challenges that aquaponics encounters, production problems can be broadly broken down into three specific themes: (1) system productivity, (2) effective value chains and (3) efficient supply chain management.

System Productivity Agricultural productivity is measured as the ratio of agricultural outputs to agricultural inputs. Traditional small-scale aquaponics systems were designed primarily to address environmental considerations such as water discharge, water inputs and nutrient recycling, but the focus in recent years has increasingly shifted towards economic feasibility in order to increase productivity for large-scale farming applications. However, this will require the productivity of aquaponics systems to be able to compete economically with independent, state-of-the-art hydroponics and aquaculture systems. If the concept of aquaponics is to be successfully applied at a large scale, the reuse of nutrients and energy must be optimized, but end markets must also be considered.

Effective Value Chains The value chains (added value) of agricultural products mainly arise from the processing of the produce such as the harvested vegetables, fruits and fish. For example, the selling price for pesto (i.e. red and green) can be more than ten times higher than that of the tomatoes, basil, olive oil and pine nuts. In addition, most processed food products have a longer shelf life, thus reducing spoilage. Evidently, fresh produce is important because nutritional values are mostly higher than those in the processed foods. However, producing fresh and high-quality produce is a real challenge and therefore a luxury in many regions of the world. Losses of nutrients during storage of fruit and vegetables are substantial if they are not canned or frozen quickly (Barrett 2007; Rickman et al. 2007). Therefore, for large-scale systems, food processing should at least be considered to balance out any fluctuations between supply and demand and reduce food waste. With respect to food waste reduction, vegetables that do not meet fresh produce standards, but are still of marketable quality, should be processed in order to reduce postharvest losses.

Although such criteria apply to all agricultural and fisheries products, value adding can substantially increase the profitability of the aquaponics farm, especially if products can reach niche markets.

Efficient Supply Chain Management In countries with well-developed transportation and refrigeration networks, fruit and vegetables can be imported from all around the world to meet consumer demands for fresh produce. But as mentioned previously, high-quality and fresh produce is a scarce commodity in many parts of the world, and the long-distance movement of goods – i.e. supply chain management – to meet high-end consumer demand is often criticized and justifiably so. Most urban dwellers around the world rely on the transport of foods over long distances to meet daily needs (Grewal and Grewal 2012). One of the major criticisms is thus the reliance on fossil fuels required to transport products over large distances (Barrett 2007). The issue of food miles directs focus on the distance that food is transported from the time of production to purchase by the end consumer (Mundler and Criner 2016). However, in terms of CO_2 emissions per tonne/km (tkm), one food mile for rail transportation (13.9 g CO_2/tkm) is not equal to one food mile of truck/road transportation, as truck transportation has more than 15 times greater environmental impact (McKinnon 2007). Therefore, transportation distance is not necessarily the only consideration, as the ecological footprint of vegetables grown on farms in rural areas is potentially less than the inputs required to grow food in greenhouses closer to urban centres.

Food miles are thus only a part of the picture. Food is transported long distances, but the greenhouse gas emissions associated with food production are dominated by the production phase (i.e. the impact of energy for heating, cooling and lighting) (Engelhaupt 2008; Weber and Matthews 2008). For example, Carlsson (1997) showed that tomatoes imported from Spain to Sweden in winter have a much lower carbon footprint than those locally grown in Sweden, since energy inputs to greenhouses in Sweden far outweigh the carbon footprint of transportation from Spain. When sourcing food, the transport of goods is not the only factor to take into consideration, as the freshness of products determines their nutritive value, taste and general appeal to consumers. By growing fresh food locally, many scholars agree that urban farming could help secure the supply of high-quality produce for urban populations of the future whilst also reducing food miles (Bon et al. 2010; dos Santos 2016; Hui 2011). Both areas will be discussed in more detail in Sect. 1.5.

From a consumer's perspective, urban aquaponics thus has advantages because of its environmental benefits due to short supply chains and since it meets consumer preferences for high-quality locally produced fresh food (Miličić et al. 2017). However, despite these advantages, there are a number of socio-economic concerns: The major issue involves urban property prices, as land is expensive and often considered too valuable for food production. Thus, purchasing urban land most likely makes it impossible to achieve a feasible expected return of investment. However, in shrinking cities, where populations are decreasing, unused space could be used for agricultural purpose (Bontje and Latten 2005; Schilling and Logan 2008) as is the case in Detroit in the United States (Mogk et al. 2010).

Additionally, there is a major issue of urban planning controls, where in many cities urban land is not designated for agricultural food production and aquaponics is seen to be a part of agriculture. Thus, in some cities aquaponic farming is not allowed. The time is ripe to engage with urban planners who need to be convinced of the benefits of urban farms, which are highly productive and produce fresh, healthy, local food in the midst of urban and suburban development.

1.5 The Future of Aquaponics

Technology has enabled agricultural productivity to grow exponentially in the last century, thus also supporting significant population growth. However, these changes also potentially undermine the capacity of ecosystems to sustain food production, to maintain freshwater and forest resources and to help regulate climate and air quality (Foley et al. 2005).

One of the most pressing challenges in innovative food production, and thus in aquaponics, is to address regulatory issues constraining the expansion of integrated technologies. A wide range of different agencies have jurisdiction over water, animal health, environmental protection and food safety, and their regulations are in some cases contradictory or are ill-suited for complex integrated systems (Joly et al. 2015). Regulations and legislation are currently one of the most confusing areas for producers and would-be entrepreneurs. Growers and investors need standards and guidelines for obtaining permits, loans and tax exemptions, yet the confusing overlap of responsibilities among regulatory agencies highlights the urgent need for better harmonization and consistent definitions. Regulatory frameworks are frequently confusing, and farm licensing as well as consumer certification remains problematic in many countries. The FAO (in 2015), the WHO (in 2017) and the EU (in 2016) all recently began harmonizing provisions for animal health/well-being and food safety within aquaponics systems and for export-import trade of aquaponic products. For instance, several countries involved in aquaponics are lobbying for explicit wording within the *Codex Alimentarius*, and a key focus within the EU, determined by the EU sponsored COST Action FA1305, the 'EU Aquaponics Hub', is currently on defining aquaponics as a clear and distinct entity. At present, regulations define production for both aquaculture and hydroponics, but have no provisions for merging of the two. This situation often creates excessive bureaucracy for producers who are required to license two separate operations or whose national legislation does not allow for co-culturing (Joly et al. 2015). The EU Aquaponics Hub, which has supported this publication (COST FA1305), defines aquaponics as 'a production system of aquatic organisms and plants where the majority (> 50%) of nutrients sustaining the optimal plant growth derives from waste originating from feeding the aquatic organisms' (see Chap. 7).

Consumer certification schemes also remain a difficult area for aquaponics producers in many parts of the world. For instance, in the United States and Australia, aquaponic products can be certified as organic, but not within the European Union.

From an economic perspective, aquaponics is in theory capable of increasing the overall value of fish farming or conventional hydroponics whilst also closing the food-water-energy cycle within a circular bio-based economy. In order to make small-scale aquaponics systems economically viable, aquaponics farmers generally have to operate in niche markets to obtain higher prices for products, so certification thus becomes very important.

The most pressing issues are whether aquaponics can become acceptable at the policy level. Food safety is a high priority for gaining public support, and although there is a much lowered pathogen risk in closed systems, thus implying less need for antimicrobials and pesticides, managing potential risks – or moreover managing perceptions of those risks, especially as they may affect food safety – is a high priority for government authorities and investors alike (Miličić et al. 2017). One concern that is often raised is the fear of pathogen transfer in sludge from fish to plants, but this is not substantiated in the literature (Chap. 6). As such, there is a need to allay any remaining food safety and biosecurity concerns through careful research and, where concerns may exist, to ascertain how it may be possible to manage these problems through improved system designs and/or regulatory frameworks.

Aquaponics is an emerging food production technology which has the ability to condense and compress production into spaces and places that would not normally be used for growing food. This not only means that it is exceptionally relevant in urban areas, where aquaponics can be placed on underutilized and unused places such as flat roofs, development sites, abandoned factories, housing estates and schools, but it provides a means both in the developed and developing world for people to take back part of the food production process by providing fresh local food to the market (van Gorcum et al. 2019). The integration of aquaponics with vertical farming and living wall technologies will, in time, most likely improve productivity by reducing the overall farming footprint with reduced land take and intensification.

The intense production methods in aquaponics rely on the knowledge of a combination of key factors which are highly suitable for use in teaching STEM (science, technology, engineering and maths) subjects in schools. Aquaponics provides the teacher and student with opportunities to explore the realm of complex systems, their design and management and a host of other subject areas, including environmental sciences, water chemistry, biology and animal welfare. Aquaponics is also being used in prisons/correctional facilities, such as at the San Francisco County Jail, to help inmates gain skills and experience in aquaculture and horticulture that they can use on their release. In the domestic context, there is a growing trend to design countertop systems that can grow herbs as well as small systems that can be located in offices, where exotic fish provide a calming effect, whilst plants, as part of living walls, similarly provide an aesthetic backdrop and clean the air.

Aquaponics is a farming technology advancing rapidly from its first exploits in the last years of the twentieth century and the first decades of the twenty-first century. But it still is an 'emerging technology and science topic' (Junge et al. 2017) which is subject to a considerable amount of 'hype'. When comparing the number of aquaculture, hydroponic and aquaponic peer-reviewed papers, aquaponic

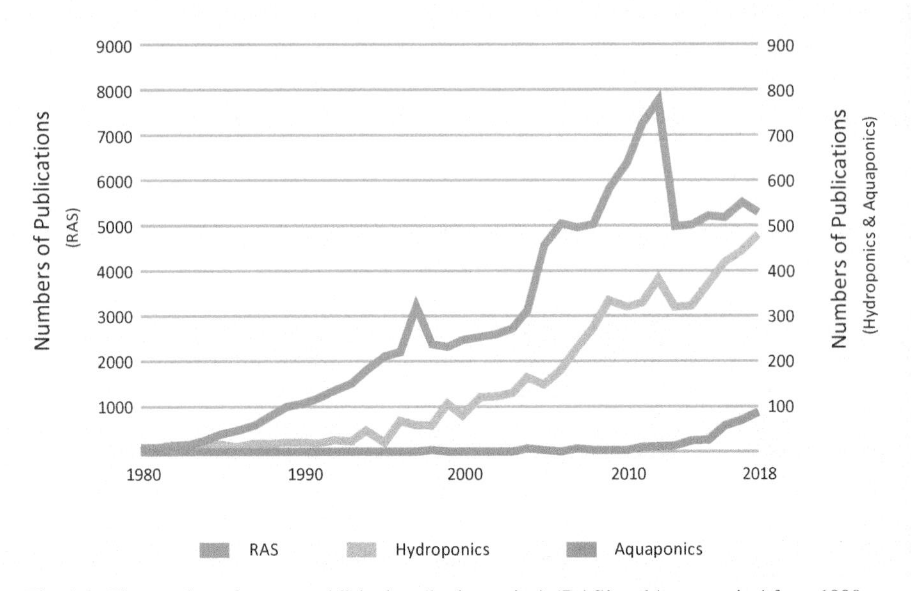

Fig. 1.3 The number of papers published on 'hydroponics', 'RAS' and 'aquaponics' from 1980 to 2018 (data were collected from the Scopus database on 30 January 2019). Please note that the scale for 'RAS' is one order of magnitude higher than that for 'hydroponics' and 'aquaponics'

papers are considerably lower (Fig. 1.3), but the numbers are rising and will continue to rise as aquaponics education, especially at university level, and general interest increases. A 'hype ratio' can be described as an indicator of the popularity of a subject in the public media relative to what is published in the academic press. This can, for example, be calculated by taking the search results in Google divided by the search results in Google Scholar. In the case of aquaponics, the hype ratio on 16 August 2016 was 1349, which is considerable when compared to the hype ratios of hydroponics (131) and recirculating aquaculture (17) (Junge et al. 2017). The sense one gets from this is that, indeed, aquaponics is an emerging technology but that there is enormous interest in the field which is likely to continue and increase over the next decades. The hype ratio, however, is likely to decline as more research is undertaken and scientific papers are published.

This book is aimed at the aquaponics researcher and practitioner, and it has been designed to discuss, explore and reveal the issues that aquaponics is addressing now and that will no doubt arise in the future. With such a broad spectrum of topics, it aims to provide a comprehensive but easily accessible overview of the rather novel scientific and commercial field of aquaponics. Apart from the production and technical side, this book has been designed to address trends in food supply and demand, as well as the various economic, environmental and social implications of this emerging technology. The book has been co-authored by numerous experts from around the world, but mostly from within the EU. Its 24 chapters cover the whole gamut of aquaponics areas and will provide a necessary textbook for all those interested in aquaponics and moving aquaponics forwards into the next decade.

References

Appelbaum S, Kotzen B (2016) Further investigations of aquaponics using brackish water resources of the Negev desert. Ecocycles 2:26. https://doi.org/10.19040/ecocycles.v2i2.53

Bajželj B, Richards KS, Allwood JM, Smith P, Dennis JS, Curmi E, Gilligan CA (2014) Importance of food-demand management for climate mitigation. Nat Clim Chang 4:924–929. https://doi.org/10.1038/nclimate2353

Barrett DM (2007) Maximizing the nutritional value of fruits and vegetables

Bernstein S (2011) Aquaponic gardening: a step-by-step guide to raising vegetables and fish together. New Society Publishers, Gabriola Island

Bon H, Parrot L, Moustier P (2010) Sustainable urban agriculture in developing countries. A review. Agron Sustain Dev 30:21–32. https://doi.org/10.1051/agro:2008062

Bontje M, Latten J (2005) Stable size, changing composition: recent migration dynamics of the Dutch large cities. Tijdschr Econ Soc Geogr 96:444–451. https://doi.org/10.1111/j.1467-9663.2005.00475.x

Brandi CA (2017) Sustainability standards and sustainable development –synergies and trade-offs of transnational governance. Sustain Dev 25:25–34. https://doi.org/10.1002/sd.1639

Carlsson A (1997) Greenhouse gas emissions in the life-cycle of carrots and tomatoes. Lund University, Lund

Conijn JG, Bindraban PS, Schröder JJ, Jongschaap REE (2018) Can our global food system meet food demand within planetary boundaries? Agric Ecosyst Environ 251:244–256. https://doi.org/10.1016/J.AGEE.2017.06.001

Delaide B, Goddek S, Gott J, Soyeurt H, Jijakli M (2016) Lettuce (*Lactuca sativa* L. var. Sucrine) growth performance in complemented aquaponic solution outperforms hydroponics. Water 8:467. https://doi.org/10.3390/w8100467

dos Santos MJPL (2016) Smart cities and urban areas–aquaponics as innovative urban agriculture. Urban For Urban Green 20:402. https://doi.org/10.1016/j.ufug.2016.10.004

Ehrlich PR, Harte J (2015) Opinion: to feed the world in 2050 will require a global revolution. Proc Natl Acad Sci U S A 112:14743–14744. https://doi.org/10.1073/pnas.1519841112

Emerenciano M, Carneiro P, Lapa M, Lapa K, Delaide B, Goddek S (2017) Mineralizacão de sólidos. Aquac Bras:21–26

Engelhaupt E (2008) Do food miles matter? Environ Sci Technol 42:3482

FAO (2009) The state of food and agriculture. FAO, Rome

FAO (2017) 2017 the state of food and agriculture leveraging food systems for inclusive rural transformation. FAO, Rome

Foley JA, Defries R, Asner GP, Barford C, Bonan G, Carpenter SR, Chapin FS, Coe MT, Daily GC, Gibbs HK, Helkowski JH, Holloway T, Howard EA, Kucharik CJ, Monfreda C, Patz JA, Prentice IC, Ramankutty N, Snyder PK (2005) Global consequences of land use. Science 309:570–574. https://doi.org/10.1126/science.1111772

Foley JA, Ramankutty N, Brauman KA, Cassidy ES, Gerber JS, Johnston M, Mueller ND, O'Connell C, Ray DK, West PC, Balzer C, Bennett EM, Carpenter SR, Hill J, Monfreda C, Polasky S, Rockström J, Sheehan J, Siebert S, Tilman D, Zaks DPM (2011) Solutions for a cultivated planet. Nature 478:337–342. https://doi.org/10.1038/nature10452

Garnett T (2011) Where are the best opportunities for reducing greenhouse gas emissions in the food system (including the food chain)? Food Policy 36:S23–S32. https://doi.org/10.1016/J.FOODPOL.2010.10.010

Goddek S (2017) Opportunities and challenges of multi-loop aquaponic systems. Wageningen University, Wageningen. https://doi.org/10.18174/412236

Goddek S, Keesman KJ (2018) The necessity of desalination technology for designing and sizing multi-loop aquaponics systems. Desalination 428:76–85. https://doi.org/10.1016/j.desal.2017.11.024

Goddek S, Vermeulen T (2018) Comparison of *Lactuca sativa* growth performance in conventional and RAS-based hydroponic systems. Aquac Int 26:1377. https://doi.org/10.1007/s10499-018-0293-8

Goddek S, Delaide B, Mankasingh U, Ragnarsdottir K, Jijakli H, Thorarinsdottir R (2015) Challenges of sustainable and commercial aquaponics. Sustainability 7:4199–4224. https://doi.org/10.3390/su7044199

Goddek S, Schmautz Z, Scott B, Delaide B, Keesman K, Wuertz S, Junge R (2016) The effect of anaerobic and aerobic fish sludge supernatant on hydroponic lettuce. Agronomy 6:37. https://doi.org/10.3390/agronomy6020037

Goddek S, Delaide BPL, Joyce A, Wuertz S, Jijakli MH, Gross A, Eding EH, Bläser I, Reuter M, Keizer LCP, Morgenstern R, Körner O, Verreth J, Keesman KJ (2018) Nutrient mineralization and organic matter reduction performance of RAS-based sludge in sequential UASB-EGSB reactors. Aquac Eng 83:10–19. https://doi.org/10.1016/J.AQUAENG.2018.07.003

Goddek S, Körner O (2019) A fully integrated simulation model of multi-loop aquaponics: a case study for system sizing in different environments. Agric Syst 171:143–154. https://doi.org/10.1016/j.agsy.2019.01.010

Godfray HCJ, Beddington JR, Crute IR, Haddad L, Lawrence D, Muir JF, Pretty J, Robinson S, Thomas SM, Toulmin C (2010) Food security: the challenge of feeding 9 billion people. Science 327:812–818. https://doi.org/10.1126/science.1185383

Graber A, Junge R (2009) Aquaponic systems: nutrient recycling from fish wastewater by vegetable production. Desalination 246:147–156

Grewal SS, Grewal PS (2012) Can cities become self-reliant in food? Cities 29:1–11. https://doi.org/10.1016/J.CITIES.2011.06.003

Hui SCM (2011) Green roof urban farming for buildings in high-density urban cities. In: World green roof conference. pp 1–9

Joly A, Junge R, Bardocz T (2015) Aquaponics business in Europe: some legal obstacles and solutions. Ecocycles 1:3–5. https://doi.org/10.19040/ecocycles.v1i2.30

Junge R, König B, Villarroel M, Komives T, Jijakli M (2017) Strategic points in aquaponics. Water 9:182. https://doi.org/10.3390/w9030182

Kahiluoto H, Kuisma M, Kuokkanen A, Mikkilä M, Linnanen L (2014) Taking planetary nutrient boundaries seriously: can we feed the people? Glob Food Sec 3:16–21. https://doi.org/10.1016/J.GFS.2013.11.002

Kloas W, Groß R, Baganz D, Graupner J, Monsees H, Schmidt U, Staaks G, Suhl J, Tschirner M, Wittstock B, Wuertz S, Zikova A, Rennert B (2015) A new concept for aquaponic systems to improve sustainability, increase productivity, and reduce environmental impacts. Aquac Environ Interact 7:179–192. https://doi.org/doi. https://doi.org/10.3354/aei00146

Kotzen B, Appelbaum S (2010) An investigation of aquaponics using brackish water resources in the Negev desert. J Appl Aquac 22:297–320. https://doi.org/10.1080/10454438.2010.527571

Manelli A (2016) New paradigms for a sustainable Well-being. Agric Agric Sci Procedia 8:617–627. https://doi.org/10.1016/J.AASPRO.2016.02.084

McKinnon A (2007) CO_2 emissions from freight transport: an analysis of UK data. Prepared for the climate change working group of the commission for integrated transport. London, England

Miličić V, Thorarinsdottir R, Santos M, Hančič M (2017) Commercial aquaponics approaching the European market: to consumers' perceptions of aquaponics products in Europe. Water 9:80. https://doi.org/10.3390/w9020080

Mogk J, Kwiatkowski S, Weindorf M (2010) Promoting urban agriculture as an alternative land use for vacant properties in the city of Detroit: benefits, problems, and proposals for a regulatory framework for successful land use integration. Law Fac Res Publ 56:1521

Monsees H, Keitel J, Kloas W, Wuertz S (2015) Potential reuse of aquacultural waste for nutrient solutions in aquaponics. In: Proc of Aquaculture Europe. Rotterdam, The Netherlands

Mundler P, Criner G (2016) Food systems: food miles. In: Encyclopedia of food and health. Elsevier, Amsterdam, pp 77–82. https://doi.org/10.1016/B978-0-12-384947-2.00325-1

Rickman JC, Barrett DM, Bruhn CM (2007) Review nutritional comparison of fresh, frozen and canned fruits and vegetables. Part 1. Vitamins C and B and phenolic compounds. J Sci Food Agric J Sci Food Agric 87:930–944. https://doi.org/10.1002/jsfa.2825

Rockström J, Steffen W, Noone K, Persson Å, Chapin FS, Lambin EF, Lenton TM, Scheffer M, Folke C, Schellnhuber HJ, Nykvist B, de Wit CA, Hughes T, van der Leeuw S, Rodhe H, Sörlin S, Snyder PK, Costanza R, Svedin U, Falkenmark M, Karlberg L, Corell RW, Fabry VJ, Hansen J, Walker B, Liverman D, Richardson K, Crutzen P, Foley JA (2009) A safe operating space for humanity. Nature 461:472–475. https://doi.org/10.1038/461472a

Schilling J, Logan J (2008) Greening the rust belt: a green infrastructure model for right sizing America's shrinking cities. J Am Plan Assoc 74:451–466. https://doi.org/10.1080/01944360802354956

Sirakov I, Lutz M, Graber A, Mathis A, Staykov Y, Smits T, Junge R (2016) Potential for combined biocontrol activity against fungal fish and plant pathogens by bacterial isolates from a model aquaponic system. Water 8:518. https://doi.org/10.3390/w8110518

Steffen W, Richardson K, Rockström J, Cornell SE, Fetzer I, Bennett EM, Biggs R, Carpenter SR, de Vries W, de Wit CA, Folke C, Gerten D, Heinke J, Mace GM, Persson LM, Ramanathan V, Reyers B, Sörlin S (2015) Planetary boundaries: guiding human development on a changing planet. Science 347(80):736

Suweis S, Carr JA, Maritan A, Rinaldo A, D'Odorico P (2015) Resilience and reactivity of global food security. Proc Natl Acad Sci U S A 112:6902–6907. https://doi.org/10.1073/pnas.1507366112

UN (2014) World's population increasingly urban with more than half living in urban areas | UN DESA | United Nations Department of Economic and Social Affairs [WWW Document]. URL http://www.un.org/en/development/desa/news/population/world-urbanization-prospects-2014.html. Accessed 19 Oct 2018

UN (2017) The sustainable development goals report. UN, New York

van Gorcum B, Goddek S, Keesman KJ (2019) Gaining market insights for aquaponically produced vegetables in Kenya. Aquac Int:1–7. https://link.springer.com/article/10.1007/s10499-019-00379-1

Van Vuuren DP, Bouwman AF, Beusen AHW (2010) Phosphorus demand for the 1970–2100 period: a scenario analysis of resource depletion. Glob Environ Chang 20:428–439. https://doi.org/10.1016/J.GLOENVCHA.2010.04.004

van Woensel L, Archer G, Panades-Estruch L, Vrscaj D (2015) Ten technologies which could change our lives – potential impacts and policy implications. European Commission, Brussels

Vermeulen T, Kamstra A (2013) The need for systems design for robust aquaponic systems in the urban environment

Weber CL, Matthews HS (2008) Food-miles and the relative climate impacts of food choices in the United States. Environ Sci Technol 42:3508–3513. https://doi.org/10.1021/es702969f

World Bank (2018) Agricultural land (% of land area) [WWW Document]. URL https://data.worldbank.org/indicator/ag.lnd.agri.zs. Accessed19 Oct 2018

Chapter 2
Aquaponics: Closing the Cycle on Limited Water, Land and Nutrient Resources

Alyssa Joyce, Simon Goddek, Benz Kotzen, and Sven Wuertz

Abstract Hydroponics initially developed in arid regions in response to freshwater shortages, while in areas with poor soil, it was viewed as an opportunity to increase productivity with fewer fertilizer inputs. In the 1950s, recirculating aquaculture also emerged in response to similar water limitations in arid regions in order to make better use of available water resources and better contain wastes. However, disposal of sludge from such systems remained problematic, thus leading to the advent of aquaponics, wherein the recycling of nutrients produced by fish as fertilizer for plants proved to be an innovative solution to waste discharge that also had economic advantages by producing a second marketable product. Aquaponics was also shown to be an adaptable and cost-effective technology given that farms could be situated in areas that are otherwise unsuitable for agriculture, for instance, on rooftops and on unused, derelict factory sites. A wide range of cost savings could be achieved through strategic placement of aquaponics sites to reduce land acquisition costs, and by also allowing farming closer to suburban and urban areas, thus reducing transportation costs to markets and hence also the fossil fuel and CO_2 footprints of production.

Keywords Aquaponics · Sustainable agriculture · Eutrophication · Soil degradation · Nutrient cycling

---

A. Joyce (✉)
Department of Marine Science, University of Gothenburg, Gothenburg, Sweden
e-mail: alyssa.joyce@gu.se

S. Goddek
Mathematical and Statistical Methods (Biometris), Wageningen University, Wageningen, The Netherlands
e-mail: simon.goddek@wur.nl; simon@goddek.nl

B. Kotzen
School of Design, University of Greenwich, London, UK
e-mail: b.kotzen@greenwich.ac.uk

S. Wuertz
Department Ecophysiology and Aquaculture, Leibniz-Institute of Freshwater Biology and Inland Fisheries, Berlin, Germany
e-mail: wuertz@igb-berlin.de

S. Goddek et al. (eds.), *Aquaponics Food Production Systems*,
https://doi.org/10.1007/978-3-030-15943-6_2

2.1 Introduction

The term ‘tipping point’ is currently being used to describe natural systems that are on the brink of significant and potentially catastrophic change (Barnosky et al. 2012). Agricultural food production systems are considered one of the key ecological services that are approaching a tipping point, as climate change increasingly generates new pest and disease risks, extreme weather phenomena and higher global temperatures. Poor land management and soil conservation practices, depletion of soil nutrients and risk of pandemics also threaten world food supplies.

Available arable land for agricultural expansion is limited, and increased agricultural productivity in the past few decades has primarily resulted from increased cropping intensity and better crop yields as opposed to expansion of the agricultural landmass (e.g. 90% of gains in crop production have been a result of increased productivity, but only 10% due to land expansion) (Alexandratos and Bruinsma 2012; Schmidhuber 2010). Global population is estimated to reach 8.3–10.9 billion people by 2050 (Bringezu et al. 2014), and this growing world population, with a corresponding increase in total as well as per capita consumption, poses a wide range of new societal challenges. The United Nations Convention to Combat Desertification (UNCCD) *Global Land Outlook Working Paper* 2017 report notes worrying trends affecting food production (Thomas et al. 2017) including land degradation, loss of biodiversity and ecosystems, and decreased resilience in response to environmental stresses, as well as a widening gulf between food production and demand. The uneven distribution of food supplies results in inadequate quantities of food, or lack of food of sufficient nutritional quality for part of the global population, while in other parts of the world overconsumption and diseases related to obesity have become increasingly common. This unbalanced juxtaposition of hunger and malnutrition in some parts of the world, with food waste and overconsumption in others, reflects complex interrelated factors that include political will, resource scarcity, land affordability, costs of energy and fertilizer, transportation infrastructure and a host of other socioeconomic factors affecting food production and distribution.

Recent re-examinations of approaches to food security have determined that a ‘water-energy-food nexus’ approach is required to effectively understand, analyse and manage interactions among global resource systems (Scott et al. 2015). The nexus approach acknowledges the interrelatedness of the resource base – land, water, energy, capital and labour – with its drivers, and encourages inter-sectoral consultations and collaborations in order to balance different resource user goals and interests. It aims to maximize overall benefits while maintaining ecosystem integrity in order to achieve food security. Sustainable food production thus requires reduced utilization of resources, in particular, water, land and fossil fuels that are limited, costly and often poorly distributed in relation to population growth, as well as recycling of existing resources such as water and nutrients within production systems to minimize waste.

In this chapter, we discuss a range of current challenges in relation to food security, focusing on resource limitations and ways that new technologies and interdisciplinary approaches such as aquaponics can help address the water-food-energy nexus in relation to the UN's goals for sustainable development. We concentrate on the need for increased nutrient recycling, reductions in water consumption and non-renewable energy, as well as increased food production on land that is marginal or unsuitable for agriculture.

2.2 Food Supply and Demand

2.2.1 Predictions

Over the last 50 years, total food supply has increased almost threefold, whereas the world's population has only increased twofold, a shift that has been accompanied by significant changes in diet related to economic prosperity (Keating et al. 2014). Over the last 25 years, the world's population increased by 90% and is expected to reach the 7.6 billion mark in the first half of 2018 (Worldometers). Estimates of increased world food demand in 2050 relative to 2010 vary between 45% and 71% depending on assumptions around biofuels and waste, but clearly there is a production gap that needs to be filled. In order to avoid a reversal in recent downward trends undernourishment, there must be reductions in food demand and/or fewer losses in food production capacity (Keating et al. 2014). An increasingly important reason for rising food demand is per capita consumption, as a result of rising per capita income, which is marked by shifts towards high protein foods, particularly meat (Ehrlich and Harte 2015b). This trend creates further pressures on the food supply chain, since animal-based production systems generally require disproportionately more resources, both in water consumption and feed inputs (Rask and Rask 2011; Ridoutt et al. 2012; Xue and Landis 2010). Even though the rate of increasing food demand has declined in recent decades, if current trajectories in population growth and dietary shifts are realistic, global demand for agricultural products will grow at 1.1–1.5% per year until 2050 (Alexandratos and Bruinsma 2012).

Population growth in urban areas has put pressure on land that has been traditionally used for soil-based crops: demands for housing and amenities continue to encroach on prime agricultural land and raise its value well beyond what farmers could make from cultivation. Close to 54% of the world's population now lives in urban areas (Esch et al. 2017), and the trend towards urbanization shows no signs of abating. Production systems that can reliably supply fresh foods close to urban centres are in demand and will increase as urbanization increases. For instance, the rise of vertical farming in urban centres such as Singapore, where land is at a premium, provides a strong hint that concentrated, highly productive farming systems will be an integral part of urban development in the future. Technological advances are increasingly making indoor farming systems economical, for instance the development of LED horticultural lights that are extremely long lasting and energy efficient has increased competitiveness of indoor farming as well as production in high latitudes.

Analysis of agrobiodiversity consistently shows that high- and middle-income countries obtain diverse foods through national or international trade, but this also implies that production and food diversity are uncoupled and thus more vulnerable to interruptions in supply lines than in low-income countries where the majority of food is produced nationally or regionally (Herrero et al. 2017). Also, as farm sizes increase, crop diversity, especially for crops belonging to highly nutritious food groups (vegetables, fruits, meat), tends to decrease in favour of cereals and legumes, which again risks limiting local and regional availability of a range of different food groups (Herrero et al. 2017).

2.3 Arable Land and Nutrients

2.3.1 Predictions

Even as more food needs to be produced, usable land for agricultural practices is inherently limited to roughly 20–30% of the world's land surface. The availability of agricultural land is decreasing, and there is a shortage of suitable land where it is most needed, i.e. particularly near population centres. Soil degradation is a major contributor to this decline and can generally be categorized in two ways: displacement (wind and water erosion) and internal soil chemical and physical deterioration (loss of nutrients and/or organic matter, salinization, acidification, pollution, compaction and waterlogging). Estimating total natural and human-induced soil degradation worldwide is fraught with difficulty given the variability in definitions, severity, timing, soil categorization, etc. However, it is generally agreed that its consequences have resulted in the loss of net primary production over large areas (Esch et al. 2017), thus restricting increases in arable and permanently cropped land to 13% in the four decades from the early 1960s to late 1990s (Bruinsma 2003). More importantly in relation to population growth during that time period, arable land per capita declined by about 40% (Conforti 2011). The term 'arable land' implies availability of adequate nutrients to support crop production. To counteract nutrient depletion, worldwide fertilizer consumption has risen from 90 kg/ha in 2002 to 135 kg in 2013 (Pocketbook 2015). Yet the increased use of fertilizers often results in excesses of nitrate and phosphates ending up in aquatic ecosystems (Bennett et al. 2001), causing algal blooms and eutrophication when decaying algal biomass consumes oxygen and limits the biodiversity of aquatic life. Large-scale nitrate and phosphate-induced environmental changes are particularly evident in watersheds and coastal zones.

Nitrogen, potassium and phosphorus are the three major nutrients essential for plant growth. Even though demand for phosphorus fertilizers continues to grow exponentially, rock phosphate reserves are limited and estimates suggest they will be depleted within 50–100 years (Cordell et al. 2011; Steen 1998; Van Vuuren et al. 2010). Additionally, anthropogenic nitrogen input is expected to drive terrestrial ecosystems towards greater phosphorous limitations, although a better understanding of the processes is critical (Deng et al. 2017; Goll et al. 2012; Zhu et al. 2016). Currently, there are no substitutes for phosphorus in agriculture, thus putting constraints on future agricultural productivity that relies on key fertilizer input of mined phosphate (Sverdrup and Ragnarsdottir 2011). The 'P-paradox', in other words, an excess of P impairing water quality, alongside its shortage as a depleting non-renewable resource, means that there must be substantial increases in recycling and efficiency of its use (Leinweber et al. 2018).

Modern intensive agricultural practices, such as the frequency and timing of tillage or no-till, application of herbicides and pesticides, and infrequent addition of organic matter containing micronutrients can alter soil structure and its microbial biodiversity such that the addition of fertilizers no longer increases productivity per hectare. Given that changes in land usage have resulted in losses of soil organic carbon estimated to be around 8%, and projected losses between 2010 and 2050 are 3.5 times that figure, it is assumed that soil water-holding capacity and nutrient losses will continue, especially in view of global warming (Esch et al. 2017). Obviously there are trade-offs between satisfying human needs and not compromising the ability of the biosphere to support life (Foley et al. 2005). However, it is clear when modelling planetary boundaries in relation to current land use practices that it is necessary to improve N and P cycling, principally by reducing both nitrogen and phosphorus emissions and runoff from agricultural land, but also by better capture and reuse (Conijn et al. 2018).

2.3.2 *Aquaponics and Nutrients*

One of the principal benefits of aquaponics is that it allows for the recycling of nutrient resources. Nutrient input into the fish component derives from feed, the composition of which depends on the target species, but feed in aquaculture typically constitutes a significant portion of input costs and can be more than half the total annual cost of production. In certain aquaponics designs, bacterial biomass can also be harnessed as feed, for instance, where biofloc production makes aquaponic systems increasingly self-contained (Pinho et al. 2017).

Wastewater from open-cage pens or raceways is often discharged into waterbodies, where it results in nutrient pollution and subsequent eutrophication. By contrast, aquaponic systems take the dissolved nutrients from uneaten fish feed and faeces, and utilizing microbes that can break down organic matter, convert the nitrogen and phosphorous into bioavailable forms for use by plants in the hydroponics unit. In order to achieve economically acceptable plant production levels, the presence of appropriate microbial assemblages reduces the need to add much of the supplemental nutrients that are routinely used in stand-alone hydroponic units. Thus aquaponics is a near-zero discharge system that offers not only economic benefit from both fish and plant production streams, but also significant reductions in both environmentally noxious discharges from aquaculture sites. It also eliminates the problem of N- and P-rich runoff from fertilizers used in soil-based agriculture. In decoupled aquaponic systems, aerobic or anaerobic bioreactors can also used to treat sludge and recover significant macro- and micronutrients in bioavailable forms for subsequent use in hydroponic production (Goddek et al. 2018) (see Chap. 8). Exciting new developments such as these, many of which are now being realized for commercial prodution, continue to refine the circular economy concept by increasingly allowing for nutrient recovery.

2.4 Pest, Weed and Disease Control

2.4.1 Predictions

It is generally recognized that control of diseases, pests and weeds is a critical component of curbing production losses that threaten food security (Keating et al. 2014). In fact, increasing the use of antibiotics, insecticides, herbicides and fungicides to cut losses and enhance productivity has allowed dramatic increases in agricultural output in the latter half of the twentieth century. However, these practices are also linked to a host of problems: pollution from persistent organic compounds in soils and irrigation water, changes in rhizobacterial and mycorrhizal activity in soils, contamination of crops and livestock, development of resistant strains, detrimental effects on pollinators and a wide range of human health risks (Bringezu et al. 2014; Ehrlich and Harte 2015a; Esch et al. 2017; FAO 2015b). Tackling pest, weed and disease control in ways that reduce the use of these substances is mentioned in virtually every call to provide food security for a growing world population.

2.4.2 Control of Pests, Weeds and Diseases

As a closed system with biosecurity measures, aquaponic systems require far fewer chemical pesticide applications in the plant component. If seed and transplant stocks are carefully handled and monitored, weed, fungal and bacterial/algal contaminants can be controlled in hydroponic units with targeted measures rather than the widespread preventive application of herbicides and fungicides prevalent in soil-based agriculture. As technology continues to advance, developments such as positive pressure greenhouses can further reduce pest problems (Mears and Both 2001). Design features to reduce pest risks can cut costs in terms of chemicals, labour, application time and equipment, especially since the land footprint of industrial-scale aquaponics systems is small, and systems are compact and tightly contained, as compared to the equivalent open production area of vegetable and fruit crops of conventional soil-based farms.

The use of RAS in aquaponic systems also prevents disease transmissions between farmed stocks and wild populations, which is a pressing concern in flow-through and open-net pen aquaculture (Read et al. 2001; Samuel-Fitwi et al. 2012). Routine antibiotic use is generally not required in the RAS component, since it is a closed system with few available vectors for disease introduction. Furthermore, the use of antimicrobials and antiparasitics is generally discouraged, as it can be detrimental to the microbiota that are crucial for converting organic and inorganic wastes into usable compounds for plant growth in the hydroponic unit (Junge et al. 2017). If disease does emerge, containment of both fish and plants from the surrounding environment makes decontamination and eradication more manageable. Although closed systems clearly do not completely alleviate all disease and pest problems (Goddek et al. 2015), proper biocontrol measures that are already practised in stand-alone RAS and hydroponics result in significant reductions of risk. These issues are discussed in further detail in subsequent chapters (for fish, see Chap. 6; for plants, further details in Chap. 14).

2.5 Water Resources

2.5.1 Predictions

In addition to requiring fertilizer applications, modern intensive agricultural practices also place high demands on water resources. Among biochemical flows (Fig. 2.1), water scarcity is now believed to be one of the most important factors constraining food production (Hoekstra et al. 2012; Porkka et al. 2016). Projected global population increases and shifts in terrestrial water availability due to climate change, demand more efficient use of water in agriculture. As noted previously, by 2050, aggregate agriculture production will need to produce 60% more food globally (Alexandratos and Bruinsma 2012), with an estimated 100% more in developing countries, based on population growth and rising expectations for standards of living (Alexandratos and Bruinsma 2012; WHO 2015). Famine in some regions of the world, as well as malnutrition and hidden hunger, indicates that the balance between food demand and availability has already reached critical levels, and that food and water security are directly linked (McNeill et al. 2017). Climate change predictions suggest reduced freshwater availability, and a corresponding decrease in agricultural yields by the end of the twenty-first century (Misra 2014).

The agriculture sector currently accounts for roughly 70% of the freshwater use worldwide, and the withdrawal rate even exceeds 90% in most of the world's least-developed countries. Water scarcity will increase in the next 25 years due to expected population growth (Connor et al. 2017; Esch et al. 2017), with the latest

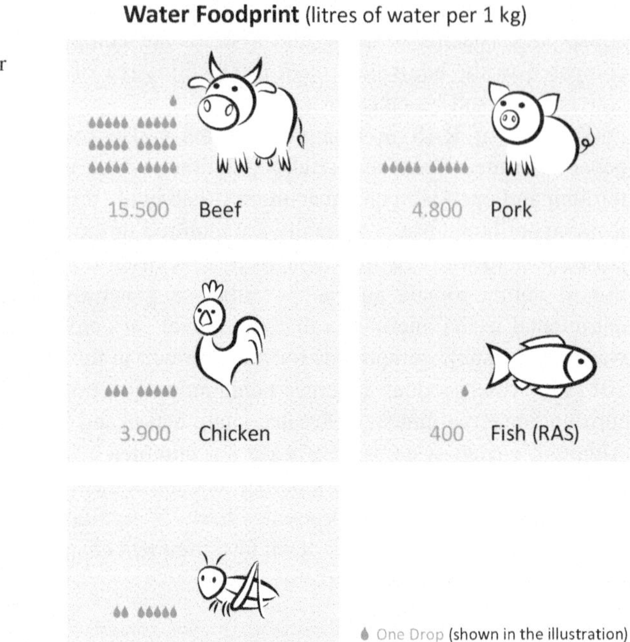

Fig. 2.1 Water footprint (L per kg). Fish in RAS systems use the least water of any food production system

modelling forecasting declining water availability in the near future for nearly all countries (Distefano and Kelly 2017). The UN predicts that the pursuit of business-as-usual practices will result in a global water deficit of 40% by 2030 (Water 2015). In this respect, as groundwater supplies for irrigation are depleted or contaminated, and arid regions experience more drought and water shortages due to climate change, water for agricultural production will become increasingly valuable (Ehrlich and Harte 2015a). Increasing scarcity of water resources compromises not only water security for human consumption but also global food production (McNeill et al. 2017). Given that water scarcity is expected even in areas that currently have relatively sufficient water resources, it is important to develop agricultural techniques with low water input requirements, and to improve ecological management of wastewater through better reuse (FAO 2015a).

The UN World Water Development Report for 2017 (Connor et al. 2017) focuses on wastewater as an untapped source of energy, nutrients and other useful by-products, with implications not only for human and environmental health but also for food and energy security as well as climate change mitigation. This report calls for appropriate and affordable technologies, along with legal and regulatory frameworks, financing mechanisms and increased social acceptability of wastewater treatment, with the goal of achieving water reuse within a circular economy. The report also points to a 2016 World Economic Forum report that lists the water crisis as the global risk of highest concern in the next 10 years.

The concept of a water footprint as a measure of humans' use of freshwater resources has been put forwards in order to inform policy development on water use. A water footprint has three components: (1) blue water, which comprises the surface and groundwater consumed while making products or lost through evaporation, (2) green water that is rainwater used especially in crop production and (3) grey water, which is water that is polluted but still within existing water quality standards (Hoekstra and Mekonnen 2012). These authors mapped water footprints of countries worldwide and found that agricultural production accounts for 92% of global freshwater use, and industrial production uses 4.4% of the total, while domestic water only 3.6%. This raises concerns about water availability and has resulted in public education efforts aimed at raising awareness about the amounts of water required to produce various types of food, as well as national vulnerabilities, especially in water-scarce countries in North Africa and the Middle East.

2.5.2 Aquaponics and Water Conservation

The economic concept of comparative productivity measures the relative amount of a resource needed to produce a unit of goods or services. Efficiency is generally construed to be higher when the requirement for resource input is lower per unit of goods and services. However, when water-use efficiency is examined in an environmental context, water quality also needs to be taken into account, because maintaining or enhancing water quality also enhances productivity (Hamdy 2007).

The growing problem of water scarcity demands improvements in water-use efficiencies especially in arid and semiarid regions, where availability of water for agriculture, and water quality of discharge, are critical factors in food production. In these regions, recirculation of water in aquaponic units can achieve remarkable water re-use efficiency of 95–99% (Dalsgaard et al. 2013). Water demand is also less than 100 L/kg of fish harvested, and water quality is maintained within the system for production of crops (Goddek et al. 2015). Obviously, such systems must be constructed and operated to minimize water losses; they must also optimize their ratios of fish water to plants, as this ratio is very important in maximizing water re-use efficiency and ensuring maximal nutrient recycling. Modelling algorithms and technical solutions are being developed to integrate improvements in individual units, and to better understand how to effectively and efficiently manage water (Vilbergsson et al. 2016). Further information is provided in Chaps. 9 and 11.

In light of soil, water and nutrient requirements, the water footprint of aquaponic systems is considerably better than traditional agriculture, where water quality and demand, along with availability of arable land, costs of fertilizers and irrigation are all constraints to expansion (Fig. 2.1).

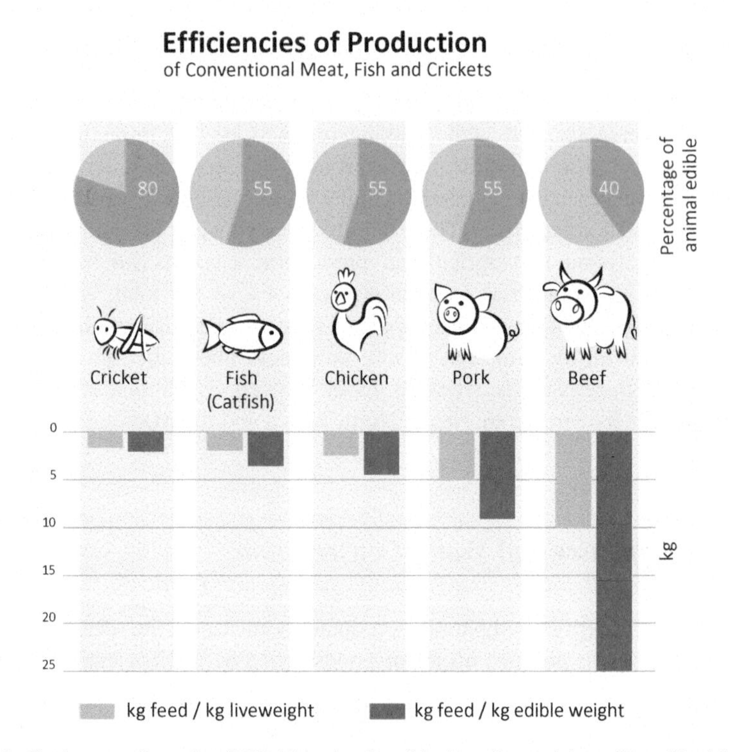

Fig. 2.2 Feed conversion ratios (FCRs) based as kg of feed per live weight and kg of feed for edible portion. Only insects, which are eaten whole in some parts of the world, have a better FCR than fish

2.6 Land Utilization

2.6.1 Predictions

Globally, land-based crops and pasture occupy approximately 33% of total available land, and expansion for agricultural uses between 2000 and 2050 is estimated to increase by 7–31% (350–1500 Mha, depending on source and underlying assumptions), most often at the expense of forests and wetlands (Bringezu et al. 2014). While there is currently still land classed as 'good' or 'marginal' that is available for rain-fed agriculture, significant portions of it are far from markets, lack infrastructure or have endemic diseases, unsuitable terrain or other conditions that limit development potential. In other cases, remaining lands are already protected, forested or developed for other uses (Alexandratos and Bruinsma 2012). By contrast, dryland ecosystems, defined in the UN's Commission on Sustainable Development as arid, semiarid and dry subhumid areas that typically have low productivity, are threatened by desertification and are therefore unsuitable for agricultural expansion but nevertheless have many millions of people living in close proximity (Economic 2007). These facts point to the need for more sustainable intensification of food production closer to markets, preferably on largely unproductive lands that may never become suitable for soil-based farming.

The two most important factors contributing to agricultural input efficiencies are considered by some experts to be (i) the location of food production in areas where climatic (and soil) conditions naturally increase efficiencies and (ii) reductions in environmental impacts of agricultural production (Michael and David 2017). There must be increases in the supply of cultivated biomass achieved through the intensification of production per hectare, accompanied by a diminished environmental burden (e.g. degradation of soil structure, nutrient losses, toxic pollution). In other words, the footprint of efficient food production must shrink while minimizing negative environmental impacts.

2.6.2 Aquaponics and Land Utilization

Aquaponic production systems are soilless and attempt to recycle essential nutrients for cultivation of both fish and plants, thereby using nutrients in organic matter from fish feed and wastes to minimize or eliminate the need for plant fertilizers. For instance, in such systems, using land to mine, process, stockpile and transport phosphate or potash-rich fertilizers becomes unnecessary, thus also eliminating the inherent cost, and cost of application, for these fertilizers.

Aquaponics production contributes not only to water usage efficiency (Sect. 2.5.2) but also to agricultural input efficiency by reducing the land footprint needed for production. Facilities for instance, can be situated on nonarable land and in suburban or urban areas closer to markets, thus reducing the carbon footprint associated with rural farms and transportation of products to city markets. With a smaller footprint, production capacity can be located in otherwise unproductive areas such as on rooftops or old factory sites, which can also reduce land acquisition costs if those areas are deemed unsuitable for housing or retail businesses. A smaller footprint for production of high-quality protein and vegetables in aquaponics can also take pressure away from clearing ecologically valuable natural and semi-natural areas for conventional agriculture.

2.7 Energy Resources

2.7.1 Predictions

As mechanization spreads globally, open-field intensive agriculture increasingly relies heavily on fossil fuels to power farm machinery and for transportation of fertilizers as well as farm products, as well as to run the equipment for processing, packaging and storage. In 2010, the OECD International Energy Agency predicted that global energy consumption would grow by up to 50% by 2035; the FAO has also estimated that 30% of global energy consumption is devoted to food production and its supply chain (FAO 2011). Greenhouse gas (GHG) emissions associated with fossil fuels (approximately 14% in lifecycle analysis) added to those from fertilizer manufacturing (16%) and nitrous oxide from average soils (44%) (Camargo et al. 2013), all contribute substantially to the environmental impacts of farming. A trend in the twenty-first century to produce crop-based biofuels (e.g. corn for ethanol) to replace fossil fuels has increased pressure on the clearing of rainforests, peatlands, savannas and grasslands for agricultural production. However, studies point to creation of a 'carbon debt' from such practices, since the overall release of CO_2 exceeds the reductions in GHGs they provide by displacing fossil fuels (Fargione et al. 2008). Arguably a similar carbon debt exists when clearing land to raise food crops via conventional agriculture that relies on fossil fuels.

In a comparative analysis of agricultural production systems, trawling fisheries and recirculating aquaculture systems (RAS) were found to emit GHGs 2–2.5 times that of non-trawling fisheries and non-RAS (pen, raceway) aquaculture. In RAS, these energy requirements relate primarily to the functioning of pumps and filters (Michael and David 2017). Similarly, greenhouse production systems can emit up to three times more GHGs than open-field crop production if energy is required to maintain heat and light within optimal ranges (ibid.). However, these GHG figures do not take into account other environmental impacts of non-RAS systems, such as eutrophication or potential pathogen transfers to wild stocks. Nor do they consider GHG from the production, transportation and application of herbicides and pesticides used in open-field cultivation, nor methane and nitrous oxide from associated livestock production, both of which have a 100-year greenhouse warming potential (GWP) 25 and 298 times that of CO_2, respectively (Camargo et al. 2013; Eggleston et al. 2006).

These sobering estimates of present and future energy consumption and GHG emissions associated with food production have prompted new modelling and approaches, for example, the UN's water-food-energy nexus approach mentioned in Sect. 2.1. The UN's Sustainable Development Goals have pinpointed the vulnerability of food production to fluctuations in energy prices as a key driver of food insecurity. This has prompted efforts to make agrifood systems 'energy smart' with an emphasis on improving energy efficiencies, increasing use of renewable energy sources and encouraging integration of food and energy production (FAO 2011).

2.7.2 Aquaponics and Energy Conservation

Technological advances in aquaponic system operations are moving towards being increasingly 'energy smart' and reducing the carbon debt from pumps, filters and heating/cooling devices by using electricity generated from renewable sources. Even in temperate latitudes, many new designs allow the energy involved in heating and cooling of fish tanks and greenhouses to be fully reintegrated, such that these systems do not require inputs beyond solar arrays or the electricity/heat generated from bacterial biogas production of aquaculture-derived sludge (Ezebuiro and Körner 2017; Goddek and Keesman 2018; Kloas et al. 2015; Yogev et al. 2016). In addition, aquaponic systems can use microbial denitrification to convert nitrous oxide to nitrogen gas if enough carbon sources from wastes are available, such that heterotrophic and facultative anaerobic bacteria can convert excess nitrates to nitrogen gas (Van Rijn et al. 2006). As noted in Sect. 2.7.1, nitrous oxide is a potent GHG and microbes already present in closed aquaponics systems can facilitate its conversion into nitrogen gas.

2.8 Summary

As the human population continues to increase, there is increasing demand for high-quality protein worldwide. Compared to meat sources, fish are widely recognized as being a particularly healthy source of protein. In relation to the world food supply, aquaculture now provides more fish protein than capture fisheries (FAO 2016). Globally, human per capita fish consumption continues to rise at an annual average rate of 3.2% (1961–2013), which is double the rate of population growth. In the period from 1974 to 2013, biologically unsustainable 'overfishing' has increased by 22%. During the same period, the catch from what are deemed to be 'fully exploited' fisheries has decreased by 26%. Aquaculture therefore provides the only possible solution for meeting increased market demand. It is now the fastest growing food sector and therefore an important component of food security (ibid.)

With the global population estimated to reach 8.3–10.9 billion people by 2050 (Bringezu et al. 2014), sustainable development of the aquaculture and agricultural sectors requires optimization in terms of production efficiency, but also reductions in utilization of limited resources, in particular, water, land and fertilizers. The benefits of aquaponics relate not just to the efficient uses of land, water and nutrient resources but also allow for increased integration of smart energy opportunities such as biogas and solar power. In this regard, aquaponics is a promising technology for producing both high-quality fish protein and vegetables in ways that can use substantially less land, less energy and less water while also minimizing chemical and fertilizer inputs that are used in conventional food production.

References

Alexandratos N, Bruinsma J (2012) World agriculture towards 2030/50: the 2012 revision, ESA Work. Paper 12-03. UN Food Agriculture Organization (FAO), Rome

Barnosky AD, Hadly EA, Bascompte J, Berlow EL, Brown JH, Fortelius M, Getz WM, Harte J, Hastings A, Marquet PA (2012) Approaching a state shift in Earth/'s biosphere. Nature 486:52–58

Bennett EM, Carpenter SR, Caraco NF (2001) Human impact on erodable phosphorus and eutrophication: a global perspective: increasing accumulation of phosphorus in soil threatens rivers, lakes, and coastal oceans with eutrophication. AIBS Bull 51:227–234

Bringezu S, Schütz H, Pengue W, OBrien M, Garcia F, Sims R (2014) Assessing global land use: balancing consumption with sustainable supply. UNEP/International Resource Panel, Nairobi/ Paris

Bruinsma J (2003) World agriculture: towards 2015/2030: an FAO perspective. Earthscan, London

Camargo GG, Ryan MR, Richard TL (2013) Energy use and greenhouse gas emissions from crop production using the farm energy analysis tool. Bioscience 63:263–273

Conforti P (2011) Looking ahead in world food and agriculture: perspectives to 2050. Food and Agriculture Organization of the United Nations (FAO), Rome

Conijn J, Bindraban P, Schröder J, Jongschaap R (2018) Can our global food system meet food demand within planetary boundaries? Agric Ecosyst Environ 251:244–256

Connor R, Renata A, Ortigara C, Koncagül E, Uhlenbrook S, Lamizana-Diallo BM, Zadeh SM, Qadir M, Kjellén M, Sjödin J (2017) The United Nations world water development report 2017. In: Wastewater: the untapped resource, The United Nations world water development report. UNESCO, Paris

Cordell D, Rosemarin A, Schröder J, Smit A (2011) Towards global phosphorus security: a systems framework for phosphorus recovery and reuse options. Chemosphere 84:747–758

Dalsgaard J, Lund I, Thorarinsdottir R, Drengstig A, Arvonen K, Pedersen PB (2013) Farming different species in RAS in Nordic countries: current status and future perspectives. Aquac Eng 53:2–13

Deng Q, Hui D, Dennis S, Reddy KC (2017) Responses of terrestrial ecosystem phosphorus cycling to nitrogen addition: a meta-analysis. Glob Ecol Biogeogr 26:713–728

Distefano T, Kelly S (2017) Are we in deep water? Water scarcity and its limits to economic growth. Ecol Econ 142:130–147

Economic UNDo (2007) Indicators of sustainable development: guidelines and methodologies. United Nations Publications, New York

Eggleston H, Buendia L, Miwa K, Ngara T, Tanabe K (2006) IPCC guidelines for national greenhouse gas inventories. Inst Glob Environ Strateg, Hayama, Japan 2:48–56

Ehrlich PR, Harte J (2015a) Food security requires a new revolution. Int J Environ Stud 72:908–920

Ehrlich PR, Harte J (2015b) Opinion: to feed the world in 2050 will require a global revolution. Proc Natl Acad Sci 112:14743–14744

Esch Svd, Brink Bt, Stehfest E, Bakkenes M, Sewell A, Bouwman A, Meijer J, Westhoek H, Berg Mvd, Born GJvd (2017) Exploring future changes in land use and land condition and the impacts on food, water, climate change and biodiversity: scenarios for the UNCCD Global Land Outlook. PBL Netherlands Environmental Assessment Agency, The Hague

Ezebuiro NC, Körner I (2017) Characterisation of anaerobic digestion substrates regarding trace elements and determination of the influence of trace elements on the hydrolysis and acidification phases during the methanisation of a maize silage-based feedstock. J Environ Chem Eng 5:341–351

FAO (2011) Energy-smart food for people and climate. Food and Agriculture Organization of the United Nations, Rome

FAO (2015a) Environmental and social management guidelines. Food and Agriculture Organization of the United Nations, Rome

FAO (2015b) Statistical pocketbook 2015. Food and Agriculture Organization of the United Nations, Rome

FAO (2016) The state of world fisheries and aquaculture 2016. Contributing to food security and nutrition for all. Food and Agriculture Organization of the United Nations, Rome, p 200

Fargione J, Hill J, Tilman D, Polasky S, Hawthorne P (2008) Land clearing and the biofuel carbon debt. Science 319:1235–1238

Foley JA, DeFries R, Asner GP, Barford C, Bonan G, Carpenter SR, Chapin FS, Coe MT, Daily GC, Gibbs HK (2005) Global consequences of land use. Science 309:570–574

Goddek S, Keesman KJ (2018) The necessity of desalination technology for designing and sizing multi-loop aquaponics systems. Desalination 428:76–85

Goddek S, Delaide B, Mankasingh U, Ragnarsdottir KV, Jijakli H, Thorarinsdottir R (2015) Challenges of sustainable and commercial aquaponics. Sustainability 7:4199–4224

Goddek S, Delaide BPL, Joyce A, Wuertz S, Jijakli MH, Gross A, Eding EH, Bläser I, Reuter M, Keizer LCP, Morgenstern R, Körner O, Verreth J, Keesman KJ (2018) Nutrient mineralization and organic matter reduction performance of RAS-based sludge in sequential UASB-EGSB reactors. Aquac Eng 83:10–19. https://doi.org/10.1016/J.AQUAENG.2018.07.003

Goll DS, Brovkin V, Parida B, Reick CH, Kattge J, Reich PB, Van Bodegom P, Niinemets Ü (2012) Nutrient limitation reduces land carbon uptake in simulations with a model of combined carbon, nitrogen and phosphorus cycling. Biogeosciences 9:3547–3569

Hamdy A (2007) Water use efficiency in irrigated agriculture: an analytical review. Water use efficiency and water productivity: WASAMED project, pp 9–19

Herrero M, Thornton PK, Power B, Bogard JR, Remans R, Fritz S, Gerber JS, Nelson G, See L, Waha K (2017) Farming and the geography of nutrient production for human use: a transdisciplinary analysis. Lancet Planetary Health 1:e33–e42

Hoekstra AY, Mekonnen MM (2012) The water footprint of humanity. Proc Natl Acad Sci 109:3232–3237

Hoekstra AY, Mekonnen MM, Chapagain AK, Mathews RE, Richter BD (2012) Global monthly water scarcity: blue water footprints versus blue water availability. PLoS One 7:e32688

Junge R, König B, Villarroel M, Komives T, Jijakli MH (2017) Strategic points in aquaponics. Water 9:182

Keating BA, Herrero M, Carberry PS, Gardner J, Cole MB (2014) Food wedges: framing the global food demand and supply challenge towards 2050. Glob Food Sec 3:125–132

Kloas W, Groß R, Baganz D, Graupner J, Monsees H, Schmidt U, Staaks G, Suhl J, Tschirner M, Wittstock B, Wuertz S, Zikova A, Rennert B (2015) A new concept for aquaponic systems to improve sustainability, increase productivity, and reduce environmental impacts. Aquac Environ Interact 7:179–192

Leinweber P, Bathmann U, Buczko U, Douhaire C, Eichler-Löbermann B, Frossard E, Ekardt F, Jarvie H, Krämer I, Kabbe C (2018) Handling the phosphorus paradox in agriculture and natural ecosystems: scarcity, necessity, and burden of P. Ambio 47:3–19

McNeill K, Macdonald K, Singh A, Binns AD (2017) Food and water security: analysis of integrated modeling platforms. Agric Water Manag 194:100–112

Mears D, Both A (2001) A positive pressure ventilation system with insect screening for tropical and subtropical greenhouse facilities. Int Symp Des Environ Control Trop Subtrop Greenh 578:125–132

Michael C, David T (2017) Comparative analysis of environmental impacts of agricultural production systems, agricultural input efficiency, and food choice. Environ Res Lett 12:064016

Misra AK (2014) Climate change and challenges of water and food security. Int J Sustain Built Environ 3:153–165

Pinho SM, Molinari D, de Mello GL, Fitzsimmons KM, Emerenciano MGC (2017) Effluent from a biofloc technology (BFT) tilapia culture on the aquaponics production of different lettuce varieties. Ecol Eng 103:146–153

Pocketbook FS (2015) World food and agriculture (2015). Food and Agriculture Organization of the United Nations, Rome

Porkka M, Gerten D, Schaphoff S, Siebert S, Kummu M (2016) Causes and trends of water scarcity in food production. Environ Res Lett 11:015001

Rask KJ, Rask N (2011) Economic development and food production–consumption balance: a growing global challenge. Food Policy 36:186–196

Read P, Fernandes T, Miller K (2001) The derivation of scientific guidelines for best environmental practice for the monitoring and regulation of marine aquaculture in Europe. J Appl Ichthyol 17:146–152

Ridoutt BG, Sanguansri P, Nolan M, Marks N (2012) Meat consumption and water scarcity: beware of generalizations. J Clean Prod 28:127e133

Samuel-Fitwi B, Wuertz S, Schroeder JP, Schulz C (2012) Sustainability assessment tools to support aquaculture development. J Clean Prod 32:183–192

Schmidhuber J (2010) FAO's long-term outlook for global agriculture–challenges, trends and drivers. International Food & Agriculture Trade Policy Council

Scott CA, Kurian M, Wescoat JL Jr (2015) The water-energy-food nexus: enhancing adaptive capacity to complex global challenges, Governing the nexus. Springer, Cham, pp 15–38

Steen I (1998) Management of a non-renewable resource. Phosphorus Potassium 217:25–31

Sverdrup HU, Ragnarsdottir KV (2011) Challenging the planetary boundaries II: assessing the sustainable global population and phosphate supply, using a systems dynamics assessment model. Appl Geochem 26:S307–S310

Thomas, R., Reed, M., Clifton, K., Appadurai, A., Mills, A., Zucca, C., Kodsi, E., Sircely, J., Haddad, F., vonHagen, C., 2017. Scaling up sustainable land management and restoration of degraded land

Van Rijn J, Tal Y, Schreier HJ (2006) Denitrification in recirculating systems: theory and applications. Aquac Eng 34:364–376

Van Vuuren DP, Bouwman AF, Beusen AH (2010) Phosphorus demand for the 1970–2100 period: a scenario analysis of resource depletion. Glob Environ Chang 20:428–439

Vilbergsson B, Oddsson GV, Unnthorsson R (2016) Taxonomy of means and ends in aquaculture production–part 2: the technical solutions of controlling solids, dissolved gasses and pH. Water 8:387

Water U (2015) Water for a sustainable world, The United Nations world water development report. United Nations Educational, Scientific and Cultural Organization, Paris

WHO (2015) Progress on sanitation and drinking water: 2015 update and MDG assessment. World Health Organization, Geneva

Xue X, Landis AE (2010) Eutrophication potential of food consumption patterns. Environ Sci Technol 44:6450–6456

Yogev U, Barnes A, Gross A (2016) Nutrients and energy balance analysis for a conceptual model of a three loops off grid, aquaponics. Water 8:589

Zhu Q, Riley W, Tang J, Koven C (2016) Multiple soil nutrient competition between plants, microbes, and mineral surfaces: model development, parameterization, and example applications in several tropical forests. Biogeosciences 13:341

Chapter 3
Recirculating Aquaculture Technologies

Carlos A. Espinal and Daniel Matulić

Abstract Recirculating aquaculture technology, which includes aquaponics, has been under development for the past 40 years from a combination of technologies derived from the wastewater treatment and aquaculture sectors. Until recently, recirculating aquaculture systems (RAS) farms have been relatively small compared with other types of modern aquaculture production. The last two decades have seen a significant increase in the development of this technology, with increased market acceptance and scale. This chapter provides a brief overview of the history, water quality control processes, new developments and ongoing challenges of RAS.

Keywords Recirculating aquaculture systems (RAS) · Wastewater treatment · Biofilter · Denitrification · Membrane technology

3.1 Introduction

Recirculating aquaculture systems (RAS) describe intensive fish production systems which use a series of water treatment steps to depurate the fish-rearing water and facilitate its reuse. RAS will generally include (1) devices to remove solid particles from the water which are composed of fish faeces, uneaten feed and bacterial flocs (Chen et al. 1994; Couturier et al. 2009), (2) nitrifying biofilters to oxidize ammonia

---

The original version of this chapter was revised: The chemical formula "NO2" has been corrected to "NO_2" on page 37. The correction to this chapter is available at https://doi.org/10.1007/978-3-030-15943-6_25

C. A. Espinal (✉)
Landing Aquaculture, Oirschot, The Netherlands
e-mail: carlos@landingaquaculture.com

D. Matulić
Department of Fisheries, Beekeeping, Game management and Special Zoology, Faculty of Agriculture, University of Zagreb, Zagreb, Croatia
e-mail: dmatulic@agr.hr

S. Goddek et al. (eds.), *Aquaponics Food Production Systems*,
https://doi.org/10.1007/978-3-030-15943-6_3

excreted by fish to nitrate (Gutierrez-Wing and Malone 2006) and (3) a number of gas exchange devices to remove dissolved carbon dioxide expelled by the fish as well as/or adding oxygen required by the fish and nitrifying bacteria (Colt and Watten 1988; Moran 2010; Summerfelt 2003; Wagner et al. 1995). In addition, RAS may also use UV irradiation for water disinfection (Sharrer et al. 2005; Summerfelt et al. 2009), ozonation and protein skimming for fine solids and microbial control (Attramadal et al. 2012a; Gonçalves and Gagnon 2011; Summerfelt and Hochheimer 1997) and denitrification systems to remove nitrate (van Rijn et al. 2006).

Modern recirculating aquaculture technology has been developing for more than 40 years, but novel technologies increasingly offer ways to change the paradigms of traditional RAS including improvements on classic processes such as solids capture, biofiltration and gas exchange. RAS has also experienced important developments in terms of scale, production capacities and market acceptance, with systems becoming progressively larger and more robust.

This chapter discusses how RAS technology has developed over the past two decades from a period of technological consolidation to a new era of industrial implementation.

3.1.1 History of RAS

The earliest scientific research on RAS conducted in Japan in the 1950s focused on biofilter design for carp production driven by the need to use locally limited water resources more productively (Murray et al. 2014). In Europe and the United States, scientists similarly attempted to adapt technologies developed for domestic wastewater treatment in order to better reuse water within recirculating systems (e.g. activated sludge processes for sewage treatment, trickling, submerged and down-flow biofilters and several mechanical filtration systems). These early efforts included primarily work on marine systems for fish and crustacean production, but were soon adopted in arid regions where the agriculture sector is restricted by water supply. In aquaculture, different solutions have been designed to maximize water use including highly intensive recirculating systems that incorporate water filtration systems such as drum filters, biological filters, protein skimmers and oxygen injection systems (Hulata and Simon 2011). Despite a strong conviction by pioneers in the industry about the commercial viability of their work, most of the early studies focused exclusively on the oxidation of toxic inorganic nitrogen wastes derived from protein metabolism. The trust in technology was reinforced by the successful operation of public as well as domestic aquaria, which generally feature over-sized treatment units to ensure crystal-clear water. Additionally, extremely low stocking densities and associated feed inputs meant that such over-engineering still made a relatively small contribution to capital and operational costs of the system compared to intensive RAS. Consequently, the changes in process dynamics associated with scale-change were unaccounted for, resulting in the under-sizing of RAS treatment units in order to minimize capital costs. As a consequence, safety margins were far too narrow or non-existent (Murray et al. 2014). Because many of the pioneering scientists had biological rather than engineering backgrounds, technical improvements were also constrained by miscommunications between scientists, designers, construction personnel and operators. The development of a standardized terminology, units of measurement and reporting formats in 1980 (EIFAC/ICES 1980), helped address the situation, though regional differences still persisted. It was not until the mid-1980s that cyclic water quality parameters became well recognized as being important in pond production, e.g. periodically measuring the concentrations of pH, oxygen, TAN (total ammonia nitrogen), NO_2 (nitrate), BOD (biochemical oxygen demand) and COD (chemical oxygen demand).

In the latter part of the last century, numerous articles were published on the early development of RAS. Rosenthal (1980) elaborated on the state of recirculation systems in Western Europe, while Bovendeur et al. (1987) developed a water recirculation system for the culture of African catfish in relation to waste production and waste removal kinetics (a design was presented for a water treatment system consisting of a primary clarifier and an aerobic fixed-film reactor that demonstrated satisfactory results for high-density culture of African catfish). This work was part of the rapid development in fish culture systems up to the mid-1990s in Northern and Western Europe (Rosenthal and Black 1993), as well as in North America (Colt 1991). New classifications, such as the classification according to how water flows through an aquaculture system, provided key insights with respect to the water quality processes that are important for fish production (Krom and van Rijn 1989). In subsequent work by van Rijn (1996), concepts were introduced focused on the biological processes underlying the treatment systems. The conclusions from this work were that incorporating methods for reducing the accumulation of sludge and nitrate resulted in more stable water quality conditions within the culture units. During this period, RAS production increased significantly in volume and species diversity (Rosenthal 1980; Verreth and Eding 1993; Martins et al. 2005). Today, more than 10 species are produced in RAS (African catfish, eel and trout as major freshwater species and turbot, seabass and sole as major marine species) (Martins et al 2010b), with RAS also becoming a crucial element in the production of larvae and juveniles of diverse species.

While maximum sustainable yields of many aquatic wild stock species have been or will soon be reached, and many species are already overfished, RAS is considered a key technology that will help the aquaculture sector meet the needs for aquatic species over the coming decades (Ebeling and Timmons 2012).

3.1.2 A Short History of Aquaponics in the Context of RAS

Aquaponics is a term that has been 'coined' in the 1970s, but in practice has ancient roots – although there are still discussions about its first occurrence. The Aztec cultivated agricultural islands known as *chinampas* (the earliest 1150–1350CE), in a system considered by some to be the first form of aquaponics for agricultural use (Fig. 3.1). In such systems, plants were raised on stable, or sometime movable and floating islands placed in lake shallows wherein nutrient rich mud could be dredged

Fig. 3.1 Chinampas (floating gardens) in Central America – artificial island construction as antecedent of aquaponic technology. (From Marzolino/Shutterstock.com)

from the chinampa canals and placed on the islands to support plant growth (Crossley 2004).

An even earlier example of aquaponics started on the other side of the world in south China and is believed to have spread within South East Asia where Chinese settlers from Yunnan settled around 5 CE. Farmers cultivated and farmed rice in paddy fields in combination with fish (FAO 2001). These polycultural farming systems existed in many Far Eastern countries to raise fish such as oriental loach (*Misgurnus anguillicaudatus*) (Tomita-Yokotani et al. 2009), swamp eel (fam. Synbranchidae), common carp (*Cyprinus carpio*) and crucian carp (*Carassius carassius*) (FAO 2004). In essence, however, these were not aquaponic systems but can be best described as early examples of integrated aquaculture systems (Gomez 2011). In the twentieth century, the first attempts to create practical, efficient and integrated fish production systems alongside vegetables were made in the 1970s with the work of Lewis and Naegel (Lewis and Wehr 1976; Naegel 1977; Lewis et al. 1978). Further early systems were designed by Waten and Busch in 1984 and Rakocy in 1989 (Palm et al. 2018).

3.2 Review of Water Quality Control in RAS

RAS are complex aquatic production systems that involve a range of physical, chemical and biological interactions (Timmons and Ebeling 2010). Understanding these interactions and the relationships between the fish in the system and the equipment used is crucial to predict any changes in water quality and system performance. There are more than 40 water quality parameters than can be used to determine water quality in aquaculture (Timmons and Ebeling 2010). Of these, only a few (as described in Sects. 3.2.1, 3.2.2, 3.2.3, 3.2.4, 3.2.5, 3.2.6 and 3.2.7) are traditionally controlled in the main recirculation processes, given that these processes can rapidly affect fish survival and are prone to change with the addition of feed to the system. Many other water quality parameters are not normally monitored or controlled because (1) water quality analytics may be expensive, (2) the pollutant to be analysed can be diluted with daily water exchange, (3) potential water sources containing them are ruled out for use or (4) because their potential negative effects have not been observed in practice. Therefore, the following water quality parameters are normally monitored in RAS.

3.2.1 Dissolved Oxygen (DO)

Dissolved oxygen (DO) is generally the most important water quality parameter in intensive aquatic systems, as low DO levels may quickly result in high stress in fish, nitrifying biofilter malfunction and indeed significant fish losses. Commonly, stocking densities, feed addition, temperature and the tolerance of the fish species to hypoxia will determine the oxygen requirements of a system. As oxygen can be transferred to water in concentrations higher than its saturation concentration under atmospheric conditions (this is called supersaturation), a range of devices and designs exist to ensure that the fish are provided with sufficient oxygen.

In RAS, DO can be controlled via aeration, addition of pure oxygen, or a combination of these. Since aeration is only capable of raising the DO concentrations to the atmospheric saturation point, the technique is generally reserved for lightly loaded systems or systems with tolerant species such as tilapia or catfish. However, aerators are also an important component of commercial RAS where the use of expensive technical oxygen is reduced by aerating water with a low dissolved oxygen content back to the saturation point before supersaturating the water with technical oxygen.

There are several types of aerators and oxygenators that can be used in RAS and these fall within two broad categories: gas-to-liquid and liquid-to-gas systems (Lekang 2013). Gas-to-liquid aerators mostly comprise diffused aeration systems where gas (air or oxygen) is transferred to the water, creating bubbles which exchange gases with the liquid medium (Fig. 3.2). Other gas-to-liquid systems include passing gases through diffusers, perforated pipes or perforated plates to

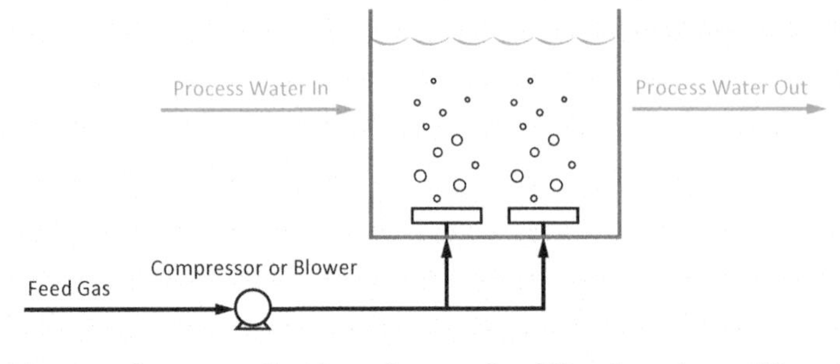

Fig. 3.2 Diagrams of two gas-to-liquid transfer examples: diffused aeration and Venturi injectors/aspirators

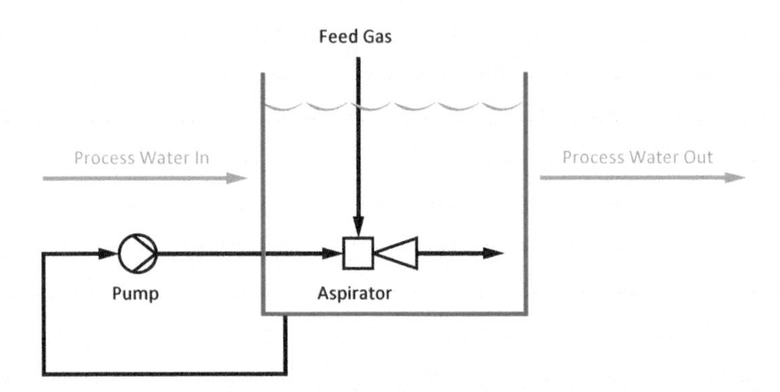

Fig. 3.3 Diagrams of two liquid-to-gas transfer examples: the packed column aerator and surface splashers in an enclosed tank. The packed column aerator allows water to trickle down an enclosed vessel, usually packed with structured media, where air is forced through using a fan or blower. Surface splashers found in pond aquaculture can also be used in enclosed atmospheres enriched with gases – normally oxygen – for gas transfer

create bubbles using Venturi injectors which create masses of small bubbles or devices which trap gas bubbles in the water stream such as the Speece Cone and the U-tube oxygenator.

Liquid-to-gas aerators are based on diffusing the water into small droplets to increase the surface area available for contact with the air, or creating an atmosphere enriched with a mixture of gases (Fig. 3.3). The packed column aerator (Colt and Bouck 1984) and the low-head oxygenators (LHOs) (Wagner et al. 1995) are examples of liquid-to-gas systems used in recirculating aquaculture. However, other liquid-to-gas systems popular in ponds and outdoor farms such as paddlewheel aerators (Fast et al. 1999) are also used in RAS.

Considerable literature is available on gas exchange theory and the fundamentals of gas transfer in water, and the reader is encouraged not only to consult aquaculture and aquaculture engineering texts, but also to refer to process engineering and wastewater treatment materials for a better understanding of these processes.

3.2.2 Ammonia

In an aqueous medium, ammonia exists in two forms: a non-ionized form (NH_3) that is toxic to fish and an ionized form (NH_4^+) that has low toxicity to fish. These two form the total ammonia nitrogen (TAN), wherein the ratio between the two forms is controlled by pH, temperature and salinity. Ammonia accumulates in the rearing water as a product of the protein metabolism of the fish (Altinok and Grizzle 2004) and can achieve toxic concentrations if left untreated. Of the 35 different types of freshwater fish that have been studied, the average acute toxicity value for ammonia is 2.79 mg NH3/l (Randall and Tsui 2002).

Ammonia has been traditionally treated in recirculation systems with nitrifying biofilters, devices that are designed to promote microbial communities that can oxidize ammonia into nitrate (NO_3). Although the use of nitrifying biofilters is not new, contemporary RAS has seen a streamlining of biofilter designs, with just a few, well-studied designs having widespread acceptance. Other highly innovative techniques to treat ammonia have been developed over the past few years, but are not widely applied commercially (examples noted below).

Ammonia is oxidized in biofilters by communities of nitrifying bacteria. Nitrifying bacteria are chemolithotrophic organisms that include species of the genera *Nitrosomonas*, *Nitrosococcus*, *Nitrospira*, *Nitrobacter* and *Nitrococcus* (Prosser 1989). These bacteria obtain their energy from the oxidation of inorganic nitrogen compounds (Mancinelli 1996) and grow slowly (replication occurs 40 times slower than for heterotrophic bacteria) so are easily outcompeted by heterotrophic bacteria if organic carbon, mostly present in biosolids suspended in the culture water, are allowed to accumulate (Grady and Lim 1980). During RAS operation, good system management greatly relies on minimizing suspended solids through adequate solids removal techniques (Fig. 3.4).

Nitrifying biofilters or biofilter reactors have been roughly classified into two main categories: suspended growth and attached growth systems (Malone and Pfeiffer 2006). In suspended growth systems, the nitrifying bacterial communities grow freely in the water, forming bacterial flocs which also harbour rich ecosystems where protozoa, ciliates, nematodes and algae are present (Manan et al. 2017). With appropriate mixing and aeration, algae, bacteria, zooplankton, feed particles and faecal matter remain suspended in the water column and naturally flocculate together, forming the particles that give biofloc culture systems their name (Browdy et al. 2012). The main disadvantage of suspended growth systems is their tendency to lose their bacterial biomass as process water flows out of the reactor, thus requiring a means to capture and return it to the system. In attached growth systems, solid forms (sand grains, stones, plastic elements) are used as substrates to retain the bacteria inside the reactor and thus, do not need a post-treatment solids capture step. Generally, attached growth systems provide more surface area for bacterial attachment than suspended growth systems, and do not produce significant solids in their outflow, which is one of the main reasons why attached growth biofilters have been so commonly used in RAS.

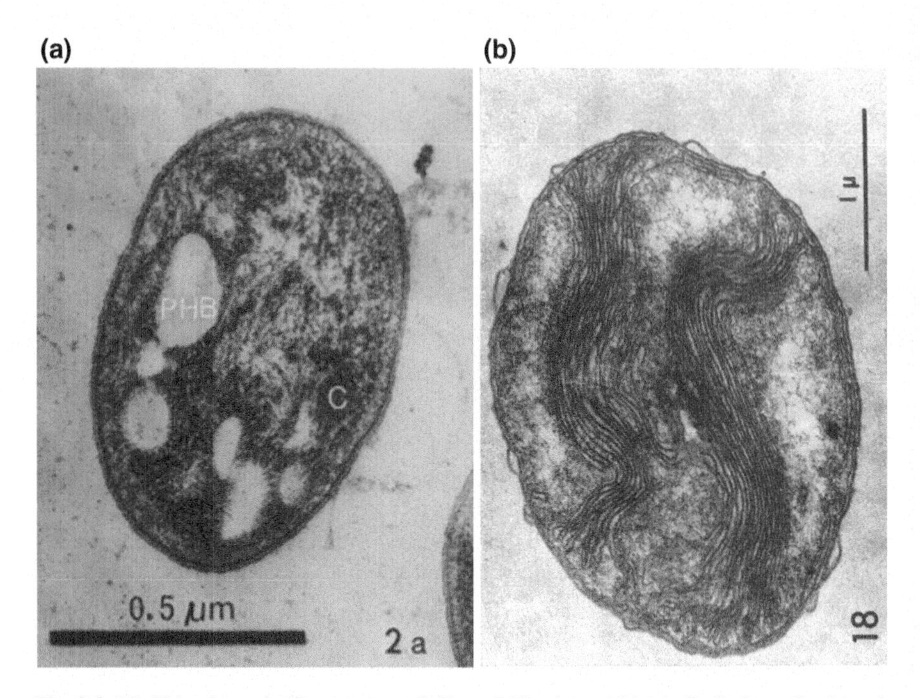

Fig. 3.4 Nitrifying bacteria *Nitrosomonas* (left), and *Nitrobacter* (right). (Left photo: Bock et al. 1983. Right photo: Murray and Watson 1965)

Efforts have been made to classify biofilters and to document their performance in order to help farmers and designers specify systems with a better degree of reliability (Drennan et al. 2006; Gutierrez-Wing and Malone 2006). In recent years, the aquaculture industry has opted for biofilter designs which have been widely studied and thus can offer predictable performance. The moving bed bioreactor (Rusten et al. 2006), the fluidized sand filter bioreactor (Summerfelt 2006) and the fixed-bed bioreactor (Emparanza 2009; Zhu and Chen 2002) are examples of attached growth biofilter designs which have become standard in modern commercial RAS. Trickling filters (Díaz et al. 2012), another popular design, have seen their popularity reduced due to their relatively high pumping requirements and relatively large sizes.

3.2.3 Biosolids

Biosolids in RAS originate from fish feed, faeces and biofilms (Timmons and Ebeling 2010) and are one of the most critical and difficult water quality parameters to control. As biosolids serve as a substrate for heterotrophic bacterial growth, an increase in their concentration may eventually result in increased oxygen consumption, poor biofilter performance (Michaud et al. 2006), increased water turbidity and even mechanical blockage of parts of the system (Becke et al. 2016; Chen et al. 1994; Couturier et al. 2009).

In RAS, biosolids are generally classified both by their size and their removal capacity by certain techniques. Of the total fraction of solids produced in a RAS, settleable solids are those generally bigger than 100 μm and that can be removed by gravity separation. Suspended solids, with sizes ranging from 100 μm to 30 μm, are those which do not settle out of suspension, but that can be removed by mechanical (i.e. sieving) means. Fine solids, with sizes of less than 30 μm, are generally those that cannot be removed by sieving, and must be controlled by other means such as physico-chemical processes, membrane filtration processes, dilution or bioclarification (Chen et al. 1994; Lee 2014; Summerfelt and Hochheimer 1997; Timmons and Ebeling 2010; Wold et al. 2014). The techniques for controlling settleable and suspended solids are well known and developed, and an extensive literature exists on the subject. For example, the use of dual-drain tanks, swirl separators, radial flow separators and settling basins is a popular means to control settleable solids (Couturier et al. 2009; Davidson and Summerfelt 2004; De Carvalho et al. 2013; Ebeling et al. 2006; Veerapen et al. 2005). Microscreen filters are the most popular method for suspended solids control (Dolan et al. 2013; Fernandes et al. 2015) and are often used in the industry to control both settleable and suspended solids with a single technique. Other popular solids capture devices are depth filters such as the bead filters (Cripps and Bergheim 2000) and rapid sand filters, which are also popular in swimming pool applications. Moreover, design guidelines to prevent the accumulation of solids in tanks, pipework, sumps and other system components are also available in the literature (Davidson and Summerfelt 2004; Lekang 2013; Wong and Piedrahita 2000). Lastly, fine solids in RAS are commonly treated by ozonation, bioclarification, foam fractionation or a combination of these techniques. The last few years in RAS development have focused on a greater understanding of how to control the fine solids fraction and to understand its effect on fish welfare and system performance.

3.2.4 Carbon Dioxide (CO_2)

In RAS, the control of dissolved gases does not stop with supplying oxygen to the fish. Other gases dissolved in the rearing water may affect fish welfare if not controlled. High dissolved carbon dioxide (CO_2) concentrations in the water inhibit the diffusion of CO_2 from the blood of fish. In fish, increased CO_2 in blood reduces the blood's pH and in turn, the affinity of haemoglobin for oxygen (Noga 2010). High CO_2 concentrations have also been associated with nephrocalcinosis, systemic granulomas and chalky deposits in organs in salmonids (Noga 2010). CO_2 in RAS originates as a product of heterotrophic respiration by fish and bacteria. As a highly soluble gas, carbon dioxide does not reach atmospheric equilibrium as easily as oxygen or nitrogen and thus, it must be put in contact with high volumes of air with a low concentration of CO_2 to ensure transfer out of water (Summerfelt 2003). As a general rule, RAS which are supplied with pure oxygen will require some form of carbon dioxide stripping, while RAS which are supplied with aeration for oxygen supplementation will not require active CO_2 stripping (Eshchar et al. 2003; Loyless and Malone 1998).

In theory, any gas transfer/aeration device open to the atmosphere will offer some form of CO_2 stripping. However, specialized carbon dioxide stripping devices require that large volumes of air are put in contact with the process water. CO_2 stripper designs have mostly focused on cascade-type devices such as cascade aerators, trickling biofilters and, more importantly, the packed column aerator (Colt and Bouck 1984; Moran 2010; Summerfelt 2003), which has become a standard piece of equipment in commercial RAS operating with pure oxygen. Although the development of packed column aeration technology has advanced over past years, most of the research done on this device has been focused on understanding its performance under different conditions (i.e. freshwater vs seawater) and design variations such as heights, packing types and ventilation rates. The effect of the hydraulic loading rate (unit flow per unit area of degasser) is known to have an effect on the efficiency of a degasser, but further research is needed to have a better understanding of this design parameter.

3.2.5 Total Gas Pressure (TGP)

Total gas pressure (TGP) is defined as the sum of the partial pressures of all the gases dissolved in an aqueous solution. The less soluble a gas is, the more 'room' it occupies in the aqueous solution and thus, the more pressure it exerts in it. Of the main atmospheric gases (nitrogen, oxygen and carbon dioxide) nitrogen is the least soluble (e.g. 2.3 times less soluble than oxygen and more than 90 times less soluble than carbon dioxide). Thus, nitrogen contributes to total gas pressure more than any other gas, but is not consumed by fish or heterotrophic bacteria, so it will accumulate in the water unless stripped. It is also important to note that oxygen will also contribute to high TGP if the gas transfer process does not allow excess gases to be displaced out of the solution. A classic example of this are ponds with photoautotrophic activity in them. Photoautotrophs (usually plant organisms that carry out photosynthesis) release oxygen into the water while a quiet water surface may not provide enough gas exchange for excess gas to escape to the atmosphere and thus, supersaturation may occur.

Fish require total gas pressures equal to atmospheric pressure. If fish breathe water with a high total gas pressure, excess gas (generally nitrogen) exits the bloodstream and forms bubbles, with often serious health effects for the fish (Noga 2010). In aquaculture this is known as gas bubble disease.

Avoiding high TGP requires careful examination of all areas in the RAS where gas transfer may occur. High-pressure oxygen injection without off-gassing (allowing excess nitrogen to be displaced out of the water) may also contribute to high TGP. In systems with fish which are very sensitive to TGP, the use of vacuum degassers is an option (Colt and Bouck 1984). However, maintaining a RAS free from areas of uncontrolled gas pressurization, using carbon dioxide strippers (which will also strip nitrogen) and dosing technical oxygen with care, is enough to keep TGP at safe levels in commercial RAS.

3.2.6 Nitrate

Nitrate (NO_3) is the end product of nitrification and commonly the last parameter to be controlled in RAS, due to its relatively low toxicity (Davidson et al. 2014; Schroeder et al. 2011; van Rijn 2013). This is mostly attributed to its low permeability at the fish gill membrane (Camargo and Alonso 2006). The toxic action of nitrate is similar to that of nitrite, affecting the capacity of oxygen-carrying molecules. The control of nitrate concentrations in RAS has traditionally been achieved by dilution, by effectively controlling the hydraulic retention time or daily exchange rate. However, the biological control of nitrate using denitrification reactors is a growing area of research and development in RAS.

Tolerance to nitrate may vary by aquatic species and life stage, with salinity having an ameliorating effect over its toxicity. It is important for RAS operators to understand the chronic effects of nitrate exposure rather than the acute effects, as acute concentrations will probably not be reached during normal RAS operation.

3.2.7 Alkalinity

Alkalinity is, in broad terms, defined as the pH buffering capacity of water (Timmons and Ebeling 2010). Alkalinity control in RAS is important as nitrification is an acid-forming process which destroys it. In addition, nitrifying bacteria require a constant supply of alkalinity. Low alkalinity in RAS will result in pH swings and nitrifying biofilter malfunction (Summerfelt et al. 2015; Colt 2006). Alkalinity addition in RAS will be determined by nitrification activity in the systems, which is in turn related to feed addition, by the alkalinity content of the make-up (daily exchange) water and by the presence of denitrifying activity, which restores alkalinity (van Rijn et al. 2006).

3.3 Developments in RAS

The last few years have seen an increase in the number and sizes of recirculating aquaculture farms, especially in Europe. With the increase in acceptance of the technology, improvements over traditional engineering approaches, innovations and new technical challenges keep emerging. The following section describes the key design and engineering trends and new challenges that recirculating aquaculture technology is facing.

3.3.1 Main Flow Oxygenation

The control of dissolved oxygen in modern RAS aims to increase the efficiency of oxygen transfer and decrease the energy requirements of this process. Increasing the oxygen transfer efficiency can be achieved by devising systems which retain oxygen gas in contact with water for longer, while a decrease in energy requirements may be achieved by the use of low-head oxygen transfer systems or using systems which do not use electricity at all, such as liquid oxygen systems connected to oxygen diffusers operating only by pressure. A defining factor of low-head oxygenators is the relatively low dissolved concentration that can be achieved compared to high-pressure systems. To overcome this limitation, low-head oxygenation devices are strategically placed to treat the full recirculating flow instead of using a smaller bypass of highly supersaturated water, thus ensuring sufficient mass transport of oxygen. Using oxygenation devices installed in the main recirculating flow generates savings in electricity consumption because the use of energy-intensive high-pressure systems that are necessary to achieve high DO concentrations in small flows is avoided. Low-head oxygenation systems may also reduce the amount of pumping systems needed, as high-pressure oxygenation systems are commonly placed on a bypass in the pipelines going to the fish tanks. In contrast, low-head oxygenation devices tend to be comparatively larger because of their need to handle larger flows and thus, their initial cost may be higher. Examples of devices that can treat the totality of the flow include the low-head oxygenator (LHO) (Wagner et al. 1995), operated by gravity as water is firstly pumped into a biofilter and a packed column (Summerfelt et al. 2004), low-head oxygen cones, variants of the Speece Cone (Ashley et al. 2008; Timmons and Losordo 1994) operated at low pressure, the deep shaft cones (Kruger Kaldnes, Norway), also a variant of the Speece cone designed to reach higher operating pressures by means of increased hydrostatic pressure resulting from placing the devices lower than the fish tanks and pump sumps, the U-tube oxygenator and its design variants such as the Farrell tube or the patented oxygen dissolver system (AquaMAOF, Israel) and the use of diffused oxygenation in deep fish tanks (Fig. 3.5).

3.3.2 Nitrifying Biofiltration Alternatives

Although nitrifying biofilters continue to be the main commercially accepted method of ammonia removal in commercial RAS, new nitrogen removal technologies have been developed over recent years. Some of these technologies consider alternative biological pathways to remove ammonia from the culture water, while others aim to

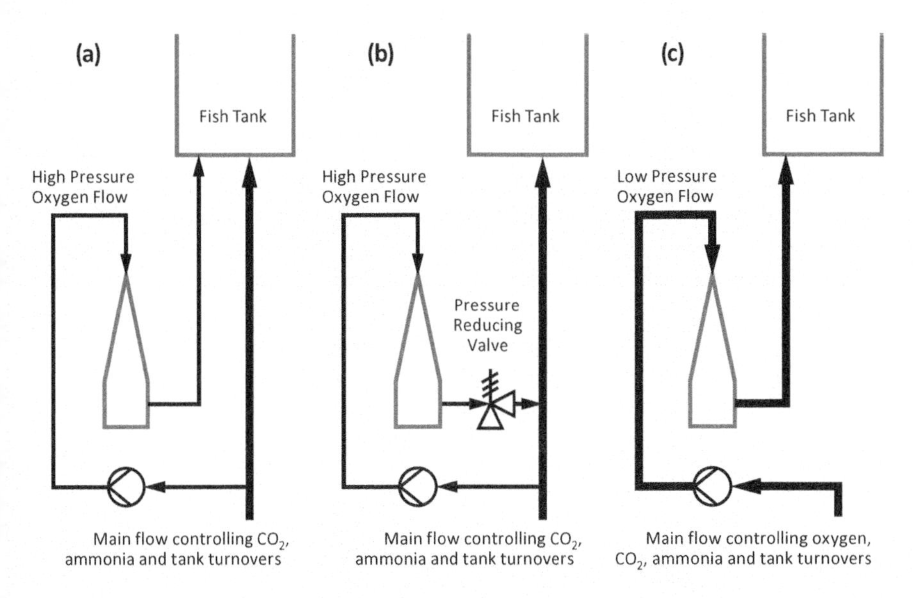

Fig. 3.5 Gas transfer alternatives for recirculating water returning into fish tanks. If the gas contacting vessel allows for pressurization, oxygen can be transferred in high concentrations in relatively small, high-pressure streams (**a**, **b**). However, oxygen at lower concentrations can be transferred into the main recirculation loop, but for this, the oxygen transfer device must be much larger to handle full flow of the system (**c**)

replace or work in parallel with nitrifying biofilters in order to reduce inherent limitations. These include large reactor sizes, susceptibility to crashing, long start-up times and poorer performance in both cold water and marine systems.

Anammox-based Processes

An alternative biological ammonia removal pathway considered for RAS is the anammox process (Tal et al. 2006), which occurs under anaerobic conditions. Anaerobic ammonia oxidation is a process which eliminates nitrogen by combining ammonia and nitrite to produce nitrogen gas (van Rijn et al. 2006). The anammox process is of interest to RAS because it allows for complete autotrophic nitrogen removal, in contrast to traditional combinations of nitrifying biofilters with heterotrophic denitrification systems requiring organic carbon addition (van Rijn et al. 2006). Moreover, in the anammox pathway, only half of the ammonia released by the fish is aerobically oxidized to nitrite (requiring oxygen), while the other half is anaerobically converted to nitrogen gas along with the nitrite produced. This may provide savings in oxygen and energy use in RAS (van Rijn et al. 2006).

Anammox reactor prototypes have been demonstrated successfully (Tal et al. 2006, 2009), while anammox activity has been suspected to occur in marine denitrification systems (Klas et al. 2006). The European FP7 project DEAMNRECIRC was also successful in creating anammox reactor prototypes for cold water and seawater aquaculture applications. However, commercial applications of the technology have as yet not been identified by the authors.

Chemical Removal of Ammonia

Ammonia removal systems based on ion exchange and electrochemical oxidation processes are being proposed as alternatives to nitrifying biofilters. Ion exchange processes rely on using adsorptive materials such as zeolites or ion-selective resins to extract dissolved ammonia from the water (Lekang 2013), while electrochemical oxidation processes convert ammonia to nitrogen gas through a number of complex oxidation reactions (Lahav et al. 2015). By comparison, ion exchange processes are suitable for waters with low concentrations of ions (i.e. freshwater), while electrochemical oxidation processes take advantage of the chloride ions present in the water to produce active chlorine species which readily react with ammonia (Lahav et al. 2015) and are thus suitable for waters with higher concentrations of chloride ions (i.e. brackish and marine waters).

Although ion exchange processes are not new, their application into RAS has been limited by their capacity to maintain performance over time: the filtering material eventually becomes 'saturated', losing its adsorptive capacity and, thus, must be regenerated. Gendel and Lahav (2013), proposed a novel approach to an ion exchange-based ammonia process in tandem with an innovative adsorbent regeneration process using electrochemical oxidation. Electrochemical oxidation of ammonia is a process which has received greater attention in recent years, and several concepts have been investigated and have been launched commercially, for example, EloxiRAS in Spain.

Factors limiting the application of these technologies into commercial RAS include, in the case of ion exchange processes, poor economic performance, difficulty to regenerate large amounts of adsorbent materials on demand (Lekang 2013), system complexity requiring the addition of chemical reagents, high electricity consumption and a high degree of suspended solids removal (Lahav et al. 2015), which is often impractical in large-scale RAS. In the case of ammonia electrooxidation processes, the production of toxic reactive species requiring active removal is their most important limitation, although their high solids control requirement, often possible only with pressurized mechanical filters, is also a challenge in RAS operating with large flows and low pressure.

3.3.3 Fine Solids Control

Fine solids are the dominant solids fraction in RAS with particles <30 μm forming more than 90% of the total suspended solids in the culture water. Recent investigations have found that more than 94% of the solids present in the culture water of a RAS are <20 μm in size or 'fine' (Fernandes et al. 2015). The accumulation of fine solids mainly occur as larger solids bypass the mechanical filters (which are not 100% efficient) and are eventually broken down by pumps, friction with surfaces and bacterial activity. Once solids sizes are reduced, traditional mechanical filtration techniques are rendered useless.

In recent years, the production, control, fish welfare effects and system performance effects of fine solids continue to be explored. The effects of fine solids on fish welfare were initially investigated through fisheries research (Chen et al. 1994). However, the direct effects of fine solids in RAS on fish welfare have not been thoroughly investigated until recently. Surprisingly, separate work on rainbow trout by Becke et al. (2016) and Fernandes et al. (2015) showed no negative welfare effects in systems with suspended solids concentrations of up to 30 mg/l in exposure trials lasting 4 and 6 weeks, respectively. Despite these findings, the indirect effects of fine solids accumulation in RAS are known (Pedersen et al. 2017) and are reported to be mostly linked to the proliferation of opportunistic microorganisms (Vadstein et al. 2004;Attramadal et al. 2014; Pedersen et al. 2017) since fine solids provide a high-surface area substrate for bacteria to colonize. Another important negative effect of fine solids accumulation is the increase in turbidity, which makes visual inspection of fish difficult and may hamper photoperiod control strategies which require light penetration in the water column to occur. Fine solids control strategies used in modern RAS include ozonation, protein skimming, floatation, cartridge filtration and membrane filtration (Couturier et al. 2009; Cripps and Bergheim 2000; Summerfelt and Hochheimer 1997; Wold et al. 2014). Protein skimmers, also known as foam fractionators, are also relatively popular fine solids control devices, especially in marine systems (Badiola et al. 2012).

3.3.4 *Ozonation*

Knowledge of ozone (O_3) application in RAS has existed since the 1970s and 1980s (Summerfelt and Hochheimer 1997). However, its application has not been as widespread as other processes such as nitrifying biofilters or mechanical filters (Badiola et al. 2012). Aside from fine solids treatment, ozone, as a powerful oxidizer, can be used in RAS to eliminate microorganisms, nitrite and humic substances (Gonçalves and Gagnon 2011). Recent years have seen an increase in knowledge about the potentials and limitations of ozone applied in both freshwater and marine RAS. Importantly, the ozone doses that can be safely achieved to improve water quality in both freshwater and seawater systems have been confirmed in several publications (Li et al. 2015; Park et al. 2013, 2015; Schroeder et al. 2011; Summerfelt 2003; Timmons and Ebeling 2010), with the conclusion that ozone doses over recommended limits (1) do not improve water quality further and (2) may cause negative welfare effects, especially in seawater systems where excessive ozonation will cause the formation of toxic residual oxidants. In coldwater RAS, ozonation requirements to achieve complete disinfection of the process flow have been determined (Summerfelt et al. 2009).

Ozonation improves microscreen filter performance and minimizes the accumulation of dissolved matter affecting the water colour (Summerfelt et al. 2009).

However, excessive ozonation may severely impact farmed fish by causing adverse effects including histopathologic tissue damage (Richardson et al. 1983; Reiser et al. 2010) and alterations in feeding behaviour (Reiser et al. 2010) as well as oxidative stress (Ritola et al. 2000, 2002; Livingstone 2003). Additionally, ozonation by-products may be harmful. Bromate is one of these and is potentially toxic. Tango and Gagnon (2003) showed that ozonated marine RAS have concentrations of bromate that are likely to impair fish health. Chronic, sublethal ozone-produced oxidants (OPO) toxicity was investigated in juvenile turbot by Reiser et al. (2011), while rainbow trout health and welfare were assessed in ozonated and non-ozonated RAS by Good et al. (2011). Raising rainbow trout to market size in ozonated RAS improved fish performance without significantly impacting their health and welfare while high OPO doses affect welfare of juvenile turbot.

3.3.5 Denitrification

In most recirculating aquaculture systems, nitrate, the end product of nitrification, tends to accumulate. Such accumulation is commonly controlled by dilution (introducing new water in the system). The control of nitrate by dilution may be a limiting factor to a RAS operation due to environmental regulations, poor availability of new water, the cost of treating the incoming and effluent water streams or the costs associated with chilling or heating the new water.

Biological nitrate removal in RAS can be achieved by facultative anaerobic bacteria using a dissimilatory pathway to convert nitrate to nitrogen gas in the presence of carbon and nitrate as electron donors (van Rijn et al. 2006). Denitrification reactors are thus biological reactors which are typically operated in anaerobic conditions and generally dosed with some type of carbon source such as ethanol, methanol, glucose, molasses, etc. Denitrification technology has been under development since the 1990s (van Rijn and Riviera 1990), but its popularity among the recirculating aquaculture industry has only increased over the past years, offering innovative denitrification reactor solutions.

One of the most notable applications of denitrification systems in aquaculture is the 'zero exchange' RAS (Yogev et al. 2016), which employ anaerobic digestion of biosolids produced in the system to produce volatile fatty acids (VFA) which are then used by denitrifiers as a carbon source. Klas at al. (2006) developed a 'single-sludge' denitrification system, where production of VFA from biosolids and denitrification occur in a single, mixed reactor. Suhr et al. (2014) developed the single-sludge concept further, adapting it for end-of-pipe treatment of fish farming effluents and adding an extra step which separates VFA production from the denitrification reactor in a hydrolysis tank. These works have provided valuable information on the possibilities of using aquacultural biosolids instead of expensive inorganic carbon sources for denitrification. Furthermore, Christianson et al. (2015) studied the effectiveness of autotrophic, sulphur-based denitrification reactors as an alternative to conventional heterotrophic denitrification reactors. Autotrophic reactors produce less biomass (solids) and can be supplied with sulphur particles, which are cheaper than conventional inorganic carbon sources.

VFAs are also the precursor component in the production of biopolymers such as Polyhydroalkanoates (PHAs), used to produce biodegradable plastics (Pittmann and Steinmetz 2013). This could hold potential for fish farms employing anaerobic activated sludge processes to be part of the 'biorefinery' concept applied to waste-water treatment plants.

3.3.6 Microbial Control

Microbial communities are important constituents of the aquatic ecosystem. In aquaculture production systems, they play significant roles in nutrient recycling, degradation of organic matter and treatment and control of disease (Zeng et al. 2017). Developing efficient, productive, biologically secure and disease-free RAS requires a thorough understanding of all life support processes from physical and chemical (gas transfer, thermal treatment, ozonation, UV irradiation, pH and salinity adjustments) to biological processes (nitrification, denitrification and aerobic heterotrophic activity). While physical and chemical processes can be controlled, biological filtration systems rely on the interaction of microbial communities with each other and their environment as a consequence of nutrient input (fish waste output) and, as such, are not as easily controlled (Schreier et al. 2010). Recent studies using molecular tools have not only allowed for evaluating microbial diversity in RAS but have also provided some insight into their activities that should lead to a better understanding of microbial community interactions. These approaches are certain to provide novel RAS process arrangements as well as insight into new processes and tools to enhance and monitor these systems (Schreier et al. 2010). Current understanding of RAS biofilter microbial diversity in both freshwater and marine systems is based on studies using 16S rRNA and functional gene-specific probes or 16S rRNA gene libraries rather than culture-based techniques (Table 3.1).

Insights into the temporal and spatial dynamics of microbiota in RAS are also still limited (Schreier et al. 2010), and potential solutions to maintain or restore beneficial microbial communities in RAS are lacking (Rurangwa and Verdegem 2015). Besides a microbial community that purifies the water, microbiota in RAS can also harbour pathogens or produce off-flavour-causing compounds (Guttman and van Rijn 2008). Given the difficulty to treat disease during operation without negatively affecting beneficial microbiota, microbial management in RAS is rather a necessity from the start-up through the whole production process. Microorganisms are introduced into RAS through different pathways: make-up water, air, animal vectors, feed, fish stocking, dirty equipment and via staff or visitors (Sharrer et al. 2005; Blancheton et al. 2013). Specific microbes can also, on the other hand, be applied intentionally to steer microbial colonization to improve system performance or animal health (Rurangwa and Verdegem 2015).

Table 3.1 Primary activities associated with RAS biofiltration units and participating microorganisms. (From Schreier et al. 2010)

Process	Reaction	Microorganism	
		Freshwater	Marine
Nitrification			
Ammonium oxidation	$NH_4^+ + 1.5O_2 \rightarrow NO_2^- + 2H^+ + H_2O$	*Nitrosomonas oligotropha*	*Nitrosomonas sp.*
			Nitrosomonas cryotolerans
			Nitrosomonas europaea
			Nitrosomonas cinnybus/nitrosa
			Nitrosococcus mobilis
Nitrite oxidation	$NO_2^- + H_2O \rightarrow NO_3^- + 2H^+ + 2e^-$	*Nitrospira spp.*	
		Nitrospira marina[a]	*Nitrospira marina*[a]
		Nitrospira moscoviensis[a]	*Nitrospira moscoviensis*[a]
Denitrification			
Autotrophic	$S_2^- + 1.6NO_3^- + 1.6H^+ \rightarrow$		*Thiomicrosporia denitrificans*
(sulfide-dependent)	$SO_4^{2-} + 0.8N_2(g) + 0.8H_2O$		*Thiothrix disciformis*[a]
			Rhodobacter litoralis[a]
			Hydrogenophaga sp.
Heterotrophic	$5CH_3COO^- + 8NO_3^- + 3H^+ \rightarrow$		*Pseudomonas fluorescens*
	$10HCO_3^- + 4N_2(g) + 4H_2O$	*Pseudomonas sp.*	*Pseudomonas stutzeri*
		Comamonas sp.	*Pseudomonas sp.*
			Paracoccus denitrificans
Dissimilatory nitrate	$NO_3^- + 2H^+ + 4H_2 \rightarrow NH_4^+ + 3H_2O$		Various *Proteobacteria* and *Firmicutes*
Reduction to ammonia (DNRA)			
Anaerobic ammonium	$NH_4^+ + NO_2^- \rightarrow N_2(g) + 2H_2O$		*Planctomycetes spp.*
oxidation (Anammox)			*Brocadia sp.*[a]
Sulfate reduction	$SO_4^{2-} + CH_3COO^- + 3H^+ \rightarrow$		*Desulfovibrio sp.,*
	$HS^- + 2HCO_3^- + 3H^+$		*Dethiosulfovibrio sp.,*
			Fusibacter sp., *Bacteroides sp.*
Sulfide oxidation	$HS^- + 2O_2 \rightarrow SO_4^{2-} + H^+$		*Thiomicrospira sp.*

(continued)

Table 3.1 (continued)

Process	Reaction	Microorganism	
		Freshwater	Marine
Methanogenesis	$4H_2 + H^+ + HCO_3^- \rightarrow CH_4(g) + 3H_2O$		Methanogenic Archaea [Mirzoyan and Gross, unpublished]

[a]Microorganisms identified solely on the basis of partial 16S rRNA gene or functional gene sequences

One of the approaches for inhibiting pathogen colonization is the use of probiotic bacteria that may compete for nutrients, produce growth inhibitors, or, quench cell-to-cell communication (quorum sensing) that allows for settling within biofilms (Defoirdt et al. 2007, 2008; Kesarcodi-Watson et al. 2008). Probiotic bacteria include *Bacillus*, *Pseudomonas* (Kesarcodi-Watson et al. 2008) and *Roseobacter* spp. (Bruhn et al. 2005), and bacteria related to these have also been identified in RAS biofilters (Schreier et al. 2010) (Table 3.1). To obtain the information needed to manage microbial stability in RAS, Rojas-Tirado et al. (2017) have identified the factors affecting changes in the bacterial dynamics in terms of their abundance and activity. Their studies show that bacterial activity was not a straightforward predictable parameter in the water phase as nitrate-N levels in identical RAS showed unexpected sudden changes/fluctuations within one of the systems. Suspended particles in RAS provide surface area that can be colonized by bacteria. More particles accumulate as the intensity of recirculation increases, thus potentially increasing the bacterial carrying capacity of the systems. Pedersen et al. (2017) explored the relationship between total particle surface area (TSA) and bacterial activity in freshwater RAS. They indicated a strong, positive, linear correlation between TSA and bacterial activity in all systems with low to moderate recirculation intensity. However, the relationship apparently ceased to exist in the systems with the highest recirculation intensity. This is likely due to the accumulation of dissolved nutrients sustaining free-living bacterial populations, and/or accumulation of suspended colloids and fine particles less than 5 μm in diameter, which were not characterized in their study but may provide significant surface area.

In RAS, various chemical compounds (mainly nitrates and organic carbon) accumulate in the rearing water. These chemical substrata regulate the ecophysiology of the bacterial communities on the biofilter and have an impact on its nitrification efficiency and reliability. Michaud et al. (2014) investigated the shift of the bacterial community structure and major taxa relative abundance in two different biological filters and concluded that the dynamics and flexibility of the bacterial community to adapt to influent water changes seemed to be linked with the biofilter performance. One of the key aspects for improving the reliability and sustainability of RAS is the appropriate management of the biofilter bacterial populations, which is directly linked to the C (carbon) availability (Avnimelech 1999). It should be noted that RAS have properties that may actually contribute to microbial stabilization, including long water retention time and a large surface area of biofilters for bacterial growth, which could potentially limit the chances of proliferation of opportunistic microbes in the rearing water (Attramadal et al. 2012a).

Attramadal et al. (2012a) compared the development of the microbial community in a RAS with moderate ozonation (to 350 mV) to that of a conventional flow-through system (FTS) for the same group of Atlantic cod, *Gadus morhua*. They found less variability in bacterial composition between replicate fish tanks of the RAS than between tanks of the FTS. The RAS had a more even microbial community structure with higher species diversity and periodically a lower fraction of opportunists. The fish in RAS performed better than their control in the FTS, despite being exposed to an apparent inferior physico-chemical water quality. While researching the effects of moderate ozonation or high-intensity UV irradiation on the microbial environment in RAS for marine fish larvae, Attramadal et al. (2012b) emphasized that a RAS for such larvae should probably not include strong disinfection because it leads to a reduction in bacterial numbers, which is likely to result in a destabilization of the microbial community. Furthermore, their results support the hypothesis of RAS as a microbial control strategy during the first feeding of fish larvae.

RAS and microbial maturation as tools for K-selection of microbial communities was the subject of the study by Attramadal et al. (2014) in which they hypothesized that fish larvae that are reared in water dominated by K-strategists (mature microbial communities) will perform better, because they are less likely to encounter opportunistic (R-selected) microbes and develop detrimental host–microbe interactions. The results of their experiment showed a high potential for increasing fish survival by using K-selection of bacteria, which is a cheap and easy method that can be used in all kinds of new or existing aquaculture systems. Small changes in the management (organic load and maturation of water) of water treatment give significantly different microbiota in fish tanks (Attramadal et al. 2016). On the other hand, humic substances (HS) are natural organic compounds, comprising a wide array of pigmented polymers of high organic weight. They are end products in the degradation of complex organic compounds and, when abundant, produce a typical brown to dark-brownish colour of the soil and water (Stevenson 1994). In a zero-discharge aquaculture system, HS-like substances were detected in the culture water as well as in the fish blood (Yamin et al. 2017a). A protective effect of HS was reported in fish exposed to toxic metal (Peuranen et al. 1994; Hammock et al. 2003) and toxic ammonia and nitrite concentrations (Meinelt et al. 2010). Furthermore, evidence was provided for their fungistatic effect against the fish pathogen, *Saprolengia parasitica* (Meinelt et al. 2007). In common carp (*Cyprinus carpio*) exposed to (a) humic-rich water and sludge from a recirculating system, (b) a synthetic humic acid and (c) a Leonardite-derived humic-rich extract, infection rates were reduced to 14.9%, 17.0% and 18.8%, respectively, as compared to a 46.8% infection rate in the control treatment (Yamin et al. 2017b). Likewise, the exposure of guppy fish (*Poecilia reticulata*), infected with the monogenea *Gyrodactylus turnbulli* and *Dactylogyrus* sp. to humic-rich culture water and feed, reduced both the infection prevalence (% of infected fish) and the infection intensity (parasites per fish) of the two parasites (Yamin et al. 2017c).

It is believed that the fundamental research in the area of microbial ecology of the nitrification/denitrification reactor systems in RAS may provide innovations which may alter and/or improve the reactor performance in RAS drastically. Up until now, the microbial community in reactors is still difficult to control (Leonard et al. 2000, 2002; Michaud et al. 2006, 2009; Schreier et al. 2010; Rojas-Tirado et al. 2017) and many of the inefficiencies of the system originate from this (Martins et al. 2010b).

3.3.7 Energy Efficiency

Economic viability of fish production in a recirculating aquaculture system depends, in part, on minimizing the energy requirements of operating such facilities. RAS require a higher technical infrastructure than open systems, thus energy costs in RAS have already been rated as major constraints which may prevent this technology from widespread application (Singh and Marsh 1996). Of all the costs associated with the electricity use in RAS, ventilation and water cooling are generally the most important. In indoor RAS, building ventilation is important to control humidity and carbon dioxide levels. Poor humidity control may result in a rapid deterioration of building structures, while atmospheric carbon dioxide accumulation will affect carbon dioxide stripping processes operating in the RAS and cause dizziness in workers. In order to keep an acceptable atmosphere inside the facilities, ventilation or air conditioning plants are widely in operation (Gehlert et al. 2018). These ventilation systems may be fitted with measures to reduce energy use. Furthermore, in order to develop an environmentally sustainable RAS, energy may be assumed as a key driving parameter, and in particular, energy can be considered an important indicator. Energetic performance analysis of the RAS has been performed by Kucuk et al. (2010) to contribute to the energy management in the RAS. In order to improve the energetic performance of the RAS, they recommended that operating conditions of the components, particularly, the pumps should be optimized and improved based on the fish production capacity of the system.

To increase the efficiency, RAS managers need guidelines and tools to optimize production. Energy audits can provide real data that can be used for decision-making. Badiola et al. (2014) investigated the total energy consumption (kWh) of a RAS cod system continuously for 14 months and identified the heat pump as a top energy consumer of rearing fish requiring high water thermal treatment. Gehlert et al. (2018) concluded that ventilation units offered a significant potential for energy savings in the RAS. Most of the time, when climate parameters in the facility stay within a desired range, air flow rates can be kept at low levels for saving energy. Additionally, energy saving measures in the RAS may include: software with energy performance data, alternative energy sources to heat the water and the use of frequency converters (Badiola et al. 2014).

3.4 Animal Welfare Issues

3.4.1 Introduction

During the last decade, fish welfare has attracted a lot of attention, and this has led to the aquaculture industry incorporating a number of husbandry practices and technologies specifically developed to improve this aspect. The neocortex, which in humans is an important part of the neural mechanism that generates the subjective experience of suffering, is lacking in fish and non-mammalian animals, and it has been argued that its absence in fish indicates that fish cannot suffer. A strong alternative view, however, is that complex animals with sophisticated behaviours, such as fish, probably have the capacity for suffering, though this may be different in degree and kind from the human experience of this state (Huntingford et al. 2006).

The UK government's Farm and Animal Welfare Committee (FAWC) has based their guidelines on the 'Five Freedoms' framework, which defines ideal states rather than specific levels of acceptable welfare (FAWC 2014). Freedom from hunger and thirst, discomfort, pain, injury, disease, fear and distress, as well as the freedom to express normal behaviour, provides us with a defined framework with which to assess welfare issues. Physical health is the most universally accepted measure of welfare and is undoubtedly a necessary requirement for good welfare. In a competitive, expanding and emerging industry, aquaculturists who incorporate welfare considerations into their daily husbandry practices can gain a competitive advantage and added price premium (Olesen et al. 2010) through improved consumer perception and confidence in their products. Grimsrud et al. (2013) provided evidence that there is a high willingness to pay, among all Norwegian households, to improve the welfare of farmed Atlantic salmon through increased resistance to diseases and salmon lice, which may imply less use of medicines and chemicals in the production process.

In intensive RAS, animal welfare is tightly connected to the performance of the systems. Over the past few years, animal welfare in the RAS has been mostly studied from the perspective of water quality and fish crowding effects on growth performance, stress bioindicators or the development of health disorders. The main goal of animal welfare research in the RAS has been to build and operate systems that maximize productivity and minimize stress and mortalities. Topics of interest have been stocking density limits (Calabrese et al. 2017), concentration limits of nitrogenous compounds in the rearing water (Davidson et al. 2014), concentration limits for dissolved carbon dioxide (Good et al. 2018), the effects of ozonation (Good et al. 2011; Reiser et al. 2011) and to a lesser extent, the accumulation of recalcitrant compounds in the RAS (van Rijn and Nussinovitch 1997) with limited water exchanges and noise (Martins et al. 2012; Davidson et al. 2017).

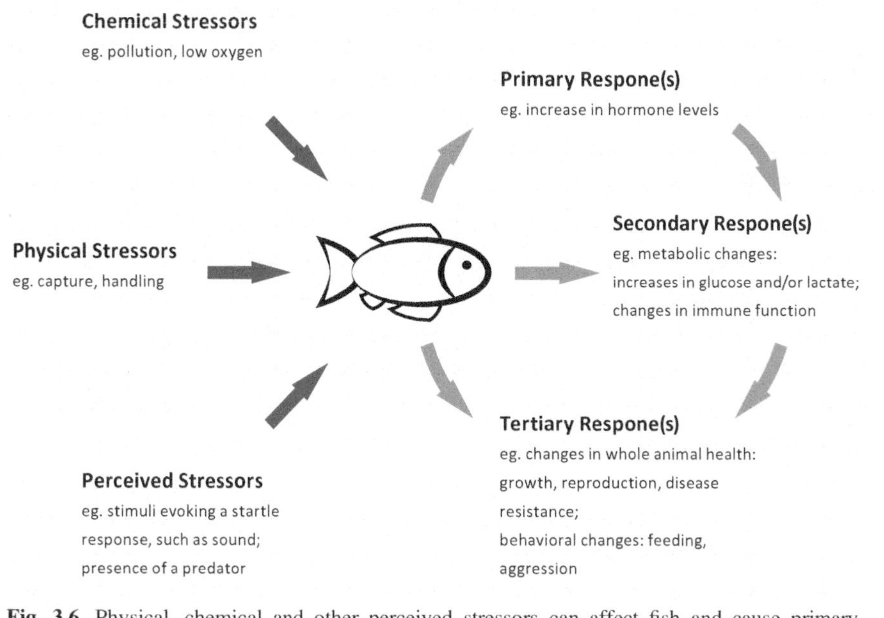

Fig. 3.6 Physical, chemical and other perceived stressors can affect fish and cause primary, secondary and/or whole-body responses. (After Barton 2002)

3.4.2 Stress

The stress response in fish is an adaptive function in the face of a perceived threat to homeostasis and stress physiology does not necessarily equate to suffering and diminished welfare (Ashley 2007) (Fig. 3.6). Stress responses serve a very important function to preserve the individual. Welfare measures in aquaculture are, therefore, largely associated with the tertiary effects of stress response that are generally indicative of prolonged, repeated or unavoidable stress (Conte 2004).

Stocking density is a pivotal factor affecting fish welfare in the aquaculture industry, especially RAS where high densities in confined environments are aimed at high productivity. Although rarely defined, stocking density is the term normally used to refer to the weight of fish per unit volume or per unit volume in unit time of water flow through the holding environment (Ellis et al. 2001). The concept of minimum space for a fish is more complex than for terrestrial species as fish utilize a three-dimensional medium (Conte 2004).

Beyond providing for the physiological needs, the FAWC (2014) recommends that fish 'need sufficient space to show most normal behaviour with minimal pain, stress and fear'. Stocking density is, therefore, an area that illustrates both the significance of species differences and the existence of a complex web of interacting factors that affect fish welfare. Calabrese et al. (2017) have researched stocking density limits for post-smolt Atlantic salmon (*Salmosalar* L.) with emphasis on production performance and welfare wherein fin damage and cataracts were observed in stocking densities of 100 kg m^{-3} and above. However, the effect of stocking density on measures of welfare varies between species. For instance, sea bass (*Dicentrarchus labrax*) showed higher stress levels at high densities, as indicated by cortisol, innate immune response and expression of stress-related genes (Vazzana et al. 2002; Gornati et al. 2004). High stocking densities in juvenile gilthead sea bream (*S. aurata*) also produce a chronic stress situation, reflected by high cortisol levels, immunosuppression and altered metabolism (Montero et al. 1999). In contrast, Arctic charr (*Salvelinus alpinus*) feed and grow well when stocked at high densities while showing a depressed food intake and growth rates at low densities (Jorgensen et al. 1993).

Diet may also play an important role in stress sensitivity. African catfish (*Clarias gariepinus*) receiving a diet with a high supplementation of ascorbic acid (vitamin C) during early development showed a lower stress sensitivity (Merchie et al. 1997). On the other hand, common carp (*Cyprinus carpio*), fed large doses of vitamin C, showed a more pronounced cortisol (a steroid hormone released with stress) increase in response to stress when compared to fish fed recommended levels of the vitamin (Dabrowska et al. 1991). Tort et al. (2004) have shown that a modified diet providing a supplementary dosage of vitamins and trace minerals to assist the immune system may help to co-reduce some of the effects of winter disease syndrome. Other common aquaculture diseases regarding animal welfare and stress are reviewed in Ashley (2007).

3.4.3 Accumulation of Substances in the Process Water

Intensive and 'zero-discharge' RAS offer significant environmental advantages. However, culturing of fish in continuously recycled water raises the question of whether substances released by the fish into the water may accumulate, resulting in decreased growth rates and impaired welfare. The existence of growth retardation in Nile tilapia (*Oreochromis niloticus*) by comparing the growth, feeding behaviour and stress response of fish cultured in the RAS with different levels of substances accumulated (TAN, NO_2-N and NO_3-N, orthophosphate-P) was investigated by Martins et al. (2010a). Results showed that the large individuals had a trend towards growth retardation in the highest accumulation RAS while small individuals, on the contrary, seem to grow better in such systems based on high levels of blood glucose as a stress indicator. A similar study done by the same author on carp embryos and larvae (Martins et al. 2011) found results that suggest that the concentration of substances (orthophosphate-P, nitrate, arsenic and copper) were likely to affect the development. Despite these findings, the authors claim that overall, the percentage of mortalities and deformities recorded in the study were relatively low compared to other studies. In both studies, the authors used systems with very limited water exchange rates with the aid of denitrification reactors (30 litres of new water per Kg of feed per day). Similarly the accumulation of hormones in coldwater salmonid RAS has been studied by Good et al. 2014, 2017). Their research in 2014 found neither a relationship between water exchange rate and hormone accumulation (except for testosterone) nor a link between hormone accumulation and precocious maturation in Atlantic salmon, but further study was suggested. Their study in 2017 focused on the use of ozonation for the reduction of hormones in the same RAS, with inconclusive results regarding the steroid hormone accumulation, but with a positive reduction of oestradiol by ozone.

On the other hand, the accumulation of humic substances in 'zero exchange' RAS has shown to have a protective effect against bacterial infections (Yamin et al. 2017a) and ectoparasites (Yamin et al. 2017b). Humic acids have also been shown to reduce ammonia and nitrite toxicity (Meinelt et al. 2010). This has implications for RAS operated with ozonation, as ozone may improve water quality while sacrificing the apparently beneficial effects of humic substances.

3.4.4 Health and Behaviour

The fundamental characteristics of good welfare are good health and absence of disease and, with respect to aquaculture, good productivity (Turnbull and Kadri 2007; Volpato et al. 2007). While the physical health of an animal is fundamental for good welfare (Ashley 2007; Duncan 2005), the fact that an animal is healthy does not necessarily mean that its welfare status is adequate. Thus, welfare is a broader and more overarching concept than the concept of health. Physiological and behavioural measures are intrinsically linked and are dependent on one another for a correct interpretation with regard to welfare (Dawkins 1998).

The behaviour of animals and in our case fish, represents a reaction to the environment as fish perceive it and behaviour is therefore a key element of fish welfare. Changes in foraging behaviour, gill ventilation activity, aggression, individual and group swimming behaviour, stereotypic and abnormal behaviours have been linked with acute and chronic stressors in aquaculture and can therefore be regarded as likely indicators of poor welfare (Martins et al. 2011). Behavioural welfare indicators have the advantage of being fast and easy to observe and therefore are good candidates for use 'on-farm'. Examples of behaviour that are commonly used as an indicators of welfare are changes in food-anticipatory behaviour, feed intake, swimming activity and ventilation rates (Huntingford et al. 2006). However, Barreto and Volpato (2004) caution the use of ventilation frequency as an indicator of stress in fish because although ventilation frequency is a very sensitivity response to disturbance, it is of limited use because it does not reflect the severity of the stimulus.

3.4.5 *Noise*

Farmed fish are cultured over long periods of time in the same tanks of the same colour(s) and the same shape and exposed to the same, potentially harmful, background noises (Martins et al. 2012). Intensive aquaculture systems and particularly recirculating systems utilize equipment such as aerators, air and water pumps, blowers and filtration systems that inadvertently increase noise levels in fish culture tanks. Sound levels and frequencies measured within intensive aquaculture systems are within the range of fish hearing, but species-specific effects of aquaculture production noise are not well defined (Davidson et al. 2009).

Bart et al. (2001) found that mean broadband sound pressure levels (SPL) differed across various intensive aquaculture systems. In his study, a sound level of 135 dB re 1μPa was measured in an earthen pond near an operating aerator, whereas large fiberglass tanks (14 m diameter) within a recirculating system had the highest SPLs of 153 dB re 1μPa.

Field and laboratory studies have shown that fish behaviour and physiology can be negatively impacted by intense sound. Terhune et al. (1990) have observed decreased growth and smoltification rates of Atlantic salmon, *Salmo salar*, in fiberglass tanks that had underwater sound levels 2–10 dB re 1μPa higher at 100–500 Hz than in concrete tanks. Therefore, chronic exposure to aquaculture production noise could cause increased stress, reduced growth rates and feed conversion efficiency and decreased survival. However, Davidson et al. (2009) found that after 5 months of noise exposure, no significant differences were identified between treatments for mean weight, length, specific growth rates, condition factor, feed conversion or survival of rainbow trout, *Oncorhynchus mykiss*. Similar findings are described by Wvysocki et al. (2007). However, these findings should not be generalized across all cultured fish species, because many species, including catfish and cyprinids, have much greater hearing sensitivity than rainbow trout and could be affected differently by noise. For instance, Papoutsoglou et al. (2008) provided initial evidence that music transmission under specific rearing conditions could have enhancing effects on *S. aurata* growth performance, at least at specific fish sizes. Moreover, the observed music effects on several aspects of fish physiology (e.g. digestive enzymes, fatty acid composition and brain neurotransmitters) imply that certain music could possibly provide even further enhancement in growth, quality, welfare and production.

3.5 Scalability Challenges in RAS

RAS are capital-intensive operations, requiring high funding expenditure on equipment, infrastructure, influent and effluent treatment systems, engineering, construction and management. Once the RAS farm is built, working capital is also needed until harvests and successful sales are achieved. Operational expenditures are also substantial and are mostly comprised of fixed costs such as rent, interest on loans, depreciation and variable costs such as fish feed, seed (fingerlings or eggs), labour, electricity, technical oxygen, pH buffers, electricity, sales/marketing, maintenance costs, etc.

When comparing productivity and economics, RAS will invariably compete with other forms of fish production and even other sources of protein production for human consumption. This competition is likely to exert a downwards pressure on the sale price of fish, which in turn must be high enough to make a RAS business profitable. As in other forms of aquaculture production, reaching higher economies of scale is generally a way to reduce the cost of production and thus obtain access to markets. Some examples of reduction in costs of production that can be achieved with larger facilities are:

1. Reduced transportation costs on bulk orders of feed, chemicals, oxygen.
2. Discounts on the purchase of larger quantities of equipment.
3. Access to industrial electricity rates.
4. Automation of farm processes such as process monitoring and control, feeding, harvesting, slaughter and processing.
5. Maximization of the use of labour: The same manpower was needed to take care of 10 tons of fish as was needed to take care of 100 tons of fish or more.

Following the increase of economies of scale in the net pen aquaculture sector, larger RAS are being developed on scales not considered a decade ago. The last decade has seen the construction of facilities with production capacities of thousands of tons per year, and this sheer size increase of RAS facilities is bringing new technical challenges, which are explored in the next section.

3.5.1 Hydrodynamics and Water Transport

Proper control of the hydrodynamic conditions in fish tanks is essential to ensure uniform water quality and adequate solids transport towards the tank outlets (Masaló 2008; Oca and Masalo 2012). Tanks which are not capable of flushing metabolites quickly enough will have less carrying capacity. Ensuring proper hydrodynamic performance in fish tanks is an important aquacultural engineering research topic which has helped the industry design and operate tanks of different shapes and sizes. However, increasing tank sizes used in commercial RAS are posing new engineering challenges to designers and operators. Recent investigations are underway to optimize the hydrodynamic characteristics in large octagonal tanks used for salmon smolt production (Gorle et al. 2018), by studying the effect of fish biomass, geometry and inlet and outlet structures in large tanks used in Norwegian smolt facilities. Similarly, Summerfelt et al. (2016) found a trend towards a decreasing feed loading rate per unit flow in modern tanks compared to tanks built more than a decade ago in Norwegian smolt facilities. A lowered feed burden effectively results in improved tank water quality as the recirculating water is treated at a faster rate, preventing the accumulation of metabolites and the depletion of oxygen in the tank even further compared to older tanks which operated at higher feed burdens. Future work will likely provide more information on the hydrodynamics of tanks with more than 1000 m^3 in volume. Other examples of enormous tanks which are currently being used are the tanks used in the RAS 2020 systems (Kruger, Denmark) or the Niri concept (Niri, Norway). Adoption of these new concepts using larger tanks will play a vital role in their profitability, as long as proper hydrodynamic conditions are achieved.

3.5.2 Stock Loss Risk

In RAS, the intensive rearing conditions can lead to sudden and catastrophic loss of fish if the system fails. Sources of system failure may include mechanical failure of pumping systems and RAS equipment, power outages, loss of oxygenation/aeration systems, hydrogen sulphide build-up and release, operational accidents and more. These risks and solutions to them need to be identified and incorporated into operational procedures.

The increasing size of RAS operations may also mean an increased risk of financial loss if catastrophic loss of fish occurs. On the other hand, risk-mitigation measures and system redundancy may also increase the cost of a RAS project and thus, designers and engineers must strike a balance between these elements.

Aside from industry and media reports, little academic research has been done on the risk of commercial RAS ventures. Badiola et al. (2012) surveyed RAS farms and analysed the main technical issues, finding that poor system design, water quality problems and mechanical problems were the main risk elements affecting the viability of RAS.

3.5.3 Economics

Debate on the economic viability of RAS focuses mostly on the high capital start-up costs of recirculating aquaculture farms and the long lead time before fish are ready to be marketed, as well as the perception that RAS farms have high operating costs. De Ionno et al. (2007) studied the commercial performance of RAS farms, concluding that economic viability increases with the scale of the operation. According to this study, farms smaller than 100 tons per year of production capacity are only marginally profitable in the Australian context where the study took place. Timmons and Ebeling (2010) also provide a case for achieving large economies of scale (in the order of magnitude of thousands of tons of production per year) which allow for the production cost reductions through vertical integration projects such as the inclusion of processing facilities, hatcheries or feed mills. Liu et al. (2016) studied the economic performance of a theoretical RAS farm with a capacity of 3300 tons per year, compared to a traditional net pen farm of the same capacity. At this scale the RAS operation reaches similar production costs compared to the net pen farm, but the higher capital investment doubles the payback period in comparison, even when the fish from the RAS farm are sold at a premium price. In the future, costly and strict licensing requiring good environmental performance may increase the viability of RAS as competitive option for the production of Atlantic salmon.

3.5.4 Fish Handling

On land-based farms, fish handling is often required for various reasons: to separate fish into weight classes, to reduce stocking densities, to transport fish across growing departments (i.e. from a nursery to an on-growing department) or to harvest fish when they are market ready. According to Lekang (2013), fish are handled most effectively with active methods such as fish pumps and also with passive methods such as the use of visual or chemical signals that allow the fish to move themselves from one place on the farm to the next.

Summerfeltet al. (2009) studied several means to crowd and harvest salmonids from large circular tanks using Cornell-type dual drains. Strategies included crowding fish with purse seines, clam-shell bar crowders and herding fish between tanks taking advantage to their innate avoidance response to carbon dioxide. Harvesting techniques included extracting the fish through the sidewall discharge port of a Cornell-type dual-drain tank or using an airlift to lift the crowded fish to a dewatering box. AquaMAOF (Israel) employs swimways and tanks sharing a common wall to passively transfer the fish through the farm, with harvesting taking place using a pescalator (Archimedes screw pump) at the end of a swimway. The RAS2020 concept from Kruger (Denmark) uses bar graders/crowders permanently installed in a donut-shaped or circular raceway tank to move and crowd the fish without the need for fish pumps.

Despite continuing developments on this topic, the increasing size of RAS farms will keep challenging designers and operators on how to handle fish safely, economically and without stress. The expanding range of designs, species under production and operation intensity of RAS farms may result in various and novel fish transport and harvest technologies.

3.6 RAS and Aquaponics

Aquaponic systems are a branch of recirculating aquaculture technology in which plant crops are included to either diversify the production of a business, to provide extra water filtration capacity, or a combination of the two.

As a branch of RAS, aquaponic systems are bound to the same physical, chemical and biological phenomena that occur in RAS. Therefore, the same fundamentals of water ecology, fluid mechanics, gas transfer, water depuration etc. apply in more or less equal terms to aquaponics with the exception of water quality control, as plants and fish may have specific and different requirements.

The fundamental economic realities of RAS and aquaponics are also related. Both technologies, are capital intensive and highly technical and are affected by economies of scale, appropriate design of the components, reliance on market conditions and the expertise of operators.

3.6.1 Welfare

In aquaponic systems, the uptake of nutrients should be maximized for the healthy production of plant biomass but without neglecting the best welfare conditions for the fish in terms of water quality (Yildiz et al. 2017). Measures to reduce the risks of the introduction or spread of diseases or infection and to increase biosecurity in aquaponics are also important. The possible impacts of allelochemicals, i.e. chemicals released by the plants, should be also taken into account. Moreover, the effect of diet digestibility, faeces particle size and settling ratio on water quality should be carefully considered. There is still a lack of knowledge regarding the relationship between the appropriate levels of minerals needed by plants, and fish metabolism, health and welfare (Yildiz et al. 2017) which requires further research.

3.6.2 Microbial Diversity and Control

As mentioned earlier in the chapter, aquaponics combines a recirculating aquaculture system with a hydroponic unit. One of its most important features is the reliance on bacteria and their metabolic products. Also, Sect. 3.2.6.discussed the importance of microbial communities and its control in RAS. Bacteria serve as the bridge that connects fish excrements, which are high in ammonium concentration, to plant fertilizer, which should be a combination of low ammonium and high nitrate (Somerville et al. 2014). As aquaponic systems can have different subunits, i.e. fish tanks, biofilter, drum filter, settler tanks and hydroponic units, each having different possible designs and different optimal conditions, the microbial communities in these components may differ considerably. This provides an interesting topic of research with the ultimate goal of improving system management processes. Schmautz et al. (2017) attempted to characterize the microbial community in different areas of aquaponic systems. They concluded that fish faeces contained a separate community dominated by bacteria of the genus *Cetobacterium*, whereas the samples from plant roots, biofilter and periphyton were more similar to each other, with more diverse bacterial communities. The biofilter samples contained large numbers of *Nitrospira* (3.9% of total community) that were found only in low numbers in the periphyton or the plant roots. On the other hand, only small percentages of *Nitrosomonadales* (0.64%) and *Nitrobacter* (0.11%) were found in the same samples. This second group of organisms are commonly tested for their presence in aquaponics systems as they are mainly held responsible for nitrification (Rurangwa and Verdegem 2015; Zou et al. 2016); *Nitrospira* has only recently been described as a total nitrifier (Daims et al. 2015), being able to directly convert ammonium to nitrate in the system. The dominance of *Nitrospira* is thus a novelty in such systems and might be correlated with a difference in the basic setup (Graber et al. 2014).

Schmautz et al. (2017) also emphasized that the increased presence of *Nitrospira* does not necessarily correlate to larger activity of these organisms in the system, as its metabolic activities were not measured. In addition, many species of bacteria and coliforms are inherently present in aquaponic recirculating biofilters carrying out transformations of organic matter and fish waste. This implies the presence of many microorganisms that can be pathogens for plants and fish, as well as for people. For this purpose, some microorganisms have been considered safety indicators for products and water quality in the system (Fox et al. 2012). Some of these safety indicators are *Escherichia coli* and *Salmonella* spp. Much needed research has thus recently been carried out in order to ascertain microbial safety of aquaponic products (Fox et al. 2012; Sirsat and Neal 2013). One future direction for the analysis of microbial activity in aquaponics has been identified by Munguia-Fragozo et al. (2015), who reviewed the *Omic* technologies for microbial community analysis. They concluded that metagenomics and metatranscriptomics analysis will be crucial in future studies of microbial diversity in aquaponic biosystems.

- From a period of technological consolidation to a new era of industrial implementation, RAS technology has considerably developed over the past two decades. The last few years have seen an increase in the number and scale of recirculating aquaculture farms. With the increase in acceptance of the technology, improvements over traditional engineering approaches, innovations and new technical challenges keep emerging.
- Aquaponics combines a recirculating aquaculture system with a hydroponic unit. RAS are complex aquatic production systems that involve a range of physical, chemical and biological interactions.
- Dissolved oxygen (DO) is generally the most important water quality parameter in intensive aquatic systems. However, addition of sufficient oxygen to the rearing water can be achieved relatively simply and thus, the control of other water parameters become more challenging.
- High dissolved carbon dioxide (CO_2) concentrations have a negative effect in fish growth. The removal of CO_2 from water to concentrations below 15 mg/L is challenging due to its high solubility and the limited efficiency of degassing equipment.
- Ammonia has been traditionally treated in recirculation systems with nitrifying biofilters. Some emerging technologies are being explored as alternatives to ammonia removal.

- Biosolids in RAS originate from fish feed, faeces and biofilms and are one of the most critical and difficult water quality parameters to control. A multi-step treatment system where solids of different sizes and removed via different mechanisms, is the most common approach.
- Ozone, as a powerful oxidizer, can be used in RAS to eliminate microorganisms, nitrite and humic substances. Ozonation improves microscreen filter performance and minimizes the accumulation of dissolved matter affecting the water colour.
- Denitrification reactors are biological reactors which are typically operated under anaerobic conditions and generally dosed with some type of carbon source such as ethanol, methanol, glucose and molasses. One of the most notable applications of denitrification systems in aquaculture is the 'zero exchange' RAS.
- In aquaculture production systems microbial communities play significant roles in nutrient recycling, degradation of organic matter and treatment and control of disease. The role of water disinfection in RAS is being challenged by the idea of using microbially mature water to control opportunistic pathogens.
- In intensive RAS, animal welfare is tightly connected to the performance of the systems. The main goal of animal welfare research in RAS has been to build and operate systems that maximize productivity and minimize stress and mortalities.

References

Altinok I, Grizzle JM (2004) Excretion of ammonia and urea by phylogenetically diverse fish species in low salinities. Aquaculture 238:499–507. https://doi.org/10.1016/j.aquaculture.2004.06.020

Ashley PJ (2007) Fish welfare: current issues in aquaculture. Appl Anim Behav Sci 104 (2–4):199–235. https://doi.org/10.1016/j.applanim.2006.09.001

Ashley KI, Mavinic DS, Hall KJ (2008) Oxygenation performance of a laboratory-scale Speece Cone hypolimnetic aerator: preliminary assessment. Can J Civ Eng 35:663–675. https://doi.org/10.1139/L08-011

Attramadal K, Salvesen I, Xue R, Øie G, Størseth TR, Vadstein O, Olsen Y (2012a) Recirculation as a possible microbial control strategy in the production of marine larvae. Aquac Eng 46:27–39. https://doi.org/10.1016/j.aquaeng.2011.10.003

Attramadal K, Øie G, Størseth TR, Morten OA, Vadstein O, Olsen Y (2012b) The effects of moderate ozonation or high intensity UV-irradiation on the microbial environment in RAS for marine larvae. Aquaculture 330–333:121–129. https://doi.org/10.1016/j.aquaculture.2011.11.042

Attramadal KJK, Truong TMH, Bakke I, Skjermo J, Olsen Y, Vadstein O (2014) RAS and microbial maturation as tools for K-selection of microbial communities improve survival in cod larvae. Aquaculture 432:483–490. https://doi.org/10.1016/j.aquaculture.2014.05.052

Attramadal K, Minniti G, Øie G, Kjørsvik E, Østensen M-A, Bakke I, Vadstein O (2016) Microbial maturation of intake water at different carrying capacities affects microbial control in rearing tanks for marine fish larvae. Aquaculture 457:68–72. https://doi.org/10.1016/j.aquaculture.2016.02.015

Avnimelech Y (1999) Carbon/nitrogen ratio as a control element in aquaculture systems. Aquaculture 176(3–4):227–235. https://doi.org/10.1016/S0044-8486(99)00085-X

Badiola M, Mendiola D, Bostock J (2012) Recirculating Aquaculture Systems (RAS) analysis: main issues on management and future challenges. Aquac Eng 51:26–35

Badiola M, Basurko OC, Gabiña G, Mendiola D (2014) Energy audits in Recirculating Aquaculture Systems (RAS): as a way forward to guarantee sustainability. AE2014 conference paper. https://doi.org/10.13140/2.1.1218.5604

Barreto RE, Volpato GL (2004) Caution for using ventilatory frequency as an indicator of stress in fish. Behav Process 66:43–51. https://doi.org/10.1016/j.beproc.2004.01.001

Bart AN, Clark J, Young J, Zohar Y (2001) Underwater ambient noise measurements in aquaculture systems: a survey. Aquac Eng 25:99–110. https://doi.org/10.1016/S0144-8609(01)00074-7

Barton B (2002) Stress in fishes: a diversity of responses with particular reference to changes in circulating corticosteroids. Integr Comp Biol 42:517–525

Becke C, Steinhagen D, Schumann M, Brinker A (2016) Physiological consequences for rainbow trout (*Oncorhynchus mykiss*) of short-term exposure to increased suspended solid load. Aquac Eng. https://doi.org/10.1016/j.aquaeng.2016.11.001

Blancheton JP, Attramadal KJK, Michaud L, Roque d'Orbcastel E, Vadstein O (2013) Insight into bacterial population in aquaculture systems and its implication. Aquac Eng 53:30–39. https://doi.org/10.1016/j.aquaeng.2012.11.009

Bock E, Sundermeyer-Klinger H, Stackerbrandt E (1983) New facultative lithoatotrophic nitrinte-oxidizing bacteria. Arch Microbiol 136:281–284. https://doi.org/10.1007/BF00425217

Bovendeur J, Eding EH, Henken AM (1987) Design and performance of a water recirculation system for high-density culture of the African catfish, *Clarias-gariepinus* (Burchell 1822). Aquaculture 63:329–353. https://doi.org/10.1016/0044-8486(87)90083-4

Browdy CL, Ray AJ, Leffler JW, Avnimelech Y (2012) Biofloc-based aquaculture systems. In: Tidwell JH (ed) Aquaculture production systems. Wiley, Oxford

Bruhn JB, Nielsen KF, Hjelm M, Hansen M, Bresciani J, Schulz S, Gram L (2005) Ecology, inhibitory activity, and morphogenesis of a marine antagonistic bacterium belonging to the *Roseobacter clade*. Appl Environ Microbiol 71:7263–7270. https://doi.org/10.1128/AEM.71.11.7263-7270.2005

Calabrese S, Nilsen TO, Kolarevic J, Ebbesson LOE, Pedrosa C, Fivelstad S, Hosfeld C, Stefansson SO, Terjesen BF, Takle H, Martins CIM, Sveier H, Mathisen F, Imsland AK, Handeland SO (2017) Stocking density limits for post-smolt Atlantic salmon (*Salmo salar* L.) with emphasis on production performance and welfare. Aquaculture 468(1):363–370. https://doi.org/10.1016/j.aquaculture.2016.10.041

Camargo JA, Alonso A (2006) Ecological and toxicological effects of inorganic nitrogen pollution in aquatic ecosystems: a global assessment. Environ Int 32:831–849. https://doi.org/10.1016/j.envint.2006.05.002

Chen S, Coffin DE, Malone RF (1994) Suspended solids control in recirculating aquaculture systems. In: Timmons MB, Losordo TM (eds) Aquaculture water reuse systems: engineering design and management. Elsevier, Amsterdam, pp 61–100

Christianson L, Lepine C, Tsukuda S, Saito K, Summerfelt S (2015) Nitrate removal effectiveness of fluidized sulfur-based autotrophic denitrification biofilters for recirculating aquaculture systems. Aquac Eng:10–18. https://doi.org/10.1016/j.aquaeng.2015.07.002

Colt J (1991) Aquacultural production systems. J Anim Sci 69:4183–4192. https://doi.org/10.2527/1991.69104183x

Colt J (2006) Water quality requirements for reuse systems. Aquac Eng 34(3):143–156. https://doi.org/10.1016/j.aquaeng.2005.08.011

Colt J, Bouck G (1984) Design of packed columns for degassing. Aquac Eng 3:251–273. https://doi.org/10.1016/0144-8609(84)90007-4

Colt JE, Watten BJ (1988) Applications of pure oxygen in fish culture. Aquac Eng 7:397–441. https://doi.org/10.1016/0144-8609(88)90003-9

Conte FS (2004) Stress and the welfare of cultured fish. Appl Anim Behav Sci 86:205–223. https://doi.org/10.1016/j.applanim.2004.02.003

Couturier M, Trofimencoff T, Buil JU, Conroy J (2009) Solids removal at a recirculating salmon-smolt farm. Aquac Eng 41:71–77. https://doi.org/10.1016/j.aquaeng.2009.05.001

Cripps SJ, Bergheim A (2000) Solids management and removal for intensive land-based aquaculture production systems. Aquac Eng 22:33–56. https://doi.org/10.1016/S0144-8609(00)00031-5

Crossley PL (2004) Sub-irrigation in wetland agriculture. Agric Hum Values 21(2/3):191–205. https://doi.org/10.1023/B:AHUM.0000029395.84972.5e. Accessed 18 Dec 2017

Dabrowska H, Dabrowski K, Meyerburgdorff K, Hanke W, Gunther KD (1991) The effect of large doses of Vitamin-C and magnesium on stress responses in Common Carp, *Cyprinus-Carpio*. Comp Biochem Physiol A 99:681–685. https://doi.org/10.1016/0300-9629(91)90150-B

Daims H, Lebedeva EV, Pjevac P, Han P, Herbold C, Albertsen M, Jehmlich N, Palatinszky M, Vierheilig J, Bulaev A, Kirkegaard RH, von Bergen M, Rattei T, Bendinger B, Nielsen PH, Wagner M (2015) Complete nitrification by *Nitrospira* bacteria. Nature 528:504–509. https://doi.org/10.1038/nature16461

Davidson J, Summerfelt S (2004) Solids flushing, mixing, and water velocity profiles within large (10 and 150 m3) circular "Cornell-type" dual-drain tanks. Aquac Eng 32:245–271. https://doi.org/10.1016/j.aquaeng.2004.03.009

Davidson J, Bebak J, Mazik P (2009) The effects of aquaculture production noise on the growth, condition factor, feed conversion, and survival of rainbow trout, *Oncorhynchus mykiss*. Aquaculture 288(3–4):337–343. https://doi.org/10.1016/j.aquaculture.2008.11.037

Davidson J, Good C, Welsh C, Summerfelt S (2014) Comparing the effects of high vs. low nitrate on the health, performance, and welfare of juvenile rainbow trout *Oncorhynchus mykiss* within water recirculating aquaculture systems. Aquac Eng 59:30–40. https://doi.org/10.1016/j.aquaeng.2014.01.003

Davidson J, Good C, Williams C, Summerfelt S (2017) Evaluating the chronic effects of nitrate on the health and performance of post-smolt Atlantic salmon *Salmo salar* in freshwater recirculation aquaculture systems. Aquac Eng 79:1–8

Dawkins MS (1998) Evolution and animal welfare. Q Rev Biol 73:305–328. https://doi.org/10.1086/420307

De Carvalho RAPLF, Lemos DEL, Tacon AGJ (2013) Performance of single-drain and dual-drain tanks in terms of water velocity profile and solids flushing for in vivo digestibility studies in juvenile shrimp. Aquac Eng 57:9–17. https://doi.org/10.1016/j.aquaeng.2013.05.004

De Ionno PN, Wines GL, Jones PL, Collins RO (2007) A bioeconomic evaluation of a commercial scale recirculating finfish growout system—An Australian perspective. Aquaculture 259(1):315–327. https://doi.org/10.1016/j.aquaculture.2006.05.047

Defoirdt T, Boon N, Sorgeloos P, Verstraete W, Bossier P (2007) Alternatives to antibiotics to control bacterial infections: luminescent vibriosis in aquaculture as an example. Trends Biotechnol 25:472–479. https://doi.org/10.1016/j.tibtech.2007.08.001

Defoirdt T, Boon N, Sorgeloos P, Verstraete W, Bossier P (2008) Quorum sensing and quorum quenching in *Vibrio harveyi*: lessons learned from in vivo work. ISME J 2:19–26. https://doi.org/10.1038/ismej.2007.92

Díaz V, Ibáñez R, Gómez P, Urtiaga AM, Ortiz I (2012) Kinetics of nitrogen compounds in a commercial marine Recirculating Aquaculture System. Aquac Eng 50:20–27. https://doi.org/10.1016/j.aquaeng.2012.03.004

Dolan E, Murphy N, O'Hehir M (2013) Factors influencing optimal micro-screen drum filter selection for recirculating aquaculture systems. Aquac Eng 56:42–50. https://doi.org/10.1016/j.aquaeng.2013.04.005

Drennan DG, Hosler KC, Francis M, Weaver D, Aneshansley E, Beckman G, Johnson CH, Cristina CM (2006) Standardized evaluation and rating of biofilters. II. Manufacturer's and user's perspective. Aquac Eng 34:403–416. https://doi.org/10.1016/j.aquaeng.2005.07.001

Duncan IJ (2005) Science-based assessment of animal welfare: farm animals. Rev Sci Tech 24:483–492

Ebeling JM, Timmons MB (2012) Recirculating aquaculture systems. In: Tidwell JH (ed) Aquaculture production systems. Wiley, Oxford

Ebeling JM, Timmons MB, Bisogni JJ (2006) Engineering analysis of the stoichiometry of photoautotrophic, autotrophic, and heterotrophic removal of ammonia-nitrogen in aquaculture systems. Aquaculture 257:346–358. https://doi.org/10.1016/j.aquaculture.2006.03.019

EIFAC/ICES (1980) World conference on flow-through and recirculation systems, Stavanger, Norway 1980 and the 1981 World Aquaculture conference, Venice, Italy

Ellis T, Scott AP, Bromage N, North B, Porter M (2001) What is stocking density? Trout News 32:35–37

Emparanza EJM (2009) Problems affecting nitrification in commercial RAS with fixed-bed biofilters for salmonids in Chile. Aquac Eng 41:91–96. https://doi.org/10.1016/j.aquaeng.2009.06.010

Eshchar M, Mozes N, Fediuk M (2003) Carbon dioxide removal rate by aeration devices in marine fish tanks. Isr J Aquac-Bamidgeh 55:79–85

FAO (2001) Integrated agriculture-aquaculture: a primer 407. ISBN 9251045992

FAO (2004) Aquaculture management and conservation service (FIMA) 2004–2018. Cultured aquatic species information programme. *Carassius carassius*. Cultured aquatic species information programme. Weimin M. Accessed 5 Jan 2018

Fast AW, Tan EC, Stevens DF, Olson JC, Qin J, Barclay DK (1999) Paddlewheel aerator oxygen transfer efficiencies at three salinities. Aquac Eng 19:99–103. https://doi.org/10.1016/S0144-8609(98)00044-2

FAWC (2014) Farm animal welfare committee. Welfare of farmed fish. Available on www.defra.gov.uk/fawc. Accessed 18 Jan 2018

Fernandes P, Pedersen L-F, Pedersen PB (2015) Microscreen effects on water quality in replicated recirculating aquaculture systems. Aquac Eng 65:17–26. https://doi.org/10.1016/j.aquaeng.2014.10.007

Fox BK, Tamaru CS, Hollyer J et al (2012) A preliminary study of microbial water quality related to food safety in recirculating aquaponic fish and vegetable production systems. College of Tropical Agriculture and Human Resources, University of Hawaii at Manoa. Food Safety and Technology

Gehlert G, Griese M, Schlachter M, Schulz C (2018) Analysis and optimisation of dynamic facility ventilation in recirculating aquaculture systems. Aquac Eng 80:1–10. https://doi.org/10.1016/j.aquaeng.2017.11.003

Gendel Y, Lahav O (2013) A novel approach for ammonia removal from fresh-water recirculated aquaculture systems, comprising ion exchange and electrochemical regeneration. Aquac Eng 52:27–38. https://doi.org/10.1016/j.aquaeng.2012.07.005

Gomez RC (2011) Integrated fish farming systems. Available at: http://www.fao.org/fileadmin/templates/FCIT/Meetings/World_Water_Day_2011/5-integrated_aquaculture.pdf. Accessed 12 Feb 2018.

Gonçalves AA, Gagnon GA (2011) Ozone application in recirculating aquaculture system: an overview. Ozone Sci Eng 33:345–367. https://doi.org/10.1080/01919512.2011.604595

Good C, Davidson J, Welsh C, Snekvik K, Summerfelt S (2011) The effects of ozonation on performance, health and welfare of rainbow trout *Oncorhynchus mykiss* in low-exchange water recirculation aquaculture systems. Aquac Eng 44:97–102. https://doi.org/10.1016/j.aquaeng.2011.04.003

Good C, Davidson J, Earley RL, Lee E, Summerfelt S (2014) The impact of water exchange rate and treatment processes on water-borne hormones in recirculation aquaculture systems containing sexually maturing Atlantic salmon *Salmo salar*. J Aquac Res Development 5:260. https://doi.org/10.4172/2155-9546.1000260

Good C, Davidson J, Earley RL, Styga J, Summerfelt S (2017) The effects of ozonation on select waterborne steroid hormones in reicirculation aquaculture systems containing sexually mature Atlantic salmon *Salmo salar*. Aquac Eng 79:9–16. https://doi.org/10.1016/j.aquaeng.2017.08.004

Good C, Davidson J, Terjesen BF, Talke H, Kolarevic J, Baeverfjord G, Summerfelt S (2018) The effects of long-term 20mg/l carbon dioxide exposure on the health and performance of Atlantic

salmon *Salmo salar* post-smolts in water recirculation aquaculture systems. Aquac Eng 81:1–9. https://doi.org/10.1016/j.aquaeng.2018.01.003
Gorle JMR, Terjesen BF, Mota VC, Summerfelt S (2018) Water velocity in commercial RAS culture tanks for Atlantic salmon smolt production. Aquac Eng 81:89–100
Gornati R, Papis E, Rimoldi S, Terova G, Saroglia M, Bernardini G (2004) Rearing density influences the expression of stress-related genes in sea bass (*Dicentrarchus labrax* L.). Gene 341:111–118. https://doi.org/10.1016/j.gene.2004.06.020
Graber A, Antenen N, Junge R (2014) The multifunctional aquaponic system at ZHAW used as research and training lab. In: Maček Jerala M, Maček MA (eds) Conference VIVUS: transmission of innovations, knowledge and practical experience into everyday practice, Strahinj, Slovenia
Grady CPL, Lim HC (1980) Biological wastewater treatment: theory and applications. Marcel Dekker, New York
Grimsrud KM, Nielsen HM, Olesen I (2013) Households' willingness-to-pay for improved fish welfare in breeding programs for farmed Atlantic salmon. Aquaculture 372–375:19–27. https://doi.org/10.1016/j.aquaculture.2012.10.009
Gutierrez-Wing MT, Malone RF (2006) Biological filters in aquaculture: trends and research directions for freshwater and marine applications. Aquac Eng 34:163–171. https://doi.org/10.1016/j.aquaeng.2005.08.003
Guttman L, Rijn J (2008) Identification of conditions underlying production of geosmin and 2-methylisoborneol in a recirculating system. Aquaculture 279:85–91. https://doi.org/10.1016/j.aquaculture.2008.03.047
Hammock D, Huang CC, Mort G, Swinehart JH (2003) The effect of humic acid on the uptake of mercury(II), cadmium(II), and zinc(II) by Chinook salmon (*Oncorhynchus tshawytscha*) eggs. Arch Environ Contam Toxicol 44:83–88. https://doi.org/10.1007/s00244-002-1261-9
Hulata G, Simon Y (2011) An overview on desert aquaculture in Israel. In: Crespi V, Lovatelli, A (eds) Aquaculture in desert and arid lands: development constraints and opportunities. FAO technical workshop. Hermosillo, Mexico. FAO fisheries and aquaculture proceedings no. 20. Rome, FAO. 6–9 July 2010, pp 85–112
Huntingford FA, Adams C, Braithwaite VA, Kadri S, Pottinger TG, Sandøe P, Turnbull JF (2006) Current issues in fish welfare. J Fish Biol 68:332–372. https://doi.org/10.1111/j.0022-1112.2006.001046.x
Jorgensen EH, Christiansen JS, Jobling M (1993) Effects of stocking density on food intake, growth performance and oxygen consumption in Arctic charr (*Salvelinus alpinus*). Aquaculture 110:191–204. https://doi.org/10.1016/0044-8486(93)90272-Z
Kesarcodi-Watson A, Kaspar H, Lategan MJ, Gibson L (2008) Probiotics in aquaculture: the need, principles and mechanisms of action and screening processes. Aquaculture 274:1–14. https://doi.org/10.1016/j.aquaculture.2007.11.019
Klas S, Mozes N, Lahav O (2006) Development of a single-sludge denitrification method for nitrate removal from RAS effluents: lab-scale results vs. model prediction. Aquaculture 259:342–353. https://doi.org/10.1016/j.aquaculture.2006.05.049
Krom MD, van Rijn J (1989) Water quality processes in fish culture systems: processes, problems, and possible solutions. In: de Pauw N, Jaspers E, Ackerfors H, Wilkins N (eds) Aquaculture-a biotechnology in progress, vol 2. European aquaculture society, Bredene, pp 1091–1111
Kucuk H, Midilli A, Özdemir A, Çakmak E, Dincer I (2010) Exergetic performance analysis of a recirculating aquaculture system. Energy Convers Manag 51(5):1033–1043. https://doi.org/10.1016/j.enconman.2009.12.007
Lahav O, Ben Asher R, Gendel Y (2015) Potential applications of indirect electrochemical ammonia oxidation within the operation of freshwater and saline-water recirculating aquaculture systems. Aquac Eng 65:55–64. https://doi.org/10.1016/j.aquaeng.2014.10.009
Lee J (2014) Separation of fine organic particles by a low-pressure hydrocyclone (LPH). Aquac Eng 63:32–38. https://doi.org/10.1016/j.aquaeng.2014.07.002
Lekang OI (2013) Aquaculture engineering, 2nd edn. Wiley, Ames

Leonard N, Blancheton JP, Guiraud JP (2000) Populations of heterotrophic bacteria in an experimental recirculating aquaculture system. Aquac Eng 22:109–120. https://doi.org/10.1016/S0144-8609(00)00035-2

Leonard N, Guiraud JP, Gasset E, Cailleres JP, Blancheton JP (2002) Bacteria and nutrients – nitrogen and carbon – in a recirculating system for sea bass production. Aquac Eng 26:111–127. https://doi.org/10.1016/S0144-8609(02)00008-0

Lewis WM, Wehr LW (1976) A fish-rearing system incorporating cages, water circulation, and sewage removal. Prog Fish Cult 38(2):78–81. https://doi.org/10.1577/1548-8659(1976)38[78:AFSICW]2.0.CO;2

Lewis WM, Yopp JH, Schramm HL Jr, Brandenburg AM (1978) Use of hydroponics to maintain quality of recirculated water in a fish culture system. Trans Am Fish Soc 107(1):92–99. https://doi.org/10.1577/1548-8659(1978)107<92:UOHTMQ>2.0.CO;2

Li X, Przybyla C, Triplet S, Liu Y, Blancheton JP (2015) Long-term effects of moderate elevation of oxidation-reduction potential on European seabass (*Dicentrarchus labrax*) in recirculating aquaculture systems. Aquac Eng 64:15–19. https://doi.org/10.1016/j.aquaeng.2014.11.006

Liu Y, Rosten T, Henriksen K, Hognes E, Summerfelt S, Vinci B (2016) Comparative economic performance and carbon footprint of two farming models for producing Atlantic salmon (*Salmo salar*): land-based closed containment system in freshwater and open net pen in seawater. Aquac Eng 71:1–12

Livingstone DR (2003) Oxidative stress in aquatic organisms in relation to pollution and aquaculture. Rev Med Vet-Toulouse 154:427–430

Loyless JC, Malone RE (1998) Evaluation of air-lift pump capabilities for water delivery, aeration, and degasification for application to recirculating aquaculture systems. Aquac Eng 18 (2):117–133. https://doi.org/10.1016/S0144-8609(98)00025-9

Malone RF, Pfeiffer TJ (2006) Rating fixed film nitrifying biofilters used in recirculating aquaculture systems. Aquac Eng 34:389–402. https://doi.org/10.1016/j.aquaeng.2005.08.007

Manan H, Moh JHZ, Kasan NA et al (2017) Identification of biofloc microscopic composition as the natural bioremediation in zero water exchange of pacific white shrimp, Penaeus vannamei, cultured in close hatchery system. Appl Water Sci 7:2437–2446. https://doi.org/10.1007/s13201-016-0421-4

Mancinelli RL (1996) The nature of nitrogen: an overview. Life Support Biosph Sci 3(1–2):17–24

Martins CIM, Eding EH, Schneider O, Rasmussen R, Olesen B, Plesner L, Verreth JAJ (2005) Recirculation aquaculture systems in Europe. CONSENSUS. Oostende, Belgium, Consensus working group. Eur Aquacult Soc 31

Martins CIM, Ochola D, Ende SSW, Eding EH, Verreth JAJ (2010a) Is growth retardation present in Nile tilapia *Oreochromis niloticus* cultured in low water exchange recirculating aquaculture systems. Aquaculture 298:43–50. https://doi.org/10.1016/j.aquaculture.2009.09.030

Martins CIM, Eding EH, Verdegem MCJ, Heinsbroek LTN, Schneider O, Blancheton JP, Roque d'Orbcastel E, Verreth JAJ (2010b) New developments in recirculating aquaculture systems in Europe: a perspective on environmental sustainability. Aquac Eng 43(3):83–93. https://doi.org/10.1016/j.aquaeng.2010.09.002

Martins CIM, Eding EH, Verreth JAJ (2011) Stressing fish in recirculating aquaculture systems (RAS): does stress induced in one group of fish affect the feeding motivation of other fish sharing the same RAS? Aquac Res 42:1378–1384. https://doi.org/10.1111/j.1365-2109.2010.02728.x

Martins CIM, Galhardo L, Noble C et al (2012) Behavioural indicators of welfare in farmed fish. Fish Physiol Biochem 38:17. https://doi.org/10.1007/s10695-011-9518-8

Masalo I (2008) Hydrodynamic characterization of aquaculture tanks and design criteria for improving self-cleaning properties. Ph.D thesis. Universitat Politechnica de Catalunya BARCELONATECH, Spain

Meinelt T, Paul A, Phan TM, Zwirnmann E, Krüger A, Wienke A, Steinberg CEW (2007) Reduction in vegetative growth of the water mold *Saprolegnia parasitica* (Coker) by humic

substances of different qualities. Aquat Toxicol 83:93–103. https://doi.org/10.1016/j.aquatox.2007.03.013
Meinelt T, Kroupova H, Stüber A, Rennert B, Wienke A, Steinberg CEW (2010) Can dissolved aquatic humic substances reduce the toxicity of ammonia and nitrite in recirculating aquaculture systems? Aquaculture 306:378–383. https://doi.org/10.1016/j.aquaculture.2010.06.007
Merchie G, Lavens P, Verreth J, Ollevier F, Nelis H, DeLeenheer A, Storch V, Sorgeloos P (1997) The effect of supplemental ascorbic acid in enriched live food for *Clarias gariepinus* larvae at start feeding. Aquaculture 151:245–258. https://doi.org/10.1016/S0044-8486(96)01505-0
Michaud L, Blancheton JP, Bruni V, Piedrahita R (2006) Effect of particulate organic carbon on heterotrophic bacterial populations and nitrification efficiency in biological filters. Aquac Eng 34:224–233. https://doi.org/10.1016/j.aquaeng.2005.07.005
Michaud L, Lo Giudice A, Troussellier M, Smedile F, Bruni V, Blancheton JP (2009) Phylogenetic characterization of the heterotrophic bacterial communities inhabiting a marine recirculating aquaculture system. J Appl Microbiol 107:1935–1946. https://doi.org/10.1111/j.1365-2672.2009.04378.x
Michaud L, Lo Giudice A, Interdonato F, Triplet S, Ying L, Blancheton JP (2014) C/N ratio-induced structural shift of bacterial communities inside lab-scale aquaculture biofilters. Aquac Eng 58:77–87. https://doi.org/10.1016/j.aquaeng.2013.11.002
Montero D, Izquierdo MS, Tort L, Robaina L, Vergara JM (1999) High stocking density produces crowding stress altering some physiological and biochemical parameters in gilthead seabream, *Sparus aurata*, juveniles. Fish Physiol Biochem 20:53–60. https://doi.org/10.1023/A:1007719928905
Moran D (2010) Carbon dioxide degassing in fresh and saline water. I: degassing performance of a cascade column. Aquac Eng 43:29–36. https://doi.org/10.1016/j.aquaeng.2010.05.001
Munguia-Fragozo P, Alatorre-Jacome O, Rico-Garcia E et al (2015) Perspective for aquaponic systems: "Omic" technologies for microbial community analysis. Biomed Res Int 2015, Article ID 480386, 10 pages. https://doi.org/10.1155/2015/480386
Murray RG, Watson SW (1965) Structure of *Nitrocystis* oceanus and comparison with *Nitrosomonas* and *Nitrobacter*. J Bacteriol 89(6):1594–1609
Murray F, Bostock J, Fletcher M (2014) Review of RAS technologies and their commercial application. Final report. Available at http://www.hie.co.uk
Naegel LCA (1977) Combined production of fish and plants in recirculating water. Aquaculture 10 (1):17–24. https://doi.org/10.1016/0044-8486(77)90029-1
Noga E (2010) Fish disease: diagnosis and treatment, 2nd edn. Wiley, Ames
Oca J, Masalo I (2012) Flow pattern in aquaculture circular tanks: influence of flow rate, water depth, and water inlet & outlet features. Aquac Eng 52:65–72
Olesen I, Alfnes F, Bensze Røra M, Kolstad K (2010) Eliciting consumers' willingness to pay for organic and welfare-labelled salmon in a non-hypothetical choice experiment. Livest Sci 127 (2–3):218–226. https://doi.org/10.1016/j.livsci.2009.10.001
Palm HW, Knaus U, Appelbaum S, Goddek S, Strauch SM, Vermeulen T, Haïssam JM, Kotzen B (2018) Towards commercial aquaponics: a review of systems, designs, scales and nomenclature. Aquac Int 26:813. https://doi.org/10.1007/s10499-018-0249-z
Papoutsoglou SE, Karakatsouli N, Batzina A, Papoutsoglou ES, Tsopelakos A (2008) Effect of Mozart's music stimulus on gilthead seabream (*Sparus aurata* L.) physiology under different light intensity in a recirculating water system. J Fish Biol 73:980–1004. https://doi.org/10.1111/j.1095-8649.2008.02001.x
Park J, Kim PK, Lim T, Daniels HV (2013) Ozonation in seawater recirculating systems for black seabream *Acanthopagrus schlegelii* (Bleeker): effects on solids, bacteria, water clarity, and color. Aquac Eng 55:1–8. https://doi.org/10.1016/j.aquaeng.2013.01.002
Park J, Kim PK, Park S, Daniels HV (2015) Effects of two different ozone doses on total residual oxidants, nitrogen compounds and nitrification rate in seawater recirculating systems for black seabream *Acanthopagrus schlegelii* (Bleeker). Aquac Eng 67:1–7. https://doi.org/10.1016/j.aquaeng.2015.05.003

Pedersen PB, von Ahnen M, Fernandes P, Naas C, Pedersen L-F, Dalsgaard J (2017) Particle surface area and bacterial activity in recirculating aquaculture systems. Aquac Eng 78A:18–23. https://doi.org/10.1016/j.aquaeng.2017.04.005

Peuranen S, Vuorinen PJ, Vuorinen M, Hollender A (1994) The effects of iron, humic acids and low pH on the gills and physiology of brown trout (*Salmo trutta*). Ann Zool Fenn 31:389–396

Pittmann T, Steinmetz H (2013) Development of a process to produce bioplastic at municipal wastewater treatment plants. Conference paper

Prosser JI (1989) Autotrophic nitrification in bacteria. Adv Microb Physiol 30:125–181

Randall DJ, Tsui TKN (2002) Ammonia toxicity in fish. Marine Poll Bull 45:17–23

Reiser S, Schroeder JP, Wuertz S, Kloas W, Hanel R (2010) Histological and physiological alterations in juvenile turbot (*Psetta maxima*, L.) exposed to sublethal concentrations of ozone-produced oxidants in ozonated seawater. Aquaculture 307:157–164. https://doi.org/10.1016/j.aquaculture.2010.07.007

Reiser S, Wuertz S, Schroeder JP, Kloas W, Hanel R (2011) Risks of seawater ozonation in recirculation aquaculture – effects of oxidative stress on animal welfare of juvenile turbot (*Psetta maxima*, L.). Aquat Toxicol 105:508–517. https://doi.org/10.1016/j.aquatox.2011.08.004

Richardson LB, Burton DT, Block RM, Stavola AM (1983) Lethal and sublethal exposure and recovery effects of ozone-produced oxidants on adult white perch (*Morone americana* Gmelin). Water Res 17:205–213. https://doi.org/10.1016/0043-1354(83)90101-X

Ritola O, Lyytikainen T, Pylkko P, Molsa H, Lindstrom-Seppa P (2000) Glutathione-dependant defence system and monooxygenase enzyme activities in Arctic charr *Salvelinus alpinus* (L.) exposed to ozone. Aquaculture 185:219–233. https://doi.org/10.1016/S0044-8486(99)00355-5

Ritola O, Livingstone DR, Peters LD, Lindstrom-Seppa P (2002) Antioxidant processes are affected in juvenile rainbow trout (*Oncorhynchus mykiss*) exposed to ozone and oxygen-supersaturated water. Aquaculture 210:1–9. https://doi.org/10.1016/S0044-8486(01)00823-7

Rojas-Tirado P, Bovbjerg Pedersen P, Pedersen L-F (2017) Bacterial activity dynamics in the water phase during start-up of recirculating aquaculture systems. Aquac Eng 78A:24–31. https://doi.org/10.1016/j.aquaeng.2016.09.004

Rosenthal H (1980) Recirculation systems in western Europe. World symposium on aquaculture in heated effluents and recirculation system, Stavanger, Institut für Kuesten- und Binnefischerei, Bundesforschungsanstalt Hamburg, BRD

Rosenthal H, Black EA (1993) Recirculation systems in aquaculture. In: Wang J-K (ed) Techniques for modem aquaculture. ASAE, St. Joseph, pp 284–294

Rurangwa E, Verdegem MCJ (2015) Microorganisms in recirculating aquaculture systems and their management. Rev Aquac 7:117–130. https://doi.org/10.1111/raq.12057

Rusten B, Eikebrokk B, Ulgenes Y, Lygren E (2006) Design and operations of the Kaldnes moving bed biofilm reactors. Aquac Eng 34:322–331. https://doi.org/10.1016/j.aquaeng.2005.04.002

Schmautz Z, Graber A, Jaenicke S, Goesmann A, Junge R, Smits THM (2017) Microbial diversity in different compartments of an aquaponics system. Arch Microbiol 199:613. https://doi.org/10.1007/s00203-016-1334-1

Schreier HJ, Mirzoyan N, Saito K (2010) Microbial diversity of biological filters in recirculating aquaculture systems. Curr Opin Biotechnol 21:318–325. https://doi.org/10.1016/j.copbio.2010.03.011

Schroeder JP, Croot PL, Von Dewitz B, Waller U, Hanel R (2011) Potential and limitations of ozone for the removal of ammonia, nitrite, and yellow substances in marine recirculating aquaculture systems. Aquac Eng 45:35–41. https://doi.org/10.1016/j.aquaeng.2011.06.001

Sharrer MJ, Summerfelt ST, Bullock GL, Gleason LE, Taeuber J (2005) Inactivation of bacteria using ultraviolet irradiation in a recirculating salmonid culture system. Aquac Eng 33:135–149. https://doi.org/10.1016/j.aquaeng.2004.12.001

Singh S, Marsh LS (1996) Modelling thermal environment of a recirculating aquaculture facility. Aquaculture 139:11–18. https://doi.org/10.1016/0044-8486(95)01164-1

Sirsat S, Neal J (2013) Microbial profile of soil-free versus insoil grown lettuce and intervention methodologies to combat pathogen surrogates and spoilage microorganisms on lettuce. Foods 2 (4):488–498. https://doi.org/10.3390/foods2040488

Somerville C, Cohen M, Pantanella E, Stankus A, Lovatelli A (2014) Small-scale aquaponic food production: integrated fish and plant farming. In: FAO fisheries and aquaculture technical paper food and agriculture organization of the United Nations, Rome, Italy, p 262

Stevenson FJ (1994) Humus Chemistry: genesis, composition, reactions, 2nd edn. Wiley, New York, p 496. https://doi.org/10.1021/ed072pA93.6

Suhr KI, Pedersen LF, Nielsen JL (2014) End-of-pipe single-sludge denitrification in pilot-scale recirculating aquaculture systems. Aquac Eng 62:28–35. https://doi.org/10.1016/j.aquaeng.2014.06.002

Summerfelt ST (2003) Ozonation and UV irradiation – An introduction and examples of current applications. Aquac Eng 28:21–36. https://doi.org/10.1016/S0144-8609(02)00069-9

Summerfelt ST (2006) Design and management of conventional fluidized-sand biofilters. Aquac Eng 34:275–302. https://doi.org/10.1016/j.aquaeng.2005.08.010

Summerfelt ST, Hochheimer JN (1997) Review of Ozone processes and applications as an oxidizing agent in Aquaculture. Progress Fish-Culturist 59:94–105. https://doi.org/10.1577/1548-8640(1997)059<0094:ROOPAA>2.3.CO;2

Summerfelt S, Davidson JW, Waldrop TB, Tsukuda S, Bebak-Williams J (2004) A partial-reuse system for coldwater aquaculture. Aquac Eng 157–181. https://doi.org/10.1016/j.cell.2013.02.048

Summerfelt ST, Sharrer MJ, Tsukuda SM, Gearheart M (2009) Process requirements for achieving full-flow disinfection of recirculating water using ozonation and UV irradiation. Aquac Eng 40:17–27. https://doi.org/10.1016/j.aquaeng.2008.10.002

Summerfelt ST, Zulke A, Kolarevic J, Megard Reiten BK, Selset R, Guiterrez X, Terjesen BF (2015) The effects of alkalinity on ammonia removal, carbon dioxide stripping, and pH in semi-commercial scale water recirculating aquaculture systems operated with moving bed bioreactors. Aquac Eng 65:46–54. https://doi.org/10.1016/j.aquaeng.2014.11.002

Summerfelt ST, Mathisen F, Buran Holan A, Terjesen BF (2016) Survey of large circular and octagonal tanks operated at Norwegian commercial smolt and post-smolt sites. Aquac Eng 74:105–110. https://doi.org/10.1016/j.aquaeng.2016.07.004

Tal Y, Watts JEM, Schreier HJ (2006) Anaerobic ammonium-oxidizing (Anammox) bacteria and associated activity in fixed-film biofilters of a marine recirculating aquaculture system. Appl Environ Microbiol 72:2896–2904. https://doi.org/10.1128/AEM.72.4.2896-2904.2006

Tal Y, Schreier HJ, Sowers KR, Stubblefield JD, Place AR, Zohar Y (2009) Environmentally sustainable land-based marine aquaculture. Aquaculture 286:28–35. https://doi.org/10.1016/j.aquaculture.2008.08.043

Tango MS, Gagnon GA (2003) Impact of ozonation on water quality in marine recirculation systems. Aquac Eng 29:125–137. https://doi.org/10.1016/S0144-8609(03)00061-X

Terhune JM, Friars GW, Bailey JK, O'Flynn FM (1990) Noise levels may influence Atlantic salmon smolting rates in tanks. J Fish Biol 37:185–197. https://doi.org/10.1111/j.1095-8649.1990.tb05939.x

Timmons MB, Ebeling JM (2010) Recirculating aquaculture, 2nd edn. Cayuga Aqua Ventures, Ithaca

Timmons MB, Losordo TM (1994) Aquaculture water reuse systems – engineering design and management. Elsevier, Amsterdam

Tomita-Yokotani K, Anilir S, Katayama N, Hashimoto H, Yamashita M (2009) Space agriculture for habitation on mars and sustainable civilization on earth. Recent Adv Space Technol:68–69. https://doi.org/10.1109/RAST.2009.5158276

Tort L, Rotllant J, Liarte C, Acerete L, Hernandez A, Ceulemans S, Coutteau P, Padros F (2004) Effects of temperature decrease on feeding rates, immune indicators and histopathological changes of gilthead sea bream *Sparus aurata* fed with an experimental diet. Aquaculture 229:55–65. https://doi.org/10.1016/S0044-8486(03)00403-4

Turnbull JF, Kadri S (2007) Safeguarding the many guises of farmed fish welfare. Dis Aquat Org 75:173–182. https://doi.org/10.3354/dao075173

Vadstein O, Mo TA, Bergh Ø (2004) Microbial interactions, prophylaxis and diseases. In: Moksness E, Kjørsvik E, Olsen Y (eds) Culture of cold-water marine fishes. Blackwell Publishing, Bath, pp 28–72. https://doi.org/10.1016/j.aquaeng.2005.04.004

van Rijn J (1996) The potential for integrated biological treatment systems in recirculating fish culture—a review. Aquaculture 139(3–4):181–201. https://doi.org/10.1016/0044-8486(95)01151-X

van Rijn J (2013) Waste treatment in recirculating aquaculture systems. Aquac Eng 53:49–56. https://doi.org/10.1016/j.aquaeng.2012.11.010

van Rijn J, Nussinovitch A (1997) An empirical model for predicting degradation of organic solids in fish culture systems based on short-term observations. Aquaculture 154(2):173–179. https://doi.org/10.1016/S0044-8486(97)00048-3

van Rijn J, Rivera G (1990) Aerobic and anaerobic biofiltration in an aquaculture unit: nitrite accumulation as a result of nitrification and denitrification. Aquac Eng 9:1–18

van Rijn J, Tal Y, Schreier HJ (2006) Denitrification in recirculating systems: theory and applications. Aquac Eng 34(3):364–376

Vazzana M, Cammarata M, Cooper EL, Parrinello N (2002) Confinement stress in sea bass (*Dicentrarchus labrax*) depresses peritoneal leukocyte cytotoxicity. Aquaculture 210:231–243. https://doi.org/10.1016/S0044-8486(01)00818-3

Veerapen JP, Lowry BJ, Couturier MF (2005) Design methodology for the swirl separator. Aquac Eng 33:21–45. https://doi.org/10.1016/j.aquaeng.2004.11.001

Verreth JAJ, Eding EH (1993) European farming industry of African catfish (*Clarias gariepinus*) facts and figures. J World Aquacult Soc 24:6–13

Volpato GL, Goncalves-de-Freitas E, Fernandes-de-Castilho M (2007) Insights into the concept of fish welfare. Dis Aquat Org 75:165–171. https://doi.org/10.3354/dao075165

Wagner EJ, Bosakowski T, Miller SA (1995) Evaluation of the absorption efficiency of the Low Head Oxygenation System. Aquac Eng 14:49–57. https://doi.org/10.1016/0144-8609(94)P4426-C

Wold P-A, Holan AB, Øie G, Attramadal K, Bakke I, Vadstein O, Leiknes TO (2014) Effects of membrane filtration on bacterial number and microbial diversity in marine recirculating aquaculture system (RAS) for Atlantic cod (*Gadus morhua* L.) production. Aquaculture 422–423:69–77. https://doi.org/10.1016/j.aquaculture.2013.11.019

Wong KB, Piedrahita RH (2000) Settling velocity characterization of aquacultural solids. Aquac Eng 21:233–246. https://doi.org/10.1016/S0144-8609(99)00033-3

Wysocki LE, Davidson JW III, Smith ME, Frankel AS, Ellison TE, Mazik PM, Popper AN, Bebak J (2007) The effects of aquaculture production noise on hearing, growth, and disease resistance of rainbow trout, *Oncorhynchus mykiss*. Aquaculture 272:687–697. https://doi.org/10.1016/j.aquaculture.2007.07.225

Yamin G, Borisover M, Cohen E, van Rijn J (2017a) Accumulation of humic-like and proteinaceous dissolved organic matter in zero-discharge aquaculture systems as revealed by fluorescence EEM spectroscopy. Water Res 108:412–421. https://doi.org/10.1016/j.watres.2016.11.028

Yamin G, Falk R, Avtalion RR, Shoshana N, Ofek T, Smirnov R, Rubenstein G, van Rijn J (2017b) The protective effect of humic-rich substances on atypical *Aeromonas salmonicida subsp. salmonicida* infection in common carp (*Cyprinus carpio* L.). J Fish Dis 40(12):1783–1790. https://doi.org/10.1111/jfd.12645

Yamin G, Zilberg D, Levy G, van Rijn J (2017c) The protective effect of humic-rich substances from monogenean parasites infecting the guppy (*Poecilia reticulata*). Aquaculture 479:487–489. https://doi.org/10.1016/j.aquaculture.2017.06.022

Yildiz HY, Robaina L, Pirhonen J, Mente E, Domínguez D, Parisi G (2017) Fish welfare in aquaponic systems: its relation to water quality with an emphasis on feed and faeces—a review. Water 9(1):13. https://doi.org/10.3390/w9010013

Yogev U, Sowers KR, Mozes N, Gross A (2016) Nitrogen and carbon balance in a novel near-zero exchange saline recirculating aquaculture system. Aquaculture 467:118–126

Zeng Q, Tian X, Wang L (2017) Genetic adaptation of microbial populations present in high-intensity catfish production systems with therapeutic oxytetracycline treatment. Sci Rep 7:17491. https://doi.org/10.1038/s41598-017-17640-3

Zhu S, Chen S (2002) The impact of temperature on nitrification rate in fixed film biofilters. Aquac Eng 26(4):221–237. https://doi.org/10.1016/S0144-8609(02)00022-5

Zou Y, Hu Z, Zhang J, Xie H, Guimbaud C, Fang Y (2016) Effects of pH on nitrogen transformations in media-based aquaponics. Bioresour Technol 210:81–87. https://doi.org/10.1016/j.biortech.2015.12.079

Chapter 4
Hydroponic Technologies

Carmelo Maucieri, Carlo Nicoletto, Erik van Os, Dieter Anseeuw, Robin Van Havermaet, and Ranka Junge

Abstract Hydroponics is a method to grow crops without soil, and as such, these systems are added to aquaculture components to create aquaponics systems. Thus, together with the recirculating aquaculture system (RAS), hydroponic production forms a key part of the aqua-agricultural system of aquaponics. Many different existing hydroponic technologies can be applied when designing aquaponics systems. This depends on the environmental and financial circumstances, the type of crop that is cultivated and the available space. This chapter provides an overview of different hydroponic types, including substrates, nutrients and nutrient solutions, and disinfection methods of the recirculating nutrient solutions.

Keywords Hydroponics · Soilless culture · Nutrients · Grow media · Aeroponics · Aquaponics · Nutrient solution

C. Maucieri (✉) · C. Nicoletto
Department of Agriculture, Food, Natural Resources, Animals and Environment, University of Padova, Campus of Agripolis, Legnaro, Italy
e-mail: carmelo.maucieri@unipd.it; carlo.nicoletto@unipd.it

E. v. Os
Wageningen University and Research, Business Unit Greenhouse Horticulture, Wageningen, The Netherlands
e-mail: erik.vanos@wur.nl

D. Anseeuw
Inagro, Roeselare, Belgium
e-mail: info@inagro.be

R. V. Havermaet
Provinciaal Proefcentrum voor de Groenteteelt Oost-Vlaanderen, Kruishoutem, Belgium
e-mail: robin@pcgroenteteelt.be

R. Junge
Institute of Natural Resource Sciences Grüental, Zurich University of Applied Sciences, Wädenswil, Switzerland
e-mail: ranka.junge@zhaw.ch

S. Goddek et al. (eds.), *Aquaponics Food Production Systems*,
https://doi.org/10.1007/978-3-030-15943-6_4

4.1 Introduction

In horticultural crop production, the definition soilless cultivation encompasses all the systems that provide plant production in soilless conditions in which the supply of water and of minerals is carried out in nutrient solutions with or without a growing medium (e.g. stone wool, peat, perlite, pumice, coconut fibre, etc.). Soilless culture systems, commonly known as hydroponic systems, can further be divided into open systems, where the surplus nutrient solution is not recycled, and closed systems, where the excess flow of nutrients from the roots is collected and recycled back into the system (Fig. 4.1).

Soilless culture systems have evolved as one possible solution to avoid soil-borne diseases that have always been a problem in the greenhouse cultivation industry.

Nowadays, soilless growing systems are common in horticultural practice in most European countries, although not in every country does this occur on a large scale. The advantages of soilless systems compared to soil grown crops are:

- Pathogen-free start with the use of substrates other than soil and/or easier control of soil-borne pathogens.
- Growth and yield are independent of the soil type/quality of the cultivated area.
- Better control of growth through a targeted supply of nutrient solution.
- The potential for reusing the nutrient solution allowing for maximizing resources.
- Increased quality of produce gained by the better control of other environmental parameters (temperature, relative humidity) and pests.

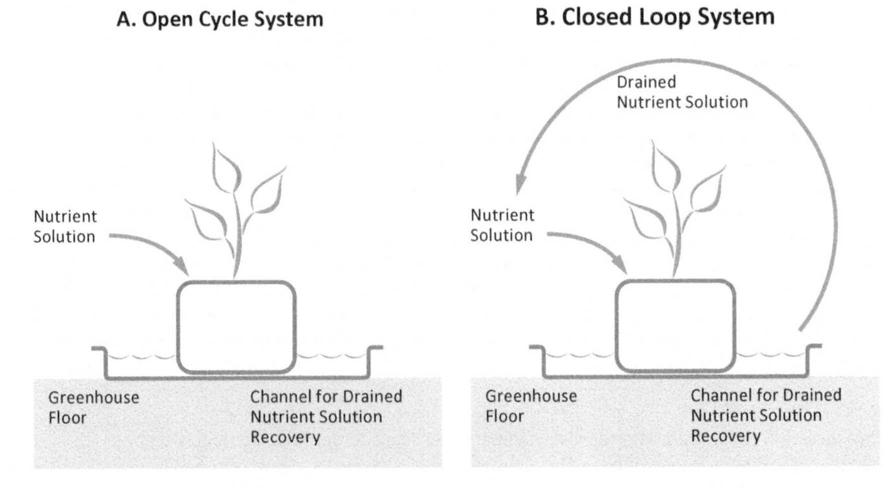

Fig. 4.1 Scheme of open cycle (**a**) and closed loop systems (**b**)

In most cases, open loop or run-to-waste systems rather than closed loop or recirculation systems are adopted, although in more and more European countries the latter are mandatory. In these open systems, the spent and/or superfluous nutrient solution is deposited into the ground and surface water bodies, or it is used in open field cultivation. However, regarding economics and environmental concerns, soilless systems should be as closed as possible, i.e. where recirculation of the nutrient solution occurs, where the substrate is reused and where more sustainable materials are used.

The advantages of closed systems are:

- A reduction in the amount of waste material.
- Less pollution of ground and surface water.
- A more efficient use of water and fertilizers.
- Increased production because of better management options.
- Lower costs because of the savings in materials and higher production.

There are also a number of disadvantages such as:

- The required high water quality.
- High investments.
- The risk of rapid dispersal of soil-borne pathogens by the recirculating nutrient solution.
- Accumulation of potential phytotoxic metabolites and organic substances in the recirculating nutrient solution.

In commercial systems, the problems of pathogen dispersal are tackled by disinfecting the water through physical, chemical and/or biological filtration techniques. However, one of the main factors that hinder the use of recirculating nutrient solution culture for greenhouse crops is the accumulation of salts in the irrigation water. Typically, there is a steady increase in electrical conductivity (EC) due to the accumulation of ions, which are not fully absorbed by the crops. This may be especially true in aquaponic (AP) settings where sodium chloride (NaCl), incorporated in the fish feed, may accumulate in the system. To amend this problem, it has been suggested that an added desalination step could improve the nutrient balance in multi-loop AP systems (Goddek and Keesman 2018).

4.2 Soilless Systems

The intense research carried out in the field of hydroponic cultivation has led to the development of a large variety of cultivation systems (Hussain et al. 2014). In practical terms all of these can also be implemented in combination with

Table 4.1 Classification of hydroponic systems according to different aspects

Characteristic	Categories	Examples
Soilless system	No substrate	NFT (nutrient film technique)
		Aeroponics
		DFT (deep flow technique)
	With substrate	Organic substrates (peat, coconut fibre, bark, wood fibre, etc.)
		Inorganic substrates (stone wool, pumice, sand, perlite, vermiculite, expanded clay)
		Synthetic substrates (polyurethane, polystyrene)
Open/closed systems	Open or run-to-waste systems	The plants are continuously fed with "fresh" solution without recovering the solution drained from the cultivation modules (Fig. 4.1a)
	Closed or recirculation systems	The drained nutrient solution is recycled and topped up with lacking nutrients to the right EC level (Fig. 4.1b)
Water supply	Continuous	NFT (nutrient film technique) DFT (deep flow technique)
	Periodical	Drip irrigation, ebb and flow, aeroponics

aquaculture; however, for this purpose, some are more suitable than others (Maucieri et al. 2018). The great variety of systems that may be used necessitates a categorization of the different soilless systems (Table 4.1).

4.2.1 Solid Substrate Systems

At the start of soilless cultivation in the 1970s, many substrates were tested (Wallach 2008; Blok et al. 2008; Verwer 1978). Many failed for reasons such as being too wet, too dry, not sustainable, too expensive and releasing of toxic substances. Several solid substrates survived: stone wool, perlite, coir (coconut fibre), peat, polyurethane foam and bark. Solid substrate systems can be divided as follows:

Fibrous Substrates These may be organic (e.g. peat, straw and coconut fibre) or inorganic (e.g. stone wool). They are characterized by the presence of fibres of different sizes, which give the substrate a high water-retention capacity (60–80%) and a modest air capacity (free porosity) (Wallach 2008). A high percentage of the retained water is easily available for the plant, which is directly reflected in the minimum volume of substrate per plant required to guarantee a sufficient water supply. In these substrates there are no obvious water and salinity gradients along the profile, and, consequently, the roots tend to grow faster, evenly and abundantly, using the entire available volume.

Granular Substrates They are generally inorganic (e.g. sand, pumice, perlite, expanded clay) and are characterized by different particle sizes and thus textures; they have high porosity and are free draining. Water-holding capacity is rather poor (10–40%), and much of the water retained is not easily available to the plant (Maher et al. 2008). Therefore, the required volume of substrate per plant is higher compared to the fibrous ones. In granular substrates, a marked gradient of moisture is observed along the profile and this causes the roots to develop mainly on the bottom of containers. Smaller particle sizes, increase in the capacity for water retention, moisture homogeneity and greater EC and a lower volume of the substrate are required for the plant.

Substrates are usually enveloped in plastic coverings (so-called grow bags or slabs) or inserted in other types of containers of various sizes and of synthetic materials.

Before planting the substrate should be saturated in order to:

- Provide adequate water and nutrients supply in the entire substrate slab.
- Achieve uniform EC and pH levels.
- Expel the presence of air and make a homogeneous wetting of the material.

It is equally important for a substrate dry phase after planting to stimulate the plants to evolve homogeneous substrate exploration by roots to obtain an abundant and well-distributed root system at the various levels and to expose the roots to air. Using a substrate for the second time by rewetting may be a problem because saturation is not possible due to drain holes in the plastic envelope. In an organic substrate (such as coir), adopting short and frequent irrigation turns, it is possible to recover the water-retention capability to use it for the second time, more easily than inert substrates (stone wool, perlite) (Perelli et al. 2009).

4.2.2 Substrates for Medium-Based Systems

A substrate is necessary for the anchorage of the roots, a support for the plant and also as a water-nutritional mechanism due to its microporosity and cation exchange capacity.

Plants grown in soilless systems are characterized by an unbalanced shoot/root ratio, demands for water, air and nutrients that are much greater than in open field conditions. In the latter case, the growth rates are slower, and the quantities of substrate are theoretically unlimited. To satisfy these requirements, it is necessary to resort to substrates which, alone or in mixture, ensure optimal and stable chemical–physical and nutritional conditions. An array of materials with different characteristics and costs can be used as substrates as illustrated in Fig. 4.2. However, as yet, there is no one substrate that can be used universally in all cultivation situations.

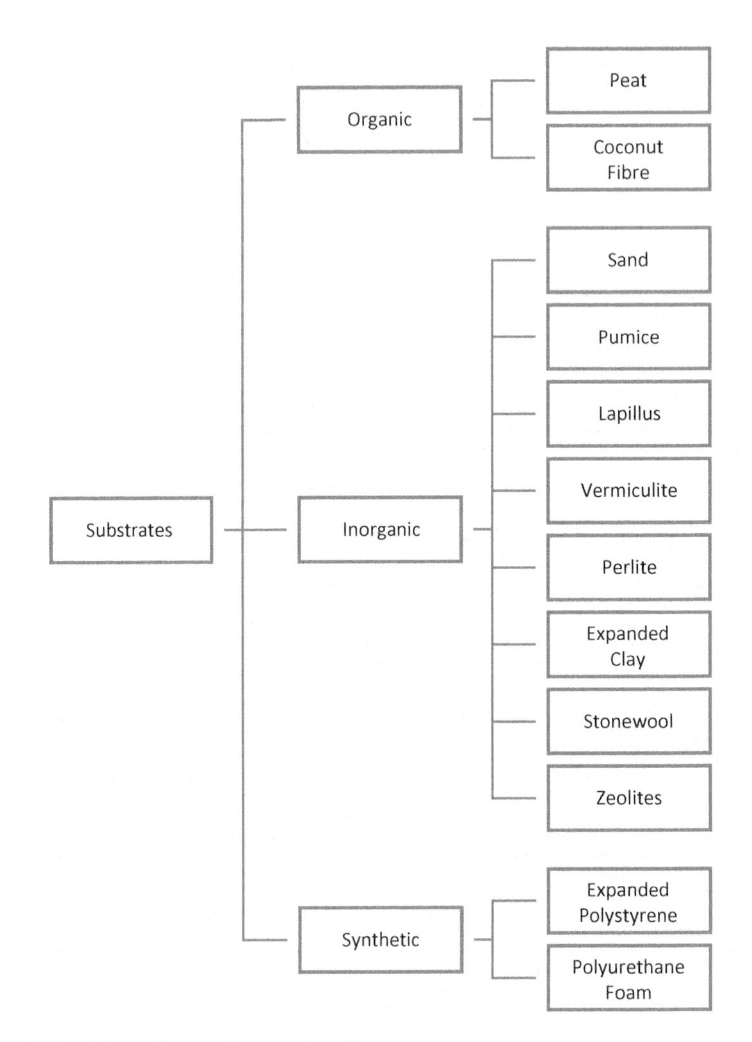

Fig. 4.2 Materials usable as substrates in soilless systems

4.2.3 Characterization of Substrates

Bulk Density (BD) BD is expressed by the dry weight of the substrate per unit of volume. It enables the anchorage of the roots and offers plant support. The optimum BD for crops in a container varies between 150 and 500 kg m^{-3} (Wallach 2008). Some substrates, because of their low BD and their looseness, as is the case of perlite (ca. 100 kg m^{-3}), polystyrene in granules (ca. 35 kg m^{-3}) and non-compressed sphagnum peat (ca. 60 kg m^{-3}), are not suitable for use alone, especially with plants that grow vertically.

Table 4.2 Main chemical–physical characteristics of peats and coconut fibre. (dm = dry matter)

	Raised bogs		Fen bogs	
Characteristics	Blond	Brown	Black	Coconut fibre (coir)
Organic matter (% dm)	94–99	94–99	55–75	94–98
Ash (% dm)	1–6	1–6	23–30	3–6
Total porosity (% vol)	84–97	88–93	55–83	94–96
Water-retention capacity (% vol)	52–82	74–88	65–75	80–85
Free porosity (% vol)	15–42	6–14	6–8	10–12
Bulk density (kg m^{-3})	60–120	140–200	320–400	65–110
CEC (meq%)	100–150	120–170	80–150	60–130
Total nitrogen (% dm)	0.5–2.5	0.5–2.5	1.5–3.5	0.5–0.6
C/N	30–80	20–75	10–35	70–80
Calcium (% dm)	<0.4	<0.4	>2	–
pH (H_2O)	3.0–4.0	3.0–5.0	5.5–7.3	5.0–6.8

Source: Enzo et al. (2001)

Porosity The ideal substrate for potted crops should have a porosity of at least 75% with variable percentages of macropores (15–35%) and micropores (40–60%) depending on the cultivated species and the environmental and crop conditions (Wallach 2008; Blok et al. 2008; Maher et al. 2008). In small-sized containers, total porosity should reach 85% of the volume (Bunt 2012). The structure should be stable over time and should resist compaction and the reduction of volume during dehydration phases.

Water-Holding Capacity Water-holding capacity ensures adequate levels of substrate moisture for crops, without having to resort to frequent irrigations. However, the water-holding capacity must not be too high in order to avoid root asphyxia and too much cooling. The water available for the plant is calculated by the difference between the quantity of water at the retention capacity and that retained at the wilting point. This should be around 30–40% of the apparent volume (Kipp et al. 2001). Finally, it must be considered that with the constant increase of the root system biomass during growth, the free porosity in the substrate is gradually reduced and the hydrological characteristics of the substrate are modified.

Cation Exchange Capacity (CEC) CEC is a measure of how many cations can be retained on substrate particle surfaces. In general, organic materials have a higher CEC and a higher buffer capability than mineral ones (Wallach 2008; Blok et al. 2008) (Table 4.2).

pH A suitable pH is required to suit the needs of the cultivated species. Substrates with a low pH are more suitable for crops in containers, as they are more easily modified towards the desired levels by adding calcium carbonate and also because they meet the needs of a wider number of species. Moreover, during cultivation the pH value tends to rise due to irrigation with water rich in carbonates. The pH may also vary in relation to the type of fertilizer used. It is more difficult to correct an alkaline substrate. This can however be achieved by adding sulphur or physiologically acid fertilizers (ammonium sulphate, potassium sulphate) or constitutionally acid fertilizers (mineral phosphate).

Electrical Conductivity (EC) Substrates should have a known nutrient content and low EC values, (see also Table 4.4). It is often preferable to use a chemically inert substrate and to add the nutrients in relation to the specific crop requirements. Particular attention has to be paid to the EC levels. High EC levels indicate the presence of ions (e.g. Na^+) that, although not being important as nutrients, can play a decisive role in the suitability of the substrate.

Health and Safety Health in the systems and safety for operatives are provided by the absence of pathogens (nematodes, fungi, insects), potentially phytotoxic substances (pesticides) and weed seeds. Some industrially produced materials (expanded clay, perlite, stone wool, vermiculite and polystyrene) guarantee high levels of sterility due to the high temperatures applied during their processing.

Sustainability Another important characteristic of a substrate is its sustainability profile. Many commonly used substrates face ecological challenges relating to their provenance, production process and/or subsequent processing and end-of-life footprint. In this regard, substrates originating from materials with a low ecological footprint (modified in an environmentally friendly way and ultimately biodegradable) are an extra characteristic to consider. Reusability of the substrate can also be an important aspect of the sustainability of a substrate.

Cost Last but not least, the substrate must be inexpensive or at least cost-effective, readily available and standardized from the chemical–physical point of view.

4.2.4 Type of Substrates

The choice of substrates ranges from products of organic or mineral origin which are present in nature and which are subjected to special processing (e.g. peat, perlite, vermiculite), to those of organic origin derived from human activities (e.g. waste or by-products of agricultural, industrial and urban activities) and of industrial origin obtained by synthesis processes (e.g. polystyrene).

4.2.4.1 Organic Materials

This category includes natural organic substrates, including residues, waste and by-products of organic nature derived from agricultural (manure, straw, etc.) or, for example, industrial, by-products of the wood industry, etc. or from urban settlements, e.g. sewage sludge, etc. These materials can be subjected to additional processing, such as extraction and maturation.

All the materials that can be used in hydroponics can also be used in AP. However, as the bacterial load in an AP solution may be higher than in conventional hydroponic solutions, it can therefore be expected that organic substrates may be prone to an increased decomposition rate, causing substrate compaction and root aeration problems. Therefore, organic materials can be considered for crops with a shorter growth cycle, whilst mineral substrates may be preferred for crops with a long growth cycle.

Peat

Peat, used alone or with other substrates, is currently the most important material of organic origin for substrate preparation. The term peat refers to a product derived from residues of bryophytes (*Sphagnum*), Cyperaceae (*Trichophorum*, *Eriophorum*, *Carex*) and others (*Calluna*, *Phragmites*, etc.) transformed in anaerobic conditions.

Raised bogs are formed in cold and very rainy environments. Rainwater, without salts, is retained on the surface by mosses and vegetable residues, creating a saturated environment. In raised bogs we can distinguish a deeper, much decomposed layer of dark colour (*brown peat*) and a slightly decomposed, shallower layer of a light colour (*blond peat*). Both of the peats are characterized by good structural stability, very low availability of nutrients and acidic pH whilst they mainly differ in their structure (Table 4.2).

Brown peats, with very small pores, have a higher water-retention capacity and less free porosity for air and have higher CEC and buffer capability. Physical characteristics vary in relation to the particle size that allows water absorption from 4 up to 15 times its own weight. Raised bogs usually satisfy the requirements needed for a good substrate. Moreover, they have constant and homogenous properties, and so they can be industrially exploited. However, the use of these peats requires pH corrections with, e.g. calcium carbonate ($CaCO_3$). Generally, for a sphagnum peat with a pH 3–4, 2 kg m^{-3} of $CaCO_3$ should be added to increase the pH for one unit. Attention must be paid to avoid the complete drying of the substrate. It should also be taken into account that peat is subjected to microbiological decomposition processes which, over time, may increase water-retention capacity and reduce the free porosity.

Fen bogs are mainly present in temperate areas (e.g. Italy and western France), where *Cyperaceae*, *Carex* and *Phragmites* are dominant. These peats are formed in the presence of stagnant water. The oxygen, salts and calcium content in the water allow for a faster decomposition and humification, compared to that which occurs in the raised bogs. This results in a very dark, brown to black peat with a higher nutrient content, in particular nitrogen and calcium, a higher pH, higher bulk density and much lower free porosity (Table 4.2). They are rather fragile in the dry state, and have a remarkable plasticity in the humid state, which confers high susceptibility to compression and deformation. The carbon/nitrogen (C/N) ratio is generally between 15 and 48 (Kuhry and Vitt 1996; Abad et al. 2002). Because of its properties, black peat is of low value and is not suitable as a substrate, but can be mixed with other materials.

It should be noted that in some countries there is a drive to reduce peat use and extraction to reduce environmental effects and various peat substitutes have been identified with varied success.

Coconut Fibre

Coconut fibre (coir) is obtained from removing the fibrous husks of coconuts and is a by-product of the copra (coconut oil production) and fibre extraction industry, and is composed almost exclusively of lignin. Before use, it is composted for 2–3 years, and then it is dehydrated and compressed. Prior to its use, it must be rehydrated by adding up to 2–4 times of its compressed volume with water. Coconut fibre possesses chemical–physical characteristics that are similar to blond peat (Table 4.2), but with the advantages of having a higher pH. It also has a lower environmental impact than peat (excessive exploitation of peat bogs) and stone wool where there are problems with disposal. This is one of the reasons why it is increasingly preferred in soilless systems (Olle et al. 2012; Fornes et al. 2003).

Wood-Based Substrates

Organic substrates which are derived from wood or its by-products, such as bark, wood chips or saw dust, are also used in global commercial plant production (Maher et al. 2008). Substrates based on these materials generally possess good air content and high saturated hydraulic conductivities. The disadvantages can include low water-retention capacities, insufficient aeration caused by microbial activity, inappropriate particle-size distribution, nutrient immobilization or negative effects due to salt and toxic compound accumulations (Dorais et al. 2006).

4.2.4.2 Inorganic Materials

This category includes natural materials (e.g. sand, pumice) and mineral products derived from industrial processes (e.g. vermiculite, perlite) (Table 4.3).

Sand

Sands are natural inorganic material with particles between 0.05 and 2.0 mm diameter, originating from the weathering of different minerals. The chemical composition of sands may vary according to origin, but in general, it is constituted by 98.0–99.5% silica (SiO_2) (Perelli et al. 2009). pH is mainly related to the carbonate content. Sands with lower calcium carbonate content and pH 6.4–7.0 are better suited as substrate material because they do not influence the solubility of phosphorus and some microelements (e.g. iron, manganese). Like all mineral-origin substrates, sands have a low CEC and low buffering capability (Table 4.3). Fine sands (0.05–0.5 mm) are the most suitable for use in hydroponic systems in mixtures 10–30% by volume with organic materials. Coarse sands (>0.5 mm) can be used in order to increase the drainage capacity of the substrate.

Pumice

Pumice comprises aluminium silicate of volcanic origin, being very light and porous, and may contain small amounts of sodium and potassium and traces of calcium, magnesium and iron depending on the place of origin. It is able to retain calcium, magnesium, potassium and phosphorus from the nutrient solutions and to gradually release these to the plant. It usually has a neutral pH, but some materials may have

Table 4.3 Main chemical–physical characteristics of inorganic substrates used in soilless systems

Substrates	Bulk density (kg m^{-3})	Total porosity (%vol)	Free porosity (%vol)	Water-retention capacity (%vol)	CEC (meq %)	EC (mS cm^{-1})	pH
Sand	1400–1600	40–50	1–20	20–40	20–25	0.10	6.4–7.9
Pumice	450–670	55–80	30–50	24–32	–	0.08–0.12	6.7–9.3
Volcanic tuffs	570–630	80–90	75–85	2–5	3–5	–	7.0–8.0
Vermiculite	80–120	70–80	25–50	30–55	80–150	0.05	6.0–7.2
Perlite	90–130	50–75	30–60	15–35	1.5–3.5	0.02–0.04	6.5–7.5
Expanded clay	300–700	40–50	30–40	5–10	3–12	0.02	4.5–9.0
Stone wool	85–90	95–97	10–15	75–80	–	0.01	7.0–7.5
Expanded Polystyrene	6–25	55	52	3	–	0.01	6.1

Source: Enzo et al. (2001)

excessively high pH, good free porosity but low water-retention capacity (Table 4.3). The structure however tends to deteriorate fairly quickly, due to the easy breaking up of the particles. Pumice, added to peat, increases the drainage and aeration of the substrate. For horticulture use, pumice particles from 2 to 10 mm in diameter are preferred (Kipp et al. 2001).

Volcanic Tuffs

Tuffs derive from volcanic eruptions, with particles ranging between 2 and 10 mm diameter. They may have a bulk density ranging between 850 and 1100 kg m^{-3} and a water-retention capacity between 15% and 25% by volume (Kipp et al. 2001).

Vermiculite

Vermiculite comprises hydrous phyllosilicates of magnesium, aluminium and iron, which in the natural state have a thin lamellar structure that retains tiny drops of water. Exfoliated vermiculite is commonly used in the horticultural industry and is characterized by a high buffer capability and CEC values similar to those of the best peats (Table 4.3), but, compared to these, it has a higher nutrient availability (5–8% potassium and 9–12% magnesium) (Perelli et al. 2009). NH_4^+ is especially strongly retained by vermiculite; the activity of the nitrifying bacteria, however, allows the recovery of part of the fixed nitrogen. Similarly, vermiculite binds over 75% of phosphate in an irreversible form, whereas it has low absorbent capacity for Cl^-, NO_3^- and SO_4^-. These characteristics should be carefully assessed when vermiculite is used as a substrate. The vermiculite structure is not very stable because of a low compression resistance and tends to deteriorate over time, reducing water drainage. It can be used alone; however, it is preferable to mix it with perlite or peat.

Perlite

Perlite comprises aluminium silicate of volcanic origin containing 75% SiO_2 and 13% Al_2O_3. The raw material is crushed, sieved, compressed and heated to 700–1000 °C. At these temperatures, the little water contained in the raw material turns into vapour by expanding the particles into small whitish-grey aggregates which, unlike vermiculite, have a closed cell structure. It is very light and possesses high free porosity even after the soaking. It contains no nutrients, has negligible CEC and is virtually neutral (Table 4.3) (Verdonk et al. 1983). pH, however, can vary easily, because the buffer capacity is insignificant. pH ought to be controlled via the quality of the irrigation water and should not fall below 5.0 in order to avoid the phytotoxic effects of the aluminium. The closed cell structure allows water to be held only on the surface and in the spaces between the agglomerations, so the water-retention capacity is variable in relation to the dimensions of the agglomerations. It is marketed in different sizes, but the most suitable for horticulture are 2–5 mm diameter. It can be used as a substrate in rooting beds, because it ensures good aeration. In mixtures with organic materials, it enhances the softness, permeability and aeration of the substrate. Perlite can be reused for several years as long as it is sterilized between uses.

Expanded Clay

Expanded clay is obtained by treating clay powder at about 700 °C. Stable aggregates are formed, and, depending on the used clay material, they have variable values with regard to CEC, pH and bulk density (Table 4.3). Expanded clay can be used in mixtures with organic materials in the amounts of about 10–35% by volume, to which it provides more aeration and drainage (Lamanna et al. 1990). Expanded clays with pH values above 7.0 are not suitable for use in soilless systems.

Stone Wool

Stone wool is the most used substrate in soilless cultivation. It originates from the fusion of aluminium, calcium and magnesium silicates and carbon coke at 1500–2000 °C. The liquefied mixture is extruded in 0.05 mm diameter strands and, after compression and addition of special resins, the material assumes a very light fibrous structure with a high porosity (Table 4.3).

Stone wool is chemically inert and, when added to a substrate, it improves its aeration and drainage and also offers an excellent anchorage for plant roots. It is used alone, as a sowing substrate and for soilless cultivation. The slabs used for the cultivation can be employed for several production cycles depending on quality, as long as the structure is able to guarantee enough porosity and oxygen availability for root systems. Usually, after several crop cycles, the greater part of substrate porosity is filled with old, dead roots, and this is due to the compaction of the substrate over time. The result is a then a reduced depth of substrate where irrigation strategies may need adaptation.

Zeolites
Zeolites comprise hydrated aluminium silicates characterized by the capacity to absorb gaseous elements; they are high in macro- and microelements, they have high absorbent power and they have high internal surface (structures with 0.5 mm pores). This substrate is of great interest as it absorbs and slowly releases K^+ and NH_4^+ ions, whilst it is not able to absorb Cl^- and Na^+, which are hazardous to plants. Zeolites are marketed in formulations which differ in the N and P content and which can be used in seed sowing, for the rooting of cuttings or during the cultivation phase (Pickering et al. 2002).

4.2.4.3 Synthetic materials

Synthetic materials include both low-density plastic materials and ion-exchange synthetic resins. These materials, called "expanded", because they are obtained by a process of dilation at high temperatures, are not yet widely used, but they possess physical properties suitable to balance the characteristics of other substrates.

Expanded Polystyrene
Expanded polystyrene is produced in granules of 4–10 mm diameter with a closed cell structure. It does not decompose, is very light and has a very high porosity but with an extremely low water-retention capacity (Table 4.3). It has no CEC and virtually zero buffer capability, so it is added to the substrate (e.g. peat) exclusively to improve its porosity and drainage. The preferred particle size is 4–5 mm (Bunt 2012).

Polyurethane Foam
Polyurethane foam is a low-density material (12–18 kg m^{-3}) with a porous structure that allows absorption of water equal to 70% of its volume. It is chemically inert, has an almost neutral pH (6.5–7.0), does not contain useful nutrients available to plants and does not undergo decomposition (Kipp et al. 2001). In the market it is possible to find it in the form of granules, rooting cubes or blocks. Like a stone wool, it can also be used for soilless cultivation.

4.2.5 Preparation of Mixed Cultivation Substrates

Mixed substrates can be useful to reduce overall substrate costs and/or to improve some characteristics of the original materials. For example, peat, vermiculite and coir can be added to increase water-retention capacity; perlite, polystyrene, coarse sand and expanded clay to increase free porosity and drainage; blonde peat to raise the acidity; higher quantities of organic material or suitable amounts of clay soil to increase CEC and buffer capability; and low decomposable substrates for increased durability and stability. The characteristics of the mixtures rarely represent the average of the components because with the mixing the structures are modified between the individual particles and consequently the relationship of physical and chemical characteristics. In general, mixtures with a low nutrient content are preferable, in order to be able to better manage cultivation. The right relationship among the different constituents of a mixture also varies with the environmental conditions in which it operates. At high temperatures it is rational to use components that possess a higher water-retention capacity and do not allow fast evaporation (e.g. peat) and, at the same time, are resilient to decomposition. In contrast, in humid environments, with low solar radiation, the components characterized by high porosity are preferred to ensure good drainage. In this case, it will be necessary to add coarse substrates such as sand, pumice, expanded clay and expanded polystyrene (Bunt 2012).

4.3 Types of Hydroponic Systems According to Water/Nutrient Distribution

4.3.1 Deep Flow Technique (DFT)

Deep flow technique (DFT), also known as deep water technique, is the cultivation of plants on floating or hanging support (rafts, panels, boards) in containers filled with 10–20 cm nutrient solution (Van Os et al. 2008) (Fig. 4.3). In AP this can be up to 30 cm. There are different forms of application that can be distinguished mainly by the depth and volume of the solution, and the methods of recirculation and oxygenation.

One of the simplest systems comprises 20–30 cm deep tanks, which can be constructed of different materials and waterproofed with polyethylene films. The tanks are equipped with floating rafts (several types are available from suppliers) that serve to support the plants above the water whilst the plants' roots penetrate the water. The system is particularly interesting as it minimizes costs and management. For example, there is a limited need for the automation of the control and correction

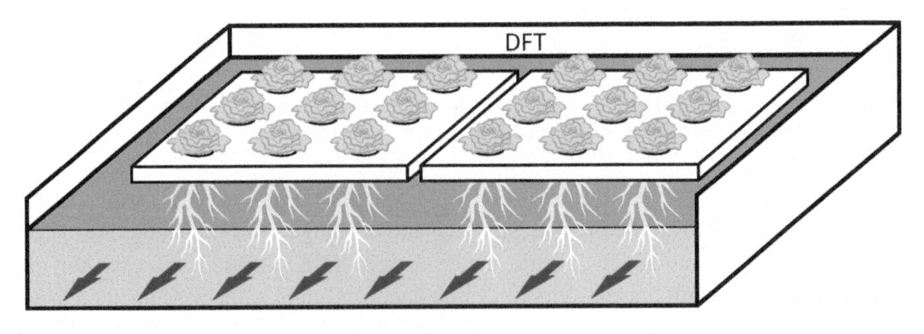

Fig. 4.3 Illustration of a DFT system with floating panels

of the nutrient solution, particularly in short duration crops such as lettuce, where the relatively high volume of solution facilitates the replenishment of the nutrient solution only at the end of each cycle, and only the oxygen content needs to be monitored periodically. Oxygen levels should be above 4–5 mg L^{-1}; otherwise, nutrient deficiencies may appear due to root systems uptake low performance. Circulation of the solution will normally add oxygen, or Venturi systems can be added which dramatically increase air into the system. This is especially important when water temperatures are greater than 23 °C, as such high temperatures may stimulate lettuce bolting.

4.3.2 Nutrient Film Technique (NFT)

The NFT technique is used ubiquitously and can be considered the classic hydroponic cultivation system, where a nutrient solution flows along and circulates in troughs with a 1–2 cm layer of water (Cooper 1979; Jensen and Collins 1985; Van Os et al. 2008) (Fig. 4.4). The recirculation of the nutrient solution and the absence of substrate represent one of the main advantages of the NFT system. An additional advantage is its great potential for automation to save on labour costs (planting & harvesting) and the opportunity to manage the optimal plant density during crop cycle. On the other hand, the lack of substrate and low water levels makes the NFT vulnerable to the failure of pumps, due to e.g. clogging or a failure in the power supply. Temperature fluctuations in the nutrient solution can cause plant stress followed by diseases.

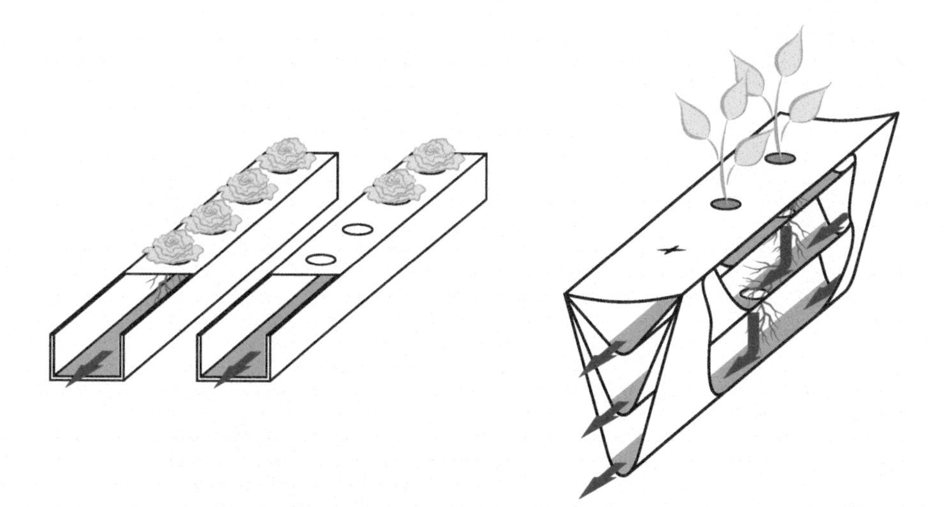

Fig. 4.4 Illustration of NFT system (left) and a multilayer NFT trough, developed and marketed by New Growing Systems (NGS), Spain (right)

The development of the root system, part of which remains suspended in air above the nutrient flow and which is exposed to an early ageing and loss of functionality, represents a major constraint as it prevents the production of long-cycle crops (over 4–5 months). Because of its high susceptibility to temperature variations, this system is not suitable for cultivation environments characterized by high levels of irradiation and temperature (e.g. southern areas of the Mediterranean basin). However, in response to these challenges, a multilayer NFT trough has been designed which allows for longer production cycles without clogging problems (NGS). It is made of a series of interconnected layers placed in a cascade, so that even in strong rooting plant species, such as tomatoes, the nutrient solution will still find its way to the roots by by-passing the root-clogged layer via a lower positioned layer.

4.3.3 Aeroponic Systems

The aeroponic technique is mainly aimed at smaller horticultural species, and has not yet been widely used due to the high investment and management costs. Plants are supported by plastic panels or by polystyrene, arranged horizontally or on inclined tops of growing boxes. These panels are supported by a structure made with inert materials (plastic, steel coated with plastic film, polystyrene boards), in order to form closed boxes where the suspended root system can develop (Fig. 4.5).

The nutrient solution is directly sprayed on the roots, which are suspended in the box in air, with static sprinklers (sprayers), inserted on pipes housed inside the box

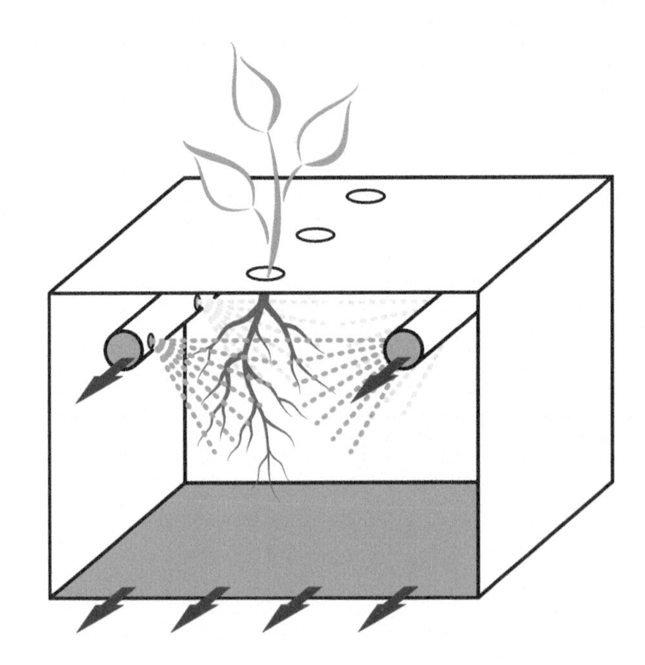

Fig. 4.5 Illustration of the aeroponics technique

module. The spray duration is from 30 to 60 s, whilst the frequency varies depending on the cultivation period, the growth stage of the plants, the species and the time of day. Some systems use vibrating plates to create micro droplets of water which form a steam which condenses on the roots. The leachate is collected on the bottom of the box modules and conveyed to the storage tank, for reuse.

4.4 Plant Physiology

4.4.1 Mechanisms of Absorption

Amongst the main mechanisms involved in plant nutrition, the most important is the absorption which, for the majority of the nutrients, takes place in ionic form following the hydrolysis of salts dissolved in the nutrient solution.

Active roots are the main organ of the plant involved in nutrient absorption. Anions and cations are absorbed from the nutrient solution, and, once inside the plant, they cause the protons (H^+) or hydroxyls (OH^-) to exit which maintains the balance between the electric charges (Haynes 1990). This process, whilst maintaining the ionic equilibrium, can cause changes in the pH of the solution in relation to the quantity and quality of the nutrients absorbed (Fig. 4.6).

The practical implications of this process for the horticulturist are two-fold: to provide adequate buffer capability to the nutrient solution (adding bicarbonates if needed) and to induce slight pH changes with the choice of fertilizer. The effect of fertilizers on the pH relates to the different chemical forms of the used compounds.

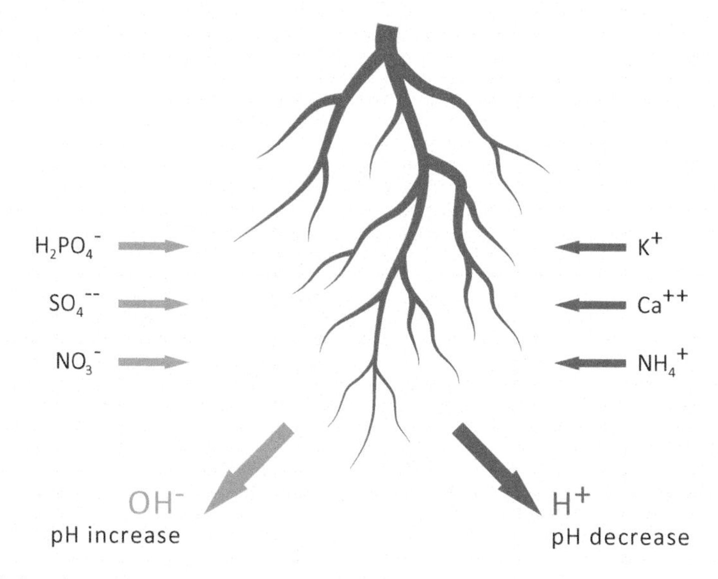

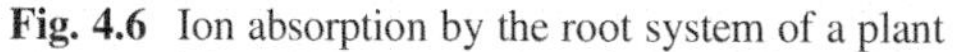

Fig. 4.6 Ion absorption by the root system of a plant

In the case of N, for example, the most commonly used form is nitric nitrogen (NO_3^-), but when the pH should be lowered, nitrogen can be supplied as ammonium nitrogen (NH_4^+). This form, when absorbed, induces the release of H^+ and consequently an acidification of the medium.

Climatic conditions, especially air and substrate temperature and relative humidity, exert a major influence on the absorption of nutrients (Pregitzer and King 2005; Masclaux-Daubresse et al. 2010; Marschner 2012; Cortella et al. 2014). In general, the best growth occurs where there are few differences between substrate and air temperature. However, persistently high temperature levels in the root system have a negative effect. Sub-optimal temperatures reduce the absorption of N (Dong et al. 2001). Whilst NH_4^+ is effectively used at optimum temperatures, at low temperatures, the bacterial oxidation is reduced, causing accumulation within the plant that can produce symptoms of toxicity and damage to the root system and the aerial biomass. Low temperatures at the root level also inhibit the assimilation of K and P, as well as the P translocation. Although the available information regarding the effect of low temperatures on the absorption of micronutrients is less clear, it appears that Mn, Zn, Cu, and Mo uptake are most affected (Tindall et al. 1990; Fageria et al. 2002).

4.4.2 Essential Nutrients, Their Role and Possible Antagonisms

The appropriate management of plant nutrition must be based on basic aspects that are influenced by uptake and use of macro, and micro-nutrients (Sonneveld and Voogt 2009). Macro-nutrients are needed in relatively large amounts, whilst micronutrients or trace elements are needed in small amounts. Furthermore, nutrient availability to the plant in the case of the soilless systems presents more or less consistent phenomena of synergy and antagonism (Fig. 4.7).

Nitrogen (N) Nitrogen is absorbed by plants to produce amino acids, proteins, enzymes and chlorophyll. The most used nitrogen forms for plant fertilization are nitrate and ammonium. Nitrates are quickly absorbed by the roots, are highly movable inside the plants and can be stored without toxic effects. Ammonium can be absorbed by plants only in low quantities and cannot be stored at high quantities because it exerts toxic effects. Quantities higher than 10 mg L^{-1} inhibit plant calcium and copper uptake, increase the shoot growth compared to root growth and result in a strong green colour of the leaves. Further excesses in ammonia concentration result in phytotoxic effects such as chlorosis along the leaves' margins. Excess in nitrogen supply causes high vegetative growth, increase of crop cycle length, strong green leaf colour, low fruit set, high content of water in the tissues, low tissue lignification and high tissue nitrate accumulation. Commonly, nitrogen deficiency is characterized by a pale green colour of the older leaves (chlorosis), reduced growth and senescence advance.

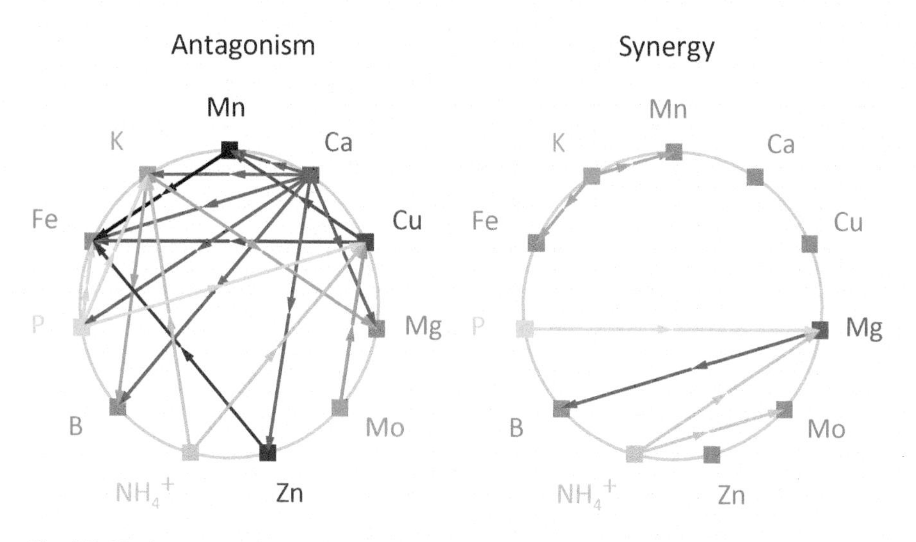

Fig. 4.7 Nutrients synergies and antagonisms amongst ions. Connected ions present synergistic or antagonistic relationship according to the direction of the arrow

Potassium (K) Potassium is fundamental for cell division and extension, protein synthesis, enzyme activation and photosynthesis and also acts as a transporter of other elements and carbohydrates through the cell membrane. It has an important role in keeping the osmotic potential of the cell in equilibrium and regulating the stomatal opening. The first signs of deficiency are manifested in the form of yellowish spots that very quickly necrotize on the margins of the older leaves. Potassium deficient plants are more susceptible to sudden temperature drops, water stress and fungal attacks (Wang et al. 2013).

Phosphorus (P) Phosphorus stimulates roots development, the rapid growth of buds and flower quantity. P is absorbed very easily and can be accumulated without damage to the plant. Its fundamental role is linked to the formation of high-energy compounds (ATP) necessary for plant metabolism. The average quantities requested by plants are rather modest (10–15% of the needs of N and K) (Le Bot et al. 1998). However, unlike what occurs in soil, P is easily leachable in soilless crops. The absorption of P appears to be reduced by low substrate temperatures (< 13 °C) or at increasing pH values (> 6.5) which can lead to deficiency symptoms (Vance et al. 2003). Under these conditions a temperature increase and/or pH reduction is more effective than additional amendments of phosphorus fertilizers. P excess can reduce or block the absorption of some other nutrients (e.g. K, Cu, Fe) (Fig. 4.7). Phosphorus deficiency manifests in a green-violet colour of the older leaves, which may follow chlorosis and necrosis in addition to the stunted growth of the vegetative apex. However, these symptoms are non-specific and make P deficiencies difficult to be identified (Uchida 2000).

Calcium (Ca) Calcium is involved in cell wall formation, membrane permeability, cell division and extension. Good availability gives the plant greater resistance to fungal attacks and bacterial infections (Liu et al. 2014). The absorption is very closely linked to the water flow between roots and aerial parts. Its movement occurs through the xylem and is therefore particularly influenced by low temperatures at the root level, by reduced water supply (drought or salinity of the solution) or by excessive relative humidity of the air. As Ca is not mobile within the plant, deficiencies start from the most recently formed parts (Adams 1991; Adams and Ho 1992; Ho et al. 1993). The main symptoms are plant growth being stunted, deformation of the margins of the younger leaves, light green or sometimes chlorotic colouring of new tissues and a stunted root system without fine roots. The deficiencies are displayed in different ways, e.g. apical rot in tomato and/or marginal browning of leaves in lettuce.

Magnesium (Mg) Magnesium is involved in the constitution of chlorophyll molecules. It is immobilized at pH values below 5.5 and enters into competition with the absorption of K and Ca (Fig. 4.7). Symptoms of deficiency are yellowing between leaf veins and internal chlorosis of the basal leaves. As Mg can be easily mobilized, magnesium-deficient plants will first break down chlorophyll in the older leaves and transport the Mg to younger leaves. Therefore, the first sign of magnesium deficiency is the interveinal chlorosis in older leaves, contrary to iron deficiency where interveinal chlorosis first appears in the youngest leaves (Sonneveld and Voogt 2009).

Sulphur (S) Sulphur is required by the plant in quantities comparable to those of phosphorus, and in order to optimize its absorption, it must be present in a 1:10 ratio with nitrogen (McCutchan et al. 2003). It is absorbed as sulphate. The deficiencies are not easily detected, as the symptoms can be confused with those of nitrogen deficiency, except that the deficiency of nitrogen begins to manifest itself from the older leaves, whilst that of sulphur from the youngest ones (Schnug and Haneklaus 2005). S nutrition has a significant role in ameliorating the damages in photosynthetic apparatus caused by Fe-deficiency (Muneer et al. 2014).

Iron (Fe) Iron is one of the most important micro-nutrients because it is key in many biological processes such as photosynthesis (Briat et al. 2015; Heuvelink and Kierkels 2016). To improve its absorption, the nutrient solution pH should be around 5.5–6.0, and the Mn content should not be allowed to become too high because the two elements subsequently enter into competition (Fig. 4.7). The optimal ratio of Fe–Mn is around 2:1 for most crops (Sonneveld and Voogt 2009). At low temperatures, the assimilation efficiency is reduced. The deficiency symptoms are characterized by interveinal chlorosis from the young leaves towards the older basal ones, and by reduced root system growth. Symptoms of deficiency are not always due to the low presence of Fe in the nutrient solution, but often they are due to the Fe unavailability for the plant. The use of chelating agents guarantees constant availability of Fe for the plant.

Chlorine (Cl) Chlorine has been recently considered a micro-nutrient, even if its content in plants (0.2–2.0% dw) is quite high. It is easily absorbed by the plant and is very mobile within it. It is involved in the photosynthetic process and the regulation of the stomata opening. Deficiencies, which are rather infrequent, occur with typical symptoms of leaves drying out, especially at the margins. Much more widespread is the damage due to an excess of Cl that leads to conspicuous plant shrinkage which is relative to the different sensitivities of different species. To avoid crop damage, it is always advisable to check the Cl content in the water used to prepare nutrient solutions and choose suitable fertilizers (e.g. K_2SO_4 rather than KCl).

Sodium (Na) Sodium, if in excess, is harmful to plants, as it is toxic and interferes with the absorption of other ions. The antagonism with K (Fig. 4.7), for example, is not always harmful because in some species (e.g. tomatoes), it improves the fruit taste, whereas in others (e.g. beans), it can reduce plant growth. Similar to Cl, it is important to know the concentration in the water used to prepare the nutrient solution (Sonneveld and Voogt 2009).

Manganese (Mn) Manganese forms part of many coenzymes and is involved in the extension of root cells and their resistance to pathogens. Its availability is controlled by the pH of the nutrient solution and by competition with other nutrients (Fig. 4.7). Symptoms of deficiency are similar to those of the Fe except for the appearance of slightly sunken areas in the interveinal areas (Uchida 2000). Corrections can be made by adding $MnSO_4$ or by lowering the pH of the nutrient solution.

Boron (B) Boron is essential for fruit setting and seed development. The absorption methods are similar to those already described for Ca with which it can compete. The pH of the nutrient solution must be below 6.0 and the optimal level seems to be between 4.5 and 5.5. Symptoms of deficiency can be detected in the new structures that appear dark green, the young leaves greatly increase their thickness and have a leathery consistency. Subsequently they can appear chlorotic and then necrotic, with rusty colouring.

Zinc (Zn) Zinc plays an important role in certain enzymatic reactions. Its absorption is strongly influenced by the pH and the P supply of the nutrient solution. pH values between 5.5 and 6.5 promote the absorption of Zn. Low temperature and high P levels reduce the amount of zinc absorbed by the plant. Zinc deficiencies occur rarely, and are represented by chlorotic spots in the interveinal areas of the leaves, very short internodes, leaf epinasty and poor growth (Gibson 2007).

Copper (Cu) Copper is involved in respiratory and photosynthetic processes. Its absorption is reduced at pH values higher than 6.5, whilst pH values lower than 5.5 may result in toxic effects (Rooney et al. 2006). High levels of ammonium and phosphorus interact with Cu reducing the availability of the latter. The excessive presence of Cu interferes with the absorption of Fe, Mn and Mo. The deficiencies are manifested by interveinal chlorosis which leads to the collapse of the leaf tissues that look like desiccated (Gibson 2007).

Molybdenum (Mo) Molybdenum is essential in protein synthesis and in nitrogen metabolism. Contrary to other micro-nutrients, it is better available at neutral pH values. Symptoms of deficiency start with chlorosis and necrosis along the main rib of old leaves, whilst the young leaves appear deformed (Gibson 2007).

4.4.3 Nutrient Management in Relation to the Requirements of Plants

Since the development of soilless horticulture systems in the 1970s (Verwer 1978; Cooper 1979), different nutrient solutions have been developed and adjusted according to the growers' preferences (Table 4.4; De Kreij et al. 1999). All mixes follow the principles of excess availability of all elements to prevent deficiencies and balance between (bivalent) cations to avoid competition between cations in plant nutrient uptake (Hoagland and Arnon 1950; Steiner 1961; Steiner 1984; Sonneveld and Voogt 2009). Commonly, the EC is allowed to rise in the root zone to a limited degree. In tomatoes, for example, the nutrient solution typically has an EC of ca. 3 dS m^{-1}, whilst in the root zone in the stone wool slabs, the EC may rise to 4–5 dS m^{-1}. However, in northern European countries, for the first irrigation of new stone wool slabs at the beginning of the production cycle, the nutrient solution may have an EC as high as 5 dS m^{-1}, saturating the stone wool substrate with ions up to an EC of 10 dS m^{-1}, which will subsequently be flushed after 2 weeks. To provide sufficient flushing of the root zone, in a typical drip-irrigation stone wool slab system, about 20–50% of the dosed water is collected as drainage water. The drainage water is then recycled, filtered, mixed with fresh water and topped up with nutrients for use in the next cycle (Van Os 1994).

In tomato production, increasing the EC can be applied to enhance lycopene synthesis (promoting the bright red coloration of the fruits), total soluble solids (TSS) and fructose and glucose content (Fanasca et al. 2006; Wu and Kubota 2008). Furthermore, tomato plants have higher absorption rates for N, P, Ca and Mg and low absorption of K during the early (vegetative) stages. Once the plants start developing fruits, leaf production is slowed down leading to a reduction in N and Ca requirements, whilst K requirement increases (e.g. Zekki et al. 1996; Silber, Bar-Tal 2008). In lettuce, on the other hand, an increased EC may promote tip-burn disease during hot growing conditions. Huett (1994) showed a significant decrease in the number of leaves with tip-burn disease per plant when the EC was dropped from 3.6 to 0.4 dS m^{-1}, as well as when the nutrient formulation K/Ca was reduced from 3.5:1 to 1.25:1. In AP the management of nutrients is more difficult than in hydroponics since they mainly depend on fish stock density, feed type and feeding rates.

4.4.4 Nutrient Solution Properties

Phosphorus is an element which occurs in forms that are strongly dependent on environmental pH. In the root zone, this element can be found as PO_4^{3-}, HPO_4^{2-} and $H_2PO_4^-$ ions, where the last two ions are the main forms of P taken up by the plants. Thus, when the pH is slightly acidic (pH 5–6), the largest amount of P is presented in a nutrient solution (De Rijck and Schrevens 1997).

Potassium, calcium and magnesium are available to plants in a wide range of pH. However, the presence of other ions may interfere in their plant availability due to the formation of compounds with different grades of solubility. At a pH above 8.3, Ca^{2+} and Mg^{2+} ions easily precipitate as carbonates by reacting with CO_3^{2-}. Also sulphate forms relatively strong complexes with Ca^{2+} and Mg^{2+} (De Rijck and Schrevens 1998). As pH increases from 2 to 9, the amount of SO_4^{2-} forming soluble complexes with Mg^{2+} as $MgSO_4$ and with K^+ as KSO_4^- increases (De Rijck and Schrevens 1999). In general, nutrient availability for plant uptake at pH above 7 may be restricted due to a precipitation of Boron, Fe^{2+}, Mn^{2+}, PO_4^{3-}, Ca^{2+} and Mg^{2+} due to insoluble and unavailable salts. The most appropriate pH values of the nutrient solution for the development of crops lie between 5.5 and 6.5 (Sonneveld and Voogt 2009).

4.4.5 Water Quality and Nutrients

The quality of the supplied water is extremely important in hydroponic and AP systems. For long-term recirculation, the chemical composition should be well known and monitored frequently to avoid an imbalance in nutrient supply but also to avoid the accumulation of certain elements leading to toxicity. De Kreij et al. (1999) made an overview of the chemical demands on water quality for hydroponic systems.

Before starting, an analysis of the water supply has to be made on the macro- and microelements. Based on the analysis, a scheme for the nutrient solution can be made. For example, if rainwater is used, special attention has to be made for Zn when collection takes place via untreated gutters. In tap water, problems may appear with Na, Ca, Mg, SO_4 and HCO_3. Furthermore, surface and bore hole water may be used which may also contain amounts of Na, Cl, K, Ca, Mg, SO_4 and Fe but also microelements as Mn, Zn, B and Cu. It should be noted that all valves and pipes should be made of synthetic materials such as PVC and PE, and not containing Ni or Cu parts.

It often happens that water supplies contain a certain amount of Ca and Mg; therefore, the contents have to be subtracted from the amount in the nutrient solution to avoid accumulation of these ions. HCO_3 has to be compensated preferably by nitric acid, about 0.5 mmol L^{-1} that can be maintained as a pH buffer in the nutrient solution. Phosphoric and sulphuric acid can also be possibly used to compensate pH, but both will rapidly give a surplus of $H_2PO_4^-$ or SO_4^{2-} in the nutrient solution. In AP systems nitric acid (HNO_3) and potassium hydroxide (KOH) can be also used to regulate pH and at same time supply macronutrients in the system (Nozzi et al. 2018).

4.4.5.1 Water Quality Management

For the formulation of nutrient solutions, simple fertilizers (granular, powder or liquid) and substances (e.g. acid compounds) that affect the pH are preferably used. The integration of the nutrient elements into the solution takes into account the optimum values of the quantities of each element. This has to be made in relation to the requirements of the species and its cultivars considering phenological phases and substrate. The calculation of nutrient supplements must be carried out considering the conditions of the water used, according to a *strict set of priorities*. On the priority scale, magnesium and sulphates are positioned at the bottom, at the same level, because they have less nutritional importance and the plants do not present damage even if their presence is abundant in the nutrient solution. This characteristic has an advantageous practical feedback as it allows an exploitation of the two elements in order to balance the nutritional composition with respect to other macronutrients whose deficiency or excess may be negative for production. As an example, we can consider a nutrient solution where an integration of only potassium or only nitrate is required. The salts to be used, in this case, are respectively potassium sulphate or magnesium nitrate. In fact, if the most common potassium nitrate or calcium nitrate were used, the levels of nitrate, in the first case, and calcium, in the second case, would automatically increase. Moreover, when the analysis of the water used shows an imbalance between cations and anions, and in order to be able to calculate a nutrient solution with the EC in equilibrium, the correction of the water values is carried out reducing the levels of magnesium and/or sulphates.

The following points provide guidelines for the formulation of nutrient solutions:

1. Definition of the species and cultivar requirements. Consideration of the cultivation environment and the characteristics of water need to be taken into account. In order to satisfy the needs of the plants in warm periods and with intense radiation, the solution must possess a lower EC and K content, which contrasts to a higher quantity of Ca. Instead, when the temperature and brightness reach sub-optimal levels, it is advisable to raise the values of the EC and K by reducing those of the Ca. It is important to note regarding cultivars that there are substantial variations, especially for the values of the NO_3^-, due to the different vegetative vigorousness of the cultivars. For tomatoes, in fact, 15 mmol L^{-1} of NO_3^- is used on average (Table 4.4), and in the case of cultivars characterized by low vegetative vigour and in certain phenological phases (e.g. fruit setting of the fourth trusses), up to 20 mM L^{-1} of NO_3^- is adopted. In case that some elements such as Na are

Table 4.4 Nutrient solutions in hydroponic cultivation of lettuce (DFT) tomato, pepper and cucumber (stone wool slabs with drip irrigation) in the Netherlands (De Kreij et al. 1999)

	pH	EC	NH_4	K	Ca	Mg	NO_3	SO_4	P	Fe	Mn	Zn	B	Cu	Mo
		dS m^{-1}	mmol L^{-1}	mmol L^{-1}	mmol L^{-1}	mmol L^{-1}	mmol L^{-1}	mmol L^{-1}	mmol L^{-1}	mmol L^{-1}	mmol L^{-1}	mmol L^{-1}	mmol L^{-1}	mmol L^{-1}	mmol L^{-1}
Lettuce (Wageningen UR)	5.9	1.7	1.0	4.4	4.5	1.8	10.6	1.5	1.5	28.1	1.5	6.4	47.0	1.0	0.7
Lettuce	5.8	1.2	0.7	4.8	2.3	0.8	8.9	0.8	1.0	35.1	4.9	3.0	18.4	0.5	0.5
Lettuce	5.8	1.2		3.0	2.5	1.0	7.5	1.0	0.5	50.0	3.7	0.6	4.8	0.5	0.01
Tomato generative	5.5	2.6–3.0	1.2	13.0	4.2	1.9	15.4	4.7	1.5	15.0	10.0	5.0	30.0	0.8	0.5
Tomato vegetative	5.5	2.6	1.2	8.3	5.7	2.7	15.4	4.7	1.5	15.0	10.0	5.0	30.0	0.8	0.5
Cucumber	5.5	3.2	1.2	10.4	6.7	2.0	23.3	1.5–2.0	1.5–2.0	15.0	10.0	5.0	25.0	0.8	0.5
Pepper	5.6	2.5–3.0	1.2	5–7	4–5	2.0	17.0	1.8–2.0	1.5–2.5	25.0	10.0	7.0	30.0	1.0	0.5
Plant propagation	5.5	2.3	1.2	6.8	4.5	3.0	16.8	2.5	1.3	25.0	10.0	5.0	35.0	1.0	0.5

Adopted and modified from Vermeulen (2016, personal communication)

present in the water, in order to reduce its effect, which is particularly negative for some crops, it will be necessary to increase the amount of NO_3^- and Ca and possibly decrease the K, keeping the EC at the same level.

2. Nutrient requirement calculations should be obtained by subtracting the values of the chemical elements of the water from the chemical elements defined above. For example, the established need for Mg of peppers (*Capsicum* sp.) is 1.5 mM L^{-1}, having the water at 0.5 mM L^{-1}, and 1.0 mM L^{-1} of Mg should be added to the water (1.5 requirement – 0.5 water supply = 1.0).
3. Choice and calculation of fertilizers and acids to be used. For example, having to provide Mg, as in the example of point 2 above, $MgSO_4$ or $Mg(NO_3)_2$ can be used. A decision will be made taking into account the collateral contribution of sulphate or nitrate as well.

4.4.6 *Comparison Between Hydroponic and Aquaponic Production*

During their life cycle, plants need several essential macro- and microelements for regular development (boron, calcium, carbon, chlorine, copper, hydrogen, iron, magnesium, manganese, molybdenum, nitrogen, oxygen, phosphorous, potassium, sulphur, zinc), usually absorbed from the nutrient solution (Bittsanszky et al. 2016). The nutrient concentration and ratio amongst them are the most important variables capable to influence plant uptake. In AP systems fish metabolic wastes contain nutrients for the plants, but it must be taken into account, especially at commercial scales, that the nutrient concentrations supplied by the fish in AP systems are significantly lower and unbalanced for most nutrients compared to hydroponic systems (Nicoletto et al. 2018). Usually, in AP, with appropriate fish stocking rates, the levels of nitrate are sufficient for good plant growth, whereas the levels of K and P are generally insufficient for maximum plant growth. Furthermore, calcium and iron could also be limited. This can reduce the crop yield and quality and so nutrient integration should be carried out to support an efficient nutrient reuse. Microbial communities play a crucial role in the nutrient dynamic of AP systems (Schmautz et al. 2017), converting ammonium to nitrate, but also contributing to the processing of particulate matter and dissolved waste in the system (Bittsanszky et al. 2016). Plant uptake of N and P represents only a fraction of the amount removed from the water (Trang and Brix 2014), indicating that microbial processes in the root zone of the plants, and in the substrate (if present) and throughout the whole system, play a major role.

The composition of fish feeds depends on the type of fish and this influences nutrient release from fish's metabolic output. Typically, fish feed contains an energy source (carbohydrates and/or lipids), essential amino acids, vitamins, as well as other organic molecules that are necessary for normal metabolism but some that the fish's cells cannot synthesize. Furthermore, it must be taken into account that a plant's nutritional requirements vary with species (Nozzi et al. 2018), variety, life cycle stage, day length and weather conditions and that recently (Parent et al. 2013; Baxter 2015), the Liebig's law (plant growth is controlled by the scarcest resource) has been superseded by complex algorithms that consider the interactions between the individual nutrients. Both these aspects do not allow a simple evaluation of the effects of changes in nutrient concentrations in hydroponic or AP systems.

The question thus arises whether it is necessary and effective to add nutrients to AP systems. As reported by Bittsanszky et al. (2016), AP systems can only be operated efficiently and thus successfully, if special care is taken through the continuous monitoring of the chemical composition of the recirculating water for adequate concentrations and ratios of nutrients and of the potentially toxic component, ammonium. The necessity to add nutrients depends on plant species and growth stage. Frequently, although fish density is optimal for nitrogen supply, the addition of P and K with mineral fertilizers, at least, should be carried out (Nicoletto et al. 2018). In contrast to, for example, lettuce, tomatoes which need to bear fruit, mature and ripen, need supplemental nutrients. In order to calculate these needs, a software can be used, such as *HydroBuddy* which is a free software (Fernandez 2016) that is used to calculate the amount of required mineral nutrient supplements.

4.5 Disinfection of the Recirculating Nutrient Solution

To minimize the risk of spreading soil-borne pathogens, disinfection of the circulating nutrient solution is required (Postma et al. 2008). Heat treatment (Runia et al. 1988) was the first method used. Van Os (2009) made an overview for the most important methods and a summary is given below. Recirculating of the nutrient solution opens possibilities to save on water and fertilizers (Van Os 1999). The big disadvantage of the recirculation of the nutrient solution is the increasing risk of spreading root-borne pathogens all over the production system. To minimize such risks, the solution should be treated before reuse. The use of pesticides for such a treatment is limited as effective pesticides are not available for all such pathogens, and if available, resistance may appear, and environmental legislation restricts discharge of water with pesticides (and nutrients) into the environment (European Parliament and European Council 2000). In addition, in AP systems, the use of pesticides exerts negative effects on fish health and cannot be carried out, even if hydroponic and AP parts of the system are in different rooms, because spraying of chemicals may enter the nutrient solution via condensation water or via direct spraying on the substrate slabs. In view of this, a biological control approach can be adopted to manage pest diseases, and this can be accessed via the EU Aquaponics Hub Fact Sheet (EU Aquaponics Hub). At the same time, similar problems can be observed for fish treatment using veterinary drugs that are not compatible with the plant's cycle.

4.5.1 Description of Disinfection Methods

Disinfection of the circulating nutrient solution should take place continuously. All drain returned (10–12 h during daytime) has to be treated within 24 h. For a greenhouse of 1000 m^2 in a substrate cultivation (stone wool, coir, perlite), a disinfection capability of about 1–3 m^3 per day is needed to disinfect an estimated needed surplus of 30% of the water supplied with drip irrigation to tomato plants during a 24-h period in summer conditions. Because of the variable return rate of drain water, a sufficiently large catchment tank for drain water is needed in which the water is stored before it is pumped to the disinfection unit. After disinfection another tank is required to store the clean water before adjusting EC and pH and blending with new water to supply to the plants. Both tanks have an average size of 5 m^3 per 1000 m^2. In a nutrient film system (NFT), about 10 m^3 per day should be disinfected daily. It is generally considered that such a capacity is uneconomical to disinfect (Ruijs 1994). DFT requires similar treatment. This is the main reason why NFT and DFT production units do not normally disinfect the nutrient solution. Disinfection is carried out either by non-chemical or chemical methods as follows:

4.5.1.1 Non-chemical Methods

In general these methods do not alter the chemical composition of the solution, and there is no build-up of residuals:

1. *Heat treatment***.** Heating the drain water to temperatures high enough to eradicate bacteria and pathogens is the most reliable method for disinfection. Each type of organism has its own lethal temperature. Non-spore-forming bacteria have lethal temperatures between 40 and 60 °C, fungi between 40 and 85 °C, nematodes between 45 and 55 °C and viruses between 80 and 95°C (Runia et al. 1988) at an exposure time of 10 s. Generally, the temperature set point of 95 °C is high enough to kill most of the organisms that are likely to cause diseases with a minimum time of 10 s. Whilst this may seem very energy intensive, it should be noted that the energy is recovered and reused with heat exchangers. Availability of a cheap energy source is of greater importance for practical application.
2. *UV radiation*. UV radiation is electromagnetic radiation with a wavelength between 200 and 400 nm. Wavelengths between 200 and 280 nm (UV-C), with an optimum at 254 nm, has a strong killing effect on micro-organisms, because it minimizes the multiplication of DNA chains. Different levels of radiation are needed for different organisms so as to achieve the same level of efficacy. Runia (1995) recommends a dose which varies from 100 mJ cm^{-2} for eliminating bacteria and fungi to 250 mJ cm^{-2} for eliminating viruses. These relatively high doses are needed to compensate for variations in water turbidity and variations in penetration of the energy into the solution due to low turbulence around the UV lamp or variations in output from the UV lamp. Zoschke et al. (2014) reviewed that UV irradiation at 185 and 254 nm offers water organic

contaminant control and disinfection. Moreover, Moriarty et al. (2018) reported that UV radiation efficiently inactivated coliforms in AP systems.

3. *Filtration*. Filtration can be used to remove any undissolved material out of the nutrient solution. Various types of filters are available relative to the range of particle sizes. Rapid sand filters are often used to remove large particles from the drain water before adding, measuring and control of EC, pH and application of new fertilizers. After passing the fertilizer unit, often a fine synthetic filter (50–80 um) is built in the water flow to remove undissolved fertilizer salts or precipitates to avoid clogging of the irrigation drippers. These synthetic filters are also used as a pretreatment for disinfection methods with heat treatment, ozone treatment or UV radiation. With reductions in filtration pore size, the flow is inhibited, so that removal of very small particles requires a combination of adequate filters and high pressure followed by frequent cleaning of the filter(s). Removal of pathogens requires relatively small pore sizes (<10 μm; so-called micro-, ultra- or nanofiltration).

4.5.1.2 Chemical Methods

1. *Ozone (O_3)*. Ozone is produced from dry air and electricity using an ozone-generator (converting $3O_2 \rightarrow 2O_3$). The ozone-enriched air is injected into the water that is being sanitized and stored for a period of 1 h. Runia (1995) concluded that an ozone supply of 10 g per hour per m^3 drain water with an exposure time of 1 h is sufficient to eliminate all pathogens, including viruses. The reduction of microbial populations in vegetable production in soilless systems managed with ozone has also been observed by Nicoletto et al. (2017). Human exposure to the ozone that vents from the system or the storage tanks should be avoided since even a short exposure time of a concentration of 0.1 mg L^{-1} of ozone may cause irritation of mucous membranes. A drawback of the use of ozone is that it reacts with iron chelate, as UV does. Consequently, higher dosages of iron are required and measures need to be taken to deal with iron deposits in the system. Recent research (Van Os 2017) with contemporary ozone installations looks promising, where complete elimination of pathogens and breakdown of remaining pesticides is achieved, with no safety problems.
2. *Hydrogen peroxide (H_2O_2)*. Hydrogen peroxide is a strong, unstable oxidizing agent that reacts to form H_2O and an $O^{\cdot}$- radical. Commercially so-called activators are added to the solution to stabilize the original solution and to increase efficacy. Activators are mostly formic acid or acetic acid, which decrease pH in the nutrient solution. Different dosages are recommended (Runia 1995) against *Pythium* spp. (0.005%), other fungi (0.01%), such as *Fusarium*, and viruses (0.05%). The 0.05% concentration is also harmful to plant roots. Hydrogen peroxide is especially helpful for cleaning the watering system, whilst the use for disinfection has been taken over by other methods. The method is considered inexpensive, but not efficient.

3. *Sodium hypochlorite (NaOCl).* Sodium hypochlorite is a compound having different commercial names (e.g. household bleach) with different concentrations but with the same chemical structure (NaOCl). It is widely used for water treatment, especially in swimming pools. The product is relatively inexpensive. When added to water, sodium hypochlorite decomposes to HOCl and $NaOH^-$ and depending on the pH to OCl^-; the latter decomposes to Cl^- and O· for strong oxidation. It reacts directly with any organic substance, and if there is enough hypochlorite, it also reacts with pathogens. Le Quillec et al. (2003) showed that the tenability of hypochlorite depends on the climatic conditions and the related decomposing reactions. High temperatures and contact with air cause rapid decomposition, at which $NaClO_3$ is formed with phytotoxic properties. Runia (1995) showed that hypochlorite is not effective for eliminating viruses. Chlorination with a concentration of 1–5 mg Cl L^{-1} and an exposure time of 2 h achieved a reduction of 90–99.9% of *Fusarium oxysporum*, but some spores survived at all concentrations. Safety measures have to be taken for safe storage and handling. Hypochlorite might work against a number of pathogens, but not all, but at the same time, Na^+ and Cl^- concentration is increased in a closed growing system which will also lead to levels which decrease productivity of the crop and at which time the nutrient solution has to be leached. Despite the above-mentioned drawbacks, the product is used and recommended by commercial operatives as a cheap and useful method.

4.5.2 Chemical Versus Non-chemical Methods

Growers prefer disinfection methods with excellent performance in combination with low costs. A good performance can be described by eliminating pathogens with a reduction of 99.9% (or a log 3 reduction) combined with a clear, understandable and controllable process. Low costs are preferably combined with low investments, low maintenance costs and no need for the grower to perform as a laboratory specialist. Heat treatment, UV radiation and ozone treatment show a good performance. However, investments in ozone treatment are very high, resulting in high annual costs. Heat treatment and UV radiation also have high annual costs, but investments are lower, whilst the eliminating process is easy to control. The latter two methods are most popular among growers, especially at nurseries larger than 1 or 2 ha. Slow sand filtration is less perfect in performance but has considerably lower annual costs. This method could be recommended for producers smaller than 1 ha and for growers with lower investment capital, as sand filters can be constructed by the grower themselves. Sodium hypochlorite and hydrogen peroxide are also cheap methods, but performance is insufficient to eliminate all pathogens. Besides it is a biocide and not a pesticide, which means by law, in the EU at least, it is legally forbidden to use it for the elimination of pathogens.

4.5.3 Biofouling and Pretreatment

Disinfection methods are not very selective between pathogens and other organic material in the solution. Therefore, pretreatment (rapid sand filters, or 50–80 um mechanical filters) of the solution before disinfection is recommended at heat treatment, UV radiation and ozone treatment. If after disinfection residuals of the chemical methods remain in the water, they may react with the biofilms which have been formed in the pipe lines of the watering systems. If the biofilm is released from the walls of the pipes, they will be transported to the drippers and cause clogging. Several oxidizing methods (sodium hypochlorite, hydrogen peroxide with activators, chlorine dioxide) are mainly used to clean pipe lines and equipment, and these create a special risk for clogging drippers over time.

References

Abad M, Noguera P, Puchades R, Maquieira A, Noguera V (2002) Physico-chemical and chemical properties of some coconut coir dusts for use as a peat substitute for containerised ornamental plants. Bioresour Technol 82:241–245

Adams P (1991) Effect of increasing the salinity of the nutrient solution with major nutrients or sodium chloride on the yield, quality and composition of tomatoes grown in rockwool. J Hortic Sci 66:201–207

Adams P, Ho LC (1992) The susceptibility of modern tomato cultivars to blossom-end rot in relation to salinity. J Hortic Sci 67:827–839

Baxter I (2015) Should we treat the ionome as a combination of individual elements, or should we be deriving novel combined traits? J Exp Bot 66:2127–2131

Bittsanszky A, Uzinger N, Gyulai G, Mathis A, Junge R, Villarroel M, Kotzen B, Komives T (2016) Nutrient supply of plants in aquaponic systems. Ecocycles 2:17–20

Blok C, de Kreij C, Baas R, Wever G (2008) Analytical methods used in soilless cultivation. In: Raviv, Lieth (eds) Soilless culture, theory and practice. Elsevier, Amsterdam, pp 245–290

Briat JF, Dubos C, Gaymard F (2015) Iron nutrition, biomass production, and plant product quality. Trends Plant Sci 20:33–40

Bunt BR (2012) Media and mixes for container-grown plants: a manual on the preparation and use of growing media for pot plants. Springer, Dordrecht

Cooper A (1979) The ABC of NFT. Grower Books, London

Cortella G, Saro O, De Angelis A, Ceccotti L, Tomasi N, Dalla Costa L, Manzocco L, Pinton R, Mimmo T, Cesco S (2014) Temperature control of nutrient solution in floating system cultivation. Appl Therm Eng 73:1055–1065

De Kreij C, Voogt W, Baas R (1999) Nutrient solutions and water quality for soilless cultures, report 196. Research station for Floriculture and Glasshouse Vegetables, Naaldwijk, p 36

De Rijck G, Schrevens E (1997) pH influenced by the elemental composition of nutrient solutions. J Plant Nutr 20:911–923

De Rijck G, Schrevens E (1998) Elemental bioavailability in nutrient solutions in relation to complexation reactions. J Plant Nutr 21:2103–2113

De Rijck G, Schrevens E (1999) Anion speciation in nutrient solutions as a function of pH. J Plant Nutr 22:269–279

Dong S, Scagel CF, Cheng L, Fuchigami LH, Rygiewicz PT (2001) Soil temperature and plant growth stage influence nitrogen uptake and amino acid concentration of apple during early spring growth. Tree Physiol 21:541–547

Dorais M, Menard C, Begin G (2006) Risk of phytotoxicity of sawdust substrates for greenhouse vegetables. In: XXVII International Horticultural Congress-IHC2006: International symposium on advances in environmental control, automation 761, pp 589–595
Enzo M, Gianquinto G, Lazzarin R, Pimpini F, Sambo P (2001) Principi tecnico-agronomici della fertirrigazione e del fuori suolo. In: Tipografia-Garbin. Padova, Italy
EU Aquaponics Hub. Plant Protection Fact Sheet. http://euaquaponicshub.com/hub/wp-content/uploads/2016/05/Plant-protection-factsheet.pdf
European Parliament and European Council (2000) Directive 2000/60/EC of the European Parliament and the Council of 23 October 2000 establishing a framework for Community action in the field of water policy (short: Water Framework Directive) Off J Eur Communities, p 327/1
Fageria NK, Baligar VC, Clark RB (2002) Micronutrients in crop production. Adv Agron 77:185–268
Fanasca S, Colla G, Maiani G, Venneria E, Rouphael Y, Azzini E, Saccardo F (2006) Changes in antioxidant content of tomato fruits in response to cultivar and nutrient solution composition. J Agric Food Chem 54:4319–4325
Fernandez D (2016) HydroBuddy v1.50: The first free Open source hydroponic nutrient calculator program Available Online [WWW document]. Sci Hydroponics. http://scienceinhydroponics.com/
Fornes F, Belda RM, Abad M, Noguera P, Puchades R, Maquieira A, Noguera V (2003) The microstructure of coconut coir dusts for use as alternatives to peat in soilless growing media. Aust J Exp Agr 43:1171–1179
Gibson JL (2007) Nutrient deficiencies in bedding plants. Ball Publishing, Batavia
Goddek S, Keesman KJ (2018) The necessity of desalination technology for designing and sizing multi-loop aquaponics systems. Desalination 428:76–85
Haynes RJ (1990) Active ion uptake and maintenance of cation-anion balance: a critical examination of their role in regulating rhizosphere pH. Plant Soil 126:247–264
Heuvelink E, Kierkels T (2016) Iron is essential for photosynthesis and respiration: iron deficiency. In Greenhouses: the international magazine for greenhouse growers 5, pp 48–49
Ho LC, Belda R, Brown M (1993) Uptake and transport of calcium and the possible causes of blossom-end rot in tomato. J Exp Bot 44:509–518
Hoagland DR, Arnon DI (1950) The water culture method for growing plants without soil, Circular 347, Agricultural Experimental station, University of California, Berkeley, CA, USA
Huett DO (1994) Growth, nutrient uptake and tipburn severity of hydroponic lettuce in response to electrical conductivity and K:Ca ratio in solution. Aust J Agric Res 45:251–267
Hussain A, Iqbal K, Aziem S, Mahato P, Negi AK (2014) A review on the science of growing crops without soil (soilless culture)-a novel alternative for growing crops. Int J Agric Crop Sci 7:833–842
Jensen MH, Collins WL (1985) Hydroponic vegetable production. Hortic Rev 7:483–558
Kipp JA, Wever G, de Kreij C (2001) International substrate manual. Elsevier, Amsterdam
Kuhry P, Vitt DH (1996) Fossil carbon/nitrogen ratios as a measure of peat decomposition. Ecology 77:271–275
Lamanna D, Castelnuovo M, D'Angelo G (1990) Compost-based media as alternative to peat on ten pot ornamentals. Acta Hortic 294:125–130
Le Bot J, Adamowicz S, Robin P (1998) Modelling plant nutrition of horticultural crops: a review. Sci Hortic 74:47–82
Le Quillec S, Fabre R, Lesourd D (2003) Phytotoxicité sur tomate et chlorate de sodium. Infos-ctifl, pp 40–43
Liu S, Hou Y, Chen X, Gao Y, Li H, Sun S (2014) Combination of fluconazole with non-antifungal agents: a promising approach to cope with resistant *Candida albicans* infections and insight into new antifungal agent discovery. Int J Antimicrob Agents 43:395–402
Maher MJ, Prasad M, Raviv M (2008) Organic soilless media components. In: Raviv, Lieth (eds) Soilless culture, theory and practice. Elsevier, Amsterdam, pp 459–504

Marschner P (2012) Rhizosphere biology. In: Marschner's mineral nutrition of higher plants, 3rd edn. Academic, Amsterdam, pp 369–388
Masclaux-Daubresse C, Daniel-Vedele F, Dechorgnat J, Chardon F, Gaufichon L, Suzuki A (2010) Nitrogen uptake, assimilation and remobilization in plants: challenges for sustainable and productive agriculture. Ann Bot 105:1141–1157
Maucieri C, Nicoletto C, Junge R, Schmautz Z, Sambo P, Borin M (2018) Hydroponic systems and water management in aquaponics: a review. Ital J Agron 13:1–11
McCutchan JH, Lewis WM, Kendall C, McGrath CC (2003) Variation in trophic shift for stable isotope ratios of carbon, nitrogen, and sulfur. Oikos 102:378–390
Moriarty MJ, Semmens K, Bissonnette GK, Jaczynski J (2018) Inactivation with UV-radiation and internalization assessment of coliforms and *Escherichia coli* in aquaponically grown lettuce. LWT 89:624–630
Muneer S, Lee BR, Kim KY, Park SH, Zhang Q, Kim TH (2014) Involvement of Sulphur nutrition in modulating iron deficiency responses in photosynthetic organelles of oilseed rape (*Brassica napus* L.). Photosynth Res 119:319–329
NGS, Multilayer. http://ngsystem.com/en/ngs/multibanda
Nicoletto C, Maucieri C, Sambo P (2017) Effects on water management and quality characteristics of ozone application in chicory forcing process: a pilot system. Agronomy 7:29
Nicoletto C, Maucieri C, Mathis A, Schmautz Z, Komives T, Sambo P, Junge R (2018) Extension of aquaponic water use for NFT baby-leaf production: mizuna and rocket salad. Agronomy 8:75
Nozzi V, Graber A, Schmautz Z, Mathis A, Junge R (2018) Nutrient management in aquaponics: comparison of three approaches for cultivating lettuce, mint and mushroom herb. Agronomy 8:27
Olle M, Ngouajio M, Siomos A (2012) Vegetable quality and productivity as influenced by growing medium: a review. Agriculture 99:399–408
Parent S-É, Parent LE, Egozcue JJ, Rozane D-E, Hernandes A, Lapointe L, Hébert-Gentile V, Naess K, Marchand S, Lafond J, Mattos D, Barlow P, Natale W (2013) The plant ionome revisited by the nutrient balance concept. Front Plant Sci 4:39
Perelli M, Graziano PL, Calzavara R (2009) Nutrire le piante. Arvan Ed., Venice.
Pickering HW, Menzies NW, Hunter MN (2002) Zeolite/rock phosphate – a novel slow release phosphorus fertiliser for potted plant production. Sci Hortic 94:333–343
Postma J, van Os EA, Bonants PJM (2008) Pathogen detection and management strategies in soilless plant growing systems. In: Raviv, Lieth (eds) Soilless culture, theory and practice. Elsevier, Amsterdam, pp 425–458
Pregitzer KS, King JS (2005) Effects of soil temperature on nutrient uptake. In: Nutrient acquisition by plants. Springer, Berlin/Heidelberg, pp 277–310
Rooney CP, Zhao FJ, McGrath SP (2006) Soil factors controlling the expression of copper toxicity to plants in a wide range of European soils. Environ Toxicol Chem 25:726–732
Ruijs MNA (1994) Economic evaluation of closed production systems in glasshouse horticulture. Acta Hortic 340:87–94
Runia WT (1995) A review of possibilities for disinfection of recirculation water from soilless culture. Acta Hortic 382:221–229
Runia WT, van Os EA, Bollen GJ (1988) Disinfection of drain water from soilless cultures by heat treatment. Neth J Agric Sci 36:231–238
Schmautz Z, Graber A, Jaenicke S, Goesmann A, Junge R, Smits TH (2017) Microbial diversity in different compartments of an aquaponics system. Arch Microbiol 199:613–620
Schnug E, Haneklaus S (2005) Sulphur deficiency symptoms in oilseed rape (*Brassica napus* L.)- the aesthetics of starvation. Phyton 45:79–95
Silber A, Bar-Tal A (2008) Nutrition of substrate-grown plants. In: Raviv, Lieth (eds) Soilless culture, theory and practice. Elsevier, Amsterdam, pp 292–342
Sonneveld C, Voogt W (2009) Plant nutrition of greenhouse crops. Springer, Dordrecht, p 403
Steiner AA (1961) A universal method for preparing nutrient solutions of a certain desired composition. Plant Soil 15:134–154

Steiner AA (1984) The universal nutrient solution. In: Proceeding 6th International congress on soilless culture. Lunteren, Netherlands, ISOSC, pp 633–649
Tindall JA, Mills HA, Radcliffe DE (1990) The effect of root zone temperature on nutrient uptake of tomato. J Plant Nutr 13:939–956
Trang NTD, Brix H (2014) Use of planted biofilters in integrated recirculating aquaculture-hydroponics systems in the Mekong Delta, Vietnam. Aquac Res 45:460–469
Uchida R (2000) Essential nutrients for plant growth: nutrient functions and deficiency symptoms. In: Silva JA, Uchida R (eds) Plant nutrient management in Hawaii's soils, approaches for tropical and subtropical agriculture. College of Tropical Agriculture and Human Resources, University of Hawaii at Manoa, Honolulu, pp 31–55
Van Os EA (1994) Closed growing systems for more efficient and environmental friendly production. Acta Hortic 361:194–200
Van Os EA (1999) Closed soilless growing systems: a sustainable solution for Dutch greenhouse horticulture. Water Sci Technol 39:105–112
Van Os EA (2009) Comparison of some chemical and non-chemical treatments to disinfect a recirculating nutrient solution. Acta Hortic 843:229–234
Van Os EA (2017) Recent advances in soilless culture in Europe. Acta Hortic 1176:1–8
Van Os EA, Gieling TH, Lieth JH (2008) Technical equipment in soilless production systems. In: Raviv, Lieth (eds) Soilless culture, theory and practice. Elsevier, Amsterdam, pp 157–207
Vance CP, Uhde-Stone C, Allan DL (2003) Phosphorus acquisition and use: critical adaptations by plants for securing a nonrenewable resource. New Phytol 157:423–447
Verdonck O, Penninck RD, De Boodt M (1983) The physical properties of different horticultural substrates. Acta Hortic 150:155–160
Verwer FLJA (1978) Research and results with horticultural crops grown in rockwool and nutrient film. Acta Hortic 82:141–148
Wallach J (2008) Physical characteristics of soilless media. In: Raviv, Lieth (eds) Soilless culture, theory and practice. Elsevier, Amsterdam, pp 41–116
Wang M, Zheng Q, Shen Q, Guo S (2013) The critical role of potassium in plant stress response. Int J Mol Sci 14:7370–7390
Wu M, Kubota C (2008) Effects of high electrical conductivity of nutrient solution and its application timing on lycopene, chlorophyll and sugar concentrations of hydroponic tomatoes during ripening. Sci Hortic 116:122–129
Zekki H, Gauthier L, Gosselin A (1996) Growth, productivity, and mineral composition of hydroponically cultivated greenhouse tomatoes, with or without nutrient solution recycling. J Am Soc Hortic Sci 121:1082–1088
Zoschke K, Börnick H, Worch E (2014) Vacuum-UV radiation at 185 nm in water treatment–a review. Water Res 52:131–145

Part II
Specific Aquaponics Technology

Chapter 5
Aquaponics: The Basics

Wilson Lennard and Simon Goddek

Abstract Aquaponics is a technology that is part of the broader integrated agri-aquaculture systems discipline which seeks to combine animal and plant culture technologies to confer advantages and conserve nutrients and other biological and economic resources. It emerged in the USA in the early 1970s and has recently seen a resurgence, especially in Europe. Whilst aquaponics broadly combines recirculating fish culture with hydroponic plant production, the application of the term aquaponic is broad and many technologies claim use of the name. Combining fish culture with aquatic-based, terrestrial plant culture via aquaponics may be better defined via its nutrient resource sharing credentials. Aquaponics applies several principles including, but not limited to, efficient water use, efficient nutrient use, lowered or negated environmental impact and the application of biological and ecological approaches to agricultural fish and plant production. Water sources are important so that the nutrients required for fish and plant production are available and balanced, and system water chemistry is paramount to optimised fish and plant production. Systems may be configured in several ways, including those that are fully recirculating and those that are decoupled. Aquaponics importantly seeks to apply methods that provide technical, biological, chemical, environmental and economic advantages.

Keywords Aquaponics · Agri-aquaculture · Aquaculture · Hydroponics · Agriculture · Fish · Plants · Nutrients · Ecology

W. Lennard (✉)
Aquaponic Solutions, Blackrock, VIC, Australia
e-mail: willennard@gmail.com

S. Goddek
Mathematical and Statistical Methods (Biometris), Wageningen University, Wageningen, The Netherlands
e-mail: simon.goddek@wur.nl; simon@goddek.nl

S. Goddek et al. (eds.), *Aquaponics Food Production Systems*,
https://doi.org/10.1007/978-3-030-15943-6_5

5.1 Introduction

Aquaponics is a technology that is a subset of a broader agricultural approach known as integrated agri-aquaculture systems (IAAS) (Gooley and Gavine 2003). This discipline consists of integrating aquaculture practices of various forms and styles (mostly fin fish farming) with plant-based agricultural production. The rationale of integrated agri-aquaculture systems is to take advantage of the resources shared between aquaculture and plant production, such as water and nutrients, to develop and achieve economically viable and environmentally more sustainable primary production practices (Gooley and Gavine 2003). In essence, both terrestrial plant and aquatic animal production systems share a common resource: water. Plants are generally consumptive of water via transpiration and release it to the surrounding gaseous environment, whereas fish are generally less consumptive of water, but their contained culture produces substantial waste water streams due to accumulated metabolic wastes. Therefore, aquaculture may be integrated within the water supply pathway of plant production in non-consumptive ways so that two crops (fish and plants) may be produced from a water source that is generally used to produce one crop (plants).

An interesting additional advantage of integrating aquaculture with the irrigation supply pathway for plant production is that aquaculture also produces waste nutrients via the dissolved and undissolved wastes produced from fish (and other aquatic animal) metabolism. Therefore, aquaculture may also produce waste nutrient streams that are suitable for, and assist, plant production by contributing to the plants nutrient requirements.

The advantages produced by integrating aquaculture with conventional terrestrial and aquatic plant production systems have been summarised by Gooley and Gavine (2003) as:

1. An increase in farm productivity and profitability without any net increase in water consumption (Chap. 2).
2. Farm diversification into higher-value crops, including high-value aquatic species.
3. Reuse of otherwise wasted on-farm resources (e.g. capture and reuse of nutrients and water).
4. Reduction of net environmental impacts of semi-intensive and intensive farming practices.
5. Net economic benefits by offsetting existing farm capital and operating expenses (Chap. 18).

Aquaponics has been said to have evolved from relatively ancient agriculture practices associated with integrating fish culture with plant production, especially those developed within the South East Asian, flooded rice paddy farming context and South American Chinampa, floating island, agriculture practices (Komives and Junge 2015). In reality, historically, fish were rarely actively added to rice paddy fields until the nineteenth century (Halwart and Gupta 2004) and were present in very low densities which would not contribute to any substantial nutritive assistance to the plants. Chinampas were traditionally built on lakes in Mexico where nutrient advantages may have been supplied via the eutrophic or semi-eutrophic lake sediments rather than directly from any designed or actively integrated fish production system (Morehart 2016; Baquedano 1993).

Modern aquaponics started in the USA in the 1970s and was co-evolved by several institutions with an interest in more sustainable farming practices. Early important work was performed by several researchers, but ultimately, the progenitor of nearly all modern aquaponics is thought to be the work performed by, and the systems produced by, James Rakocy and his team at the University of the Virgin Islands (UVI) starting in the early 1980s (Lennard 2017).

Aquaponics is now considered a new and emerging industry with a relevant place in the broader, global agricultural production context and there are a number of variations of the technology of integrating fish culture with aquatic plant culture that are collectively defined under the aquaponics banner or name (Knaus and Palm 2017). Therefore, aquaponics seeks to integrate *aqua*culture animal production with hydro*ponic* plant production using various methods to share water and nutrient resources between the major production components to produce commercial and saleable fish and plant products.

5.2 A Definition of Aquaponics

Aquaponics fits into the broader definition of integrated agri-aquaculture systems (IAAS). However, IAAS applies many different aquatic animal and plant production technologies in many contexts, whereas aquaponics is far more tightly associated with integrating tank-based fish culture technologies (e.g. recirculating aquaculture systems; RAS) with aquatic or hydroponic plant culture technologies (Lennard 2017). RAS technologies apply conserved and standard methods for the culture of fish in tanks with applied filtration to control and alter the water chemistry to make it suitable for fish (i.e. fast and efficient solid fish wastes removal, efficient, bacteria-mediated conversion of potentially toxic dissolved fish waste ammonia to less toxic nitrate and oxygen maintenance via assisted aeration or directly injected oxygen gas) (Timmons et al. 2002). Hydroponics and substrate culture technologies apply conserved and standard methods for the culture of edible terrestrial plants within aquatic environments (i.e. the plants gain access to the nutrients required for growth via a water-based delivery method) (Resh 2013).

The association of aquaponics with standard RAS aquaculture and hydroponics/substrate culture means that aquaponics is often defined simply as "... the combination of fish production (aquaculture) and soil-less plant cultivation hydroponics under coupled or decoupled water circulation" (Knaus and Palm 2017). This broad definition places an emphasis on the integration of hardware, equipment or technologies and places little, if any, emphasis on any other aspects of the method.

Because aquaponics is a relatively new industrial-scale technology that applies different methods and approaches, the applied definition appears very broad. Some define aquaponics within a recirculating context only (Cerozi and Fitzsimmons

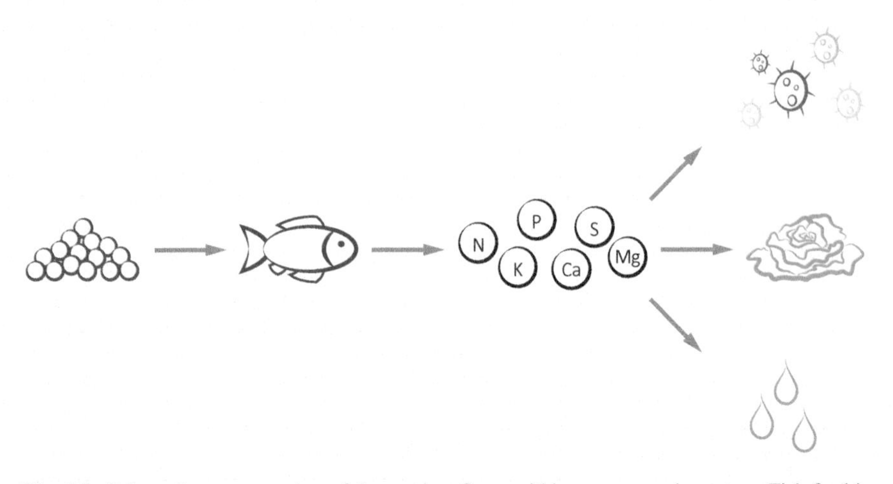

Fig. 5.1 Schematic representation of the nutrient flows within an aquaponic system. Fish feed is the major nutrient entry point. The fish eat the feed, use what nutrients they need, release the rest as waste and this waste is then partitioned between the microbes, plants and system water. (adapted from Lennard 2017)

2017), some concentrate on approaches that do not return the water from the plants to the fish (Delaide et al. 2016) and others include both recirculating and decoupled methods (Knaus and Palm 2017). Further still, some researchers are including the use of aquaculture effluents irrigated to soil-based crop production under the aquaponic title (Palm et al. 2018). Historically, aquaponics, as the breakdown of the word (*aqua*culture and hydro*ponics*) suggests, was defined as only concerning aquaculture and hydroponic plant production (Rakocy and Hargreaves 1993), so current attempts at associations with soil-based culture seem incongruous.

Whilst aquaponic systems do integrate tank-based aquaculture technologies with hydroponic plant culture technologies, aquaponic systems work by supplying nutrients to, and partitioning nutrients between, the production inhabitants (fish and plants) and the inhabitants that perform biological and chemical services that assist the production inhabitant outcome (microflora) (Fig. 5.1) (Lennard 2017). Therefore, is aquaponics more a system associated with nutrient supply, dynamics and partitioning rather than one associated with the technology, equipment or hardware applied?

Over the past decades, the definition of aquaponics has included a similar theme, with subtle variations. The broadest definition has generally been provided in the scientific publications of Rakocy and his UVI team, for example:

> *Aquaponics is the combined culture of fish and plants in closed recirculating systems.*
> – Rakocy et al. (2004a, b)

This early definition was based on the assumption that one-loop, fully recirculating systems, consisting of a recirculating aquaculture component and a hydroponic component, represented all aquaponic systems, which at the time, they did. Graber and Junge (2009) expanded the definition, due to changes and developments in the approach, as follows:

> *Aquaponic is a special form of recirculating aquaculture systems (RAS), namely a polyculture consisting of fish tanks (aquaculture) and plants that are cultivated in the same water circle (hydroponic).*
> – Graber and Junge (2009)

Recent developments and methods ask for a reconsideration of this standpoint. In recent years a shifting of the focus of aquaponics towards a production system that tackles both ecological responsibility and economic sustainability has been present. Kloas et al. (2015) and Suhl et al. (2016) were one of the first to address this economic consideration:

> *[...] a unique and innovative double recirculating aquaponic system was developed as a prerequisite for a high productivity comparable to professional stand-alone fish/plant facilities.*
> – Suhl et al. (2016)

The definition issue, or clarifying "what can be defined as aquaponics", has been a point of discussion over the past years. One of the main areas of development has been that of multi-loop (or decoupled) aquaponic systems that aim at providing additional fertilisers to the plants in order to expose them to an optimal nutrient concentration (Goddek 2017). There should be no opposition between the ideologies of fully recirculating and multi-loop aquaponic methodologies, both have their respective places and applications within the appropriate industrial context and a single common driving force of both should be that the technology, whilst being nutrient and water efficient, also needs to be economically competitive to establish itself in the market. In order to replace conventional practices, more than an ideology needs to be offered to potential clients/users – i.e. technical and economic feasibility.

The European COST sponsored Aquaponics Hub (COST FA1305 2017) applies the definition "*...a production system of aquatic organisms and plants where the majority (> 50%) of nutrients sustaining the optimal plant growth derives from waste originating from feeding the aquatic organisms*", which clearly places an emphasis on the nutrient sharing aspect of the technology.

It must also be stated that the proportion of fish to plants should remain at a level that supports a core prospect of aquaponics; that plants are grown using fish wastes. A system containing one fish and several hectares of hydroponic plant cultivation, for example, should not be considered as aquaponics, simply because that one fish effectively contributes nothing to the nutrient requirements of the plants. Since the labelling of aquaponic products plays an increasingly important role in consumer choice, we want to encourage a discussion by redefining aquaponics based on these multiple developments of the technology. Even though we advocate closing the nutrient cycle to the highest possible degree in the context of best practicable means, a potential definition should also take all developments into consideration.

Therefore, the definition should contain as a minimum, the requirement for a majority of aquaculture-derived nutrients for the plants. A new definition may therefore be represented as:

> *Aquaponics is defined as an integrated multi-trophic, aquatic food production approach comprising at least a recirculating aquaculture system (RAS) and a connected hydroponic unit, whereby the water for culture is shared in some configuration between the two units. Not less than 50% of the nutrients provided to the plants should be fish waste derived.*

Nutrient-based definitions are open and non-judgemental of the applied technology choice, or even the proportions of each component (fish and plants), as long as fish culture and some form of aquatic (hydroponic or substrate culture) plant production technology is utilised. However, it also focuses the definition on the nutrient dynamics and nutrient sharing aspects of the methods applied and therefore ensures, to at least some extent, that the advantages often associated with aquaponics (water saving, nutrient efficiency, lowered environmental impact, sustainability) are present in some proportion.

The nutrient association definition applied to aquaponics will always be a source of further contention among those who practise it. This is supported by the fact that the name aquaponics is applied to a vast array of different technologies with different nutrient supply motivations and usage outcomes: from system designs and methods that expect, if not demand, that the vast majority of the nutrients required to grow the plants arise from the fish wastes (in some cases, greater than 90%; Lennard 2017) to designs that share plant nutrient supplies between fish wastes and more substantial external additions (e.g. approximately 50:50 fish waste to external supplementation – as many modern, European decoupled aquaponic system designs do; COST FA1305 2017) to those designs that add so few fish that no discernible nutrient supply from the fish wastes to the plants is present (Lennard 2017).

The name aquaponics, until relatively recently (i.e. the last 3–5 years), has been universally applied to coupled and fully recirculating system designs that seek to supply as much of the required plant nutrition from the fish wastes as possible (Rakocy and Hargreaves 1993; Lennard 2017) (Fig. 5.2).

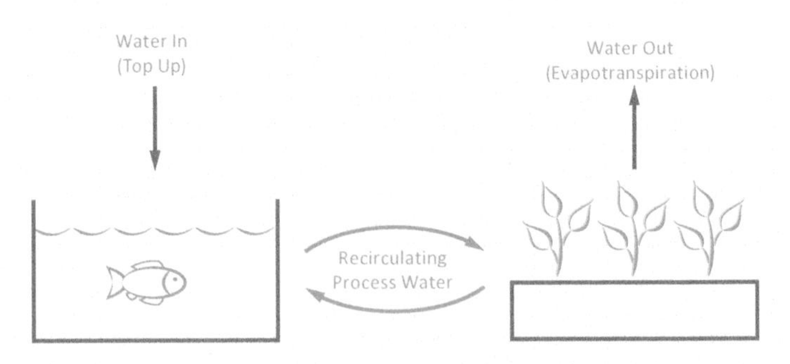

Fig. 5.2 Simplified scheme of the main water flows within a coupled aquaponic system. The nutrient concentrations in the process water are equally distributed throughout the whole system

However, decoupled approaches now represent a proportion of the systems being researched or commercially applied, especially in Europe, and in current practice do not supply plant nutrient requirements from the fish wastes to the same extent as fully recirculating systems do (Lennard 2017; Goddek and Keesman 2018). For example, Goddek and Keesman (2018) state that for 3 examples of current European decoupled aquaponic system designs, the relative addition requirements for external hydroponic-derived nutrients are 40–60% (NerBreen), 60% (Tilamur) and 38.1% (IGB Berlin). Because these decoupled designs are based on integrating existing RAS and hydroponic/substrate culture technologies, they are regarded as aquaponic in nature (Delaide et al. 2016) (Fig. 5.3) (see Chap. 8).

The definition of aquaponics is now being expanded beyond ecological, water and nutrient efficiency drivers and optimisation to also include economic drivers (Goddek and Körner 2019; Goddek and Keesman 2018; Goddek 2017; Kloas et al. 2015; Reyes Lastiri et al. 2016; Yogev et al. 2016) (Chap. 8). The benefits of such an approach are that a positive economic outcome from aquaponics technology is as important as its biological, chemical, engineering, ecological and sustainable credentials and therefore, the economic outcome should play a role within the overall definition (Chap. 8).

Many advantages are often associated with aquaponics, especially in terms of its water-use efficiency, its nutrient use efficiency, its sustainable nature, its ability to produce two crops from the one input source (fish feed) and its lowered environmental impact (Timmons, et al., 2002; Buzby and Lian-shin 2014; Wongkiew et al. 2017; Roosta and Hamidpour 2011; Suhl et al. 2016). These advantages are regularly quoted and applied by commercial aquaponic operators and are used as a marketing and price regulation pathway for the products (fish and plants), and therefore, the use of the name "aquaponics" directly and immediately associates that the products labelled as such have been produced with methods that contain or utilise the advantages listed. However, there is no formal regulation of the industry that dictates that the use of the word (aquaponics) only occurs when the advantages are apparent and present within the technology and methods applied. If the above advantages are assigned to aquaponics as a technology, then surely the technology

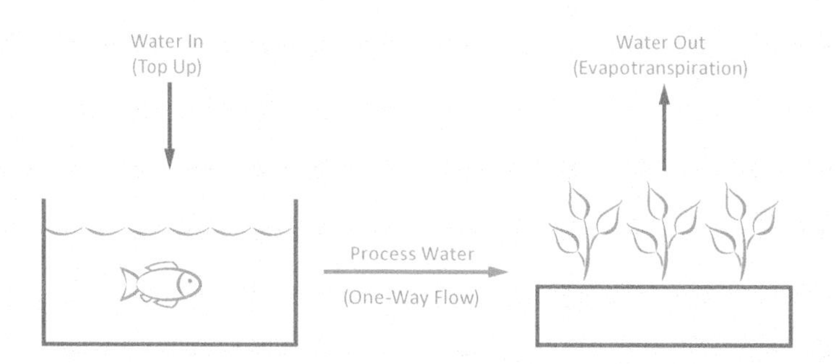

Fig. 5.3 Simplified scheme of the main water flows within a decoupled aquaponic system. The nutrient concentrations in each component may be separately tailored to the individual component requirement

should provide the prescribed advantages, and if the technology does not provide the advantages, then the word should not be applied (Lennard 2017).

Because aquaponics may be defined either in terms of its hardware equipment integration aspect (RAS with hydroponics), its nutrient sharing or partitioning properties or its ability to provide important advantages, there is still a wide spectrum of possible applications of the name to many different technical approaches that utilise different methods and demand different outcomes. Therefore, it appears that the actual definition of aquaponics is still unresolved.

It appears therefore that very important questions are yet to be answered: what is aquaponics and how is it defined?

This would suggest that one very important aspect for the aquaponic industry to consider is the development of a truthful and agreed-upon definition. The broader aquaponics industry will continue to be full of disagreement if a definition is not agreed upon, and more importantly, consumers of the products produced within aquaponic systems will become more and more confused about what aquaponics actually is – a state of affairs that will not assist the growth and evolution of the industry.

5.3 General Principles

Even though the definition of aquaponics has not been entirely resolved, there are some general principles that are associated with the broad range of aquaponic methods and technologies.

Using the nutrients added to the aquaponic system as optimally and efficiently as possible to produce the two main products of the enterprise (i.e. fish and plant biomass) is an important and shared first principle associated with the technology (Rakocy and Hargreaves 1993; Delaide et al. 2016; Knaus and Palm 2017). There is no use in adding nutrients (which possess an inherent cost in terms of money, time and value) to a system to watch a high percentage of those nutrients are partitioned into processes, requirements or outcomes that are not directly associated with the fish and plants produced, or any intermediary life forms that may assist nutrient access by the fish and plants (i.e. microorganisms – bacteria, fungi, etc.) (Lennard 2017). Therefore, probably the most important general principle associated with aquaponics is to use the applied nutrients as efficiently as possible to achieve the optimised production of both fish and plants.

This same argument may also be applied to the water requirement of the aquaponic system in question; again, the water added to the system should be utilised principally by the fish and plants and used as efficiently as possible and not allowed to leak to processes, life forms or outcomes that are not directly associated with fish and plant production or may impact on the surrounding environment (Lennard 2017).

In real terms, the efficient use of nutrients and water leads to several design principles that are broadly applied to the aquaponic method:

1. The most important principle of aquaponics is to use the wastes produced by the fish as a principle nutrient source for the plants. In fact, this is the entire idea of aquaponics and so should be a first order driver for the method. Aquaponics was historically envisaged as a system to grow plants using fish aquaculture wastes so that those aquaculture wastes had less environmental impact and were seen as a positive and profitable commodity, rather than a troublesome waste product with an associated cost to meet environmental legislative requirements (Rakocy and Hargreaves 1993; Love et al. 2015a, b).
2. The system design should encourage the use of fish keeping and plant culturing technologies that do not inherently uptake or destructively utilise the water or nutrient resources added. For example, fish keeping components based on using earthen ponds are discouraged, because the earthen pond has the ability to use and make unavailable water and nutrient resources to the associated fish and plants, thus lowering the water and nutrient use efficiency of the system. Similarly, hydroponic plant culturing methods should not use media that uptakes excessive amounts of nutrients or water and renders them unavailable to the plants (Lennard 2017).
3. The system design should not waste nutrients or water via the production of external waste streams. Principally, if water and nutrients leave the system via a waste stream, then that water and those nutrients are not being used for fish or plant production, and therefore, that water and those nutrients are being wasted, and the system is not as efficient as possible. In addition, the production of a waste streams can have a potential environmental impact. If waste water and nutrients do leave the aquaponic system, they should be used in alternate, exterior-to-system plant production technologies so water and nutrients are not wasted, contribute to the overall production of edible or saleable biomass and do not present a broader environmental impact potential (Tyson et al. 2011).
4. The system should be designed to lower or ideally, completely negate, direct environmental impact from water or nutrients. A first order goal of aquaponics is to use the wastes produced by the fish as a nutrient source for the plants so as to negate the release of those nutrients directly to the surrounding environment where they may cause impacts (Tyson et al. 2011).
5. Aquaponic system designs should ideally lend themselves to being located within environmentally controlled structures and situations (e.g. greenhouses, fish rooms). This allows the potential to achieve the best productive rates of fish and plants from the system. Most aquaponic designs are relatively high in terms of capital costs and ongoing costs of production, and therefore, the ability to house the system in the perfect environment enhances profit potentials that financially justify the high capital and costs of production (Lennard 2017).

The above outlined principles of design directly associate with a set of general principles that are often, but not always, applied to the aquaponic production environment. These general principles relate to how the system operates and how nutrients are portioned among the system and its inhabitants.

The basic premise of aquaponics, in a nutrient dynamic context, is that fish are fed fish feed, fish metabolise and utilise the nutrients in the fish feed, fish release wastes based on the substances in the fish feed they do not utilise (including elements), microflora access those fish metabolic wastes and use small amounts of them, but transform the rest, and the plants then access and remove those microflora transformed, fish metabolic wastes as nutrient sources and, to some extent, clean the water medium of those wastes and counteract any associated accumulation (Rakocy and Hargreaves 1993; Love et al. 2015a, b).

Because earthen-based fish production systems remove nutrients themselves, aquaponics generally utilises what are known as recirculating aquaculture system (RAS) principles for the fish production component (Rakocy and Hargreaves 1993; Timmons et al. 2002). Fish are kept in tanks made of materials that do not remove nutrients from the water (plastic, fibreglass, concrete, etc.), the water is filtered to treat or remove the metabolic waste products of the fish (solids and dissolved ammonia gases) and the water (and associated nutrients) is then directed to a plant culturing component whereby the plants use the fish wastes as part of their nutrient resource (Timmons et al. 2002). As for the fish, earthen-based plant culturing components are not used because the soils involved remove nutrients and may not necessarily make them fully available to the plants. In addition, hydroponic plant culturing techniques do not use soil and are cleaner than soil-based systems and allow some passive control of the microorganism mixtures present.

Plants cultured in conventional hydroponics require the addition of what are known as mineral fertilisers: nutrients that are present in their basal, ionic forms (e.g. nitrate, phosphate, potassium, calcium, etc. as ions) (Resh 2013). Conversely, recirculating aquaculture systems must apply regular (daily) water exchanges to control the accumulation of fish waste metabolites (Timmons et al. 2002). Aquaponics seeks to combine the two separate enterprises to produce an outcome that achieves the best of the two technologies while negating the worst (Goddek et al. 2015).

Plants require a suite of macro and micro elements for optimal and efficient growth. In aquaponics, the majority of these nutrients arise from the fish wastes (Rakocy and Hargreaves 1993; Lennard 2017; COST FA1305 2017). However, fish feeds (the major source of aquaponic system nutrients) do not contain all the nutrients required for optimised plant growth, and therefore, external nutrition, to varying extents, is required.

Standard hydroponics and substrate culture add nutrients to the water in forms that are directly plant-available (i.e. ionic, inorganic forms produced via designed salt variety additions) (Resh 2013). A proportion of the wastes released by fish are in forms that are directly plant-available (e.g. ammonia) but potentially toxic to the fish (Timmons et al. 2002). These dissolved, ionic fish waste metabolites, like ammonia, are transformed by ubiquitous bacterial species that replace hydrogen ions with oxygen ions, the product from ammonia being nitrate, which is far less toxic to the fish and the preferred nitrogen source for the plants (Lennard 2017). Other nutrients appropriate to plant uptake are bound in the solid fraction of the fish waste as organic compounds and require further treatment via microbial interaction to render the nutrients available to plant uptake (Goddek et al. 2015). Therefore, aquaponic systems require a suite of microflora to be present to perform these transformations.

The key to optimised aquaponic integration is determining the ratio between fish waste output (as directly influenced by fish feed addition) and plant nutrient utilisation (Rakocy and Hargreaves 1993; Lennard and Leonard 2006; Goddek et al. 2015). Various rules of thumb and models have been developed in an attempt to define this balance. Rakocy et al. (2006) developed an approach that matches the plant growing area requirement with the daily fish feed input and called it "The Aquaponic Feeding Rate Ratio". The feeding rate ratio is set between 60 and 100 grams of fish feed added per day, per square meter of plant growing area (60–100 $g/m^2/day$). This feeding rate ratio was developed using *Tilapia* spp. fish eating a standard, 32% protein commercial diet (Rakocy and Hargreaves 1993). In addition, the aquaponic system this ratio is particular to (known as the University of the Virgin Islands Aquaponic System – UVI System) does not utilise the solid fish waste fraction, is over-supplied with nitrogen and requires in-system, passive de-nitrification to control the nitrogen accumulation rate (Lennard 2017). Others have determined alternate ratios based on different fish and plant combinations, tested in different specific conditions (e.g. Endut et al. 2010 – 15–42 $g/m^2/day$ for African catfish, *Clarias gariepinus* and water spinach plants, *Ipomoea aquatica*).

The UVI feeding rate ratio was developed by Rakocy and his team as an approximate approach; hence why it is stated as a range (Rakocy and Hargreaves 1993). The UVI ratio tries to account for the fact that different plants require different nutrient amounts and mixtures and therefore a "generic" aquaponic design approach is a difficult prospect. Lennard (2017) has developed an alternate approach that seeks to directly match individual fish waste nutrient production rates (based on the fish feed utilised and the fish conversion and utilisation of that feed) with specific plant nutrient uptake rates so that exacting fish to plant ratio matching for any fish or plant species chosen may be realised and accounted for in the aquaponic system design. He matches this design approach with a specific management approach that also utilises all the nutrients available within the fish solid waste fraction (via aerobic remineralisation of the fish solid wastes) and only adds the nutrients required by the chosen plant species for culture that are missing from the fish waste production fractions. Therefore, this substantially lowers the associated feeding rate ratio (e.g. less than 11 $g/m^2/day$ for some leafy green varieties as a UVI equivalent) and allows any fish species to be specifically and exactly matched to any plant species chosen (Lennard 2017). Similarly, Goddek et al. (2016) have proposed models that allow more exacting fish to plant component ratio determination for decoupled aquaponic systems.

The general principles of efficient nutrient use, low and efficient water use, low or negated environmental impact, ability to be located away from traditional soil resources and sustainability of resource use are the general principles applied to aquaponic system design and configuration and their ongoing application should be encouraged within the field and industry.

5.4 Water Sources

Water is the key medium used in aquaponic systems because it is shared between the two major components of the system (fish and plant components), it is the major carrier of the nutrient resources within the system and it sets the overall chemical environment the fish and plants are cultured within. Therefore, it is a vital ingredient that may have a substantial influence over the system.

In an aquaponic system, water-based environment context, the source of water and what that source water contains chemically, physically and biologically are a major influence over the system because it sets a baseline for what is required to be added to the system by the various inputs of the system. These inputs, in turn, effect and set the environment that the fish and plants are cultured within. For example, some of the major inputs in terms of nutrients to any aquaponic system include, but are not limited to, the fish feed (a primary nutrient resource for the system), the buffers applied (which assist to control and set the pH values associated with both the fish and plant components) and any external nutrient additions or supplementations required to meet the nutrient needs of the fish and plants (Lennard 2017).

Fish feeds are designed to provide the nutrition required for fish growth and health and therefore contain nutrient mixtures and quantities primarily to aid the fish being cultured (Timmons et al. 2002; Rakocy et al. 2006). Plants, on the other hand, have different nutrient requirements to fish, and fish feeds rarely, if ever, meet the total nutrient requirements of the plants (Rakocy et al. 2006). Because of this, aquaponic systems that culture fish and plants solely using fish feed-derived nutrient resources may efficiently and optimally produce fish, but they rarely do so for the plants. The best aquaponic system designs recognise that the ultimate outcome is to produce both fish and plants at optimal and efficient growth rates and therefore, also recognise that some form of additional nutrition is required to meet the total plant nutrient requirement (Rakocy et al. 2006; Suhl et al. 2016).

Classical, fully recirculating aquaponic systems generally rely on fish feeds (after the fish have consumed that feed, metabolised it and utilised the nutrients within it) as the major nutrient source for the plants and supplement any missing nutrients required by the plants via some form of buffering regime (Rakocy et al. 2006) or via additional nutrient supplementation (e.g. adding chelated nutrient forms directly to the culture water or by adding nutrients via foliar sprays) (Roosta and Hamidpour 2011).

The best example of this classical recirculating aquaponic approach is the UVI (University of the Virgin Islands) aquaponic system developed by Dr. James Rakocy and his UVI team (Rakocy and Hargreaves 1993; Rakocy et al. 2006). The UVI design principally adds nutrients for both fish and plant culture via fish feed additions. However, fish feeds do not contain enough calcium (Ca^{+}) and potassium (K^{+}) for optimal plant culture. The bacteria-mediated conversion of fish waste-dissolved ammonia to nitrate causes system-wide production of hydrogen ions within the water column, and the proliferation of these hydrogen ions results in a constant fall in the system water pH towards acid. The buffering regime employed adds the missing calcium and potassium by adding basic salts (often salts based on carbonate, bicarbonate or hydroxyl ions paired with calcium or potassium) to the system that assist to control the system water pH at a level that meets both the shared pH environmental requirements of the fish and the plants, whilst providing the additional calcium and potassium the plants require (Rakocy et al. 2006). In addition, the UVI system adds another major nutrient for plant growth that is not available in standard fish feeds, iron (Fe), via regular and controlled iron chelate additions. Therefore, the potassium, calcium and iron the plants require that are not found in the fish feed are available via these two additional nutrient supply mechanisms (Rakocy et al. 2006).

Decoupled aquaponic designs adopt an approach to culture the fish and plants in a way whereby the water is used by the fish and the fish waste nutrients are supplied to the plants, without recirculation of the water back to the fish (Karimanzira et al. 2016). Decoupled designs therefore allow more flexibility in customising the water chemistry, after fish use, for optimised plant production because supplementation of the nutrients not present in the fish feed (and fish waste) may be achieved with no concerns of the water returning to the fish (Goddek et al. 2016). This means decoupled designs potentially may apply more exacting nutrient mixtures and strengths to the culture water, post fish use, for plant culture, and this may be achieved with more exacting and intense nutrient supplementation.

In both cases (recirculating and decoupled aquaponic system designs), an understanding of the chemical quality of the source water is vital so that as close to optimal nutrient concentrations for the plants may be achieved. If, for example, the source water contains calcium (a case often seen when ground water resources are utilised), this will affect and change the buffering regime applied to recirculating aquaponic designs and the extent of the nutrient supplementation applied to a decoupled design because the calcium present in the source water will offset any required supplementation required for plant calcium needs (Lennard 2017). Or, if the source water contains elevated sodium (Na^+) concentrations (again, often seen with ground water resources and a nutrient plants do not use and which can accumulate in system waters), it is important to know how much is present so management methods may be applied to avoid potential plant nutrient toxicity (Rakocy et al. 2006). The chemical nature of the source water, therefore, is vital to overall aquaponic system health and management.

Ultimately, because source water chemistry can affect aquaponic system nutrient management and because aquaponic operators like to have the ability to manipulate aquaponic water and nutrient chemistry to a high degree, a water source with little, if any, associated water chemistry is highly desirable (Lennard 2017). In this sense, rainwater or water treated for chemical removal (e.g. reverse osmosis) is the best source water for aquaponics in a water chemistry context (Rakocy et al. 2004a, b; Lennard 2017). Ground waters are also suitable, but it must be ensured that they do not contain chemicals or salts in concentrations that are too high to be practical (e.g. high magnesium or iron concentrations) or contain chemical species that are not used by the fish or plants (e.g. high sodium concentrations) (Lennard 2017). River waters may also be suitable as aquaponic source water, but as for other water sources, they should be tested for chemical presence and concentrations. Town water sources (i.e. water reticulated and supplied for domestic and consumptive purposes) are broadly applied in aquaponics (Love et al. 2015a, b) and are also acceptable if they contain acceptable nutrient, salt or chemical concentrations. In the case of town- or municipal-supplied water resources, it should be noted that many supplies have some form of sterilisation applied to make the water drinkable for humans. If this source of water is to be used for aquaponics, then it is important to ensure that any chemicals that may be applied to achieve sterilisation (e.g. chlorine, chloramine, etc.) are not present in concentrations that could harm the fish, plants or microorganisms within the aquaponic system (Lennard 2017).

The chemistry associated with source water is not the only factor that needs consideration when supplying source water for aquaponic use. Many natural waters may also contain microbial and other microorganisms that may affect the overall ecological health of the aquaponic system or present a discernible human health risk. Rainwaters rarely contain microbes themselves; however, the vessels or tanks the rainwater may be stored within may contain or allow microbial proliferation. Ground waters are usually good in terms of microbial presence but may also contain high microbial loads, especially if sourced from areas associated with animal farming or human waste treatment. River waters may also contain high microbial loads due to farming or human waste treatment outflows and again should be checked via detailed microbial analysis (Lennard 2017).

Because the chemical and microbial nature of the source water used in aquaponic systems can have potential effects on system water chemistry and microbiology, it is recommended that any applied water source be sterilised and treated for chemical removal (e.g. reverse osmosis, distillation, etc.) before being used in an aquaponic system (Lennard 2017). If sterilisation is universally applied, the chance of introducing any foreign and unwanted microbes to the system is substantially lowered. If water treatment and filtration is applied, any chemicals, salts, unwanted nutrients, pesticides, herbicides, etc. will be removed and therefore cannot contribute negatively to the system.

A clean water source, free of microbes, salts, nutrients and other chemicals allows the aquaponic operator to manipulate the system water to contain the nutrient mixture and strength they require without the fear that any external influences may affect the operation of the system or the health and strength of the fish and plants and is a vital requirement for any commercial aquaponic operation.

5.5 Water Quality Requirements

Aquaponics represents an effort to control water quality so that all the present life forms (fish, plants and microbes) are being cultured in as close to ideal water chemistry conditions as possible (Goddek et al. 2015). If water chemistry can be matched to the requirements of these three sets of important life forms, efficiency and optimisation of growth and health of all may be aspired to (Lennard 2017).

Optimisation is important to commercial aquaponic production because it is only through optimisation that commercial success (i.e. financial profitability) may be realised. Therefore, water chemistry and water quality requirements within the aquaponic system are pivotal to the ultimate commercial and economic success of the enterprise (Goddek et al. 2015).

There is currently disagreement within the broader aquaponic industry and community in terms of what represents good or acceptable water quality within aquaponic systems. It appears that it is universally accepted that the natural water chemistry requirements of the individual life form subsets (fish, plants and microbes) are broadly agreed upon (Rakocy and Hargreaves 1993; Rakocy et al. 2006; Goddek et al. 2015; Delaide et al. 2016; Lennard 2017). However, the presence of a broad range of approaches, methods and technology choices that are called aquaponics and the background or history of the associated, stand-alone technologies of recirculating aquaculture systems (RAS) and hydroponic plant culture (including substrate culture) appears to lead to disagreements among operators, scientists and designers. For example, taking only one single water chemistry parameter into consideration, pH, some argue that the pH requirements of hydroponically cultured plants are very different to the pH requirements of RAS-cultured freshwater fish species (Suhl et al. 2016). The hydroponic industry generally applies pH settings between 4.5 and 6.0 for water-based plant culture (Resh 2013), whereas the RAS industry typically applies pH settings between 7.0 and 8.0 (Timmons et al. 2002) to meet the requirements of the fish and the microbes present (which perform important transformations of potentially toxic fish waste metabolites to less toxic forms). The argument, therefore, is that any pH set point is a compromise between the requirements of the plants, the fish and the microbes and that therefore an optimal pH for all life forms is not achievable which leads to suboptimal plant production (Suhl et al. 2016). Others argue, however, that a closer scrutiny of the complexities of the nutrient dynamics of plant nutrient uptake may elucidate a different opinion (Lennard 2017).

Hydroponic (and substrate culture) systems feed nutrients to the plants in their basal, ionic forms by adding nutrient salts to the water that dissociate to release the available nutrient ions (Resh 2013). Research has demonstrated that these ionic nutrient forms exist in a window of availability to the plant, based on the available system water pH. Therefore, in a standard hydroponic context, with no present microbial flora (i.e. sterilised – as most hydroponic systems are), it is important to set the pH of the system water to a level that makes the mixture of ionic nutrients the plant requires as available as possible (Resh 2013). Within any hydroponic system, this is a compromise itself, because as any ionic nutrient availability chart demonstrates (see Fig. 5.4), different ionic nutrient forms are the most available at differing pHs (Resh 2013). It is this standard ionic nutrient availability association that the hydroponic industry uses as its primer for pH set points and explains why the desired hydroponic operational pH is somewhere between 4.5 and 6.0 (an acid environment) in sterilised hydroponic and substrate culture systems.

Alternatively, RAS applies a water pH set point based on what is natural for the fish being cultured and the microbes treating and converting the fish waste products

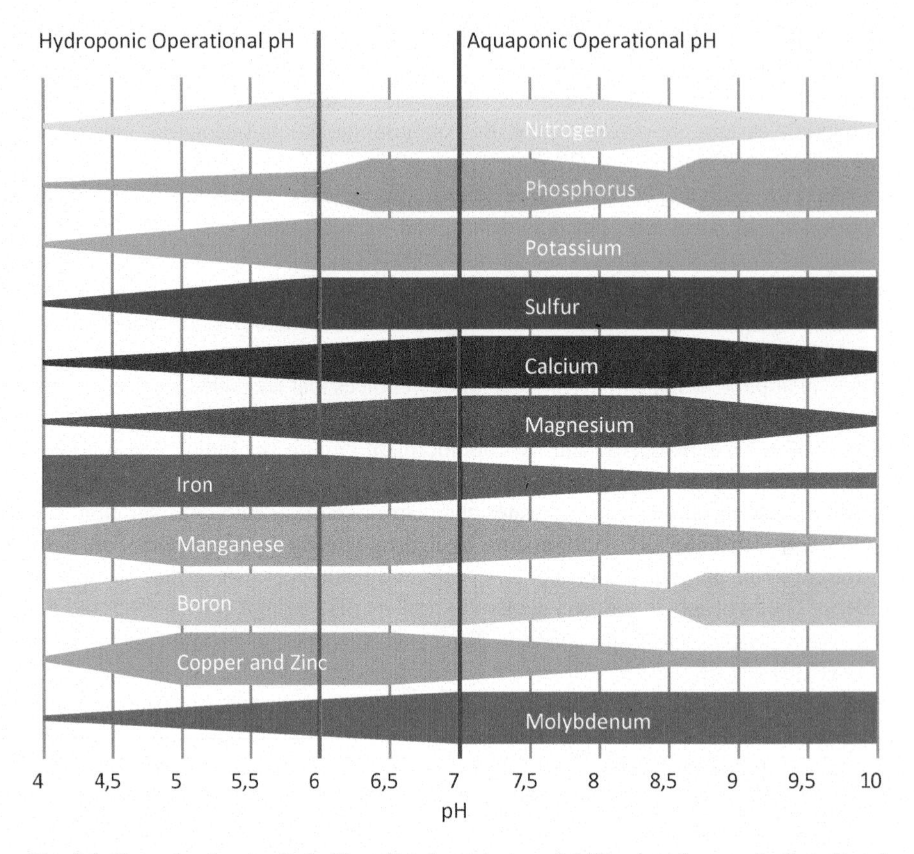

Fig. 5.4 Example of a standard pH mediated, nutrient availability chart for aquatically cultured plants. Red line represents a normal operating pH for a hydroponic system; the blue line that for an aquaponic system

(Timmons et al. 2002; Goddek et al. 2015; Suhl et al. 2016). In natural freshwater environments, most fish species require an environmental pH (i.e. water pH) that closely matches the internal pH of the fish, which is often close to a pH of 7.4 (Lennard 2017). In addition, the major microbes associated with dissolved metabolite transformation in RAS culture (the nitrification bacteria of several species) also require a pH around 7.5 for optimal ammonia transformation to nitrate (Goddek et al. 2015; Suhl et al. 2016). Therefore, RAS operators apply a pH set point of approximately 7.5 to RAS freshwater fish culture.

There is an obvious difference between a pH of 5.5 (an average for standard, sterilised, hydroponic plant culture) and a pH of 7.5 (an average standard for RAS fish culture). Therefore, it is argued broadly that pH represents one of the largest water quality compromises present in aquaponic science (Goddek et al. 2015; Suhl et al. 2016). Advocates of decoupled aquaponic designs often cite this difference in optimal pH requirement as an argument for the decoupled design approach, stating that fully recirculating designs must find a pH compromise when decoupled designs have the luxury of applying different water pH set points to the fish and plant components (Suhl et al. 2016; Goddek et al. 2016). However, what this argument ignores is that aquaponic systems, as opposed to hydroponic systems, are not sterile and employ ecological aquatic techniques that encourage a diverse population of microflora to be present within the aquaponic system (Eck 2017; Lennard 2017). This results in a broad variety of present microbes, many of which form intricate and complex associations with the plants, especially the plant roots, within the aquaponic system (Lennard 2017). It is well known and established in plant physiology that many microbes, associated with the soil medium and matrix, closely associate with plant roots and that many of these microbes assist plants to access and uptake vital nutrients (Vimal et al. 2017). It is also known that some of these microbes produce organic molecules that directly further assist plant growth, assist plant immunity development and assist to outcompete plant (especially root) pathogens (Vimal et al. 2017; Srivastava et al. 2017). In essence, these microbes assist plants in many ways that are simply not present in the sterilised environment applied in standard hydroponic culture.

With these diverse microbes present, the plants gain access to nutrients in many ways that are not possible in systems that rely on aquatic pH settings alone to enable plant nutrient access (e.g. standard hydroponics and substrate culture). Many of these microbes operate at broad pH levels, just like other soil-based microbes, such as the nitrification bacteria (pH of 6.5–8.0, Timmons et al. 2002). Therefore, with these microbes present in aquaponic systems, the pH set point may be raised above what is normally applied in hydroponic or substrate culture techniques (i.e. pH of 4.5–6.0) while advanced and efficient plant growth is still present (Lennard 2017). This is evidenced in the work of several aquaponic researchers who have demonstrated better plant growth rates in aquaponics than in standard hydroponics (Nichols and Lennard 2010).

Other water quality requirements in aquaponic systems relate to physical/chemical parameters and more specifically, plant nutrient requirement parameters. In terms of physical/chemical requirements, plants, fish and microbes share many commonalities. Dissolved oxygen (DO) is vital to fish, plant roots and microflora and must be maintained in aquaponic systems (Rakocy and Hargreaves 1993; Rakocy et al. 2006). Plant roots and microflora generally require relatively lower DO concentrations than most fish; plant roots and microbes can survive with DO below 3 mg/L (Goto et al. 1996), whereas most fish require above 5 mg/L (Timmons et al. 2002). Therefore, if the DO concentration within the aquaponic system is set and maintained for the fish requirement, the plant and microbe requirement is also met (Lennard 2017). Different fish species require different DO concentrations: warm water fish (e.g. *Tilapia* spp., barramundi) can generally tolerate lower DO concentrations than cool water fish species (e.g. salmonids like rainbow trout and arctic char); because the fish DO requirement is almost always greater than the plant roots and microfloral requirement, DO should be set for the specific fish species being cultured (Lennard 2017).

Water-carbon dioxide (CO_2) concentrations, like that for DO, are generally set by the fish because the plant roots and microbes can tolerate higher concentrations than the fish. Carbon dioxide concentrations are important to optimal fish health and growth and are often ignored in aquaponic designs. Parameters and set points for CO_2 concentrations should be the same as for the same fish species cultured in fish-only, RAS systems and in general, should be kept below 20 mg/L (Masser et al. 1992).

Water temperature is important to all the present life forms within an aquaponic system. Fish and plant species should be matched as closely as possible for water temperature requirements (e.g. *Tilapia* spp. of fish like 25° C plus, and plants like basil thrive in this relatively high water temperature; lettuce varieties like cooler water, and therefore, a better matched fish candidate is rainbow trout) (Lennard 2017). However, as for other water physical and chemistry parameters, meeting the fish's requirement for water temperature is paramount because the microbes have the ability to undergo specific species selection based on the ambient conditions (e.g. nitrification bacterial species differentiation occurs at different water temperatures and the species that matches best to the particular water temperature will dominate the nitrification bacterial biomass of the system) and many plants can grow very well at a broader range of water temperatures (Lennard 2017). Matching the water temperature, and maintaining it within plus or minus 2° C (i.e. a high-level temperature control) to the fish, is an important requirement in aquaponics because when water temperature is correct and does not deviate from the ideal average, the fish achieve efficient and optimised metabolism and eat and convert feed efficiently, leading to better fish growth rates and stable and predictable waste load releases, which assists plant culture (Timmons et al. 2002).

Maintaining water clarity (low turbidity) is another important parameter in aquaponic culture (Rakocy et al. 2006). Most water turbidity is due to suspended solids loads that have not been adequately filtered, and these solids may affect fish by adhering to their gills, which may lower potential oxygen transfer rates and ammonia release rates (Timmons et al. 2002). Suspended solids loads less than 30 mg/L are recommended for aquaponically cultured fish (Masser et al. 1992; Timmons et al. 2002). High suspended solids loads also affect plant roots because they have the ability to adhere to the roots which may cause nutrient uptake inefficiency, but more commonly provides increased potential for pathogenic organism colonisation, which leads to poor root health and ultimate plant death (Rakocy et al. 2006). These suspended solids also encourage the prevalence of heterotrophic bacteria (species that break down and metabolise organic carbon) which, if allowed to dominate systems, may outcompete other required species, such as nitrification bacteria.

Electrical conductivity (EC) is a measure often applied in hydroponics to gain an understanding of the amount of total nutrient present in the water. It, however, cannot provide information on the nutrient mix, the presence or absence of individual nutrient species or the amount of individual nutrient species present (Resh 2013). It is not often applied in aquaponics because it only measures the presence of ionic (charged) nutrient forms, and it has been argued that aquaponics is an organic nutrient supply method, and therefore, EC is not a relevant measure (Hallam 2017). However, plants generally only source ionic forms of nutrients, and therefore,

EC can be used as a general tool or guide to the total amount of plant-available nutrient in an aquaponic system (Lennard 2017).

For fully recirculating aquaponic systems, in terms of physical and chemical parameters, it is the fish that are more exacting in their requirements, and therefore, if systems are managed to maintain the requirements of the fish, the plants and microbes are having their requirements more than satisfied (Lennard 2017). The difference when it comes to the plants, however, is their requirement for the correct mixture and strength of nutrients to be present to allow optimised nutrient access and uptake (whether stand-alone or microbial-assisted) which leads to efficient and fast growth. Decoupled aquaponic systems may therefore be more attractive because of the perception that they allow more exacting nutrient delivery to the plants (Goddek et al. 2016). Fish feed and, therefore, fish waste do not contain the correct mixture of nutrients to meet the plant requirements (Rakocy et al. 2006). Therefore, the aquaponic system design must account for those missing nutrients and supplement them. Fully recirculating aquaponic systems generally supplement nutrients by adding them in the salt species used to manage the daily pH buffering regime; the basic portion of the salt adjusts the pH and the positive portion of the salt allows the supplementation of missing plant nutrients (e.g. potassium, calcium, magnesium) (Rakocy et al. 2006). Decoupled aquaponic designs take the waste water and associated solid wastes from the fish component and adjust the water to contain the nutrients required for plant production by adding nutrients in different forms (Goddek et al. 2016). These nutrient additions are generally based on using standard hydroponic salt species that do not necessarily provide any pH adjustment outcome (e.g. calcium phosphate, calcium sulphate, potassium phosphate, etc.).

The pathway to efficient plant growth in aquaponic systems is to provide an aquatic nutrient profile that provides all the nutrients the plant requires (mixture) at the strengths required (concentration) (Lennard 2017). In fully recirculating aquaponic designs, or decoupled aquaponic designs that do not apply sterilisation methods, there appears to be less of a requirement to meet the nutrient concentrations or strengths applied in standard hydroponics, because the ecological nature of the system associates many diverse microflora with the plant roots and these microflora assist plant nutrient access (Lennard 2017). For decoupled, or other, aquaponic designs that apply sterilisation to the plant component and follow a standard hydroponic analogue approach, there appears to be a requirement to try and approach standard hydroponic nutrient concentrations (Suhl et al. 2016; Karimanzira et al. 2016). The compromise, however, with the decoupled approach is that it leads to external supplementation ratios far beyond those of fully recirculating aquaponic designs; European decoupled designs currently average 50% or greater external nutrient additions (COST FA1305 2017; Goddek and Keesman 2018), while the UVI method supplies less than 20%, and other systems may supply less than 10% external nutrient supplementation (Lennard 2017).

No matter the method, all aquaponic systems should strive to supply the plants with the nutrition required for optimised growth so as to provide the enterprise with the greatest chance of financial viability. In this context, the nutrient content and strength of the water being delivered to the plants is very important and regular nutrient testing of the water should be employed so that nutrient mixture and strength may be maintained and managed as a very important water quality requirement.

5.6 Applicable Fish Culture Technologies

In aquaponics, the aquaculture portion of the integration equation is broadly applied in a tank-based context, where the fish are kept in tanks, the water is filtered via mechanical (solids removal) and biological (ammonia transformation to nitrate) mechanisms and dissolved oxygen is maintained, either via aeration or direct oxygen injection (Rakocy et al. 2006; Lennard 2017).

As has been argued in Sect. 5.0 (Introduction) of this chapter, historical examples of chinampas (Somerville et al. 2014) and Asian rice paddy farming (Halwart and Gupta 2004) as early iterations of aquaponics are unfounded and inappropriate examples of aquaponic principles, because modern aquaponics relies on designed additions of fish and fish feeds to supply a designed level of nutrition to the plants, and therefore, these historical examples cannot be considered in any way similar (Lennard 2017).

The above historical examples, which rely on soil-based plant culturing systems, lead to the question of what aquaculture technologies are suitable for aquaponic integration. Soil-based, extensive, freshwater pond aquaculture of fin fish is the largest culturing method applied to produce freshwater fish for human consumption (Boyd and Tucker 2012). A pond approach relies on the earthen base of the pond, and the associated microflora present in that soil, to treat and remediate the wastes produced from fish culture so the fish are not living in water that has a potential to be toxic to them (Boyd and Tucker 2012). Because this system relies on the inherent treatment capacity of the earthen pond itself, fish densities are relatively low compared to other aquaculture methods. Because the fish densities are low (and therefore the associated feeding rates are low) and the pond itself treats and uptakes the waste nutrients produced by the fish, pond waters exhibit extremely low water nutrient concentrations. These pond system aquatic nutrient concentrations are so low that they are often inappropriate as nutrient sources for substantial, commercial aquatic plant production methods (Lennard 2017). Therefore, ponds are not an appropriate aquaculture method to be integrated with hydroponics in terms of acceptable plant production rates.

Similarly, raceway fin fish culture methods (as regularly applied for freshwater Salmonid production), which supply very large volumes of water at high turnover rates, or low residence times, through controlled raceway fish culture tanks, are not appropriate for aquaponic integration because the high water turnover rates do not allow adequate nutrient accumulations to meet plant nutrient requirements (Rakocy and Hargreaves 1993).

The most appropriate fish culture technologies to apply within an aquaponic integration context are those that culture fish in tanks and allow a level of fish waste accumulation (plant nutrient accumulation) that has the potential to lead to water nutrient concentrations that are applicable for efficient hydroponic plant production (Rakocy et al. 2006). Recirculating Aquaculture System (RAS) principles are broadly applied to aquaponics because they provide the methodologies to successfully keep and grow the fish, in controlled volumes of water, with low daily water replacement rates, that allow fish waste (plant nutrient) accumulations that approach those required to efficiently hydroponically culture the plants (Rakocy and Hargreaves 1993; Lennard 2017). The complexities and design requirements of RAS are discussed in Chap. 3 of this book. Suffice to say that RAS fish culture is the only real appropriate method to apply for fish culturing components in an aquaponic context and as discussed above, soil-based aquaculture systems, such as extensive pond systems and raceway culture systems, cannot provide the nutrient requirements of the plants and therefore should not be considered.

5.7 Nutrient Sources

The major input to any aquaponic system are the nutrients added because aquaponic systems are designed to efficiently partition the nutrients added to them to the three important forms of life present: the fish and plants (which are the main products of the system) and the microflora (which assist to make the added nutrients available to the fish and plants) (Lennard 2017).

In classical, fully recirculating aquaponic designs, one of the key design drivers is to use the main nutrient input source, the fish feed, as efficiently as possible and therefore fully recirculating designs strive to supply as many of the nutrients required for the plants from the fish feed (Lennard 2017). Decoupled designs, on the other hand, place an emphasis on optimised plant growth by directly comparing the nutrient mixtures and strengths applied in standard hydroponics and substrate culture and trying to replicate those within the aquaponic context and therefore do not strive to supply as many of the nutrients required for the plants from the fish feed and utilise substantial external nutrient supplementations to achieve the required plant growth rates (Delaide et al. 2016). This means that a different emphasis is placed on the origin of the nutrients added, based on the technical design approach, and this, therefore, affects the main nutrient supply source of the aquaponic system; for fully recirculating designs, the major plant nutrient source is fish feed (via fish waste production), and for decoupled designs the major nutrient supply source for the plants is external supplements (e.g. nutrient salts) (Lennard 2017).

Fully recirculating aquaponic designs, such as the UVI aquaponic system model, rely on the fish feed as the major nutrient source for the system (Rakocy et al. 2006). The fish feed is added to the fish, which eat it, metabolise it and use the nutrients from it as required and then produce a waste stream (both solids and dissolved). This waste stream from the fish becomes the major nutrient source for the plants, and hence, the fish feed is the major nutrient source for the plants. The UVI system provides approximately 80% or more of the nutrients required to grow the plants from the fish feed alone (Lennard 2017). The remaining nutrients required for plant growth, because the fish feed does not contain them in the amounts required, are added via a nutrient supplementation method that provides the dual role of supplementing the additional nutrients and controlling the system aquatic pH (Rakocy et al. 2006). This dual role approach is referred to as "buffering" and the supplement is referred to as "buffer". For the UVI model, the two important plant nutrients identified as lacking in fish feed and which require supplementation are potassium (K) and calcium (Ca) and are supplemented daily via the buffering regime. In addition, plant-required iron (Fe) is also lacking in the fish feed and is supplemented in a chelated form via direct addition to the system water at a frequency measured in weeks (i.e. every 2–4 weeks based on weekly aquatic iron analysis) (Rakocy et al. 2006).

Other fully recirculating aquaponic design approaches or methods place an even higher emphasis on providing nutrients via the fish feed. Lennard (2017) has developed a method for fully recirculating systems that supplies greater than 90% of the nutrients required for plant growth from the fish feed added. The increase in the efficiency of nutrients supplied via the fish feed of this method when compared to the UVI method is that this approach remineralises the solid fish wastes (via external, bacteria-mediated biodigestion) and adds these nutrients back into the aquaponic system for plant utilisation, whereas the UVI method sends the majority of the solid fish wastes to an external waste stream (Rakocy et al. 2006; Lennard 2017). This approach also adds nutrients deficient in the fish feed for plant growth via a buffering regime; however, this regime is far more exacting and allows greater manipulation of nutrient strengths and mixtures than the UVI approach (Lennard 2017).

Therefore, the major nutrient addition pathways for most fully recirculating aquaponic system designs are the fish feed (major route), buffer external supplementation for added potassium and calcium (minor route) and direct supplementation of iron chelate (minor route).

Decoupled aquaponic system designs, such as those being widely adopted currently in Europe, rely on a mixture of fish feed nutrients and active, external supplementation to provide the nutrients required for plant growth (Suhl et al. 2016). Because decoupled designs do not return water from the plant component to the fish component, it is possible to customise the nutrient profile within the water specifically to the plant requirement (Goddek et al. 2016). Therefore, decoupled aquaponic designs almost always rely on substantial external nutrient supplementation to meet the plant requirement and place far less emphasis on providing as much nutrition as possible for the plants from the fish wastes. In addition, the amount of external supplementation is substantial when compared to fully recirculating approaches (Lennard 2017) with external fractions regularly 50% or more of the total plant nutrient requirement or greater (Goddek 2017). The external nutrients supplemented to decoupled aquaponic systems are most often hydroponic nutrient salt analogues or derivatives (Delaide et al. 2016; Karimanzira et al. 2016). This reliance on a source of substantial additional nutrients other than those that arise from fish wastes (fish feeds) that are hydroponic salt in nature, for plant supply of the European decoupled approach, has even directly affected the definition of aquaponics that the European aquaponics community currently applies, with the EU COST, EU Aquaponics Hub, defining aquaponics as "...a production system of aquatic organisms and plants where the majority (> 50%) of nutrients sustaining the plants derives from wastes originating from feeding the aquatic organisms" (Goddek 2017; COST FA 1305, 2017) compared to Lennard (2017) who defines aquaponics as requiring at least 80% nutrient supply from fish wastes. Some also argue whether a method that relies on 50% of the nutrients required for plant growth originating from external sources other than fish feeds is actually aquaponic in nature or rather, a hydroponic method with some fish integrated or added (Lennard 2017)?

Another proposed supply source for nutrients to aquaponic systems is that of external nutrient supplementation via the application of foliar plant sprays (Tyson et al. 2008; Roosta and Hamidpour 2011; Roosta and Hamidpour 2013; Roosta 2014). These foliar sprays are again, an aquatic delivery of standard hydroponic nutrient salts or derivatives. The difference is that in the decoupled examples above, the nutrient salts are added directly to the culture water and are therefore accessed by the plants via root uptake (Resh 2013), whereas foliar sprays, as the name implies, add dissolved nutrient salts to the plant leaves and uptake is achieved via plant leaf stomatal or cuticle access (Fernandez et al. 2013).

There are therefore several major nutrient sources applied in aquaponics: fish feeds, buffering systems (via basic, pH-adjusting salt species added to the water column), nutrient salt additions (hydroponic nutrient salts added to the water column) and foliar sprays (hydroponic nutrient salts added to the leaf surface). All of which supply nutrients to the aquaponic system for the health and growth of the fish and plants that are cultured.

5.8 Aquaponics as an Ecological Approach

Aquaponics, until recently, has been dominated by fully recirculating (or coupled) design approaches that share and recirculate the water resource constantly between the two major components (fish and plant culture) (Rakocy et al. 2006; Lennard 2017). In addition, the low to medium technology approaches historically applied to aquaponics have driven a desire to remove costly components so as to increase the potential of a positive economic outcome. One of the filtration components almost always applied to standard RAS and hydroponics/substrate culture technologies, that of aquatic sterilisation, has regularly not been included by aquaponic designers.

Sterilisation in a RAS and hydroponic/substrate culture context is universally applied because the high densities of either the fish or plants cultured usually attract pressure from aquatic, pathogenic organisms that substantially lower overall production rates (Van Os 1999; Timmons et al. 2002). The major reason for this increased aquatic pest pressure in both technologies is that each concentrates on providing minimal biotic, ecological resources and therefore allows considerable "ecological space" within the system water for biotic colonisation. In these "open" biological conditions, pest and pathogenic species proliferate and tend to colonise quickly to take advantage of the species present (i.e. fish and plants) (Lennard 2017). In this context, sterilisation or disinfection of the culture water has historically be seen as an engineered approach to counteract the issue (Van Os 1999; Timmons et al. 2002). This means that both RAS and hydroponic/substrate culture industries adopt a sterilisation approach to control pathogenic organisms within the associated culture water.

Aquaponics has always placed an emphasis on the importance of the associated microbiology to perform important biological services. In all the coupled aquaponic designs of Rakocy and his UVI team, a biological filter was not included because they demonstrated that the raft culture, hydroponic component provided more than enough surface area to support the nitrifying bacteria colony size to treat all the ammonia produced by the fish as a dissolved waste product and convert it to nitrate (Rakocy et al. 2006, 2011). Rakocy and his team therefore did not advocate applied sterilisation of the system water because it may have affected the nitrifying bacterial colonies. This historical UVI/Rakocy perspective dictated aquaponic system design into the future. Other advantages of not including aquatic sterilisation to aquaponic systems were identified and discussed, especially in the context of assistive plant microbiota (Savidov 2005; Goddek et al. 2016).

The current thinking in aquaponic research and industry is that not applying any form of aquatic sterilisation or disinfection allows the system water to develop a complex aquatic ecology that consists of many different microbiological life forms (Goddek et al. 2016; Lennard 2017). This produces a situation similar to a natural ecosystem whereby a high diversity of microflora interacts with each other and the other associated life forms within the system (i.e. fish and plants). The proposed outcome is that this diversity leads to a situation in which no single pathogenic organism can dominate due to the presence of all of the other microflora and can therefore not cause devastating effects to fish or plant production. It has been demonstrated that aquaponic systems contain a high diversity of microflora (Eck 2017) and via the proposed ecological diversity mechanism outlined above, assistance to both fish and plant health and growth is provided via this microbial diversity (Lennard 2017).

The non-sterilised, ecologically diverse approach to aquaponics has been historically applied to coupled or fully recirculating aquaponic designs (Rakocy et al. 2006), whereas a sterilised, hydroponic analogy has been proposed for some decoupled aquaponic design approaches (Monsees et al. 2016; Priva 2009; Goddek 2017). However, it appears more decoupled designers are now applying principles that take an ecological, non-sterilised approach into consideration (Goddek et al. 2016; Suhl et al. 2016; Karimanzira et al. 2016) and therefore acknowledge there is a positive effect associated with a diverse aquaponic microflora (Goddek et al. 2016; Lennard 2017).

5.9 Advantages of Aquaponics

Because there are two separate, existing, analogous technologies that produce fish and plants at high rates (RAS fish culture and hydroponic/substrate culture plant production), a reason for their integration seems pertinent. RAS produces fish at productive rates in terms of individual biomass gain, for the feed weight added, that rivals, if not betters, other aquaculture methods (Lennard 2017). In addition, the high fish densities that RAS allows lead to higher collective biomass gains (Rakocy et al. 2006; Lennard 2017). Hydroponics and substrate culture possess, within a controlled environment context, advanced production rates of plants that better most other agriculture and horticulture methods (Resh 2013). Therefore, initially, there is a requirement for aquaponics to produce fish and plants at rates that equal these two separate productive technologies; if not, then any loss of productive effort counts against any integration argument. If the productive rate of the fish and plants in an aquaponic system can equal, or better, the RAS and hydroponic industries, then a further case may be made for other advantages that may occur due to the integration process.

Standard hydroponics or substrate culture has been directly compared with aquaponics in terms of the plant growth rates of the two technologies. Lennard (2005) compared aquaponic system lettuce production to a hydroponic control in several replicated laboratory experiments. He demonstrated that aquaponic lettuce production was statistically lower in aquaponics (4.10 kg/m^2) when compared to hydroponics (6.52 kg/m^2) when a standard approach to media bed aquaponic system design and management was applied. However, he then performed a series of experiments that isolated specific parameters of the design (e.g. reciprocal vs constant hydroponic subunit water delivery, applied water flow rate to the hydroponic subunit and comparing different hydroponic subunits) or comparing specific management drivers (e.g. buffering methodologies and species and the overall starting nutrient concentrations) to achieve optimisation and then demonstrated that aquaponics (5.77 kg/m^2) was statistically identical to hydroponic lettuce production (5.46 kg/m^2) after optimisation of the aquaponic system based on the improvements suggested by his earlier experiments, the result suggesting that improvements to coupled or fully recirculating aquaponic designs can equal standard hydroponic plant production rates. Lennard (2005) also demonstrated fish survival, SGR, FCR and growth rates equal to those exhibited in standard RAS and extensive pond aquaculture for the fish species tested (Australian Murray Cod).

Pantanella et al. (2010) also demonstrated statistically similar lettuce production results within high fish density (5.7 kg/m^2 lettuce production) and low fish density (5.6 kg/m^2 lettuce production) aquaponic systems compared to a standard hydroponic control (6.0 kg/m^2).

Lennard (Nichols and Lennard 2010) demonstrated statistically equal or better results for all lettuce varieties and almost all herb variety production tested in a nutrient film technique (NFT) aquaponic system when compared to a hydroponic NFT system within the same greenhouse.

Delaide et al. (2016) compared nutrient-complemented RAS production water (nutrients were added to match one water nutrient mixture and strength example by Rakocy – denoted as an aquaponic analogue), fully nutrient-complemented RAS production water (RAS production water with added hydroponic nutrient salts to meet a water nutrient mixture and strength as used for standard hydroponics – denoted as a decoupled analogue) and a hydroponic control (standard hydroponic nutrient solution) in terms of plant growth rate and showed the aquaponic water analogue equalled the hydroponic control and the decoupled analogue water bettered the hydroponic control. However, it must be noted that these were not fully operating aquaponic systems containing fish (and the associated, full and active microbial content) that were compared, but simply water removed from an operating RAS and complemented, then compared to a hydroponic control water.

Rakocy and his UVI team have demonstrated with several studies that *Tilapia* spp. fish growth rates equal industry standards set by standard aquaculture production practices (Rakocy and Hargreaves 1993; Rakocy et al. 2004a, b, 2006, 2011).

These and other studies have demonstrated that aquaponics, no matter the configuration (coupled and decoupled), has the potential to produce plant production rates equal to, or better than, standard hydroponics and fish production rates of a similar standard to RAS. Therefore, the above discussed requirement for aquaponics to equal its industry analogues (RAS and hydroponics) seems to have been adequately proven, and therefore, the other advantages of aquaponics should be considered.

Efficient water use is regularly attributed to aquaponics. Lennard (2005) stated that the water savings associated with an optimised aquaponic test system (laboratory) were 90% or greater when compared to a standard RAS aquaculture control system where water was exchanged to control nitrate accumulations, whereas the plants in the aquaponics performed the same requirement. Therefore, he demonstrated that aquaponics provides a substantial water-saving benefit compared to standard RAS aquaculture. Interestingly, this 90% water-saving figure has subsequently been stated broadly within the global aquaponics community in a plant use context (e.g. aquaponics uses 90% less water than soil-based plant production (Graber and Junge 2009)) – an example of how scientific argument can be incorrectly adopted by nonscientific industry participants.

McMurtry (1990) demonstrated a water consumption rate in his aquaponic system of approximately 1% of that required in a similar pond culture system. Rakocy (1989) has demonstrated similar 1% water consumption rates compared to pond-based aquaculture. Rakocy and Hargreaves (1993) stated that the daily water replacement rate for the UVI aquaponic system was approximately 1.5% of the total system volume and Love et al. (2015a, b) stated an approximate 1% of system volume water loss rate per day for their aquaponic research system.

The comparison of aquaponics to RAS can realise substantial water savings and aquaponics uses small amounts of replacement water on a daily basis. A well-designed aquaponic system will seek to use water as efficiently as possible and therefore only replace that water lost via plant evapotranspiration (Lennard 2017). In fact, it has been proposed that water may even be recovered from that lost due to plant evapotranspiration via employing some form of air water content harvesting scheme or technology (Kalantari et al. 2017). Coupled aquaponic systems appear to provide a greater potential to conserve and lower water use (Lennard 2017). If the nutrient dynamics between fish production and plant use can be balanced, the only water loss is via plant evapotranspiration, and because the water is integrally shared between the fish and plant components, daily makeup water volumes simply represent all the water lost from the system plants (Lennard 2017). Decoupled aquaponic designs present a more difficult proposition because the two components are not integrally linked and the daily water use of the fish component does not match the daily water use of the plant component (Goddek et al. 2016; Goddek and Keesman 2018). Therefore, water use and replacement rates for aquaponic systems are not completely resolved and probably never will be due to the broad differences in system design approaches.

Efficient nutrient utilisation is assigned to the aquaponic method and cited as an advantage of the aquaponic approach (Rakocy et al. 2006; Blidariu and Grozea 2011; Suhl et al. 2016; Goddek et al. 2015). This is generally because standard RAS aquaculture utilises the nutrients within the fish feed to grow the fish, with the remainder being sent to waste. Fish metabolise much of the feed they are fed, but only utilise approximately 25–35% of the nutrients added (Timmons et al. 2002; Lennard 2017). This means up to 75% of the nutrients added to fish-only RAS are wasted and not utilised. Aquaponics seeks to use the nutrients wasted in RAS for plant production, and therefore, aquaponics is said to use the nutrients added more efficiently because two crops are produced from the one input source (Rakocy and Hargreaves 1993; Timmons et al. 2002; Rakocy et al. 2006; Lennard 2017). The extent of fish waste nutrient use does differ between the various aquaponic methods. The fully recirculating UVI model does not utilise the majority of the solid fish wastes generated in the fish component and sends them to waste (Rakocy et al. 2006), the fully recirculating Lennard model takes this a step further by using all the wastes generated by the fish component (dissolved wastes directly and solids via external microbial remineralisation with main system replacement) (Lennard 2017). Many decoupled approaches also attempt to utilise all the wastes generated by the fish component, via direct use of dissolved wastes and again, via external microbial remineralisation with main system replacement (Goddek et al. 2016; Goddek and Keesman 2018). All of these methods and approaches demonstrate that a primary driver for the aquaponic method is to utilise as many of the added nutrients as possible and therefore attempt to use the added nutrients as efficiently as possible.

Independence from soil has been cited as an advantage of the aquaponic method (Blidariu and Grozea 2011; Love et al. 2015a, b). The advantage perceived is that because soil is not required, the aquaponic system or facility may be located where the operator chooses, rather than where suitable soil is present (Love et al. 2015a, b). Therefore, the aquaponic method is independent of location based on soil availability, which is an advantage over soil-based agriculture.

It has been argued that aquaponics provides an advantage by mimicking natural systems (Blidariu and Grozea 2011; Love et al. 2014). This is supported by the ecological nature of the aquaponic approach/method, as outlined in Sect. 5.7 above, with the associated advantages related to diverse and dense microfloral communities (Lennard 2017).

Aquaculture has a potential direct environmental impact due to the release of nutrient-rich waste waters to the surrounding environment – generally, aquatic environments (Boyd and Tucker 2012). Some hydroponic methods may also possess this potential. However, aquaponics can exhibit lowered or negated direct environmental impact from nutrient-rich waste streams because the main waste-generating component (i.e. the fish) is integrated with a nutrient use component (i.e. the plants) (Rakocy et al. 2006; Blidariu and Grozea 2011; Goddek et al. 2015; Lennard 2017). However, some aquaponic methods do produce wastes (e.g. the UVI model); but these are generally treated and reused for other agricultural practices at the aquaponic facility site (Timmons et al. 2002; Rakocy et al. 2006). Many aquaponic methods rely on the use of standard aquaculture feeds, which contain varying concentrations of sodium, usually via the use of fish meal or fish oil as an ingredient (Timmons et al. 2002). Sodium is not utilised by plants and therefore may accumulate over time in aquaponic systems, which may lead to a requirement for some form of water replacement so sodium does not accumulate to concentrations that affect the plants (Lennard 2017). However, it has been reported that some species of lettuce had the ability to take up sodium, when being exposed to aquaculture water (Goddek and Vermeulen 2018).

Coupled or fully recirculating aquaponic systems integrally share the water resource between the two main components (fish and plant). Because of this fully connected and recirculating aquatic nature, coupled aquaponic systems exhibit a self-controlling mechanism in terms of an inability to safely apply herbicides and pesticides to the plants; if they are applied, their presence may negatively affect the fish (Blidariu and Grozea 2011). Fully recirculating advocates see this inability to applying pesticides and herbicides as an advantage, the argument being that it guarantees a spray-free product (Blidariu and Grozea 2011). Decoupled aquaponics advocates also seek to not apply herbicides or pesticides; however, due to the fact that the water is not recirculated back to the fish from the plants, the ability to apply pesticides and herbicides to the plants is present (Goddek 2017). Therefore, the application or lack of application of pesticides and herbicides to the plant component of aquaponic designs is seen differently by groups who advocate for different design approaches.

There is a perception that the presence of both fish and plants in the same aquatic system provides positive synergistic effects to fish and plant health (Blidariu and Grozea 2011). This has been indirectly demonstrated by the ability of aquaponics in some studies to produce plant growth rates greater than those seen in standard hydroponics (Nichols and Lennard 2010; Delaide et al. 2016). However, no direct causal link has been established between the presence of both fish and plants and any positive outcome to fish or plant health.

References

Baquedano E (1993) Aztec Inca & Maya. A Dorling Kindersley Book, Singapore
Blidariu F, Grozea A (2011) Increasing the economic efficiency and sustainability of indoor fish farming by means of aquaponics – review. Anim Sci Biotechnol 44(2):1–8
Boyd CE, Tucker CS (2012) Pond aquaculture water quality management. Springer Science and Business Media
Buzby KM, Lian-shin L (2014) Scaling aquaponic systems: balancing plant uptake with fish output. Aquac Eng 63:39–44
Cerozi BS, Fitzsimmons K (2017) Phosphorous dynamics modelling and mass balance in an aquaponics system. Agric Syst 153:94–100
COST FA1305 (2017) The EU aquaponics hub – realising sustainable integrated fish and vegetable production for the EU. https://euaquaponicshub.com
Delaide B, Goddek S, Gott J, Soyeurt H, Haissam Jijakli M (2016) Lettuce (*Lactuca sativa* L. var. Sucrine) (2016). Growth performance in complemented aquaponic solution outperforms hydroponics. Water 8(10):467
Eck M (2017) Taxonomic characterisation of bacteria communities from water of diversified aquaponic systems. Thesis for the partial fulfillment of a Masters Degree. Université de Liège, Liège
Endut A, Jusoh A, Ali N, Wan Nik WB, Hassan A (2010) A study on the optimal hydraulic loading rate and plant ratios in recirculation aquaponic system. Technol 101:1511–1517
Fernandez V, Sotiropoulos T, Brown P (2013) Foliar fertilization: scientific principles and field practices, 1st edn. International Fertilizer Industry Association (IFA), Paris
Goddek S (2017) Opportunities and challenges of multi-loop aquaponic systems. Wageningen University, Wageningen
Goddek S, Keesman KJ (2018) The necessity of desalination technology for designing and sizing multi-loop aquaponics systems. Desalination 428:76–85
Goddek S, Körner O (2019) A fully integrated simulation model of multi-loop aquaponics: a case study for system sizing in different environments. Agric Syst 171:143–154
Goddek S, Vermeulen T (2018) Comparison of *Lactuca sativa* growth performance in conventional and RAS-based hydroponic systems. Aquac Int 26:1–10. https://doi.org/10.1007/s10499-018-0293-8
Goddek S, Delaide B, Mankasingh U, Vala Ragnarsdottir K, Jijakli H, Thorarinsdottir R (2015) Challenges of sustainable and commercial aquaponics. Sustainability 7(4):4199–4224
Goddek S, Espinal CA, Delaide B, Jikali MH, Schmautz Z, Wuertz S, Keesman J (2016) Navigating towards decoupled aquaponic systems: a system dynamics design approach. Water 8(7):303. 1–29
Gooley GJ, Gavine FM (2003) Integrated agri-aquaculture systems: a resource handbook for Australian industry development. RIRDC Publication No. 03/012
Goto E, Both AJ, Albright LD, Langhans RW, Leed AR (1996) Effect of dissolved oxygen concentration on lettuce growth in floating hydroponics. Proceedings of the International Symposium in Plant Production in Closed Systems. Acta Hortic 440:205–210
Graber A, Junge R (2009) Aquaponic systems: nutrient recycling from fish wastewater by vegetable production. Desalination 246:147–156
Hallam M (2017) EC. Murray Hallam's practical aquaponics. https://aquaponics.net.au/ec/
Halwart M, Gupta MV (eds) (2004) Culture of fish in rice fields. FAO and The WorldFish Center, Penang
Kalantari F, Tahir OM, Lahijani AM, Kalantari S (2017) A review of vertical farming technology: a guide for implementation. Adv Eng Forum 24:76–91
Karimanzira D, Keesman KJ, Kloas W, Baganz D, Raushenbach T (2016) Dynamic modelling of the INAPRO aquaponic system. Aquac Eng 75:29–45
Kloas W, Groß R, Baganz D, Graupner J, Monsees H, Schmidt U, Staaks G, Suhl J, Tschirner M, Wittstock B, Wuertz S, Zikova A, Rennert B (2015) A new concept for aquaponic systems to

improve sustainability, increase productivity, and reduce environmental impacts. Aquac Environ Interact 7:179–192. https://doi.org/10.3354/aei00146

Knaus U, Palm HW (2017) Effects of the fish species choice on vegetables in aquaponics under spring-summer conditions in northern Germany (Mecklenburg Western Pomerania). Aquaculture 473:62–73

Komives T, Junge R (2015) Editorial: on the "aquaponic corner" section of the journal. Ecocycles 1 (2):1–2

Lennard WA (2005) Aquaponic integration of Murray Cod (*Maccullochella peelii peelii*) aquaculture and lettuce (*Lactuca sativa*) hydroponics. Thesis (Ph.D.). RMIT University, 2005

Lennard W (2017) Commercial aquaponic systems: integrating recirculating fish culture with hydroponic plant production. In press

Lennard WA, Leonard BV (2006) A comparison of three different hydroponic sub-systems (gravel bed, floating and nutrient film technique) in an aquaponic test system. Aquac Int 14:539–550

Love DC, Fry JP, Genello G, Hill ES, Frederick JA, Li X, Semmens K (2014) An international survey of aquaponics practitioners. PLoS One 9(7):E102662

Love DC, Fry JP, Li X, Hill ES, Genello L, Semmens K, Thompson RE (2015a) Commercial aquaponics production and profitability: findings from an international survey. Aquaculture 435:67–74

Love DC, Uhl MS, Genello L (2015b) Energy and water use of a small-scale raft aquaponics system in Baltimore, Maryland, United States. Aquac Eng 68:19–27

Masser MP, Rakocy J, Losordo TM (1992) Recirculating aquaculture tank production systems. SRAC Publication No. 452. Southern Regional Aquaculture Center. USA

McMurtry M (1990) Performance of an integrated aquaculture-olericulture system as influenced by component ratio. PhD. dissertation, North Carolina State University, Raleigh, North Carolina, USA

Monsees H, Kloas W, Wuertz S (2016) Comparison of coupled and decoupled aquaponics - Implications for future system design. Abstract from aquaculture Europe, 2016. Edinburgh, Scotland

Morehart CT (2016) Chinampa agriculture, surplus production and political change at Xaltocan, Mexico. Anc Mesoam 27(1):183–196

Nichols MA, Lennard W (2010) Aquaponics in New Zealand. Practical Hydroponics and Greenhouses, 115: 46–51

Palm HW et al (2018) Towards commercial aquaponics: a review of systems, designs, scales and nomenclature. Aquac Int 26(3):813–842

Pantanella E, Cardarelli M, Colla G, Rea A, Marcucci A (2010) Aquaponic vs hydroponic: production and quality of lettuce crop. Acta Hortic 927:887–893

Priva (2009) Eindrapport project EcoFutura, visteelt in de glastuinbouw. Priva B.V., Aqua-Terra Nova B.V., Green Q Group B.V., Groen Agro Control. www.ecofutura.nl

Rakocy JE (1989) Vegetable hydroponics and fish culture, a productive interface. World Aquacult 20:42–47

Rakocy JE, Hargreaves JA (1993) Integration of vegetable hydroponics with fish culture: a review. In: Wang J (ed) Techniques for modern aquaculture. American Society of Agricultural Engineers, St Joseph

Rakocy JE, Bailey DS, Shultz RC, Thoman ES (2004a) Update on tilapia and vegetable production in the UVI aquaponic system. In: New Dimensions on Farmed Tilapia: Proceedings of the Sixth International Symposium on Tilapia in Aquaculture, Manila, pp 676–690

Rakocy JE, Shultz RC, Bailey DS, Thoman ES (2004b) Aquaponic integration of Tilapia and Basil: Comparing a batch and staggered cropping system. Acta Hortic 648:63–69

Rakocy JE, Masser MP, Losordo TM (2006) Recirculating aquaculture tank production systems: aquaponics – integrating fish and plant culture. SRAC Publication No. 454. Southern Regional Aquaculture Center. USA

Rakocy JE, Bailey DS, Shultz RC, Danaher JJ (2011) A commercial scale aquaponic system developed at the University of the Virgin Islands. Proceedings of the 9th International Symposium on Tilapia in Aquaculture

Resh HM (2013) Hydroponic food production, 7th edn. CRC Press, Boca Raton

Reyes Lastiri D, Slinkert T, Cappon HJ, Baganz D, Staaks G, Keesman KJ (2016) Model of an aquaponic system for minimised water, energy and nitrogen requirements. Water Sci Technol 74:1. https://doi.org/10.2166/wst.2016.127
Roosta HR (2014) Effects of foliar spray of K on mint, radish, parsley and coriander plants in aquaponic system. J Plant Nutr 37(14):2236–2254
Roosta HR, Hamidpour M (2011) Effects of foliar application of some macro- and micro-nutrients on tomato plants in aquaponic and hydroponic systems. Sci Hortic 129:396–402
Roosta HR, Hamidpour M (2013) Mineral nutrient content of tomato plants in aquaponic and hydroponic systems: Effect of foliar application of some macro-and micro-nutrients. J Plant Nutr 36(13):2070–2083
Savidov N (2005) Comparative study of aquaponically and hydroponically grown plants in model system. In: Evaluation and development of aquaponics production and product market capabilities in Alberta. Chapter 3.2., Phase II, pp 21–31
Somerville C, Cohen M, Pantanella E, Stankus A, Lovatelli A (2014) Small-scale aquaponic food production: integrated fish and plant farming. FAO Fisheries and Aquaculture Technical Paper No. 589
Srivastava JK, Chandra H, Kalra SJ, Mishra P, Khan H, Yadav P (2017) Plant–microbe interaction in aquatic system and their role in the management of water quality: a review. Appl Water Sci 7:1079–1090
Suhl J, Dannehl D, Kloas W, Baganz D, Jobs S, Schiebe G, Schmidt U (2016) Advanced Aquaponics: evaluation of intensive tomato production in aquaponics vs conventional hydroponics. Agric Water Manag 178:335–344
Timmons MB, Ebeling JM, Wheaton FW, Summerfelt ST, Vinci BJ (2002) Recirculating aquaculture systems, 2nd edn. Cayuga Aqua Ventures, Ithaca
Tyson RV, Simonne EH, Treadwell DD, Davis M, White JM (2008) Effect of water pH on yield and nutritional status of greenhouse cucumber grown in recirculating hydroponics. J Plant Nutr 31 (11):2018–2030
Tyson RV, Treadwell DD, Simonne EH (2011) Opportunities and challenges to sustainability in aquaponic systems. Hort Technol 21(1):6–13
Van Os E (1999) Design of sustainable hydroponic systems in relation to environment-friendly disinfection methods. Acta Hortic 548:197–205
Vimal SR, Singh JS, Arora NK, Singh S (2017) Soil-plant-microbe interactions in stressed agriculture management: a review. Pedosphere 27(2):177–192
Wongkiew S, Zhen H, Chandran K, Woo Lee J, Khanal SK (2017) Nitrogen transformations in aquaponic systems: a review. Aquac Eng 76:9–19
Yogev U, Barnes A, Gross A (2016) Nutrients and energy balance analysis for a conceptual model of a three loops off grid, aquaponics. Water 8:589. https://doi.org/10.3390/W8120589

Chapter 6
Bacterial Relationships in Aquaponics: New Research Directions

Alyssa Joyce, Mike Timmons, Simon Goddek, and Timea Pentz

Abstract The growth rates and welfare of fish and the quality of plant production in aquaponics system rely on the composition and health of the system's microbiota. The overall productivity depends on technical specifications for water quality and its movement amongst components of the system, including a wide range of parameters including factors such as pH and flow rates which ensure that microbial components can act effectively in nitrification and remineralization processes. In this chapter, we explore current research examining the role of microbial communities in three units of an aquaponics system: (1) the recirculating aquaculture system (RAS) for fish production which includes biofiltration systems for denitrification; (2) the hydroponics units for plant production; and (3) biofilters and bioreactors, including sludge digester systems (SDS) involved in microbial decomposition and recovery/remineralization of solid wastes. In the various sub-disciplines related to each of these components, there is existing literature about microbial communities and their importance within each system (e.g. recirculating aquaculture systems (RAS), hydroponics, biofilters and digesters), but there is currently limited work examining interactions between these components in aquaponics system, thus making it an important area for further research.

Keywords Microbiota · Aquaponics · Biofilters · Bioreactors · RAS · Hydroponics · Metagenomics

A. Joyce (✉)
Department of Marine Science, University of Gothenburg, Gothenburg, Sweden
e-mail: alyssa.joyce@gu.se

M. Timmons
Biological & Environmental Engineering, Cornell University, Ithaca, NY, USA
e-mail: mbt3@cornell.edu

S. Goddek
Mathematical and Statistical Methods (Biometris), Wageningen University, Wageningen, The Netherlands
e-mail: simon.goddek@wur.nl; simon@goddek.nl

T. Pentz
Eat & Shine VOF, Velp, The Netherlands
e-mail: timea.pentz@eatandshine.nl

S. Goddek et al. (eds.), *Aquaponics Food Production Systems*,
https://doi.org/10.1007/978-3-030-15943-6_6

6.1 Introduction

Recirculating water in the aquaculture portion of an aquaponics system contains both particulate and dissolved organic matter (POM, DOM) which enter the system primarily via fish feed; the portion of feed that is not eaten or metabolized by fish remains as waste in the recirculating aquaculture system (RAS) water, either in dissolved form (e.g. ammonia) or as suspended or settled solids (e.g. sludge). Once the majority of sludge is removed by mechanical separation, the remaining dissolved organic matter must still be removed from a RAS system. Such processes rely on microbiota in various biofilters in order to maintain water quality for the fish and to convert inorganic/organic wastes into forms of bioavailable nutrients for the plants. Microbial communities in aquaponics system include bacteria, archaea, fungi, viruses and protists in assemblages that fluctuate in composition based on an ebb and flow of nutrients and changes in environmental conditions such as pH, light and oxygen. Microbial communities play a significant role in denitrification and mineralization processes (see Chap. 10) and thus have key roles in the overall productivity of the system, including fish welfare and plant health.

The challenges within any aquaponics system are to control inputs – water, fingerlings, feed, plantlets – and their associated microbiota to maximize the benefits of organic matter and its breakdown into bioavailable forms for target organisms. Given that optimal environmental growth parameters and nutrients differ for fish and plants (see Chap. 8), various separation and aeration systems, and biofilters containing relevant microbial assemblages, must be situated at strategic points in the water supply in order to help maintain nutrient levels, pH and dissolved oxygen (DO) levels within desired ranges for both target fish and plant species. Indeed, water quality parameters, including temperature, DO, electrical conductivity, redox potential, nutrient levels, carbon dioxide, lighting, feed and flow rates, all affect the behaviour and composition of microbial communities within an aquaponics system (Junge et al. 2017). In this regard, it is important to refine setup and operation so that each unit contributes adequate quantities of bioavailable forms of nutrients to its successor, rather than enabling proliferation of pathogens or opportunistic microbes that can consume the bulk of macronutrients needed downstream.

Various techniques for the analysis of microbial communities can yield important information about changes in community structure and function over time in different aquaponic configurations. By correlating these changes with nutrient bioavailability and operational parameters, it is possible to reduce over- or under-production of essential nutrients or the production of noxious by-products. For instance, maximizing recovery of beneficial plant nutrients from waste organic matter in the fish component depends primarily on the ability of microbiota to facilitate breakdown of nutrients within a series of biofilters and sludge digesters, whose performance is based on a range of operational parameters such as flow rates, residence time and pH (Van Rijn 2013). Since not all aquaponics system include sludge digesters, we will address this aspect in more detail in the latter half of this review whilst referring the reader to Chap. 3 for more details on solid separation techniques and Chaps. 7 and 8 for discussions on coupled vs decoupled aquaponics system. If we consider here only dissolved and suspended particulates in the water (and not sludge), all aquaponics system employ a range of different biofilters that expose the attached microorganisms to organic matter passing through the filter and provide an appropriate substrate and sufficient surface area for microbial attachment and formation of biofilms. Degradation of this organic matter provides energy to the microbial communities, which in turn release macronutrients (e.g. nitrate, orthophosphate) and micronutrients (e.g. iron, zinc, copper) back to the system in usable forms (Blancheton et al. 2013; Schreier et al. 2010; Vilbergsson et al. 2016a).

There is considerable agricultural research on the role of microbiota in plant rooting, growth and health. The preponderance of this research focuses on soil-based systems; however, research on hydroponics has also increased in recent years (Bartelme et al. 2018). The microbiota in aquaculture have also been similarly well-characterized, where the role of microbes in fish health and digestion has received considerable attention as researchers attempt to better characterize the role of gut health on nutrient assimilation. Given the importance of biofiltration in RAS systems, bacteria involved in the nitrification process for RAS have also been comparatively well-studied and thus are not be addressed here (see Chaps. 10 and 12). However, there has been comparatively limited research on microbes in aquaponics system, especially the crucial interactions of microbiota amongst various compartments of the system. This lack of research currently limits the scope and productivity of such systems, where there is considerable potential for enhancement with pre- and probiotics, as well as other opportunities to improve the health of aquaponics system through a better understanding, and thus better ability to control, the vast set of uncharacterized microbiota that affect system health and performance.

As such, this chapter focuses primarily on recent studies that reveal how and where microbial communities determine productivity within compartments, whilst also highlighting the relatively small number of studies linking those microbial communities to interactions amongst components and overall system productivity. We attempt to identify gaps where further knowledge about microbial communities could address operational challenges and provide important insights for enhancing efficiency and reliability.

6.2 Tools for Studying Microbial Communities

New technologies for studying how microbial communities change over time, and which groups of organisms predominate under particular environmental conditions, have increasingly offered opportunities to anticipate adverse outcomes within system components and thus lead to the design of better sensors and tests for the effective monitoring of microbial communities in fish or plant cultures. For instance, various ‘omics’ technologies – metagenomics, metatranscriptomics, community proteomics, metabolomics – are increasingly enabling researchers to study the diversity of microbiota in RAS, biofilters, hydroponics and sludge digestor systems where sampling includes whole microbial assemblages instead of a given genome. Analysis of prokaryotic diversity in particular, has been helped enormously in recent decades by metagenomic and metatranscriptomic techniques. In particular, amplification and sequence analysis of the 16S rRNA gene, based on intraspecific conservation of neutral gene sequences flanking ribosomal operons in bacterial DNA, has been considered the 'gold standard' for taxonomic classification and identification of bacterial species. Such data is also used in microbiology to track epidemics and geographical distributions and study bacterial populations and phylogenies (Bouchet et al. 2008). The methodology can be labour-intensive and expensive, but recent automated systems, whilst not necessarily discriminatory at the species and strain level, offer opportunities for application in aquaponics settings (Schmautz et al. 2017). Recent reviews summarize applications of 16S rRNA as they pertain to RAS (Martínez-Porchas and Vargas-Albores 2017; Munguia-Fragozo et al. 2015; Rurangwa and Verdegem 2015). Advances in metagenomics of microbes other than bacteria found in RAS and hydroponics rely on similar methodologies but use 18S (eukaryotes), 26S (fungi) and 16S in combination with 26S (yeasts) rRNA clone libraries to characterize these microbiota (Martínez-Porchas and Vargas-Albores 2017). Detailed rRNA libraries, for instance, have also been used in hydroponics to characterize microbial communities in the rhizosphere (Oburger and Schmidt 2016). Such libraries can be particularly useful in aquaponics, given that they can examine assemblage of microorganisms such as bacteria, archaea, protozoans and fungi and provide feedback on changes within the system.

The development of automated next-generation sequencing (NGS) has also enabled data analysis of genomes from population samples (metagenomics) that can be used to characterize microbiota, reveal temporal-spatial phylogenetic changes and trace pathogens. Applications in RAS include tracking certain bacterial strains amongst cultured fish and eliminating populations that carry virulent strains, whilst preserving carriers of other strains (review: (Bayliss et al. 2017). Metagenomic approaches can be culture- and amplification-independent, which allows previously unculturable species to become known and investigated for their possible effects (Martínez-Porchas and Vargas-Albores 2017). Next-generation sequencing techniques are commonly used in plant microbiology along with follow-up metatranscriptomics analyses. An excellent example is the first whole-plant study of microbial communities in the rhizosphere, wherein root exudates were shown to correlate with developmental stages (Knief 2014).

Proteomics is most useful when studying a particular bacterial species or strain under specific environmental conditions in order to describe its pathogenicity or possible role in symbiosis. Nevertheless, there are advances in community proteomics that build on prior metagenomic studies and use various biochemical techniques to identify, for example, secreted proteins associated with commensal or symbiotic microbial communities, and further possibilities abound as the capability of NGS technologies advance rapidly (review: (Knief et al. 2011).

Metabolomics characterizes the functions of genes, but the techniques are not organism-specific or sequence-dependent and thus can reveal the wide range of metabolites that are end-products of cellular biochemistry in organisms, tissues, cells or cell compartment (depending on which samples are analysed). Nevertheless, knowledge about the metabolome of microbial communities under particular environmental conditions (microcosms) reveals a great deal about the biogeochemical cycling of nutrients and the effects of perturbations. Such knowledge characterizes various metabolic pathways and the range of metabolites present in samples. Subsequent biochemical and statistical analyses can point to physiological states that can in turn be correlated with environmental parameters which may not be evident from genomic or proteomic approaches. Nevertheless, combining metabolomics with gene function studies has tremendous potential in furthering aquaponics research; see review (van Dam and Bouwmeester 2016).

6.3 Biosecurity Considerations for Food Safety and Pathogen Control

6.3.1 Food Safety

Good food safety and ensuring animal welfare are high priorities in gaining public support for aquaponics. One of the most frequent issues raised by food safety experts in relation to aquaponics is the potential risk of contamination with human pathogens when using fish effluent as fertilizer for plants (Chalmers 2004; Schmautz et al. 2017). A recent literature search to determine zoonotic risks in aquaponics concluded that pathogens in contaminated intake water, or pathogens in components of feeds originating with warm-blooded animals, can become associated with fish gut microbiota, which, even if not detrimental to the fish themselves, can potentially be passed up the food chain to humans (Antaki and Jay-Russell 2015). The mechanisms of introduction of pathogens to an aquaponics system are thus of concern, with the likeliest source of faecal coliforms or other pathogenic bacteria stemming from feed inputs to fish. From a biological perspective, there are potential risks of these pathogens proliferating either in biofilters, or, in one-loop systems by introducing airborne pathogens from open plant components back to the fish tanks. Although biosecurity risks are low in the relatively closed environmental space of an aquaponics system – as compared for instance to open pond aquaculture – and are even lower in decoupled aquaponics system wherein portions of the system can be isolated, there is still a perception that fish sludge could be potentially dangerous when applied to plants for human consumption. *Escherichia coli* (*E. coli*) is a human enteric pathogen causing foodborne illnesses that has been a key concern regarding the use of animal waste as fertilizer in agriculture or aquaculture, e.g. integrated pig-fish systems (Dang and Dalsgaard 2012). However, it is generally not considered to present a risk in fish-plant aquaponics. For instance, Moriarty et al. (2018)

previously demonstrated that UV-radiation treatment can successfully reduce *E. coli* but also noted that the coliforms detected in the aquaponics system were at background levels and did not proliferate in the fish raceways or in the hydroponically grown lettuce within the experimental system, and thus did not present a health risk. There is limited research on these aspects, but a few preliminary studies have found very low risks of coliform contamination, for instance, by showing no difference in coliform levels from sterilized and non-sterilized RAS water treatments applied to plants (Pantanella et al. 2015). Even though there is a potential risk of internalization of microbes within plant leaves, and thus their transmission to the consumed portions of some edible leafy plants grown in aquaponics, other studies have come to similar conclusions that the risks are minimal of introducing potentially dangerous human pathogens (Elumalai et al. 2017).

However, managing risks, or more importantly managing the perceptions of those risks, remains a high priority for government authorities and aquaponics investors. It is assumed that the quality control of feed inputs and careful handling of fish/fish wastes can limit most of these potential concerns (Fox et al. 2012). Indeed, no known human health incidents have to our knowledge currently been reported in relation to aquaponics system, and this may be a function of the fact that RAS facilities and hydroponic greenhouses typically have good biosecurity measures, including hygiene and quarantine practices that are stringently observed. Recommended microbiological practices for biosecurity have been evaluated for different aquaculture production systems and recommendations formulated into Hazard Analysis Critical Control Points guidelines, an international system for controlling food safety (Orriss and Whitehead 2000). However, there is still a need for better scientific documentation of risks for pathogen transfers to humans, and direct research into management in this area of aquaponics production.

6.3.2 Fish and Plant Pathogens

There is existing discipline-specific literature in aquaculture, hydroponics and bioengineering that can help inform and enhance microbial performance in aquaponics. For instance, microbial communities serve a wide range of important functions in fish health, including playing a key role in the digestibility and assimilation of feed, as well as immunodulation, and these functions as well as the role of probiotics in enhancing aquaculture systems are well-reviewed (Akhter et al. 2015). The role of microbes in RAS systems specifically is also well covered, including microbial management of biofilters, as well as research into pathogen control, as well as various techniques to control off-flavours deriving from RAS systems (Rurangwa and Verdegem 2015). Likewise, the microbes in the rhizosphere of plants are important for rooting and plant growth (Dessaux et al. 2016) but also for controlling the spread of pathogens in hydroponic plant production; these areas are well explored in a recent review by Bartelme et al. (2018). However, there is still a very limited understanding of linkages in the microbiome amongst the compartments of aquaponics system, knowledge that is crucial for maximizing productivity and reducing pathogen transfer.

The proliferation of opportunistic pathogens that are dangerous to fish or plant health are important considerations in the economics of aquaponics operations, given that any use of antibiotics or disinfectants can have a potentially detrimental effect on biofilter function, as well as destabilizing microbial relationships in other compartments of the system. Disinfection protocols commonly used in RAS include treating water with ultraviolet light (Elumalai et al. 2017), which, combined with ozone (and usually a combination of both), comprises a first-line abiotic approach to maintaining water quality. Fish eggs/larvae are also often quarantined before being introduced, and any intake water treated, thus reducing direct potential sources of fish pathogen entry to the system.

Incoming water to RAS is also typically allowed to 'mature' in biofilters before being fed into the recirculating system. Experiments, for instance, have shown that inoculating a pre-biofilter with a mixture of nitrifying bacteria, and 'feeding' it with organic matter until bacterial populations match the carrying capacity of the fish tanks, means that the rearing tank water is less likely to be unstable and overtaken by opportunistic bacteria (Attramadal et al. 2016; Rurangwa and Verdegem 2015). However, should pathogens become problematic, the use of high-dose UV, ozone, chemical or antibiotic treatments can sometimes be necessary, although such use is generally disruptive to other compartments of the system, especially the biofilters (Blancheton et al. 2013). Indeed, depending on the dose and location within the system, non-selective treatments for pathogens can actually favour proliferation of opportunists. For instance, high levels of ozone treatment not only kills bacteria, protists and viruses but also oxidizes DOM and affects aggregation of POM, thereby exerting selection pressure on bacterial populations (ibid.).

A detailed discussion of plant pathogens in aquaponics system and their control is included in Chap. 14 and thus is not reiterated here. However, it is worth noting that *Bacillus* species are routinely used as commercial probiotics in aquaculture, and there is growing evidence that similar *Bacillus* species are also effective for plants, that are already available in some commercial hydroponics probiotics solutions (Shafi et al. 2017). A recent study has extended such studies on *Bacillus* to include experimentation in aquaponics system (Cerozi and Fitzsimmons 2016b). The location where the probiotics are introduced – in the fish, plant or biofilters – may be important, but it is not clear from existing work whether the addition of probiotics in the fish component, with potential benefits for the fish, also has better effects on plant growth and health relative to the addition of similar levels of probiotics directly to the hydroponics compartment.

In addition to standard application probiotics, there are a variety of innovative techniques for biocontrol that may in the future become increasingly valuable for reducing the presence and proliferation of harmful microbes. In one recent study, bacterial isolates were selected from an established aquaponics system based on their ability to exert inhibitory effects on both fish and plant fungal pathogens. The goal was to culture these isolates as inocula that could subsequently act as biological controls for diseases within that aquaponics system (Sirakov et al. 2016). For instance, Sirakov et al. demonstrated that a *Pseudomonas* sp. that they isolated was effective as a biocontrol for the pathogenic fungi *Saprolegnia parasitica* of fish and *Pythium ultimum* of plants. The researchers also reported in vitro inhibition of a variety of other bacterial isolates from the different aquaponics compartments, but without testing their in vivo effects. The potential for using such isolates as biological controls is not new, but applications of NGS techniques can now reveal more about interactions of such isolates with each other and with potential pathogens, thus making it possible to optimize the effectiveness of delivery. Use of other 'omics' techniques could help reveal overall community structure and associated metabolic functions, and begin elucidating which organisms and functions are most beneficial. In future, such techniques might allow selection for 'helper strains' within microbial communities, or the identification of exudates that have anti-microbial effects (Massart et al. 2015).

6.4 Microbial Equilibrium and Enhancement in Aquaponics Units

Productivity in aquaponics system involves monitoring and managing environmental parameters in order to provide each component, whether microbial, animal or plant, with optimal growth conditions. Whilst this is not always possible given trade-offs in requirements, one of the key goals of aquaponics revolves around the concept of homeostasis, wherein maintaining stability of the system involves adjusting operational parameters to minimize unnecessary perturbations that cause stress within a unit, or detrimental effects on other components. With ever-changing microbial assemblages, homeostasis never implies a permanent state of equilibrium, but rather a goal of achieving as much stability as possible, particularly within water quality parameters.

A RAS coupled to a hydroponics system will be ever-changing, but within this configuration, the RAS component remains relatively stable, particularly in decoupled systems (Goddek and Körner 2019). The hydroponics system, on the other hand, tends to be more erratic in water quality since the plant crops are often harvested in batch modes, and rarely in synchrony with fish production.

During the initial start-up phase of any aquaponics system, water quality – particularly with regard to microbial communities in biofilters – is a concern, and in order to minimize proliferation of opportunistic bacteria, a routine practice has been to allow microbial maturation of intake water before its introduction into the RAS, adding fish only after the capacity of the biofilters matches the carrying capacity of rearing tanks at a particular stocking density (Blancheton et al. 2013). A similar practice is observed in hydroponics where at least a portion of recycled water is used to inoculate a new crop, given that mature microbial communities take time to develop and introducing all new water results in long lag times. Such practices lead to greater stability in culture conditions and greater productivity. For instance, improved performance in RAS systems has been noted when the pre-intake filter is supplied with pulverized fish food to develop microbial communities more similar to those in the rearing tanks (Attramadal et al. 2014).

6.5 Bacterial Roles in Nutrient Cycling and Bioavailability

Considerable research has been conducted to characterize heterotrophic and autotrophic bacteria in RAS systems and to better understand their roles in maintaining water quality and cycling of nutrients (for reviews, see Blancheton et al. (2013); Schreier et al. (2010). Non-pathogenic heterotrophs, typically dominated by *Alphaproteobacteria* and *Gammaproteobacteria*, tend to thrive in biofilters, and their contributions to transformations of nitrogen are fairly well understood because nitrogen cycling (NC) has been of paramount importance in developing recirculating culture systems (Timmons and Ebeling 2013). It has long been recognized that the bacterial transformation of the ammonia excreted by fish in a RAS system must be matched with excretion rates, because excess ammonia quickly becomes toxic for fish (see Chap. 9). Therefore in freshwater and marine RAS, the functional roles of microbial communities in NC dynamics – nitrification, denitrification, ammonification, anaerobic ammonium oxidation and dissimilatory nitrate reduction – have received considerable research attention and are well described in recent reviews (Rurangwa and Verdegem 2015; Schreier et al. 2010). There are far fewer studies of nitrogen transformations in aquaponics, but a recent review (Wongkiew et al. 2017) provides a summary along with discussion of nitrogen utilization efficiency, which is a prime consideration for plant growth in hydroponics.

After nitrogen, the second most essential macronutrient in aquaponics is phosphorus, which is not a limiting factor for fish that acquire it from feed, but is crucial for plants in hydroponics. However, the forms of phosphate in fish wastes are not immediately bioavailable for plants. Plants must have adequate quantities of inorganic ionic orthophosphate ($H_2PO_4^-$ and HPO_4^{2-} = Pi) (Becquer et al. 2014), as this is the only bioavailable form for uptake and assimilation. Inorganic phosphate binds to calcium above pH 7.0, so aquaponics system must be careful to maintain pH conditions near pH 7.0. As pH values rise above 7.0, various insoluble forms of calcium phosphate can end up as precipitates in sludge (Becquer et al. 2014; Siebielec et al. 2014). Hence, RAS losses of P are primarily through removal of sludge from the system (Van Rijn 2013). However, somewhere in the aquaponics system, particulate matter must be captured and allowed to mineralize in order to provide sufficient supplies of usable nutrients for crops in the hydroponics unit. The mineralization step will also release other macro- and micronutrients so that there are fewer deficiencies, thus reducing the need for supplementation in the hydroponics compartment. Given that world supplies of phosphate-rich fertilizers are dwindling and supplementation with P is increasingly costly, efforts are being made to maximize the recovery of P from RAS sludge (Goddek et al. 2016b; Monsees et al. 2017).

The bioavailability of macro and micronutrients is currently poorly understood. Previous research (Cerozi and Fitzsimmons 2016a) suggests that the availability of nutrients becomes compromised as pH is reduced below 7.0 and has resulted in coupled hydroponics system for leafy greens being operated around pH 6.0. However, recent research comparing aquaponic conditions and pH 7.0 to hydroponic conditions of pH 5.8 showed no difference in productivity (Anderson et al. 2017a, b). In these studies, hydroponic conditions at pH 7.0 reduced productivity by ~ 22% compared to hydroponic pH 5.8. Initially, the hypothesis was that the differences in productivity could be ascribed to the microbiota of the aquaponic water, but subsequent research dismissed that theory (Wielgosz et al. 2017).

In RAS where C:N ratios increase due to availability of organic matter, denitrifying bacteria, especially *Pseudomonas* sp., use carbon as an electron donor in anoxic conditions, to produce N_2 at the expense of nitrate (Schreier et al. 2010; Wongkiew et al. 2017). Biofloc systems are sometimes used to augment feed for fish (Crab et al. 2012; Martínez-Córdova et al. 2015), and biofloc is increasingly being used in aquaponics system, especially in Asia (Feng et al. 2016; Kim et al. 2017; Li et al. 2018). When biofloc is used in aquaponics (da Rocha et al. 2017; Pinho et al. 2017), nutrient cycling becomes even more complex given that DO, temperature and pH influence whether heterotrophic (carbon-utilizing) microbial communities predominate over autotrophic denitrifiers that are capable of reducing sulphide to sulphate (Schreier et al. 2010). Heterotrophs tend to have a higher growth rate than autotrophs in the presence of adequate sources of carbon (Michaud et al. 2009); therefore, manipulating feed type or regimes, or adding an organic carbon source directly, whilst monitoring dissolved oxygen levels, can help keep populations equilibrated whilst still providing hydroponics with N in a usable form (Vilbergsson et al. 2016a).

In hydroponics system, nutrient cycling has been less well-studied since inorganic compounds containing the required balance of nutrients are typically added in order to ensure proper plant growth. However, high nutrient concentrations, especially in humid warm environments such as greenhouses, easily facilitate growth of microbial communities, especially phytopathogens such as fungi (*Fusarium*) and oomycota (*Phytophthora*, *Pythium* sp.), that can quickly spread in circulating water and may result in die-offs (Lee and Lee 2015). Recent efforts to better understand hydroponic rhizobacteria and their beneficial effects in promoting plant growth (but also for inhibiting pathogen proliferation) have utilized various 'omics' techniques to analyse microbial communities and their interactions with root systems (Lee and Lee 2015).

For instance, when probiotic bacteria such as *Bacillus*, are present, they were shown to enhance P availability and also appear to have an added plant growth-promoting effect in a tilapia-lettuce system (Cerozi and Fitzsimmons 2016b). In aquaponics system, the addition of probiotics to fish feed and RAS water, as well as to the hydroponic water supply, deserves further experimentation, since microbial communities can have multiple modulatory effects on plant physiology. For example, the microbial communities (bacteria, fungi, oomycetes) of four food crops were analysed by metagenomic sequencing when maintained in a constant nutrient film hydroponics system where pH and nutrient concentrations were allowed to fluctuate naturally throughout the plants' life cycles (Sheridan et al. 2017). The authors concluded that treatment with a commercial mixture of plant growth-promoting microbes (PGPMs), in this case bacteria, mycorrhizae and fungi, appeared to confer greater stability and similarities in community composition after 12–14 weeks than in controls. They suggest that this could be attributed to root exudates, which purportedly favour and even control the development of microbial communities appropriate to successive plant developmental stages. Given the known effects of PGPMs in soil-based crop production, and the few studies that are available for soilless systems, further investigation is warranted to determine how to enhance PGPMs and to improve their effects in aquaponics system (Bartelme et al. 2018). If hydroponic cultures are more stable and plant growth is more robust with PGPMs, then the goal should be to characterize microbial communities in aquaponics via metagenomics and correlate them with optimal macro- and micronutrient availability via metabolomics and proteomics.

6.6 Suspended Solids and Sludge

The parameters for operating aquaponics at a given scale – including water volume, temperature, feed and flow rates, pH, fish and crop ages and densities – all affect the temporal and spatial distribution of the microbial communities that develop within its compartments, for reviews: RAS (Blancheton et al. 2013); hydroponics (Lee and Lee 2015).

In addition to controlling dissolved oxygen, carbon dioxide levels and pH in aquaponics, it is also essential to control the accumulation of solids in the RAS system as fine suspended particles can adhere to gills, cause abrasion and respiratory distress and increase susceptibility to disease (Yildiz et al. 2017). More relevant, the particulate organic matter (POM) must be quickly and effectively removed from RAS systems, or else excessive heterotrophic growth will cause almost all unit processes to fail. RAS feeding rates must be carefully managed to minimize solids loading on the system (e.g. avoid over-feeding and minimize feeding costs). The biophysical properties of feed – particle size, nutrient content, digestibility, sensory appeal, density and settling rate – determine ingestion and assimilation rates, which in turn have an impact on solids build-up and thus water quality. Although water quality is frequently studied in the context of nutrient cycling (see Chap. 9), it is also important to obtain a better understanding of the composition of microbial communities and changes in these based on feed composition, particulate loading and how this influences the growth of heterotrophic and autotrophic bacterial communities.

Various features of RAS system designs have been developed specifically to deal with solids (Timmons and Ebeling 2013); see also review: (Vilbergsson et al. 2016b). For instance, some biofilters function to keep substantial portions of wastes suspended in order to facilitate degradation, whilst others mechanically filter through screens or granular media. Still others rely on sedimentation to simply collect and remove sludge. However, such methods are not particularly effective at recovering nutrients within the sludge and making it bioavailable for plant use. Historically, this sludge has been handled in bioreactors for its methanogenic value or dewatered to be used as fertilizer for soil-based crops, but various newer designs have attempted to improve recovery for use in the hydroponic component. Improving recovery of this sludge is an important area of investigation given that a significant portion of the essential macro- and micronutrients required for plant growth are bound to the particulate organic matter, which, if discarded, is lost from the system. By adding an additional sludge recycling loop to aquaponics system, solid wastes can be converted into dissolved nutrients for reuse by plants rather than being discarded (Goddek et al. 2018). Digesters or remineralizing bioreactors are one way of accomplishing this, however one of the key areas that is currently under-developed includes knowledge of how microbial communities within these sludge digesters can be enhanced (e.g. through addition of microbes) or better utilized (e.g. through better engineered design of linked reactors) to recover nutrients into bioavailable forms for plants. Even though the actual microbial communities within sludge digesters have not been well researched for aquaponics, there is considerable literature on the microbiota of sludge digesters for sewage and animal wastes in agriculture, including fish effluent, that can provide further insight into ideal designs for sludge recovery in aquaponics system. Current research on the incorporation of sludge into aquaponics system involves remineralization in digesters situated between the RAS and hydroponic unit (Goddek et al. 2016a, 2018). Within aerobic or anaerobic bioreactors, environmental conditions that are favourable for waste degradation can effectively break down this sludge into bioavailable nutrients, which can subsequently be delivered to hydroponics system without the presence of soil (Monsees et al. 2017). Many one-loop aquaponics system already include aerobic (Rakocy et al. 2004) and anaerobic (Yogev et al. 2016) digesters to transform nutrients that are trapped in the fish sludge and make them bioavailable for plants. The ability to decouple these has a number of advantages that are further discussed in Chap. 8 and appears to lead to higher growth rates (Goddek and Vermeulen 2018). However, despite the many advances, the actual technology to accomplish this remains challenging. For example, some heterotrophic denitrifying bacteria cultured in anoxic or even aerobic conditions with sludge from RAS will use nitrate as an electron receptor and oxidized carbon sources for energy, while storing excess P as polyphosphate along with divalent metal ions such as Ca^{+2} or Cu^{+2}. When stressed at alkaline pH, these bacteria degrade polyphosphate and release orthophosphate, which is the necessary form for assimilation of phosphate by plants (Van Rijn et al. 2006). Inserting remineralization bioreactor units, such as those in Goddek et al. (2018), could provide a way to better recover P for hydroponics. Similar methods have, for instance, been used with trout sludge from a RAS that were treated for nitrate and P content in excess of allowable disposal limits (Goddek et al. 2015). However, the microbial communities involved in these processes are sensitive to culture conditions such as C:N ratios, oxygenation, metal ions and pH, so nitrites and other noxious intermediates can accumulate. Despite a vast literature on digesters of various organic wastes, primarily anaerobic for biogas production (Ibrahim et al. 2016), there is far less research on treating RAS wastes (Van Rijn

2013), and in the case of aquaponics system, even less available research about the relationship between nutrient bioavailability and crop growth in hydroponics system (Möller and Müller 2012). At this time, more studies of RAS sludge bioreactors could provide important insights into culture conditions for microbial populations that produce favourable results, for instance, on P recovery, and its introduction into hydroponics units.

One of the current challenges in efforts to assess the recovery of P from sludge arises when comparing trials of anaerobic and aerobic digesters for their efficacy (Goddek et al. 2016b; Monsees et al. 2017). Although both studies used similar sludge composition initially, the results were quite different. In one study (Monsees et al. 2017), measures of various soluble nutrients in aerobic treatments resulted in a 330% increase in P concentration and a 16% decrease in nitrate concentration compared to minor increases in P and a 97% decrease in nitrate in anaerobic treatments. By contrast, results from a similar study (Goddek et al. 2016b) showed that growth of lettuce plants in a hydroponic unit was superior using anaerobic supernatant, even though both anaerobic and aerobic treatments only resulted in slightly better nitrate recovery from anaerobic conditions and almost complete loss of PO_4 from both treatments (Goddek et al. 2016b). Obviously, factors such as feed composition and rates, the suspension versus settling of solids, pH (maintained at 7 ± 1 with $CaOH_2$ in the former and variable 8.2–8.65 in the latter), sampling and fish strains differed in these two studies. Nevertheless, the contrasting results for PO_4 and NO_3 indicate the need for further research to optimize nutrient recovery, with the addition of a metagenomics approach to characterize microbial communities so as to better understand their role in these processes.

6.7 Conclusions

Formerly the domain of small-scale producers, technological advances are increasingly moving aquaponics into larger-scale commercial production by focusing on improved macro- and micronutrient recovery whilst providing technical innovations to reduce water and energy requirements. However, scaling up of aquaponics to an industrial scale requires a much better understanding and maintenance of microbial assemblages, and the implementation of strong biocontrol measures that favour the health and well-being of both fish and crops, whilst still meeting food safety standards for human consumption. Further research on biocontrol of microbial pathogens in aquaponics, including potential human, fish and plant pathogens are needed, in light of the sensitivity of such systems to perturbation, and the fact that the use of chemicals and antibiotics can have profound effects on microbial populations, fish and plant physiologies, as well as overall system operation. Elucidating microbial interactions can improve the productivity of aquaponics system given the crucial roles of microbes in converting organic matter into usable forms that can allow fish and plants to thrive.

References

Akhter N, Wu B, Memon AM, Mohsin M (2015) Probiotics and prebiotics associated with aquaculture: a review. Fish Shellfish Immunol 45:733–741

Anderson T, de Villiers D, Timmons M (2017a) Growth and tissue elemental composition response of butterhead lettuce (*Lactuca sativa*, cv. Flandria) to hydroponic and aquaponic conditions. Horticulturae 3:43

Anderson TS, Martini MR, de Villiers D, Timmons MB (2017b) Growth and tissue elemental composition response of Butterhead lettuce (*Lactuca sativa*, cv. Flandria) to hydroponic conditions at different pH and alkalinity. Horticulturae 3:41

Antaki ET, Jay-Russell M (2015) Potential zoonotic risks in aquaponics. IAFP, Portland

Attramadal KJK, Truong TMH, Bakke I, Skjermo J, Olsen Y, Vadstein O (2014) RAS and microbial maturation as tools for K-selection of microbial communities improve survival in cod larvae. Aquaculture 432:483–490

Attramadal KJ, Minniti G, Øie G, Kjørsvik E, Østensen M-A, Bakke I, Vadstein O (2016) Microbial maturation of intake water at different carrying capacities affects microbial control in rearing tanks for marine fish larvae. Aquaculture 457:68–72

Bartelme RP, Oyserman BO, Blom JE, Sepulveda-Villet OJ, Newton RJ (2018) Stripping away the soil: plant growth promoting microbiology opportunities in aquaponics. Front Microbiol 9:8

Bayliss SC, Verner-Jeffreys DW, Bartie KL, Aanensen DM, Sheppard SK, Adams A, Feil EJ (2017) The promise of whole genome pathogen sequencing for the molecular epidemiology of emerging aquaculture pathogens. Front Microbiol 8:121

Becquer A, Trap J, Irshad U, Ali MA, Claude P (2014) From soil to plant, the journey of P through trophic relationships and ectomycorrhizal association. Front Plant Sci 5:548

Blancheton J, Attramadal K, Michaud L, d'Orbcastel ER, Vadstein O (2013) Insight into bacterial population in aquaculture systems and its implication. Aquac Eng 53:30–39

Bouchet V, Huot H, Goldstein R (2008) Molecular genetic basis of ribotyping. Clin Microbiol Rev 21:262–273

Cerozi BD, Fitzsimmons K (2016a) The effect of pH on phosphorus availability and speciation in an aquaponics nutrient solution. Bioresour Technol 219:778–781

Cerozi BD, Fitzsimmons K (2016b) Use of Bacillus spp. to enhance phosphorus availability and serve as a plant growth promoter in aquaponics systems. Sci Hortic 211:277–282

Chalmers GA (2004) Aquaponics and food safety. Lethbridge, Alberta

Crab R, Defoirdt T, Bossier P, Verstraete W (2012) Biofloc technology in aquaculture: beneficial effects and future challenges. Aquaculture 356:351–356

da Rocha A, Biazzetti Filho M, Stech M, Paz da Silva R (2017) Lettuce production in aquaponic and biofloc systems with silver catfish *Rhamdia quelen*. Bol Inst Pesca 43:64

Dang STT, Dalsgaard A (2012) *Escherichia coli* contamination of fish raised in integrated pig-fish aquaculture systems in Vietnam. J Food Prot 75:1317–1319

Dessaux Y, Grandclément C, Faure D (2016) Engineering the rhizosphere. Trends Plant Sci 21:266–278

Elumalai SD, Shaw AM, Pattillo DA, Currey CJ, Rosentrater KA, Xie K (2017) Influence of UV treatment on the food safety status of a model aquaponics system. Water 9:27

Feng J, Li F, Zhou X, Xu C, Fang F (2016) Nutrient removal ability and economical benefit of a rice-fish co-culture system in aquaculture pond. Ecol Eng 94:315–319

Fox BK, Tamaru CS, Hollyer J, Castro LF, Fonseca JM, Jay-Russell M, Low T (2012) A preliminary study of microbial water quality related to food safety in recirculating aquaponic fish and vegetable production systems. College of Tropical Agriculture and Human Resources, Honolulu

Goddek S, Körner O (2019) A fully integrated simulation model of multi-loop aquaponics: a case study for system sizing in different environments. Agric Syst 171:143

Goddek S, Vermeulen T (2018) Comparison of *Lactuca sativa* growth performance in conventional and RAS-based hydroponics system. Aquac Int 26:1377. https://doi.org/10.1007/s10499-018-0293-8

Goddek S, Delaide B, Mankasingh U, Ragnarsdottir KV, Jijakli H, Thorarinsdottir R (2015) Challenges of sustainable and commercial aquaponics. Sustainability 7:4199–4224

Goddek S, Espinal CA, Delaide B, Jijakli MH, Schmautz Z, Wuertz S, Keesman KJ (2016a) Navigating towards decoupled aquaponics system: a system dynamics design approach. Water 8:303

Goddek S, Schmautz Z, Scott B, Delaide B, Keesman KJ, Wuertz S, Junge R (2016b) The effect of anaerobic and aerobic fish sludge supernatant on hydroponic lettuce. Agronomy-Basel 6:37

Goddek S, Delaide BP, Joyce A, Wuertz S, Jijakli MH, Gross A, Eding EH, Bläser I, Reuter M, Keizer LP (2018) Nutrient mineralization and organic matter reduction performance of RAS-based sludge in sequential UASB-EGSB reactors. Aquac Eng 83:10–19

Ibrahim MH, Quaik S, Ismail SA (2016) An introduction to anaerobic digestion of organic wastes, prospects of organic waste management and the significance of earthworms. Springer, Cham, pp 23–44

Junge R, König B, Villarroel M, Komives T, Haïssam Jijakli M (2017) Strategic points in aquaponics. Water 9:182

Kim SK, Jang IK, Lim HJ (2017) Inland aquaponics system using biofloc technology. Google Patents

Knief C (2014) Analysis of plant microbe interactions in the era of next generation sequencing technologies. Front Plant Sci 5:216

Knief C, Delmotte N, Vorholt JA (2011) Bacterial adaptation to life in association with plants–a proteomic perspective from culture to in situ conditions. Proteomics 11:3086–3105

Lee S, Lee J (2015) Beneficial bacteria and fungi in hydroponics system: types and characteristics of hydroponic food production methods. Sci Hortic 195:206–215

Li G, Tao L, Li X-l, Peng L, Song C-f, Dai L-l, Wu Y-z, Xie L (2018) Design and performance of a novel rice hydroponic biofilter in a pond-scale aquaponic recirculating system. Ecol Eng 125:1–10

Martínez-Córdova LR, Emerenciano M, Miranda-Baeza A, Martínez-Porchas M (2015) Microbial-based systems for aquaculture of fish and shrimp: an updated review. Rev Aquac 7:131–148

Martínez-Porchas M, Vargas-Albores F (2017) Microbial metagenomics in aquaculture: a potential tool for a deeper insight into the activity. Rev Aquac 9:42–56

Massart S, Martinez-Medina M, Jijakli MH (2015) Biological control in the microbiome era: challenges and opportunities. Biol Control 89:98–108

Michaud L, Lo Giudice A, Troussellier M, Smedile F, Bruni V, Blancheton J-P (2009) Phylogenetic characterization of the heterotrophic bacterial communities inhabiting a marine recirculating aquaculture system. J Appl Microbiol 107:1935–1946

Möller K, Müller T (2012) Effects of anaerobic digestion on digestate nutrient availability and crop growth: a review. Eng Life Sci 12:242–257

Monsees H, Keitel J, Paul M, Kloas W, Wuertz S (2017) Potential of aquacultural sludge treatment for aquaponics: evaluation of nutrient mobilization under aerobic and anaerobic conditions. Aquac Environ Interact 9:9–18

Moriarty MJ, Semmens K, Bissonnette GK, Jaczynski J (2018) Inactivation with UV-radiation and internalization assessment of coliforms and *Escherichia coli* in aquaponically grown lettuce. LWT 89:624–630

Munguia-Fragozo P, Alatorre-Jacome O, Rico-Garcia E, Torres-Pacheco I, Cruz-Hernandez A, Ocampo-Velazquez RV, Garcia-Trejo JF, Guevara-Gonzalez RG (2015) Perspective for

aquaponics system: "omic" technologies for microbial community analysis. Biomed Res Int 2015:480386
Oburger E, Schmidt H (2016) New methods to unravel rhizosphere processes. Trends Plant Sci 21:243–255
Orriss GD, Whitehead AJ (2000) Hazard analysis and critical control point (HACCP) as a part of an overall quality assurance system in international food trade. Food Control 11:345–351
Pantanella E, Cardarelli M, Di Mattia E, Colla G (2015) Aquaponics and food safety: effects of UV sterilization on total coliforms and lettuce production. In: Carlile WR (ed) International conference and exhibition on soilless culture, pp 71–76
Pinho SM, Molinari D, de Mello GL, Fitzsimmons KM, Coelho Emerenciano MG (2017) Effluent from a biofloc technology (BFT) tilapia culture on the aquaponics production of different lettuce varieties. Ecol Eng 103:146–153
Rakocy JE, Bailey DS, Shultz RC, Thoman ES (2004) Update on tilapia and vegetable production in the UVI aquaponics system. New dimensions on farmed tilapia. Proceedings from the 6th international symposium on tilapia in aquaculture 000, 1–15
Rurangwa E, Verdegem MC (2015) Microorganisms in recirculating aquaculture systems and their management. Rev Aquac 7:117–130
Schmautz Z, Graber A, Jaenicke S, Goesmann A, Junge R, Smits THM (2017) Microbial diversity in different compartments of an aquaponics system. Arch Microbiol 199:613–620
Schreier HJ, Mirzoyan N, Saito K (2010) Microbial diversity of biological filters in recirculating aquaculture systems. Curr Opin Biotechnol 21:318–325
Shafi J, Tian H, Ji M (2017) Bacillus species as versatile weapons for plant pathogens: a review. Biotechnol Equip 31:446–459
Sheridan C, Depuydt P, De Ro M, Petit C, Van Gysegem E, Delaere P, Dixon M, Stasiak M, Aciksöz SB, Frossard E (2017) Microbial community dynamics and response to plant growth-promoting microorganisms in the rhizosphere of four common food crops cultivated in hydroponics. Microb Ecol 73:378–393
Siebielec G, Ukalska-Jaruga A, Kidd P (2014) Bioavailability of trace elements in soils amended with high-phosphate materials. In: Selim HM (ed) Phosphate in soils: interaction with micronutrients, radionuclides and heavy metals bioavailability of trace elements in soils amended with high-phosphate materials. CRC Press/Taylor & Francis Group, Boca Raton, pp 237–268
Sirakov I, Lutz M, Graber A, Mathis A, Staykov Y, Smits TH, Junge R (2016) Potential for combined biocontrol activity against fungal fish and plant pathogens by bacterial isolates from a model aquaponics system. Water 8:518
Timmons MB, Ebeling JM (2013) Recirculating aquaculture. Ithaca Publishing Company, Ithaca. 788 p
van Dam NM, Bouwmeester HJ (2016) Metabolomics in the rhizosphere: tapping into belowground chemical communication. Trends Plant Sci 21:256–265
Van Rijn J (2013) Waste treatment in recirculating aquaculture systems. Aquac Eng 53:49–56
Van Rijn J, Tal Y, Schreier HJ (2006) Denitrification in recirculating systems: theory and applications. Aquac Eng 34:364–376
Vilbergsson B, Oddsson GV, Unnthorsson R (2016a) Taxonomy of means and ends in aquaculture production-part 3: the technical solutions of controlling n compounds, organic matter, p compounds, metals, temperature and preventing disease. Water 8:506
Vilbergsson B, Oddsson GV, Unnthorsson R (2016b) Taxonomy of means and ends in aquaculture production – Part 2: The technical solutions of controlling solids, dissolved gasses and pH. Water 8:387
Wielgosz ZJ, Anderson TS, Timmons MB (2017) Microbial effects on the production of aquaponically grown lettuce. Horticulturae 3:46
Wongkiew S, Hu Z, Chandran K, Lee JW, Khanal SK (2017) Nitrogen transformations in aquaponics system: a review. Aquac Eng 76:9–19

Yildiz HY, Robaina L, Pirhonen J, Mente E, Domínguez D, Parisi G (2017) Fish welfare in aquaponics system: its relation to water quality with an emphasis on feed and faeces-a review. Water 9:13

Yogev U, Barnes A, Gross A (2016) Nutrients and energy balance analysis for a conceptual model of a three loops off grid, aquaponics. Water 8:589

Chapter 7
Coupled Aquaponics Systems

Harry W. Palm, Ulrich Knaus, Samuel Appelbaum, Sebastian M. Strauch, and Benz Kotzen

Abstract Coupled aquaponics is the archetype form of aquaponics. The technical complexity increases with the scale of production and required water treatment, e.g. filtration, UV light for microbial control, automatic controlled feeding, computerization and biosecurity. Upscaling is realized through multiunit systems that allow staggered fish production, parallel cultivation of different plants and application of several hydroponic subsystems. The main task of coupled aquaponics is the purification of aquaculture process water through integration of plants which add economic benefits when selecting suitable species like herbs, medicinal plants or ornamentals. Thus, coupled aquaponics with closed water recirculation systems has a particular role to fulfil.

Under fully closed recirculation of nutrient enriched water, the symbiotic community of fish, plants and bacteria can result in higher yields compared with stand-alone fish production and/or plant cultivation. Fish and plant choices are highly diverse and only limited by water quality parameters, strongly influenced by fish feed, the plant cultivation area and component ratios that are often not ideal. Carps, tilapia and catfish are most commonly used, though more sensitive fish species and crayfish have been applied. Polyponics and additional fertilizers are methods to improve plant quality in the case of growth deficiencies, boosting plant production and increasing total yield.

The original version of this chapter was revised: The correct DOI "https://doi.org/10.1016/j.aquaculture.2018.03.021" has been updated for author Palm et al. (2019) in references list on page 197. The correction to this chapter is available at https://doi.org/10.1007/978-3-030-15943-6_25

H. W. Palm (✉) · U. Knaus (✉) · S. M. Strauch
Faculty of Agricultural and Environmental Sciences, Department of Aquaculture and Sea-Ranching, University of Rostock, Rostock, Germany
e-mail: harry.palm@uni-rostock.de; ulrich.knaus@uni-rostock.de

S. Appelbaum
French Associates Institute for Agriculture and Biotechnology of Drylands, Jacob Blaustein Institutes for Desert Research, Ben-Gurion University of the Negev, Beersheba, Israel
e-mail: sappl@bgu.ac.il

B. Kotzen
School of Design, University of Greenwich, London, UK
e-mail: b.kotzen@greenwich.ac.uk

S. Goddek et al. (eds.), *Aquaponics Food Production Systems*,
https://doi.org/10.1007/978-3-030-15943-6_7

The main advantages of coupled aquaponics are in the most efficient use of resources such as feed for nutrient input, phosphorous, water and energy as well as in an increase of fish welfare. The multivariate system design approach allows coupled aquaponics to be installed in all geographic regions, from the high latitudes to arid and desert regions, with specific adaptation to the local environmental conditions. This chapter provides an overview of the historical development, general system design, upscaling, saline and brackish water systems, fish and plant choices as well as management issues of coupled aquaponics especially in Europe.

Keywords Coupled aquaponics · Fish and plant choice · Nutrient cycles · Polyponic systems · Functions

7.1 Introduction

The combination of fish and plant cultivation in coupled aquaponics dates back to the first design by Naegel (1977) in Germany, using a 2000 L hobby scale system (Fig. 7.1) located in a controlled environment greenhouse. This system was developed in order to verify the use of nutrients from fish waste water under fully controlled water recirculating conditions intended for plant production including a dual sludge system (aerobic/anaerobic wastewater treatment). Naegel based his concept on the open pond aquaponic system of the South Carolina Agricultural Experiment Station, in the USA, where excess nutrients from the fishponds, stocked with channel catfish (*Ictalurus punctatus*), were eliminated by the hydroponic production of water chestnuts (*Eleocharis dulcis*) (Loyacano and Grosvenor 1973). By including nitrification and denitrification tanks to increase the nitrate concentration inside his system, Naegel (1977) attempted a complete oxidation of all nitrogenous compounds, reaching nitrate concentrations of 1200 mg/L, and demonstrating the effectiveness of the nitrification step. Although the system was stocked at a low density (20 kg/m^3 each) using tilapia (*Tilapia mossambica*) and carp (*Cyprinus carpio*), the tomatoes (*Lycopersicon esculentum*) and iceberg lettuce (*Lactuca scariola*) grew well and produced harvestable yield. These first research results led to the concept of coupled aquaponic systems, in which the plants eliminate the waste produced by the fish, creating adequate growth, demonstrating highly efficient water use in both units. The principle of coupled aquaponics was first described by Huy Tran at the World Aquaculture Conference in 2015 (Tran 2015).

Coupled aquaponic systems do not necessarily use mechanical particulate filtering in the classical sense and keep consistent nutrient flow between the aquaculture and hydroponic units. The main challenge is how to manage the faecal load in the coupled aquaponic system where the plants absorb the nutrients and particulate waste can be removed from the system by filter presses or geotextiles.

The development of modern agriculture, human population growth and shrinking resources worldwide, has promoted the development of coupled aquaponic systems. Since fish farming is considerably more efficient in protein production and water use compared with other farmed animals and since closed systems are largely site-independent, coupled aquaponic systems have been able to develop worldwide (Graber

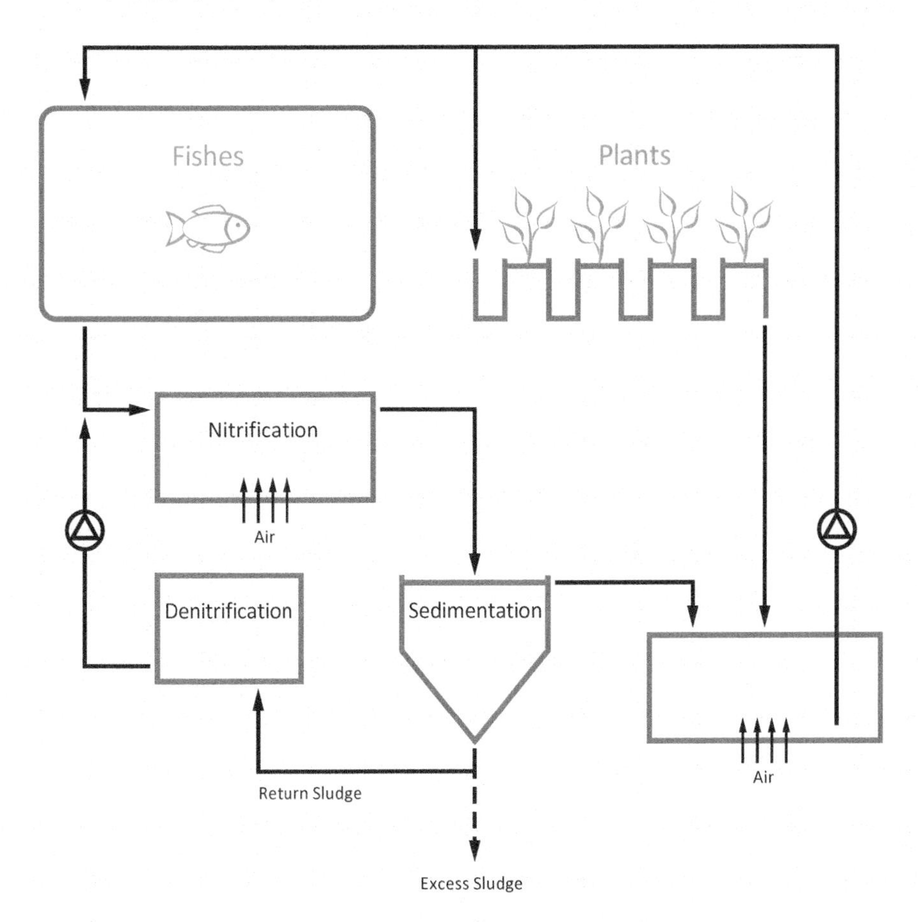

Fig. 7.1 Diagram of the first system by Naegel (1977) growing tilapia and common carp in combination with lettuce and tomatoes in a closed recirculation system

and Junge 2009), under arid conditions (Kotzen and Appelbaum 2010; Appelbaum and Kotzen 2016) and even in urban settings (König et al. 2016). Most described systems belong to domestic, small-scale and semi-commercial installations (Palm et al. 2018) that are driven by hobby aquarists, enthusiasts or smaller start-up companies. New research results, summarized in this chapter, demonstrate both the potentials and constraints regarding the continued development of these systems into commercial aquaponics, being capable of making a significant contribution to future food production.

7.2 Historical Development of Coupled Aquaponics

Most original research efforts on coupled aquaponic systems took place in the USA with an increasing presence in the EU partly initiated by COST Action FA1305, The EU Aquaponics Hub and in other European research centres. Nowadays, fully recirculating aquaponic system designs almost completely dominate the American aquaponics industry, with estimates that over 90% of the existing aquaponic systems in the USA are of a fully recirculating design (Lennard, pers. comm.). The first American coupled aquaponics research was undertaken at the Illinois Fisheries and Aquaculture Center (formerly the SIU Cooperative Fisheries Research Laboratory) and the Department of Zoology, focusing on coupled aquaponic systems stocked with channel catfish (*Ictalurus punctatus*) in combination with tomatoes (*Lycopersicon esculentum*) (Lewis et al. 1978). The authors noted that an optimal plant growth is only possible when all the essential macro- and micronutrients are available in the process water, and thus nutrient supplementation is required in the event of nutrient deficiencies. The authors also demonstrated a deficiency in plant-available iron, constraining plant growth, which could be solved through iron-chelate supplementation. Other early studies in the USA focused on analysing technological functionality and the quality of the harvested channel catfish and tomatoes (Lewis et al. 1978; Sutton and Lewis 1982). Laboratory-scale aquaponic systems examined parameters, such as resource efficiencies with regard to materials, costs, water and energy consumption, and examined the use of other fish species such as *Tilapia* spp. in the US Virgin Islands (UVI) (Watten and Busch 1984). Dr. James Rakocy at the UVI developed the first commercial coupled aquaponic system, a raft system that combined the production of Nile tilapia (*Oreochromis niloticus*) and lettuce (*Lactuca sativa*), and later investigated the production of further plant species (Rakocy 1989, 2012; Rakocy et al. 2000, 2003, 2004, 2006, 2011). This medium scale commercial installation took advantage of the local climate where greenhouses were not necessary and the market conditions of the Virgin Islands to generate profit. The UVI aquaponic system was subsequently adopted in different countries with respect to the respective needs of different plants and the appropriateness of the technology, e.g. in Canada by Savidov (2005) and in Saudi Arabia by Al-Hafedh et al. (2008). This is the case in Europe as well, where coupled aquaponic systems have evolved from the original UVI design, e.g. the vertical aquaponic system at the Aquaponics Research Lab., University of Greenwich (Khandaker and Kotzen 2018). Several other research departments investigated the technological feasibility of closed – or 'coupled' – aquaponics production using various fish and plant species as well as hydroponic subsystems to increase yields and reducing different emission parameters (Graber and Junge 2009). For example, at Rostock University (Germany), the research focused on the stability of backyard systems (Palm et al. 2014a), combining different fish species, African catfish (*Clarias gariepinus*) and Nile tilapia (*Oreochromis niloticus*), with different plants (Palm et al. 2014b, 2015). In 2015, a modern experimental semi-commercial scale aquaponic system, the 'FishGlassHouse', was built on the campus of the University of Rostock (Palm et al. 2016). However, the system was designed allowing both coupled and decoupled operations. Other notable facilities were built at the Zürich University of Applied Sciences (ZHAW) at Waedenswil in Switzerland (Graber and Junge 2009; Graber et al. 2014), both coupled and decoupled research facility of the Icelandic company Svinna-verkfraedi Ltd. (Thorarinsdottir 2014; Thorarinsdottir et al. 2015), the cold water aquaponic system NIBIO Landvik at Grimstad (Skar et al. 2015; Thorarinsdottir et al. 2015), the PAFF Box (Plant And Fish Farming

Box) one loop aquaponic system at Gembloux Agro-Bio Tech – University of Liège, in Gembloux, Belgium (Delaide et al. 2017), the combined living wall and vertical farming aquaponic system at the University of Greenwich (Khandaker and Kotzen 2018), as well as the research-domestic coupled aquaponic system (changed from decoupled to coupled in 2018, Morgenstern and Dapprich 2018, pers. comm.) at the South Westphalia University of Applied Sciences, i.GREEN Institute for Green Technology & Rural Development.

7.3 Coupled Aquaponics: General System Design

The coupled aquaponics principle combines three classes of organisms: (1) aquatic organisms, (2) bacteria and (3) plants that benefit from each other in a closed recirculated water body. The water serves as a medium of nutrient transport, mainly from dissolved fish waste, which is converted into nutrients for plant growth by bacteria. These bacteria (e.g. *Nitrosomonas* spec., *Nitrobacter* spec.) oxidize ammonium to nitrite and finally to nitrate. Therefore, it is necessary for the bacteria to receive substantial amounts of ammonium and nitrite to stabilize colony growth and the quantity of nitrate production. Consequently, in a coupled aquaponic system, volumes are critically important, *i*) the aquaculture unit following the principles of recirculating aquaculture systems (RAS), *ii*) the bacterial growth substrate and *iii*) the space for the plant units and the amount of plants to be cultivated. Together, they form the aquaponics unit (Fig. 7.2).

The specific biological-chemical components of the process water have particular importance for coupled aquaponic systems. With food or uneaten feed particles, the organic fish waste and the bacteria inside the process water, an emulsion of nutrients combined with enzymes and digestive bacteria support the growth of fish and plants. There is evidence that compared to stand-alone systems such as aquaculture (fish) and hydroponics (plants), the growth of aquatic organisms and crops in a coupled aquaponics can be similar or even higher. Rakocy (1989) described a slightly higher

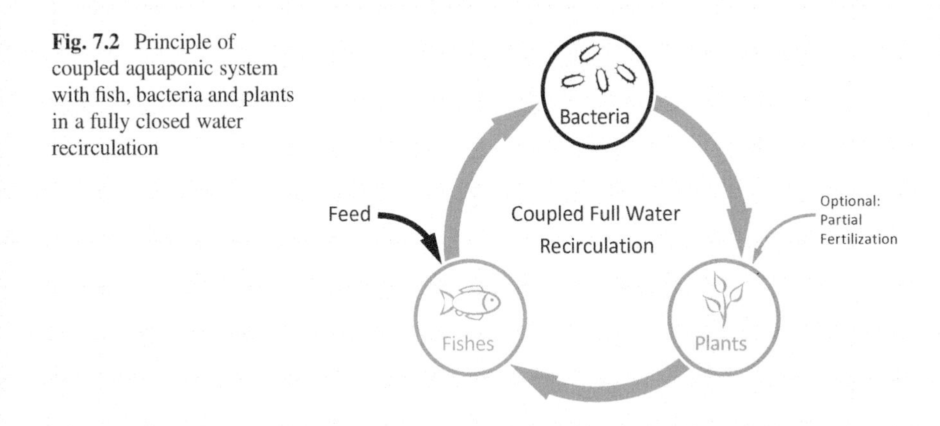

Fig. 7.2 Principle of coupled aquaponic system with fish, bacteria and plants in a fully closed water recirculation

yield of tilapia (*Tilapia nilotica*, 46.8 kg) in coupled aquaponics in contrast to stand-alone fish culture (41.6 kg) and slight increases in Summer Bibb lettuce yield (385.1 kg) compared to vegetable hydroponic production (380.1 kg). Knaus et al. (2018b) recorded that aquaponics increased biomass growth of *O. basilicum*, apparently due to increased leaf generation of the plants (3550 leaves in aquaponics) compared to conventional hydroponics (2393 leaves). Delaide et al. (2016) demonstrated that aquaponic and hydroponic treatments of lettuce exhibited similar plant growth, whereas the shoot weight of the complemented aquaponic solution with nutrients performed best. Similar observations have been made by Goddek and Vermeulen (2018). Lehmonen and Sireeni (2017) observed an increased root weight, leaf area and leaf colour in Batavia salad (*Lactuca sativa* var. *capitata*) and iceberg lettuce (*L. sativa*) with aquaponics process water from *C. gariepinus* combined with additional fertilizer. Certain plants such as lettuce (*Lactuca sativa*), cucumbers (*Cucumis sativus*) or tomatoes (*Solanum lycopersicum*) can consume nutrients faster, and as a result flower earlier in aquaponics compared with hydroponics (Savidov 2005). Also, Saha et al. (2016) reported a higher plant biomass yield in *O. basilicum* in combination with crayfish *Procambarus* spp. and a low start-up fertilization of the aquaponic system.

The basic system design of coupled aquaponics consists of one or more fish tanks, a sedimentation unit or clarifier, substrates for the growth of bacteria or suitable biofilters and a hydroponic unit for plant growth (Fig. 7.3). These units are connected by pipes to form a closed water cycle. Often, after the mechanical filtration and the biofilter, a pump sump is used (one pump or one loop system) which, as the deepest point of the system, pumps the water back to the fish tanks from where it flows by gravity to the hydroponic unit.

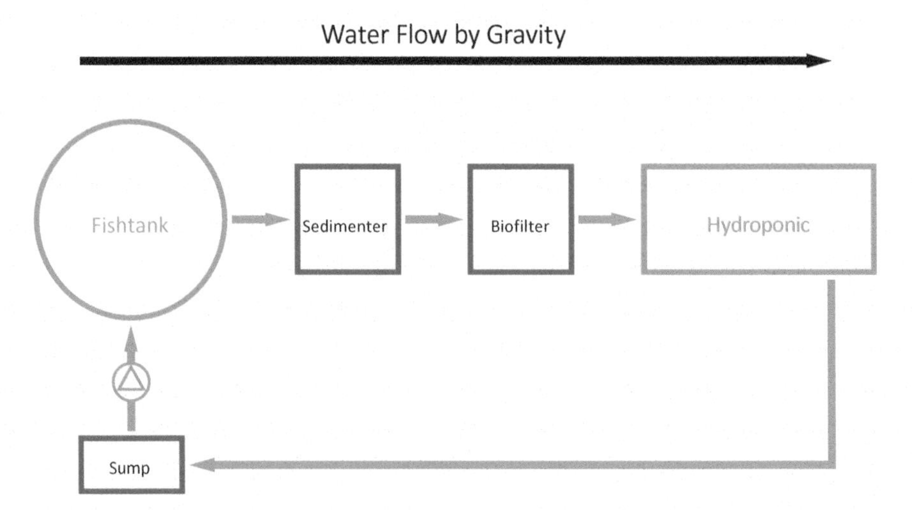

Fig. 7.3 Basic technical system design of a coupled aquaponic system with fish tank, sedimenter, biofilter, hydroponic unit and a sump where the water is pumped or airlifted back to the fish tanks and flows by gravity along the components

Coupled aquaponic systems are used in different scales. The closed-loop principle can be used in domestic systems (mini/hobby/backyard-coupled), demonstration units (e.g. living walls coupled), commercial aquaponics and aquaponics farming (with soil) ranging from small/semi-commercial to large-scale systems (Palm et al. 2018). A recent development in aquaponics has included partial fertilization, which is dependent on the tolerance of the fish species. This, however, can result in a short-term nutrient peak in the system but can be compensated through the nutrient retention by the plants. In coupled aquaponics, an optimal ratio of the production area (or fish volumes) of the aquaculture unit with the resulting feed demand as well as an adequate amount of plants to be cultured in the hydroponic unit (plant production area) must be achieved. (For discussions on the role of evapotranspiration and solar radiation within the systems, see Chaps. 8 and 11). For gravel aquaponics, Rakocy (2012) as a first attempt suggested 'component ratio principles', with a fish-rearing volume of 1 m^3 of fish tank volume to 2 m^3 hydroponic media of 3 to 6 cm pea gravel as a rule of thumb. Ultimately, the amount of fish determines the yield of crops in coupled aquaponics. Additionally, the technical conditions of the fish-rearing unit must be adapted according to the needs of the cultivated aquatic species.

7.4 Aquaculture Unit

The fish-rearing tanks (size, numbers and design) are selected depending on the scale of production and fish species in use. Rakocy et al. (2006) used four large fish-rearing tanks for the commercial production of *O. niloticus* in the UVI aquaponic system (USA). With the production of omnivorous or piscivorous fish species, such as *C. gariepinus*, several tanks should be used due to the sorting of the size classes and staggered production (Palm et al. 2016). Fish tanks should be designed so that the solids that settle at the bottom of the tanks can effectively be removed through an effluent at the bottom. This solid waste removal is the first crucial water treatment step in coupled aquaponics as is the case in aquaculture and decoupled aquaponics. The waste originates from uneaten feed, fish faeces, bacterial biomass and flocculants produced during aquaculture production, increasing BOD and reducing water quality and oxygen availability with respect to both the aquaculture and hydroponic units. In aquaculture, the solid waste consists to a large extent of organic carbon, which is used by heterotrophic bacteria to produce energy through oxygen consumption. The better the solid waste removal, the better the general performance of the system for both fish and plants, i.e. with optimal oxygenation levels and no accumulation of particles in the rhizosphere inhibiting nutrient uptake, and with round or oval tanks proving to be particularly efficient (Knaus et al. 2015).

Fish production in coupled aquaponics in the FishGlassHouse in Germany was tested at different scales in order to ascertain cost effectiveness. This was done effectively as extensive (max. 50 kg, 35 fish m^{-3}) or intensive (max. 200 kg, 140 fish m^{-3}) African catfish production. The semi-intensive production (max. 100 kg, 70 fish m^{-3}) cannot be recommended due to a negative cost benefit balance. In the semi-intensive production mode, system maintenance, labour and feed input were as much as under intensive production but with reduced fish and plant biomass output, and any economic gains in the aquaculture unit did not pay off (Palm et al. 2017). This resulted from the high biochemical oxygen demands (BOD), high denitrification because of the reduced oxygen availability, relatively high water exchange rates, predominantly anaerobic mineralization with distinct precipitation, low P and K-levels as well as a low pH-values with much less fish output compared with the intensive conditions. In contrast, the extensive fish production allowed higher oxygen availability with less water exchange rates and better nutrient availability for plant growth. Thus, under the above conditions, a RAS fish production unit for coupled aquaponics therefore either functions under extensive or intensive fish production conditions, and intermediate conditions should be avoided.

7.4.1 Filtration

Clarifiers, sometimes also called sedimenters or swirl separators (also see Chap. 3), are the most frequently used devices for the removal of solid waste in coupled aquaponics (Rakocy et al. 2006; Nelson and Pade 2007; Danaher et al. 2013, Fig. 7.4). Larger particulate matters must be removed from the system to avoid anoxic zones with denitrifying effects or the development of H_2S. Most clarifiers use lamella or plate inserts to assist in solids removal. Conical bottoms support sludge concentration at the bottom during operation and cleaning, whereas flat bottoms require large quantities of water to flush out and remove the sludge. During operation, the solids sink to the bottom of the clarifier to form sludge. Depending on the feed input and retention time, this sludge can build up to form relatively thick layers. The microbial activity inside the sludge layers gradually shifts towards anaerobic conditions, stimulating microbial denitrification. This process reduces plant available nitrate and should be avoided, especially if the process water is to be used for hydroponic plant production. Consequently, denitrification can be counterproductive in coupled aquaponics.

The density of the solid waste removed by the clarifier is rather low, compared with other technologies, maintenance is time-consuming, and cleaning the clarifier with freshwater is responsible for the main water loss of the entire system. The required amount of water is affected by its general design, the bottom shape and the accessibility of the PVC baffles to flushing water (Fig. 7.4a, b). Increasing fish stocking densities require higher quantities of water exchange (every day in the week under intensive conditions) to maintain optimal water quality for fish production, which can result in the loss of large amounts of process water, also losing substantial amounts of nutrients required for plant growth. Furthermore, replacement with freshwater introduces calcium and magnesium carbonates which may then precipitate with phosphates. Therefore, the use of such manually operated clarifiers makes predictions on process water composition with respect to optimal plant growth nearly impossible (Palm et al. 2019). It would be more effective to follow

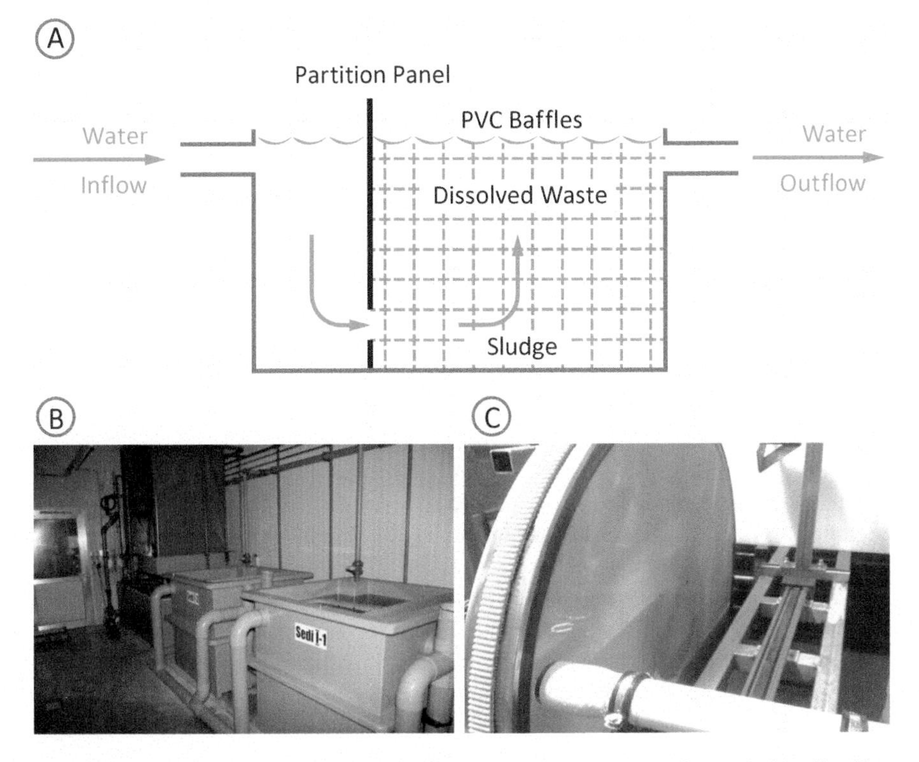

Fig. 7.4 Principle of aquaponic filtration with a sedimenter (**a–b**) and (**c**) disc-filter (PAL-Aquakultur GmbH, Abtshagen, Germany) of commercial African catfish (*Clarias gariepinus*) RAS in the FishGlassHouse (Rostock University, Germany)

Naegle's (1977) example of separating aerobic and anaerobic sludge and gaseous nitrogen discharge with a dual sludge system.

More effective solid waste removal can be achieved by automatic drum- or disk-filters which provide mechanical barriers that hold back solids, which are then removed through rinsing. New developments aim to reduce the use of rinse water through vacuum cleaning technologies, allowing the concentration of total solids in the sludge up to 18% (Dr. Günther Scheibe, PAL-Aquakultur GmbH, Germany, personal communication, Fig. 7.4c). Such effective waste removal has a positive influence on the sludge composition, improving effluent water control in order to better meet the horticultural requirements. Another option is the application of multiple clarifiers (sedimenters) or sludge-removal components in a row.

Biofilters are another essential part of RAS, as they convert ammonia nitrogen via microbial oxidation to nitrate (nitrification). Even though plant roots and the system itself provide surfaces for nitrifying bacteria, the capability to control the water quality is limited. Systems that do not have biofiltration are restricted to mini or hobby installations with low feed inputs. As soon as the biomass of fish and the feed input increases, additional biofilter capacity is required to maintain adequate water quality for fish culture and to provide sufficient nitrate quantities for plant growth.

For domestic and small-scale aquaponics, plant media (gravel or expanded clay for example) can suffice as effective biofilters. However, due to the high potential for clogging and thus the requirement for regular manual cleaning and maintenance, these methods are not suitable for larger-scale commercial aquaponics (Palm et al. 2018). Additionally, Knaus and Palm (2017a) demonstrated that the use of a simple biofilter in a bypass already increased the possible daily feed input in a backyard-coupled aquaponic system by approximately 25%. Modern biofilters that are used in intensive RAS are effective in providing sufficient nitrification capacity for fish and plant production. Because of increased investment costs, such components are more applicable in medium- and larger-scale commercial aquaponic systems.

7.4.1.1 Hydroponics in Coupled Aquaponics

In coupled aquaponics, a wide range of hydroponic subsystems can be used (also see Chap. 4) depending on the scale of operation (Palm et al. 2018). Unless labour has no significant impact on the yield (or profit) and the system is not too large, different hydroponic subsystems can be used at the same time. This is common in domestic and demonstration aquaponics that often use media bed substrate systems (sand, gravel, perlite, etc.) in ebb and flow troughs, DWC channels (deep water culture or raft systems) and even often self-made nutrient film channels (NFT). Most labour-intensive are media substrate beds (sand/gravel) in ebb and flow troughs, which can clog due to the deposition of detritus and often need to be washed (Rakocy et al. 2006). Due to the handling of the substrates, these systems are usually limited in size. On the other hand, DWC hydroponic subsystems require less labour and are less prone to maintenance, allowing them to be adopted for larger planting areas. For this reason, DWC subsystems are mainly found in domestic to small/semi-commercial systems, however, not usually in large-scale aquaponic systems. For larger commercial aquaponic production, the proportion of labour and maintenance in the DWC system is still seen to be too high. Even the use of water resources and energy for pumping are also unfavourable for large-scale systems.

If closed aquaponic systems are designed for profit-oriented production, the use of labour must decrease whilst the production area must increase. This is only possible by streamlining fish production combined with the application of easy-to-use hydroponic subsystems. The nutrient film technique (NFT) can, at present, be considered the most efficient hydroponic system, combining low labour with large plant cultivation areas and a good ratio of water, energy and investment costs. However, not all aquaponic plants grow well in NFT systems and thus it is necessary to find the right plant choice for each hydroponic subsystem, which in turn correlates with the nutrient supply of a specific fish species integrated in a specific hydroponic subsystem design. For coupled aquaponics, the sometimes higher particle load in the water can be problematic by clogging drips, pipes and valves in NFT installations. Hence, large aquaponic systems have to contain professional water management with effective mechanical filtration to avoid recirculation blockages. When the continuous supply of water is ensured through the pipes, the NFT system can be used in all types of coupled aquaponic systems, but is most recommended for production under small/semi-commercial systems and large-scale systems (Palm et al. 2018).

7.5 Scaling Coupled Aquaponic Systems

Typical coupled aquaponic system range from small to medium scale and larger sized systems (Palm et al. 2018). Upscaling remains one of the future challenges because it requires careful testing of the possible fish and plant combinations. Optimal unit sizes can be repeated to form multiunit systems, independent of the scale of production. According to Palm et al. (2018), the range of aquaponic systems were categorized into (1) mini, (2) hobby, (3) domestic and backyard, (4) small/semi-commercial and (5) large(r)-scale systems, as described below:

Mini installations (Fig. 7.5) usually consist of a small fish reservoir such as a fish tank or aquarium on which the plants grow on the surface or within a small hydroponic bed. Conventional aquarium filters, aeration and pumps are usually used. Mini systems are usually 2 m^2 or less in size (Palm et al. 2018). These small aquaponic systems can be used in the home with only few plants for home consumption and planted with plants such as tomatoes, herbs or ornamentals. Such systems add new values to human living space by adding 'nature' back into the family life area which is especially popular in big cities. Some mini systems consist of only a plant vase and one or more fish without filter and pump. However, these systems are only short-term to operate because a regulated filtration is missing.

Hobby aquaponic systems are categorized to reach a maximum size of 10 m^2 (Palm et al. 2018). With a higher fish stocking density, more feed and aeration, a mechanical sedimentation unit (sedimenter/clarifier) is necessary (Fig. 7.6). The sedimenter removes particulate matter –'sludge' such as faeces and uneaten feed

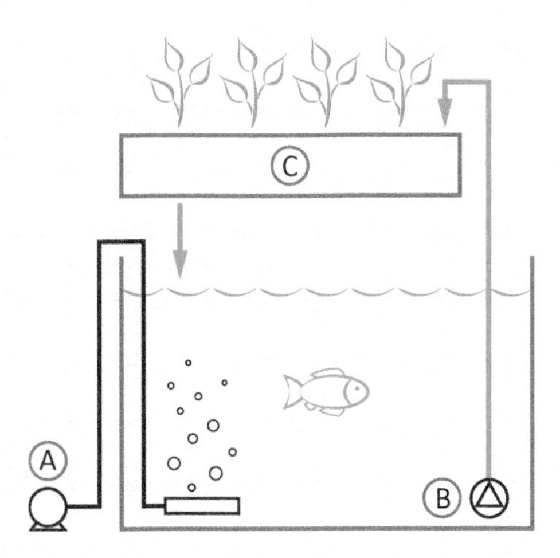

Fig. 7.5 Principle of a domestic coupled mini aquaponic system (< 2 m^2, after Palm et al. 2018) with aeration (**a**) and a pump (**b**), the hydroponics (**c**) act like a biofilter

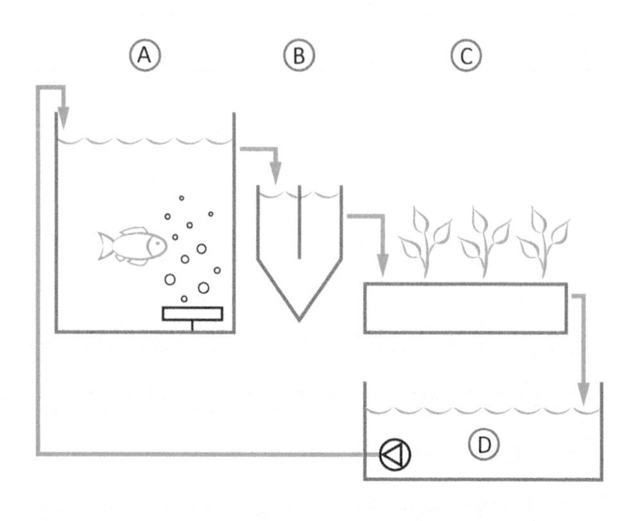

Fig. 7.6 Principle of a coupled domestic hobby aquaponic system (2–10 m^2 after Palm et al. 2018) with (**a**) fish tank and aeration, (**b**) sedimenter or clarifier altered after Nelson and Pade (2007), (**c**) hydroponics bed, e.g. gravel with different crops which acts like a biofilter and (**d**) a sump with the pump

from the system without using energy. The water flows by gravity from the fish tank to the sedimenter and then through the hydroponic tanks and then drops into a sump from where a pump or air lift pumps the water back to the fish tanks. In hobby installations, the plant beds act as a natural microbial filter and often media bed substrates such as sand, fine gravel or perlite are used. Hobby aquaponic systems are more the category of gimmicks that do not target food production. They rather enjoy the functionality of the integrated system. Hobby systems, as the name implies, are usually installed by hobbyists who are interested in growing a variety of aquatic organisms and plants for their own use and for 'fun'.

Domestic/backyard aquaponics has the purpose of external home use production of fish and plants characterized as having a maximum production area of 50 m^2 (Palm et al. 2018). These systems are built by enthusiasts. The construction is technically differentiated with a higher fish production, additional aeration and a higher feed input. The coupled aquaponics principle is applied with the use of one single pump which recirculates the water from a sump (lowest point) to higher standing fish tanks and then by gravity via sedimenter and a biofilter (with aeration and bacteria substrates) to the hydroponic units (Fig. 7.7).

For biofiltration, conventional bed filters can also be used as described in Palm et al. (2014a, b, 2015). In backyard aquaponics, hydroponics could consist alone or together of raft or DWC (deep water culture) troughs, substrate subsystems such as coarse gravel/sand ebb and flow boxes or nutrient film technique (NFT) channels. In the northern hemisphere, in outside installations, production is limited to the spring, summer and early autumn periods because of the weather conditions. With this scale of operation, fish and plants can be produced for private consumption (and production can be extended through small greenhouse production), but direct sales in small quantities are also possible.

Small and semi-commercial scale aquaponic systems are characterized by being up to 100 m^2 (Palm et al. 2018) with production focused on the retail market. More tanks, often with a higher stocking density, additional filters and water treatment

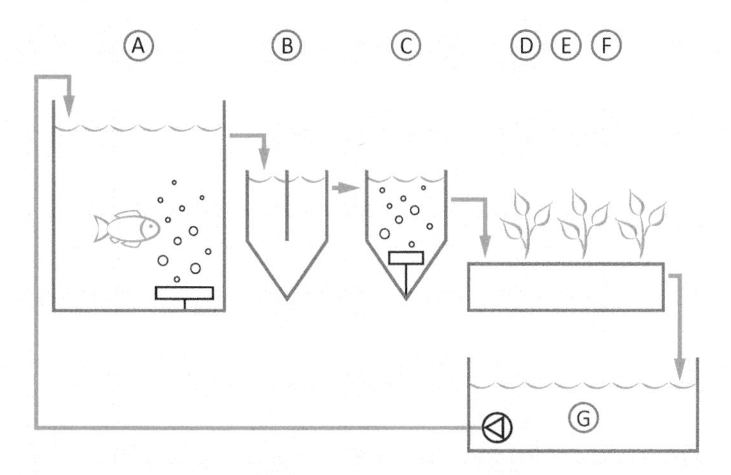

Fig. 7.7 Principle of a coupled domestic backyard aquaponic system, 10–50 m^2 (from Palm et al. 2018) with (**a**) fish tank and aeration, (**b**) sedimenter or clarifier altered after Nelson and Pade (2007), (**c**) biofilter with substrates and aeration, (**d**) hydroponic unit which could consists of combined raft or DWC channels, (**e**) gravel or sand media substrate system, (**f**) nutrient film technique NFT-channels and (**g**) a sump with one pump

systems and a larger hydroponic area with more diverse designs characterize these systems.

Large(r)-scale commercial operations above 100 m^2 (Palm et al. 2018) and reaching many thousands of square metres reach the highest complexity and require careful planning of the water flow and treatment systems (Fig. 7.8). General components are multiple fish tanks, designed as intensive recirculation aquaculture systems (RAS), a water transfer point or a sump allowing water exchange between the fish and plants, and commercial plant production units (aquaponics *s.s./s.l.*). As fish production is meant for intensive stocking densities, components such as additional filtration with the help of drum filters, oxygen supply, UV light treatments for microbial control, automatic controlled feeding and computerization including automatic water quality control classify these systems.

These systems have a multiunit design capable of upscaling under fully closed water recirculation which also allows for staggered production, parallel cultivation of different plants that require different hydroponic subsystems and better control of the different units in the case of disease outbreak and plant pest control.

7.6 Saline/Brackish Water Aquaponics

A relatively new field of research is the evaluation of different salinities of the process water for plant growth. Since freshwater worldwide is in continuously increasing demand and at high prices, some attention has been given to the use of saline/brackish water resources for agriculture, aquaculture and also aquaponics. The use of brackish water is significant as many countries such as Israel have

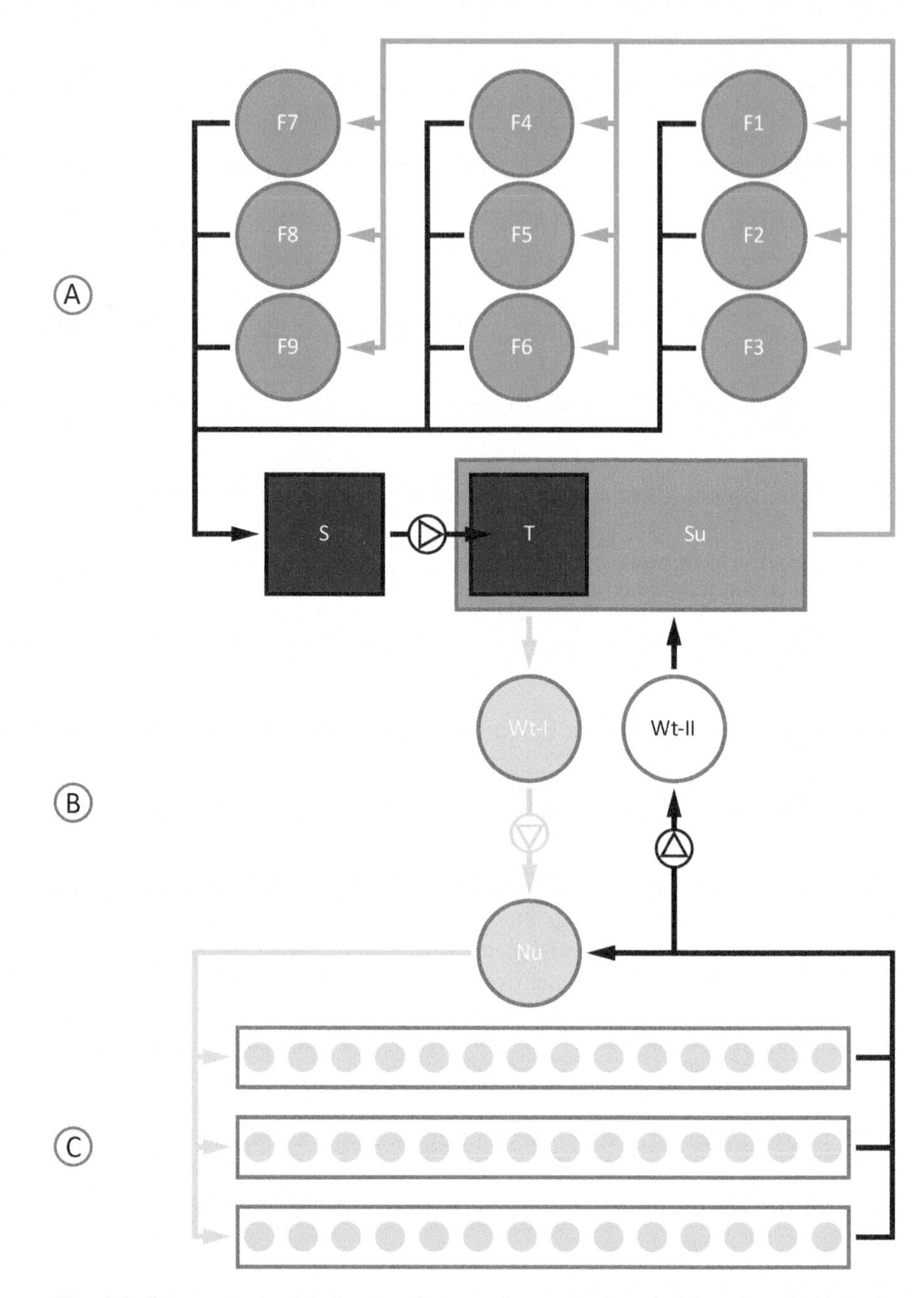

Fig. 7.8 Schema (supervision) of a large-scale aquaponics module adopted after the FishGlassHouse at University of Rostock (Germany) (1000 m^2 total production area, Palm et al. 2018) with (**a**) independent aquaculture unit, (**b**) the water transfer system and (**c**) the independent hydroponic unit; F1-F9 fish tanks, S) sedimenter, P-I pump one (biofilter pump), P-II pump two (aquaculture recirculation pump), T) trickling filter, Su) sump. In the middle, nutrient water transfer system with Wt-I) water transfer tank from the aquaculture unit, P-III) pump three, which pumps the nutrient rich water from aquaculture to C) hydroponics unit on the right with Nu) nutrient tank and an independent hydroponic recirculation system and planting tables (or NFT); P-IV) pump four, which pumps the nutrient low water from the hydroponic unit back to Wt-II) water transfer tank two and to the aquaculture unit for coupled (or decoupled if not used) aquaponic conditions

underground brackish water resources, and more than half the world's underground water is saline. Whilst the amount of underground saline water is only estimated as 0.93% of world's total water resources at 12,870,000 km^3, this is more than the underground freshwater reserves (10,530,000 km^3), which makes up 30.1% of all freshwater reserves (USGS).

The first published research on the use of brackish water in aquaponics was carried out in 2008–2009 in the Negev Desert of Israel (Kotzen and Appelbaum 2010). The authors studied the potential for brackish water aquaponics that could utilize the estimated 200–300 billion m^3 located 550–1000 metres underground in the region. This and additional studies used up to 4708–6800 μS/cm (4000–8000 μS/cm = moderately saline, Kotzen and Appelbaum 2010; Appelbaum and Kotzen 2016) in coupled aquaponic systems with *Tilapia* sp. (red strain of Nile tilapia *Oreochromis niloticus* x blue tilapia *O. aureus* hybrids), combined with deep water culture floating raft and gravel systems. The systems were mirrored with potable water systems as a control. A wide range of herbs and vegetables were grown, with very good and comparative results in both brackish and freshwater systems. In both systems fish health and growth were as good as plant growth of leeks (*Allium ampeloprasum*), celery (*Apium graveolens*) (Fig. 7.9), kohlrabi (*Brassica oleracea v. gongylodes*), cabbage (*Brassica oleracea v. capitata*), lettuce (*Lactuca sativa*), cauliflower (*Brassica oleracea v. botrytis*), Swiss chard (*Beta vulgaris vulgaris*), spring onion (*Allium fistulosum*), basil (*Ocimum basilicum*) and water cress (*Nasturtium officinale*) (Kotzen and Appelbaum 2010; Appelbaum and Kotzen 2016).

A 'mission report' by van der Heijden et al. (2014) on integrating agriculture and aquaculture with brackish water in Egypt suggests that red tilapia (probably red strains of *Oreochromis mossambicus*) has high potential combined with vegetables such as peas, tomatoes and garlic that can tolerate low to moderate salinity. Plants that are known to have saline tolerance include the cabbage family (Brassicas), such

Fig. 7.9 Mature celery plant grown in brackish water

as cabbage (*Brassica oleracea*), broccoli (*Brassica oleracea italica*), kale (*Brassica oleracea var. sabellica*), Beta family, such as *Beta vulgaris* (beetroot), perpetual spinach (*Beta vulgaris subsp. Vulgaris*), and bell peppers (*Capsicum annuum*) and tomatoes (*Solanum lycopersicum*). An obvious plant candidate for brackish water aquaponics is marsh samphire (*Salicornia europaea*) and potentially other 'strand vegetables' such as sea kale (*Crambe maritima*), sea aster (*Tripolium pannonicum*) and sea purslane (*Atriplex portulacoides*). Gunning (2016) noted that in the most arid regions of the word the cultivation of halophytes as an alternative to conventional crops is gaining significant popularity and *Salicornia europea* is becoming increasingly popular on the menus of restaurants and the counters of fishmongers and health-food stores across the country. This is similarly the case across the UK and the EU where most of the produce is exported from Israel and now also Egypt. A distinct advantage of growing marsh samphire is that it is a 'cut and come again' crop which means it can be harvested at intervals of around 1 month. In its natural environment along saline estuaries *Salicornia europaea* grows along a saline gradient from saline through brackish (Davy et al. 2001). In trials by Gunning (2016), plants were grown from seed, whereas Kotzen grew his trial plants from cut stems bought at the supermarket fish counter. Further studies under saline conditions were performed by Nozzi et al. (2016), who studied the effects of dinoflagellate (*Amyloodinum ocellatum*) infection in sea bass (*Dicentrarchus labrax*) at different salinity levels. Pantanella (2012) studied the growth of the halophyte *Salsola soda* (salt cabbage) in combination with the flathead grey mullet (*Mugil cephalus*) under marine conditions of increasing salt contents on an experimental farm at the University of Tuscia (Italy). Marine water resources have also been successful used in coupled aquaponics with the production of European sea bass (*Dicentrarchus labrax*) and salt-tolerant plants (halophytes) such as *Salicornia dolichostachya*, *Plantago coronopus* and *Tripolium pannonicum* in an inner land marine recirculating aquaculture system (Waller et al. 2015).

7.7 Fish and Plant Choices

7.7.1 Fish Production

In larger scale commercial aquaponics fish and plant production need to meet market demands. Fish production allows species variation, according to the respective system design and local markets. Fish choice also depends on their impact onto the system. Problematic coupled aquaponics fish production due to inadequate nutrient concentrations, negatively affecting fish health, can be avoided. If coupled aquaponic systems have balanced fish to plant ratios, toxic nutrients will be absorbed by the plants that are cleaning the water. Since acceptance of toxic substances is species dependent, fish species choice has a decisive influence on the economic success. Therefore, it is important to find the right combination and ratio between the fish and the plants, especially of those fish species with less water polluting activities and plants with high nutrient retention capacity.

The benefits of having a particular fish family in coupled aquaponic systems are not clearly understood with respect to their specific needs in terms of water quality and acceptable nutrient loads. Naegel (1977) found there was no notable negative impact on the fish and fish growth in his use of tilapia (*Tilapia mossambica*) and common carp (*Cyprinus carpio*). The channel catfish (*Ictalurus punctatus*) was also used by Lewis et al. (1978) and Sutton and Lewis (1982) in the USA. It was demonstrated that the quality of the aquaponics water readily met the demands of the different fish species, especially through the use of 'easy-to-produce' fish species such as the blue tilapia (*Oreochromis aureus*, formerly *Sarotherodon aurea*) in Watten and Busch (1984); Nile tilapia (*Oreochromis niloticus*), which was often used in studies with different plant species as a model fish species (Rakocy 1989; Rakocy et al. 2003, 2004; Al-Hafedh et al. 2008; Rakocy 2012; Villarroel et al. 2011; Simeonidou et al. 2012; Palm et al. 2014a, 2014b; Diem et al. 2017); and also tilapia hybrids-red strain (*Oreochromis niloticus x* blue tilapia *O. aureus* hybrids), that were investigated in arid desert environments (Kotzen and Appelbaum 2010; Appelbaum and Kotzen 2016).

There has been an expansion in the types of fish species used in aquaponics, at least in Europe, which is based on the use of indigenous fish species as well as those that have a higher consumer acceptance. This includes African catfish (*Clarias gariepinus*) which was grown successfully under coupled aquaponic conditions by Palm et al. (2014b), Knaus and Palm (2017a) and Baßmann et al. (2017) in northern Germany. The advantage of *C. gariepinus* is a higher acceptance of adverse water parameters such as ammonium and nitrate, as well as there is no need for additional oxygen supply due to their special air breathing physiology. Good growth rates of *C. gariepinus* under coupled aquaponic conditions were further described in Italy by Pantanella (2012) and in Malaysia by Endut et al. (2009). An expansion of African catfish production under coupled aquaponics can be expected, due to unproblematic production and management, high product quality and increasing market demand in many parts of the world.

In Europe, other fish species with high market potential and economic value have recently become the focus in aquaponic production, with particular emphasis on piscivorous species such as the European pikeperch 'zander' (*Sander lucioperca*). Pikeperch production, a fish species that is relatively sensitive to water parameters, was tested in Romania in coupled aquaponics. Blidariu et al. (2013a, b) showed significantly higher P_2O_5 (phosphorous pentoxide) and nitrate levels in lettuce (*Lactuca sativa*) using pikeperch compared to the conventional production, suggesting that the production of pikeperch in coupled aquaponics is possible without negative effects on fish growth by nutrient toxicity. The Cyprinidae (Cypriniformes) such as carp have been commonly used in coupled aquaponics and have generally shown better growth with reduced stocking densities and minimal aquaponic process water flow rates (efficient water use) during experiments in India. The optimal stocking density of koi carp (*Cyprinus carpio* var. koi) was at 1.4 kg/m (Hussain et al. 2014), and the best weight gain and yield of *Beta vulgaris* var. *bengalensis* (spinach) was found with a water flow rate of 1.5 L/min (Hussain et al. 2015). Good fish growth and plant yield of water spinach (*Ipomoea aquatica*) with a maximum percentage of nutrient removal (NO_3-N, PO_4-P, and K) was reported at a minimum water flow rate of 0.8 L/min with polycultured koi carp (*Cyprinus carpio* var. koi) and gold fish (*Carassius auratus*) by Nuwansi et al. (2016). It is interesting to note that plant growth and nutrient removal in koi (*Cyprinus carpio* var. koi) and gold fish (*Carassius auratus*) production (Hussain et al. 2014, 2015) with *Beta vulgaris* var. *bengalensis* (spinach) and water spinach (*Ipomoea aquatica*) increased linearly with a decrease in process water flow between 0.8 L/min and 1.5 L/min. These results suggest that for cyprinid fish culture, lower water flow is recommended as this has no negative impacts on fish growth. In contrast, however, Shete et al. (2016) described a higher flow rate of 500 L h^{-1} (approx. 8 L/min) for common carp and mint (*Mentha arvensis*) production, indicating the need for different water flow rates for different plant species. Another cyprinid, the tench (*Tinca tinca*), was successfully tested by Lobillo et al. (2014) in Spain and showed high fish survival rates (99.32%) at low stocking densities of 0.68 kg m^{-3} without solids removal devices and good lettuce survival rates (98%). Overall, members of the Cyprinidae family highly contribute to the worldwide aquaculture production (FAO 2017); most likely this would also be true under aquaponic conditions and productivity, but the economic situation should be tested for each country separately.

Other aquatic organisms such as shrimp and crayfish have been introduced into coupled aquaponic production. Mariscal-Lagarda et al. (2012) investigated the influence of white shrimp process water (*Litopenaeus vannamei*) on the growth of tomatoes (*Lycopersicon esculentum*) and found good yields in aquaponics with a twofold water sparing effect under integrated production. Another study compared the combined semi-intensive aquaponic production of freshwater prawns (*Macrobrachium rosenbergii* – the Malaysian shrimp) with basil (*Ocimum basilicum*) versus traditional hydroponic plant cultivation with a nutrient solution (Ronzón-Ortega et al. 2012). However, basil production in aquaponics was initially less effective (25% survival), but with increasing biomass of the prawns, the plant biomass also increased so that the authors came to a positive conclusion with the production of basil with *M. rosenbergii.* Sace and Fitzsimmons (2013) reported a better plant growth in lettuce (*Lactuca sativa*), Chinese cabbage (*Brassica rapa pekinensis*) and pakchoi (*Brassica rapa*) with *M. rosenbergii* in polyculture with the Nile tilapia (*O. niloticus*). The cultivation with prawns stabilized the system in terms of the chemical-physical parameters, which in turn improved plant growth, although due to an increased pH, nutrient deficiencies occurred in the Chinese cabbage and lettuce. In general, these studies demonstrate that shrimp production under aquaponic conditions is possible and can even exert a stabilizing effect on the closed loop – or coupled aquaponic principle.

7.7.2 Plant Production

The cultivation of many species of plants, herbs, fruiting crops and leafy vegetables have been described in coupled aquaponics. In many cases, the nutrient content of the aquaponics process water was sufficient for good plant growth. A review by

Thorarinsdottir et al. (2015) summarized information on plant production under aquaponic production conditions from various sources. Lettuce (*Lactuca sativa*) was the main cultivated plant in aquaponics and was often used in different variations such as crisphead lettuce (iceberg), butterhead lettuce (bibb in the USA), romaine lettuce and loose leaf lettuce under lower night (3–12 °C) and higher day temperatures (17–28 °C) (Somerville et al. 2014). Many experiments were carried out with lettuce in aquaponics (e.g. Rakocy 1989) or as a comparison of lettuce growth between aquaponics, hydroponics and complemented aquaponics (Delaide et al. 2016). Romaine lettuce (*Lactuca sativa longifolia* cv. Jericho) was also investigated by Seawright et al. (1998) with good growth results similar to stand-alone hydroponics and an increasing accumulation of K, Mg, Mn, P, Na and Zn with increasing fish biomass of Nile tilapia (*Oreochromis niloticus*). Fe and Cu concentrations were not affected. Lettuce yield was insignificant with different stocking densities of fish (151 g, 377 g, 902 g, 1804 g) and plant biomass between 3040 g (151 g fish) and 3780 g (902 g fish). Lettuce was also cultivated, e.g. by Lennard and Leonard (2006) with Murray Cod (*Maccullochella peelii peelii*), and by Lorena et al. (2008) with the sturgeon 'bester' (hybrid of *Huso huso* female and *Acipenser ruthenus* male) and by Pantanella (2012) with Nile tilapia (*O. niloticus*). As a warm water crop, basil (*Ocimum basilicum*) was reported as a good herb for cultivation under coupled aquaponics and was reported as the most planted crop by 81% of respondents in findings of an international survey (Love et al. 2015). Rakocy et al. (2003) investigated basil with comparable yields under batch and staggered production (2.0; 1.8 kg/m^2) in contrast to field cultivation with a comparatively low yield (0.6 kg/m^2). Somerville et al. (2014) described basil as one of the most popular herbs for aquaponics, especially in large-scale systems due to its relatively fast growth and good economic value. Different cultivars of basil can be grown under higher temperatures between 20 and 25 °C in media beds, NFT (nutrient film technique) and DWC (deep water culture) hydroponic systems. Basil grown in gravel media beds can reach 2.5-fold higher yield combined with tilapia juveniles (*O. niloticus*, 0.30 g) in contrast to *C. gariepinus* (0.12 g) (Knaus and Palm 2017a).

Tomatoes (*Lycopersicon esculentum*) were described by Somerville et al. (2014) as an 'excellent summer fruiting vegetable' in aquaponics and can cope with full sun exposure and temperatures below 40 °C depending on tomato type. However, economic sustainability in coupled aquaponics is disputed due to the reduced competitiveness of aquaponics tomato production compared to high-engineered conventional hydroponic production in greenhouses in, e.g. the Netherlands Improvement Centre of DLV GreenQ in Bleiswijk with tomato yield of 100.6 kg m^{-2} (Hortidaily 2015), or even higher (Heuvelink 2018). Earlier investigations focused on the cultivation of this plant mostly compared to field production. Lewis et al. (1978) reported nearly double the crop of tomatoes under aquaponics compared to field production and the iron deficiency which occurred was fixed by using ethylene diamine tetra-acetic acid. Tomatoes were also produced in different aquaponic systems over the last decades, by Sutton and Lewis (1982) with good plant yields at water temperatures up to 28 °C combined with Channel catfish (*Ictalurus punctatus*), by Watten and Busch (1984) combined with tilapia

(*Sarotherodon aurea*) and a calculated total marketable tomato fruit yield of 9.6 kg/m^{-2}, approximately 20% of recorded yields for decoupled aquaponics (47 kg/m^2/y, Geelen 2016). McMurtry et al. (1993) combined hybrid tilapia (*Oreochromis mossambicus* x *Oreochromis niloticus*) with tomatoes in associated sand biofilters which showed optimal 'plant yield/high total plant yield' of 1:1.5 tank/biofilter ratio (sand filter bed) and McMurtry et al. (1997) with increasing total plant fruit yield with increasing biofilter/tank ratio. It must be stated that the production of tomatoes is possible under coupled aquaponics. Following the principle of soilless plant cultivation in aquaponics sensu stricto after Palm et al. (2018), it is advantageous to partially fertilize certain nutrients such as phosphorous, potassium or magnesium to increase yields (see challenges below).

The cultivation of further plant species is also possible and testing of new crops is continuously being reported. In the UK, Kotzen and Khandaker have tested exotic Asian vegetables, with particular success with bitter gourd, otherwise known as kerala or bitter melon (*Momordica charantia*) (Kotzen pers. comm.). Taro (*Colocasia esculenta*) is another species which is readily grown with reported success both for its large 'elephant ear' like leaves as well as its roots (Kotzen pers. comm.). Somerville et al. (2014) noted that crops such as cauliflower, eggplant, peppers, beans, peas, cabbage, broccoli, Swiss chard and parsley have the potential for cultivation under aquaponics. But there are many more (e.g. celery, broccoli, kohlrabi, chillies, etc.) including plants that prefer to have wet root conditions, including water spinach (*Ipomoea aquatica*) and mint (*Menta* sp.) as well as some halophytic plants, such as marsh samphire (*Salicornia europaea*).

Ornamental plants can also be cultivated, alone or together with other crops (intercropping), e.g. *Hedera helix* (common ivy) grown at the University of Rostock by Palm & Knaus in a coupled aquaponic system. The trials used 50% less nutrients that would be normally supplied to the plants under normal nursery conditions with a 94.3% success rate (Fig. 7.10).

Fig. 7.10 Three quality categories of ivy (*Hedera helix*), grown in a coupled aquaponic system indicating the quality that the nursery trade requires (**a**) very good and directly marketable, (**b**) good and marketable and (**c**) not of high enough quality

Besides the chosen plant and variant, there are two major obstacles that concern aquaponics plant production under the two suggested states of fish production, extensive and intensive. Under extensive conditions, nutrient availability inside the process water is much lower than under commercial plant production, nutrients such as K, P and Fe are deficient, and the conductivity is between 1000 and 1500 µS / cm, which is much less than applied under regular hydroponic production of commercial plants regularly between 3000 and 4000 µS / cm. Plants that are deficient in some nutrients can show signs of leaf necroses and have less chlorophyll compared with optimally fertilized plants. Consequently, selective addition of some nutrients increases plant quality that is required to produce competitive products.

In conclusion, commercial plant production of coupled aquaponics under intensive fish production has the difficulty to compete with regular plant production and commercial hydroponics at a large scale. The non-optimal and according to Palm et al. (2019) unpredictable composition of nutrients caused by the fish production process must compete against optimal nutrient conditions found in hydroponic systems. There is no doubt that solutions need to be developed allowing optimal plant growth whilst at the same time providing the water quality required for the fish.

7.7.3 *Fish and Plant Combination Options*

Combining fish and plants in closed aquaponics can generate better plant growth (Knaus et al. 2018b) combined with benefits for fish welfare (Baßmann et al. 2017). Inside the process water, large variations in micronutrients and macronutrients may occur with negative effects on plant nutritional needs (Palm et al. 2019). A general analysis of coupled aquaponic systems has shown that there are low nutrient levels within the systems (Bittsanszky et al. 2016) in comparison with hydroponic nutrient solutions (Edaroyati et al. 2017). Plants do not tolerate an under or oversupply of nutrients without effects on growth and quality, and the daily feed input of the aquaponic system needs to be adjusted to the plant's nutrient needs. This can be achieved by regulating the stocking density of the fish as well as altering the fish feed. Somerville et al. (2014) categorized plants in aquaponics according to their nutrient requirements as follows:

1. Plants with low nutrient requirements (e.g. basil, *Ocimum basilicum*)
2. Plants with medium nutritional requirements (e.g. cauliflower, *Brassica oleracea var. Botrytis*)
3. Plants with high nutrient requirements such as fruiting species (e.g. strawberries, *Fragaria* spec.).

Not all plants can be cultured in all hydroponic subsystems with the same yield. The plant choice depends on the hydroponic subsystem if conventional soilless aquaponic systems (e.g. DWC, NFT, ebb and flow; aquaponics sensu stricto' – *s.s.* – in the narrow sense) are used. Under aquaponics farming ('aquaponics sensu lato' – *s.l.* – in a broader sense, Palm et al. 2018), the use of inert soil or with addition of fertilizer applies gardening techniques from horticulture, increasing the possible range of species.

Under hydroponic conditions, the component structures of the subsystems have a decisive influence on plant growth parameters. According to Love et al. (2015), most aquaponic producers used raft and media bed systems and to a smaller amount NFT and vertical towers. Lennard and Leonard (2006) studied the growth of Green oak lettuce (*Lactuca sativa*) and recorded the relationship Gravel bed > Floating raft > NFT in terms of biomass development and yield in combination with the Murray Cod (*Maccullochella peelii peelii*) in Australia. Knaus & Palm (2016–2017, unpublished data) have tested different hydroponic subsystems such as NFT, floating raft and gravel substrate on the growth of different plants in the FishGlassHouse in a decoupled aquaponic experimental design, requiring subsequent testing under coupled conditions. With increasing production density of African catfish (*C. gariepinus*, approx. 20–168 kg/m^3), most of the cultured crops such as cucumbers (*Cucumis sativus*), basil (*Ocimum basilicum*) and pak choi (*Brassica rapa chinensis*) tended to grow better, in contrast to Lennard and Leonard (2006), in gravel and NFT aquaponics (GRAVEL > NFT > RAFT; Wermter 2016; Pribbernow 2016; Lorenzen 2017), and Moroccan mint 'spearmint' (*Mentha spicata*) showed the opposite growth performance (RAFT = NFT > GRAVEL) with highest leaf numbers in NFT (Zimmermann 2017). This demonstrates an advantage of gravel conditions and can be used figuratively also in conventional plant pots with soil substrate under coupled aquaponic conditions. This type of aquaponics was designated as 'horticulture – aquaponics (*s.l.*)' due to the use of substrates from the horticultural sector (soil, coco fibre, peat, etc.) (see Palm et al. 2018). This involves all plant cultivation techniques that allow plants to grow in pots, whereby the substrate in the pot itself may be considered equivalent to a classical gravel substrate for aquaponics. Research by Knaus & Palm (unpublished data) showed variance in the quality of commonly grown vegetables and thus their suitability for growing in this type of aquaponics with soil (Fig. 7.11, Table 7.1). In this type of aquaponics, beans, lambs lettuce and radish did well.

The plant choice (species and strain) and especially the hydroponic subsystem and/or substrate, including peat, peat substitutes, coco fibre, composts, clay, etc. or a mix of them (see Somerville et al. 2014), has a significant impact on the economic success of the venture. The efficiency of some substrates must be tested in media bed hydroponic sub-units (e.g. the use of sand (McMurtry et al. 1990, 1997), gravel (Lennard and Leonard 2004) and perlite (Tyson et al. 2008). The use of other media bed substrates such as volcanic gravels or rock (tuff/tufa), limestone gravel, river bed gravel, pumice stone, recycled plastics, organic substrates such as coconut fibre, sawdust, peat moss and rice trunk have been described by Somerville et al. (2014). Qualitative comparative studies with recommendations, however, are very rare and subject of future research.

7.7.4 Polyponics

The combination of different aquatic organisms in a single aquaponic system can increase total yields. First applied by Naegel (1977), this multispecies production principle was coined from the term polyculture combined with aquaponics in

Fig. 7.11 Experiments with a variety of commonly grown vegetables, under winter conditions in winter 2016/2017 in the FishGlassHouse (University of Rostock, Germany)

Table 7.1 Recommendation for the use of gardening plants in aquaponic farming with the use of 50% of the regular fertilizer in pots with soil

Name	Lat. Name	Possible for aquaponics	Mark	Nutrient regime
Beans	*Phaseolus vulgaris*	Yes	1	Extensive
Peas	*Pisum sativum*	No	2	Intensive
Beet	*Beta vulgaris*	No	2	Both
Tomatoes	*Solanum lycopersicum*	No	2.3	Both
Lamb's lettuce	*Valerianella locusta*	Yes	1	Both
Radish	*Raphanus sativus*	Yes	1	Both
Wheat	*Triticum aestivum*	No	2	Both
Lettuce	*Lactuca sativa*	Yes	1	Intensive

coupled systems as 'polyponic' (polyculture + aquaponics) by Knaus and Palm (2017b). Like IMTA (integrated multitrophic aquaculture), polyponics expands the diversity of the production systems. Using multiple species in one system has both advantages and disadvantages as (a) diversification allows the producer to respond to local market demands but (b) on the other hand, focus is spread across a number of products, which requires greater skill and better management. Published information on polyponics is scarce. However, Sace and Fitzsimmons (2013) reported better plant growth of lettuce, Chinese cabbage and pakchoi in polyculture with freshwater shrimp (*Macrobrachium rosenbergii*) and Nile tilapia (*O. niloticus*) in coupled aquaponics. Alberts-Hubatsch et al. (2017) described the cultivation of noble crayfish (*Astacus astacus*), hybrid striped bass (*Morone saxatilis* x *M. chrysops*), microalgae (*Nannochloropsis limnetica*) and watercress (*Nasturtium officinale*), where crayfish growth was higher than expected, feeding on watercress roots, fish faeces and a pikeperch-designed diet.

Initial investigations at the University of Rostock showed differences in plant growth in two identical 25m^2 backyard-coupled aquaponic units with the production of African catfish (*Clarias gariepinus*) and Nile tilapia (*Oreochromis niloticus*, Palm et al. 2014b). The plant yields of lettuce (*Lactuca sativa*) and cucumber fruits (*Cucumis sativus*) were significantly better in combination with *O. niloticus*. This effect was also seen by Knaus and Palm (2017a) with a 2.5-fold higher yield in basil (*Ocimum basilicum*) and two times more biomass of parsley (*Petroselinum crispum*) combined with *O. niloticus*. Another comparison between *O. niloticus* and common carp (*Cyprinus carpio*) showed a twofold higher gross biomass per plant (g plant^{-1}) of tomatoes (*Solanum lycopersicum*) with tilapia and a slightly increased gross biomass of cucumbers (*Cucumis sativus*) with carp, however, with higher cucumber fruit weight in the *O. niloticus* aquaponic unit (Knaus and Palm 2017b). The yield of mint (*Mentha* x *piperita*) was approximately 1.8 times higher in the tilapia unit, but parsley was 2.4 times higher combined with the carp (Knaus et al. 2018a). The results of these experiments followed the order of plant growth: *O. niloticus* > *C. carpio* > *C. gariepinus*, whilst fish growth showed a reverse order with: *C. gariepinus* > *O. niloticus* > *C. carpio*.

According to these results, the fish choice influences the plant yield and a combination of different fish species and their respective growth performance allows adjustment of a coupled aquaponics to optimal fish and plant yields. During consecutive experiments (*O. niloticus* only, *C. gariepinus* only), a higher basil (*O. basilicum*) biomass yield of 20.44% (Plant Growth Difference – PGD) was observed for *O. niloticus* in contrast to the basil yield with *C. gariepinus* (Knaus et al. 2018b). Thus, *O. niloticus* can be used to increase the plant yield in a general *C. gariepinus* system. This so-called boost effect by tilapia enhances the overall system production output and compensates *i*) poorer plant growth with high fish growth of *C. gariepinus* as well as *ii*) poorer fish growth in *O. niloticus* with a boost to the plant yield. A first commercial polyponic farm has opened in Bali, Indonesia, producing tilapia combined with Asian catfish (*Clarias batrachus*) and conventional farm products.

7.8 System Planning and Management Issues

Coupled aquaponics depends on the nutrients that are provided from the fish units, either a commercial intensive RAS or tanks stocked under extensive conditions in smaller operations. The fish density in the latter is often about 15–20 kg/m^3 (tilapia, carp), but extensive African catfish production can be higher up to 50 kg/m^3. Such different stocking densities have a significant influence on nutrient fluxes and nutrient availability for the plants, the requirement of water quality control and adjustment as well as appropriate management practices.

The process water quality with respect to nutrient concentrations is primarily dependent on the composition of the feed and the respective turnover rates of the fish. The difference between feed input and feed nutrients, assimilating inside the fish or lost through maintenance of the system, equals the maximum potential of

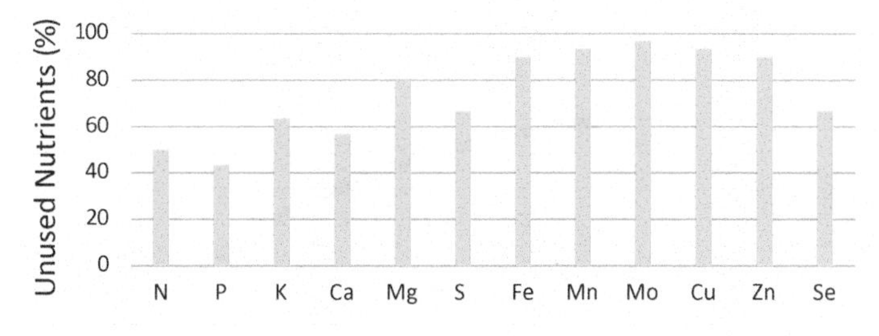

Fig. 7.12 Unused nutrients in African catfish aquaculture that are potentially available for aquaponic plant production (original data)

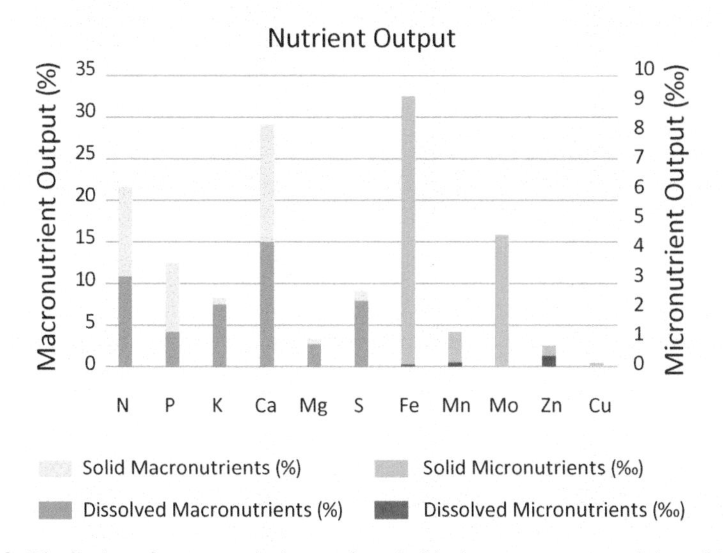

Fig. 7.13 Distribution of macro- and micronutrients inside the process water and the solids. (Data from Strauch et al. (2018))

plant available nutrients from aquaculture. As noted above, the nutrient concentrations should be adjusted to levels, which allow the plants to grow effectively. However, not all fish species are able to withstand such conditions. Consequently, resilient fish species such as the African catfish, tilapia or carp are preferred aquaponic candidates. At the University of Rostock, whole catfish and its standard diet as output and input values were analysed to identify the turnover rates of the macronutrients N, P, K, Ca, Mg and S and the micronutrients Fe, Mn, Mo, Cu, Zn and Se. With the exception of P, more than 50% of the feed nutrients given to the fish are not retained in its body and can be considered potentially available as plant nutrients (Strauch et al. 2018; Fig. 7.12). However, these nutrients are not equally distributed inside the process water and the sediments. Especially macronutrients (N, P, K) accumulate in the process water as well as inside the solid fraction whilst the micronutrients, such as iron, disappear in the solid fraction separated by the clarifier. Figure 7.13 shows the nutrient output per clarifier cleaning after 6 days of

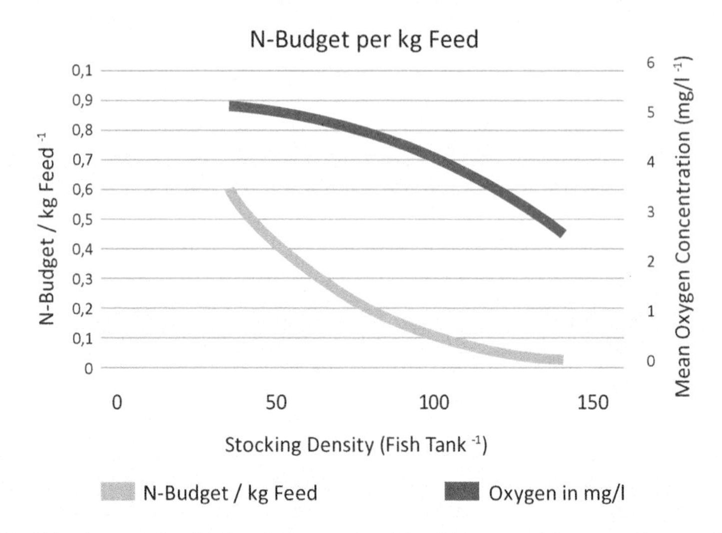

Fig. 7.14 N-budget per kg feed and oxygen level in African catfish aquaculture under three different stocking densities (original data)

sludge collection in an intensive African catfish RAS. The proportions of plant essential nutrients that are bound in the solids relative to the respective amounts that appear dissolved are significant: N = 48%, P = 61%, K = 10%, Ca = 48%, Mg = 16%, S = 11%, Fe = 99%, Mn = 86%, Mo = 100%, Zn = 48% and Cu = 55%.

One key management factor is the availability of oxygen inside the system, which is crucial to keep the concentration of plant available nitrate in the process water high. Conventional clarifiers that are applied in many RAS remove carbon-rich solid wastes from the recirculation but will leave them in contact with the process water until the next cleaning interval of the sedimentation tank. During this time, the carbon-rich organic matter is utilized as a source of energy by denitrifying bacteria, accounting for significant losses of nitrate. It outgasses as nitrogen into the atmosphere and is lost. Under intensive production conditions, large quantities of organic sludge will accumulate inside the sedimentation tanks, with consequences for maintenance, replacement with freshwater and subsequently for the nutrient composition inside the process water. Figure 7.14 illustrates the nutrient concentrations in the holding tanks of African catfish RAS under three different stocking densities (extensive: 35 fish / tank, semi-intensive: 70 fish / tank, intensive: 140 fish/ tank). The higher the stocking density and the lower the resulting oxygen content inside the system, the lower is the plant available nitrate per kg feed inside the system.

In general, with increasing fish intensity, the availability of oxygen inside the system decreases because of the consumption of the oxygen by the fish and aerobic sludge digestion inside the clarifier and the hydroponic subsystems. Oxygen levels can be maintained at higher levels, but this requires additional investment for oxygen monitoring and control. This issue is of tremendous importance for coupled aquaponics, right from the beginning of the planning phase of the systems because the different scenarios are decisive for the planned fish production, the resulting quality of the process water for the plant production units, and consequently for economic sustainability. Four principals of coupled aquaponic production systems with management consequences in terms of system design, maintenance procedures and nutrient availability for plant growth, with transitions between them, can be defined as follows:

- Extensive production, oxygen resilient fish (e.g. tilapia, carp), no oxygen control, O_2 above 6 mg/L, little water use with high nutrient concentrations, small investment, low BOD, high nitrate per kg feed.
- Intensive production, oxygen resilient fish (e.g. African catfish), no oxygen control, O_2 below 6 mg/L, high water use, medium investment, high BOD, low nitrate per kg feed, high nutrient concentrations.
- Extensive production, oxygen demanding fish (e.g. Trout), oxygen control, O_2 above 6–8 mg/L, high water use, medium investment, low BOD, high nitrate per kg feed, low nutrient concentrations.
- Intensive production, oxygen demanding fish (e.g. Trout, pikeperch), oxygen control, O_2 above 6–8 mg/L, high water use, high investment, low BOD, medium nitrate per kg feed.

In addition to the stocking density and the average amount of oxygen inside the system, the plant production regime, i.e. batch or staggered cultivation, has consequences for the plant available nutrients inside the process water (Palm et al. 2019). This is the case especially with fast growing fish, where the feed increase during the production cycle can be so rapid that there needs to be a higher water exchange rate and thus nutrient dilution can increase, with consequences for the nutrient composition and management.

The same oxic or anoxic processes that occur in the RAS as a part of the coupled aquaponic system also occur inside the hydroponic subsystems. Therefore, oxygen availability and possibly aeration of the plant water can be crucial in order to optimize the water quality for good plant growth. The oxygen allows the heterotrophic bacteria to convert organic bound nutrients to the dissolved phase (i.e. protein nitrogen into ammonia) and the nitrifying bacteria to convert the ammonia into nitrate. The availability of oxygen in the water also reduces anoxic microbial metabolism (i.e. nitrate- and/or sulphate-reducing bacteria, Comeau 2008), processes which can have tremendous effects on the reduction of nutrient concentrations. The aeration of the roots also has the advantage that water and nutrients are transported to the root surface, and that particles that settle on the root surface are removed (Somerville et al. 2014).

7.9 Some Advantages and Disadvantages of Coupled Aquaponics

The following discussion reveals a number of key pros and challenges of coupled aquaponics as follows:

Pro: Coupled aquaponic systems have many food production benefits, especially saving resources under different production scales and over a wide range of geographical regions. The main purpose of this production principle is the most efficient and sustainable use of scarce resources such as feed, water, phosphorous as a limited plant nutrient and energy. Whilst, aquaculture and hydroponics (as stand-alone), in comparison to aquaponics are more competitive, coupled aquaponics may have the edge in terms of sustainability and thus a justification of these systems especially when seen in the context of, for example, climate change, diminishing resources, scenarios that might change our vision of sustainable agriculture in future.

Pro: Small-scale and backyard-coupled aquaponics are meant to support local and community-based food production by households and farmers. They are not able to stem high investment costs and require simple and efficient technologies. This applies for tested fish and plant combinations in coupled aquaponics.

Pro: The plants in contemporary coupled aquaponics have the similar role in treating waste as constructed wetlands do in the removal of waste from water (Fig. 7.15). The plants in the hydroponic unit in coupled aquaponics therefore fulfil the task of purifying the water and can be considered a 'biological advanced unit of water purification' in order to reduce the environmental impact of aquaculture.

Challenge: It has been widely accepted that using only fish feed as the input for plant nutrition is often qualitatively and quantitatively insufficient in comparison to conventional agriculture production systems (e.g. N-P-K hydroponics manure) (Goddek et al. 2016), limiting the growth of certain crops in coupled aquaponics.

Pro: Coupled aquaponic systems have a positive influence on fish welfare. Most recent studies demonstrate that in combination with cucumber and basil, the agonistic behaviour of African catfish (*C. gariepinus*) was reduced (Baßmann et al. 2017, 2018). More importantly, comparing injuries and behavioural patterns with the control, aquaponics with high basil density influenced African catfish even more positively. Plants release substances into the process water like phosphatases (Tarafdar and Claassen 1988; Tarafdar et al. 2001) that are able to hydrolyse biochemical phosphate compounds around the root area and exude organic acids (Bais et al. 2004). Additionally, microorganisms on the root surfaces play an important role through the excretion of organic substances increasing the solubilization of minerals making them available for plant nutrition. It is evident that the environment of the rhizosphere, the 'root exudate', consists of many organic compounds such as organic acid anions, phytosiderophores, sugars, vitamins, amino acids, purines, nucleosides, inorganic ions, gaseous molecules, enzymes and root border cells (Dakora and Phillips

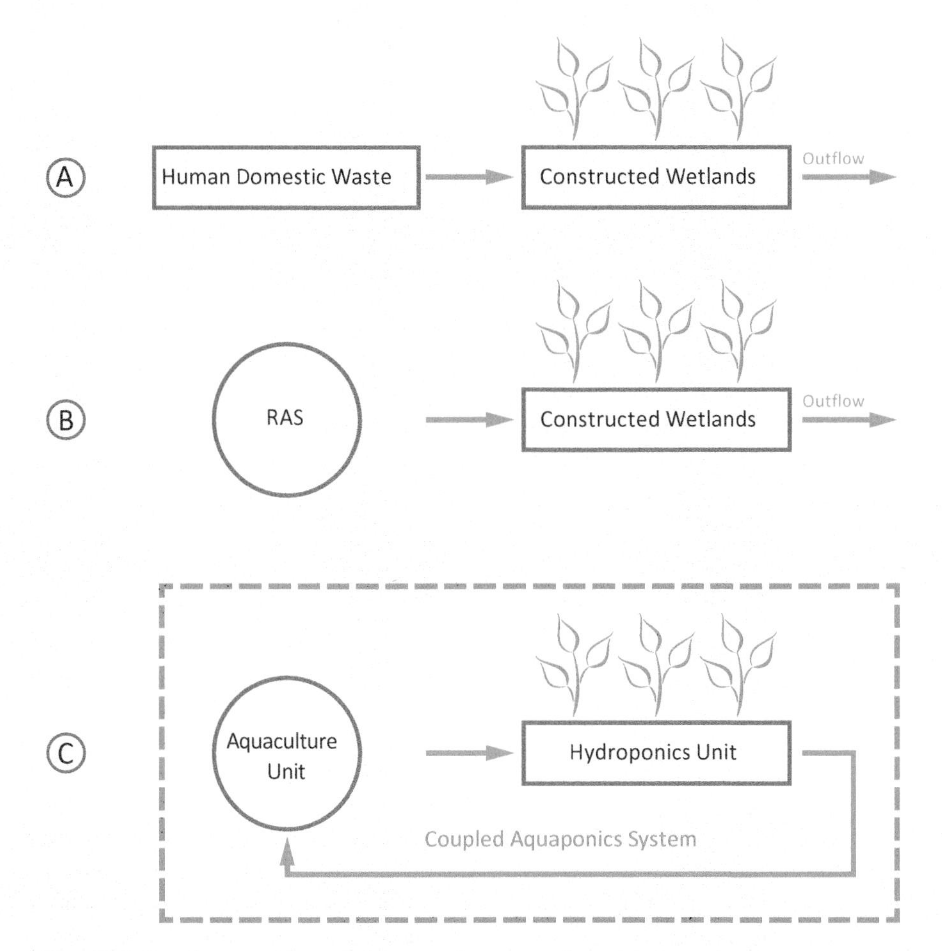

Fig. 7.15 Development of coupled aquaponic systems from (**a**) domestic waste constructed wetlands (CW) and (**b**) CW in combination with recirculating aquaculture systems (RAS) to (**c**) hydroponic units in coupled aquaponic systems

2002), which may influence the health of aquatic organisms in coupled aquaponic systems. This symbiotic relationship is not available in either pure aquaculture or decoupled aquaponics. However, considerable research still needs to be undertaken to understand the responsible factors for better fish welfare.

Pro: Aquaponics can be considered as an optimized form of the conventional agricultural production especially in those areas where production factors caused by the environmental conditions are particularly challenging, e.g. in deserts or highly populated urban areas (cities). Coupled aquaponic systems can be easily adjusted to the local conditions, in terms of system design and scale of operation.

Challenge: Coupled aquaponic also show disadvantages, due to often unsuitable component ratio conditions of the fish and plant production. In order to avoid consequences for fish welfare, coupled aquaponic systems must balance the feed input, stocking density as well as size of the water treatment units and hydroponics. So far knowledge of component ratios in coupled aquaponics is still limited, and modelling to overcome this problem is at the beginning. Rakocy (2012) suggested 57 g of feed/day per square meter of lettuce growing area and a composite ratio of 1 m^3 of fish-rearing tank to 2 m^3 of pea gravel that allows a production of 60 kg / m^3 tilapia. Based on the UVI-system, the size ratios themselves were perceived as a disadvantage since a relatively large ratio of plant growing area to fish surface area of at least 7:3 must be achieved for adequate plant production. On the other hand, system designs of coupled systems are highly variable, often not comparable, and the experiences made cannot be easily transferred to another system or location. Consequently, far more research data is needed in order to identify the best possible production ratios finally also enabling upscaling of coupled aquaponic systems through multiplying optimal designed basic modules (also see Chap. 11).

Challenge: Adverse water quality parameters have been stated to negatively affect fish health. As Yavuzcan Yildiz et al. (2017) pointed out, nutrient retention of plants should be maximized to avoid negative effects of water quality on fish welfare. It is important to select adequate fish species that can accept higher nutrient loads, such as the African catfish (*C. gariepinus*) or the Nile tilapia (*O. niloticus*,). More sensible species such as the Zander or pikeperch (*Sander lucioperca*) might be also applied in aquaponics because they prefer nutrient enriched or eutrophic water bodies with higher turbidity (Jeppesen et al. 2000; Keskinen and Marjomäki 2003; [see Sect. 7.7.1. Fish production]). So far, there is scant data allowing precise statements on fish welfare impairments. With plants generally needing high potassium concentrations between 230 and 400 mg/L inside the process water, 200–400 mg/L potassium showed no negative influence on African catfish welfare (Presas Basalo 2017). Similarly, 40 and 80 mg/L ortho-P in the rearing water had no negative impact on growth performance, feed efficiency and welfare traits of juvenile African catfish (Strauch et al. 2019).

Challenge: Another issue is the potential transmission of diseases in terms of food safety, to people through the consumption of plants that have been in contact with fish waste. In general, the occurrence of zoonoses is minor because closed aquaponics are fully controlled systems. However, germs can accumulate in the process water of the system components or in the fish gut. *Escherichia coli* and *Salmonella* spp. (zoonotic enteric bacteria) were identified as indicators of faecal contamination and microbial water quality, however, they were detected in aquaponics only in very small quantities (Munguia-Fragozo et al. 2015). Another comparison of smooth-textured leafy greens between aquaponics, hydroponics and soil-based production showed no significant differences in aerobic plate counts (APC, aerobic bacteria), Enterobacteriaceae, non-pathogenic *E. coli* and *Listeria*, suggesting a comparable contamination level with pathogens (Barnhart et al. 2015). *Listeria* spp. was most frequent (40%) in hydroponics with de-rooted plants (aquaponic plants with roots 0%, aquaponic plants without roots <10%), but not necessarily the harmful *L. monocytogenes* species. It was suggested that the source of the bacteria may be due to the lack of hygiene management, with little relevance to aquaponics as such. Another infectious bacterium, *Fusobacteria* (*Cetobacterium*), was detected by Schmautz et al. (2017) in the fish faeces with a high prevalence of up to 75%. Representatives of *Fusobacteria* are responsible for human diseases (hospital germ, abscesses, infections), reproducing in biofilms or as part of the fish intestines. Human infections with *Fusobacteria* from aquaponics have not yet been recorded but may be possible by neglecting the required hygiene protocols.

In general, there is rather little information about diseases caused by the consumption of fish and plants originating from coupled aquaponic systems. In Wilson (2005), Dr. J.E. Rakocy stated that there was no recorded human disease outbreak in 25 years of coupled aquaponic production. However, a washing procedure of the plant products should be used to reduce the number of bacteria as a precaution. A chlorine bath (100 ppm) followed by a potable water rinse was recommended by Chalmers (2004). If this methodology is used and the contact of the plants or plant products with the recirculating process water is avoided, the likelihood of contamination with human pathogenic bacteria can be strongly reduced. This is a necessary precaution not only for coupled but also for all other forms of aquaponics.

References

Alberts-Hubatsch H, Ende S, Schuhn A, von der Marwitz C, Wirtz A, Fuchs V, Henjes J, Slater M (2017) Integration of hybrid striped bass *Morone saxatilis* x *M. chrysops*, noble crayfish *Astacus Astacus*, watercress *Nasturtium officinale* and microalge *Nannochloropsis limnetica* in an experimental aquaponic system. Dubrovnik meeting abstract, EAS 2017 Oostende, Belgium

Al-Hafedh YS, Alam A, Beltagi MS (2008) Food production and water conservation in a recirculating aquaponic system in Saudi Arabia at different ratios of fish feed to plants. J World Aquacult Soc 39(4):510–520. https://doi.org/10.1111/j.1749-7345.2008.00181.x

Appelbaum S, Kotzen B (2016) Further investigations of aquaponics using brackish water resources of the Negev desert. Ecocycles Sci J Eur Ecocycles Soc 2(2):26–35. https://doi.org/10.19040/ecocycles.v2i2s.53

Bais HP, Park SW, Weir TL, Callaway RM, Vivanco JM (2004) How plants communicate using the underground information superhighway. Trends Plant Sci 9(1):26–32

Barnhart C, Hayes L, Ringle D (2015) Food safety hazards associated with smooth textured leafy greens produced in aquaponic, hydroponic, and soil-based systems with and without roots in retail. University of Minnesota Aquaponics. The University of Minnesota Aquaponics. 17 p

Baßmann B, Brenner M, Palm HW (2017) Stress and welfare of African catfish (*Clarias gariepinus* Burchell, 1822) in a coupled aquaponic system. Water 9(7):504. https://doi.org/10.3390/w9070504

Baßmann B, Harbach H, Weißbach S, Palm HW (2018) Effect of plant density in coupled aquaponics on the welfare status of African catfish (*Clarias gariepinus*). J World Aquacult Soc. (in press)

Bittsanszky A, Uzinger N, Gyulai G, Mathis A, Villarroel M, Kotzen B, Komives T (2016) Nutrient supply of plants in aquaponic systems. Ecocycles Sci J Eur Ecocycles Soc 2(2):17–20. https://doi.org/10.19040/ecocycles.v2i2.57

Blidariu F, Drasovean A, Grozea A (2013a) Evaluation of phosphorus level in green lettuce conventional grown under natural conditions and aquaponic system. Bull Univ Agric Sci Vet Med Cluj-Napoca Anim Sci Biotechnol:128–135
Blidariu F, Alexandru D, Adrian G, Isidora R, Dacian L (2013b) Evolution of nitrate level in green lettuce conventional grown under natural conditions and aquaponic system. Sci Pap Anim Sci Biotechnol 46(1):244–250
Chalmers GA (2004) Aquaponics and food safety. Alberta April, Lethbridge, 77 p
Comeau Y (2008) Microbial metabolism. In: Biological wastewater treatment: principles, modeling and design, Cap, 2. IWA/Cambridge University Press, London, pp 9–32
Dakora FD, Phillips DA (2002) Root exudates as mediators of mineral acquisition in low-nutrient environments. In: Food security in nutrient-stressed environments: exploiting plants' genetic capabilities. Springer, Dordrecht, pp 201–213
Danaher JJ, Shultz RC, Rakocy JE, Bailey DS (2013) Alternative solids removal for warm water recirculating raft aquaponic systems. J World Aquacult Soc 44(3):374–383. https://doi.org/10.1111/jwas.12040
Davy AJ, Bishop GF, Costa CSB (2001) Salicornia L. (*Salicornia pusilla* J. Woods, *S. ramosissima* J. Woods, *S. europaea* L., *S. obscura* PW ball & Tutin, *S. nitens* PW ball & Tutin, *S. fragilis* PW ball & Tutin and *S. dolichostachya* moss). J Ecol 89(4):681–707
Delaide B, Goddek S, Gott J, Soyeurt H, Jijakli MH (2016) Lettuce (*Lactuca sativa* L. var. Sucrine) growth performance in complemented aquaponic solution outperforms hydroponics. Water 8 (10):467
Delaide B, Delhaye G, Dermience M, Gott J, Soyeurt H, Jijakli MH (2017) Plant and fish production performance, nutrient mass balances, energy and water use of the PAFF box, a small-scale aquaponic system. Aquac Eng 78:130–139
Diem TNT, Konnerup D, Brix H (2017) Effects of recirculation rates on water quality and *Oreochromis niloticus* growth in aquaponic systems. Aquac Eng 78:95–104. https://doi.org/10.1016/j.aquaeng.2017.05.002
Edaroyati MP, Aishah HS, Al-Tawaha AM (2017) Requirements for inserting intercropping in aquaponics system for sustainability in agricultural production system. Agron Res 15 (5):2048–2067. https://doi.org/10.15159/AR.17.070
Endut A, Jusoh A, Ali N, Wan Nik WNS, Hassan A (2009) Effect of flow rate on water quality parameters and plant growth of water spinach (*Ipomoea aquatica*) in an aquaponic recirculating system. Desalin Water Treat 5:19–28. https://doi.org/10.5004/dwt.2009.559
FAO (2017) In: Subasinghe R (ed) World aquaculture 2015: a brief overview, FAO fisheries and aquaculture circular no. 1140. FAO, Rome
Geelen C (2016) Dynamic model of an INAPRO demonstration aquaponic system. Thesis biobased chemistry and technology. Wageningen University Agrotechnology and Food Sciences. 50 p
Goddek S, Vermeulen T (2018) Comparison of Lactuca sativa growth performance in conventional and RAS-based hydroponic systems. Aquac Int:1–10. https://doi.org/10.1007/s10499-018-0293-8
Goddek S, Espinal C, Delaide B, Jijakli M, Schmautz Z, Wuertz S, Keesman K (2016) Navigating towards decoupled aquaponic systems: a system dynamics design approach. Water 8:303. https://doi.org/10.3390/w8070303
Graber A, Junge R (2009) Aquaponic systems: nutrient recycling from fish wastewater by vegetable production. Desalination 246(1–3):147–156
Graber A, Antenen N, Junge R (2014) The multifunctional aquaponic system at ZHAW used as research and training lab. In: Conference VIVUS: transmission of innovations, knowledge and practical experience into everyday practice. Strahinj: Biotehniški center, Naklo. http://www.bc-naklo.si/uploads/media/29-Graber-Antenen-Junge-Z.pdf
Gunning D (2016) Cultivating *Salicornia europaea* (Marsh Samphire), Irish Sea Fisheries Board, http://www.bim.ie/media/bim/content/news,and,events/BIM,Cultivating,,Salicornia,europaea,-,Marsh,Samphire.pdf

Heuvelink E (2018) Tomatoes. In: Heuvelink E (ed) Crop production science in horticulture. Wageningen University/CABI, Wallingford, 388 p

Hortidaily (2015) Dutch start third trial on tomatoes with 100% LED-lighting: 100,6 Kg/m2 at the Improvement Centre in Bleiswijk. What's next? https://www.hortidaily.com/article/6022598/kg-m-at-the-improvement-centre-in-bleiswijk-what-s-next/

Hussain T, Verma AK, Tiwari VK, Prakash C, Rathore G, Shete AP, Nuwansi KKT (2014) Optimizing koi carp, *Cyprinus carpio* var. koi (Linnaeus, 1758), stocking density and nutrient recycling with spinach in an aquaponic system. J World Aquacult Soc 45(6):652–661. https://doi.org/10.1111/jwas.12159

Hussain T, Verma AK, Tiwari VK, Prakash C, Rathore G, Shete AP, Saharan N (2015) Effect of water flow rates on growth of *Cyprinus carpio* var. koi (*Cyprinus carpio* L., 1758) and spinach plant in aquaponic system. Aquac Int 23(1):369–384. https://doi.org/10.1007/s10499-014-9821-3

Jeppesen E, Jensen JP, SØndergaard M, Lauridsen T, Landkildehus F (2000) Trophic structure, species richness and biodiversity in Danish lakes: changes along a phosphorus gradient. Freshw Biol 45(2):201–218

Keskinen T, Marjomäki TJ (2003) Growth of pikeperch in relation to lake characteristics: total phosphorus, water colour, lake area and depth. J Fish Biol 63(5):1274–1282

Khandaker M, Kotzen B (2018) The potential for combining living wall and vertical farming systems with aquaponics with special emphasis on substrates. Aquac Res 49(4):1454–1468

Knaus U, Palm HW (2017a) Effects of fish biology on ebb and flow aquaponical cultured herbs in northern Germany (Mecklenburg Western Pomerania). Aquaculture 466:51–63. https://doi.org/10.1016/j.aquaculture.2016.09.025

Knaus U, Palm HW (2017b) Effects of the fish species choice on vegetables in aquaponics under spring-summer conditions in northern Germany (Mecklenburg Western Pomerania). Aquaculture 473:62–73. https://doi.org/10.1016/j.aquaculture.2017.01.020

Knaus U, Segade Á, Robaina L (2015) Training School 1 – Aquaponic trials: improving water quality and plant production through fish management and diet. 25–29 May 2015, Universidad de Las Palmas de Gran Canaria, Spain. Cost Action FA1305, The EU Aquaponics Hub – Realising Sustainable Integrated Fish and Vegetable Production for the EU

Knaus U, Appelbaum S, Palm HW (2018a) Significant factors affecting the economic sustainability of closed backyard aquaponics systems. Part IV: autumn herbs and polyponics. AACL Bioflux 11(6):1760–1775

Knaus U, Appelbaum S, Castro C, Sireeni J, Palm HW (2018b) Growth performance of basil in a small-scale aquaponic system with the production of tilapia (*Oreochromis niloticus*) and African catfish (*Clarias gariepinus*). (In preparation)

König B, Junge R, Bittsanszky A, Villarroel M, Komives T (2016) On the sustainability of aquaponics. Ecocycles Sci J Eur Ecocycles Soc 2(1):26–32. https://doi.org/10.19040/ecocycles.v2i1.50

Kotzen B, Appelbaum S (2010) An investigation of aquaponics using brackish water resources in the Negev Desert. J Appl Aquac 22(4):297–320. https://doi.org/10.1080/10454438.2010.527571

Lehmonen R, Sireeni J (2017) Comparison of plant growth and quality in hydroponic and aquaponic systems. Bachelor's thesis, University of Jyväskylä, Jyväskylä, Finland. 27 p

Lennard WA, Leonard BV (2004) A comparison of reciprocating flow versus constant flow in an integrated, gravel bed, aquaponic test system. Aquac Int 12(6):539–553. https://doi.org/10.1007/s10499-004-8528-2

Lennard WA, Leonard BV (2006) A comparison of three different hydroponic sub-systems (gravel bed, floating and nutrient film technique) in an aquaponic test system. Aquac Int 14(6):539–550. https://doi.org/10.1007/s10499-006-9053-2

Lewis WM, Yopp JH, Schramm HL Jr, Brandenburg AM (1978) Use of hydroponics to maintain quality of recirculated water in a fish culture system. Trans Am Fish Soc 107(1):92–99. https://doi.org/10.1577/1548-8659(1978)107<92:UOHTMQ>2.0.CO;2

Lobillo JR, Fernández-Cabanás VM, Carmona E, Candón FJL (2014) Manejo básico y resultados preliminares de crecimiento y supervivencia de tencas (*Tinca tinca* L.) y lechugas (*Lactuca sativa* L.) en un prototipo acuapónico. ITEA 110(2):142–159. https://doi.org/10.12706/itea.2014.009

Lorena S, Cristea V, Oprea L (2008) Nutrients dynamic in an aquaponic recirculating system for sturgeon and lettuce (*Lactuca sativa*) production. Lucrări Ştiinţifice-Zootehnie şi Biotehnologii, Universitatea de Ştiinţe Agricole şi Medicină Veterinară a Banatului Timişoara 41(2):137–143

Lorenzen L (2017) Vergleich des Wachstums von chinesischem Blätterkohl (*Brassica rapa chinensis*) in drei verschiedenen Hydroponiksubsystemen unter aquaponischen Bedingungen. Agrar- und Umweltwissenschaftliche Fakultät, Lehrstuhl Aquakultur und Sea-Ranching, Universität Rostock. Masterarbeit. 63 p. [in German]

Love DC, Fry JP, Li X, Hill ES, Genello L, Semmens K, Thompson RE (2015) Commercial aquaponics production and profitability: findings from an international survey. Aquaculture 435:67–74. https://doi.org/10.1016/j.aquaculture.2014.09.023

Loyacano HA, Grosvenor RB (1973) Effects of Chinese waterchestnut in floating rafts on production of channel catfish in plastic pools. Proc Annu Conf Southeast Assoc Game Fish Comm 27:471–473

Mariscal-Lagarda MM, Páez-Osuna F, Esquer-Méndez JL, Guerrero-Monroy I, del Vivar AR, Félix-Gastelum R (2012) Integrated culture of white shrimp (*Litopenaeus vannamei*) and tomato (*Lycopersicon esculentum* mill) with low salinity groundwater: management and production. Aquaculture 366:76–84. https://doi.org/10.1016/j.aquaculture.2012.09.003

McMurtry MR, Nelson PV, Sanders DC, Hodges L (1990) Sand culture of vegetables using recirculated aquacultural effluents. Appl Agric Res 5(4):280–284

McMurtry MR, Sanders DC, Patterson RP, Nash A (1993) Yield of tomato irrigated with recirculating aquacultural water. J Prod Agric 6(3):428–432. https://doi.org/10.2134/jpa1993.0428

McMurtry MR, Sanders DC, Cure JD, Hodson RG, Haning BC, Amand PCS (1997) Efficiency of water use of an integrated fish/vegetable co-culture system. J World Aquacult Soc 28 (4):420–428. https://doi.org/10.1111/j.1749-7345.1997.tb00290.x

Munguia-Fragozo P, Alatorre-Jacome O, Rico-Garcia E, Torres-Pacheco I, Cruz-Hernandez A, Ocampo-Velazquez RV, Garcia-Trejo JF, Guevara-Gonzalez RG (2015) Perspective for aquaponic systems: “omic” technologies for microbial community analysis. BioMed Res Int 2015: 480386. Hindawi Publishing Corporation. BioMed Research International. 2015, 10 pages. https://doi.org/10.1155/2015/480386

Naegel LC (1977) Combined production of fish and plants in recirculating water. Aquaculture 10 (1):17–24. https://doi.org/10.1016/0044-8486(77)90029-1

Nelson RL, Pade JS (2007) Aquaponic equipment the clarifier. Aquaponics J 4(47):30–31

Nozzi V, Strofaldi S, Piquer IF, Di Crescenzo D, Olivotto I, Carnevali O (2016) *Amyloodinum ocellatum* in *Dicentrarchus labrax*: study of infection in salt water and freshwater aquaponics. Fish Shellfish Immunol 57:179–185. https://doi.org/10.1016/j.fsi.2016.07.036

Nuwansi KKT, Verma AK, Prakash C, Tiwari VK, Chandrakant MH, Shete AP, Prabhath GPWA (2016) Effect of water flow rate on polyculture of koi carp (*Cyprinus carpio* var. koi) and goldfish (*Carassius auratus*) with water spinach (*Ipomoea aquatica*) in recirculating aquaponic system. Aquac Int 24(1):385–393. https://doi.org/10.1007/s10499-015-9932-5

Palm HW, Seidemann R, Wehofsky S, Knaus U (2014a) Significant factors influencing the economic sustainability of closed aquaponic systems. part I: system design, chemo-physical parameters and general aspects. AACL Bioflux 7(1):20–32

Palm HW, Bissa K, Knaus U (2014b) Significant factors affecting the economic sustainability of closed aquaponic systems. part II: fish and plant growth. AACL Bioflux. 7(3):162–175

Palm HW, Nievel M, Knaus U (2015) Significant factors affecting the economic sustainability of closed aquaponic systems. part III: plant units. AACL Bioflux 8(1):89–106

Palm HW, Strauch S, Knaus U, Wasenitz B (2016) Das FischGlasHaus – eine Innovationsinitiative zur energie- und nährstoffeffizienten Produktion unterschiedlicher Fisch- und Pflanzenarten in Mecklenburg-Vorpommern (“Aquaponik in MV”). Fischerei & Fischmarkt in Mecklenburg-Vorpommern 1/2016–16:38–47 [in German]

Palm HW, Wasenitz B, Knaus U, Bischoff A, Strauch SM (2017) Two years of aquaponics research in the FishGlassHouse – lessons learned. Dubrovnik meeting abstract, Aquaculture Europe 2017, EAS Oostende Belgium

Palm HW, Knaus U, Appelbaum S, Goddek S, Strauch SM, Vermeulen T, Jijakli MH, Kotzen B (2018) Towards commercial aquaponics: a review of systems, designs, scales and nomenclature. Aquac Int 26(3):813–842

Palm HW, Knaus U, Wasenitz B, Bischoff-Lang AA, Strauch SM (2019) Proportional up scaling of African catfish (*Clarias gariepinus* Burchell, 1822) commercial recirculating aquaculture systems disproportionally affects nutrient dynamics. Aquac Int 26(3):813–842. https://doi.org/10.1016/j.aquaculture.2018.03.021

Pantanella E (2012) Nutrition and quality of aquaponic systems. Ph.D. thesis. Università degli Studi della Tuscia. Viterbo, Italy. 124 p

Presas Basalo F (2017) Does water potassium concentration in aquaponics affect the performance of African catfish *Clarias gariepinus* (Burchell, 1822)? Master's thesis, Department of Animal Sciences Aquaculture and Fisheries Group, Wageningen University and University of Rostock Department of Aquaculture and Sea-Ranching. 75 p

Pribbernow M (2016) Vergleich des Wachstums von Basilikum (*Ocimum basilicum* L.) in drei verschiedenen Hydroponik-Subsystemen unter aquaponischer Produktion. Agrar- und Umweltwissenschaftliche Fakultät, Lehrstuhl Aquakultur und Sea-Ranching, Universität Rostock. Masterarbeit. 98 p. [in German]

Rakocy JE (1989) Hydroponic lettuce production in a recirculating fish culture system. In: Island perspectives. Vol. 3. agricultural experiment station, University of the Virgin Islands. pp 5–10

Rakocy JE (2012) Chapter 14: aquaponics – integrating fish and plant culture. In: Tidwell JH (ed) Aquaculture production systems. Wiley-Blackwell, Oxford, pp 344–386

Rakocy JE, Shultz RC, Bailey DS (2000) Commercial aquaponics for the Caribbean: proceedings of the Gulf and Caribbean Fisheries Institute [Proc. Gulf Caribb. Fish. Inst.]. no. 51, pp 353–364

Rakocy J, Shultz RC, Bailey DS, Thoman ES (2003) Aquaponic production of tilapia and basil: comparing a batch and staggered cropping system. In: South Pacific soilless culture conference-SPSCC 648. pp 63–69

Rakocy JE, Bailey DS, Shultz RC, Thoman ES (2004) Update on tilapia and vegetable production in the UVI aquaponic system. In: New dimensions on farmed Tilapia: proceedings of the sixth international symposium on Tilapia in Aquaculture, held September. pp 12–16

Rakocy JE, Masser MP, Losordo TM (2006) Recirculating aquaculture tank production systems: aquaponics-integrating fish and plant culture. SRAC Publication 454:1–16

Rakocy JE, Bailey DS, Shultz RC, Danaher JJ (2011) A commercial-scale aquaponic system developed at the University of the Virgin Islands. In: Proceedings of the 9th international symposium on Tilapia in Aquaculture. pp 336–343

Ronzón-Ortega M, Hernández-Vergara MP, Pérez-Rostro CI (2012) Hydroponic and aquaponic production of sweet basil (*Ocimum basilicum*) and giant river prawn (*Macrobrachium rosenbergii*). Trop Subtrop Agroecosyst 15(2):63–71

Sace CF, Fitzsimmons KM (2013) Vegetable production in a recirculating aquaponic system using Nile tilapia (*Oreochromis niloticus*) with and without freshwater prawn (*Macrobrachium rosenbergii*). Acad J Agric Res 1(12):236–250

Saha S, Monroe A, Day MR (2016) Growth, yield, plant quality and nutrition of basil (*Ocimum basilicum* L.) under soilless agricultural systems. Ann Agric Sci 61(2):181–186

Savidov N (2005) Evaluation of aquaponics technology in Alberta, Canada. Aquaponics Journal 37:20–25

Schmautz Z, Graber A, Jaenicke S, Goesmann A, Junge R, Smits TH (2017) Microbial diversity in different compartments of an aquaponics system. Arch Microbiol 199(4):613–620

Seawright DE, Stickney RR, Walker RB (1998) Nutrient dynamics in integrated aquaculture–hydroponics systems. Aquaculture 160(3–4):215–237. https://doi.org/10.1016/S0044-8486(97)00168-3

Shete AP, Verma AK, Chadha NK, Prakash C, Peter RM, Ahmad I, Nuwansi KKT (2016) Optimization of hydraulic loading rate in aquaponic system with common carp (*Cyprinus carpio*) and mint (*Mentha arvensis*). Aquac Eng 72:53–57. https://doi.org/10.1016/j.aquaeng.2016.04.004

Simeonidou M, Paschos I, Gouva E, Kolygas M, Perdikaris C (2012) Performance of a small-scale modular aquaponic system. AACL Bioflux. 5(4):182–188

Skar SLG, Liltved H, Drengstig A, Homme JM, Kledal PR, Paulsen H, Björnsdottir R, Oddson S, Savidov N (2015) Aquaponics NOMA (Nordic Marin) – new innovations for sustainable aquaculture in the Nordic countries. Nordic Innovation Publication 2015:06, 108 p

Somerville C, Cohen M, Pantanella E, Stankus A, Lovatelli A (2014) Small-scale aquaponic food production. Integrated fish and plant farming, FAO fisheries and aquaculture technical paper no. 589. FAO, Rome. (262 pp)

Strauch SM, Wenzel LC, Bischoff A, Dellwig O, Klein J, Schüch A, Wasenitz B, Palm HW (2018) Commercial African catfish (*Clarias gariepinus*) recirculating aquaculture systems: assessment of element and energy pathways with special focus on the phosphorus cycle. Sustainability 2018 (10):1805. https://doi.org/10.3390/su10061805

Strauch SM, Bischoff AA, Bahr J, Baßmann B, Oster M, Wasenitz B, Palm HW (2019, submitted) Effects of ortho-phosphate on growth performance, welfare and product quality of juvenile African catfish (*Clarias gariepinus*). Fishes 4:3. https://doi.org/10.3390/fishes4010003

Sutton RJ, Lewis WM (1982) Further observations on a fish production system that incorporates hydroponically grown plants. Progress Fish Cult 44(1):55–59. https://doi.org/10.1577/1548-8659(1982)44[55,FOOAFP]2.0.CO;2

Tarafdar JC, Claassen N (1988) Organic phosphorus compounds as a phosphorus source for higher plants through the activity of phosphatases produced by plant roots and microorganisms. Biol Fertil Soils 5(4):308–312

Tarafdar JC, Yadav RS, Meena SC (2001) Comparative efficiency of acid phosphatase originated from plant and fungal sources. J Plant Nutr Soil Sci 164(3):279–282

Thorarinsdottir RI (2014) Implementing commercial aquaponics in Europe – first results from the Ecoinnovation project EcoPonics. Aquaculture Europe 14, October 14–17 2014, San Sebastian, Spain

Thorarinsdottir RI, Kledal PR, Skar SLG, Sustaeta F, Ragnarsdottir KV, Mankasingh U, Pantanella E, van de Ven R, Shultz RC (2015) Aquaponics guidelines. 64 p

Tran H (2015) Aquaponics-coupled and decoupled systems and the water quality needs of each. World Aquaculture 2015 – Meeting Abstract Jeju, Korea, May 27, 2015. World Aquaculture Society PO Box 397 Sorrento, LA 70778–0397 (USA)

Tyson RV, Simonne EH, Treadwell DD, White JM, Simonne A (2008) Reconciling pH for ammonia biofiltration and cucumber yield in a recirculating aquaponic system with perlite biofilters. Hortscience 43(3):719–724

USGS. The USGS Water Science School. https://water.usgs.gov/edu/gallery/global-water-volume.html

Van der Heijden PGM, Roest CWJ, Farrag F, ElWageih H, Sadek S, Hartgers EM, Nysingh SL (2014) Integrated agri-aquaculture with brackish waters in Egypt: mission report (March 9–March 17, 2014) (No. 2526). Alterra Wageningen UR

Villarroel M, Rodriguez Alvariño JM, Duran Altisent JM (2011) Aquaponics: integrating fish feeding rates and ion waste production for strawberry hydroponics. Span J Agric Res 9 (2):537–545

Waller U, Buhmann AK, Ernst A, Hanke V, Kulakowski A, Wecker B, Orellana J, Papenbrock J (2015) Integrated multi-trophic aquaculture in a zero-exchange recirculation aquaculture system for marine fish and hydroponic halophyte production. Aquac Int 23(6):1473–1489. https://doi.org/10.1007/s10499-015-9898-3

Watten BJ, Busch RL (1984) Tropical production of tilapia (*Sarotherodon aurea*) and tomatoes (*Lycopersicon esculentum*) in a small-scale recirculating water system. Aquaculture 41 (3):271–283. https://doi.org/10.1016/0044-8486(84)90290-4

Wermter L (2016) Comparison of three different hydroponic sub-systems of *Cucumis sativus* L. grown in an aquaponic system. Department of Agronomy. Faculty of Agricultural and Environmental Sciences (AUF), Professorship of Crop production and Aquaculture and Sea-Ranching. University of Rostock, Germany. Masterthesis. 40 p

Wilson G (2005) Greenhouse aquaponics proves superior to inorganic hydroponics. Aquaponics J 39(4):14–17

Yavuzcan Yildiz H, Robaina L, Pirhonen J, Mente E, Domínguez D, Parisi G (2017) Fish welfare in aquaponic systems: its relation to water quality with an emphasis on feed and faeces-a review. Water 9(1):13

Zimmermann J (2017) Vergleich des Wachstums von Marokkanischer Minze (*Mentha spicata*) in drei verschiedenen Hydroponik Subsystemen unter aquaponischer Produktion. Agrar- und Umweltwissenschaftliche Fakultät, Lehrstuhl Aquakultur und Sea-Ranching, Universität Rostock. Masterarbeit. 71 p. [in German]

Chapter 8
Decoupled Aquaponics Systems

Simon Goddek, Alyssa Joyce, Sven Wuertz, Oliver Körner, Ingo Bläser, Michael Reuter, and Karel J. Keesman

Abstract Traditional aquaponics systems were arranged in a single process loop that directs nutrient-rich water from fish to the plants and back. Given the differing specific nutrient and environmental requirements of plants and fish, such systems presented a compromise to the ideal conditions for rearing of both, thus reducing the efficiency and productivity of such coupled systems. More recently, designs that allow for decoupling of units provide for a more finely tuned regulation of the process water in each of the respective units while also allowing for better recycling of nutrients from sludge. Suspended solids from the fish (e.g. faeces and uneaten feed) need to be removed from the process water before water can be directed to plants in order to prevent clogging of hydroponic systems, a step that represents a significant loss of total nutrients, most importantly phosphorus. The reuse of sludge and mobilization of nutrients contained within that sludge present a number of engineering challenges that, if addressed creatively, can dramatically increase the efficiency and sustainability of aquaponics systems. One solution is to separate, or when there are pathogens or production problems, to isolate components of the system, thus maximizing overall control and efficiency of each component, while

S. Goddek (✉) · K. J. Keesman
Mathematical and Statistical Methods (Biometris), Wageningen University, Wageningen, The Netherlands
e-mail: simon.goddek@wur.nl; simon@goddek.nl

A. Joyce
Department of Marine Science, University of Gothenburg, Gothenburg, Sweden
e-mail: alyssa.joyce@gu.se

S. Wuertz
Department Ecophysiology and Aquaculture, Leibniz-Institute of Freshwater Biology and Inland Fisheries, Berlin, Germany
e-mail: wuertz@igb-berlin.de

O. Körner
Leibniz-Institute of Vegetable and Ornamental Crops (IGZ), Grossbeeren, Germany
e-mail: koerner@igzev.de

I. Bläser · M. Reuter
aquaponik manufaktur GmbH, Issum, Germany
e-mail: ingo.blaeser@aquaponik-manufaktur.de

S. Goddek et al. (eds.), *Aquaponics Food Production Systems*,
https://doi.org/10.1007/978-3-030-15943-6_8

reducing compromises between the conditions and species-specific requirements of each subsystem. Another potential innovation that is made possible by the decoupling of units involves introducing additional loops wherein bioreactors can be used to treat sludge. An additional distillation loop can ensure increased nutrient concentrations to the hydroponics unit while, at the same time, reducing adverse effects on fish health from high nutrient levels in the RAS unit. Several studies have documented the aerobic and anaerobic digestion performance of bioreactors for treating sludge, but the benefits of the digestate on plant growth are not well-researched. Both remineralization and distillation components consequently have a high unexplored potential to improve decoupled aquaponics systems.

Keywords Decoupled aquaponics · Multi-loop aquaponics · System dynamics · System design · Anaerobic digestion · Desalination

8.1 Introduction

As discussed in Chaps. 5 and 7, single-loop aquaponics systems are well-researched, but such systems have a suboptimal overall efficiency (Goddek et al. 2016; Goddek and Keesman 2018). As aquaponics scales up to industrial-level production, there has been emphasis on increasing the economic viability of such systems. One of the best opportunities to optimize production in terms of harvest yield can be accomplished by uncoupling the components within an aquaponics system to ensure optimal growth conditions for both fish and plants. Decoupled systems differ from coupled systems insomuch as they separate the water and nutrient loops of both the aquaculture and hydroponics unit from each another and thus provide a control of the water chemistry in both systems. Figure 8.1 provides a schematic overview of a traditional coupled system (A), a decoupled two-loop system (B), and a decoupled multi-loop system (C). However, there is considerable debate whether decoupled aquaponics systems are economically advantageous over more traditional systems, given that they require more infrastructure. In order to answer that question, it is necessary to consider different system designs in order to identify their strengths and weaknesses.

The concept of a coupled one-loop aquaponics system as shown in Fig. 8.1a can be regarded as the traditional basis of all aquaponics systems in which water recirculates freely between the aquaculture and hydroponics units, while nutrient-rich sludge is discharged. One of the key drawbacks of such systems is that it is necessary to make trade-offs in the rearing conditions of both subsystems in terms of pH, temperature, and nutrient concentrations (Table 8.1).

In contrast, decoupled or two-loop aquaponics systems separate the aquaculture and aquaponics units from each other (Fig. 8.1b). Here, the sizing of the hydroponic unit is a critical aspect, because ideally it needs to assimilate the nutrients provided by the fish unit directly or via sludge mineralization (e.g. extracting nutrients from

Fig. 8.1 The evolution of aquaponics systems. (**a**) shows a traditional one-loop aquaponics system, (**b**) a simple decoupled aquaponics system, and (**c**) a decoupled multi-loop aquaponics system. The blue font stands for water input, output, and flows and the red for waste products

A. Coupled (One-Loop) System

Input
Fingerlings
Fish Feed
Top-Up Water
Plant Material
Fertilizer (optional)
Process Water
Output
Fish
Nutrient-Rich Sludge
Crops
Biowaste
Water (Evapotranspiration)

B. Decoupled (Two-Loop) System

Input
Fingerlings
Fish Feed
Top-Up Water
Plant Material
Fertilizer (obligatory)
Process Water
Output
Fish
Nutrient-Rich Sludge
Water Bleed-Off
Crops
Biowaste
Water (Evapotranspiration)

C. Decoupled (Multi-Loop) System

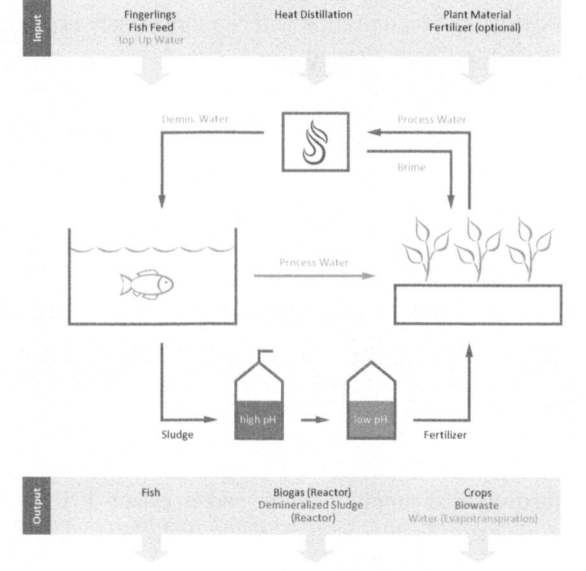

the sludge and providing it to the plants in a soluble form). Indeed, both the plant area size and environmental conditions (e.g. surface, leaf area index, relative humidity, solar radiation, etc.) determine the amount of water that can be evapotranspired and are the main factors determining the rate of RAS water replacement. The water sent from the RAS to the hydroponic unit is consequently replaced by clean water which reduces nutrient concentrations and thus improves water quality (Monsees et al. 2017a, b). The amount of water that can be replaced depends on evapotranspiration rate of plants that is controlled by net radiation, temperature, wind velocity, relative humidity, and crop species. Notably, there is a seasonal dependency, with more water evaporated in the warmer, sunnier seasons which is also when plant growth rates are highest. This approach has been suggested by Goddek et al. (2015) and Kloas et al. (2015) as an approach for improving the design of one-loop systems and better utilizing capacity to assure optimal plant growth performance. The concept has been adopted, inter alia, by *ECF* in Berlin, Germany, and the now bankrupt *UrbanFarmers* in The Hague, Netherlands.

Despite potential benefits, initial experiments with a decoupled single-loop design met with serious drawbacks. This resulted from the high amounts of additional nutrients that were needed to be added to the hydroponic loop given that the process water flowing from the RAS to the hydroponics loop is purely evapotranspiration dependent (Goddek et al. 2016; Kloas et al. 2015; Reyes Lastiri et al. 2016). Nutrients also tended to accumulate in the RAS systems when evapotranspiration rates were lower, and could reach critical levels, thus requiring periodic bleeding off of water (Goddek 2017).

Overcoming these drawbacks required the implementation of additional loops to reduce the amount of waste produced in the system (Goddek and Körner 2019). Such multi-loop systems are outlined in Fig. 8.1c and enhance the two-loop approach (8.1b) with two units that will be more closely explored in the next two subchapters as well as Chaps. 10 and 11:

1. Efficient nutrient mineralization and mobilization, using a two-stage anaerobic reactor system to reduce the discharge of nutrients from the system via fish sludge
2. Thermal distillation/desalination technology to concentrate the nutrient solution in the hydroponics unit in order to reduce the need for additional fertilizers

Such approaches have been partly implemented by various aquaponics producers such as the Spanish company NerBreen (Fig. 8.1) (Goddek and Keesman 2018) as well as Kikaboni AgriVentures Ltd. in Nairobi, Kenya, (van Gorcum et al. 2019) (Fig. 8.2).

In terms of economic advantages (Goddek and Körner 2019; Delaide et al. 2016), optimizing growth conditions in each respective loop of decoupled aquaponics systems has inherent advantages for both plants and fish (Karimanzira et al. 2016; Kloas et al. 2015) by reducing waste discharge as well as improving nutrient recovery and supply (Goddek and Keesman 2018; Karimanzira et al. 2017; Yogev et al. 2016). In their work, Delaide et al. (2016), Goddek and Vermeulen (2018), and Woodcock (pers. Comm.) show that decoupled aquaponics systems achieve better growth performance than their respective one-loop aquaponics and hydroponics control groups. Despite this, there are various problems that still need to be resolved,

Fig. 8.2 Pictures of existing multi-loop system in (1) Spain (NerBreen) and (2) Kenya (Kikaboni AgriVentures Ltd.). Whereas the NerBreen System is located in a controlled environment, the Kikaboni System is using a semi-open foil-tunnel system

including technical issues such as system scaling, parameter optimization, and engineering choices for greenhouse technologies for different regional scenarios. In the rest of this chapter, we will focus on some of the current developments to provide an overview of ongoing challenges, as well as promising developments in the field.

8.2 Mineralization Loop

In RAS, solid and nutrient-rich sludge must be removed from the system to maintain water quality. By adding an additional sludge recycling loop, accumulating RAS wastes can be converted into dissolved nutrients for reuse by plants rather than discarded (Emerenciano et al. 2017). Within bioreactors, microorganisms can break down this sludge into bioavailable nutrients, which can subsequently be delivered to plants (Delaide et al. 2018; Goddek et al. 2018; Monsees et al. 2017a, b). Many one-loop aquaponics systems already include aerobic (Rakocy et al. 2004) and anaerobic (Yogev et al. 2016) digesters to transform nutrients that are trapped in the fish sludge and make them bioavailable for plants. However, integrating such a system into a coupled one-loop aquaponics system has several disadvantages:

1. The dilution factor for nutrient-rich effluents is much higher when discharging them to a single-loop system in relation to discharging them to the hydroponics unit only. Effectively, nutrients diluted by entering in contact with large volumes of fish rearing water.
2. Fish are unnecessarily exposed to the mineralization reactor's effluents; e.g. the effluents of anaerobic reactors can include volatile fatty acids (VFAs) and ammonia that might potentially harm the fish; such reactors also represent an additional source for potential introduction of pathogens.
3. Around 90% of the nutrients trapped in the sludge can be recovered when RAS sludge is maintained at a pH of 4 (Jung and Lovitt 2011). Such a low pH is not possible when operating bioreactors at a pH around 7 (Goddek et al. 2018), which is the usual trade-off pH value within one-loop aquaponics systems.

With respect to pH, Fig. 8.3 shows the approximate pH values of the respective process water flows in a multi-loop aquaponics system (e.g. as presented in Fig. 8.1c). Figure 8.3 also shows the impact of mineralization reactors on the performance of the system as a whole, based on the anaerobic reactors proposed by Goddek et al. (2018). Such a system represents only one possible solution for treating sludge, with alternative approaches discussed in Chap. 10. The decrease in pH of the process water flowing from the RAS subsystem into the hydroponics loop as shown in Fig. 8.3 demonstrates acidification in the nutrient concentration loop (i.e. demineralized water has a pH of 7). Thus, the effluent has a lower pH than the RAS outlet, which reduces the need to adjust the pH for optimal plant growth conditions.

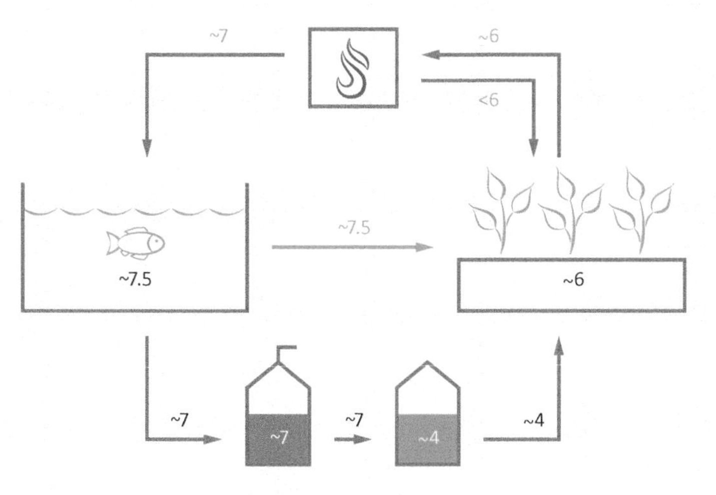

Fig. 8.3 Approximate pH of the water within the different system components as well as the process water. The '~' indicated an approximation

Table 8.1 Overview of optimal growth conditions for fish and plants and preferred operational conditions for sludge nutrient recycling treatment

Subsystem	Species/ function	pH	Temperature (°C)	Nitrate (NO_3) (mg/L)
Recirculating aquaculture system (RAS)	*Oreochromis niloticus* (Nile tilapia)	7–9 (Ross 2000)	27–30 (El-Sayed 2006)	<100–200 (Dalsgaard et al. 2013)
	Oncorhynchus mykiss (rainbow trout)	6.5–8.5 (FAO 2005)	15 (Coghlan and Ringler 2005)	<40 (Davidson et al. 2011; Schrader et al. 2013)
Hydroponics	*Lactuca sativa* (lettuce)	5.5–6.5 (Resh 2012)	21–25 (Resh 2012)	730 (Resh 2012)
	Lycopersicon esculentum (tomato)	6.3–6.5 (Resh 2002)	18–24 (Resh 2002)	666 (Sonneveld and Voogt 2009)
Anaerobic reactor	Methanogenesis	6.8–7-4 (de Lemos Chernicharo 2007)	30–35 (Alvarez and Lidén 2008; de Lemos Chernicharo 2007)	–
	Sludge mobilization	4.0 (Jung and Lovitt 2011)	n/a	–

The two-stage reactor system works as follows:

- *In the first stage* (pH around 7 to provide optimal conditions for methanogenesis; Table 8.1), the organic matter is broken down to sustain a high degree of methane production (i.e. carbon removal). Mirzoyan and Gross (2013) reported a total suspended solids reduction of around 90%, using upflow anaerobic sludge

blanket reactor technology. This has the benefit that (1) biogas is harvested as a renewable energy source and (2) fewer VFAs are produced in the second stage. The sludge retention time in the first stage should be several months, before removing the accumulated nutrients in the sludge (e.g. calcium phosphate aggregation) within the second stage.
- *In the second stage*, nutrients in suspended solids are effectively mobilized and become available for plant uptake. This mobilization is the most effective in a low-pH environment (Goddek et al. 2018; Jung and Lovitt 2011). Once the pH of acidic reactors is decreased, it usually remains stable; thus less pH regulation is required in the hydroponic unit.

The effluents that are rich in nutrients may require some post-treatment depending on the amount of measured total suspended solids and VFAs. However, it is important to keep in mind that ammonia can stimulate plant growth, e.g. leafy greens, when it accounts for 5–25% of the total nitrogen concentration (Jones 2005). However, fruit vegetables such as tomatoes or sweet peppers are particularly sensitive to ammonia in the nutrient solution. An aerobic post-effluent treatment or a well-aerated hydroponics sump would be required in systems growing those types of crops.

8.2.1 Determining Water and Nutrient Flows

For system sizing (Sect. 8.4), the amount of water flowing from the RAS system via the reactor(s) to the hydroponics unit (Q_{MIN}) needs to be known (Eq. 8.1):

$$Q_{\mathrm{MIN}}\ (\mathrm{kg/day}) = \frac{n_{\mathrm{feed}} \times k_{\mathrm{sludge}}}{\pi_{\mathrm{sludge}}} \tag{8.1}$$

where n_{feed} is the amount of fish feed in kg, k_{sludge} is the proportion coefficient of fish feed ending up as sludge, and π_{sludge} is the proportion of total solids (i.e. sludge) in the sludge water flow entering the mineralization loop.

The sludge concentration can be increased by adding a gravity separation device prior to the bioreactors, directing the 'clear' supernatant back to the RAS system. This formula can also be used to get an input for sizing the reactor based on the hydraulic retention time (Chap. 10). Between 20 and 40% of the fish feed ends up as total suspended solids in the RAS-derived sludge (Timmons and Ebeling 2013). As an example, it has been found that tilapia sludge contains around 55% of nutrients that were added to the system via feed (Neto and Ostrensky 2013; Yavuzcan Yildiz et al. 2017) which represents a valuable resource for crop growth.

The main nutrients that can be recovered via a mineralization process are N and P. As P (one of the major components of sludge) is the most valuable macronutrient in terms of cost and availability for crop production, it should be the first element to be optimized in the aquaponic system.

The mineralization rate of the mineralization loop is calculated as follows:

$$\text{Mineralization}\ (\text{g/day}) = (n_{\text{feed}} \times 1000) \times \pi_{\text{feed}} \times \pi_{\text{sludge}} \times \eta_{\text{min}} \qquad (8.2)$$

where n_{feed} is the feed input to the system (in kg); $\pi_{\text{feed}}$is the proportion of the nutrient in the feed formulation;$\pi_{\text{sludge}}$is the proportion of a specific feed-derived element ending up in the sludge; and $\eta_{\text{min}}$is the mineralization and mobilization efficiency of the reactor system.

The last step would be to determine the concentration of the respective element in the effluent of the mineralization loop:

$$\text{Nutrient concentration}\ (\text{mg/L}) = \frac{\text{Mineralization} \times 1000}{Q_{\text{MIN}}} \qquad (8.3)$$

Example 8.1

Our RAS system is fed with 10 kg of fish feed per day. We assume that 25% of the fed feed ends up as sludge. In our system, we use a Radial Flow Settler (RFS) to concentrate the sludge to 1% dry matter. Consequently, the flow from the RAS to HP via the mineralization loop is calculated as follows:

$$Q_{\text{MIN}}\ (\text{kg/day}) = \frac{10\text{kg} \times 0.25}{0.01} = 250 \approx 250\text{kg/day}$$

We decide to size our system on P. The P content of our feed (in most cases provided by the feed manufacturer) is 1.5% and 55% of it ends up in the sludge (Neto and Ostrensky 2013). We assume that our reactors achieve a mineralization efficiency of 90% for this element. Therefore, the grams of P transferred to the hydroponics unit each day can be determined:

$$\text{Mineralization}\ (\text{g/day}) = (10\text{kg} \times 1000) \times 0.55 \times 0.015 \times 0.9 = 74.25$$

The concentration of the effluent is consequently:

$$\text{Nutrient concentration}\ (\text{mg/L}) = \frac{74{,}25\text{g} \times 1000}{250\text{L}} = 297\ \text{mg/L}$$

This concentration of P in the effluent in the example box above is approximately six times higher than in most hydroponics nutrient solutions. The research of Goddek et al. (2018) underpins this theoretical number, and they report that their RAS sludge contained 150 and 200 mg/L of P for two independent systems, respectively (1% TSS sludge), with a fish feed P content of 0.83% in dry matter feed for the latter (200 mg/L).

8.3 Distillation/Desalination Loop

In decoupled aquaponics systems, there is a one-way flow from the RAS to the hydroponics unit. In practice, plants take up water supplied by RAS, which in turn is topped up with fresh (i.e. tap or rain) water. The necessary outflow from the RAS unit is equal to the difference between the water leaving the HP system via plants (and via the distillation unit) and the water entering the hydroponics unit from the mineralization reactor, if the system includes a reactor (Fig. 8.4). A simplified summary is that the long-term water flux requirement from RAS to HP is equal to the crop water consumption by evapotranspiration and plant water storage in the plant biomass.

However, in terms of mass balances, the amount of nutrients leaving the hydroponics system via the plants needs to be replaced to assure a constant equilibrium. This poses a dilemma, as the maximum tolerable nutrient concentration in RAS is much lower than what is necessary in HP. The high nutrient flows ($\rho_{RAS} \times Q_{RAS}$) for HP can thus not be accomplished by the low RAS nutrient concentrations. Instead, without a distillation/desalination loop, the nutrient concentration would increase in the RAS while decreasing in the hydroponics system. A possible remedy is to discharge RAS water (and thus also nutrients) to decrease the nutrient concentration there and add fertilizer to the hydroponics nutrient solution. In terms of environmental and economic impact, this solution is less satisfying and does not serve the aim of a closed loop combined production.

The implementation of a distillation unit as shown in Fig. 8.3 represents a potential solution for this dilemma. Such distillation technologies (e.g. thermal

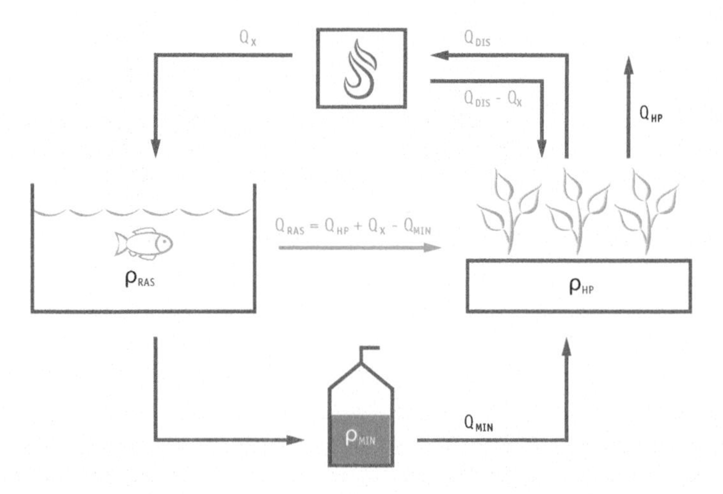

Fig. 8.4 Scheme of water fluxes and different concentrations of nutrients in a decoupled aquaponics system, where Q, flow volume in L; ρ, nutrient concentration in mg/L; RAS, recirculating aquaculture system; MIN, mineralization reactor; DIS, distillation unit; and X, unknown/flexible flow parameter

membrane distillation) have the potential to separate dissolved salts and nutrients from water (Shahzad et al. 2017; Subramani and Jacangelo 2015). In the context of multi-loop aquaponics systems, and as an alternative to additional fertilization and water bleed-off with corresponding extra costs, this technology could not only provide fresh water to the system but also achieve desired nutrient concentrations for the respective subsystems (Goddek and Keesman 2018).

For the implementation (i.e. sizing) of such a distillation unit, simple mass balance equations can be used. The remaining system, however, must be sized beforehand (either via rules of thumb or via mass balance equations; see Sect. 8.5), because the nutrients that enter the system should be in equilibrium with the bioavailable nutrients taken up by the crop (***Note:*** the sweet spot of decoupled systems is its flexibility. Consequently, one can also oversize the hydroponics part of the system although that will necessitate the use of more fertilizer). The easiest way to estimate nutrient uptake is to use the assumption that nutrients are taken up/absorbed much the same as dissolved ions in irrigation water (i.e. no element-specific chemical, biological or physical resistances). Consequently, to maintain equilibrium, all nutrients taken up by the crop as contained in the nutrient solution need to be added back to the hydroponics system (Eq. 8.4).

$$\phi_{RAS} + \phi_{MIN} - \phi_{HP} = 0 \tag{8.4}$$

where ϕ_{RAS} is the nutrient flow from the RAS system to the hydroponics system, ϕ_{MIN} is the nutrient flow from the mineralization unit to the hydroponics system and ϕ_{HP} is the nutrient plant uptake. For this equation, it is assumed that the distillation system has an efficiency of close to 100%. Thus, Q_{DIS} goes back to the hydroponics subsystem.

Consequently:

$$(\rho_{HP} \times Q_{HP}) = (\rho_{RAS} \times Q_{RAS}) + (\rho_{MIN} \times Q_{MIN}) \tag{8.5}$$

where Q is the flow volume in L, and $\rho$is the nutrient concentration in mg/L.

As stated above, the flow from RAS to the hydroponics unit is the difference of the sum of the water flows leaving the hydroponics system (i.e. $Q_{HP} + Q_X$) and the inflow from the bioreactor (Q_{MIN}), i.e. $Q_{RAS} = Q_{HP} + Q_X - Q_{MIN}$, which leads us to the following equation:

$$\begin{aligned}(\rho_{HP} \times Q_{HP}) = (\rho_{RAS} \times Q_{HP}) + (\rho_{RAS} \times Q_X) - (\rho_{RAS} \times Q_{MIN}) \\ + (\rho_{MIN} \times Q_{MIN})\end{aligned} \tag{8.6}$$

The targeted variable is the distillation flow (Q_X) that is required to maintain the nutrient concentration equilibrium in the hydroponics system. For this, Eq. 8.6 is solved for $Q_X$in the following steps:

$$(\rho_{RAS} \times Q_X) = (\rho_{HP} \times Q_{HP}) - (\rho_{MIN} \times Q_{MIN}) - (\rho_{RAS} \times Q_{HP}) + (\rho_{RAS} \times Q_{MIN}) \tag{8.7}$$

$$Q_X = \frac{\rho_{HP} \times Q_{HP}}{\rho_{RAS}} - \frac{\rho_{MIN} \times Q_{MIN}}{\rho_{RAS}} - Q_{HP} + Q_{MIN} \tag{8.8}$$

Note that the distillation flow $Q_X$is highly dynamic and depends on the evapotranspiration rate of the plants, which is climate-dependent. The dynamic outcome, however, can be used for sizing the distillation unit. To calculate the required inflow into the distillation unit, the following formula can be used:

$$Q_{DIS} = Q_X \times \frac{100}{\eta_{DIS}} \tag{8.9}$$

where Q is the flow volume in L and η the demineralization efficiency of the used device (in %).

Distillation technology can hence drastically reduce the water and environmental (i.e. fertilizer usage) footprint of multi-loop aquaponics systems. However, aquaponics systems become even more complex when considering their implementation. Even though this additional loop might not make any sense for small-scale systems, it has the potential to take larger commercial systems to a new level. Yet, one has to consider that thermal distillation technology requires high amounts of thermal energy and might not be economically reasonable everywhere. Regions with high global solar radiation levels or geothermal energy sources might be the most suitable for this technology. The economical sustainability of such systems is consequently also location dependent.

Another point to bear in mind is the high temperature of distilled water and brine from the distillation unit. Depending on the environmental conditions and the fish species used, the hot distillation water could be used to heat up the RAS water; the brine, however, needs to cool down before re-entering the HP subsystem.

8.4 Sizing Multi-loop Systems

Sizing an aquaponics system requires balancing the nutrient input and -output. Here, we basically apply the same principle as sizing a one-loop system. Yet, this approach is a bit more complicated, but will be fully illustrated with the aid of an example.

Figure 8.5 illustrates the mass balance diagram for our system approach. In the optimal situation, the system has only one input and output. However, in practice, one will have to add additional nutrients to the hydroponics part to optimize plant growth. This model can be used to size the system, e.g. based on phosphorus, which is a non-renewable resource (Chap. 2). The input to the system (m_{feed}) is the fraction

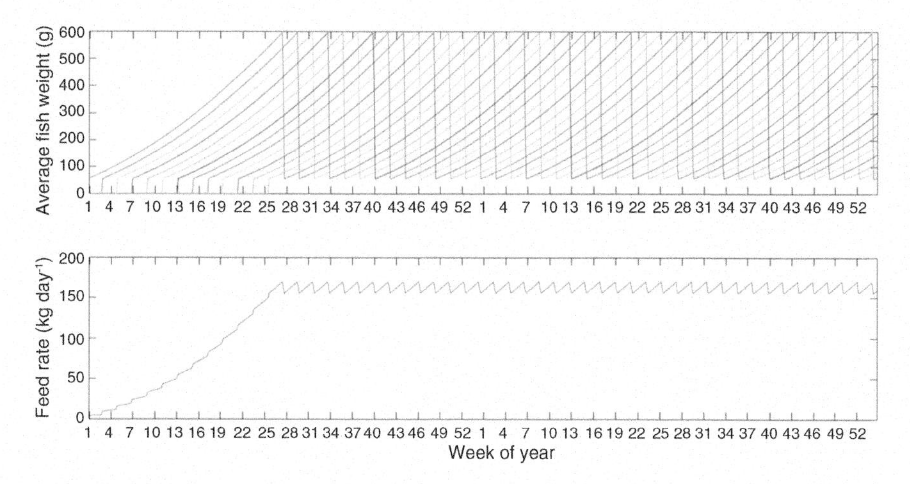

Fig. 8.6 Example of biomass balance for tilapia reared in 13 tanks in cohorts with a total volume (including biofilter and sump) of 482.000 L at a max. Total biomass of 80 t for a period of 2 years including start-up phase with average fish weight (**a**) (each line represents one tank/cohort) and the daily total feed rate (**b**) (data taken from Goddek and Körner 2019)

These are the nutrients that finally accumulate in the RAS system and can be taken up by the plants.

8.4.3 *Plant Uptake*

Table 8.3 gives an overview of the crop-specific evapotranspiration (ET_c) rates that are linked to global radiation. One mm of ET per square meter equals 1 L. For simple sizing, one should take the annual daily average (see next section).

8.4.4 *Balancing the Subsystems*

Balancing the loops is necessary for sizing the system. The input should be equal to the output (Fig. 8.5). In a decoupled aquaponics system incorporating a bioreactor unit, we have two nutrient inflow streams: (1) the fraction of the feed that is excreted to the RAS system in a soluble form and (2) the fraction of the nutrients in the fish sludge that the bioreactor(s) manage to mineralize and mobilize. The major outflow stream (apart from the periodic removal of demineralized sludge) of nutrients is the nutrient uptake of the plants. The differential Eq. 8.12. expresses this balance:

Table 8.3 Overview of outside global radiation levels subarctic, temperate maritime, and arid conditions (based on Goddek and Körner 2019) and their respective crop evapotranspiration (ET_c, *mm day-1*) rates for *lettuce* and *tomato* grown in a controlled greenhouse environment of 20°C and 80% relative humidity. Lettuce was cultivated with continuous planting year round; tomato was planted in January and removed in December (Faroe Islands and the Netherlands) or July and June (Namibia)

	Faroe Islands			The Netherlands			Namibia		
	Global radiation	ET_c/lettuce	ET_c/tomato	Global radiation	ET_c/lettuce	ET_c/tomato	Global radiation	ET_c/lettuce	ET_c/tomato
Month	mol m^{-2} day^{-1}	kg m^{-2} day^{-1}		mol m^{-2} day^{-1}	kg m^{-2} day^{-1}		mol m^{-2} day^{-1}	kg m^{-2} day^{-1}	
January	1.4	0.78	0.52	4.5	0.78	0.53	54.2	2.74	4.55
February	5.2	0.85	1.38	9.1	0.93	1.40	53.7	2.70	4.47
March	13.7	1.20	2.12	17.0	1.28	2.14	51.2	2.42	3.96
April	30.6	1.90	3.05	27.9	1.82	2.90	40.2	3.05	5.38
May	39.2	2.29	3.57	32.2	2.40	3.74	30.0	2.70	4.59
June	39.6	2.33	3.60	36.6	2.52	3.91	30.5	2.28	3.80
July	34.5	2.17	3.37	36.4	2.54	3.94	32.1	2.40	1.76
August	21.3	1.67	2.73	31.7	2.28	3.51	37.5	2.61	3.92
September	13.2	1.20	2.04	23.1	1.75	2.77	43.2	2.00	3.02
October	6.0	0.91	1.77	13.3	1.17	1.94	51.6	2.07	3.09
November	2.1	0.78	1.60	6.2	0.87	1.62	57.9	2.30	3.58
December	0.4	0.79	1.66	3.5	0.77	1.52	59.8	2.44	3.95
Average	17.3	1.41	2.28	20.6	1.59	2.50	45.6	2.47	3.83

$$\text{Mineralization (Eq.8.2)} + \underline{m_{\text{feed}}} = \frac{\underline{Q_{\text{HP}}} \times \rho_{\text{HP}}}{1000} \tag{8.12}$$

$$\left(\underline{n_{\text{feed}}} \times 1000 \times \pi_{\text{feed}} \times \pi_{\text{sludge}} \times \eta_{\text{min}}\right) + \underline{m_{\text{feed}}} = \frac{\underline{Q_{\text{HP}}} \times \rho_{\text{HP}}}{1000} \tag{8.13}$$

where $\underline{n_{\text{feed}}}$ is the average feed (in kg) entering the RAS system, $\pi_{\text{feed}}$is the proportion of the nutrient in the feed formulation, π_{sludge} is the proportion of a specific feed-derived element ending up in the sludge, and η_{min} is the mineralization and mobilization efficiency of the reactor system, $\underline{m_{\text{feed}}}$ is the average amount of a nutrient that the fish defecate in a dissolved form, $\underline{Q_{\text{HP}}}$ is the average total evapotranspiration, and $\rho_{\text{HP}}$is the target (i.e. optimal) nutrient concentration for a specific nutrient in the hydroponic subsystem.

However, to be able to determine the required area, there are two variables that need to be redefined in order to solve this equation. Equation 8.14 shows how to calculate the soluble nutrient excretion. In Eq. 8.15, we show that the average total evapotranspiration is a product of the area and the plant-specific evapotranspiration rate (here shown as an average) per m^2.

$$\underline{m_{\text{feed}}} = \underline{n_{\text{feed}}} \times \pi_{\text{feed}} \times \eta_{\text{excr}} \tag{8.14}$$

where η_{excr} represents the fraction of the nutrient excreted by the fish in a soluble form.

$$\underline{Q_{\text{HP}}} = A \times \underline{ET_{\text{c}}} \tag{8.15}$$

where $\underline{Q_{\text{HP}}}$ represents the average total evapotranspiration (in L), A the area, and $\underline{ET_{\text{c}}}$ the average crop-specific evapotranspiration in mm/m^2 (i.e. L/m^2).

Solving Eq. 8.13 by incorporating Eqs. 8.14 and 8.15 to find A, we are able to calculate the required plant area with respect to the average feed input (Eq. 8.15).

$$A = \frac{\left(\underline{n_{\text{feed}}} \times 1000 \times \pi_{\text{feed}} \times \eta_{\text{excr}} \times 1000\right) + \left(\underline{n_{\text{feed}}} \times 1000 \times \pi_{\text{feed}} \times \pi_{\text{sludge}} \times \eta_{\text{min}} \times 1000\right)}{\underline{ET_{\text{c}}} \times \rho_{\text{HP}}} \tag{8.16}$$

Example 8.2
For this example, we want to size (i.e. balance) the system with respect to P. We assume that the RAS component of our system requires an average daily feed input of 150 kg. The manufacturer reports the P content of the fish feed to be 1%. We estimate the P ending up in the sludge to be 55% and the P that fish excrete in a soluble form to be 17%. The bioreactors perform quite well and mineralize around 85% of the P.

On the output side, we calculated the average crop-specific evapotranspiration rate for lettuce (by, e.g. using the FAO Penman-Monteith equation). At our location, it is around 1.3 mm/day (i.e. 1.3 L/day). The optimal P composition of the nutrient solution is reported to be 50 mg/L (Resh 2013). Finding the area of plant cultivation needed to uptake the P produced by the system is then solved by:

$$A = \frac{\left(\underline{n_{\text{feed}}} \times 1000 \times \pi_{\text{feed}} \times \eta_{\text{excr}} \times 1000\right) + \left(\underline{n_{\text{feed}}} \times 1000 \times \pi_{\text{feed}} \times \pi_{\text{sludge}} \times \eta_{\text{min}} \times 1000\right)}{\underline{ET_{\text{c}}} \times \rho_{\text{HP}}}$$

$$A = \frac{(150000 \times 0.01 \times 0.17 \times 1000) + (150000 \times 0.01 \times 0.55 \times 0.85 \times 1000)}{1.3 \times 50}$$

$$= \frac{255000 + 701250}{65}$$

$$= 14711\text{m}^2 = 1.47\text{ha}$$

The example above shows that the majority of the P in the hydroponics unit originates from the bioreactors. Thus, implementation of a bioreactor within a decoupled system has a very high impact on P sustainability. By contrast, in order to size simple one-loop aquaponics systems, a rule of thumb is usually applied. For leafy plants approx. 40–50 g and for fruity plants approx. 50–80 g of feed is required per m^2 cultivation area (FAO 2014). When looking at the feed input in the given example above = 150 kg, and dividing it by 45 (the average of the leafy plant approximation), the proposed cultivation area is around 3750 m^2. Leaving out the sludge mineralization, our example would suggest a cultivation area of 3333 m^2 when sizing the system on P.

8.4.5 *Role of the Distillation Unit*

The role of the distillation unit is to keep the nutrient concentration of the RAS system and the hydroponics system at their respective desired levels. Since nutrient accumulation and the corresponding specific nutrient density are dynamic in RAS systems (i.e. depending on the ET_c rates) that depend on the Q_{HP} and Q_X flow (Fig. 8.5), the size of the distillation unit cannot be determined using a differential

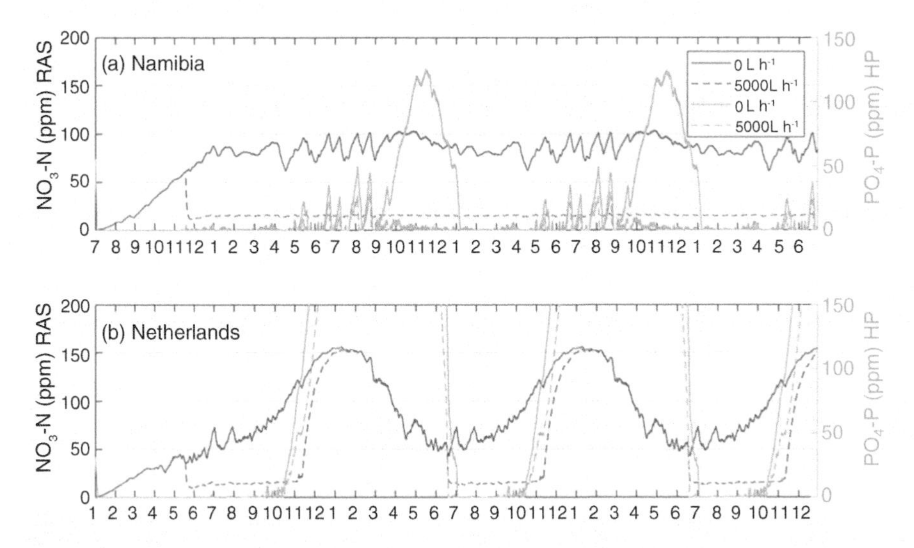

Fig. 8.7 Simulations comparing NO3-N concentration in the RAS water system on the impact of distillation flows (no, solid line; 5000 L h-1, dashed line) on hydroponics (yellow, ---) and RAS (blue, ---) nutrient solution concentrations in (**a**) Namibia and (**b**) the Netherlands, i.e. in low and high latitudes (Namibia 22.6°S and the Netherlands, 52.1°N, respectively) within a 36-month period (including the system run-up phase) using local climate data and climate-adjusted greenhouses as model input

equation. Thus, a time series model is required to determine the nutrient concentration in the RAS over time. The nutrient concentration at a specific time is necessary to be able to execute mass balance equations within the system (Sect. 8.3).

For the system to be balanced (i.e. input = output), we can give a general guideline on the required capacity of such a distillation unit. The objective is to avoid nutrient accumulation in the RAS system. Figure 8.7a, b shows the impact of distillation flows on the hydroponics and RAS nutrient solution *without a mineralization loop in two different latitudes*. Both systems have the same feed input (in average 158.6 kg day^{-1}; see Fig. 8.6). However, by taking the environmental conditions and climate-adjusted greenhouses into account, the necessary and optimal hydroponic area differs between geographical locations (see Chap. 11). Hydroponic systems with low potential evaporation rates, as are common in locations at high latitudes (i.e. far from the equator) would need larger cultivation areas than places closer to the equator. At the same time, a higher annual variation in irradiation and thus transpiration is common in these regions, thus a higher demand on seasonal variability on water and nutrients is present (see Fig. 8.7). In greenhouse cultivation, however, supplementary lighting may be necessary, and in countries such as Norway, vegetable cultivation without supplementary lighting hardly takes place. In addition, the total crop leaf surface makes a difference; crops with a high leaf area per unit ground area (i.e. leaf area index) transpire more than crops with smaller leaf areas, and a distinct difference can be seen between tomato and lettuce crops. All of these factors need to be considered when planning and sizing the aquaponic system.

In the following we provide an overview of the optimized hydroponic area size for the above described aquaponics systems: The cultivation area for monocultures simulated with scenarios in steps of 250 m^2 to find the fitting area of either lettuce or tomatoes in order to balance the system appropriately was *without supplementary lighting* (for lettuce or tomato, respectively):

- 17.000 m^2 or 11.750 m^2 for Faroe Islands
- 15.500 m^2 or 11.000 m^2 for the Netherlands
- 8750 m^2 or 6500 m^2 for Namibia

Even though the size of the systems differs, the average annual nutrient uptake is similar. However, when integrating a digester system, we have to take the additional nutrient source into consideration (Fig. 8.1c). Changing one component inevitably leads to imbalances of the system, yet the system must aim to provide optimal nutrients to both RAS and HP. For example, NO3-N in RAS must be below a certain threshold <200 mg L^{-1} for, e.g. tilapia, while PO4-P in HP should be as close as possible to the recommended concentration of 50 mg L^{-1} for good-quality plant cultivation. Thus, simulation studies help determine sizing of components in a decoupled closed multi-loop aquaponics system in order to achieve optimal nutrient supplies for both fish and plants. For that purpose, Goddek and Körner (2019) created a numeric aquaponics simulator.

However, planning an aquaponic system involves some basic system understanding in order to reach a balance that minimizes the unwanted peaks in nutrient demand and supply. Since the driving force for nutrient dynamics is the evapotranspiration of the crop (ET_c in the HP system), that is largely driven by microclimate and absorbed light. In a perfectly balanced system, this would be a fully automated and controlled (see Sect. 8.5 Monitoring and Control) environment with 24-h lighting. Plants need a certain dark period of about 4–6 h, so the best-balanced system is realistically to carry out aquaponics in closed plant factories solely with artificial light sources. This, however, demands high electrical input and investment costs and is only feasible with very high product prices. Therefore, we recommend greenhouse production with supplementary lighting (if necessary and if it pays off) as a practical and economically feasible way of building an aquaponic unit. Placing both plants and fish in the same physical construction results in additional synergies including reduced heating and increased plant growth through elevated CO_2 (Körner et al. 2017).

In addition to these technical issues, plant cultivation procedures (the practical horticultural part of the system) have to be adjusted to the needs of aquaponics such that there is a constant crop nutrient demand (assuming same climate and light) as shown in Table 8.3. Cultivation of lettuce and other leafy greens are carried out continuously (Körner et al. 2018), while larger crops, like fruit vegetables such as tomato, cucumber, or sweet pepper, are usually sown in winter, and the first harvest is often in late winter/early spring followed by removal of plants and another crop sown for harvest in winter again. Without interplanting, i.e. either various crop types in the same system or batches of fruit vegetables planted throughout the year in order to sustain nutrient demand, periods of low nutrient demand and high nutrient levels

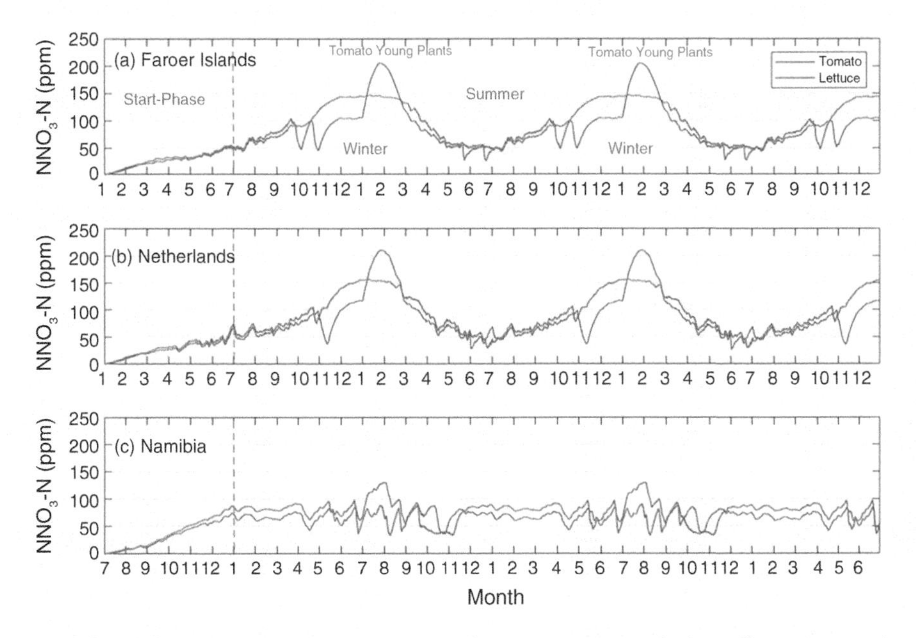

Fig. 8.8 NO3-N in RAS combined with HP growing tomato or lettuce in three climate zones and decreasing latitudes (Faroe Islands 62.0°N, the Netherlands 52.1°N, Namibia 22.6°S) with optimized area for hydroponics (see above) in a 36-month simulation using local climate data and climate-adjusted greenhouses as model input

will occur. Based on Goddek and Körner (2019), we show the variation of NO_3-N in RAS for tomato (often not adjusted in aquaponics) and lettuce when no supplementary lighting is used for three climate zones (Faroe Islands, the Netherlands, and Namibia) (Fig. 8.8). System balance is achievable by increasing the daily light integral (i.e. sum of mol light received during a 24-h period) with dynamic supplementary lighting control (Körner et al. 2006).

Applying distillation/desalination technologies can contribute to significant reductions in nutrient levels in the RAS while adjusting levels in the HP system closer to optima, i.e. the unit concentrates nutrients to levels required by plants. Figure 8.9 illustrates the effect of a desalination unit on RAS NO_3-N concentration when applying between 0 and 5000 L h^{-1} and system^{-1}. It is obvious that with increasing desalination flux, the NO_3-N concentration in the RAS system is decreasing. The unit, however, is controlled by the demand of PO_4 in the HP system. Peaks need to be avoided and, as stated above, this can be achieved by creating a stable climatic environment with dynamic light controls. It is obvious that in climate regions with fewer annual differences in solar radiation, there is less variation in ET_c and the complete system is more stable. Installing lamps and keeping a daily light integral of at least 10 mol m^{-2} can compensate for seasonal variations. Interplanting and mixed crop production help level the peak resulting from the traditional tomato cultivation protocol with young plants in winter when both

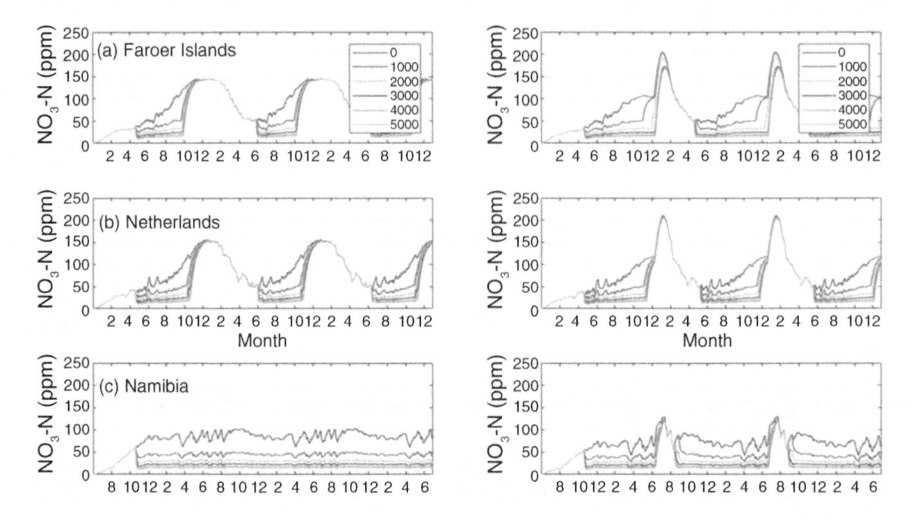

Fig. 8.9 NO3-N in RAS combined with HP with tomato (right) or lettuce (left) with desalination between 0 and 5000 L h^{-1} supply in three climate zones and decreasing latitudes (Faroe Islands 62.0°N, the Netherlands 52.1°N, Namibia 22.6°S) with adjusted area for HP (see above) in a 36-month simulation using local climate data and climate-adjusted greenhouses as model input

climate (low radiation) and cultivation (small plants, low potential ET_c) contribute to nutrient accumulation.

8.5 Monitoring and Control

In classical feedback control, like PI or PID (Proportional-Integral-Derivative) control, the controlled variables (CV) are directly measured, compared with a setpoint, and subsequently fed back to the process via a feedback control law.

In Fig. 8.10, the signals, without the time argument, are denoted by a small letter, where y is the controlled variable (CV) which is compared with the reference (setpoint) signal r. The tracking error $\boldsymbol{\varepsilon}$ (i.e. $r - y$) is fed into the controller, either in hardware or software, from which the control input u, also known as the manipulated variable (MV), is generated. The input u directly affects the process (P) from which an output (y) results. The sampled output is subsequently compared with r, which closes the loop. In practice, this loop continues until the controller is switch off. There exists extensive literature on feedback control (Doyle et al. 1992; Morris 2001; Ogata 2010), and this has been a subject of research for many years, starting with the works of Bode (1930) and Nyquist (1932).

In RAS, typical CVs are temperature, pH, and dissolved oxygen (DO) concentration, for which reliable sensors exist. Consequently, feedback control of these water quality parameters can be easily realized. However, in practice, most often, the input and output signals are disturbed by noise processes, such as

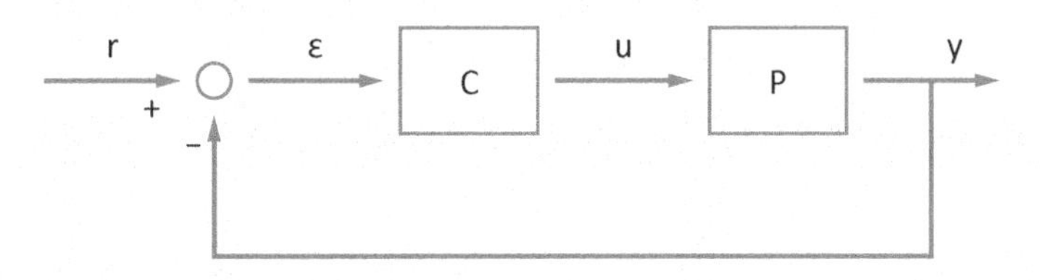

Fig. 8.10 Feedback control with controller (C) and process (*P*). r reference signal, eps tracking error, u input signal, y output sigma

unknown random inputs and measurement noise. Moreover, the overall process (*P*) may change over time as a result of growth, maturation, senescence, etc. Fish feed is another input into the RAS and its effect on fish growth cannot be directly seen or measured. For these parameters, model-based controllers (e.g. feedforward, model predictive, and optimal control) are typically introduced to predict the response of a change in the control input. However, fish feed is commonly added on the basis of values found in tables or recipes, but this rule-based control may need some adjustment in real practice to act as a feedback controller. Fish behaviours in RAS are a classical feedback control measure as fish react physiologically to environmental changes with variations in movement, location, receptiveness to feed, etc.

Hydroponic production usually takes place in protected environments such as greenhouses or plant factories where both the root and aerial environment need to be controlled. On-off controllers that predictively model optimal aerial environments have been proven superior in experimental research, but commercialization has been slow, whereas feedback controllers are standard in most climate-controlled greenhouses. However, the actuator varies with the type of controller with heating valves and vents typically feedback-controlled but lighting usually having an ON-OFF mechanism and only a few being dimmable. Controllers that rely on sensor or data input can respond to fast growth in a protected environment and result in high-quality produce with high market prices that improves its cost benefits. Many commercial greenhouses still have the classical centrally located sensor hanging 1–2 m above the crop and covering several hundred square meters is still in use, but multiple wireless sensors covering smaller areas are being introduced although much of the detailed data cannot be used because rather large climatic zones are controlled by the same actuators. Advances in sensor technology (e.g. microclimate temperature sensors, image processors, real-time gas-exchange or chlorophyll fluorescence measurements) connected to modelling software could use decision-support systems and become automated control systems.

In typical bioreactor systems, temperature, pH, dissolved oxygen in aerated systems, and gas fluxes in anaerobic systems are continuously measured and adjusted with available temperature, pH, and dissolved oxygen controllers. In addition to this, both hydraulic (HRT) and sludge retention (SRT) times are also frequently set by controlling (waste)water flows and biomass waste flows, respectively.

8.6 Economic Impact

Technologies that generate less profit, but are better for the environment usually only get implemented when the operators either receive an incentive in the form of subsidies or policies force them to do so. In the case of one-loop aquaponics systems, the appeal lies in the novel technology and the system's approach to sustainable resource use rather than its economic potential. However, recent publications provide evidence for production gains: leafy greens grow better in decoupled environments than in sterile hydroponic systems (Delaide et al. 2016; Goddek and Vermeulen 2018) and lettuce in decoupled aquaponics systems had a growth advantage of approximately 40% compared to state-of-the-art hydroponic approaches.

Even though higher growth rates can be expected, multi-loop aquaponics systems are still far more complex than hydroponics systems and significant initial investments are required for implementation. Most geographic locations require a high-tech greenhouse to control environmental conditions (i.e. a relative humidity of 80%, constant temperatures of around 20°C). Renewable energy sources can be used for cooling and heating, but currently such systems are only profitable when setting up on a large scale (i.e. > 1 ha) where good market conditions prevail.

8.7 Environmental Impact

Based on Example 8.2, there is evidence that treating sludge in digesters can have a beneficial impact on nutrient reutilization, especially phosphorus. Bioreactor systems, such as a sequential two-stage UASB reactor system, can increase the phosphorus recycling efficiency up to 300% (Chap. 10). Previously, in Chap. 2, we discussed the phosphorus paradox in relation to both phosphate scarcity and problems with eutrophication. Bioreactors have significant advantages for increased nutrient recovery from sludge, thus helping to close the nutrient cycling loop within aquaponics systems. However, further research is needed to refine such systems to optimize the bioavailability of specific nutrients. Figures 8.11, 8.12, and 8.13 show the input, output, and waste streams of stand-alone aquaculture and hydroponics systems compared with a decoupled aquaponics system. It can be seen that the decoupled approach constitutes a promising agricultural concept for a waste reduction and recycling system.

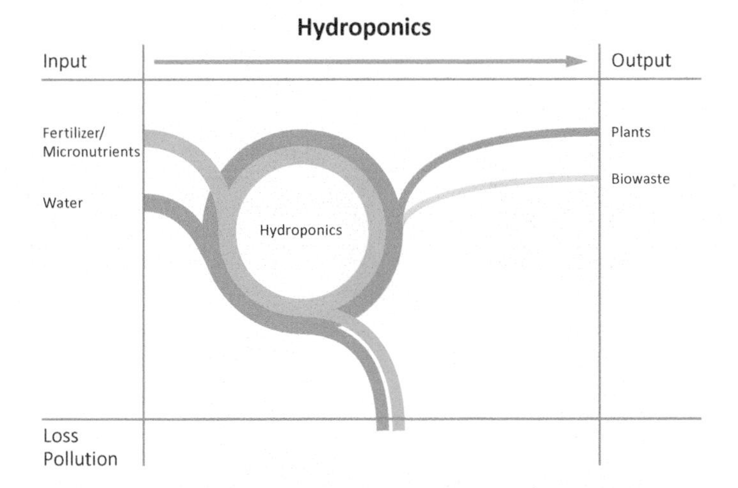

Fig. 8.11 Input, output, and loss streams in a stand-alone hydroponic system

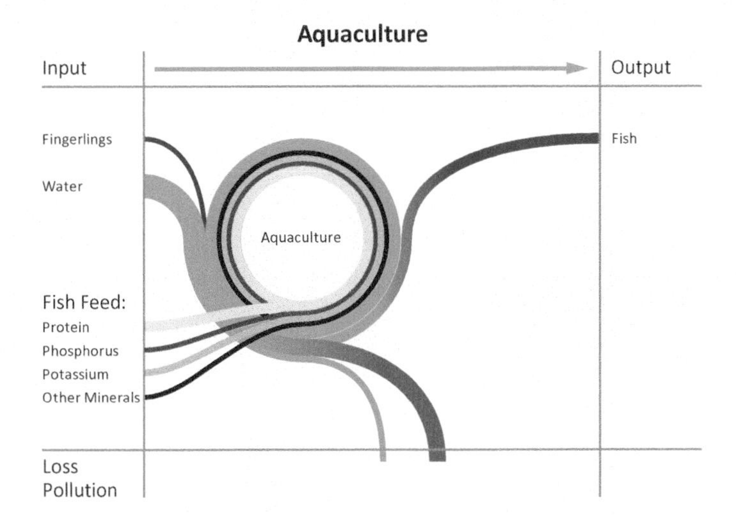

Fig. 8.12 Input, output, and loss streams in a stand-alone aquaculture system

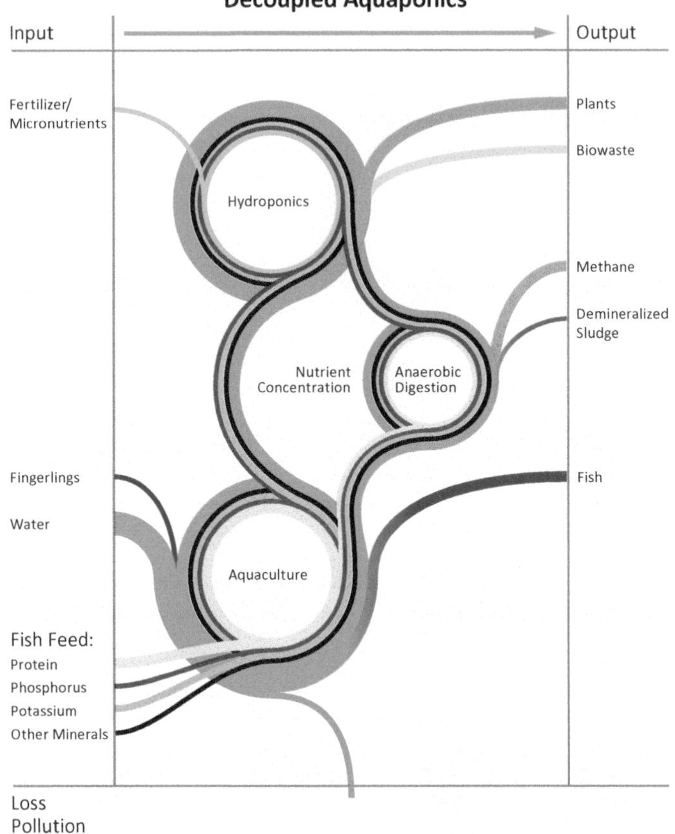

Fig. 8.13 Input, output, and loss streams in a decoupled multi-loop aquaponics system comprising an anaerobic reactor system

References

Alvarez R, Lidén G (2008) The effect of temperature variation on biomethanation at high altitude. Bioresour Technol 99:7278–7284. https://doi.org/10.1016/j.biortech.2007.12.055

Bode HW (1930) A method of impedance correction. Bell Syst Tech J 9:794–835. https://doi.org/10.1002/j.1538-7305.1930.tb02326.x

Coghlan SM, Ringler NH (2005) Temperature-dependent effects of rainbow trout on growth of Atlantic Salmon Parr. J Great Lakes Res 31:386–396. https://doi.org/10.1016/S0380-1330(05)70270-7

Dalsgaard J, Lund I, Thorarinsdottir R, Drengstig A, Arvonen K, Pedersen PB (2013) Farming different species in RAS in Nordic countries: current status and future perspectives. Aquac Eng 53:2–13

Davidson J, Good C, Welsh C, Summerfelt ST (2011) Abnormal swimming behavior and increased deformities in rainbow trout Oncorhynchus mykiss cultured in low exchange water recirculating aquaculture systems. Aquac Eng 45:109–117

de Lemos Chernicharo CA (2007) Anaerobic reactors, 4th edn. IWA Publishing, New Delhi
Delaide B, Goddek S, Gott J, Soyeurt H, Jijakli M (2016) Lettuce (Lactuca sativa L. var. Sucrine) growth performance in complemented aquaponic solution outperforms hydroponics. Water 8:467. https://doi.org/10.3390/w8100467
Delaide B, Goddek S, Keesman KJ, Jijakli MH (2018) Biotechnologie, agronomie, société et environnement = Biotechnology, agronomy, society and environment: BASE. https://popups.uliege.be:443/1780-4507 22, 12
Doyle JC, Francis BA, Tannenbaum A (1992) Feedback control theory. Macmillan Pub. Co
El-Sayed A-FM (2006) Tilapia culture. CABI Publishing, Oxfordshire
Emerenciano M, Carneiro P, Lapa M, Lapa K, Delaide B, Goddek S (2017) Mineralizacão de sólidos. Aquac Bras: 21–26
FAO (2005) Cultured aquatic species information programme. Oncorhynchus mykiss. [WWW Document]. FAO Fish. Aquac. Dep. [online]. URL http://www.fao.org/fishery/culturedspecies/Oncorhynchus_mykiss/en. Accessed 21 Sep 2015
FAO (2014) Small-scale aquaponic food production: integrated fish and plant farming. FAO, Rome
Goddek S (2017) Opportunities and challenges of multi-loop aquaponic systems. Wageningen University, Wageningen. https://doi.org/10.18174/412236
Goddek S, Keesman KJ (2018) The necessity of desalination technology for designing and sizing multi-loop aquaponics systems. Desalination 428:76–85. https://doi.org/10.1016/j.desal.2017.11.024
Goddek S, Körner O (2019) A fully integrated simulation model of multi-loop aquaponics: a case study for system sizing in different environments. Agric Syst 171:143–154
Goddek S, Vermeulen T (2018) Comparison of Lactuca sativa growth performance in conventional and RAS-based hydroponic systems. Aquac Int 26:1–10. https://doi.org/10.1007/s10499-018-0293-8
Goddek S, Delaide B, Mankasingh U, Ragnarsdottir K, Jijakli H, Thorarinsdottir R (2015) Challenges of sustainable and commercial aquaponics. Sustainability 7:4199–4224. https://doi.org/10.3390/su7044199
Goddek S, Espinal C, Delaide B, Schmautz Z, Jijakli H (2016) Navigating towards decoupled Aquaponic systems (DAPS): a system dynamics design approach. Water 8:7
Goddek S, Delaide BPL, Joyce A, Wuertz S, Jijakli MH, Gross A, Eding EH, Bläser I, Reuter M, Keizer LCP, Morgenstern R, Körner O, Verreth J, Keesman KJ (2018) Nutrient mineralization and organic matter reduction performance of RAS-based sludge in sequential UASB-EGSB reactors. Aquac Eng 83:10–19. https://doi.org/10.1016/J.AQUAENG.2018.07.003
Jones BJ (2005) Hydroponics – a practical guide for the soilless grower, 2nd edn. CRC Press, Boca Raton
Jung IS, Lovitt RW (2011) Leaching techniques to remove metals and potentially hazardous nutrients from trout farm sludge. Water Res 45:5977–5986. https://doi.org/10.1016/j.watres.2011.08.062
Karimanzira D, Keesman KJ, Kloas W, Baganz D, Rauschenbach T (2016) Dynamic modeling of the INAPRO aquaponic system. Aquac Eng 75:29–45. https://doi.org/10.1016/j.aquaeng.2016.10.004
Karimanzira D, Keesman K, Kloas W, Baganz D, Rauschenbach T (2017) Efficient and economical way of operating a recirculation aquaculture system in an aquaponics farm. Aquac Econ Manag 21:470–486. https://doi.org/10.1080/13657305.2016.1259368
Kloas W, Groß R, Baganz D, Graupner J, Monsees H, Schmidt U, Staaks G, Suhl J, Tschirner M, Wittstock B, Wuertz S, Zikova A, Rennert B (2015) A new concept for aquaponic systems to improve sustainability, increase productivity, and reduce environmental impacts. Aquac Environ Interact 7:179–192. https://doi.org/10.3354/aei00146
Körner O, Andreassen AU, Aaslyng JM (2006) Simulating dynamic control of supplementary lighting. Acta Hortic 711:151–156. https://doi.org/10.17660/ActaHortic.2006.711.17

Körner O, Gutzmann E, Kledal PR (2017) A dynamic model simulating the symbiotic effects in aquaponic systems. Acta Hortic 1170:309–316. https://doi.org/10.17660/ActaHortic.2017.1170.37

Körner O, Pedersen JS, Jaegerholm J (2018) Simulating lettuce production in multi layer moving gutter systems. Acta Hortic 1227:283–290

Lupatsch I, Kissil GW (1998) Predicting aquaculture waste from gilthead seabream (*Sparus aurata*) culture using a nutritional approach. Aquat Living Resour 11:265–268. https://doi.org/10.1016/S0990-7440(98)80010-7

Mirzoyan N, Gross A (2013) Use of UASB reactors for brackish aquaculture sludge digestion under different conditions. Water Res 47:2843–2850. https://doi.org/10.1016/j.watres.2013.02.050

Monsees H, Keitel J, Paul M, Kloas W, Wuertz S (2017a) Potential of aquacultural sludge treatment for aquaponics: evaluation of nutrient mobilization under aerobic and anaerobic conditions. Aquac Environ Interact 9:9–18. https://doi.org/10.3354/aei00205

Monsees H, Kloas W, Wuertz S, Cao X, Zhang L, Liu X (2017b) Decoupled systems on trial: eliminating bottlenecks to improve aquaponic processes. PLoS One 12:e0183056. https://doi.org/10.1371/journal.pone.0183056

Morris KA (2001) Introduction to feedback control. Harcourt/Academic Press, San Diego

Neto RM, Ostrensky A (2013) Nutrient load estimation in the waste of Nile tilapia Oreochromis niloticus (L.) reared in cages in tropical climate conditions. Aquac Res 46:1309–1322. https://doi.org/10.1111/are.12280

Nyquist H (1932) Regeneration theory. Bell Syst Tech J 11:126–147. https://doi.org/10.1002/j.1538-7305.1932.tb02344.x

Ogata K (2010) Modern control engineering, 5th edn. Pearson, Delhi

Rakocy JE, Shultz RC, Bailey DS, Thoman ES (2004) Aquaponic production of tilapia and basil: comparing a batch and staggered cropping system. Acta Hortic 648:63–69

Resh HM (2002) Hydroponic tomatoes. CRC Press, Boca Raton

Resh HM (2012) Hydroponic food production: a definitive guidebook for the advanced home gardener and the commercial hydroponic grower. CRC Press, Boca Raton

Resh HM (2013) Hydroponic food production: a definite guidebook for the advanced home gardener and the commercial hydroponic grower, 7th edn. CRC Press, Hoboken

Reyes Lastiri D, Slinkert T, Cappon HJ, Baganz D, Staaks G, Keesman KJ (2016) Model of an aquaponic system for minimised water, energy and nitrogen requirements. Water Sci Technol. wst2016127 74:30–37. https://doi.org/10.2166/wst.2016.127

Ross LG (2000) Environmental physiology and energetics. In: McAndrew BJ (ed) Tilapias: biology and exploitation. Springer, Dordrecht, pp 89–128

Schrader KK, Davidson JW, Summerfelt ST (2013) Evaluation of the impact of nitrate-nitrogen levels in recirculating aquaculture systems on concentrations of the off-flavor compounds geosmin and 2-methylisoborneol in water and rainbow trout (Oncorhynchus mykiss). Aquac Eng 57:126–130. https://doi.org/10.1016/j.aquaeng.2013.07.002

Shahzad MW, Burhan M, Ang L, Ng KC (2017) Energy-water-environment nexus underpinning future desalination sustainability. Desalination 413:52–64. https://doi.org/10.1016/j.desal.2017.03.009

Sonneveld C, Voogt W (2009) Nutrient management in substrate systems. In: Plant nutrition of greenhouse crops. Springer, Dordrecht, pp 277–312. https://doi.org/10.1007/978-90-481-2532-6_13

Subramani A, Jacangelo JG (2015) Emerging desalination technologies for water treatment: a critical review. Water Res 75:164–187. https://doi.org/10.1016/j.watres.2015.02.032

Timmons MB, Ebeling JM (2013) Recirculating aquaculture, 3rd edn. Ithaca Publishing Company LLC, Ithaca

van Gorcum B, Goddek S, Keesman KJ (2019) Gaining market insights for aquaponically produced vegetables in Kenya. Aquac Int:1–7. https://link.springer.com/article/10.1007/s10499-019-00379-1

Yavuzcan Yildiz H, Robaina L, Pirhonen J, Mente E, Domínguez D, Parisi G (2017) Fish welfare in aquaponic systems: its relation to water quality with an emphasis on feed and faeces—a review. Water 9:13. https://doi.org/10.3390/w9010013

Yogev U, Barnes A, Gross A (2016) Nutrients and energy balance analysis for a conceptual model of a three loops off grid. Aquaponics Water 8:589. https://doi.org/10.3390/W8120589

Chapter 9
Nutrient Cycling in Aquaponics Systems

Mathilde Eck, Oliver Körner, and M. Haïssam Jijakli

Abstract In aquaponics, nutrients originate mainly from the fish feed and water inputs in the system. A substantial part of the feed is ingested by the fish and either used for growth and metabolism or excreted as soluble and solid faeces, while the rest of any uneaten feed decays in the tanks. While the soluble excretions are readily available for the plants, the solid faeces need to be mineralised by microorganisms in order for its nutrient content to be available for plant uptake. It is thus more challenging to control the available nutrient concentrations in aquaponics than in hydroponics. Furthermore, many factors, amongst others pH, temperature and light intensity, influence the nutrient availability and plant uptake. Until today, most studies have focused on the nitrogen and phosphorus cycles. However, to ensure good crop yields, it is necessary to provide the plants with sufficient levels of all key nutrients. It is therefore essential to better understand and control nutrient cycles in aquaponics.

Keywords Aquaponics · Nutrient cycling · Solubilisation · Microbiological processes

9.1 Introduction

Aquaponic systems offer various advantages when it comes to producing food in an innovative and sustainable way. Besides the synergistic effects of increased aerial CO_2 concentration for greenhouse crops and decreased total heat energy consumption when cultivating fish and crops in the same space (Körner et al. 2017),

M. Eck (✉) · M. H. Jijakli
Integrated and Urban Plant Pathology Laboratory, Université de Liège, Agro-Bio Tech, Gembloux,, Belgium
e-mail: mathilde.eck@uliege.be; mh.jijakli@uliege.be

O. Körner
Leibniz-Institute of Vegetable and Ornamental Crops (IGZ), Grossbeeren, Germany
e-mail: koerner@igzev.de

S. Goddek et al. (eds.), *Aquaponics Food Production Systems*,
https://doi.org/10.1007/978-3-030-15943-6_9

aquaponics has two main advantages for nutrient cycling. First, the combination of a recirculating aquaculture system with hydroponic production avoids the discharge of aquaculture effluents enriched in dissolved nitrogen and phosphorus into already polluted groundwater (Buzby and Lin 2014; Guangzhi 2001; van Rijn 2013), and second, it allows for the fertilisation of the soilless crops with what can be considered an organic solution (Goddek et al. 2015; Schneider et al. 2004; Yogev et al. 2016) instead of using fertilisers of mineral origin made from depleting natural resources (Schmautz et al. 2016; Chap. 2). Furthermore, aquaponics yields comparable plant growth as compared with conventional hydroponics despite the lower concentrations of most nutrients in the aquaculture water (Graber and Junge 2009; Bittsanszky et al. 2016; Delaide et al. 2016), and production can be even better than in soil (Rakocy et al. 2004). Increased CO_2 concentrations in the aerial environment and changes in the biomes of the root zone are thought to be main reasons for this. In addition, the mineral content and the nutritional quality of tomatoes grown aquaponically have been reported to be equivalent or superior to the mineral content of conventionally grown ones (Schmautz et al. 2016).

Despite having two attractive assets (i.e. the recycling of aquaculture effluents and the use of organic fertilisers), the use of aquaculture effluents increases the challenge of monitoring the nutrients within the solution. Indeed, it is harder to control the composition of a solution where the nutrients originate from a biological degradation of organic matter than to follow the evolution of the nutrients' concentration in a precisely dosed hydroponic solution based on mineral compounds (Bittsanszky et al. 2016; Timmons and Ebeling 2013). Moreover, a plant's nutritional needs vary during the growth period in accordance with physiological stages, and it is necessary to meet these needs to maximise yields (Bugbee 2004; Zekki et al. 1996; Chap. 4).

In order to recycle aquaculture effluents to produce plant biomass, it is necessary to optimise the recycling rates of phosphorus and nitrogen (Goddek et al. 2016; Graber and Junge 2009; Chap. 1). Several factors can influence this, such as the fish species, fish density, water temperature, the type of plants and the microbial community (ibid.). Therefore, it is of prime importance to understand the functioning of the nutrient cycles in aquaponics (Seawright et al. 1998). This chapter aims at explaining the origins of the nutrients in an aquaponic system, describing the nutrient cycles and analysing the causes of nutrient losses.

9.2 Origin of Nutrients

The major sources of nutrients in an aquaponic system are the fish feed and the water added (containing Mg, Ca, S) (see Sect. 9.3.2.) into the system (Delaide et al. 2017; Schmautz et al. 2016) as further elaborated in Chap. 13. With respect to fish feed, there are two main types: fishmeal-based and plant-based feed. Fishmeal is the classic type of feed used in aquaculture where lipids and proteins rely on fish meal and fish oil (Geay et al. 2011). However, for some time now, concerns regarding the sustainability of such feed have been raised and attention drawn towards plant-based diets (Boyd 2015; Davidson et al. 2013; Hua and Bureau 2012; Tacon and Metian 2008). A meta-analysis conducted by Hua and Bureau (2012) revealed that the use of plant proteins in fish feed can influence the growth of fish if incorporated in high proportions. Indeed, plant proteins can have an impact on the digestibility and levels of anti-nutritional factors of the feed. In particular, phosphorus originating from plants and thus in the form of phytates does not benefit, for example, salmon, trout and several other fish species (Timmons and Ebeling 2013). It is not surprising that this observation is highly dependent on the fish species and on the quality of the ingredients (Hua and Bureau 2012). However, little is known of the impact of varying fish feed composition on crop yields (Yildiz et al. 2017).

Classical fish feed is composed of 6–8 macro ingredients and contains 6–8% organic nitrogen, 1.2% organic phosphorus and 40–45% organic carbon (Timmons and Ebeling 2013) with around 25% protein for herbivorous or omnivorous fish and around 55% protein for carnivorous fish (Boyd 2015). Lipids can be fish or plant based as well (Boyd 2015).

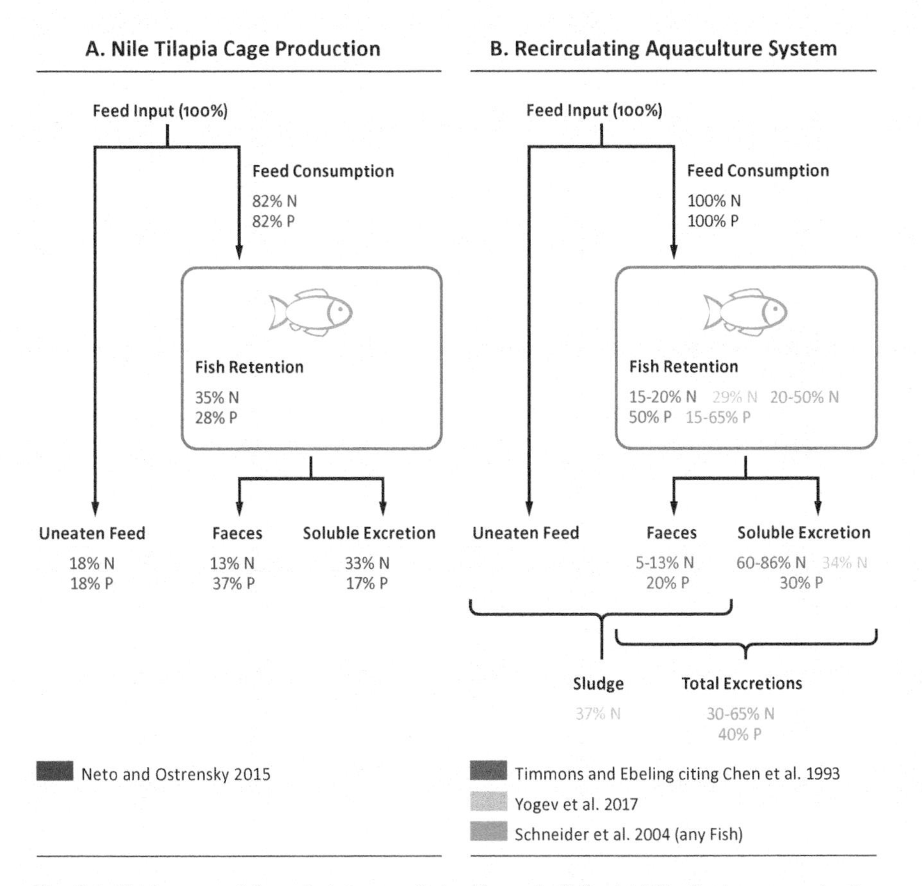

Fig. 9.1 Environmental flow of nitrogen and phosphorus in % for (**a**) Nile tilapia cage production (after Neto and Ostrensky 2015) and (**b**) RAS production (from a variety of sources)

Once fish feed is added into the system, a substantial part of it is eaten by the fish and either used for growth and metabolism or excreted as soluble and solid faeces, while the rest of the given feed decays in the tanks (Goddek et al. 2015; Schneider et al. 2004) (Fig. 9.1). In this case, the feed leftovers and metabolic products are partly dissolved in the aquaponic water, thus enabling the plants to uptake nutrients directly from the aquaponic solution (Schmautz et al. 2016).

In most cultivation systems (Chaps. 7 and 8), nutrients can be added to complement the aquaponic solution and ensure a better matching with the plants' needs (Goddek et al. 2015). Indeed, even when the system is coupled, it is possible to add iron or potassium (which are often lacking) without harming the fish (Schmautz et al. 2016).

9.2.1 Fish Feed Leftovers and Fish Faeces

Ideally, all the given feed should be consumed by the fish (Fig. 9.1). However, a small part (less than 5% (Yogev et al. 2016)) is often left to decompose in the system and contributes to the nutrient load of the water (Losordo et al. 1998; Roosta and Hamidpour 2013; Schmautz et al. 2016), thus consuming dissolved oxygen and releasing carbon dioxide and ammonia (Losordo et al. 1998), amongst other things. The composition of fish feed leftovers depends on the composition of the feed.

Logically enough, the composition of fish faeces depends on the fish's diet which also has an impact on the water quality (Buzby and Lin 2014; Goddek et al. 2015). However, the nutrient retention in fish biomass is highly dependent on fish species, feeding levels, feed composition, fish size and system temperature (Schneider et al. 2004). At higher temperatures, for example, fish metabolism is accelerated and thus results in more nutrients contained in the solid fraction of the faeces (Turcios and Papenbrock 2014). The proportion of excreted nutrients also depends on the quality and digestibility of the diet (Buzby and Lin 2014). The digestibility of the fish feed, the size of the faeces and the settling ratio should be carefully considered to ensure a good balance in the system and to maximise crop yields (Yildiz et al. 2017). Indeed, while it is a priority that fish feed should carefully be chosen to suit fish needs, the feed components could also be selected to suit plant's requirements when it makes no difference to fish (Goddek et al. 2015; Licamele 2009; Seawright et al. 1998).

9.3 Microbiological Processes

9.3.1 Solubilisation

Solubilisation consists of the breaking down of the complex organic molecules composing fish waste and feed leftovers into nutrients in the form of ionic minerals which plants can absorb (Goddek et al. 2015; Somerville et al. 2014). In both aquaculture (Sugita et al. 2005; Turcios and Papenbrock 2014) and aquaponics, solubilisation is conducted mainly by heterotrophic bacteria (van Rijn 2013; Chap. 6) which have not yet been fully identified (Goddek et al. 2015). Some studies have started deciphering the complexity of these bacteria communities (Schmautz et al. 2017). In current aquaculture, the most commonly observed bacteria are *Rhizobium* sp., *Flavobacterium* sp., *Sphingobacterium* sp., *Comamonas* sp., *Acinetobacter* sp., *Aeromonas* sp. and *Pseudomonas* sp. (Munguia-Fragozo et al. 2015; Sugita et al. 2005). An example of the major role of bacteria in aquaponics could be the transformation of insoluble phytates into phosphorus (P) made available for plant uptake through the production of phytases which are particularly present in *γ-proteobacteria* (Jorquera et al. 2008). (More research needs to be done in this area). Other nutrients than P can also be trapped as solids and evacuated from the system with the sludge. Efforts are thus being made to remineralise this sludge with UASB-EGSB reactors in order to reinject nutrients into the aquaponic system (Delaide 2017; Goddek et al. 2016; Chap. 10). Furthermore, different minerals are not released at the same rate, depending on the composition of the feed (Letelier-Gordo et al. 2015), thus leading to more complicated monitoring of their concentration in the aquaponic solution (Seawright et al. 1998).

9.3.2 Nitrification

The main nitrogen source in an aquaponic system is the fish feed and the proteins it contains (Goddek et al. 2015; Ru et al. 2017; Wongkiew et al. 2017; Yildiz et al. 2017). Ideally, 100% of this feed should be eaten by the fish. However, is has been observed that fish only use about 30% of the nitrogen contained in the given feed (Rafiee and Saad 2005). The ingested feed is partly used for assimilation and metabolism (Wongkiew et al. 2017), while the rest is excreted either through the gills or as urine and faeces (Ru et al. 2017). The nitrogen which is excreted through the gills is mainly in the form of ammonia, NH_3 (Wongkiew et al. 2017; Yildiz et al. 2017), while urine and faeces are composed of organic nitrogen (Wongkiew et al. 2017) which is transformed into ammonia by proteases and deaminases (Sugita et al. 2005). In general, the fish excrete nitrogen under the form of TAN, i.e. NH_3 and NH_4^+. The balance between NH_3 and NH_4^+ depends mostly on the pH and temperature. Ammonia is the major waste produced by fish catabolism of the feed proteins (Yildiz et al. 2017).

Nitrification is a two-step process during which the ammonia NH_3 or ammonium NH_4^+ excreted by the fish is transformed first into nitrite NO_2^- and then into nitrate NO_3^- by specific aerobic chemosynthetic autotrophic bacteria. A high availability of dissolved oxygen is required as nitrification consumes oxygen (Carsiotis and Khanna 1989; Madigan and Martinko 2007; Shoda 2014). The first step of this transformation is carried out by ammonia-oxidising bacteria (AOB) such as *Nitrosomonas*, *Nitrosococcus*, *Nitrosospira*, *Nitrosolobus* and *Nitrosovibrio*. The second step is conducted by nitrite-oxidising bacteria (NOB) such as *Nitrobacter*, *Nitrococcus*, *Nitrospira* and *Nitrospina* (Rurangwa and Verdegem 2013; Timmons and Ebeling 2013; Wongkiew et al. 2017). *Nitrospira* is currently deduced to be a complete nitrifier, i.e. to be involved in the production of both nitrite and nitrate (Daims et al. 2015). The same bacteria can be found both in aquaculture and aquaponic systems (Wongkiew et al. 2017). These bacteria are mainly found in biofilms fixed to the media composing the biofilter but can also be observed in the other compartments of the system (Timmons and Ebeling 2013).

Nitrification is of prime importance in aquaponics as ammonia and nitrite are quite toxic for fish: 0.02–0.07 mg/L of ammonia–nitrogen are sufficient to observe damage in warm water fish, and nitrite–nitrogen should be kept under 1 mg/L (Losordo et al. 1998; Timmons and Ebeling 2013). Ammonia affects the central nervous system of the fish (Randall and Tsui 2002; Timmons and Ebeling 2013), while nitrite induces problems with oxygen fixation (Losordo et al. 1998). Nitrate–nitrogen is, however, tolerated by the fish up to 150–300 mg/L (Goddek et al. 2015; Graber and Junge 2009; Yildiz et al. 2017).

Nitrification mostly takes place in biofilters (Losordo et al. 1998; Timmons and Ebeling 2013). Therefore, when starting a system, it is recommended to run the system without fish in order to allow the slowly growing population of nitrifying bacteria to establish (Timmons and Ebeling 2013; Wongkiew et al. 2017). It is also necessary to avoid, as far as possible, the presence of organic matter in the biofilters in order to prevent the growth of highly competitive heterotrophic bacteria (Timmons and Ebeling 2013). Alternatively, commercial mixes of nitrifying bacteria can be added to the system, prior to stocking, to hasten the colonisation process (Kuhn et al. 2010). Nevertheless, small aquaponic systems without biofilter also exist. In these systems, nitrifying bacteria form biofilms of the available surfaces (e.g. hydroponic compartment walls, inert media when using the media bed technique) (Somerville et al. 2014).

9.4 Mass Balance: What Happens to Nutrients once They Enter into the Aquaponic System?

9.4.1 Context

The functioning of aquaponic systems is based on a dynamic equilibrium of the nutrient cycles (Somerville et al. 2014). It is therefore necessary to understand these cycles in order to optimise the management of the systems. Plants growing hydroponically have specific requirements, which should be met during their various growing stages (Resh 2013). Therefore, nutrient concentrations in the different compartments of the system must be closely monitored, and nutrients should be supplemented to prevent deficiencies (Resh 2013; Seawright et al. 1998) either in the system water or via foliar application (Roosta and Hamidpour 2011).

According to Delaide et al. (2016), in some cases, supplementing an aquaponic solution with mineral nutrients in order to reach the same nutrient concentrations as in hydroponics could lead to higher yields than those achieved in hydroponics. The first step to take towards a balanced system is the correct design and relative sizing of the compartments (Buzby and Lin 2014). If the hydroponic compartment is too small compared to the fish tanks, then the nutrients will accumulate in the water and could reach toxic levels. The feed rate ratio (i.e. the amount of fish feed in the system based on the plant-growing surface and the plant type) is often used for the first sizing of the system (Rakocy et al. 2006; Somerville et al. 2014). However, according to Seawright et al. (1998), it is not possible to reach a plant/fish ratio which will enable an optimal match of plants' needs if only fish feed is used as an input. To make sure that the system is well balanced and functions properly, monitoring methods are usually based on the nitrogen cycle (Cerozi and Fitzsimmons 2017; Somerville et al. 2014), but to ensure optimal functioning of the system, it is necessary to monitor more closely the balance of the other macronutrients (N, P, K, Ca, Mg, S) and micronutrients (Fe, Zn, B, Mn, Mo, Cu) (Resh 2013; Somerville et al. 2014; Sonneveld and Voogt 2009) as well. Recent studies (Delaide et al. (2017), Schmautz et al. (2015, 2016)) have started tackling this topic. Schmautz et al. (2015, 2016) compared the impact of three different hydroponic layouts (i.e. nutrient film technique (NFT), floating raft and drip irrigation) on the nutrient uptake of aquaponic tomatoes. Drip irrigation was the system which produced slightly better yields with tomatoes. The mineral content of the fruits (P, K, Ca, Mg) was equivalent to the conventional values even though the iron and zinc contents were higher. The leaves however had lower levels of P, K, S, Ca, Mg, Fe, Cu and Zn than in conventional agriculture. Delaide et al. (2016) followed the cycles of macro- and micronutrients in a coupled aquaponic system. They observed that K, P, Fe, Cu, Zn, Mn and Mo were lacking in their aquaponic solution, while N, Ca, B and Na were quickly accumulated. Graber and Junge (2009) noted that their aquaponic solution contained three times less nitrogen and ten times less phosphorus than a hydroponic solution. As for potassium (K), it was 45 times lower compared to hydroponics. Nevertheless, they obtained yields as similar yields even though the quality was poorer due to a lack of potassium (K).

Factors Influencing the Nutrient Cycles

Light intensity, root zone temperature, air temperature, nutrient availability, growth stage and growth rate all influence a plant's nutrient uptake (Buzby and Lin 2014). Experiments conducted by Schmautz et al. (2016) and Lennard and Leonard (2006) showed that the hydroponic method could also play a role in a plant's nutrient uptake capacity, and it is therefore necessary to match the growing system to the type of vegetables being grown. NFT and DWC (deep water culture – raft) are thus suitable for leafy greens, whereas drip irrigation on rockwool slabs is more suitable for fruity vegetables (Resh 2013).

9.4.2 Macronutrient Cycles

Carbon (C)

Carbon is provided to the fish via the feed (Timmons and Ebeling 2013) and to the plants via CO_2 fixation. Fish can use 22% of the carbon contained in the fish feed for biomass increase and metabolism. The rest of the ingested carbon is either expirated under the form of CO_2 (52%) or excreted in a dissolved (0.7–3%) and solid (25%) form (Timmons and Ebeling 2013). The expirated CO_2 can be used by plants for their own carbon source as well (Körner et al. 2017). The uneaten part of the feed carbon is left to decompose in the system. The type of carbohydrates found in fish feed (e.g. starch or non-starch polysaccharides) can also influence the digestibility of the feed and the biodegradability of the waste in an aquaculture or aquaponic system (Meriac et al. 2014).

Nitrogen (N)

Nitrogen is absorbed by the plants either in the nitrate or ammonium form (Sonneveld and Voogt 2009; Xu et al. 2012) depending on the concentration and plant's physiology (Fink and Feller 1998 cited by Wongkiew et al. 2017). Associations between plants and microorganisms should not be overlooked as plants affect the presence of the microorganisms in aquaponics, and microorganisms can play a significant role in the nitrogen uptake capacity of plants (Wongkiew et al. 2017). The uptake of nitrogen by plants is affected by the ambient carbon dioxide concentration (Zhang et al. 2008 cited by Wongkiew et al. 2017).

Phosphorus (P)

Phosphorus is one of the essential elements for plant growth and can be absorbed under its ionic orthophosphate form ($H_2PO_4^-$, HPO_4^{2-}, PO_4^{3-}) (Prabhu et al. 2007; Resh 2013). Little is known about the dynamics of phosphorus in aquaponics. The main input of phosphorus in the system is the fish feed (Cerozi and Fitzsimmons 2017; Delaide et al. 2017; Schmautz et al. 2015), and in un-supplemented systems (Chap. 7), phosphorus tends to be limiting and thus can impede plant growth (Graber and Junge 2009; Seawright et al. 1998). According to Rafiee and Saad (2005), fish can use up to 15% of the phosphorus contained in the feed. In a system growing lettuce, Cerozi and Fitzsimmons (2017) noticed that the amount of phosphorus provided by the fish feed can be sufficient or insufficient depending on the growth stage. Up to 100% of phosphorus present in the fish water can be recycled in the plant biomass, depending on the design of the system. Graber and Junge (2009) observed a 50% recycling, while Schmautz et al. (2015) reported that 32% of the phosphorus could be found in the fruit and 28% in the leaves. The solubility of phosphorus depends on the pH, and a higher pH will foster the precipitation of phosphorus, thus rendering it unavailable for the plants (Yildiz et al. 2017). Phosphorus can precipitate as struvite (magnesium ammonium phosphate) (Le Corre et al. 2005) and/or hydroxyapatite (Cerozi and

Fitzsimmons 2017; Goddek et al. 2015). These insoluble complexes are removed via solid fish sludge from the system. Schneider et al. (2004) reported that 30–65% of the phosphorus contained in the fish feed remains unavailable to plants as it is fixed in the solid excretions which are then removed through mechanical filtration. Yogev et al. (2016) estimated that this loss can be up to 85%. One option to prevent this massive loss of P via solid sludge is to add a digestion compartment to the aquaponic system. During aerobic or anaerobic digestion, the P is released into the digestate and could be re-introduced into the circulating water (Goddek et al. 2016).

Potassium (K)
Delaide et al. (2017) found that the major source of K in their system was the fish feed. Fish can use up to 7% of the K contained in the fish feed (Rafiee and Saad 2005). However, potassium is not necessary for fish which leads to a low potassium composition of the fish feed and to even lower levels of potassium available for the plants (Graber and Junge 2009; Seawright et al. 1998; Suhl et al. 2016). To supply potassium, a KOH pH buffer is often used as the pH often decreases in aquaponics due to nitrification (Graber and Junge 2009). In an aquaponic system planted with tomatoes, potassium accumulated mainly in the fruits (Schmautz et al. 2016).

Magnesium (Mg), Calcium (Ca) and Sulphur (S)
The main source for Mg, Ca and S is tap water which facilitates the absorption by the plants as the nutrients are already available (Delaide et al. 2017). Calcium is however present in insufficient levels in aquaponics (Schmautz et al. 2015; Seawright et al. 1998) and is added under the form of calcium hydroxide $Ca(OH)_2$ (Timmons and Ebeling 2013). According to Rafiee and Saad (2005), fish can use on average 26.8% of the calcium and 20.3% of the magnesium present in the feed. Sulphur is often at low levels in aquaponic systems (Graber and Junge 2009; Seawright et al. 1998).

9.4.3 Micronutrient Cycles

Iron (Fe), manganese (Mn) and zinc (Zn) derive mainly from the fish feed, while boron (B) and copper (Cu) derive from the tap water (Delaide et al. 2017). In aquaponics, key micronutrients are often present but at too low levels (Delaide et al. 2017), and supplementation from external sources of nutrients is then necessary (Chap. 8). Iron deficiencies occur very often in aquaponics (Schmautz et al. 2015; Seawright et al. 1998; Fitzsimmons and Posadas; 1997 cited by Licamele 2009), mostly because of the non-availability of the ferric ion form. This deficiency can be solved by the use of bacterial siderophore (i.e. organic iron-chelating compounds) produced by genera such as *Bacillus* or *Pseudomonas* (Bartelme et al. 2018) or by iron supplementation with chemical chelated iron to avoid iron precipitation.

9.4.4 Nutrient Losses

Reducing nutrient loss is a constant challenge facing aquaponics practitioners. Nutrient loss occurs in several ways, e.g. the settlement of the sludge (37% of faeces and 18% of uneaten feed) (Neto and Ostrensky 2015), water losses, denitrification, ammonia volatilisation, etc. (Wongkiew et al. 2017). As an example, Rafiee and Saad (2005) note that 24% of the iron, 86% of the manganese, 47% of the zinc, 22% of the copper, 16% of the calcium, 89% of the magnesium, 6% of the nitrogen, 6% of the potassium and 18% of the phosphorus contained in the fish feed were contained in the sludge. The sludge can hold up to 40% of the nutrients present in the feed input (Yogev et al. 2016).

Denitrification can lead to a loss of 25–60% of the nitrogen (Hu et al. 2015; Zou et al. 2016). Denitrification is also linked to anoxic conditions (Madigan and Martinko 2007; van Lier et al. 2008) and low carbon levels and is responsible for the transformation of nitrate into nitrite, nitric oxide (NO), nitrous oxide (N_2O) and eventually nitrogen gas (N_2) with flows into the atmosphere (ibid.). Denitrification is conducted by several facultative heterotrophic bacteria such as *Achromobacter*, *Aerobacter*, *Acinetobacter*, *Bacillus*, *Brevibacterium*, *Flavobacterium*, *Pseudomonas*, *Proteus* and *Micrococcus* sp. (Gentile et al. 2007; Michaud et al. 2006; Wongkiew et al. 2017). Some bacteria can perform both nitrification and denitrification if dissolved oxygen levels are below 0.3 mg/L (Fitzgerald et al. 2015; Wongkiew et al. 2017). The loss of nitrogen can also occur via anaerobic ammonium oxidation (ANAMMOX), i.e. the oxidation of ammonium into dinitrogen gas in the presence of nitrite (Hu et al. 2011).

Another important loss of nitrogen which should be available for the plants is the consumption of the nitrogen by the heterotrophic aerobic bacteria present in the aquaponic systems. Indeed, the nitrogen used by these bacteria is lost to nitrifying bacteria, and nitrification is thus impeded (Blancheton et al. 2013). These bacteria are particularly present when the C/N ratio increases as they are more competitive and more able to colonise the media than the autotrophic nitrifying bacteria (Blancheton et al. 2013; Wongkiew et al. 2017).

9.4.5 Nutrient Balance Systems Dynamics

The nutrient concentration of the two major subsystems in an aquaponic system, i.e. the fish tanks (aquaculture) and the hydroponic solution, needs to be balanced for each of their needs. In closed aquaponic systems, nutrients are transported from fish to plants more or less directly through filters (usually a drum filter or a settler and then a biofilter) for nitrification. However, the nutrient needs of crops and the supplied nutrients from the aquaponic subsystem are not in balance. In multi-loop and decoupled systems (Chap. 8), it is easier to provide optimal conditions for both, fish and plant sections. Through modelling of the systems (Chap. 11), the optimal size of the hydroponic area to the fish tanks, biofilters and other equipment can be calculated (Goddek and Körner 2019). This is particularly important with decoupled multi-loop systems which comprise various types of equipment. For example, UASB (upflow anaerobic sludge blanket) (Goddek et al. 2018) or desalination units (Goddek and Keesman 2018) need to be sized carefully as discussed in Chap. 8. The basic mismatch of nutrients supplied by the fish environment and the needs of the crops needs to be rectified and balanced. For the purpose of up-concentrating the nutrients, Goddek and Keesman (2018) have described an appropriate desalination unit (Chap. 8). This approach, however, only solves part of the problem, as the perfectly balanced system is driven by a non-dynamic evaporation rate achievable only in closed chambers and perfectly working plants. The reality, however, is that the evapotranspiration of the crop (ET_c) in greenhouse-based aquaponics systems is highly dependent on multiple factors such as physical climate and biological variables. ET_c is calculated per area of the ground surface covered by the crop and is calculated for different levels in the canopy (z) by integrating irradiative net fluxes, boundary layer resistance, stomata resistance and the vapour pressure deficit, in the canopy (Körner et al. 2007) using the Penman–Monteith equation. This equation, nevertheless, only calculates the water flux through the crop. Nutrient uptake can either be calculated simply by assuming that all diluted nutrients in the water are taken up by the crop. In reality though, uptake of nutrients is a highly complicated matter. Different nutrients have different states, changing with parameters such as pH. Meanwhile nutrient availability to the plants strongly depends on pH and the relationships of nutrients to each other (e.g. K/Ca availability). In addition, the microbiome in the root zone plays an important role (Orozco-Mosque et al. 2018) which is not yet implemented in models although some models differentiate between phloem and xylem pathways. Thus, the vast amount of nutrients is not modelled in detail for aquaponic nutrient balancing and sizing of systems. The easiest way to estimate nutrient uptake is the assumption that nutrients are taken up/absorbed as dissolved in irrigation water and apply to the above-explained ET_c calculation approach and assuming that no element-specific chemical, biological or physical resistances exist. Consequentially, to maintain equilibrium, all nutrients taken up by the crop as contained in the nutrient solution need to be added back to the hydroponic system.

9.5 Conclusions

9.5.1 Current Drawbacks of Nutrient Cycling in Aquaponics

In hydroponics, the nutrient solution is accurately determined and the nutrient input into the system is well understood and controlled. This makes it relatively easy to adapt the nutrient solution for each plant species and for each growth stage. In aquaponics, according to the definition (Palm et al. 2018), the nutrients have to originate at least at 50% from uneaten fish feed, fish solid faeces and fish soluble excretions, thus making the monitoring of the nutrient concentrations available for plant uptake more difficult. A second drawback is the loss of nutrients through several pathways such as sludge removal, water renewal or denitrification. Sludge removal induces a loss of nutrients as several key nutrients such as phosphorus often precipitate and are then trapped in the evacuated solid sludge. Water renewal, which has to take place even if in small proportions, also adds to the loss of nutrients from the aquaponic circuit. Finally, denitrification happens because of the presence of denitrifying bacteria and conditions favourable to their metabolisms.

9.5.2 How to Improve Nutrient Cycling?

To conclude, nutrient cycling still needs to be improved in order to optimise plant growth in aquaponics. Several options are therefore currently explored in Chap. 8. To avoid losing the nutrients captured in the sludge, sludge remineralisation units have been developed (Chap. 10). The aim of these units is to extract the nutrients captured in solid form in the sludge and to reinject these into the system under a form which the plants can absorb (Delaide 2017). A further technique to reduce nutrient loss would be to foster plant uptake through the concentration of the aquaponic solution (i.e. the removal of a fraction of the water to keep the same amount of nutrients but in a lesser water volume). Such a concentration could be achieved via the addition of a desalination unit as part of the aquaponic system (Goddek and Körner 2019; Goddek and Keesman 2018). Finally, the use of decoupled/multi-loop systems enables optimal living and growing conditions for all fish, plants and microorganisms. While some research has been undertaken in this field, more research should be conducted to better understand nutrient cycling in aquaponics. Indeed, more information concerning the exact cycles of each macronutrient (what form, how it can be transformed or not by microorganisms, how it is taken up by plants in aquaponics) or the influence of the plant and fish species and water parameters on the nutrient cycles could greatly help the understanding of aquaponic systems.

References

Bartelme RP, Oyserman BO, Blom JE, Sepulveda-Villet OJ, Newton RJ (2018) Stripping away the soil: plant growth promoting microbiology opportunities in aquaponics. Front Microbiol 9(8). https://doi.org/10.3389/fmicb.2018.00008

Bittsanszky A, Uzinger N, Gyulai G, Mathis A, Junge R, Villarroel M, Kotzen B, Komives T (2016) Nutrient supply of plants in aquaponic systems. Ecocycles 2(1720). https://doi.org/10.19040/ecocycles.v2i2.57

Blancheton JP, Attramadal KJK, Michaud L, d'Orbcastel ER, Vadstein O (2013) Insight into bacterial population in aquaculture systems and its implication. Aquac Eng 53:30–39

Boyd CE (2015) Overview of aquaculture feeds: global impacts of ingredient use. Feed Feed Pract Aquac 3–25. https://doi.org/10.1016/B978-0-08-100506-4.00001-5

Bugbee B (2004) Nutrient management in recirculating hydroponic culture. Acta Hortic 648:99–112. https://doi.org/10.17660/ActaHortic.2004.648.12

Buzby KM, Lin LS (2014) Scaling aquaponic systems: balancing plant uptake with fish output. Aquac Eng 63:39–44. https://doi.org/10.1016/j.aquaeng.2014.09.002

Carsiotis M, Khanna S (1989) Genetic engineering of microbial nitrification. United States Environmental Protection Agency, Risk Reduction Engineering Laboratory, Cincinnati

Cerozi BS, Fitzsimmons K (2017) Phosphorus dynamics modeling and mass balance in an aquaponics system. Agric Syst 153:94–100. https://doi.org/10.1016/j.agsy.2017.01.020

Daims H, Lebedeva EV, Pjevac P, Han P, Herbold C, Albertsen M, Jehmlich N, Palatinszky M, Vierheilig J, Bulaev A, Kirkegaard RH, Von Bergen M, Rattei T, Bendinger B, Nielsen H, Wagner M (2015) Complete nitrification by Nitrospira bacteria. Nature 528:504. https://doi.org/10.1038/nature16461

Davidson J, Good C, Barrows FT, Welsh C, Kenney PB, Summerfelt ST (2013) Comparing the effects of feeding a grain- or a fish meal-based diet on water quality, waste production, and rainbow trout *Oncorhynchus mykiss* performance within low exchange water recirculating aquaculture systems. Aquac Eng 52:45–57. https://doi.org/10.1016/J.AQUAENG.2012.08.001

Delaide B(2017) A study on the mineral elements available in aquaponics, their impact on lettuce productivity and the potential improvement of their availability. PhD thesis. Gembloux Agro-Bio Tech, University of Liege

Delaide B, Goddek S, Gott J, Soyeurt H, Haissam Jijakli M, Lalman J, Junge R (2016) Lettuce (*Lactuca sativa* L. var. Sucrine) growth performance in complemented aquaponic solution outperforms hydroponics. Water 8. https://doi.org/10.3390/w8100467

Delaide B, Delhaye G, Dermience M, Gott J, Soyeurt H, Jijakli MH (2017) Plant and fish production performance, nutrient mass balances, energy and water use of the PAFF box, a small-scale aquaponic system. Aquac Eng 78:130–139. https://doi.org/10.1016/j.aquaeng.2017.06.002

Fitzgerald CM, Camejo P, Oshlag JZ, Noguera DR (2015) Ammonia-oxidizing microbial communities in reactors with efficient nitrification at low-dissolved oxygen. Water Res 70:38–51. https://doi.org/10.1016/J.WATRES.2014.11.041

Geay F, Ferraresso S, Zambonino-Infante JL, Bargelloni L, Quentel C, Vandeputte M, Kaushik S, Cahu CL, Mazurais D (2011) Effects of the total replacement of fish-based diet with plant-based diet on the hepatic transcriptome of two European sea bass (*Dicentrarchus labrax*) half-sibfamilies showing different growth rates with the plant-based diet. BMC Genomics 12:522. https://doi.org/10.1186/1471-2164-12-522

Gentile ME, Lynn Nyman J, Criddle CS (2007) Correlation of patterns of denitrification instability in replicated bioreactor communities with shifts in the relative abundance and the denitrification patterns of specific populations. ISME J 1:714–728. https://doi.org/10.1038/ismej.2007.87

Goddek S, Keesman KJ (2018) The necessity of desalination technology for designing and sizing multi-loop aquaponics systems. Desalination 428:76–85. https://doi.org/10.1016/J.DESAL.2017.11.024

Goddek S, Körner O (2019) A fully integrated simulation model of multi-loop aquaponics: a case study for system sizing in different environments. Agric Syst 171:143

Goddek S, Delaide B, Mankasingh U, Ragnarsdottir KV, Jijakli H, Thorarinsdottir R (2015) Challenges of sustainable and commercial aquaponics. Sustainability 7:4199–4224. https://doi.org/10.3390/su7044199

Goddek S, Schmautz Z, Scott B, Delaide B, Keesman K, Wuertz S, Junge R (2016) The effect of anaerobic and aerobic fish sludge supernatant on hydroponic lettuce. Agronomy 6:37. https://doi.org/10.3390/agronomy6020037

Goddek S, Delaide BPL, Joyce A, Wuertz S, Jijakli MH, Gross A, Eding EH, Bläser I, Reuter M, Keizer LCP et al (2018) Nutrient mineralization and organic matter reduction performance of RAS-based sludge in sequential UASB-EGSB reactors. Aquac Eng 83:10–19

Graber A, Junge R (2009) Aquaponic systems: nutrient recycling from fish wastewater by vegetable production. Desalination 246:147–156. https://doi.org/10.1016/j.desal.2008.03.048

Guangzhi G (2001) Mass balance and water quality in aquaculture tanks. The United Nations University, Fisheries Training Programme, Reyjavik

Hu B, Shen L, Xu X, Zheng P (2011) Anaerobic ammonium oxidation (anammox) in different natural ecosystems. Biochem Soc Trans 39:1811–1816. https://doi.org/10.1042/BST20110711

Hu Z, Lee JW, Chandran K, Kim S, Brotto AC, Khanal SK (2015) Effect of plant species on nitrogen recovery in aquaponics. Bioresources 188:92–98. https://doi.org/10.1016/j.biortech.2015.01.013

Hua K, Bureau DP (2012) Exploring the possibility of quantifying the effects of plant protein ingredients in fish feeds using meta-analysis and nutritional model simulation-based approaches. Aquaculture 356–357:284–301. https://doi.org/10.1016/J.AQUACULTURE.2012.05.003

Jorquera M, Martínez O, Maruyama F, Marschner P, de la Luz Mora M (2008) Current and future biotechnological applications of bacterial Phytases and phytase-producing Bacteria. Microbes Environ 23:182–191. https://doi.org/10.1264/jsme2.23.182

Körner O, Aaslyng JM, Andreassen AU, Holst N (2007) Modelling microclimate for dynamic greenhouse climate control. Hortscience 42:272–279

Körner O, Gutzmann E, Kledal PR (2017) A dynamic model simulating the symbiotic effects in aquaponic systems. Acta Hortic 1170:309–316

Kuhn DD, Drahos DD, Marsh L, Flick GJ (2010) Evaluation of nitrifying bacteria product to improve nitrification efficacy in recirculating aquaculture systems. Aquac Eng 43:78–82. https://doi.org/10.1016/J.AQUAENG.2010.07.001

Le Corre KS, Valsami-Jones E, Hobbs P, Parsons SA (2005) Impact of calcium on struvite crystal size, shape and purity. J Cryst Growth 283:514–522. https://doi.org/10.1016/J.JCRYSGRO.2005.06.012

Lennard WA, Leonard BV (2006) A comparison of three different hydroponic sub-systems (gravel bed, floating and nutrient film technique) in an Aquaponic test system. Aquac Int 14:539–550. https://doi.org/10.1007/s10499-006-9053-2

Letelier-Gordo CO, Dalsgaard J, Suhr KI, Ekmann KS, Pedersen PB (2015) Reducing the dietary protein:energy (P:E) ratio changes solubilization and fermentation of rainbow trout (*Oncorhynchus mykiss*) faeces. Aquac Eng 66:22–29

Licamele J (2009) Biomass production and nutrient dynamics in an aquaponics system. PhD thesis. Department of Agriculture and biosystems engineering, University of Arizona

Losordo TM, Masser MP, Rakocy J (1998) Recirculating aquaculture tank production systems: an overview of critical considerations. SRAC No. 451, pp 18–31

Madigan MT, Martinko JM (2007) Biologie des micro-organismes, 11th edn. Pearson Education France, Paris

Meriac A, Eding EH, Schrama J, Kamstra A, Verreth JAJ (2014) Dietary carbohydrate composition can change waste production and biofilter load in recirculating aquaculture systems. Aquaculture 420–421:254–261. https://doi.org/10.1016/j.aquaculture.2013.11.018

Michaud L, Blancheton JP, Bruni V, Piedrahita R (2006) Effect of particulate organic carbon on heterotrophic bacterial populations and nitrification efficiency in biological filters. Aquac Eng 34:224–233. https://doi.org/10.1016/j.aquaeng.2005.07.005

Munguia-Fragozo P, Alatorre-Jacome O, Rico-Garcia E, Torres-Pacheco I, Cruz-Hernandez A, Ocampo-Velazquez RV, Garcia-Trejo JF, Guevara-Gonzalez RG (2015) Perspective for aquaponic systems: "omic" technologies for microbial community analysis. Biomed Res Int 2015:1. https://doi.org/10.1155/2015/480386

Neto MR, Ostrensky A (2015) Nutrient load estimation in the waste of Nile tilapia *Oreochromis niloticus* (L.) reared in cages in tropical climate conditions. Aquac Res 46:1309–1322. https://doi.org/10.1111/are.12280

Orozco-Mosque MC, Rocha-Granados MC, Glick BR, Santoyo G (2018) Microbiome engineering to improve biocontrol and plant growth-promoting mechanism. Microbiol Res. In Press

Palm HW, Knaus U, Appelbaum S, Goddek S, Strauch SM, Vermeulen T, Haïssam Jijakli M, Kotzen B (2018) Towards commercial aquaponics: a review of systems, designs, scales and nomenclature. Aquac Int 26:813–842

Prabhu AS, Fageria NK, Berni RF, Rodrigues FA (2007) Phosphorus and plant disease. In: Datnoff LE, Elmer WH, Huber DM (eds) Mineral nutrition and plant disease. The American Phytopathological Society, St. Paul, pp 45–55

Rafiee G, Saad CR (2005) Nutrient cycle and sludge production during different stages of red tilapia (Oreochromis sp.) growth in a recirculating aquaculture system. Aquaculture 244:109–118. https://doi.org/10.1016/J.AQUACULTURE.2004.10.029

Rakocy JE, Shultz RC, Bailey DS, Thoman ES (2004) Aquaponic production of tilapia and basil: comparing a batch and staggered cropping system. Acta Hortic 648:63–69. https://doi.org/10.17660/ActaHortic.2004.648.8

Rakocy JE, Masser MP, Losordo TM (2006) Recirculating aquaculture tank production systems: aquaponics- integrating fish and plant culture. SRAC Publ South Reg Aquac Cent 16. https://doi.org/454

Randall D, Tsui TK (2002) Ammonia toxicity in fish. Mar Pollut Bull 45:17–23. https://doi.org/10.1016/S0025-326X(02)00227-8

Resh HM (2013) Hydroponic food production: a definitive guidebook for the advanced home gardener and the commercial hydroponic grower, 7th edn. CRC Press, Boca Raton

Roosta HR, Hamidpour M (2011) Effects of foliar application of some macro- and micro-nutrients on tomato plants in aquaponic and hydroponic systems. Sci Hortic 129:396–402. https://doi.org/10.1016/J.SCIENTA.2011.04.006

Roosta HR, Hamidpour M (2013) Mineral nutrient content of tomato plants in Aquaponic and hydroponic systems: effect of foliar application of some macro- and micro-nutrients. J Plant Nutr 36:2070–2083. https://doi.org/10.1080/01904167.2013.821707

Ru D, Liu J, Hu Z, Zou Y, Jiang L, Cheng X, Lv Z (2017) Improvement of aquaponic performance through micro- and macro-nutrient addition. Environ Sci Pollut Res 24:16328–16335. https://doi.org/10.1007/s11356-017-9273-1

Rurangwa E, Verdegem MCJ (2013) Microorganisms in recirculating aquaculture systems and their management. Rev Aquac 7:117–130. https://doi.org/10.1111/raq.12057

Schmautz Z, Graber A, Mathis A, Griessler Bulc T, Junge R (2015) Tomato production in aquaponic system: mass balance and nutrient recycling (abstract)

Schmautz Z, Loeu F, Liebisch F, Graber A, Mathis A, Bulc TG, Junge R (2016) Tomato productivity and quality in aquaponics: comparison of three hydroponic methods. Water 8:1–22. https://doi.org/10.3390/w8110533

Schmautz Z, Graber A, Jaenicke S, Goesmann A, Junge R, Smits THM (2017) Microbial diversity in different compartments of an aquaponics system. Arch Microbiol 199:613. https://doi.org/10.1007/s00203-016-1334-1

Schneider O, Sereti V, Eding EH, Verreth JAJ (2004) Analysis of nutrient flows in integrated intensive aquaculture systems. Aquac Eng 32:379–401. https://doi.org/10.1016/j.aquaeng.2004.09.001

Seawright DE, Stickney RR, Walker RB (1998) Nutrient dynamics in integrated aquaculture–hydroponics systems. Aquaculture 160:215–237

Shoda M (2014) Heterotrophic nitrification and aerobic denitrification by *Alcaligenes faecalis*. J Biosci Bioeng 117:737–741. https://doi.org/10.5772/68052

Somerville C, Stankus A, Lovatelli A (2014) Small-scale aquaponic food production. Integrated fish and plant farming. Food and Agriculture Organisation of the United Nations, Rome

Sonneveld C, Voogt W (2009) Plant nutrition of greenhouse crops. Springer, Dordrecht/Heidelberg/London/New York

Sugita H, Nakamura H, Shimada T (2005) Microbial communities associated with filter materials in recirculating aquaculture systems of freshwater fish. Aquaculture 243:403. https://doi.org/10.1016/j.aquaculture.2004.09.028

Suhl J, Dannehl D, Kloas W, Baganz D, Jobs S, Scheibe G, Schmidt U (2016) Advanced aquaponics: evaluation of intensive tomato production in aquaponics vs. conventional hydroponics. Agric Water Manag 178:335–344. https://doi.org/10.1016/j.agwat.2016.10.013

Tacon AGJ, Metian M (2008) Global overview on the use of fish meal and fish oil in industrially compounded aquafeeds: trends and future prospects. Aquaculture 285:146–158. https://doi.org/10.1016/j.aquaculture.2008.08.015

Timmons MB, Ebeling JM (2013) Recirculating aquaculture. Ithaca Publishing, New York

Turcios AE, Papenbrock J (2014) Sustainable treatment of aquaculture effluents—what can we learn from the past for the future? Sustainability 6:836–856. https://doi.org/10.3390/su6020836

van Lier JB, Mahmoud N, Zeeman G (2008) In: Henze M, van Loosdrecht MCM, Ekama GA, Brdjanovic D (eds) Anaerobic wastewater treatment, in: biological wastewater treatment: principles, modelling and design. IWA Publishing, London. ISBN: 9781843391883

van Rijn J (2013) Waste treatment in recirculating aquaculture systems. Aquac Eng 53:49–56. https://doi.org/10.1016/J.AQUAENG.2012.11.010

Wongkiew S, Hu Z, Chandran K, Lee JW, Khanal SK (2017) Nitrogen transformations in aquaponic systems: a review. Aquac Eng 76:9–19. https://doi.org/10.1016/j.aquaeng.2017.01.004

Xu G, Fan X, Miller AJ (2012) Plant nitrogen assimilation and use efficiency. Annu Rev Plant Biol 63:153–182. https://doi.org/10.1146/annurev-arplant-042811-105532

Yildiz HY, Robaina L, Pirhonen J, Mente E, Domínguez D, Parisi G (2017) Fish welfare in Aquaponic systems: its relation to water quality with an emphasis on feed and Faeces—a review. Water 9. https://doi.org/10.3390/w9010013

Yogev U, Barnes A, Gross A (2016) Nutrients and energy balance analysis for a conceptual model of a three loops off grid, aquaponics. Water 8. https://doi.org/10.3390/w8120589

Zekki H, Gauthier L, Gosselin A (1996) Growth, productivity, and mineral composition of hydroponically cultivated greenhouse tomatoes, with or without nutrient solution recycling. J Am Soc Hortic Sci 121:1082–1088

Zou Y, Hu Z, Zhang J, Xie H, Guimbaud C, Fang Y (2016) Effects of pH on nitrogen transformations in media-based aquaponics. Bioresour Technol 210:81–87. https://doi.org/10.1016/J.BIORTECH.2015.12.079

Chapter 10
Aerobic and Anaerobic Treatments for Aquaponic Sludge Reduction and Mineralisation

Boris Delaide, Hendrik Monsees, Amit Gross, and Simon Goddek

Abstract Recirculating aquaculture systems, as part of aquaponic units, are effective in producing aquatic animals with a minimal water consumption through effective treatment stages. Nevertheless, the concentrated sludge produced after the solid filtration stage, comprising organic matter and valuable nutrients, is most often discarded. One of the latest developments in aquaponic technology aims to reduce this potential negative environmental impact and to increase the nutrient recycling by treating the sludge on-site. For this purpose, microbial aerobic and anaerobic treatments, dealt with either individually or in a combined approach, provide very promising opportunities to simultaneously reduce the organic waste as well as to recover valuable nutrients such as phosphorus. Anaerobic sludge treatments additionally offer the possibility of energy production since a by-product of this process is biogas, i.e. mainly methane. By applying these additional treatment steps in aquaponic units, the water and nutrient recycling efficiency is improved and the dependency on external fertiliser can be reduced, thereby enhancing the sustainability of the system in terms of resource utilisation. Overall, this can pave the way for the economic improvement of aquaponic systems because costs for waste disposal and fertiliser acquisition are decreased.

B. Delaide (✉)
Developonics asbl, Brussels, Belgium
e-mail: boris@developonics.com; borisdelaide@hotmail.com

H. Monsees
Leibniz-Institute of Freshwater Ecology and Inland Fisheries, Berlin, Germany
e-mail: h.monsees@igb-berlin.de

A. Gross
Department of Environmental Hydrology and Microbiology, Zuckerberg Institute for Water Research, Blaustein Institutes for Desert Research, Ben-Gurion University of the Negev, Beersheba, Israel
e-mail: amgross@bgu.ac.il

S. Goddek
Mathematical and Statistical Methods (Biometris), Wageningen University, Wageningen, The Netherlands
e-mail: simon.goddek@wur.nl; simon@goddek.nl

S. Goddek et al. (eds.), *Aquaponics Food Production Systems*,
https://doi.org/10.1007/978-3-030-15943-6_10

Keywords Sludge recycling · Phosphorus · Microbial sludge conversion · Mass balance · Nutrient recycling

10.1 Introduction

The concept of aquaponics is associated with being a sustainable production system, as it re-utilises recirculating aquaculture system (RAS) wastewater enriched in macronutrients (i.e. nitrogen (N), phosphorus (P), potassium (K), calcium (Ca), magnesium (Mg) and sulphur (S)) and micronutrients (i.e. iron (Fe), manganese (Mn), zinc (Zn), copper (Cu), boron (B) and molybdenum (Mo)) to fertilise the plants (Graber and Junge 2009; Licamele 2009; Nichols and Savidov 2012; Turcios and Papenbrock 2014). A much debated question is whether this concept can match its own ambition of being a quasi-closed-loop system, as high amounts of the nutrients that enter the system are wasted by discharging the nutrient-rich fish sludge (Endut et al. 2010; Naylor et al. 1999; Neto and Ostrensky 2013). Indeed, to maintain a good water quality in a RAS and aquaponic systems, the water has to be constantly filtrated for solid removal. The two main techniques for solid filtration are to retain the particles in a mesh (i.e. mesh filtration as drum filters) and to allow the particles to decant in clarifiers. In most conventional plants, sludge is recovered out of these mechanical filtration devices and is discharged as sewage. In the best cases, the sludge is dried and applied as fertiliser on land fields (Brod et al. 2017). Notably, up to 50% (in dry matter) of the feed ingested is excreted as solids by fish (Chen et al. 1997), and most of the nutrients that enter aquaponic systems via fish feed accumulate in these solids and so in the sludge (Neto and Ostrensky 2013; Schneider et al. 2005). Hence, effective solid filtration removes, for example, more than 80% of the valuable P (Monsees et al. 2017) that could otherwise be used for plant production. Therefore, recycling these valuable nutrients for aquaponic applications is of major importance. Developing an appropriate sludge treatment able to mineralise the nutrients contained in sludge for re-using them in the hydroponic unit seems to be a necessary process for contributing to close the nutrient loop to a higher degree and thus lowering the environmental impact of aquaponic systems (Goddek et al. 2015; Goddek and Keesman 2018; Goddek and Körner 2019).

It has been shown in experimental studies that complemented aquaponic nutrient solution (i.e. after addition of lacking nutrients) promotes plant growth compared to hydroponics (Delaide et al. 2016; Ru et al. 2017; Saha et al. 2016). Hence, sludge mineralisation is also a promising way to improve the aquaponic system performance as the nutrients recovered are used to complement the aquaponic solution. In addition, on-site mineralisation units can also increase the self-sufficiency of aquaponic facilities, especially with respect to finite resources as P which is essential for plant growth. P is produced by mining activities, whereby the deposits are not equally distributed around the world. In addition, its price has risen by up to 800% within the last decade (McGill 2012). Thus, mineralisation units applied in aquaponic systems are also likely to increase its future economic success and stability.

Sludge treatment in aquaponics needs to be approached differently than has been done in the past. Indeed, in conventional wastewater treatment, the main objective is to obtain a decontaminated and clean effluent. The treatment performances are expressed in terms of removal of contaminants (e.g. solids, nitrogen (N), phosphorus (P), etc.) out of the wastewater and by quantifying the effluents with respect to the achieved quality (Techobanoglous et al. 2014). Using this conventional approach, several studies have provided quantitative evidence that a consistent proportion of chemical oxygen demand (COD) and total suspended solids (TSS) can be removed by digesting the RAS wastewater under aerobic, anaerobic and sequential aerobic–anaerobic conditions (Goddek et al. 2018; Chowdhury et al. 2010; Mirzoyan et al. 2010; Van Rijn 2013). However, in aquaponic systems, the wastewater from fish is considered to be a valuable fertiliser source. Within a closed-loop approach, the solid part discharged needs to be minimised (i.e. organic reduction maximised), and the nutrient content in the effluents needs to be maximised (i.e. nutrient mineralisation maximised). Therefore, the wastewater treatment performance in aquaponics no longer needs to be expressed in terms of contaminants removal but in terms of its contaminant reduction and nutrient mineralisation ability.

A few studies have demonstrated the functional capability of digesting fish sludge with aerobic and anaerobic treatments for organic reduction purposes (Goddek et al. 2018; van Rijn et al. 1995). With anaerobic treatment in bioreactor, high dry matter (i.e. TSS) reduction performance (e.g. higher than 90%) can be achieved while methane can also be produced (van Lier et al. 2008; Mirzoyan and Gross 2013; Yogev et al. 2016).

The aerobic treatment of sludge is also a very effective way to reduce organic matter, which is oxidised to CO_2 during respiration (see Eq. 10.1). For example, reduction rates of 90% (here determined as suspended solids, COD and BOD reduction) were reported from a water resource recovery facility (Seo et al. 2017). Aerobic processes are faster than anaerobic, but they can be more expensive (Chen et al. 1997) as a constant aeration of the sludge–water mixture requires energy-intensive pumps or motors. Moreover, significant fractions of the nutrients are converted to microbial biomass and do not stay dissolved in the water.

Although these studies have shown the organic reduction potential of fish sludge, only a few authors have examined the release of specific nutrients (e.g. for N and P) from fish sludge. Most of these studies were for short in vitro batch experiments (Conroy and Couturier 2010; Monsees et al. 2017; Stewart et al. 2006) and from an operating RAS (Yogev et al. 2016), rather than an aquaponic setup. While discussed to some extent in theory (Goddek et al. 2016; Yogev et al. 2017), the research has to start now to systematically investigate the organic reduction and nutrient mineralisation performance of fish sludge for both aerobic and anaerobic reactors and its effect on the water composition and plant growth. Therefore, this chapter aims to give an overview on the diverse fish sludge treatments that can be integrated into aquaponic setups to achieve organic reduction and nutrient mineralisation. Some design approaches will be highlighted. The nutrient mass balance approach in the context of aquaponic sludge treatment will be discussed, and specific methodology to quantify the sludge treatment performance will be developed.

10.2 Wastewater Treatment Implementation in Aquaponics

In aquaponics, the wastewater charged with solids (i.e. the sludge) is a valuable source of nutrients, and appropriate treatments need to be carried out. The treatment goals differ from conventional wastewater treatment because in aquaponics solids and water conservation is of interest. Moreover, regardless of the wastewater treatment applied, its aim should be to reduce solids and at the same time mineralise its nutrients. In other words, the aim is to obtain a solid-free effluent but rich in solubilised nutrients (i.e. anions and cations) that can be reinserted into the water loop in a coupled setup (Fig. 10.1a) or directly into the hydroponic grow beds in a decoupled setup (Fig. 10.1b). Fish sludge solids are mainly composed of degradable organic matter so that the solid reduction can be called organic reduction. Indeed, the complex organic molecules (e.g. proteins, lipids, carbohydrates, etc.) are principally composed of carbon and will be successively reduced to lower molecular weight compounds until the ultimate gaseous forms of CO_2 and CH_4 (in the case of anaerobic fermentation). During this degradation process, the macronutrients (i.e. N, P, K, Ca, Mg and S) and micronutrients (i.e. Fe, Mn, Zn, Cu, B and Mo) that were bound to the organic molecules are released into the water in their ionic forms. This phenomenon is called nutrient leaching or nutrient mineralisation. It can be assumed that when high organic reduction is achieved, high nutrient mineralisation would also be achieved. On the one hand, sludge contains a proportion of undissolved minerals, and on the other hand, some macro- and micronutrients are released during the mineralisation process. These can quickly precipitate together and form insoluble minerals. The state between ions and precipitated minerals of most of the macro- and micronutrients is pH dependent. The most well-known minerals that precipitate in bioreactors are calcium phosphate, calcium sulphate, calcium carbonate, pyrite and struvite (Peng et al. 2018; Zhang et al. 2016). Conroy and Couturier (2010) observed that Ca and P were released in anaerobic reactor when the pH dropped under 6. They showed that the release corresponded exactly to the mineralisation of calcium phosphate. Goddek et al. (2018) also observed the solubilisation of P, Ca and other macronutrients in upflow anaerobic sludge blanket reactor (UASB) that turned acidic. Jung and Lovitt (2011) reported a 90% nutrient mobilisation of aquaculture-derived sludge at a very low pH value of 4. In this condition, all the macro- and micronutrients were solubilised. There is thus an antagonism between organic reduction and nutrient mineralisation. Indeed, organic reduction is maximal when the microorganisms are active for degrading the organic compounds, and this happens at pH in a range of 6–8. Because nutrient leaching occurs at pH below 6, for optimal organic reduction and nutrient mineralisation, the most effective would be to divide the process in two steps, i.e. an organic reduction step at pH close to neutral and a nutrient leaching step under acidic conditions. To our knowledge, no operation using this two-step approach has been yet reported. This opens a new field in wastewater treatment and more research for implementation in aquaponics is needed.

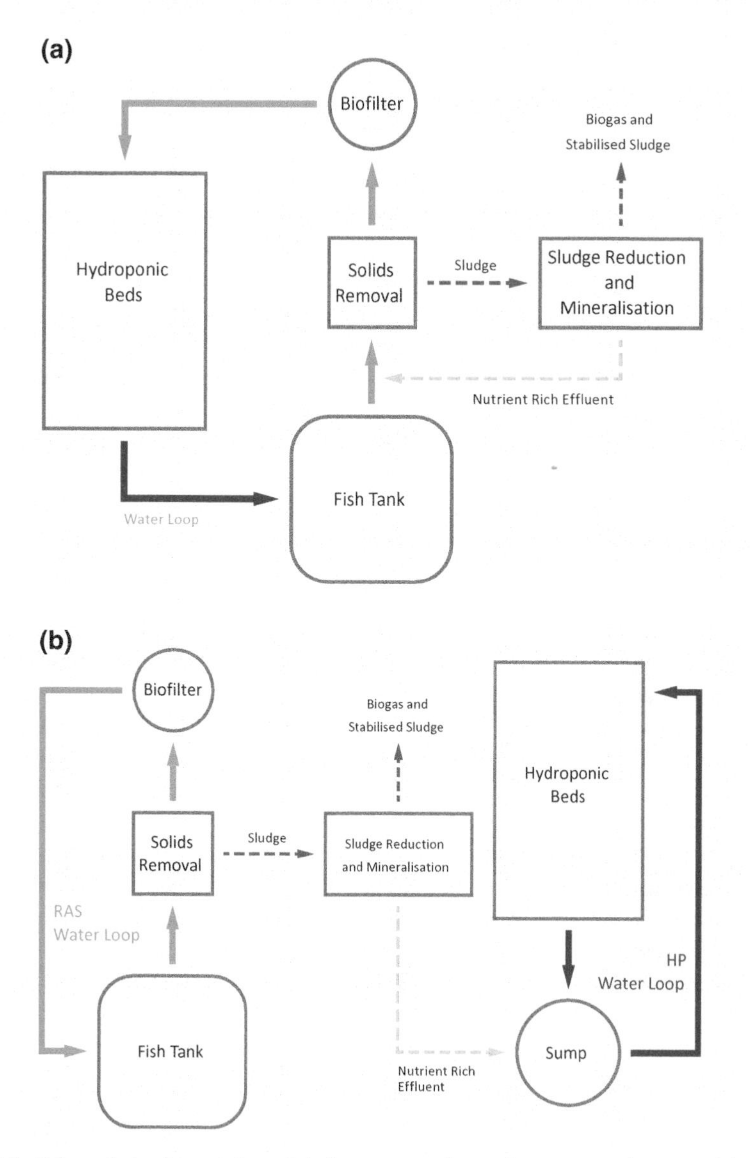

Fig. 10.1 Schematic implementation of sludge treatment in one loop aquaponic system (**a**) and in decoupled aquaponic system (**b**)

The choice of feed is also important in this context. In animal-based feeds where a major ingredient fraction is still based on animal sources (e.g. fishmeal, bone meal), bound phosphate, e.g. as apatite (derived from bone meal), is easily available under acidic conditions, whereas plant-based feeds contain phytate as a major phosphate source. Phytate in contrast to, e.g. apatite requires enzymatic (phytase) conversion (Kumar et al. 2012), and so the phosphate is not as easily available.

10.3 Aerobic Treatments

Aerobic treatment enhances the oxidation of the sludge by supporting its contact with oxygen. In this case, the oxidation of the organic matter is driven mainly by the respiration of heterotrophic microorganisms. CO_2, the end product of respiration, is released as is shown in Eq. (10.1).

$$C_6H_{12}O_6 + 6\,O_2 \rightarrow 6\,CO_2 + 6\,H_2O + \text{energy} \tag{10.1}$$

This process in aerobic reactors is mainly achieved by injecting air into the sludge–water mixture with air blowers connected to diffusers and propellers. Air injection also ensures a proper mixing of the sludge.

During this oxidative process, the macro- and micronutrients bound to the organic matter are released. This process is called aerobic mineralisation. Therefore, further nutrients can be recycled during the mineralisation process, whereas some nutrients, e.g. sodium and chloride, can also exceed their threshold for hydroponic application and must be monitored carefully before application (Rakocy et al. 2007). Aerobic mineralisation of organic matter, derived from the solid removal unit (e.g. clarifier or drum filter) in RAS, is an easy way to recycle nutrients for subsequent aquaponic application.

Moreover, during the aerobic digestion process, the pH drops and promotes the mineralisation of bound minerals trapped in the sludge. For example, Monsees et al. (2017) showed that P was released from RAS sludge due to this pH shift. This decrease in pH is mainly driven by respiration and to a lower extent probably by nitrification.

Due to a constant supply of oxygen via aeration of the mineralisation chamber and the abundance of organic matter, heterotrophic microorganisms find ideal conditions to grow. This results in an increase of respiration and the release of CO_2 that dissolves in water. CO_2 forms carbonic acid which dissociates and thereby lowers the pH of the process water as illustrated in the following equation:

$$CO_{2(g)} + 2\,H_2O \rightarrow H_3O^+ + HCO_3^- \tag{10.2}$$

RAS-derived wastewater often contains NH_4^+ and additionally is characterised by a neutral pH of around 7, because the pH in RAS needs to be kept at that level to ensure optimal microbial conversion of NH_4^+ to NO_3^- within the biofilter (i.e. nitrification). The nitrification process can contribute to the decrease in pH in aerobic reactors in the starting phase by releasing protons to the process water as can been seen in the following equation:

$$NH_4^+ + 2\,O_2 \rightarrow NO_3^- + 2\,H^+ + H_2O + \text{energy} \tag{10.3}$$

This is at least valid for the starting phase where the pH is still above 6. At a pH $\leq$ 6, nitrification might significantly slow down or even cease (Ebeling et al. 2006). However, this does not represent a problem for the mineralisation unit.

The general decrease of the pH in the aerobic mineralisation unit in the ongoing process is the main driver of the release of nutrients present under the form of precipitated minerals as calcium phosphates. Monsees et al. (2017) noted that around 50% of the phosphate in the sludge was acid soluble, derived from a tilapia RAS where a standard feed containing fishmeal was applied. Here, around 80% of the phosphate within the RAS was lost by the cleaning of the decanter and the discarding of the sludge–water mixture. Considering this fact, the big potential of mineralisation units for aquaponic applications becomes clears.

The advantages of aerobic mineralisation are the low maintenance with no need for skilled personnel and no subsequent reoxygenation. The enriched water can be used directly for plant fertilisation, ideally managed by an online system for the adequate preparation of the nutrient solution. A disadvantage compared to anaerobic mineralisation is that no methane is produced (Chen et al. 1997) and, as already mentioned, the higher energy demand due to the need for constant aeration.

10.3.1 Aerobic Mineralisation Units

A design example of an aerobic mineralisation unit is presented in Fig. 10.2. The inlet is connected to the solid removal unit via a valve, which allows discontinuous refilling of the mineralisation chamber with a mixture of sludge and water. The

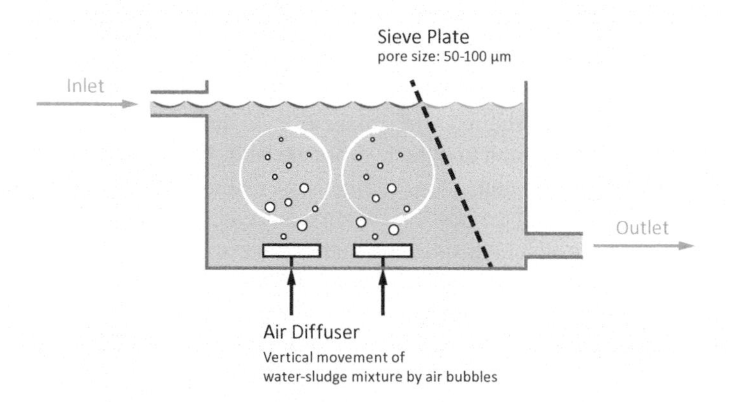

Fig. 10.2 Schematic example of an aerobic mineralisation unit operated in a batch mode. Mineralisation chamber (brown) is separated from the outlet chamber (blue) by a sieve plate that is covered by a solid cover plate during the mineralisation process (strong aeration) to prevent clogging and formation of fine particles. Organic-rich water from a clarifier or drum filter enters the mineralisation unit via the inlet. After a mineralisation cycle is completed, nutrient-rich, solid-free water exits the mineralisation unit via the outlet and is either directly transferred to the hydroponic unit or kept in a storage tank until needed

mineralisation chamber is aerated via compressed air to promote the respiration of heterotrophic bacteria and to keep anaerobic denitrification processes as minimal as possible. To prevent organic material from leaving the mineralisation chamber, a sieve plate could serve as a barrier. Ideally, a second, impermeable cover plate should be used to cover the sieve during the mineralisation process (during aeration). This should prevent the sieve plate from clogging as during the heavy aeration the organic material would be constantly moved against the sieve plate. Before transferring the nutrient-rich water from the mineralisation chamber to the hydroponic unit, aeration is stopped to allow the particles to settle. Subsequently, the cover plate is removed, and the nutrient-enriched water can pass through the sieve plate and leave the mineralisation chamber via the outlet as suggested in Fig. 10.2. Finally, the cover plate is put in place again, mineralisation chamber is refilled with RAS-derived sludge–water mixture, and the mineralisation process starts again (i.e. batch process).

The mineralisation unit should have at least twice the volume of the clarifier to allow for a continuous mineralisation. One mineralisation cycle can last for up to 5–30 days depending on the system, organic load and required nutrient profile and has to be elaborated for each individual system. For systems including a drum filter, as it is the case in most modern RAS, the mineralisation unit size has to be adjusted according to the daily or weekly sludge outflow of the drum filter. Since that has not been tested in an experimental setup so far, specific recommendations are not currently possible.

10.3.2 Implementation

An example of the implementation of an aerobic mineralisation unit into a decoupled aquaponic system is presented in Fig. 10.3. Since no pre- and post-treatment (e.g. re-oxygenation) is required, the mineralisation unit can be directly placed between the solid removal unit and the hydroponic beds. By installing a valve before and after the mineralisation unit, a discontinuous operation and nutrient delivery to the hydroponic unit on-demand are possible, but in many cases, an additional storage tank would be required. Ideally, after directing nutrient-rich water to the hydroponic unit, the displaced water is replaced with new sludge and water from the solid removal unit. Depending on the volume of the mineralisation unit, it is important to note that refilling with new sludge–water mixture can lead to an increase in pH again, and thus the mineralisation process could be interrupted. By increasing the size of the mineralisation unit, this effect would be buffered. In the study by Rakocy et al. (2007) investigating liquid organic waste from two aquaculture systems, a retention time of 29 days for aerobic mineralisation resulted in a substantial mineralisation success. Nevertheless, this also depends on the TS content within the mineralisation chamber, on the feed applied to the RAS, on the temperature and on the nutrient requirements of the plants that are produced within the hydroponic unit.

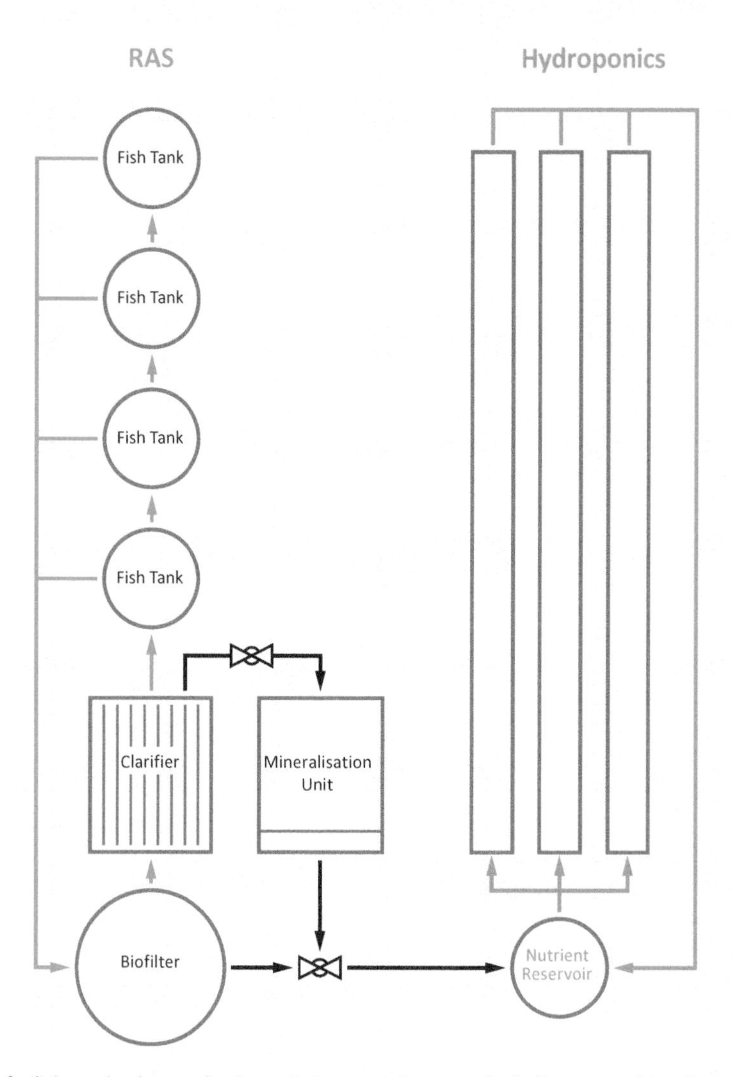

Fig. 10.3 Schematic picture of a decoupled aquaponic system including an aerobic mineralisation unit. Water can be transferred to the nutrient reservoir either from the RAS water loop or directly from the mineralisation unit

10.4 Anaerobic Treatments

Anaerobic digestion (AD) has long been used for the stabilisation and reduction of sludge mass process, mainly because of the simplicity of operation, relatively low costs and production of biogas as potential energy source. General stoichiometric representation of anaerobic digestion can be described as follows:

$$CnHaOb + (n - a/4 - b/2) \cdot H_2O \rightarrow (n/2 - a/8 + b/4) \cdot CO_2 + (n/2 + a/8 - b/4) \cdot CH4 \quad (10.4)$$

Equation 10.4 Biogas general mass balance (Marchaim 1992).

And the theoretical methane concentration can be calculated as follows:

$$[CH_4] = 0.5 + (a/4 + b/2)/2n \quad (10.5)$$

Equation 10.5 Theoretical expected methane concentration in the biogas (Marchaim 1992).

The ultimate products from AD are mostly inorganic material (e.g. minerals), slightly degraded organic compounds and biogas which is typically composed of >55% methane (CH_4) and carbon dioxide (CO_2), with only small levels (<1%) of hydrogen sulphide (H_2S) and total ammonia nitrogen ($NH_3^+/NH4^+$) (Appels et al. 2008).

During the process of AD, the organic sludge undergoes considerable changes in its physical, chemical and biological properties and schematically can be divided into four stages (Fig. 10.4). The first stage is hydrolysis, where complex organic matter such as lipids, polysaccharides, proteins and nucleic acids degrade into soluble organic substances (sugars, amino acids and fatty acids). This step is generally considered rate-limiting (Deublein and Steinhauser 2010). In the second acidogenesis step, the monomers formed in the first step split further, and volatile fatty acids (VFA) are produced by acidogenic (fermentative) bacteria along with ammonia, CO_2, H_2S and other by-products. The third step is acetogenesis, where the VFA and alcohols are further digested by acetogens to produce mainly acetic acid as well as CO_2 and H_2. This conversion is controlled to a large extent by the partial pressure of H_2 in the mixture. The last step is methanogenesis where methane is mainly produced by two groups of methanogenic bacteria: acetotrophic archaea, which split acetate into methane and CO_2, and hydrogenotrophic archaea, which use hydrogen as an electron donor and carbon dioxide as electron acceptor to produce methane (Appels et al. 2008).

Various factors such as sludge pH, salinity, mineral composition, temperature, loading rate, hydraulic retention time (HRT), carbon-to-nitrogen (C/N) ratio and volatile fatty acid content influence the digestibility of the sludge and the biogas production (Khalid et al. 2011).

Anaerobic sludge treatment from RAS began about 30 years ago with reports on sludge from freshwater RAS (Lanari and Franci 1998) followed by reports on marine (Arbiv and van Rijn 1995; Klas et al. 2006; McDermott et al. 2001) and brackish water operations (Gebauer and Eikebrokk 2006; Mirzoyan et al. 2008). Recently, the use of UASB (Fig. 10.5) for AD of RAS sludge followed by biogas production as an alternative source of energy was suggested (Mirzoyan et al. 2010). The reactor is made of a tank, part of which is filled with an anaerobic granular sludge blanket containing the active microorganism species. Sludge flows upwards through a 'microbial blanket' where it is degraded by the anaerobic microorganisms and

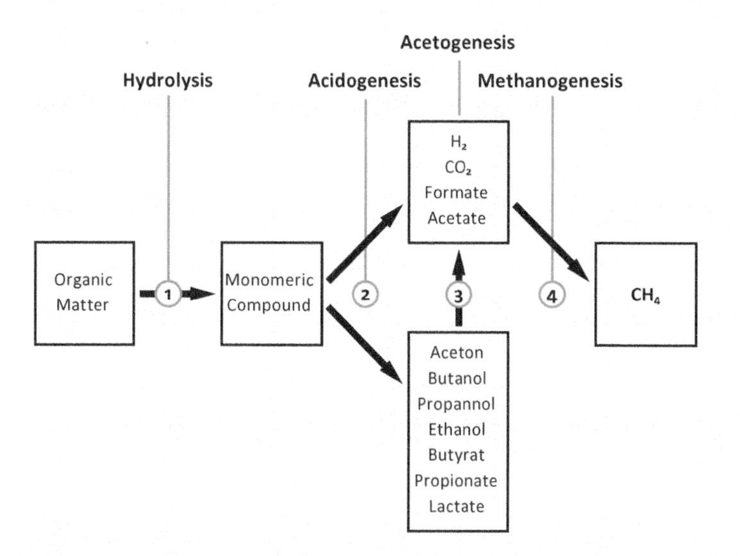

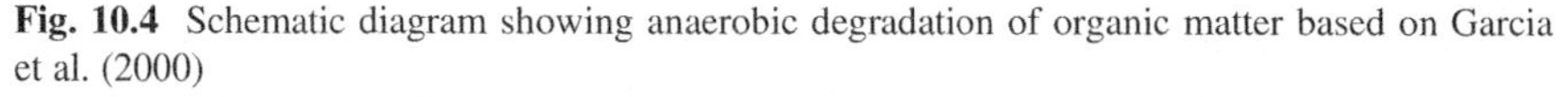

Fig. 10.4 Schematic diagram showing anaerobic degradation of organic matter based on Garcia et al. (2000)

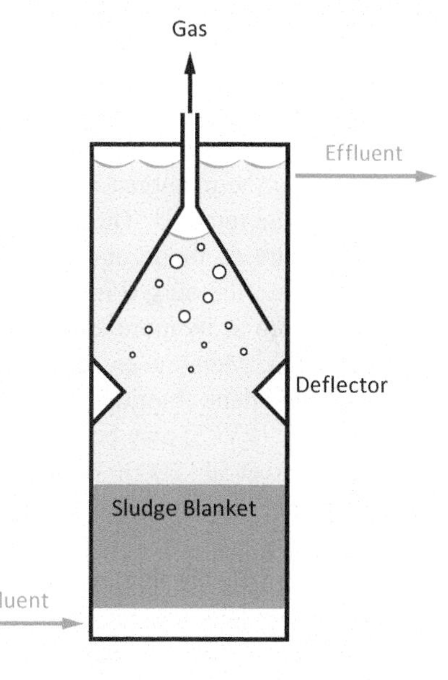

Fig. 10.5 Scheme of an upflow anaerobic sludge blanket reactor (UASB)

biogas is produced. An inverted cone settler at the top of the digester allows gas–liquid separation. When the biogas is released from the floc, it is oriented into the cone by the deflectors to be collected. A slow mixing in the reactor results from the upwards flow coupled with the natural movement of the microbial flocs that are attached to biogas bubbles. At some point, the floc leaves the gas bubble and settles back down allowing for the effluent to be free from TSS, which can then be recycled back to the system or released. The main advantages of the UASB are the low operational costs and simplicity of operation while providing high (>92%) solid-removal efficiency for wastes with low (1–3%) TSS content (Marchaim 1992; Yogev et al. 2017).

Two recent case studies demonstrated the use of UASB as a treatment for solids in pilot scale marine and saline RAS, which provide an example of the potential advantages of this unit in aquaponics (Tal et al. 2009; Yogev et al. 2017). A detailed look at the carbon balance suggested that about 50% of the introduced carbon (from feed) was removed by fish assimilation and respiration, 10% was removed by aerobic biodegradation in the nitrification bioreactor and 10% was removed in the denitrification reactor (Yogev et al. 2017). Therefore, overall about 25% carbon was introduced into the UASB reactor of which 12.5% was converted to methane, 7.5% to CO_2 and the rest (~5%) remained as nondegradable carbon in the UASB. In summary, it was demonstrated that the use of UASB allowed better water recirculation (>99%), smaller (<8%) production of sludge when compared with typical RAS that do not have on-site solid treatment, and recovery of energy that can account for 12% of the overall energy demand of the RAS. It should be noted that using UASB in aquaponics will also allow significant recovery of up to 50% more nutrients such as nitrogen, phosphorus and potassium since they are released into the water as a result of solid biodegradation (Goddek et al. 2018).

The anaerobic membrane bioreactor (AnMBR) is a more advanced technology. The main process consists of using a special membrane to separate the solids from the liquid instead of using a decanting process as in UASB. The sludge fermentation occurs in a simple anaerobic tank and the effluents leave it through the membrane. Depending on the membrane pore size (going down to 0.1–0.5 μm) even microorganisms can be retained. There are two types of membrane bioreactor design: one uses a side-stream mode outside the tank, and the other has the membrane unit submerged into the tank (Fig. 10.6), the latter being more favourable in AnMBR application due to its more compact configuration and lower energy consumption (Chang 2014). Membranes of different materials such as ceramic or polymeric (e.g. polyvinylidene fluoride (PVDF), polyethylene, polyethersulfone (PES), polyvinyl chloride (PVC)) may be configured as plate and frame, hollow fibre or tubular units (Gander et al. 2000; Huang et al. 2010). AnMBR has several significant advantages over typical biological reactors such as the UASB, namely, decoupling of (long) sludge retention time (SRT) and (short) hydraulic residence time (HRT), hence enabling the problem of the AD process's slow kinetics to be overcome; very high effluent quality in which most nutrients remain; and removal of pathogens and a small footprint (Judd and Judd 2008). In addition, efficient biogas production in the AnMBR can possibly result in a net energy balance.

While this technology deserves a lot of attention and research, it should be noted that since it is a fairly new technology, there are still several significant drawbacks that must be addressed before AnMBR would be adopted by the aquaculture industry. These are the high operational costs due to membrane maintenance to prevent biofouling, regular membrane exchange and high CO_2 fraction (30–50%) in the biogas which limits its utilisation and contributes to greenhouse gas (GHG)

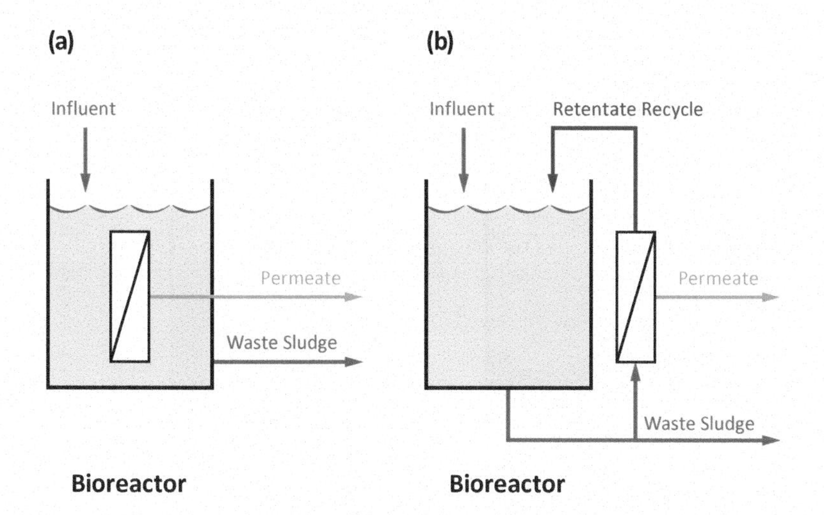

Fig. 10.6 (**a**) Side-stream MBR with a separate filtration unit with the retained fluid recycled back to the bioreactor; (**b**) submerged MBR: filtration unit integrated into the bioreactor. (Gander et al. 2000)

emission (Cui et al. 2003). On a positive note, in the near future, new biofouling prevention techniques will be developed while the membrane cost will certainly drop with the broader use of this technology. The combination of a UASB with a membrane reactor to filter the UASB effluent has been successfully studied to remove organic carbon and nitrogen (An et al. 2009). This combination seems a promising option for aquaponics for safe and sanitary use of UASB effluents.

10.4.1 Implementation

One possible solution of implementing anaerobic reactors is in a sequential manner (see also Chap. 8). A 'high pH–low pH' combination allows for harvesting methane (and thus reducing carbon) in the first high pH step and mobilising nutrients in the decarbonised sludge in a subsequent low pH environment. The advantage of this method is that the carbon reduction under high pH conditions results in less VFAs, which can occur during the low pH second step (Fig. 10.7). This approach also allows for co-digestion of green vegetative matter (i.e. from any harvesting of plants, there will be waste vegetative matter which could be put through such a digester) to increase both biogas production and nutrient recovery from the overall scheme.

Another technical integration possibility has been presented by Ayre et al. (2017). They propose to discharge the effluent of a high-pH anaerobic digester to an algal culture pond. Within that pond, algae are grown, whose biomass can be used for animal–aquaculture feed or biofertilisation (Fig. 10.8). More detailed information on this approach can be found in Chap. 11.

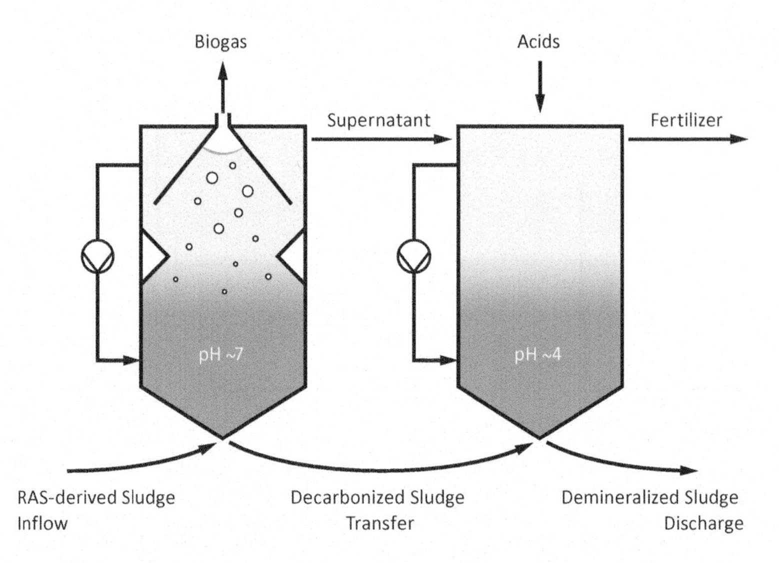

Fig. 10.7 Two-stage anaerobic system. In the first stage (high pH), the carbon will be removed from the sludge as biogas, whereas the low pH in the second stage allows nutrients that are trapped in the sludge do dissolve in the water. Usually, volatile fatty acids (VFA) would form in low pH environments. The removal of the carbon source in the first stage, however, limits VFA production in such a sequential setup

10.5 Methodology to Quantify the Sludge Reduction and Mineralisation Performance

To determine the digestion of aquaponic sludge treatment in aerobic and anaerobic bioreactors, a specific methodology needs to be followed. A methodology adapted for aquaponic sludge treatment purposes is presented in this chapter. Specific equations have been developed to precisely quantify their performance (Delaide et al. 2018), and these should be used to evaluate the performance of the treatment applied in a specific aquaponic plant.

In order to evaluate the treatment's performance, a mass balance approach needs to be achieved. It requires that TSS, COD and nutrient masses are determined for the all reactor inputs (i.e. fresh sludge) and outputs (i.e. effluents). The reactor content also needs to be sampled at the beginning and at the end of the studied period. The input, output and content of the reactors have to be perfectly mixed for sampling. Reactor input and output should basically be sampled every time the reactors are fed with fresh sludge.

Then, reactor sludge reduction performance (η) can be formulated as follows:

$$\eta_S = 100\%(1 - (\Delta S + S_{out})/S_{in})) \tag{10.6}$$

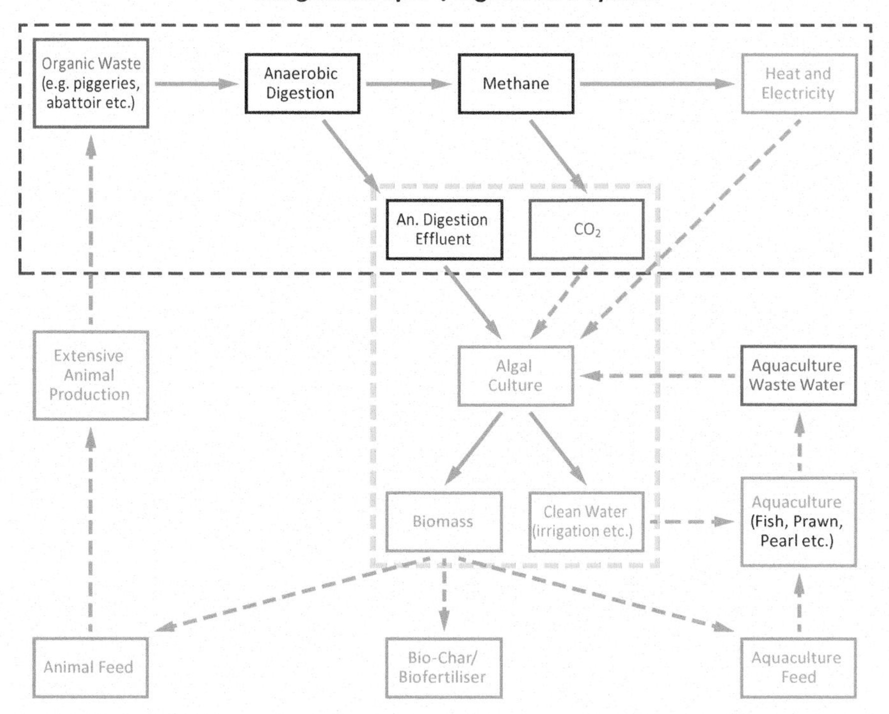

Fig. 10.8 Anaerobic digestion system integrated with aquaculture and algal culture based on Ayre et al. (2017)

where ΔS is the sludge inside the reactor at the end of the studied period minus the one at the beginning of the period, S_{out} is the total sludge that left the reactor in the outflow, and S_{in} is the total sludge that entered the reactor via inflow.

For organic reduction, the sludge (i.e. the term S) can be characterised by the dry mass of sludge (i.e. TSS) or the mass of oxygen needed to oxidise the sludge (i.e. COD). Thus, for COD and TSS reduction performances, the smaller the accumulation and the smaller the quantity in the outflow, the higher the reduction performance (i.e. high percentage) and so the less solids discharged out of the loop.

Based on the same mass balance, the nutrient mineralisation performance of the treatment (ζ), i.e. the conversion into soluble ions of the macro- and micronutrients present in the sludge under undissolved forms, the following formula can be used:

$$\zeta_N = 100\%((DN_{out} - DN_{in})/(TN_{in} - DN_{in})) \quad (10.7)$$

where ζ is the recovery of N nutrient at the end of the studied period in percent, DN_{out} is the total mass of dissolved nutrient in the outflow, DN_{in} is the total mass of dissolved nutrient in the inflow and TN_{in} is the total mass of dissolved plus undissolved nutrients in the inflow.

Thus, similar to the organic reduction performances, the smaller the accumulation inside the reactor and undissolved nutrient content in the outflow, the higher the mineralisation performance (i.e. high percentage) and so the dissolved nutrient recovered in the effluent (or outflow) for aquaponic crop fertilisation (see Example 10.1). The presented mass balance equations are used in the example box.

Example 10.1
The digestion performance of a 250-L anaerobic bioreactor has been evaluated for an 8-week period. It was fed once a day with 25 L of fresh sludge coming from a tilapia RAS system, and the equivalent supernatant volume (or output) was removed from the bioreactor. The fresh sludge (input) had a TSS of 10 g dry mass (DM) per litre or 1%, and the supernatant (output) had a TSS of 1 gDM/L or 0.1%. The TSS inside the bioreactor at the beginning and at the end of the period was 20 gDM/L. Consequently, the total DM inputs, outputs and inside the bioreactor during the evaluated period are calculated as follows:

DM in = 0.01 kg/Ld × 25 L × 7 days × 8 weeks = 14 kg
DM out = 0.001 kg/Ld × 25 L × 7 days × 8 weeks = 1.4 kg
DM to = DM tf = 250 L × 0.02 kg/L = 5 kg

The TSS reduction performance (η_{TSS}) of the bioreactor can then be calculated as follows:

$$\eta_{TSS} = 100\%(1 - ((5 - 5) + 1.4)/14)) = 90\%$$

The bioreactor P mineralisation performance can be evaluated knowing that the fresh sludge (input) had a concentration of dissolved P of 15 mg/L and a total P content of 90 mg/L. The concentration of dissolved P in the supernatant (output) was 20 mg/L. Consequently, the total P content in the input, the total dissolved P in the inputs and outputs during the evaluated period are calculated as follows:

TP in = 0.090 g/Ld × 25 L × 7 days × 8 weeks = 126 g
DP in = 0.015 g/Ld × 25 L × 7 days × 8 weeks = 21 g
DP out = 0.020 g/Ld × 25 L × 7 days × 8 weeks = 28 g

The P mineralisation performance (ζ_P) of the bioreactor can then be calculated as follows:

$$\zeta_P = 100\%((28 - 21)/(126 - 21)) = 6.67\%$$

10.6 Conclusions

Fish sludge treatment for reduction and nutrient recovery is in an early phase of implementation. Further research and improvements are needed and will see the day with the increased concern of circular economy. Indeed, fish sludge needs to be considered more as a valuable source instead of a disposable waste.

References

An Y, Yang F, Wong FS, Chua HC (2009) Effect of recirculation ratio on simultaneous methanogenesis and nitrogen removal using a combined up-flow anaerobic sludge blanket–membrane bioreactor. Environ Eng Sci 26:1047–1053. https://doi.org/10.1089/ees.2007.0317

Appels L, Baeyens J, Degrève J, Dewil R (2008) Principles and potential of the anaerobic digestion of waste-activated sludge. Prog Energy Combust Sci 34:755. https://doi.org/10.1016/j.pecs.2008.06.002

Arbiv R, van Rijn J (1995) Performance of a treatment system for inorganic nitrogen removal in intensive aquaculture systems. Aquac Eng 14:189. https://doi.org/10.1016/0144-8609(94)P4435-E

Ayre JM, Moheimani NR, Borowitzka MA (2017) Growth of microalgae on undiluted anaerobic digestate of piggery effluent with high ammonium concentrations. Algal Res 24:218–226. https://doi.org/10.1016/j.algal.2017.03.023

Brod E, Oppen J, Kristoffersen AØ, Haraldsen TK, Krogstad T (2017) Drying or anaerobic digestion of fish sludge: nitrogen fertilisation effects and logistics. Ambio 46:852. https://doi.org/10.1007/s13280-017-0927-5

Chang S (2014) Anaerobic membrane bioreactors (AnMBR) for wastewater treatment. Adv Chem Eng Sci 4:56. https://doi.org/10.4236/aces.2014.41008

Chen SL, Coffin DE, Malone RF (1997) Sludge production and management for recirculating aquacultural systems. J World Aquac Soc 28:303–315. https://doi.org/10.1111/j.1749-7345.1997.tb00278.x

Chowdhury P, Viraraghavan T, Srinivasan A (2010) Biological treatment processes for fish processing wastewater–a review. Bioresour Technol 101:439–449. https://doi.org/10.1016/j.biortech.2009.08.065

Conroy J, Couturier M (2010) Dissolution of minerals during hydrolysis of fish waste solids. Aquaculture 298:220–225. https://doi.org/10.1016/j.aquaculture.2009.11.013

Cui ZF, Chang S, Fane AG (2003) The use of gas bubbling to enhance membrane processes. J Memb Sci. https://doi.org/10.1016/S0376-7388(03)00246-1

Delaide B, Goddek S, Gott J, Soyeurt H, Jijakli MH (2016) Lettuce (*Lactuca sativa* L. var. Sucrine) growth performance in complemented aquaponic solution outperforms hydroponics. Water (Switzerland) 8. https://doi.org/10.3390/w8100467

Delaide B, Goddek S, Keesman KJ, Jijakli MH (2018) A methodology to quantify the aerobic and anaerobic sludge digestion performance for nutrient recycling in aquaponics. Biotechnol Agron Soc Environ 22

Deublein D, Steinhauser A (2010) Biogas from waste and renewable resources: an introduction. In: Biogas from waste and renewable resources: an introduction, 2nd edn. https://doi.org/10.1002/9783527632794

Ebeling JM, Timmons MB, Bisogni JJ (2006) Engineering analysis of the stoichiometry of photoautotrophic, autotrophic, and heterotrophic removal of ammonia-nitrogen in aquaculture systems. Aquaculture 257:346. https://doi.org/10.1016/j.aquaculture.2006.03.019

Endut A, Jusoh A, Ali N, Wan Nik WB, Hassan A (2010) A study on the optimal hydraulic loading rate and plant ratios in recirculation aquaponic system. Bioresour Technol 101:1511–1517. https://doi.org/10.1016/j.biortech.2009.09.040

Gander M, Je B, Judd S (2000) Aerobic MBRs for domestic wastewater treatment: a review with cost considerations. Sep Purif Technol 18:119–130

Garcia J-L, Patel BKC, Ollivier B (2000) Taxonomic, phylogenetic, and ecological diversity of methanogenic archaea. Anaerobe 6:205. https://doi.org/10.1006/anae.2000.0345

Gebauer R, Eikebrokk B (2006) Mesophilic anaerobic treatment of sludge from salmon smolt hatching. Bioresour Technol 97:2389–2401. https://doi.org/10.1016/j.biortech.2005.10.008

Goddek S, Keesman KJ (2018) The necessity of desalination technology for designing and sizing multi-loop aquaponics systems. Desalination 428:76–85. https://doi.org/10.1016/j.desal.2017.11.024

Goddek S, Körner O (2019) A fully integrated simulation model of multi-loop aquaponics: a case study for system sizing in different environments. Agric Syst 171:143

Goddek S, Delaide B, Mankasingh U, Ragnarsdottir K, Jijakli H, Thorarinsdottir R (2015) Challenges of sustainable and commercial aquaponics. Sustainability 7:4199–4224. https://doi.org/10.3390/su7044199

Goddek S, Espinal CA, Delaide B, Jijakli MH, Schmautz Z, Wuertz S, Keesman KJ (2016) Navigating towards decoupled aquaponic systems: a system dynamics design approach. Water (Switzerland) 8. https://doi.org/10.3390/W8070303

Goddek S, Delaide B, Oyce A, Wuertz S, Jijakli MH, Gross A, Eding EH, Bläser I, Keizer LCP, Morgenstern R, Körner O, Verreth J, Keesman KJ (2018) Nutrient mineralisation and organic matter reduction performance of RAS-based sludge in sequential UASB-EGSB reactors. Aquac Eng 83:10. https://doi.org/10.1016/J.AQUAENG.2018.07.003

Graber A, Junge R (2009) Aquaponic systems: nutrient recycling from fish wastewater by vegetable production. Desalination 246:147–156

Huang X, Xiao K, Shen Y (2010) Recent advances in membrane bioreactor technology for wastewater treatment in China. Front Environ Sci Eng China 4:245. https://doi.org/10.1007/s11783-010-0240-z

Judd S, Judd C (2008) The MBR book: principles and applications of membrane bioreactors in water and wastewater treatment. Elsevier. https://doi.org/10.1016/B978-185617481-7/50005-2

Jung IS, Lovitt RW (2011) Leaching techniques to remove metals and potentially hazardous nutrients from trout farm sludge. Water Res 45:5977–5986. https://doi.org/10.1016/j.watres.2011.08.062

Khalid A, Arshad M, Anjum M, Mahmood T, Dawson L (2011) The anaerobic digestion of solid organic waste. Waste Manag 31:1737. https://doi.org/10.1016/j.wasman.2011.03.021

Klas S, Mozes N, Lahav O (2006) Development of a single-sludge denitrification method for nitrate removal from RAS effluents: lab-scale results vs. model prediction. Aquaculture 259:342. https://doi.org/10.1016/j.aquaculture.2006.05.049

Kumar V, Sinha AK, Makkar HPS, De Boeck G, Becker K (2012) Phytate and phytase in fish nutrition. J Anim Physiol Anim Nutr (Berl) 96:335. https://doi.org/10.1111/j.1439-0396.2011.01169.x

Lanari D, Franci C (1998) Biogas production from solid wastes removed from fish farm effluents. Aquat Living Resour 11:289–295. https://doi.org/10.1016/S0990-7440(98)80014-4

Licamele JD (2009) Biomass production and nutrient dynamics in an aquaponics system. The University of Arizona

Marchaim U (1992) Biogas processes for sustainable development. FAO Agricultural Services Bulletin 95. Food and Agriculture Organization of the United Nations

McDermott BL, Chalmers AD, Goodwin JAS (2001) Ultrasonication as a pre-treatment method for the enhancement of the psychrophilic anaerobic digestion of aquaculture effluents. Environ Technol (United Kingdom) 22:823. https://doi.org/10.1080/095933322086180317

McGill SM (2012) 'Peak' phosphorus? The implications of phosphate scarcity for sustainable investors. J Sustain Financ Invest. https://doi.org/10.1080/20430795.2012.742635

Mirzoyan N, Gross A (2013) Use of UASB reactors for brackish aquaculture sludge digestion under different conditions. Water Res 47:2843–2850. https://doi.org/10.1016/j.watres.2013.02.050

Mirzoyan N, Parnes S, Singer A, Tal Y, Sowers K, Gross A (2008) Quality of brackish aquaculture sludge and its suitability for anaerobic digestion and methane production in an upflow anaerobic sludge blanket (UASB) reactor. Aquaculture 279:35–41. https://doi.org/10.1016/j.aquaculture.2008.04.008

Mirzoyan N, Tal Y, Gross A (2010) Anaerobic digestion of sludge from intensive recirculating aquaculture systems: review. Aquaculture 306:1–6. https://doi.org/10.1016/j.aquaculture.2010.05.028

Monsees H, Keitel J, Paul M, Kloas W, Wuertz S (2017) Potential of aquacultural sludge treatment for aquaponics: evaluation of nutrient mobilization under aerobic and anaerobic conditions. Aquac Environ Interact 9:9–18. https://doi.org/10.3354/aei00205

Naylor SJ, Moccia RD, Durant GM (1999) The chemical composition of settleable solid fish waste (manure) from commercial rainbow trout farms in Ontario, Canada. North Am J Aquac 61:21–26

Neto RM, Ostrensky A (2013) Nutrient load estimation in the waste of Nile tilapia *Oreochromisniloticus* (L.) reared in cages in tropical climate conditions. Aquac Res 46:1309–1322. https://doi.org/10.1111/are.12280

Nichols MA, Savidov NA (2012) Aquaponics: a nutrient and water efficient production system. Acta Hortic:129–132

Peng L, Dai H, Wu Y, Peng Y, Lu X (2018) A comprehensive review of phosphorus recovery from wastewater by crystallization processes. Chemosphere 197:768. https://doi.org/10.1016/j.chemosphere.2018.01.098

Rakocy JE, Bailey DS, Shultz RC, Danaher JJ (2007) Preliminary evaluation of organic waste from two aquaculture systems as a source of inorganic nutrients for hydroponics. Acta Hortic 742:201–208

Ru D, Liu J, Hu Z, Zou Y, Jiang L, Cheng X, Lv Z (2017) Improvement of aquaponic performance through micro- and macro-nutrient addition. Environ Sci Pollut Res 24:16328. https://doi.org/10.1007/s11356-017-9273-1

Saha S, Monroe A, Day MR (2016) Growth, yield, plant quality and nutrition of basil (*Ocimumbasilicum* L.) under soilless agricultural systems. Ann Agric Sci 61:181–186. https://doi.org/10.1016/j.aoas.2016.10.001

Schneider O, Sereti V, Eding EH, Verreth JAJ (2005) Analysis of nutrient flows in integrated intensive aquaculture systems. Aquac Eng 32:379–401. https://doi.org/10.1016/j.aquaeng.2004.09.001

Seo KW, Choi YS, Gu MB, Kwon EE, Tsang YF, Rinklebe J, Park C (2017) Pilot-scale investigation of sludge reduction in aerobic digestion system with endospore-forming bacteria. Chemosphere 186:202–208. https://doi.org/10.1016/j.chemosphere.2017.07.150

Stewart NT, Boardman GD, Helfrich LA (2006) Characterization of nutrient leaching rates from settled rainbow trout (*Oncorhynchus mykiss*) sludge. Aquac Eng 35:191–198. https://doi.org/10.1016/j.aquaeng.2006.01.004

Tal Y, Schreier HJ, Sowers KR, Stubblefield JD, Place AR, Zohar Y (2009) Environmentally sustainable land-based marine aquaculture. Aquaculture 286:28–35. https://doi.org/10.1016/j.aquaculture.2008.08.043

Techobanoglous G, Burton FL, Stensel HD (2014) Wastewater engineering: treatment and reuse, 5th edn. Metcalf and Eddy. https://doi.org/10.1016/0309-1708(80)90067-6

Turcios AE, Papenbrock J (2014) Sustainable treatment of aquaculture effluents-what can we learn from the past for the future? Sustain 6:836–856

Van Lier JB, Mahmoud N, Zeeman G (2008) Anaerobic wastewater treatment, Biological wastewater treatment: principles, modelling and design. https://doi.org/10.1021/es00154a002

Van Rijn J (2013) Waste treatment in recirculating aquaculture systems. Aquac Eng 53:49–56. https://doi.org/10.1016/j.aquaeng.2012.11.010

van Rijn J, Fonarev N, Berkowitz B (1995) Anaerobic treatment of intensive fish culture effluents: digestion of fish feed and release of volatile fatty acids. Aquaculture 133:9–20. https://doi.org/10.1016/0044-8486(94)00385-2

Yogev U, Barnes A, Gross A (2016) Nutrients and energy balance analysis for a conceptual model of a three loops off grid, Aquaponics. Water 8:589. https://doi.org/10.3390/W8120589

Yogev U, Sowers KR, Mozes N, Gross A (2017) Nitrogen and carbon balance in a novel near-zero water exchange saline recirculating aquaculture system. Aquaculture 467:118–126. https://doi.org/10.1016/j.aquaculture.2016.04.029

Zhang X, Hu J, Spanjers H, van Lier JB (2016) Struvite crystallization under a marine/brackish aquaculture condition. Bioresour Technol 218:1151. https://doi.org/10.1016/j.biortech.2016.07.088

Chapter 11
Aquaponics Systems Modelling

Karel J. Keesman, Oliver Körner, Kai Wagner, Jan Urban, Divas Karimanzira, Thomas Rauschenbach, and Simon Goddek

Abstract Mathematical models can take very different forms and very different levels of complexity. A systematic way to postulate, calibrate and validate, as provided by systems theory, can therefore be very helpful. In this chapter, dynamic systems modelling of aquaponic (AP) systems, from a systems theoretical perspective, is considered and demonstrated to each of the subsystems of the AP system, such as fish tanks, anaerobic digester and hydroponic (HP) greenhouse. It further shows the links between the subsystems, so that in principle a complete AP systems model can be built and integrated into daily practice with respect to management and control of AP systems. The main challenge is to choose an appropriate model complexity that meets the experimental data for estimation of parameters and states and allows us to answer questions related to the modelling objective, such as simulation, experiment design, prediction and control.

K. J. Keesman (✉) · S. Goddek
Mathematical and Statistical Methods (Biometris), Wageningen University, Wageningen, The Netherlands
e-mail: karel.keesman@wur.nl; simon.goddek@wur.nl; simon@goddek.nl

O. Körner
Leibniz-Institute of Vegetable and Ornamental Crops (IGZ), Grossbeeren, Germany
e-mail: koerner@igzev.de

K. Wagner
Institut für physikalische Prozesstechnik, University of Applied Sciences Saarbrücken, Saarbrücken, Germany
e-mail: kai.wagner@htwsaar.de

J. Urban
Laboratory of Signal and Image Processing, Institute of Complex Systems, South Bohemian Research Center of Aquaculture and Biodiversity of Hydrocenoses, Faculty of Fisheries and Protection of Waters, University of South Bohemia in Ceske Budejovice, Nove Hrady, Czech Republic
e-mail: urbanj@frov.jcu.cz

D. Karimanzira · T. Rauschenbach
Fraunhofer IOSB-AST, Ilmenau, Germany
e-mail: divas.karimanzira@iosb-ast.fraunhofer.de; Thomas.Rauschenbach@iosb-ast.fraunhofer.de

S. Goddek et al. (eds.), *Aquaponics Food Production Systems*,
https://doi.org/10.1007/978-3-030-15943-6_11

Keywords Modelling · Recirculating aquaculture system · Anaerobic digestion · Hydroponic greenhouse · Multi-loop aquaponic system · Tools

11.1 Introduction

In general, mathematical models can take very different forms depending on the system under study, which may range from social, economic and environmental to mechanical and electrical systems. Typically, the internal mechanisms of social, economic or environmental systems are not very well known or understood and often only small data sets are available, while the prior knowledge of mechanical and electrical systems is at a high level, and experiments can easily be done. Apart from this, the model form also strongly depends on the final objective of the modelling procedure. For instance, a model for process design or simulation should contain much more detail than a model used for studying different long-term scenarios.

In particular, for a wide range of applications (e.g. Keesman 2011), models are developed to:

- Obtain or enlarge insight in different phenomena, for example, recovering physical or economic relationships.
- Analyse process behaviour using simulation tools, for example, process training of operators or weather forecasts.
- Estimate state variables that cannot be easily measured in real time on the basis of available measurements, for instance, online process information.
- Control, for instance, in the internal model control or model-based predictive control concept or to manage processes.

A critical step in the modelling of any system is to find a mathematical model which adequately describes the actual situation or state. Firstly, the system boundaries and the system variables have to be specified. Then relationships between these variables have to be specified on the basis of prior knowledge, and assumptions about the uncertainties in the model have to be made. Combining this information defines the model structure. Still the model may contain some unknown or incompletely known coefficients, the model parameters, which in case of time-varying behaviour define an additional set of system variables. For a general introduction to mathematical modelling we refer to, for instance, Sinha and Kuszta (1983), Willems and Polderman (1998) and Zeigler et al. (2000).

In this chapter, the modelling of an aquaponic (food) production (AP) system will be described. Figure 11.1 shows a typical example of an AP system, i.e. the so-called decoupled three-loop aquaponic system. As a result of basic principles modelling, using conservation laws and constitutive relationships, mathematical models of all kinds of AP systems are usually represented as a set of ordinary or partial differential equations. These mathematical models are commonly used for design, estimation and control. In each of these specific modelling objectives, we distinguish between analysis and synthesis.

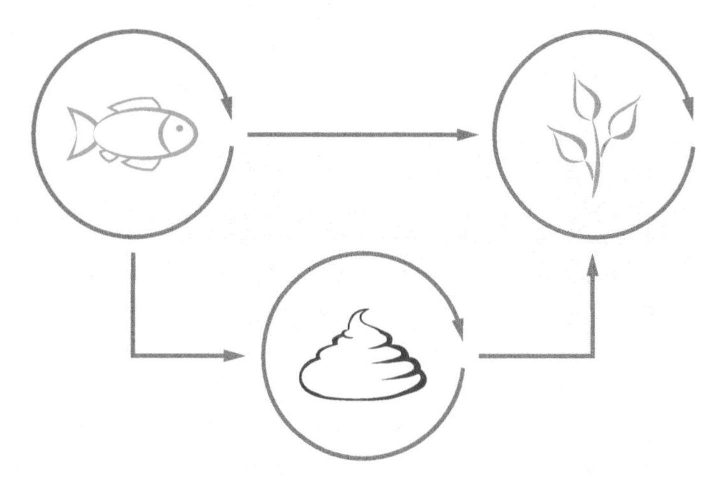

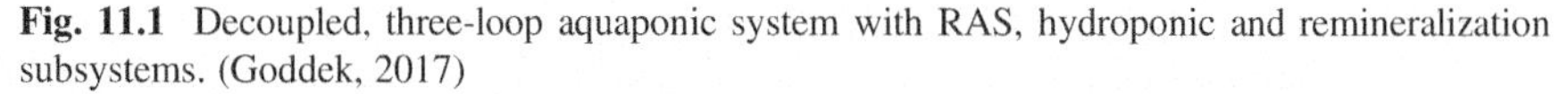

Fig. 11.1 Decoupled, three-loop aquaponic system with RAS, hydroponic and remineralization subsystems. (Goddek, 2017)

The outline of the chapter is as follows. In Sect. 11.1 some background on mathematical systems modelling is presented. Sections 11.2, 11.3, 11.4 and 11.5 describe the modelling of a recirculating aquaculture system (RAS), anaerobic digestion, hydroponic (HP) greenhouse and a multi-loop AP system, respectively. In Sect. 11.6 modelling tools are introduced and illustrated with some examples. The chapter concludes with a Discussion and Conclusions section.

11.2 Background

Many definitions of a system are available, ranging from loose descriptions to strict mathematical formulations. In what follows, a *system* is considered to be an object in which different variables interact at all kinds of time and space scales and that produces observable signals. These types of systems are also called *open systems*. A graphical representation of a general open system (***S***) with vector-valued input and output signals is represented in Fig. 11.2. Thus, multiple inputs or outputs are combined in one single arrow. So, the system variables may be scalars or vectors. In addition, they can be continuous or discrete functions of time. It is important to stress that the arrows in Fig. 11.2 represent signal flows and thus not necessarily physical flows.

It is also possible to connect systems into a network, as in an AP system, with parallel, feedback and feedforward paths. Figure 11.3 presents an example of such a network.

For controller/management analysis and synthesis, it is often convenient to connect the system (S) to the controller or management strategy (C), as in Fig. 11.4. Most often the input to the controller or management strategy is the external steering signal of the controlled system, and the output of the system is the observed system's behaviour.

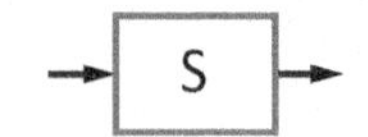

Fig. 11.2 General open system representation

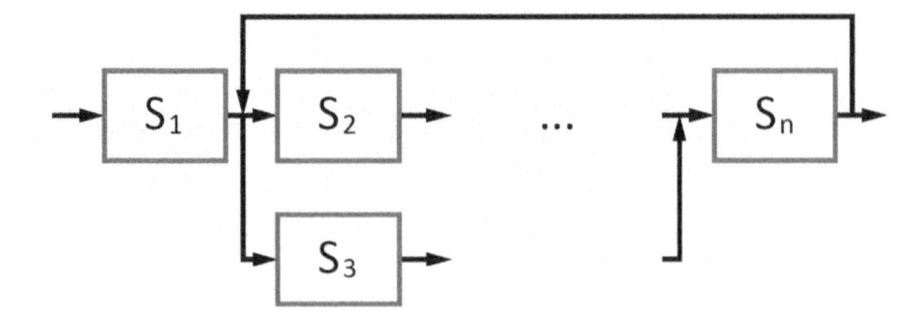

Fig. 11.3 Open system network representation

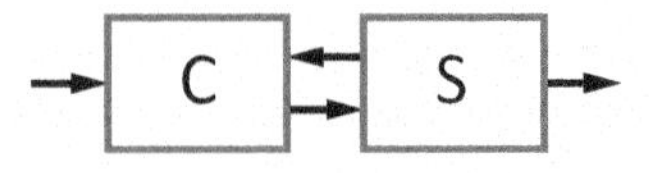

Fig. 11.4 Controlled system

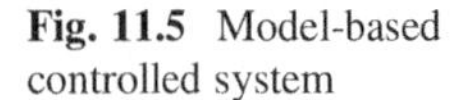

Fig. 11.5 Model-based controlled system

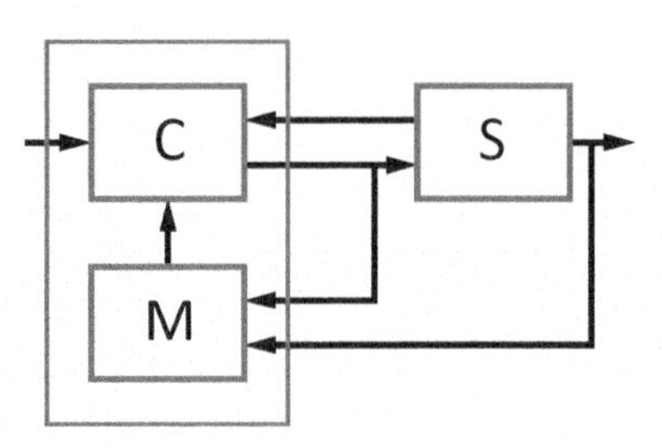

Finally, to emphasize the incorporation of a mathematical model (M) into the controller structure or management strategy, the following model-based controlled system representation is introduced (Fig. 11.5).

For now, it suffices to present the block diagram representation. In subsequent sections, the modelling of AP systems will be worked out in more detail.

In systems theory the basic structure of a mathematical model (M) is schematically represented as in Fig. 11.6. In Fig. 11.6, x is the so-called state of the system, u the control input, y the output, w the disturbance input and v the output noise. In general, each of these variables is vector-valued.

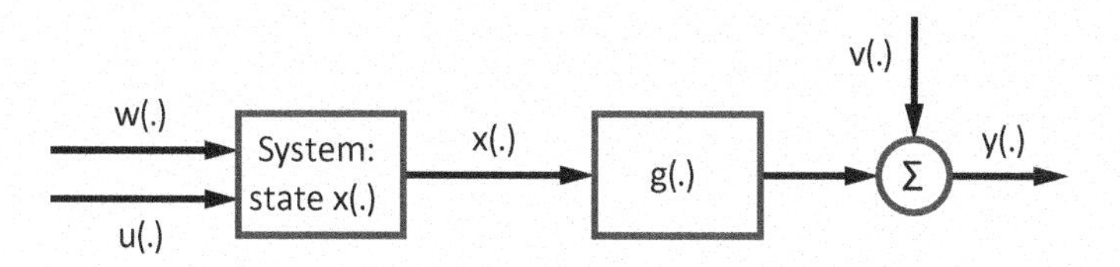

Fig. 11.6 Basic structure of mathematical model (M)

In continuous time, the following set of equations describes a general dynamic model (M), with parameter vector p, in what is called *state-space* form:

$$\frac{dx(t)}{dt} = f(t, x(t), u(t), w(t); p), \qquad x(0) = x_0$$
$$y(t) = g(t, x(t), u(t); p) + v(t), \qquad t \in \mathfrak{R}^+ \tag{11.1}$$

where the first equation describes the nonlinear and time-varying dynamics of the system in terms of state variables (x) and the second one expresses the algebraic relationship between u, x and y. This state-space model representation has been a starting point for many software implementations for design, control and estimation. In what follows, however, only deterministic models, thus without the stochastic vectors v and w, are considered. Let us illustrate this theory on a fish tank system.

Example: Fish Tank System

Consider the following fish tank, which is a typical example of the general system presented in Fig. 11.7.

Let us start with specifying our prior knowledge of the internal system mechanisms. The following mass balance can be defined in terms of the volume of the storage tank (V), also called the *state* of the system, inflows $u(t)$ and outflows $y(t)$:

$$\frac{dV(t)}{dt} = u(t) - y(t) \tag{11.2}$$

Suppose there is a level controller (LC) that keeps the outflow proportional to the volume in the tank. This can be enforced by implementing the following proportional control law,

$$Y(t) = KV(t) \tag{11.3}$$

with K a real, positive constant. Hence, after substituting Eq. (11.3) into (11.2), we obtain the following differential equation

$$\frac{dV(t)}{dt} + KV(t) = u(t) \tag{11.4}$$

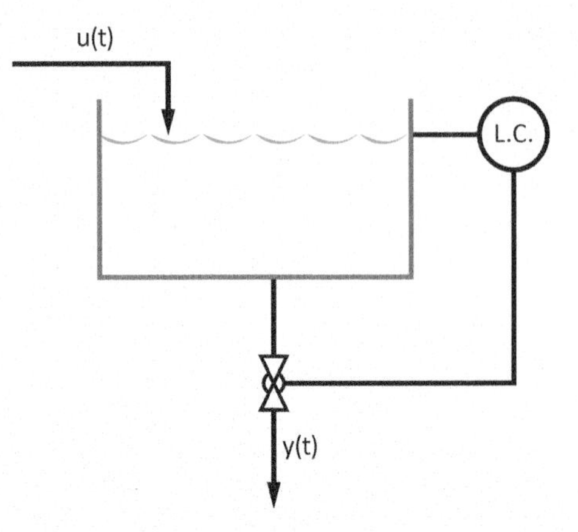

Fig. 11.7 Fish tank with volume-controlled flow using level controller (LC)

For this specific linear differential equation with constant coefficients, an analytical solution exists and is given by

$$y(t) = y(0)e^{-Kt} + \int_0^t Ke^{-K(t-s)}u(s)ds \tag{11.5}$$

under the assumption that $u(t) = 0$ for $t < 0$. From this example it is clear that applying first principles – mass conservation in this case – directly leads to an ordinary differential equation. In state-space format, the model can be represented as

$$\begin{aligned} \frac{dx(t)}{dt} &= -Kx(t) + u(t) \\ y(t) &= Kx(t) \end{aligned} \tag{11.6}$$

With x volume, u flow input and K controller gain. Thus, in terms of the general state-space Eq. (11.1), $f(t, x(t), u(t); p) \equiv -Kx(t) + u(t)$ and $g(t, x(t), u(t); p) \equiv Kx(t)$.

For two volume-controlled fish tanks in series with volume V_1 and V_2, and controller gain K_1 and K_2, respectively, two mass balances can be formulated, i.e.

$$\begin{aligned} \frac{dV_1(t)}{dt} &= -K_1V_1(t) + u(t) \\ \frac{dV_2(t)}{dt} &= K_1V_1(t) - K_2V_2(t) \end{aligned} \tag{11.7}$$

In vector-matrix form, and for physical outflow $y(t)$, we can write:

$$\begin{aligned} \frac{d}{dt}\begin{bmatrix} V_1(t) \\ V_2(t) \end{bmatrix} &= \begin{bmatrix} -K_1 & 0 \\ K_1 & -K_2 \end{bmatrix}\begin{bmatrix} V_1(t) \\ V_2(t) \end{bmatrix} + \begin{bmatrix} 1 \\ 0 \end{bmatrix}u(t) \\ y(t) &= K_2V_2(t) \end{aligned} \tag{11.8}$$

And thus, with $x_1 = V_1$,$x_2 = V_2$:$f(t, x(t), u(t); p) \equiv \begin{bmatrix} -K_1 x_1(t) + u(t) \\ K_1 x_1(t) - K_2 x_2(t) \end{bmatrix}$ and $g(t, x(t), u(t); p) \equiv K_2 x_2(t)$.

In the next sections, each of the subsystems of the AP system (Fig. 11.1) will be described in more detail.

11.3 RAS Modelling

Global fish aquaculture reached 50 million tons in 2014 (FAO 2016). Given the growing human population, there is a growing demand for fish proteins. Sustainable growth of aquaculture requires novel (bio)technologies such as recirculating aquaculture systems (RAS). RAS have a low water consumption (Orellana 2014) and allow for a recycling of excretory products (Waller et al. 2015). RAS provide suitable living conditions for fish, as a result of a multistep water treatment, such as particle separation, nitrification (biofiltration), gas exchange and temperature control. Dissolved and particulate excretory products can be transferred to secondary treatment such as plant (Waller et al. 2015) or algae production in integrated aqua-agriculture (IAAC) systems. IAAC systems are sustainable alternatives to conventional aquaculture systems and in particular are a promising expansion to RAS. In RAS it would be necessary to circulate the process water which has special implications for the process technology in both, the RAS and the algae/plant system. To combine RAS and algae/plant system, a deep understanding of the interaction between fish and water treatment is prerequisite and can be derived from dynamic modelling. The metabolism in fish follows a daily pattern which is well represented by the gastric evacuation rate (Richie et al. 2004). Particle separation, biofiltration and gas exchange are subjected to the same pattern. For design purposes the characterization of the basic components of a RAS treatment system should be investigated through simulation models. These simulation models are highly complex. Available numerical models for RAS capture only a small part of the complexity and consider only a part of the components with corresponding mechanisms. Hence, in this chapter, only a small part of a dynamic RAS model will be presented, i.e. nitrification-based biofiltration. The conversion of toxic ammonia into nitrate is a central process in the water treatment process in RAS. In the following, the dynamic modelling of the mass balance of ammonia excretion of fish and the conversion of ammonia into nitrate will be demonstrated as well as the transfer of the nutrient into an aquaponic system. With this it is possible not only to engineer a RAS but also to integrate fish production into an IAAC system based on valid parameters.

11.3.1 Dynamic Model of Nitrification-Based Biofiltration in RAS

The model is subdivided into a fish model for European seabass, *Dicentrarchus labrax*, a model describing the time- dependenting excretion of ammonia, and a nitrification model (Fig. 11.8). The fish excretion pattern is introduced into the model through the input vector u (Eq. 11.15), similar to the approach used by Wik et al. (2009). The complexity of the fish model is kept low to be able to explain its method of implementation. Nonetheless, a short introduction into modelling fish is presented in Sect. 11.3.2. Four basic aspects important to describe the nutrient flow in RAS (Badiola et al. 2012) are:

1. The flow Q, which is the total process water flow per unit time through the RAS, determines the mass transfer of all dissolved and particulate matter, including ammonia and nitrate.
2. The excretion of the fish input ammonia to the RAS process water and is depicted by the product of matrix B and vector u (Eq. 11.15).
3. The ammonia conversion into nitrate, taking place in the nitrification, is depicted in the nitrification vector n (Eq. 11.15).
4. The nutrient transfer from the RAS to a connected HP system is depicted in vector u (Eq. 11.15). Other important aspects of the RAS process chain such as solid removal, dissolved oxygen concentration and carbon dioxide concentration are not considered here. Hints for modelling these can be found in Sects. 3.1.1 and 3.2.2 of this book.

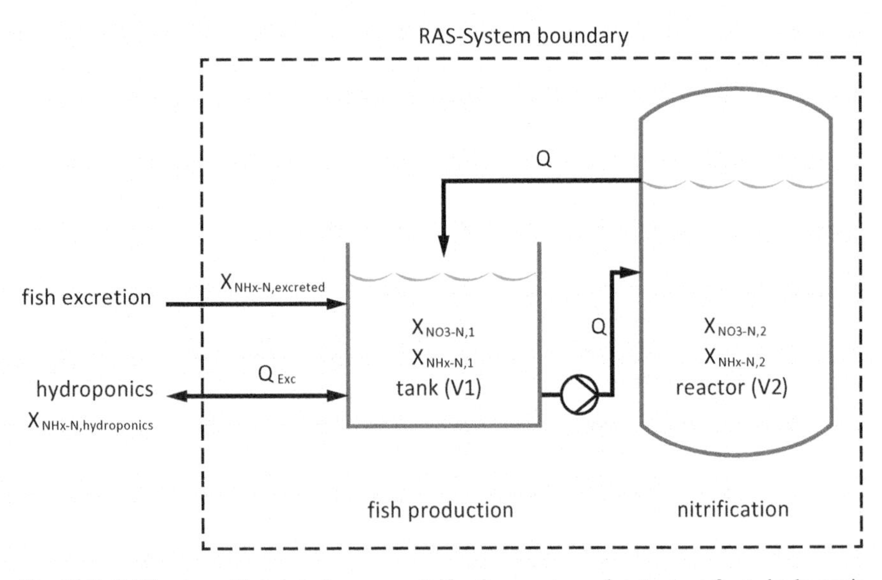

Fig. 11.8 RAS setup with fish tank, pump, nitrification reactor and water transfer to hydroponic system

11.3.2 Fish

A variety of models in the scientific literature predict growth and feed intake of different aquatic species. The models describe growth as weight gain per day, as percentage growth increment or as specific growth rate based on an exponential growth model. Models are often valid for specific life stages. Feed consumption, biomass and gender are influencing the model output as well as the environmental conditions such as temperature, oxygen level and nutrient concentration (Lugert et al. 2014). Careful research is needed to identify the correct model used for the specific application. Commercial RAS that consists of several cohorts of fish in different life stages require the modelling to incorporate cohorts into the model (Fig. 8.6) (Halamachi and Simon 2005). The excretory mass flow for the European seabass (*Dicentrarchus labrax*) can be estimated with algorithms published by Lupatsch and Kissil (1998).

Here the net nitrogen mass flow into the process water is estimated from the feed composition (protein content), the amount of given feed and the nitrogen retained in body tissues through the growth (weight increment) of fish. The faecal nitrogen losses are not included in the model, but excretion rate is corrected assuming a share of 0.25 and 0.75 of nitrogen excretion for faecal loss and ammonia excretion, respectively. The nitrogen input through the feeding of fish is estimated from the protein content and the average relative nitrogen content of proteins which is assumed to be 0.16. The protein content of seabass tissue is reported at around 0.17 g proteins g^{-1} seabass (Lupatsch et al. 2003). For a fish gaining body weight by consuming a given amount of feed, the nitrogen excretion ($X_{N,excreted}$, g) can be calculated from Eq. (11.9). It is assumed that the feed (X_{feed}) contains 0.5 g protein g^{-1} fish. It is further assumed that the feed conversion rate equals 1, i.e. 1 g of feed consumption is resulting in 1 g of body weight increase (Fig. 11.9):

$$X_{N,\mathrm{excreted}} = X_{\mathrm{feed}} * 0.16 * 0.75 * (0.5 - 0.17) \tag{11.9}$$

Dissolved ammonia excreted via the gills of fish follows similar daily pattern as the gastric evacuation rate (GER). GER is described for cold water and warm water fish by He and Wurtsbaugh (1993) and Richie et al. (2004), respectively. The excretory pattern can be well simulated with a sine function. The ammonia excretion can be calculated from Eq. (11.10):

$$X_{\mathrm{NH}_x-N,\mathrm{excreted}} = X_{N,\mathrm{excreted}}\,[g] * \left(\sin\left(\frac{2\pi}{1440}\right) + 1\right) \tag{11.10}$$

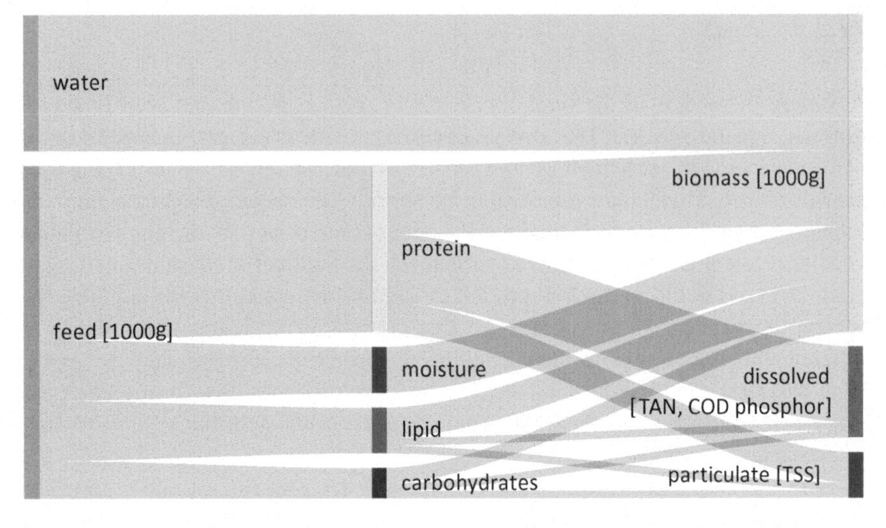

Fig. 11.9 Representation of the mass flows (Sankey chart) of feed ingredients and excretory products for a fish consuming 1000 g of feed assuming an FCR of 1

11.3.3 RAS

A variety of models describing RAS having different levels of complexity can be found in the literature. Very complex models are available for specific aspects, such as the interaction of soluble gases and alkalinity (Colt 2013) or the description of the microbial community (Henze et al. 2002). More practical models for the mass balance of RAS are published by Sánchez-Romero et al. (2016), Pagand et al. (2000), Wik et al. (2009) and Weatherley et al. (1993). All models provide information on excretory mass flows and/or nutrient flows in dependence of time and location in the process chain. Such models provide a base for the simulation of the coupling of RAS and HP. The most important dissolved matter in RAS modelling is total ammonia nitrogen (TAN). Besides TAN the chemical (COD) and biological (BOD) oxygen demand, the total suspended solids (TSS) and dissolved oxygen concentration need to be considered. However, different notations in the scientific literature make it sometimes hard to read, to convert and to implement the information into models. In the following, notations as recommended by Corominas et al. (2010) will be used. TAN will be rewritten as $X_{NHx\text{-}N}$ and nitrate nitrogen will be expressed as $X_{NO3\text{-}N}$.

11.3.4 Model Example

The model as it is described in the following is only valid for the RAS presented in Fig. 11.8. Other possible process chains for RAS are discussed in Sect. 11.3 of this chapter. For the mathematical depiction of physical systems, the following assumptions were made:

(a) Density of water is assumed to be constant.
(b) Tank and reactor are assumed to be well mixed.
(c) Tank and reactor volume are assumed to be constant.
(d) Process water flow is always greater than zero.

The assumption of a well-mixed tank and reactor leads to a mass balance equation for continuous stirred-tank reactor (CSTR) as described by Drayer and Howard (2014) in Eq. (11.11). It must be mentioned that diffusive processes can usually be neglected in RAS calculations because of a typically high process water flow rate. For a multi-tank RAS, the following holds:

$$\begin{aligned} \text{Accumulation} &= \text{Inflow} - \text{outflow} + \text{generation} - \text{reduction} \\ V_i \dot{x}_i &= Q_{\text{in}} x_{i,\text{in}} - Q_{\text{out}} x_{i,\text{out}} + x_{i,\text{gen}} - x_{i,\text{red}} \\ j &= \begin{cases} n, & i = 1 \\ i-1, & i \neq 1 \end{cases} \end{aligned} \tag{11.11}$$

In the above given equation n represents the number of tanks in the System, $\dot{x}_i$ is the change of concentration of a given substrate x in a volume given by $V_{i.}$. The process water flow into the tank or reactor is represented by Q_{in}. V_i is the volume of the component where the process water flow Q_{in} is entering in. The process water flow Q_{in} came from a component having the volume V_j.

The conversion of $X_{NHx\text{-}N}$ into $X_{NO3\text{-}N}$ in nitrifying biofilters takes place on the surface area A [m^2] available on the bio-carriers in the nitrification reactor (Rusten 2006). The available bioactive surface in the nitrification is calculated by multiplying the volume of the reactor with the volume-specific active surface of the bio-carriers A_S [$m^2 \cdot m^{-3}$]. The total bioactive surface is calculated (Eq. 11.12) from the relative filling f_{bc} of the nitrification reactor which usually is 0.6 (for details, see Rusten 2006).

$$A = V_{\text{nitrification}} * A_S * f_{bc} \tag{11.12}$$

The total daily TAN microbial conversion μ_{max} [$g\ d^{-1}$] (nitrification) was calculated by multiplying the specific TAN conversion (nitrification) rate, $NHx_{conversion\text{-}rate}$ [$g\ m^{-2}\ d^{-1}$], with the total active surface area, A [m^2], of the bio-carriers. Values for TAN conversion in different types of nitrifying biofilters can be found in literature. For moving bed biofilm reactors (MBBR), values are reported by Rusten (2006). This rate is valid for certain process conditions, and it is assumed that the bacteria biofilm is fully developed over the whole.

$$\mu_{mm} = A^{*}NHx_{Conversion-rate} \tag{11.13}$$

The total mass of NH_x converted into NO_3-N can subsequently be calculated with a Monod kinetic (Eq. 11.14). For this the $NH_{x\text{-}N}$ concentration, $X_{NHx\text{-}N,2}$ $[g \cdot l^{-1}]$, in the volume of the nitrification reactor (MBBR) V_2, is needed.

$$\begin{aligned} \frac{d}{dt}X_{NH_x-N,2} &= -\mu_{max} * \left(\frac{X_{NH_x-N,2}}{K_s + X_{NH_x-N,2}}\right) * \frac{1}{V_2} \quad \text{with} \quad K_s = \frac{\mu_{max}}{2} \\ \frac{d}{dt}X_{NO_s-N,2} &= +\mu_{max} * \left(\frac{X_{NH_x-N,2}}{K_s + X_{NH_x-N,2}}\right) * \frac{1}{V_2} \quad \text{with} \quad K_s = \frac{\mu_{max}}{2} \end{aligned} \tag{11.14}$$

Given Eqs. (11.9, 11.10, 11.11, 11.12, 11.13 and 11.14), the following state-space model (combining fish-nitrification) results

$$\frac{dx(t)}{dt} = \boldsymbol{A}^{*}\boldsymbol{X} + \boldsymbol{B}^{*}\boldsymbol{u} + \boldsymbol{n}$$

$$\boldsymbol{X} = \begin{bmatrix} X_{NH_x-N,1} \\ X_{NH_x-N,2} \\ X_{NO_3-N,1} \\ X_{NO_3-N,2} \end{bmatrix} \quad \boldsymbol{u} = \begin{bmatrix} X_{NH_x-N,\text{excreted}} \\ \boldsymbol{0} \\ \boldsymbol{Q}_{\mathbf{Exc}}{}^{*}X_{NH_x-N,\text{hydroponics}} \\ \boldsymbol{0} \end{bmatrix} \quad \boldsymbol{n} = \begin{bmatrix} \boldsymbol{0} \\ -\dfrac{\boldsymbol{\mu_{max}} * [\boldsymbol{X}]_2}{\boldsymbol{K_s} + [\boldsymbol{X}]_2} * \dfrac{1}{\boldsymbol{V_2}} \\ \boldsymbol{0} \\ +\dfrac{\boldsymbol{\mu_{max}} * [\boldsymbol{X}]_2}{\boldsymbol{K_s} + [\boldsymbol{X}]_2} * \dfrac{1}{\boldsymbol{V_2}} \end{bmatrix}$$

$$A = \begin{bmatrix} -\dfrac{Q}{V_1} - \dfrac{Q_{Exc}}{V_1} & \dfrac{Q}{V_1} & 0 & 0 \\ \dfrac{Q}{V_2} & -\dfrac{Q}{V_2} & 0 & 0 \\ 0 & 0 & -\dfrac{Q}{V_1} - \dfrac{Q_{Exc}}{V_1} & \dfrac{Q}{V_1} \\ 0 & 0 & \dfrac{Q}{V_2} & -\dfrac{Q}{V_1} \end{bmatrix}$$

$$\times B = \begin{bmatrix} \dfrac{1}{V_1} & 0 & 0 & 0 \\ 0 & \dfrac{1}{V_2} & 0 & 0 \\ 0 & 0 & \dfrac{1}{V_1} & 0 \\ 0 & 0 & 0 & \dfrac{1}{V_2} \end{bmatrix} \tag{11.15}$$

Example

In this example, a theoretical RAS with V_reactor = 1300 l and V_tank = 6000 l is simulated.

All simulations had a daily feed input of 2000 g/day with 500 g protein/kg feed (Eq. 11.8). The daily TAN excretion was assumed to be a sine curve (Eq. 11.9). Active surface of the bio-carriers A_S is 300 [m^2 m^{-3}], and the relative filling of the reactor f_{bc} is 0.6. Specific TAN conversion rate, $NHx_{conversion\text{-}rate}$, is 1.2 [g m^{-2} $^{-}$d], and the biofilm is supposed to be fully developed (Eqs. 11.11 and 11.12). The state-space representation (Eq. 11.14) was implemented in MATLAB Simulink. The Example showcases the importance of mass flow for nutrient concentrations in coupled systems (Fig. 11.10 and 11.11).

11.4 Modelling Anaerobic Digestion

Anaerobic digestion (AD) of organic material is a process that involves the sequential steps of hydrolysis, acidogenesis, acetogenesis and methanogenesis (Batstone et al. 2002). The anaerobic digestion of a mixture of proteins, carbohydrates and lipids is visualized in Figure 11.11. Most often, hydrolysis is considered as the rate-limiting step in the anaerobic digestion of complex organic matter (Pavlostathis and Giraldo-Gomez 1991). Thus, increasing the hydrolysis reaction rate will most likely lead to a

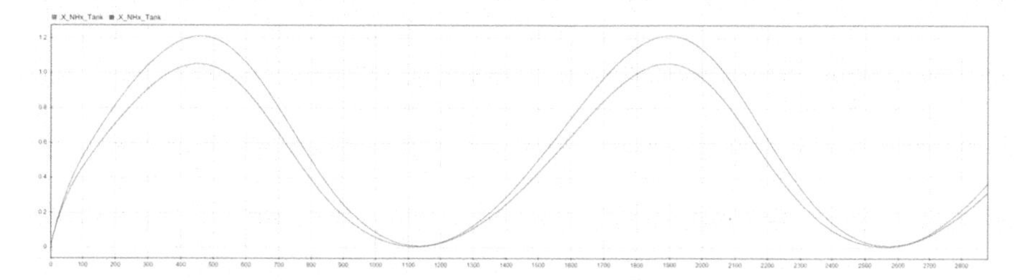

Fig. 11.10 Simulation of TAN ($X_{NHx\text{-}N,1}$) in [mg/l] over 2 days = 2880 min with Q = 300 l/min (blue) and Q = 200 l/min (orange)

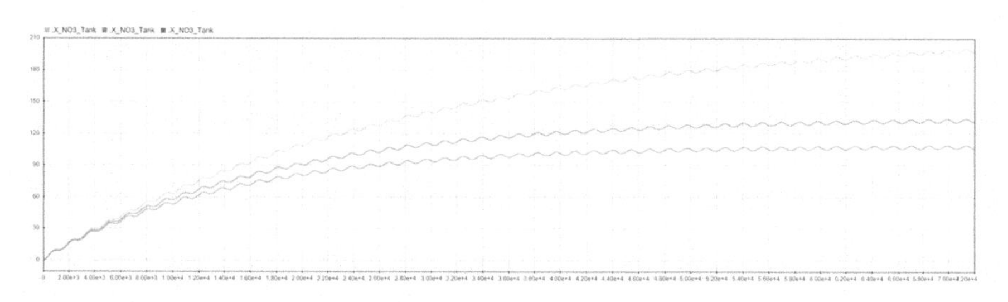

Fig. 11.11 Simulation of nitrate-N ($X_{NO3\text{-}N,1}$) in [mg/l] over 50 days = 72,000 min with Q_{Exc} = 300 l/day (yellow), Q_{Exc} = 480 l/day (orange) and Q_{Exc} = 600 l/day (blue)

higher anaerobic digestion reaction rate. However, increasing the reaction rates needs further understanding of the related process. Further understanding can be obtained via experimentation and/or mathematical modelling. As there are many factors influencing, for instance, the hydrolysis process, such as ammonia concentration; temperature; substrate composition; particle size; pH; intermediates; degree of hydrolysis; i.e. the potential of hydrolysable content; and residence time, it is almost impossible to evaluate the total effect of the factors on the hydrolysis reaction rate through experimentation. Mathematical modelling could therefore be an alternative, but as a result of all the uncertainties in model formulation, rate coefficients and initial conditions, no unique answers can be expected. But, a mathematical modelling framework would allow sensitivity and uncertainty analyses to facilitate the modelling process. As mentioned before, hydrolysis is just one of the steps in anaerobic digestion. Consequently, understanding and optimization of the full anaerobic digestion process needs connections from hydrolysis to the other processes taking place during anaerobic digestion and interactions between all these steps.

The well-known and widely used ADM1 (anaerobic digestion model #1) is a structured model including disintegration and hydrolysis, acidogenesis, acetogenesis and methanogenesis steps. Disintegration and hydrolysis are two extracellular steps. In the disintegration step, composite particulate substrates are converted into inert material, particulate carbohydrates, protein and lipids. Subsequently, the enzymatic hydrolysis step decomposes particulate carbohydrates, protein and lipids to monosaccharides, amino acids and long-chain fatty acids (LCFA), respectively (Batstone et al. 2002) (see Fig. 11.12).

ADM1 is a mathematical model that describes the biological processes and physicochemical processes of anaerobic digestion as a set of differential and algebraic equations (DAE). The model contains 26 dynamic state variables in terms of concentrations, 19 biochemical kinetic processes, 3 gas-liquid transfer kinetic processes and 8 implicit algebraic variables for each process unit. As an alternative, Galí et al. (2009) described the anaerobic process as a set of differential equations with 32 dynamic state variables in terms of concentrations and an additional 6 acid-base kinetic processes per process unit. For an overview of the modelling of anaerobic digestion processes, we refer to Ficara et al. (2012). However, in what follows and for some first insights into the AD process, we will present a simple nutrient-balance model of AD in a sequencing batch reactor (SBR).

11.4.1 Nutrient Mineralization

The nutrient mineralization can be calculated using the following equation (Delaide et al. 2018):

$$\mathrm{NR} = 100\% \times \left(\frac{\mathrm{DN_{out}} - \mathrm{DN_{in}}}{\mathrm{TN_{in}} - \mathrm{DN_{in}}} \right) \tag{11.15a}$$

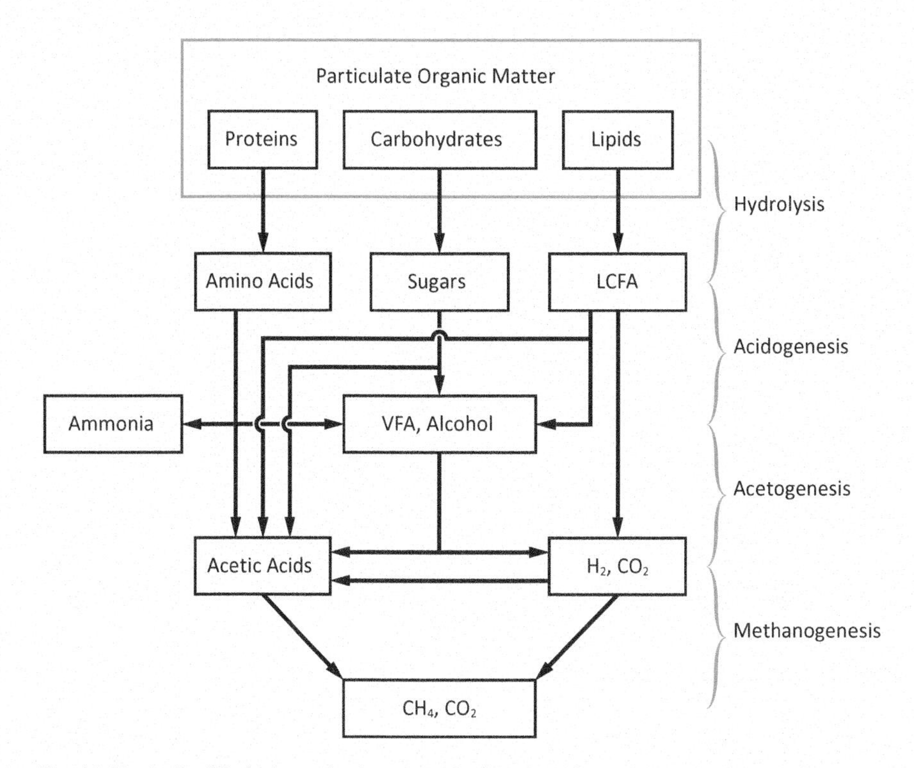

Fig. 11.12 A simplified scheme for the anaerobic digestion of complex particulate organic matter. (based on El-Mashad 2003)

where NR is the nutrient recovery at the end of the experiment in percent, DN_{out} is the total mass of dissolved nutrient in the outflow, DN_{in} is the total mass of dissolved nutrient in the inflow and TN_{in} is the total mass of dissolved plus undissolved nutrients in the inflow (see also Fig. 11.13).

11.4.2 Organic Reduction

The organic reduction performance of the reactor can be calculated using the following equation:

$$\eta_{OM} = 1 - \frac{\Delta OM + T_{OM\ out}}{T_{OM\ in}} \tag{11.15b}$$

where ΔOM is the organic matter (i.e. COD, TS, TSS, etc.) inside the reactor at the end of the experiment minus the one at the beginning of the experiment, $T_{OM\ out}$ is the total OM outflow and $T_{OM\ in}$ is the total OM inflow (see also Fig. 11.14).

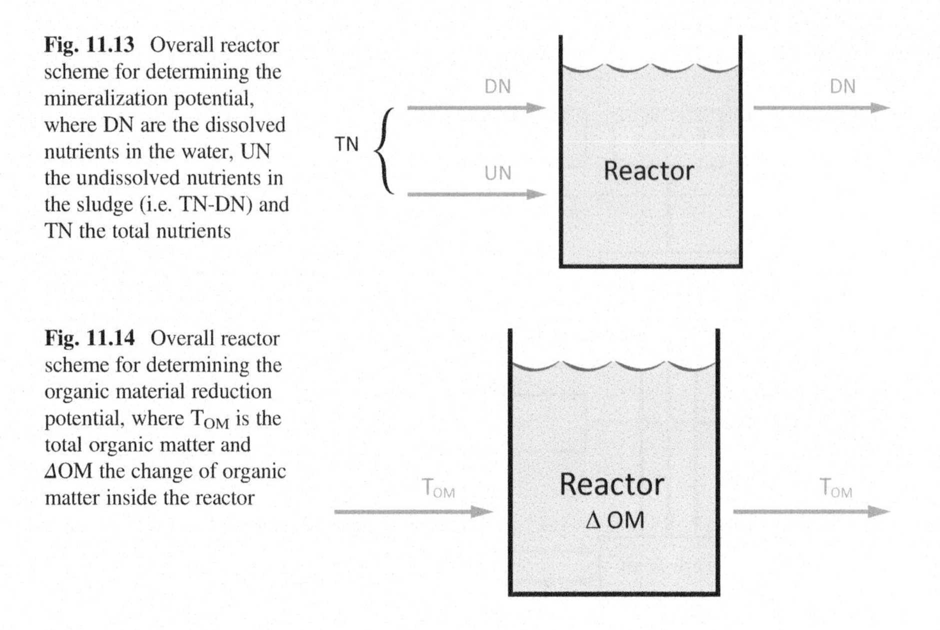

Fig. 11.13 Overall reactor scheme for determining the mineralization potential, where DN are the dissolved nutrients in the water, UN the undissolved nutrients in the sludge (i.e. TN-DN) and TN the total nutrients

Fig. 11.14 Overall reactor scheme for determining the organic material reduction potential, where T_{OM} is the total organic matter and ΔOM the change of organic matter inside the reactor

11.5 HP Greenhouse Modelling

The crop water use and nutrient uptake is a central subsystem of aquaponics. The HP part is complex, as pure uptake of water and dissolved nutrients do not simply follow a rather simple linear relationship as, e.g. fish growth. To create a full-functional model, a complete greenhouse simulator is needed. This involves sub-model systems of greenhouse physics including climate controllers and crop biology covering interactive processes with biological and physical stressors.

However, from the HP point of view, greenhouse climate is the main driver for the complete aquaponic system, including, next to the nutrient balances, feedback loops of heat produced by the fish and additional CO_2 supplied to the plants as reported by Körner et al. (2017) (Fig. 11.15).

In this model, the fish culture produces heat through metabolic processes. The amount of heat produced by the fish is directly calculated from oxygen consumption that is a function of temperature and a constant for heat production for one unit oxygen consumed (i.e. 13608 J g^{-1} fish). Heat from breakdown of organic matter (Q_{bio}), e.g. faeces and feed remain, is also contributing to the heat balance. Energy supply to the water system can then be calculated by heat production through the fish calculated from an average oxygen consumption rate ($f_{O2,Twb}$). Additional heat production can then be calculated by biological breakdown of faeces (Fig. 11.16).

CO_2 production from the aquatic subsystem (d_{CO_2}, g h^{-1}), i.e. delivery to the aerial environment (d, g h^{-1}), can be calculated for the given water temperature (T_{H2O}, K) from oxygen delivery to the system (d_{O_2}, g h^{-1}) at water base temperature ($T_{H2O,b}$, K) and the Q_{10} value of fish respiration ($Q_{10,R}$). The following relationships are used:

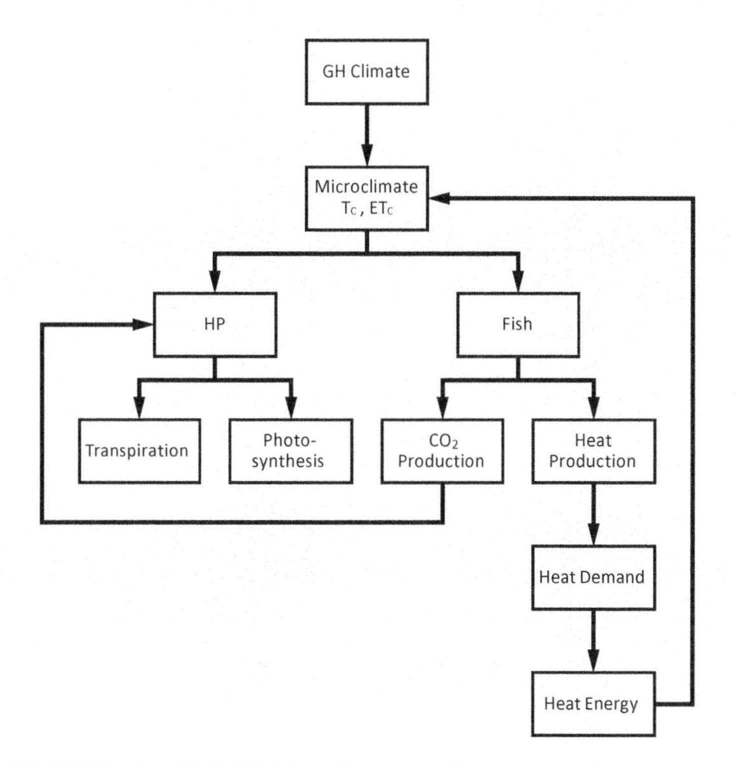

Fig. 11.15 Additional symbiotic behaviour of an aquaponic system

$$d_{O_2} = f_{fish}\, f_{O_2}\, w_{O_2}$$
$$d_{CO_2} = \frac{[CO_2]}{[O_2]}\, d_{O_2}\, Q_{10,R}{}^{\left(T_{H2O} - T_{H2O,b}\right)/10} \tag{11.16}$$

with feed amount for the fish (f_{fish}, g h^{-1}), oxygen consumption rate at base temperature (f_{O2}, kg $[O_2]$ kg^{-1} [feed]), fraction of feed loss w_{O2} (‾) and mass balance of O_2/CO_2 (‾).

To calculate the basis of aquaponics, i.e. the process flow (indicated with arrows, →) greenhouse macroclimate → microclimate → evapotranspiration → nutrients uptake, various greenhouse simulators that were developed in the past can be used and combined with aquaculture to an aquaponic system. All greenhouse models include a crop growth model. The model quality, however, can vary a lot from simple empirical regression models, e.g. Boote and Jones (1987), via deterministic models, e.g. Heuvelink (1996), to functional structural plant models (FSPM), e.g. Buck-Sorlin et al. (2011). As current crop growth and development models are inaccurate and have limited predictive power (Poorter et al. 2013), models are occasionally employed in crop management, but then mainly for planning issues in greenhouse simulators, e.g. Vanthoor (2011) and Körner and Hansen (2011). Prediction accuracy is jeopardized by many sources of uncertainty, such as modelling

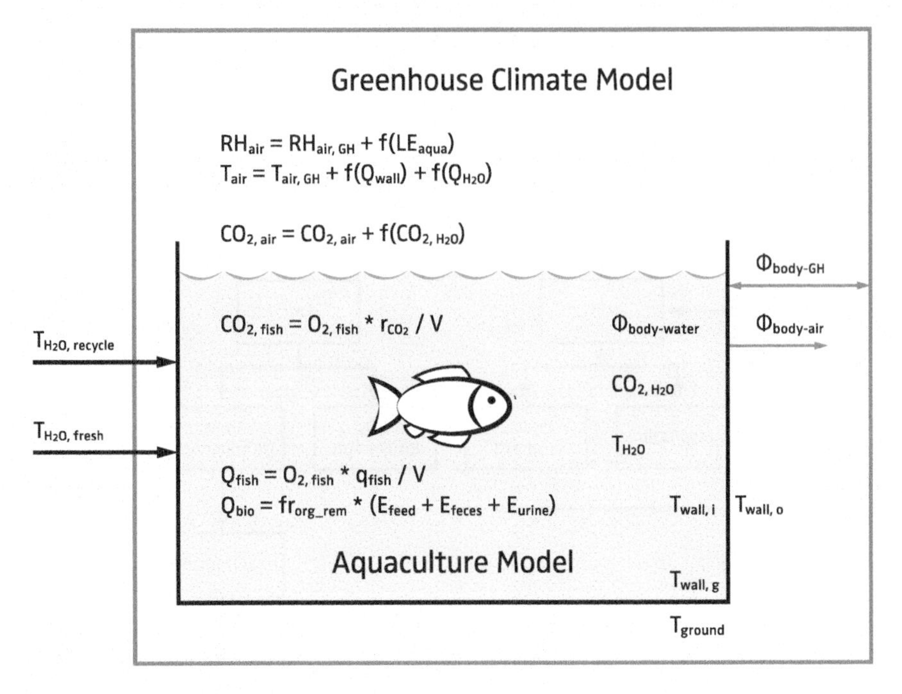

Fig. 11.16 Aquaculture system implemented in the greenhouse with humidity, temperature and CO2 concentrations of the air (RH_{air} T_{air}, $CO_{2,air}$), heat from (Q) fish environment (fish), biological breakdown (bio) and heat fluxes (ϕ), taken from Körner et al. (2017)

errors, variability between plants, variability between greenhouses and uncertain external climate conditions. As for predictions, accuracy also varies strongly per situation. However, online feeding of sensor information into the plant model can make plant model predictions considerably more reliable and useful to the grower.

Greenhouse simulators were developed in and for several places, e.g. the Virtual Grower (Frantz et al. 2010), KASPRO (De Zwart 1996), Greenergy Energy Audit Tool (Körner et al. 2008), The Virtual Greenhouse (Körner and Hansen 2011), The Adaptive Greenhouse (Vanthoor 2011), Hortex (Rath 1992, 2011) and the integrated aquaponic greenhouse model (Goddek and Körner 2019). At a research level, some models (i.e. simulation models in combination with certain greenhouse technologies) have been developed that potentially can be used to optimize investments and structural modifications to the production unit and production process. However, most systems entail closed software environments that can only be used by the developers, and many of them only exist in a research mode and lack further development and acceptance from the industry. However, there is yet no common basis for model sharing and collaborative model development. As a result, most modellers and modelling teams work in isolation developing their own models and codes. A shortcoming of that procedure is that greenhouse simulation models are developed in parallel in disparate research environments, which fail in cooperative growth and development.

All HP greenhouse model simulators are a compilation of sub-models that depend on the aim to integrate the interaction of plants and greenhouse equipment. The general two-part differentiation in greenhouse models and also in control and planning is the shoot and the root environment. Rather complicated and differentiated model approaches have been done for the greenhouse climate (Bot 1993; de Zwart 1996), and greenhouse crop growth has been intensively modelled in the 1990s for the main greenhouse crops such as tomatoes (Heuvelink 1996), cucumber (Marcelis 1994) and lettuce (Liebig and Alscher 1993). However, in order to calculate the water and nutrient uptake of crops, the microclimate, i.e. the climate close to and on the plant organs, needs to be known (Challa and Bakker 1999). This is an ongoing issue in greenhouse modelling, as microclimate variables, such as the central leaf temperature, are highly variable and dependent on many parameters and variables. One version of a leaf temperature model used in a crop canopy for crop temperature (T_c) integrated over vertical layers (z) by Körner et al. (2007) integrating absorbed irradiative net fluxes ($R_{n,a}$, Wm^{-2}), boundary layer and stomata resistances (r_b and r_s, respectively, sm^{-1}) and vapour pressure deficit at the leaf surface (VPD_s, Pa) in the canopy is shown here, i.e.

$$T_c(z) - T_a = \frac{\frac{1}{\rho_a c_p}(r_b(z) + r_s(z))R_{n,a}(z) - \frac{1}{\gamma}VPD_s(z)}{1 + \frac{\delta}{\gamma} + \frac{r_s(z)}{r_b(z)} + \frac{1}{\rho_a c_p/4\sigma T_a^3}(r_b(z) + r_s(z))} \tag{11.17}$$

with greenhouse air temperature (T_a, K), vapour pressure air density (ρ_a, g m^{-2}), Stefan-Boltzmann constant (σ, Wm^{-2} K^{-4}), specific heat capacity of the air (c_p, J g^{-1} K^{-1}), the psychrometric constant (γ, Pa K^{-1}) and the slope between saturated vapour pressure and greenhouse air temperature (δ, Pa K^{-1}).

Leaf temperature is the central part of the microclimate model, it has feedback loops to several input variables and especially stomata resistance (often also used as its reciprocal, the conductance), and the calculation needs several simulation steps for equilibrium. For HP, as part of the aquaponic system, however, modelling water and nutrient fluxes is most important. All water and nutrient balances in a closed multi-loop system are controlled based on the evapotranspiration rate of the crop ET_c (Chap. 8). Commonly ET_c is calculated as latent heat of evaporation, i.e. in energy terms (λE, Wm^{-2}), and can be in accordance to leaf temperature expressed in different canopy layers z

$$\lambda E(z) = \left(\frac{\frac{\delta}{\gamma}R_{n,a}(z) + \frac{\rho_a c_p}{\gamma}\left(\frac{1}{r_b(z)} + \frac{1}{\rho_a c_p/4\sigma T_a^3}\right)VPD_s(z)}{1 + \frac{\delta}{\gamma} + \frac{r_s(z)}{r_b(z)} + \frac{1}{\rho_a c_p/4\sigma T_a^3}(r_s(z) + r_b(z))} \right) \tag{11.18}$$

To calculate ET_c (L m^{-2}), λE needs to be multiplied with the constant L_w (heat of vaporization of water; 2454·103 J kg^{-1}) and the specific weight of water (9.789 kN·m^3 at 20 °C).

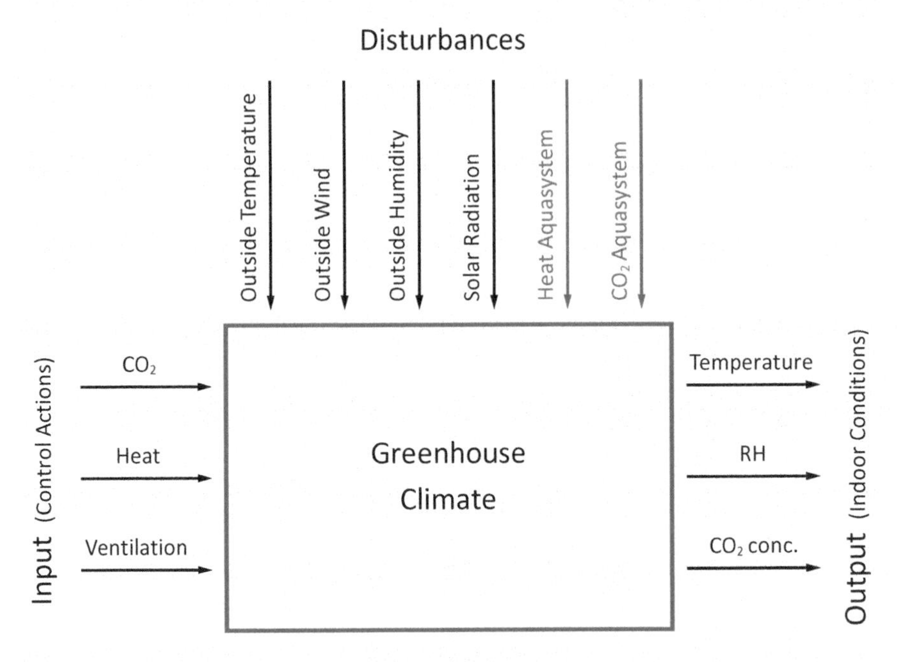

Fig. 11.17 Input-output system of a greenhouse

Equation (11.18), however, only calculates the water flux through the crop, while the easiest way to estimate nutrient uptake is the assumption that nutrients are taken up/absorbed as dissolved in irrigation water and assuming that no element specific chemical, biological or physical resistances exist. In reality uptake of nutrients is a highly complicated matter. Consequently, to maintain equilibrium, all nutrients taken up by the crop as contained in the nutrient solution need to be added back to the hydroponics system (see Chap. 8). However, Eq. (11.18) only calculates the potential ETc, while too high potential levels can result in a higher transpiration than plants can handle, and then potential water loss may exceed water uptake. For that, the simple assumption of nutrient uptake is not satisfying. As described in Chap. 10, the different nutrients can have different states and change states with, e.g. pH, while the plant availability strongly depends on pH and the relation of nutrients to each other. In addition, the microbiome in the root zone plays an important role, which is not implemented in models yet. Some models, however, differentiate between phloem and xylem pathways. The vast amount of nutrients, however, is not modelled in detail for aquaponics nutrient balancing and sizing of systems, while the easiest way to estimate nutrient uptake is the assumption that nutrients are taken up/absorbed as dissolved in irrigation water and apply the above explained ETc calculation approach.

For control purposes the greenhouse is typically considered as a black box, where outside climate conditions determine the disturbance inputs, CO_2 supply, heating and ventilation are the control inputs, and the greenhouse macro- and microclimate define the output of the system (Fig. 11.17).

To control the greenhouse, the actions are directed to minimize the fast impacts of the disturbances, i.e. being ahead of expected changes by smart control. For that, control actions such as feedback and feedforward are used (Chap. 8). The best control, however, can be achieved when using a complete greenhouse model and combine it with weather forecast (Körner and Van Straten, 2008) attaining a model-based optimal greenhouse climate control, as worked out by Van Ooteghem (2007).

11.6 Multi-loop Aquaponic Modelling

Traditional aquaponic designs comprise of aquaculture and hydroponic units involving recirculating water between both subsystems (Körner et al. 2017; Graber and Junge 2009). In such one-loop aquaponic systems, it is necessary to make trade-offs between the conditions of both subsystems in terms of pH, temperature and nutrient concentrations, as fish and plants share one ecosystem (Goddek et al. 2015). By contrast, decoupled double-loop aquaponic systems separate the RAS and hydroponic units from one another, creating detached ecosystems with inherent advantages for both plants and fish. Recently, there has been an increased interest in closing the loop in terms of nutrients as well as increasing the input-output efficiency. For that reason, remineralization (Goddek 2017; Emerenciano et al. 2017; Goddek et al. 2018; Yogev et al. 2016) and desalination loops (Goddek and Keesman 2018) have been incorporated into the overall system design. Such systems are called decoupled multi-loop aquaponic system (Goddek et al. 2016).

Sizing the respective subsystems is fundamental of having a functioning check-and-balance system. For sizing one-loop systems, a simple rule of thumb is generally used, determining the hydroponic cultivation area based on the daily feed input to the RAS (Knaus and Palm 2017; Licamele 2009). The higher degree of complexity of multi-loop systems does not allow this approach anymore, as it comes with inherent risks for making false assumptions for each subsystem. There is a growing body of literature that examines mass balances for aquaponic systems (Körner et al. 2017; Goddek et al. 2016; Reyes Lastiri et al. 2016; Karimanzira et al. 2016). While some research has been carried out in developing numerical models for one- and multi-loop aquaponic systems, no single study exists that integrates a multi-loop aquaponic model with a complemented full-scale deterministic greenhouse model. This is particularly relevant for sizing the system, since plant growth and nutrient uptake are location dependent with crop transpiration as major driver. In concrete terms, this means that the climate within a greenhouse – which is highly dependent on the external weather conditions – has a high impact on plant growth given environmental factors such as relative humidity (RH), light irradiation, temperature, carbon dioxide (CO_2) levels, etc. that were incorporated in greenhouse microclimate modelling (Körner et al. 2007; Janka et al. 2018).

11.7 Modelling Tools

In aquaponics, flow charts or stock and flow diagrams (SFD) and causal loop diagrams (CLDs) are commonly used to illustrate the functionality of the aquaponic system. In the following, flow chart and CLDs will be described.

11.7.1 Flow Charts

To get a systemic understanding of the aquaponics, flow charts with the most important components of the aquaponics are a good tool to show how material flows in the system. This can help, for example, in finding missing components and unbalanced flows and mainly influencing determinants of the subprocesses. Figure 11.18 shows a simple flow chart in aquaponics. In the flow chart, fish food and water are added to the fish tank, where the feed is taken by the fish for growth, the water is enriched with the fish waste and the nutrient-enriched water is added to the hydroponics system to produce plant biomass. From the flow chart, a CLD shown in Fig. 11.19 can be easily constructed.

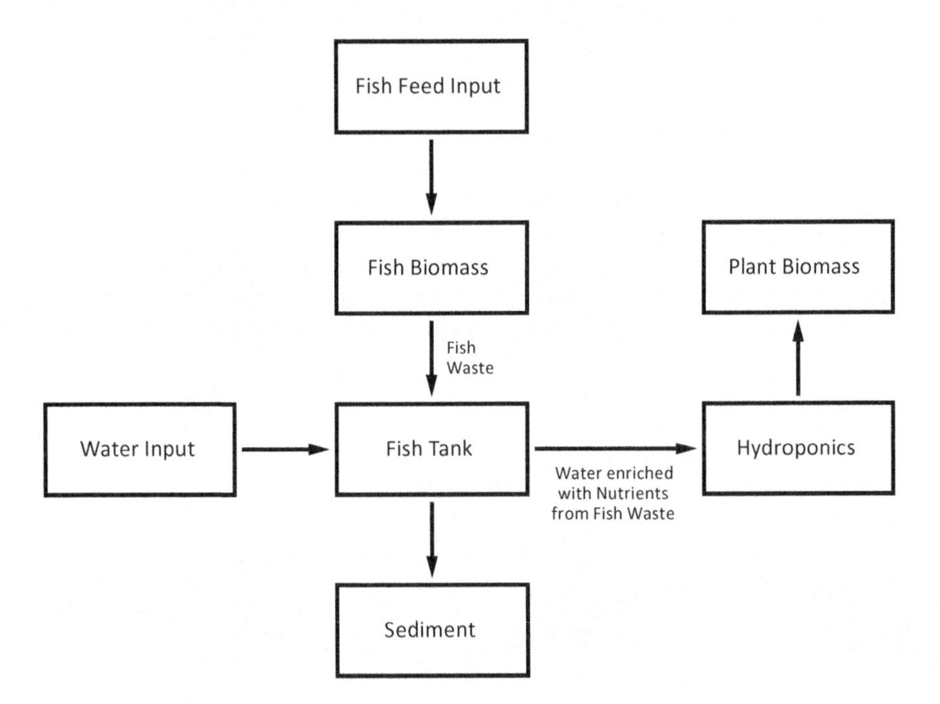

Fig. 11.18 Example of a flow chart in aquaponics (only RAS and HP exchange)

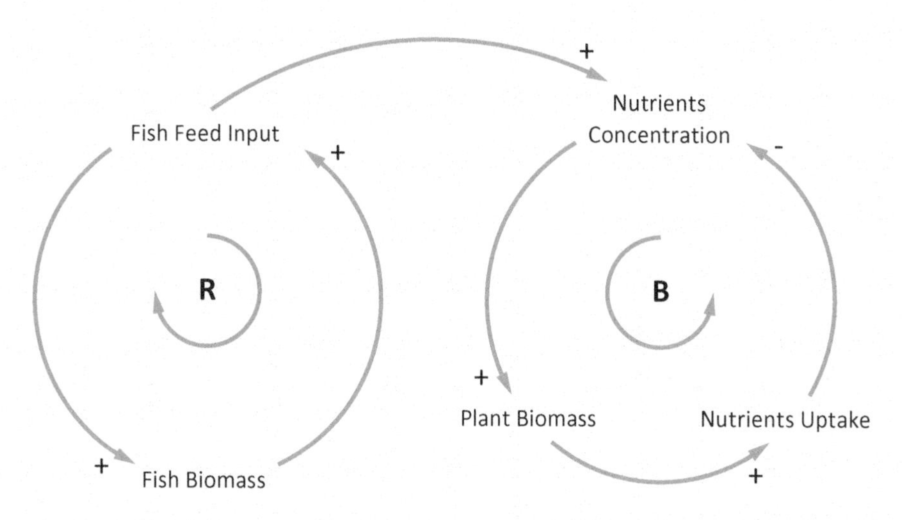

Fig. 11.19 Causal loop diagram (CLD) illustrating examples of a reinforcing and a balanced loop within aquaponic systems. The reinforcing loop (R) is one in which an action produces a result which influences more of the same action and consequently resulting in growth or decline, where as a balancing loop (B) attempts to bring things to a desired state and keep them there (e.g. temperature regulation in the house)

11.7.2 Causal Loop Diagrams

Causal loop diagrams (CLDs) are a tool to show the feedback structure of a system (Sterman 2000). These diagrams can create a foundation for understanding complex systems by visualizing the interconnection of different variables within a system. When drawing a CLD, variables are pictured as nodes. These nodes are connected by edges, which form a connection between two variables accordingly. Figure 11.19 shows that such edges can be marked as either positive or negative. This depends on the relation of the variables to one another. When both variables change into the same direction, then one can speak of a positive causal link. A negative causal link thus causes a change in opposite directions. When connecting two nodes from both sides, one creates a closed cycle that can have two characteristics: (1) a *reinforcing loop* that describes a causal relationship, creating exponential growth or collapse within the loop or (2) a *balancing loop* in which the causal influences keep the system in an equilibrium. Figure 11.19 shows an example of both types of loops.

Let us illustrate this (Fig. 11.20) for the flow chart of Fig. 11.18.

It is obvious that CLD and SFD are very useful for system understanding, when the model does not require numerical accuracy. If numerical accuracy is required, the process should be studied further with a system dynamic tool diagram (SDTD) and modelled in dynamic system simulation software. For example, the CLD in Fig. 11.20 can be augmented with differential equations to a SDTD (Fig. 11.21).

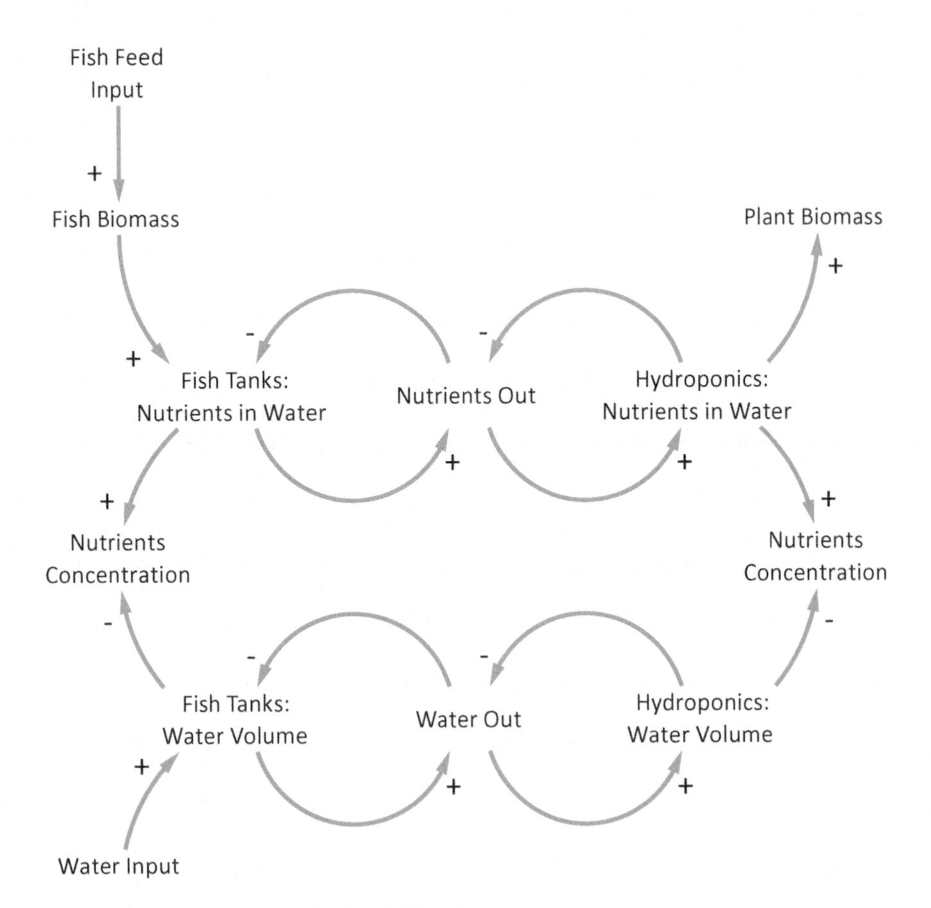

Fig. 11.20 Example CLD for RAS and HP exchange

From the SDTD, we can now see how the differential equations for the nutrients balance in the tank look like. We know that the nutrient flow out of the fish tank (M_{xfout}) must be the water flow (Q_{fout}) times the concentration in the out stream (C_{xf}):

$$M_{xfout} == C_{xf}\ Q_{fout}$$

Assuming a stirred tank gives the nutrient concentration of the fink tank to:

$$C_{\mathrm{xf}} = M_{\mathrm{xf}}/V_f$$

The differential equations of the RAS part can be derived to:

$$dV_f/dt = Q_{\mathrm{fin}} - Q_{\mathrm{fout}}$$
$$dM_{\mathrm{xf}}/dt = M_{\mathrm{xfin}} - M_{\mathrm{xfout}},$$

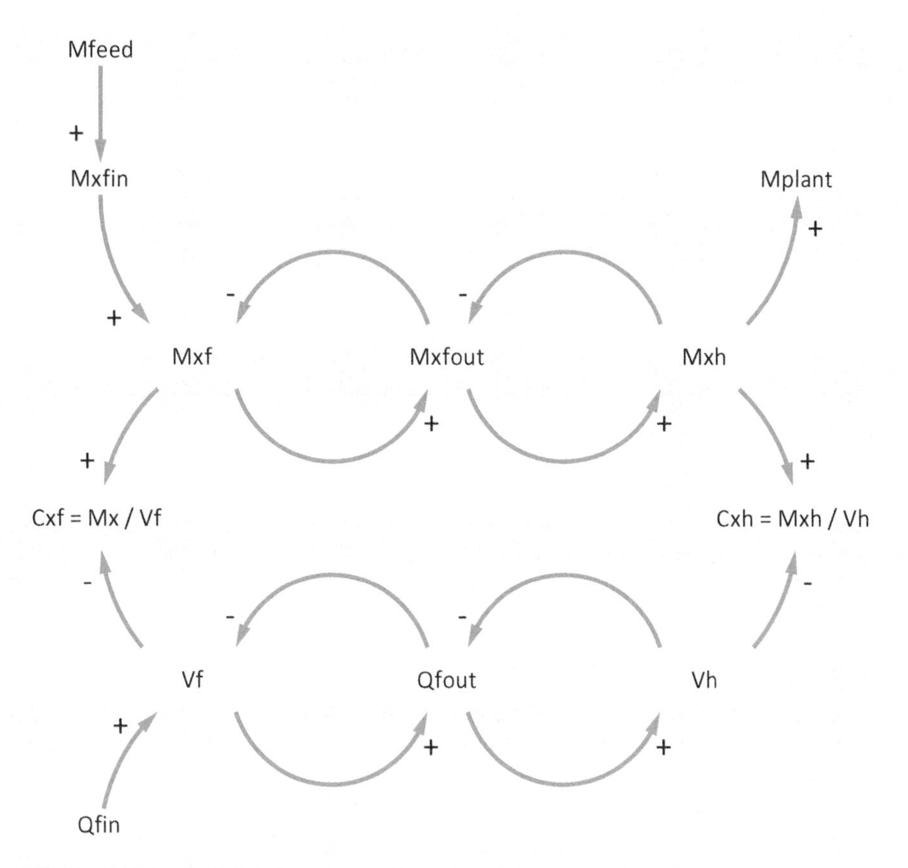

Fig. 11.21 Example SDTD for RAS and HP exchange

and for the concentration

$$dc/dt = (Q_{\mathrm{fin}} C_{\mathrm{xfin}} - Q_{\mathrm{fout}} C_{\mathrm{xf}})/V_f$$

11.7.3 Software

In addition to basic computer languages, such as Fortran, C++ and Python, for fast computation and fully user-specific implementation, all kinds of advanced software tools are available. These advanced software tools offer a variety of environments, concepts and options. We can model state variables, differential equations, connections and loops. In addition, we can use the model for simulations, stability analysis, optimization and control.

The main reasons for modelling of a system are to understand and control it. Therefore, the model helps to predict the system dynamics or behaviour. The software applications could allow us to do three consequent tasks: (a) the modelling itself, (b) the simulations of the model(s) and (c) optimization of the model and/or simulation.

Mathematica software is for functional analysis of mathematically described problems (Wolfram 1991). The concept is based on the LISP approach (McCarthy and Levin 1965.), a very effective functional programming language. The syntax is reasonably simple, and this software is popular in mathematics, physics and systems biology. Especially, the Ndsolve module helps to solve ordinary differential equations, plot the solution and find specific values.

Very similar tools for solving ODEs are offered by the *Maple*. This software is very powerful; between its features belong boundary problems solution, exact solutions and mathematical approximations. *Copasi* (complex pathway simulator) is a software tool for simulation and analysis of biochemical networks via ordinary differential equations.

SageMath is a free open-source mathematics software system. The software is Python-based and facilitates the simulation of ODE models. *Data2Dynamics* software is a collection of numerical methods for quantitative dynamic modelling and is a comprehensive model and data description language. The software allows the analysis of noise, calibration and uncertainty predictions and has libraries of biological models.

Probably the best simulation language is Simula (probably not in use anymore) and *Simula 67*, considered at the beginning as a package for Algol 60. These were the first fully object-oriented languages, introducing classes, inheritance, subclasses, garbage collector and others. In the beginning of the twenty-first century, the creators Ole-Johan Dahl and Kristen Nygaard were awarded the IEEE John von Neumann Medal and the A. M. Turing Award (Dahl and Nygaard 1966).

The idea behind Simula was that objects have life; they start to exist, do their being and cease. The objects are defined as general classes (template code), and each instance of such object has a 'life' in the simulation. The language was quite difficult to learn. However, it offered the possibility to model processes object by object and run simulation of their lives. The simulation runs on the basis of discrete events, and it is possible to simulate objects in co-routine. More tasks can start, run, detach, resume and complete in overlapping time periods in quasiparallel processes. Today's hardware allows us modelling and simulation in fully parallel threads. However, many of the Simula concepts were already used for development of other languages, namely, Java, C/C++/C# and persistent objects libraries like DOL (Soukup and Machacek 2014). Current successor of Simula is *BETA*, extending and featuring the possibilities of inheritance in concepts of nested (sub)classes (with nested local time) and patterns (Madsen et al. 1993).

It is always an option to use any of the object-oriented languages and specific libraries and program all the necessary code for a specific model. On the other hand, already existing graphical programming environments allow to design and link the structure of the modelled system from libraries of objects (signal generator, sum, integrator, etc.), parametrize them and run the simulation in virtual time.

Another popular software for simulation is MathWorks *Simulink*, describing itself as a model-based design tool. The environment allows to combine and parametrize predefined blocks (from wide range of libraries) and diagrams into subsystems. The programming is done using graphical blocks and their connections into functional parts with feedback loops. The environment is widely used for control, automation and signal processing. Another possibility is to integrate own code from MathWorks *Matlab* language or use various toolboxes (Jablonsky et al. 2016). One of them, *SensSB*, is focused on the sensitivity analysis and allows to import other models using the Systems Biology Markup Language. For just the visualization of existing models in Simulink, it is also possible to use the very quick models viewer *DiffPlug*. *PottersWheel* supports the modelling of time-dependent dynamic systems, parameter calibration, analysis and prediction. Interesting tool is the experimental design for model verification.

For the modelling and analysis of system dynamics, similar strategy is used by the *Stella* Architect isee software application, where the model is composed of blocks, which are connected by relations. Stella allows modelling and simulation of very different types of applications, ranging from medical needs through building construction to the airplanes. Stella is sometimes marketed as *iThink* software. *Powersim* software was designed originally for economic purposes. However, it developed into more sophisticated tool, including electronic, solar power or drug treatment simulations. The former developer of *Powersim* is currently producing a similar software for more complex tasks *Dynaplan Smia*. *Vensim* is a system for modelling big data relationships of real systems. The power of *Vensim* is that it allows causal tracking, sensitivity analysis, calibration and intensive simulation. However, the software is also capable of dealing with wide range of simple and complex real systems (Hassan et al. 2016). *True-World* system dynamics software facilitates complex multibody dynamic simulations in discrete and continuous time. The modelling basically starts from balances.

Completely different approaches to modelling and simulation are cellular automata or agent-based modelling approaches, popularized by Stephen Wolfram (Wolfram 1991) as a new kind of science. The approach is sometimes also called the game of life. The modelling is implemented via interactions of autonomous individuals (Macal and North 2005). The simulations show emergent behaviour and therefore are very popular in systems biology for population dynamics. A simple tool for basic (and advanced, as well) agent-based modelling and simulation is the *NetLogo* software, where simple descriptions and parametrization create powerful models. The software allows visualization of the time development and noise induction (Stys et al. 2015). The application is written in Java, which sometimes limits the available memory. Probably the biggest effort in the multi-agent modelling was done by development of *Wolfram* software, which is the continuation of popular Mathematica, with extended tools for modelling and simulation. It has put Simulink-like modelling in a more attractive suit and also creates possibilities of agent-based modelling and much more tools for other mathematical disciplines (multivariate statistics, data mining, global optimization).

AnyLogic is very interesting software for problems of flow – information, money, traffic, logistic and mining. The simulation solves the problem of optimal flow in the designed system with minimal effort and maximal efficiency. The used concepts are system dynamics and agent-based and discrete-event modelling. It also offers hybrids between different modelling concepts. The software is useful, for example, in the epidemic spread simulations (Emrich et al. 2007).

Another agent-based modelling tool is *Insight Maker*, for simulation of population interacting in geographic or network space. The software supports graphical model construction, usage of multiple paradigms, embedded scripting and optimization toolset (Fortmann-Roe 2014).

For the modelling itself, description of the state variables, solution of the ODEs, parametrization and time dependency analysis, the first group of software, from *Mathematica* to *Matlab*, could be used without hesitation. They represent powerful tools for the modelling purposes. In case of more complex analysis, such as related to big data, simulation, noise induction, optimization, sensitivity and stochastics, more advanced tools are required, with object-oriented approach, involving also higher induction in the programming languages syntax.

11.8 Discussion and Conclusions

Aquaponics are complex technical and biological systems. For example, possible explanations for fish not growing properly can be small food rations, adverse water quality, technical problems causing stress, etc. Due to the inherently slow biology, scientific investigations of the validity of these explanations would be tedious and require several experimental trials to get all important factors and their interactions, demanding a lot of facilities, expertise, research time and financial assets. Therefore, the issue of modelling aquaponic systems was addressed in this chapter. In aquaponics, modelling is required for different objectives: (i) insight/understanding, (ii) analysis, (iii) estimation and (iv) management and control. For all these objectives, appropriate models are required. For example, to achieve objectives (ii) and (iii), an empirical approach can be utilized which uses statistical models to analyse data from previous experimental trials with the objective of extracting as much information as possible without conducting new experiments. Statistical models can reveal the most important factors affecting fish and crop production in the aquaponic systems. Future experiments could concentrate on these factors, thus making the utilization of costly research assets more effective.

The complexity of aquaponic systems, due to their feedback character and the interactions between RAS and hydroponic system, water treatment and fish growth, implies that in order to fulfil objectives (i) and (iv), i.e. to understand or optimize a plant (configuration, size, fish, feed, flows, etc.) with respect to cost, stability, robustness and water quality, non-trivial theoretical models of most of the system components described in this chapter are required. The advantage of these theoretical models presented over statistical models is their stronger ability to analyse the process underlying the aquaponics and the possibility to model the time aspect (dynamics). Statistical models just confirm or refute a hypothesis and to what extent variables covary but give no evidence of the underlying processes. On the other hand, theoretical models allow us to simulate the processes according to a hypothesis, compare simulated with observed data, evaluate both the hypothesis and the model and make adaptions. The validity of statistical models may not be beyond the operational range they were trained for, whereas theoretical models can be defined and used for a wide range of environments, provided that the models are validated for these ranges before application. For example, the multiple regression model used to assess relationships between fish growth with *Oreochromis niloticus* as fish species and environmental variables in an aquaponics facility in Germany cannot be easily applied to Spain with *Cyprinus carpio*, whereas a theoretical model describing the underlying processes (e.g. fish behaviour, aquaculture, freshwater ecology) as mathematical equations can be adjusted relatively easily because the fish and ecological process underlying that model are basically the same for the two sites.

Nevertheless, theoretical models also require some parameters such as reaction constants and substance settling velocity in settling tank to be determined. This is achieved commonly based on empirical study of one facility or very few facilities or in most cases from previously published studies (secondary sources). Studies based on secondary sources have limitations imposed by the given structure and amount of the available data, which is not existent when the data come from an experimental setup designed ad hoc for the study. However, estimating model parameters using experimental data from one aquaponics facility only can have problems regarding generalizability and replication of the results due to particular conditions present in the study. The data scarcity sometimes imposes strong restrictions to models that limit their practicality. The development of studies for parameter estimation with primary data that use a larger number of aquaponics facilities than earlier studies does help to overcome the present limitations and provide better and reliable results. This, however, is not an easy challenge for aquaponics researchers.

Simulation of aquaponics with the mathematical models under a wide range of management conditions will improve the understanding of aquaponics, verify different aquaponics configurations and point the way to the most promising strategies for improving aquaponics facilities. Again, this can lead to a more efficient way of conducting experiments.

Some modelling tools were also presented in this chapter. Traditionally, stock and flow diagrams (SFD) have been used for understanding processes as support tools for quantitative analysis. They are used to comprehend the flow and fluxes of quantities but lack the ability to illustrate the information associated to the flow and fluxes. Causal loop diagram (CLD) can be used to transfer complex SFD system into understandable simplified feedback structures. Together, the SFDs and the CLDs fully define the differential equation system. If only a simple qualitative understanding of the system is required, then CLD and SFD may be enough, but if the answer requires a numerical accuracy, then the problem can be investigated further with system dynamic tool diagrams (SDTD) and subsequently be modelled in a software tool for numerical simulation.

References

Badiola M, Mendiola D, Bostock J (2012) Recirculating Aquaculture Systems (RAS) analysis: main issues on management. Aquac Eng 51:26–35. https://doi.org/10.1016/j.aquaeng.2012.07.004

Batstone DJ, Keller J, Angelidaki I, Kalyuzhnyi SV, Pavlostathis SG, Rozzi A, Sanders WTM, Siegrist H, Vavilin VA (2002) The IWA Anaerobic Digestion Model no 1 (ADM1). Water Sci Technol 45:65–73

Boote KJ, Jones JW (1987) Equations to define canopy photosynthesis from given quantum efficiency, maximum leaf rate, light extinction, leaf area index, and photon flux density. In: Biggins J (ed) Progress in photosynthesis research. Martinus Nijhoff, Dordrecht, pp 415–418

Bot GPA (1993) Physical modelling of greenhouse climate. In: Hashimoto Y, Bot GPA, Day W, Tantau HJ, Nonami H (eds) The computerized greenhouse. Academic Press, San Diego, pp 51–74

Buck-Sorlin G, De Visser PHB, Henke M, Sarlikioti V, Ven der Heijden G, Marcelis LFM, Vos J (2011) Towards a functional–structural plant model of cut-rose: simulation of light environment, light absorption, photosynthesis and interference with the plant structure. Ann Bot 108:1121–1134

Challa H, Bakker M (1999) Potential production within the greenhouse environment. In: Stanhill G, Enoch HZ (eds) Ecosystems of the world 20 – Greenhouse ecosystems. Elsevier, pp 333–347

Colt JEK (2013) Impact of aeration and alkalinity on the water quality and product quality of transported tilapia—a simulation study. *Aquac Eng*:46–58

Corominas L, Riegler L, Takács I (2010) New framework for standardized notation in wastewater. J Int Assoc Water Pollut Res 61(4):S841–S857

Dahl O-J, Nygaard K (1966) SIMULA: an ALGOL-based simulation language. Commun ACM 9 (9):671–678. https://doi.org/10.1145/365813.365819

de Zwart HF (1996) Analyzing energy-saving options in greenhouse cultivation using a simulation model. Wageningen Agricultural University, Wageningen, p 236

Delaide B, Goddek S, Keesman, KJ, Jijakli MH (2018). A methodology to quantify the aerobic and anaerobic sludge digestion performance for nutrient recycling in aquaponics. https://popups.uliege.be:443/1780-4507 22, 12

Drayer GE, Howard AM (2014) Modeling and simulation of an aquatic habitat for bioregenerative life support research. *Acta Astronaut* 93:S.138–S.147. https://doi.org/10.1016/j.actaastro.2013.07.013

El-Mashad H (2003) Solar Thermophilic Anaerobic Reactor (STAR) for renewable energy production PhD thesis Wageningen University. ISBN: 9058089533-238

Emerenciano M, Carneiro P, Lapa M, Lapa K, Delaide B, Goddek S (2017) Mineralizacão de sólidos. Aquac Bras 21–26

Emrich S, Suslov S, Judex F (2007) Fully agent based. Modellings of epidemic spread using anylogic. In: Proceedings of the EUROSIM

FAO (2016) *The State of World Fisheries and Aquaculture 2016. Contributing to food security and nutrition for all.* Food and Agriculture Organization of the United Nations, Rome

Ficara E, Hassam S, Allegrini A, Leva A, Malpei F, Ferretti G (2012) Anaerobic digestion models: a comparative study. In: Proceedings of the 7th Vienna international conference on mathematical modelling 2012, p 1052

Fortmann-Roe S (2014) Insight maker: a general-purpose tool for web-based modeling & simulation. *Simul Model Pract Theory* 47:28–45

Frantz JM, Hand B, Buckingham L, Ghose S (2010) Virtual grower: software to calculate heating costs of greenhouse production in the United States. HortTechnology 20:778–785

Galí A, Benabdallah T, Astals S, Mata-Alvarez J (2009) Modified version of ADM1 model for agro-waste application. Bioresour Technol 100(11):2783–2790

Goddek S (2017) Opportunities and challenges of multi-loop aquaponic systems. Wageningen University. https://doi.org/10.18174/412236

Goddek S, Delaide BPL, Joyce A, Wuertz S, Jijakli MH, Gross A, Eding EH, Bläser I, Reuter M, Keizer LCP, Morgenstern R, Körner O, Verreth J, Keesman KJ (2018) Nutrient mineralization and organic matter reduction performance of RAS-based sludge in sequential UASB-EGSB reactors. Aquac Eng 83:10–19. ISSN: 0144-8609

Goddek S, Keesman KJ (2018) The necessity of desalination technology for designing and sizing multi-loop aquaponics systems. Desalination 428:76–85. https://doi.org/10.1016/j.desal.2017.11.024

Goddek S, Körner O (2019) A fully integrated simulation model of multi-loop aquaponics: A case study for system sizing in different environments. Agric Syst

Goddek S, Delaide B, Mankasingh U, Ragnarsdottir K, Jijakli H, Thorarinsdottir R (2015) Challenges of sustainable and commercial aquaponics. Sustainability 7:4199–4224. https://doi.org/10.3390/su7044199

Goddek S, Espinal CA, Delaide B, Jijakli MH, Schmautz Z, Wuertz S, Keesman KJ (2016) Navigating towards decoupled aquaponic systems: a system dynamics design approach. Water (Switzerland) 8:303. https://doi.org/10.3390/W8070303

Graber A, Junge R (2009) Aquaponic systems: nutrient recycling from fish wastewater by vegetable production. Desalination 246:147–156

Halamachi I, Simon Y (2005) A novel computer simulation model for design and management of re-circulating aquaculture systems. Aquac Eng 32(3–4):S443–S464. https://doi.org/10.1016/j.aquaeng.2004.09.010

Hassan J et al (2016) Transient accumulation of NO2-and N2O during denitrification explained by assuming cell diversification by stochastic transcription of denitrification genes. PLoS Comput Biol 11(1):e1004621

He E, Wurtsbaugh W (1993) An empirical model of gastric evacuation rates for fish and an analysis of digestion in piscivorous brown trout. Trans Am Fish Soc *122(5)*:S.717–S.730

Henze M, Willi G, Takashi M, Mark L (2002) Activated sludge models ASM1, ASM2, ASM2d AND ASM3. IWA Publishing in its Scientific and Technical Report series, UK. ISBN: 1-900222-24-8

Heuvelink E (1996) Tomato growth and yield: quantitative analysis and synthesis. Department of Horticulture. Wageningen Agricultural University, Wageningen, The Netherlands, p 326

Jablonsky J, Papacek S, Hagemann M (2016) Different strategies of metabolic regulation in cyanobacteria: from transcriptional to biochemical control. Sci Rep 6:33024

Janka E, Körner O, Rosenqvist E, Ottosen CO (2018) Simulation of PSII-operating efficiency from chlorophyll fluorescence in response to light and temperature in chrysanthemum (Dendranthema grandiflora) using a multilayer leaf model. Photosynthetica 56:633–640

Karimanzira D, Keesman KJ, Kloas W, Baganz D, Rauschenbach T (2016) Dynamic modeling of the INAPRO aquaponic system. Aquac Eng 75:29–45. https://doi.org/10.1016/j.aquaeng.2016.10.004

Keesman KJ (2011) System identification: an introduction. Springer, London

Knaus U, Palm HW (2017) Effects of fish biology on ebb and flow aquaponical cultured herbs in northern Germany (Mecklenburg Western Pomerania). Aquaculture 466:51–63. https://doi.org/10.1016/j.aquaculture.2016.09.025

Körner O, Hansen JB (2011) An on-line tool for optimising greenhouse crop production. Acta Hortic 957:147–154

Körner O, Van Straten G (2008) Decision support for dynamic greenhouse climate control strategies. Comput Electron Agric 60:18–30

Körner O, Aaslyng JM, Andreassen AU, Holst N (2007) Modelling microclimate for dynamic greenhouse climate control. HortScience 42:272–279

Körner O, Warner D, Tzilivakis J, Eveleens-Clark B, Heuvelink E (2008) Decision support for optimising energy consumption in European greenhouses. Acta Hortic 801:803–810

Körner O, Gutzmann E, Kledal PR (2017) A dynamic model simulating the symbiotic effects in aquaponic systems. Acta Hortic 1170:309–316

Licamele JD (2009) Biomass production and nutrient dynamics in an aquaponics system. The University of Arizona

Liebig HP, Alscher G (1993) Combination of growth models for optimized CO_2- and temperature-control of lettuce. Acta Hortic 328:155–162

Lugert V, Thaller G, Tetens J, Schulz C, Krieter J (2014) A review on fish growth calculation: multiple functions in fish production and their specific application. Rev Aquac 8(1):30–42

Lupatsch I, Kissil GW (1998) Predicting aquaculture waste from gilthead seabream (*Sparus aurata*) culture using nutritional approach. Aquat Living Resour 11(4):265–268. https://doi.org/10.1016/S0990-7440(98)80010-7

Lupatsch I, Kissil GW, Sklan D (2003) Comparison of energy and protein efficiency among three fish species gilthead sea bream (Sparus aurata), European sea bass (Dicentrarchus labrax) and white grouper (Epinephelus aeneus): energy expenditure for protein and lipid deposition. *Aquaculture*:175–189

Macal CM, North MJ (2005) Tutorial on agent-based modeling and simulation. In: Simulation conference, 2005 proceedings of the winter. IEEE

Madsen LO, Møller-Pedersen B, Nygaard K (1993) Object-oriented programming in the BETA programming language. Addison-Wesley. ISBN 0-201-62430-3

Marcelis LFM (1994) Fruit growth and dry matter partitioning in cucumber. Department of Horticulture. Wageningen Agricultural University, Wageningen, p 173

McCarthy J, Levin MI (1965) LISP 1.5 programmer's manual. MIT Press, Cambridge, MA

Orellana JUW (2014) Culture of yellowtail kingfish (Seriola lalandi) in a marine recirculating aquaculture system (RAS) with artificial seawater. Aquac Eng:20–28

Pagand P, Blancheton JP, Casellas C (2000) A model for predicting the quantities of dissolved inorganic nitrogen released in effluents from a sea bass (Dicentrarchus labrax) recirculating water system. *Aquac Eng* 22(1–2):S137–S153

Pavlostathis SG, Giraldo-gomez E (1991) Kinetics of anaerobic treatment: A critical review. Crit Rev Environ Control 21:411–490

Poorter H, Anten NP, Marcelis LFM (2013) Physiological mechanisms in plant growth models: do we need a supra-cellular systems biology approach. Plant Cell Environ 36:1673–1690

Rath T (1992) Einsatz wissensbasierter Systeme zur Modellierung und Darstellung von gartenbautechnischem Fachwissen am Beispiel des hybriden Expertensystems HORTEX. University of Hannover, Germany

Rath T (2011) Softwaresystem zur Planung von Heizanlagen von Gewächshäusern. Fachgebiet Biosystem- und Gartenbautechnik. Leibniz University Hannover, Germany

Reyes Lastiri D, Slinkert T, Cappon HJ, Baganz D, Staaks G, Keesman KJ, (2016) Model of an aquaponic system for minimised water, energy and nitrogen requirements. Water Sci Technol. wst2016127. https://doi.org/10.2166/wst.2016.127

Richie M, Haley D, Oetker M (2004) Effect of feeding frequency on gastric evacuation and the return of appetite in tilapia Oreochromis niloticus (L.). Aquaculture 234(1–4):S657–S673. https://doi.org/10.1016/j.aquaculture.2003.12.012

Rusten BE (2006) Design and operations of the Kaldnes moving bed biofilm reactors. Aquac Eng:322–331

Sánchez-Romero A, Miranda-Baeza A, Rivas-Vega M (2016) Development of a model to simulate nitrogen dynamics in an integrated shrimp–macroalgae culture system with zerowater exchange. *J World Aquacult Soc 47(1):129–138*

Sinha NK, Kuszta B (1983) Modelling and identification of dynamic systems. Von-Nostrand Reinhold, New York

Soukup J, Macháček P (2014) Serialization and persistent objects. Springer. https://doi.org/10.1007/978-3-642-39323-5

Sterman J (2000) Business dynamics: systems thinking and modeling for a complex world. McGraw Hill, Boston

Štys D, Stys D Jr, Pecenkova J, Stys KM, Chkalova M, Kouba P, Pautsina A, Durniev D, Nahlık T, Cısa P (2015) 5iD Viewer-observation of fish school behaviour in labyrinths and use of semantic and syntactic entropy for school structure definition. World Acad Sci Eng Technol Int J Comput Electr Autom Control Inf Eng 9(1):281–285

van Ooteghem RJC (2007) Optimal control design for a solar greenhouse. Wageningen University, Wageningen, p 304

Vanthoor B (2011) A model-based greenhouse design method. Wageningen University, Wageningen, p 307

Waller U, Buhmann AK, Ernst A et al (2015) Integrated multi-trophic aquaculture in a zero-exchange recirculation aquaculture system for marine fish and hydroponic halophyte production. *Aquac Int* 23:1473

Weatherley LR, Hill RG, Macmillan KJ (1993) Process modelling of an intensive aquaculture system. *Aquac Eng*:215–230

Wik TEI, Lindén BT, Wramner PI (2009) Integrated dynamic aquaculture and wastewater treatment modelling for recirculating aquaculture systems. *Aquaculture* 287(3/4):361–370

Willems JC, Polderman JW (1998) Introduction to mathematical systems theory: a behavioral approach. Springer. ISBN: 978-1-4757-2953-5

Wolfram S (1991) Mathematica: a system for doing mathematics by computer. Wolfram Research, Champagne

Yogev U, Barnes A, Gross A (2016) Nutrients and energy balance analysis for a conceptual model of a three loops off grid, aquaponics. Water 8:589. https://doi.org/10.3390/W8120589

Zeigler BP, Praehofer H, Kim TG (2000) Theory of modeling and simulation, 2nd edn. Elsevier, London

Chapter 12
Aquaponics: Alternative Types and Approaches

Benz Kotzen, Maurício Gustavo Coelho Emerenciano, Navid Moheimani, and Gavin M. Burnell

Abstract Whilst aquaponics may be considered in the mid-stage of development, there are a number of allied, novel methods of food production that are aligning alongside aquaponics and also which can be merged with aquaponics to deliver food efficiently and productively. These technologies include algaeponics, aeroponics, aeroaquaponics, maraponics, haloponics, biofloc technology and vertical aquaponics. Although some of these systems have undergone many years of trials and research, in most cases, much more scientific research is required to understand intrinsic processes within the systems, efficiency, design aspects, etc., apart from the capacity, capabilities and benefits of conjoining these systems with aquaponics.

Keywords Aquaponics alternatives · Algaeponics · Aeroponics · Aquaeroponics · Biofloc Technologies · Digeponics · Haloponics · Maraponics · Vermiponics · Vertical aquaponics

---

B. Kotzen (✉)
School of Design, University of Greenwich, London, UK
e-mail: b.kotzen@greenwich.ac.uk

M. G. C. Emerenciano
Santa Catarina State University (UDESC), Aquaculture Laboratory (LAQ), Laguna, SC, Brazil

CSIRO Agriculture and Food, Aquaculture Program, Bribie Island Research Centre, Bribie Island, QLD, Woorim, Australia
e-mail: mauricio.emerenciano@csiro.au

N. Moheimani
Algae R&D Centre Director, School of Veterinary and Life Sciences, Murdoch University, Murdoch, WA, Australia
e-mail: n.moheimani@murdoch.edu.au

G. M. Burnell
School of Biological, Earth and Environmental Sciences, University College Cork, Cork, Ireland
e-mail: g.burnell@ucc.ie

S. Goddek et al. (eds.), *Aquaponics Food Production Systems*,
https://doi.org/10.1007/978-3-030-15943-6_12

12.1 Introduction

This chapter discusses a number of key allied and alternative technologies that either expand or have the potential to expand the functionality/productivity of aquaponic systems or are associated/stand-alone technologies that can be linked to aquaponics. The creation and development of these systems have at their core the ability, amongst other things, to increase production, reduce waste and energy and in most cases reduce water usage. Unlike aquaponics, which may be seen to be in a mid/teenage stage of development, the novel approaches discussed below are in their infancy. This, however, does not mean that they are not technologies valuable in their own right and have the potential to deliver future food, efficiently and sustainably. The methods discussed below include aeroponics, aeroaquaponics, algaeponics, biofloc technology for aquaponics, maraponics and haloponics and vertical aquaponics.

12.2 Aeroponics

12.2.1 Background

The US National Aeronautics and Space Administration (NASA) describes aeroponics as *the process of growing plants suspended in air without soil or media* providing *clean, efficient, and rapid food production.* NASA furthermore notes that *crops can be planted and harvested year-round without interruption, and without contamination from soil, pesticides, and residue* and that *aeroponic systems also reduce water usage by 98%, fertilizer usage by 60% percent, and eliminate pesticide usage altogether. Plants grown in aeroponic systems have been shown to absorb more minerals and vitamins, making the plants healthier and potentially more nutritious* (NASA Spinoff). Other advantages of aeroponics are seen to be that:

- The growing environment can be kept clean and sterile.
- This reduces the chances for plant diseases and the spread of infection.
- Seedlings do not stretch or wilt during root formation.
- Seedlings are easily removed for transplanting without transplant shock.
- Seedling growth is accelerated, which leads to increased crop cycles and thus more produce per annum.

For Weathers and Zobel (1992), aeroponics is defined as *the culture of whole plants and/or tissues with their roots or the whole tissue fed by an air/water fog (as opposed to immersion in/on water, soil, nutrient agar or other substrates).* For them, plants that are grown only partially with their roots in air and part in nutrient solutions or are grown for part of the time in air and part of the time in nutrient solution are grown through a process of aero-hydroponics and not areoponics.

Aeroponic systems thus function by spraying or misting the root zone area with nutrient solution. The roots of the plants are thus suspended in air and are subjected to a continuous or intermittent/periodic spray/misting of nutrient-rich water droplets, in the form of droplets or very fine mists, with droplet sizes from 5 to 50 μm (microns). It is usual to find 'hobby/domestic' kit with spray droplet sizes of 30–80 μm. Ultrasonic or dry-fog atomizers produce a droplet size <5 μm, but these require compressed air and very fine nozzles, or it may be possible to use ultrasonic transducers to produce these mists.

In aeroponics, as with hydroponics, nutrient supply can be optimized and in a comparison between hydroponics and aeroponics, Hikosaka et al. (2014) note that no difference was found between growth and harvest quality in lettuce using dry-fog aeroponics. However, there was a significant increase in root respiration rates and photosynthesis rates of leaves. They also note that this system also uses less water and that it can be more efficient and easier to manage than conventional hydroponics (Hikosaka et al. 2014). In a review paper on modern plant cultivation technologies in agriculture under controlled environments, Lakhiar et al. (2018) note that aeroponics 'is considered the best plant growing method for food security and sustainable development'.

12.2.2 Origin of Aeroponics

Richard J. Stoner II is considered the father of aeroponics. The NASA review of aquaponics (Clawson et al. 2000) notes that the origin of aquaponics is largely in the study of root morphology, but originates in nature, e.g. with plants, for example, orchids growing in tropical areas where mists occur naturally. Clawson et al. (2000) note the development of aeroponics from B. T. Barker, who 'succeeded in growing apple trees with a spray', and F. W. Went, who in 1957 grew tomatoes and coffee plants in mists and termed the process 'aeroponics'. With regard to the study of root morphology, Carter in 1942 used aeroponics as a way of investigating pineapple roots, and Klotz in 1944 investigated the roots of avocado and citrus, and then numerous others including Hubick and Robertson; Barak, Soffer, and Burger; Yurgalevitch and Janes; and Dutoit, Weathers, and Briggs all undertook various experiments in aeroponics (refer to Clawson et al. 2000 for details).

12.2.3 Aeroponics Growing Issues

Clawson et al. (2000) report the tests by Tibbits et al. (1994) that continuous misting can 'contribute to fungal and bacterial growth in the vicinity of or on the plants', and furthermore some researchers have found that due to fine droplets and with continuous fogging systems, there can be difficulties 'in delivering nutrients to all the plants where there is a high density of plants'. In this respect it has been shown that misting at intervals delivers a healthier system and healthier roots compared to continuous fogging and hydroponic techniques. Using intervals also makes the plants more resistant to any interruptions in misting, conditioning the plants to thrive longer on lower moisture levels, with a likely reduction in pathogen levels. For effective misting, 'droplet size and velocity are also important aeroponic parameters. The root's mist collection efficiency depends on its filament size, drop size, and velocity' (Clawson et al. 2000).

12.2.4 Combining Aquaponics and Aeroponics

Whilst a number of entrepreneurs and keen hobbyists are promoting combining aquaponics with aeroponics, there are a number of issues that need to be solved if considering this combo-technology for future farming. One issue that needs to be resolved is a name for this system, and it is suggested here that we call this combo-system 'aquaeroponics'.

Whilst there are numerous videos and discussion threads on the web, on combining aeroponics and aquaponics, the field is void of scientific literature. The web-based discussions raise the issues of clogging of mist sprayers and the need for fine filtration of aquaponic solutions. Another issue with aquaeroponics is the potential for pathogens to grow in the airy wet environment and research will be required to ascertain this. One solution to solving the problem of misters is to use ultrasonic vibration to create the mists but this does not solve any problems there may be with the growth of pathogens.

12.3 Algaeponics

12.3.1 Background

Microalgae are unicellular photoautotrophs (ranging from 0.2 μm up to 100 μm) and are classified in various taxonomic groups. Microalgae can be found in most environments but are mostly found in aquatic environments. Phytoplankton are responsible for over 45% of world's primary production as well as generating over 50% of atmospheric O_2. In general, there is no major difference in photosynthesis of microalgae and higher plants (Deppeler et al. 2018). However, due to their smaller size and the reduction in a number of internally competitive physiological organelles, microalgae can grow much faster than higher plants (Moheimani et al. 2015). Microalgae can also grow under limited nutrient conditions and have the ability to adapt to a wider range of environmental conditions (Gordon and Polle 2007). Most importantly, microalgal culture does not compete with food crop production regarding arable land and freshwater (Moheimani et al. 2015). Furthermore, microalgae can efficiently utilize inorganic nutrients from waste effluents (Ayre et al. 2017). In general, microalgal biomass contains up to 50% carbon making them a perfect candidate for bioremediating atmospheric CO_2 (Moheimani et al. 2012).

The increase in extensive worldwide agriculture and animal farming has resulted in significant increases in biologically available nitrogen and phosphorus entering the terrestrial biosphere (Galloway et al. 2004). Crop and animal farming and sewage systems contribute significant amounts to these nutrient loads (Schoumans et al. 2014). The infiltration of these nutrients into water streams can cause massive environmental issues such as harmful algal blooms and mass fish mortality. For instance, in the USA, nutrient pollution from agriculture is acknowledged as one of the major sources of eutrophication (Sharpley et al. 2008). Controlling the flow of nutrients from farming operations into the surrounding environment results in both technical and economic challenges that must be overcome to reduce such effects. There have been various successful processes developed to treat waste effluent with high organic loads. However, almost all of these methods are not very effective in removing inorganic elements from water. Furthermore, some of these methods are rather expensive to operate. One simple method for treating organic waste is anaerobic digestion (AD). The AD process is well understood and when operated efficiently, it can convert over 90% of the wastewater organic matters to bio-methane and CO_2 (Parkin and Owen 1986). The methane can be used to generate electricity and the generated heat can be used for various additional purposes. However, the AD process results in creating an anaerobic digestion effluent (ADE) which is very rich in inorganic phosphate and nitrogen as well as high COD (carbon oxygen demand). In certain locations, this effluent can be treated using microalgae and macroalgae (Ayre et al. 2017).

12.3.2 *Algal Growth Systems*

Since the United Nations committee recommended that conventional agricultural crops be supplemented with high-protein foods of unconventional origin, microalgae have become natural candidates (Richmond and Becker 1986). The first microalgal cultivation was achieved though in 1890 by culturing *Chlorella vulgaris* (Borowitzka 1999). Due to the fact that microalgae normally divide at a certain time of the day, the term cyclostat was developed in order to introduce a light/dark (circadian) cycle to the culture (Chisholm and Brand 1981). The large-scale culturing of microalgae and the partial use of its biomass especially as a base for certain products such as lipids was probably started seriously as early as 1953 with the aim of producing food from a large-scale culture of *Chlorella* (Borowitzka 1999). Typically, algae can be cultured in liquid using open ponds (Borowitzka and Moheimani 2013), closed photobioreactors (Moheimani et al. 2011), or a combination of these systems. Alga can also be cultured as biofilms (Wijihastuti et al. 2017).

Closed Photobioreactors (after Moheimani et al. 2011): Closed algal cultures (photobioreactors) are not exposed to the atmosphere but are covered with a transparent material or contained within transparent tubing. Photobioreactors have the distinct advantage of preventing evaporation. Closed and semi-closed photobioreactors are mainly used for producing high-value algal products. Due to the overall cost of operating expenditure (OPEX) and capital expenditure (CAPEX), closed photobioreactors are less economical than open systems. On the other hand, there is less contamination and less CO_2 losses, and by creating reproducible cultivation conditions and flexibility in technical design, this makes them a good substitute for open ponds. Some of the closed systems' weaknesses can be overcome by (a) reducing the light path, (b) solving shear (turbulence) complexity, reducing oxygen concentration, and (c) a temperature control system. Closed photobioreactors are mainly divided into (a) carboys, (b) tubular, (c) airlift and (d) plate photobioreactors.

Open Ponds (after Borowitzka and Moheimani 2013): Open ponds are most commonly used for large-scale outdoor microalgal cultivation. Major algal commercial production is based in open channels (raceways) which are less expensive, easier to build and operate when compared to closed photobioreactors. In addition, the growth of microalgae meets less difficulties in open than closed cultivation systems. However just a few species of microalgae (e.g. *Dunaliella salina*, *Spirulina* sp., *Chlorella* sp.) have been grown successfully in open ponds. Commercial microalgal production costs are high, approximated to be between 4 and 20 $\$US/g^{-1}$. Large-scale outdoor open pond commercial microalgal culture has developed over the last 70 years, and both still (unstirred) and agitated ponds have been developed and have been used on a commercial basis. The very large unstirred open ponds are simply constructed from natural water ponds with open beds that are usually less than 0.5 m in depth. In some smaller ponds the surface may be lined with plastic lining sheets. Unstirred open ponds represent the most economical and least technical of all commercial culture methods and have been commercially used for *Dunaliella salina* β-carotene production in Australia. Such ponds are mainly limited to growing microalgae which are capable of surviving in poor conditions or have a competitive advantage that allows them to outgrow contaminants such as protozoa, unwanted microalgae, viruses, and bacteria. Agitated ponds on the other hand have the advantage of a mixing regime. Most agitated ponds are either (a) circular ponds with rotating agitators or (b) single or joined raceway ponds.

Circular cultivation ponds have primarily been used for the large-scale cultivation of microalgae especially in South East Asia. The circular ponds up to 45 m in diameter and usually 0.3–0.7 m in depth are uncovered, but there are some examples which are covered by glass domes. The low shear stresses that are required for microalgae production are produced in these systems particularly in the centre of the pond, and this is a distinct advantage of these kinds of systems. Some disadvantages include expensive concrete structures, inefficient land use with large footprints, difficulties in controlling the movement of the agitating device and the added cost in supplying CO_2.

Paddlewheel-driven raceways are the most common commercial microalgal cultivation system. Raceways are usually constructed in either a single channel or as linked channels. Raceways are usually shallow (0.15 to 0.25 m deep), are constructed in a loop and normally cover an area of approximately 0.5 to 1.5 ha. Raceways are mostly used and recommended for the major commercial culturing of three species of microalgae including *Chlorella*, *Spirulina* and *Dunaliella*. A high risk of contamination and low productivity, resulting mainly from poor mixing regimes and light penetration, are the main disadvantages of these open systems. In raceways, biomass concentrations of up to 1000 mg dry weight.L^{-1} and productivities of 20 g dry weight.m^{-2}.d^{-1} have been shown to be possible.

The price of microalgal production makes economic achievement highly dependent on the marketing of expensive and exclusive products, for which demand is naturally restricted. Raceways are also the most used cultivation system used for treating wastewater (Parks and Craggs 2010).

Solid Cultivation (after Wijihastuti et al. 2017): An alternative microalgal cultivation method is immobilizing the cells in a polymer matrix or attaching them to the surface of a solid support (biofilm). In general, the biomass yield of such biomass cultures are at least 99% more concentrated than liquid-based cultures. Dewatering is one of the most expensive and energy-intensive parts of any mass algal production. The main advantage of biofilm growth is the potential of reducing the dewatering process and the related energy consumption and thus costs. Biofilm cultivation can also increase cellular light capture, reduce environmental stress (e.g. pH, salinity, metal toxicity, very high irradiance), reduce the cost of production and reduce nutrient consumption. Solid-based cultivation methods can be used for treating wastewater (nutrient and metal removal). There are three main methods for biofilm cultivation: (a) 100% directly submerged in medium, (b) partially submerged in medium and (c) using a porous substrate to deliver the nutrients and moisture from the medium to the cells.

12.3.3 Algal Growth Nutrient Requirements

A number of inhibitory physical, chemical and biological factors can inhibit high microalgal production. These are described in Table 12.1.

A basic knowledge of the critical growth limitations is probably the most essential factor before applying any microalgae to any process. Light is by far the most important limiting factor affecting the growth of any alga. Temperature is also a critical factor for mass algal production (Moheimani and Parlevliet 2013). However, these variables are difficult to control (Moheimani and Parlevliet 2013). Next to light and temperature, nutrients are the most important limiting factor affecting the growth of any alga (Moheimani and Borowitzka 2007) and each microalgal species tends to have its own optimum nutrient requirements. The most important nutrients are nitrogen, phosphorus and carbon (Oswald 1988). Most

Table 12.1 Limits to growth and productivity of microalgae (Moheimani and Borowitzka 2007)

Abiotic factors	Light (quality, quantity)
	Temperature
	Nutrient concentration
	O_2
	CO_2 and pH
	Salinity
	Toxic chemicals
Biotic factors	Pathogens (bacteria, fungi, viruses)
	Competition by other algae
Operational factors	Shear produced by mixing
	Dilution rate
	Depth
	Harvest frequency
	Addition of bicarbonate

algae respond to N-limitation by increasing their lipid content (Moheimani 2016). For example, Shifrin and Chisholm (1981) reported that in 20 to 30 species of microalgae that they examined, the algae increased their lipid content under N-deprivation. Phosphorus is also an important nutrient required for microalgal growth as it plays an essential role in cell metabolism and regulation, being involved in the production of enzymes, phospholipids and energy-supplying compounds (Smith 1983). Brown and Button's (1979) studies on green alga *Selenastrum capricornutum* showed an apparent growth limitation when the phosphate concentration of the medium was lower than 10 nM. CO_2 is also a critical nutrient for achieving high algal productivity (Moheimani 2016). For example, if additional CO_2 is not added to the algal culture, the average productivity can be reduced by up to 80% (Moheimani 2016). However, the addition of CO_2 to algal ponds is rather costly (Moheimani 2016). The most economical way for introducing CO_2 to a culture media is the direct transfer of the gas into the media by bubbling through sintered porous stones or using pipes under submerged plastic sheets as CO_2 injectors (Moheimani 2016). Unfortunately, in all of these methods there is still high loss of CO_2 to the atmosphere because of the short retention time of the gas bubbles in the algal suspension.

Although adding N, P and C is critical, other nutrients also affect microalgal growth and metabolism. A lack of other nutrients, such as manganese (Mn) and various other cations (Mg^{2+}, K^+ and Ca^{2+}), is also known to reduce algal growth (Droop 1973). Trace elements are also critical for microalgal growth and some microalga also require vitamins for their growth (Croft et al. 2005). One effective and inexpensive way of supplying nutrients is by combining algal culture and wastewater treatment which is discussed immediately below.

12.3.4 Algae and Wastewater Treatment

With an increase in environmental deterioration and a greater necessity to generate alternative food and energy sources, there is the impetus to explore the feasibility of biological wastewater treatments coupled with resource recovery. Microalgal wastewater treatments have been particularly attractive, due to algal photosynthetic activities, where light is transferred into profitable biomass. Under certain conditions, wastewater-grown microalgal biomass can be equivalent or superior in biomass production to higher plant species. Thus, the process can transform a waste product into useful products (e.g. animal feed, aquaculture feed, bio-fertilizer and bioenergy). Thus, the waste effluent is no longer a negative waste product, but it becomes a valuable substrate for producing important substances and successful microalgal wastewater bioremediation has been reported for over half a century (Oswald and Gotass 1957; Delrue et al. 2016). Algal phytoremediation indeed provides an environmentally favourable solution for the treatment of wastewater as it can utilize organic and inorganic nutrients efficiently (Nwoba et al. 2017). Microalgal cultures hold an enormous potential for the later steps of wastewater treatment, especially for reducing 'N', 'P' and 'COD' (Nwoba et al. 2016). Moreover, the added ability of microalgae to grow via different nutritional conditions such as photoautotrophic, mixotrophic and heterotrophic conditions also enhances its capabilities in removing various different types of pollutants and chemicals from aqueous matrices. The ability of microalgae in sequestrating carbon (CO_2) allows CO_2 bioremediation. The synchronized algal-bacteria relationship established is also ideally synergetic for the bioremediation of wastewater (Munoz and Guieysse 2006). Through photosynthesis, microalgae provide oxygen required by aerobic bacteria for the mineralization of organic matter as well as the oxidation of NH_4^+ (Munoz and Guieysse 2006). In return, the bacteria supply carbon dioxide for the growth of microalgae, significantly reducing the amount of oxygen required for the overall wastewater treatment process (Delrue et al. 2016). In general, waste effluents with low carbon to nitrogen ratios are fundamentally suited to the growth of photosynthetic organisms. Most importantly, the microalgal domestic and agricultural wastewater treatments is an attractive option since the technology is relatively easy and they require very low energy compared to the standard of effluent treatment. Optimization of microalgal wastewater treatment in large-scale raceway ponds is appealing since it combines the effective treatment of a harmful waste product and the production of potentially valuable protein-rich algal biomass. Figure 12.1 summarizes a closed loop system for treating any organic waste by combination of anaerobic digestion and algal cultivation.

12.3.5 Algae and Aquaponics

Microalgae in aquaculture and in aquaponic systems is most often seen to be a nuisance as they can restrict water flows by clogging up pipes, consume oxygen,

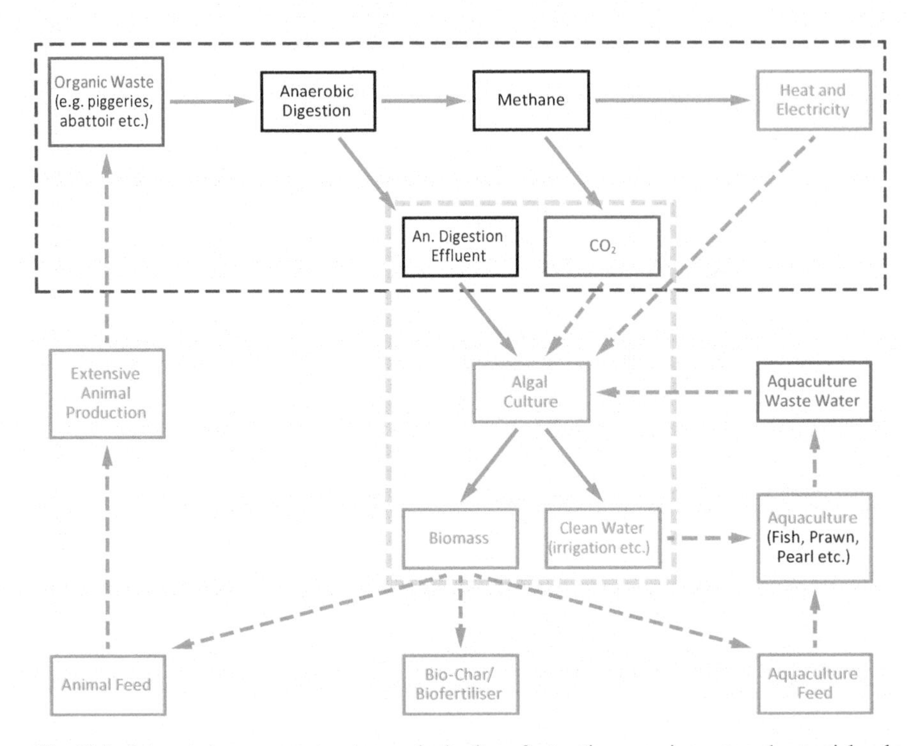

Fig. 12.1 Integrated process system to use algal culture for treating organic waste and potential end users. (The process is designed based on information from Ayre et al. 2017 and Moheimani et al. 2018)

may attract insects, reduce water quality and when decomposing can deplete oxygen. However, an experiment by Addy et al. (2017) shows that algae can improve water quality in an aquaponic system, help control pH drops related to the nitrification process, generate dissolved oxygen in the system, 'produce polyunsaturated fatty acids as a value-added fish feed and add diversity and improve resilience to the system'. One of the 'holy grails' of aquaponics is to produce at least part of the food that is fed to the fish as part of the system and it is here that research is required in producing algae that could be grown with part of the aquaponics water, most probably in a separate loop, which can then be fed as part of the diet to the fish.

12.4 Maraponics and Haloponics

Although freshwater aquaponics is the most widely described and practiced aquaponic technique, resources of freshwater for food production (agriculture and aquaculture) are becoming increasingly limited and soil salinity is progressively increasing in many parts of the world (Turcios and Papenbrock 2014). This has led to an increased interest and/or move towards alternative water sources (e.g. brackish to highly saline water as well as seawater) and the use of euryhaline or saltwater fish, halophytic plants, seaweed and low salt-tolerant glycophytes (Joesting et al. 2016). It is interesting to note that whilst the amount of saline in underground water is only estimated as 0.93% of world's total water resources at 12,870,000 km^3, this is more than the underground freshwater reserves (10,530,000 km^3) which makes up 30.1% of all freshwater reserves (Appelbaum and Kotzen 2016).

The use of saline water in aquaponics is a relatively new development and as with most new developments the terms used to describe the range/hierarchy of types needs to be established on a firm footing. In its short history, the term maraponics (i.e. marine aquaponics) has been coined for seawater aquaponics (SA), in other words, systems that use seawater as well as brackish water (Gunning et al. 2016). These systems are mainly located on-land, in coastal locations and in the case of SA, close to a seawater source. But there are fish as well as plants that grow and can be used in aquaponic units where water salinity levels vary. Thus whilst it makes etymological sense to use the term 'maraponics' for seawater aquaponics, it makes less sense to term brackish water aquaponics using this term. We thus suggest that a new term needs to be added to the aquaponic lexicon and this is 'haloponics', deriving from the Latin word *halo* meaning salt and combining this with suffix ponics. Thus maraponics is an on-land integrated multitrophic aquaculture (IMTA) system combining the aquacultural production of marine fish, marine crustaceans, marine molluscs, etc. with the hydroponic production of marine aquatic plants (e.g. marine seaweeds, marine algae and seawater halophytes) using oceanic strength seawater (approximately 35,000 ppm [35 g/L]). However aquaponic systems utilizing saline water below oceanic levels in a range of salinities should be termed haloponics (slightly saline water –1000 to 3000 ppm [1–3 g/L], moderately saline 3000–10,000 ppm [3–10 g/L] and high salinity 10,000–35,000 ppm [10–35 g/L]). These systems are also on-land IMTA systems combining aquacultural production with the hydroponic production of aquatic plants, but both the fish and plants are adapted to or grow well in what may be termed brackish water.

Although the concept of maraponics is very new, an interest in on-land seaweed-based integrated mariculture began to appear in the 1970s, starting from a laboratory-scale and then expanding to outdoor pilot-scale trials. In some of the earliest experimental studies, Langton et al. (1977) successfully demonstrated the growth of the red seaweed, *Hypnea musciformis*, cultured in tanks with shellfish culture effluent. Alternatively, crops that would usually be classed as glycophytes, such as the common tomato (*Lycopersicon esculentum*), the cherry tomato (*Lycopersicon esculentum* var. Cerasiforme) and basil (*Ocimum basilicum*), can achieve remarkably successful production levels at up to 4 g/L (4000 ppm) salinity and are often referred to as having low-moderate levels of salt tolerance (not to be confused with true halophytes, which are resistant to high salinities). Other crops that are tolerant of low-moderate salinities include turnip, radish, lettuce, sweet potato, broad bean, corn, cabbage, spinach, asparagus, beets, squash, broccoli and cucumber (Kotzen and Appelbaum 2010; Appelbaum and Kotzen 2016). For example, Dufault et al. (2001) and Dufault and Korkmaz (2000) experimented with shrimp waste (shrimp faecal matter and decomposed feed) as a fertilizer for broccoli (*Brassica oleracea italica*) and bell pepper (*Capsicum annuum*) production, respectively. Although their studies did not use maraponic techniques, they involved plants that are commonly grown using aquaponic (freshwater) techniques. Therefore, due to their salinity tolerance levels, these crops have enormous potential as candidate species for production in haloponic systems using low to medium salinities.

Recently, a number of studies have shown that halophytes can be successfully irrigated with aquacultural wastewater from marine systems using hydroponic techniques or as part of a recirculating aquaculture system (RAS). Waller et al. (2015) demonstrated the feasibility of nutrient recycling from a saltwater (16 psu salinity [16,000 ppm]) RAS for European sea bass (*D. labrax*) through the hydroponic production of three halophytic plants: *Tripolium pannonicum* (sea aster), *Plantago coronopus* (buck's horn plantain) and *Salicornia dolichostachya* (long spiked glasswort).

The majority of the maraponic work conducted so far involves the integration of two trophic levels – plants/algae and fish. However, an example of a system incorporating more than two trophic levels can be seen in an experiment conducted by Neori et al. (2000), who designed a small system for the intensive land-based culture of Japanese abalone (*Haliotis discus hannai*), seaweeds (*Ulva lactuca* and *Gracilaria conferta*) and pellet-fed gilthead bream (*Sparus aurata*). This system consisted of unfiltered seawater (2400 L/day) pumped to two abalone tanks and drained through a fish tank and finally through a seaweed filtration/production unit before being discharged back to the sea. Filter feeding molluscs could also be used in such a system. Kotzen and Appelbaum (2010) and Appelbaum and Kotzen (2016) compared the growth of common vegetables using potable water and moderately saline water (4187–6813 ppm) and found that basil (*Ocimum basilicum*), celery (*Apium graveolens*), leeks (*Allium ampeloprasum porrum*), lettuce (*Lactuca sativa* – various types), Swiss chard (*Beta vulgaris. 'cicla'*), spring onions (*Allium cepa*) and watercress (*Nasturtium officinale*) performed extremely well.

Maraponics (SAs) and haloponics offer a number of advantages over traditional crop and fish production methods. Because they use saline water (marine to brackish), there is a reduced dependence on freshwater, which in some parts of the world has become a very limited resource. It is typically practiced in a controlled environment (e.g. a greenhouse; controlled flow-rate tanks) giving better opportunities for intensive production. Many maraponic and haloponic systems are closed RAS with organic and/or mechanical biofilters and subsequently, water reuse is high, wastewater pollution is vastly reduced or eliminated, and contaminants are removed or treated. Even systems that are not RAS can significantly reduce the excess nutrients in the wastewater prior to discharge. Additionally, the occurrence of contaminants in non-RAS maraponic and haloponic systems can be reduced or eliminated through the use of water containing low levels of naturally occurring contaminants and the use of alternatives aquafeeds that do not contain dioxins or PCDs (e.g. novel feeds made from macroalgae). This improvement in water quality reduces the potential for disease occurrence and the need for antibiotic use is therefore vastly reduced. Due to their versatile configuration and low water requirements, maraponics and haloponics can be successfully implemented in a wide variety of settings, from fertile coastal areas to arid deserts (Kotzen and Appelbaum 2010), as well as in urban or peri-urban settlements. Another potential benefit is that many of the species that are suitable for these systems have a high commercial value. For example, the euryhaline European sea bass (*Dicentrarchus labrax*) and gilthead sea bream (*Sparus aurata*) can fetch a market price of €9/kg and €6/kg, respectively. Additionally, edible halophytes tend to have a high market price, with sea-agretti (*Salsola soda*), for example, having a market price of €4–€4.5/kg and marsh samphire (*Salicornia europaea*) selling at €18/kg in supermarkets.

The evidence is therefore compelling. Maraponics and haloponics provide a dynamic and rapidly growing field that has the potential to provide a number of services to communities, many of which are explored elsewhere in this publication.

12.5 Vertical Aquaponics

12.5.1 Introduction

Whilst aquaponics can be seen as part of a global solution to increase food production in more sustainable and productive ways and where growing more food in urban areas is now recognized as part of the solution to food security and a global food crisis (Konig et al. 2016), aquaponic systems can themselves become more productive and sustainable by adopting alternative growing technologies and learning from emerging technologies such as vertical farming and living walls [LWs] (Khandaker and Kotzen 2018). Additionally by being space-efficient, they can be better integrated into urban areas.

In the developed world most aquaponic systems are placed in greenhouses in order to control temperature; in northern Europe and North America for example, winter temperatures are too cold in winter and in Mediterranean areas such as Spain, Italy, Portugal, Greece and Israel, summer temperatures are too warm. There are of course many additional advantages in growing food in controlled greenhouses, such as the ability to regulate relative humidity and control air movement, to quarantine fish as well as plants from diseases as well as pests and potentially being able to add CO_2, to aid plant growth. However, growing produce in a greenhouse can readily raise costs through (a) the capital costs of the greenhouse (a broad estimate of US $350/m^2 Arnold 2017) and (b) allied infrastructure such as microclimate controls which include heating and cooling systems and lighting. On top of the initial infrastructure costs, there are also the specific greenhouse production costs which include the energy/power supply for heating and cooling as well as lighting.

Most aquaponic systems such as the University of the Virgin Island (UVI) system (Fig. 12.1), designed by Dr. James Rakocy and his colleagues, use horizontal grow tanks or beds, emulating traditional land-based arable growing patterns to produce vegetables (Khandaker and Kotzen 2018). In other words, the system relies on horizontal rows/arrays of plants usually elevated to around waist level so that plant-related management tasks can be readily undertaken. Parallel developments in living wall and vertical farming technologies have arisen at almost the same time that aquaponics has evolved and are similarly in the adolescent stage of development. Similarly as in aquaponics, as more people become involved there is a concomitant increase in systems and technological development to increase productivity and reduce costs. The coupling of vertical growing systems (vertical farming systems and living walls) rather than horizontal beds to the fish and filtration tanks is potentially one key way of increasing productivity as it should be possible to increase the number of vegetables grown compared to the numbers produced in typical horizontal bed aquaponics. The UVI aquaponic systems (Fig. 12.2) produce approximately 32 plants per square metre (Al-Hafedh et al. 2008), depending on the species and cultivar that is grown, but as Khandaker and Kotzen (2018) note, approximately 96 plants can be grown per square metre 'using back-to-back elements of the *Terapia Urbana*[1] LW system which is more than three times the density

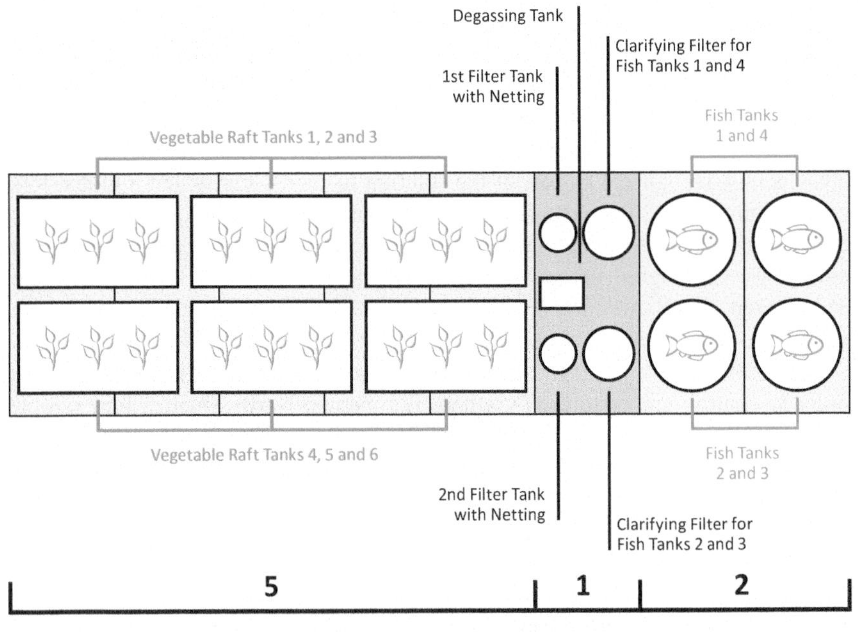

Fig. 12.2 Schematic diagram of a typical UVI system illustrating the ratio of fish tanks/filters/plant growing tanks which is 2:1:5. This shows that the greatest area is subsumed by the plants and it is in this area that space savings may be considered. (Khandaker and Kotzen 2018)

[1]Terapia Urbana S.L. produces a felt pocket type of living wall in Seville, Spain.

compared to the UVI horizontal growing system'. A conservative estimate should at least double the maximum amount grown in horizontal beds to 64 plants/m^2. In an experiment with lettuce (*Lactuca sativa* L. cv. 'Little Gem') using horizontal beds and planted columns, planted at similar densities, Touliatos et al. (2016) suggest that the 'Vertical Farming System (VFS) presents an attractive alternative to horizontal hydroponic growth systems (and) that further increases in yield could be achieved by incorporating artificial lighting in the VFS'.

Vertical Farming Systems (VFS)

Before we discuss the specific requirements for vertical systems we need to discuss the types of systems that are available. In VFS there are three main generic types (Fig. 12.3):

1. Stacked Horizontal Beds: Instead of only having one horizontal grow bed, the beds are stacked like shelves in tiers. This arrangement means that in a greenhouse, only the upper bed will be facing direct natural light and supplementary light needs to be provided at all levels. This is usually provided from directly under the grow bed above. In principle this could mean that the growing beds could be stacked as high as the greenhouse allows, but of course growing things at

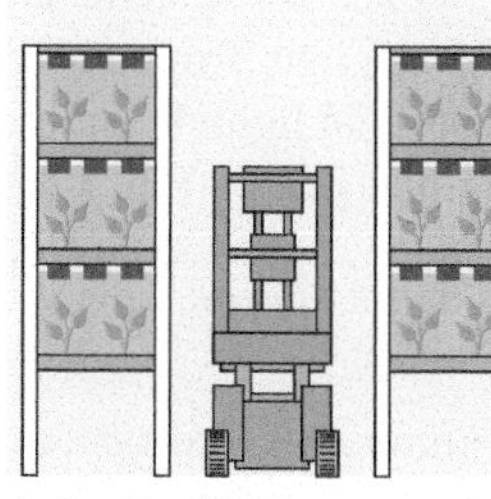

A. Stacked System
(Elevation View)

B1. Vertical System
Zipgrow Type
(Plan View)

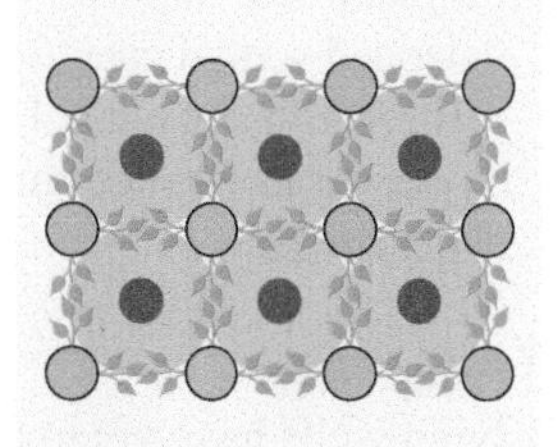

B2. Vertical System
Stacking Type
(Plan View)

C1. Stacked Tier System
with moving Troughs
(Elevation View)

C2. Stepped Tier System
with rigid Troughs
(Elevation View)

Fig. 12.3 Vertical farming systems and their lighting arrangements

height means greater difficulty in the system's management including planting, maintenance and harvesting, requiring scissor lifts and additional energy to pump nutrient-rich water to all levels. According to Bright Agrotech (Storey 2015), up to four tiers is profitable and anything above that is unprofitable. Storey (2015) further notes that labour increases by 25% at the second, third and fourth levels when a scissor lift is required (Fig. 12.3, Illustration A).

2. Vertical Tower Systems (VTS): Vertical Tower Systems comprise systems which grow plants in vertical arrays within a container or series of stacked modules. Depending on the system, plants are grown facing one direction or if, for example, they are planted in a tube-like form, then they can be arranged facing any direction. An example of a vertical array system, where plants are grown facing in a single direction is the ZipGrow™ which are either hung or supported in rows (Fig. 12.3, Illustration B1). The rows between are approximately 0.5 metres (20 inches). Growing in a more three-dimensional way occurs with stacked systems or in tubular systems which allow more plants to be grown, but lighting is more complex (Fig. 12.3, Illustration B2).
3. Stepped Tiers: These systems contain rigid or moving plant troughs. The Sky Greens VFS in Singapore uses a rotating trough system which moves the troughs upwards and into the light. Additional natural light is more significant towards the top and less so at the bottom (Fig. 12.3, Illustration C1). Other tier systems are stepped so that each tier has an unobstructed interface with the light from above, whether this is natural light from the greenhouse roof or artificial light. But these systems have to be quite low in order for people to reach the plants (Fig. 12.3, Illustration C2).

Living Walls

Living walls have yet to be used in aquaponics except in a number of trial systems such as at the University of Greenwich, London (Khandaker and Kotzen 2018). Whereas most VFS use nutrient film technique (NFT) grow channels or encapsulated mineral wool blocks, LWs sometimes also use soil type substrates in pots or troughs, which provide the rooting medium. Whilst this is fine for growing ornamental plants as well as vegetables and herbs, when coupled with fish tanks, any addition of soil to the system may complicate the microbial character of the system and be detrimental to the fish. This is however unknown and requires research. Experiments undertaken at the University of Greenwich (Khandaker and Kotzen 2018) indicate that from a number of single, inert substrates tested (including hydroleica, perlite, straw, sphagnum moss, mineral wool and coconut fibre), coconut fibre and then mineral wool were superior in terms of root penetration and root growth in lettuce (*Lactuca sativa*).

Vertical v. Horizontal: Factors to Be Considered

There are four key aspects which need to be taken into account when comparing the benefits (productivity and sustainability) of vertical growing, compared to horizontal growing. These are (1) space, (2) lighting, (3) energy and (4) life cycle costs.

1. *Space*

The benefits of being able to grow produce vertically, back to back, need to be balanced with the amount of space that is required to provide an even spread of lighting as well as the row space required for management and maintenance. The width of a row in hydroponic systems varies. As noted the standard ZipGrow™ system is approximately 0.5 metres, whereas the usual row width for growing tomatoes and cucumbers hydroponically varies from 0.9 to 1.2 metres (Badgery-Parker and James 2010). Growing smaller plants such as lettuce and herbs such as basil, may allow for narrower rows, but of course row width must ensure that produce is not compromised by moving items such as trolleys and scissor lifts. A key issue with growing vertically is the conflict that occurs between having fixed rows and fixed lighting, which needs to be located in the rows between the planting facades. These lights will impede people movements and thus either the lights need to be (i) part of the growing structure or (ii) retractable or movable, so that workers can readily undertake tasks, or (iii) the planting structures are movable and the lights remain static.

2. *Lighting*

Greenhouse production of vegetables and other plants rely on specific spatial arrangements which allow for planting, management through growth and then harvesting. The spatial arrangement will depend on the types of plants and the types of mechanization that is installed. Additionally, growing efficiently relies on the supplement of additional light of different types, which have their own pros and cons. In general what these lights do is provide specific wavelengths for plant growth and for fruit or flower production. Whereas it is relatively simple and more common to evenly light plants grown horizontally, it is more of a challenge to evenly light a vertical surface.

With regard to types of lighting, many producers have moved to or are tempted to install LEDs (light-emitting diodes), due to their long lifespan, up to 50,000 hours or more (Gupta 2017), their low power requirements and their recent reduction in cost. Virsile et al. in Gupta (2017) note that *most applications of LED lighting in greenhouses choose the combinations of red and blue wavelengths with high photon efficiency* but that green and white light *containing substantial amounts of green wavelengths has a positive physiological impact on plants*. However, *the combination of blue and red lights creates a purplish-grey image*, and this hampers the visual evaluation of plant health. The type of wavelengths chosen is complex and can have benefits at different stages in the plant's life and even according to the cultivars of, for example, lettuce. Red-leafed lettuces, for example, respond to blue LED lighting, increasing their pigmentation (Virsile et al. in Gupta 2017). Additionally, blue LED lighting can improve the nutritional quality of green vegetables, reducing nitrate content, increasing antioxidants and phenolic and other beneficial compounds. The light spectra also affect taste, shape and texture (Virsile et al. in Gupta 2017). The costs of LEDs have dropped significantly and as the efficacy of LEDs has increased so the break-even return time on investment has decreased (Bugbee in Gupta 2017).

Other lighting of course exists and this includes fluorescent lighting, metal halide (MH) lighting and high pressure sodium (HPS) lighting. The type of lighting that is used in vertical farming and with living walls varies considerably depending on the scale and location. Compact fluorescent lamps (CFLs) are relatively thin and can easily fit into small spaces, but they require an inductive ballast to regulate current through the tubes. CFLs use only 20–30% of an incandescent bulb and they last six to eight times longer but they are almost 50% less efficient than LEDs. They are by far the cheapest of the three major types of grow lights. HPS grow light technology is over 75 years old and is well established for growing under glass, but they produce a lot of heat and are thus not suitable for vertical farming and living walls, where light needs to be delivered quite close to the plants. The heat produced by LED grow lights, on the other hand, is minimal. The cost however is higher than other two types, and eye protection is needed for longer-term exposure to LEDs as the long-term exposure to the light spectra can be damaging to the eyes. The arrangement of VFS units will dictate the lighting arrangement but on the whole these are lit by LEDs. The method of lighting living walls will depend on the height of the wall. The taller the wall the more difficult it is to apply an even spread across the surface, although it should be noted that the number of lights used should be no different to those used in horizontal grow beds and if the wall is tall then the lights may need to be staggered. As most living walls are located for aesthetic purposes, lighting needs to be kept as far as possible, out of the way and the lighting has to not only provide adequate light for plant growth and health, but also so that the plants look good (Fig. 12.4).

Fig. 12.4 A 4-metre-tall, 5-metre-long living wall can be adequately lit with six high-efficiency discharge lamps. Note these were chosen not only to provide adequate light for growth but also so that the plants in the living wall would look good. (University of Greenwich Living Wall. Source: Benz Kotzen)

The advances in LED technology, where lighting frequencies and intensity can be engineered to suit individual species and cultivars as well as their various life cycles means that LEDs will become the technology of choice in the near future. This will additionally be enhanced by reductions in costs.

3. *Energy*

More energy for lighting is likely to be required for VFS as well as LWs as even natural lighting cannot be achieved over vertical surfaces. Additionally more pumping power for irrigation will be required and this will be relative to the height of the VFS or LWs.

4. *Comparative Life Cycle Analysis (LCA)*

Whilst there are numerous studies undertaken on life cycle analysis of aquaponics and various aspects of aquaponic systems, there are no comparative studies that compare vertical versus horizontal aquaponics. This has yet to be done. We are getting to a point where vertical aquaponics is likely to warrant further testing and research and in time vertical aquaponics, which couples vertical farming systems or living wall systems with the fish tanks and filtration units, is likely to become more mainstream, as long as these can be profitable and sustainable.

12.6 Biofloc Technology (BFT) Applied for Aquaponics

12.6.1 Introduction

Biofloc technology (BFT) is considered the new ‘blue revolution’ in aquaculture (Stokstad 2010) since nutrients can be continuously recycled and reused in the culture medium, benefited by the in situ microorganism production and by the minimum or zero water exchange (Avnimelech 2015). These approaches might face some serious challenges in the sector such as competition for land and water and the effluents discharged to the environment which contain excess of organic matter, nitrogenous compounds and other toxic metabolites.

BFT was first developed in the early 1970s by the Aquacop team at Ifremer-COP (French Research Institute for Exploitation of the Sea, Oceanic Center of the Pacific) with different shrimp penaeid species including *Litopenaeus vannamei*, *L. stylirostris* and *Penaeus monodon* (Emerenciano et al. 2011). In the same period, Ralston Purina (a private US company) in connection with Aquacop applied the technology both in Crystal River (USA) and Tahiti, which lead to the greater understanding of the benefits of biofloc for shrimp culture. Several other studies enabled a comprehensive approach to BFT and researched the inter-relationships between water, animals and bacteria, comparing BFT with an ‘external rumen’ but now applied for shrimp. In the 1980s and at the beginning of the 1990s, both Israel and the USA (Waddell Mariculture Center) started R&D in BFT with tilapia and Pacific white shrimp *L. vannamei*, respectively, in which environmental concerns, water limitation and land costs were the main causative agents that promoted the research (Emerenciano et al. 2013).

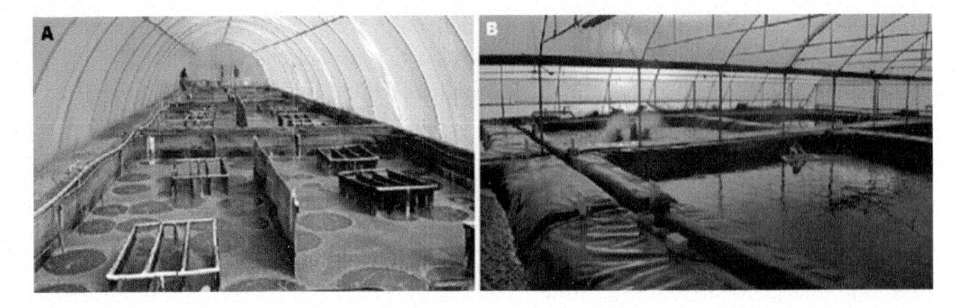

Fig. 12.5 Biofloc technology (BFT) applied for marine shrimp culture in Brazil (**a**) and for tilapia culture in Mexico (**b**) (Source: EMA-FURG, Brazil and Maurício G. C. Emerenciano)

The first commercial BFT operations and probably the most famous commenced in the 1980s at the 'Sopomer' farm in Tahiti, French Polynesia, and in the early 2000s at the Belize Aquaculture farm or 'BAL', located in Belize, Central America. The yields obtained using 1000 m^2 concrete tanks and 1.6 ha lined grow-out ponds were approximately 20–25 ton/ha/year with two crops at Sopomer and 11–26 ton/ha/cycle at BAL, respectively. More recently, BFT has been successfully expanded in large-scale shrimp farming in Asia, in South and Central America as well as in small-scale greenhouses in the USA, Europe and other areas. At least in one phase (e.g. nursery phase) BFT has been used with great success in México, Brazil, Ecuador and Peru. For commercial-scale tilapia culture, farms in Mexico, Colombia and Israel are using BFT with productions around 7 to 30 kg/m^3 (Avnimelech 2015) (Fig. 12.5b). Additionally, this technology has been used (e.g. in Brazil and Colombia) to produce tilapia juveniles (~30 g) for further stock in cages or earthen ponds (Durigon et al. 2017). BFT has mainly been applied to shrimp culture and to some extent with tilapia. Other species have been tested and show promise, as noted for silver catfish (*Rhamdia quelen*) (Poli et al., 2015), carp (Zhao et al., 2014), piracanjuba (*Brycon orbignyanus*) (Sgnaulin et al., 2018), cachama (*Colossoma macropomum*) (Poleo et al., 2011) and other crustacean species such as *Macrobrachium rosenbergii* (Crab et al., 2010), *Farfantepenaeus brasiliensis* (Emerenciano et al., 2012), *F. paulensis* (Ballester et al., 2010), *Penaeus semisulcatus* (Megahed, 2010), *L. stylirostris* (Emerenciano et al., 2011) and *P. monodon* (Arnold et al., 2006). The interest in BFT is evident by the increasing number of universities and research centres carrying out research particularly in the key fields of grow-out management, nutrition, reproduction, microbial ecology, biotechnology and economics.

12.6.2 How does BFT Work?

Microorganisms play a key role in BFT systems (Martinez-Cordoba et al. 2015). The maintenance of water quality, mainly by the control of the bacterial community over autotrophic microorganisms, is achieved using a high carbon to nitrogen ratio (C:N) since nitrogenous by-products can be easily taken up by heterotrophic bacteria. In the beginning of the culture cycles a high carbon to nitrogen ratio is required to guarantee optimum heterotrophic bacteria growth, using this energy for its maintenance and growth (Avnimelech 2015). Additionally, other microorganism groups are crucial in BFT systems. The chemoautotrophic bacterial community (i.e. nitrifying bacteria) stabilizes after approximately 20–40 days and might be responsible for two-thirds of the ammonia assimilation in the system (Emerenciano et al. 2017). Thus, the addition of external carbon should be reduced and alkalinity consumed by the microorganisms must be replaced by different carbonate/bicarbonate sources (Furtado et al. 2011). The stability of zero or minimal water exchange depends on the dynamic interaction amongst communities of bacteria, microalgae, fungi, protozoans, nematodes, rotifers, etc. that will occur naturally (Martinez-Cordoba et al. 2017). The aggregates (bioflocs) are a rich protein-lipid natural source of food that become available 24 h per day due to a complex interaction between organic matter, physical substrate and large range of microorganisms (Kuhn and Boardman 2008; Ray et al. 2010). The natural productivity in a form of microorganisms' production plays three major roles in the tanks, raceways or lined ponds: (1) in the maintenance of water quality, by the uptake of nitrogen compounds generating in situ microbial protein; (2) in nutrition, increasing culture feasibility by reducing feed conversion ratios and a decrease in feed costs; and (3) in competition with pathogens (Emerenciano et al. 2013).

Regarding the water quality for the culture organisms, besides oxygen, excess of particulate organic matter and toxic nitrogen compounds are the major concern in the biofloc systems. In this context, three pathways occur for the removal of ammonia nitrogen: at a lesser rate (1) photoautotrophic removal by algae and at a higher rate (2) heterotrophic bacterial conversion of ammonia nitrogen directly to microbial biomass and (3) autotrophic bacterial conversion from ammonia to nitrate (Martinez-Cordoba et al. 2015). The nitrate available in the systems plus other minor and major nutrients accumulated over the cycle could be used as substrate for plant growth in aquaponic systems (Pinho et al. 2017).

12.6.3 BFT in Aquaponics

The application of BFT in aquaponic systems is relatively new, although Rakocy (2012) mentions a commercial pilot-scale project with tilapia. Table 12.2 summarizes key recent studies that have used BFT in aquaponic systems.

Overall, the results demonstrate that biofloc technology can be used and integrated in a fish or shrimp-plant production. BFT when compared to other conventional aquaculture systems (such as RAS) actually improved the plant and fish yields and promoted better plant visual quality (Pinho et al. 2017), but not in all cases (Rahman 2010; Pinho 2018). Pinho et al. (2017) observed that lettuce yields with the BFT system were greater compared to the clear-water recirculation system (Fig. 12.6). This is possibly due to the higher nutrient availability provided by the

Table 12.2 Recent studies around the world applying the BFT in aquaponic systems for different aquatic and plant species

Aquatic species	Plant species	Main results	References
Tilapia	Lettuce	Biofloc technology did not improve lettuce production as compared to conventional hydroponic solution	Rahman (2010)
Tilapia	Lettuce	Yield and visual quality of lettuce was improved using BFT as compared to clear-water recirculation system	Pinho et al. (2017)
Tilapia (nursery)	Lettuce	Plant performance (lettuce) using tilapia in a nursery phase (1–30 g) was negatively influenced by biofloc wastewater as compared to RAS wastewater after two plant cycles (13 days each). Plant visual aspects were better in RAS as compared to BFT	Pinho (2018)
Tilapia	Lettuce	The presence of filtering elements (mechanical filter and biological filter) positively affected the lettuce production in aquaponic systems as compared to treatment without filters using BFT	Barbosa (2017)
Tilapia	Lettuce	Low salinity (3 ppt) can be performed in aquaponics using BFT. Visual and performance parameters indicated that the purple variety had better performance than the smooth and crisped varieties	Lenz et al. (2017)
Silver catfish	Lettuce	The use of bioflocs in the aquaponic system may improve the productivity of lettuce in an integrated culture with silver catfish	Rocha et al. (2017)
Litopenaeus vannamei	*Sarcocornia ambigua*	The performance of marine shrimp *L. vannamei* was not affected by the *S. ambigua* integrated aquaponics production and also improve the use of nutrients (e.g. nitrogen) in the culture system	Pinheiro et al. (2017)

Fig. 12.6 Experimental aquaponics greenhouse comparing biofloc technology and RAS wastewater at Santa Catarina State University (UDESC), Brazil. (Source: Pinho et al. 2017)

higher microbial activity. However, this trend was not observed in the study by Rahman (2010), who compared effluent from fish culture in a BFT system to a conventional hydroponic solution in a lettuce production. In addition, Pinho

Fig. 12.7 High salinity halophyte *Sarcocornia ambigua* aquaponics production integrated with Pacific white shrimp *Litopenaeus vannamei* successfully applying biofloc technology at Santa Catarina Federal University (UFSC), Brazil. (Source: LCM-UFSC, Brazil)

Fig. 12.8 Aquaponics lettuce production integrated with tilapia using biofloc technology (left) and accumulation of suspended solids in lettuce roots (right). Barbosa (2017)

(2018) in a recent study observed that productive performance of lettuce in aquaponic system using tilapia in a nursery phase (1–30 g) was negatively influenced by biofloc wastewater as compared to RAS wastewater over 46 days. The variation in results identifies the need for additional studies in this area.

BFT can be used with low salinity water, e.g. with some varieties of lettuce (Lenz et al. 2017), and higher salinity waters can be used, e.g. with halophyte plant species such as *Sarcocornia ambigua* co-culture with Pacific white shrimp *Litopenaeus vannamei* (Pinheiro et al. 2017) (Fig. 12.7). Silver catfish *Rhamdia quelen* has also shown good potential for the integration of aquaponics with BFT (Rocha et al. 2017).

With BFT, the concentration of solids can severely affect the roots and impact nutrient absorption and oxygen availability. As a result, yields can be affected but also the visual quality of the plants (e.g. lettuces) which is an important criterion for consumers. With this in mind, solids management is an important subject for further studies where the impact of solids (particulate fraction and also dissolved fraction) in aquaponic systems when applying BFT is considered (Fig. 12.8). In addition, economic studies need to be undertaken to compare the costs involved of the various aquaculture and plant growing systems and to identify appropriateness relative to different locations and conditions.

12.7 Digeponics

Anaerobic processing of purposely cultivated biomass, as well as residual plant material from agricultural activity, for biogas production is a well-established method. The bacterially indigestible digestate is returned to the fields as a fertilizer and for building humus. Whilst this process is widespread in agriculture, the application of this technology in horticulture is relatively new. Stoknes et al. (2016) claim that within the 'Food to waste to food' (F2W2F) project, an efficient method for the utilization of digestate as substrate and fertilizer has been developed for the first time. The research team coined the term 'digeponics' for this circular system. Digeponics, in contrast to aquaponics, replaces the aquaculture part with an anaerobic digester, or, when comparing it to a three loop aquaponic system that includes an anaerobic, the aquaculture part is removed from the system, leaving two main loops, the digestion loop and the horticultural loop.

The required organic input that is provided in the form of the fish food to an aquaponic system is replaced with food waste from human food production for digeponics. The varying composition of nutrients in the input stream opposed to the well-known, constant and probably nutritionally optimized nutrient stream resulting from the fish feed will most likely call for a more strict nutrient analysis and management regime than that required in aquaponics.

The produced biogas, which mainly contains methane and carbon dioxide, can be utilized within the facility for electricity and heat production. The resulting carbon dioxide-rich exhaust gas can be used as a fertilizer directly in the greenhouse reducing emissions in comparison to classical biogas plants used in agriculture.

Since the 'fresh and untreated digestate in anaerobic liquid slurry (contains) plant toxic substances, a very high electrical conductivity (EC) and chemical oxygen demand (COD)' (Stoknes et al. 2016), it has to be treated to make it suitable for plant fertilization. Several methods of moderation have been examined within the F2W2F project. The relatively high EC of the digestate and the operational flexibility of a digester fed with low-cost food waste alleviate some of the tight coupling issues often attributed to coupled aquaponic systems (see Chap. 7). Thus digeponics may serve as an interesting alternative to aquaponics in situations where the aquaculture part represents a challenge. With respect to a three loop aquaponic system that already comprises a loop with an anaerobic digester, the inclusion of a food waste stream for organic input might represent an interesting future direction. The methane yield of aquaculture sludge is rather limited. A targeted inclusion of residual agricultural biomass with the aim of methane yield optimization could enhance overall performance.

12.8 Vermiponics and Aquaponics

It would be remiss in this chapter not to mention earthworms and their introduction into aquaponics, and thus this chapter concludes with a brief résumé of these detritivore invertebrates and their abilities to convert organic waste into fertilizer. It is said that worms and the way that they digest matter were of interest to Aristotle and Charles Darwin as well as the philosophers Pascal and Thoreau (Adhikary 2012) and they were protected by law under Cleopatra. Earthworms are valued in agriculture and horticulture as they are 'vital to soil health because they transport nutrients and minerals from below to the surface via their waste, and their tunnels aerate the ground' (National Geographic).

Modern vermiculture is attributed to Mary Appelhof, who in the early 1970s and 1980s produced a number of publications on composting with worms. Contemporary vermicomposting occurs on large and small scales with the objective of getting rid of organic waste and producing fertilizer in the forms of compost and 'worm tea'. Worm tea can be produced by soaking worm casts or by leaching the nutrients from the compost through wetting or natural wetting leachate from precipitation.

Vermiponics uses the worm casts of mainly red wriggler worms also known as tiger worms (*Eisenia fetida*) or (*E. foetida*) to provide nutrients in a hydroponic system. When worms are introduced into an aquaponic system, we suggest that the system is termed 'vermi-aquaponics' to differentiate the systems. It is thus the introduction of worms into the growing beds of the plant parts of an aquaponic system. It should be noted that vermi-aquaponics is in its infancy and mainly practiced by hobbyists and in research laboratories. The worms are introduced mainly into the plant growing media, usually gravel beds, where they can help to break down any solid waste from the fish and any detritus from the plants and additionally provide additional nutrients for the plants, and they can also be fed to carnivore fish. In most instances the beds are of a flood and drain type, so that the worms are not constantly under water.

Acknowledgements The authors thank National Council for Scientific and Technological Development-CNPq (Project Number 455349/2012-6) and Scientific and Technological Research Foundation of Santa Catarina State-FAPESC (Project Number 2013TR3406 and 2015TR453).

References

Addy MM, Kabir F, Zhang R, Lu Q, Deng X, Current D, Griffith R, Ma Y, Zhou W, Chen P, Ruan R (2017) Co-cultivation of microalgae in aquaponic systems. Bioresour Technol 245 (2017):27–34

Adhikary S (2012) Vermicompost, the story of organic gold: a review. Agricult Sci 3(7). https://doi.org/10.4236/as.2012.37110

Al-Hafedh YS, Alam A, Beltagi MS (2008) Food production and water conservation in a recirculating Aquaponic system in Saudi Arabia at different ratios of fish feed to plants. J World Aquacult Soc 39(4):510–520

Appelbaum S, Kotzen B (2016) Further investigations of Aquaponics using brackish water resources of the Negev Desert. Ecocycles Scientific J Eur Ecocycl Soc 2(2):26–35. ISSN 2416-2140 https://doi.org/10.19040/ecocycles.v2i2.53

Arnold J (2017) Greenhouse business: start-up costs, profits and labor, 19 April 2017. https://blog.brightagrotech.com/author/jason-arnold

Arnold SJ, Sellars MJ, Crocos PJ, Coman GJ (2006) An evaluation of stocking density on the intensive production of juvenile brown tiger shrimp (*Penaeus esculentus*). Aquaculture 256 (1):174–179

Avnimelech Y (2015) Biofloc technology: a practical guide book, 3rd edn. The World Aquaculture Society, Baton Rouge

Ayre JM, Moheimani NR, Borowitzka MA (2017) Growth of microalgae on undiluted anaerobic digestate of piggery effluent with high ammonium concentrations. Algal Res 24:218–226

Badgery-Parker J, James L (2010) Commercial greenhouse cucumber production. NSW Agriculture, Orange

Ballester ELC, Abreu PC, Cavalli RO, Emerenciano M, De Abreu L, Wasielesky W Jr (2010) Effect of practical diets with different protein levels on the performance of *Farfantepenaeus paulensis* juveniles nursed in a zero exchange suspended microbial flocs intensive system. Aquac Nutr 16 (2):163–172

Barbosa M (2017) Biofloc technology: do filtering elements might affects lettuce aquaponics production integrated with tilapia? A thesis presented at animal science postgraduate program, Santa Catarina State University (Degree of Master of Science), Chapecó, Santa Catarina, Brazil, December 2017

Borowitzka MA (1999) Commercial production of microalgae: ponds, tanks, tubes and fermenters. J Biotechnol 70(1–3):313–321

Borowitzka MA, Moheimani NR (2013) Open pond culture systems. In: Algae for biofuels and energy. Springer, Dordrecht

Brown EJ, Button DK (1979) Phosphate-limited growth kinetics of *Selenastrum capricornutum* (CHLOROPHYCEAE). J Phycol 15(3):305

Bugbee B (2017) Economics of LED lighting. In: Gupta D (ed) Light emitting diodes for agriculture – smart lighting. Springer, Singapore

Chisholm SW, Brand LE (1981) Persistence of cell division phasing in marine phytoplankton in continuous light after entrainment to light: dark cycles. J Exp Mar Biol Ecol 51(2–3):107–118

Clawson JM, Hoehn A, Stodieck LS, Todd P, Stoner RJ (2000) NASA – review of aeroponics – aeroponics for spaceflight plant growth. Aeroponics DIY. https://aeroponicsdiy.com/nasa-review-of-aeroponics

Crab R, Chielens B, Wille M, Bossier P, Verstraete W (2010) The effect of different carbon sources on the nutritional value of biofloc, a feed for *Macrobrachium rosenbergii* postlarvae. Aquac Res 41:559–567

Croft MT, Lawrence AD, Raux-Deery E, Warren MJ, Smith AG (2005) Algae acquire vitamin B 12 through a symbiotic relationship with bacteria. Nature 438(7064):90

Delrue F, Álvarez-Díaz PD, Fon-Sing S, Fleury G, Sassi JF (2016) The environmental biorefinery: using microalgae to remediate wastewater, a win-win paradigm. Energies 9(3):132

Deppeler S, Petrou K, Schulz KG, Westwood K, Pearce I, McKinlay J, Davidson A (2018) Ocean acidification of a coastal Antarctic marine microbial community reveals a critical threshold for CO_2 tolerance in phytoplankton productivity. Biogeosciences 15(1):209–231

Droop MR (1973) Some thoughts on nutrient limitation in algae. J Phycol 9(3):264–272

Dufault RJ, Korkmaz A (2000) Potential of biosolids from shrimp aquaculture as a fertiliser in bell pepper production. Compost Sci Util 3:310–319

Dufault RJ, Korkmaz A, Ward B (2001) Potential of biosolids from shrimp aquaculture as a fertiliser for broccoli production. Compost Sci Util 9:107–114

Durigon EG, Sgnaulin T, Pinho SM, Brol J, Emerenciano MGC (2017) Bioflocos e seus benefícios nutricionais na pré-engorda de tilápias. Aquacult Brasil 8:50–54

Emerenciano M, Cuzon G, Goguenheim J, Gaxiola G (2011) Floc contribution on spawning performance of blue shrimp *Litopenaeus stylirostris*. Aquac Res 44(1):75–85

Emerenciano MGC, Ballester ELC, Cavalli RO, Wasielesky W (2012) Biofloc technology application as a food source in a limited water exchange nursery system for pink shrimp *Farfantepenaeus brasiliensis* (Latreille, 1817). Aquac Res 43(3):447–457

Emerenciano M, Gaxiola G, Cuzon G (2013) Biofloc technology (BFT): a review for aquaculture application and animal food industry. In: Biomass now-cultivation and utilization. InTech, Rijeka

Emerenciano MGC, Martínez-Córdova LR, Martínez-Porchas M, Miranda-Baeza A (2017) Biofloc technology (BFT): a tool for water quality management in aquaculture, water quality, In: Hlanganani Tutu (ed). InTech

Furtado PS, Poersch LH, Wasielesky W (2011) Effect of calcium hydroxide, carbonate and sodium bicarbonate on water quality and zootechnical performance of shrimp Litopenaeus vannamei reared in bio-flocs technology (BFT) systems. Aquaculture 321:130–135

Galloway JN, Dentener FJ, Capone DG, Boyer EW, Howarth RW, Seitzinger SP, Asner GP (2004) Nitrogen cycles: past, present, and future. Biogeochemistry 70(2):153–226

Gordon JM, Polle JE (2007) Ultra-high bioproductivity from algae. Appl Microbiol Biotechnol 76 (5):969–975

Gunning D, Maguire J, Burnell G (2016) The development of sustainable saltwater-based food production systems: a review of established and novel concepts. Water 8(12):598. https://doi.org/10.3390/w8120598

Gupta D (ed) (2017) Light emitting diodes for agriculture – smart lighting. Springer, Singapore

Hikosaka Y, Kanechi M, Uno Y (2014) A novel aeroponic technique using dry-fog spray fertigation to grow leaf lettuce (*Lactuca sativa* L. var. crispa) with water-saving hydroponics. Adv Hortic Sci 28(4):184–189

Joesting HM, Blaylock R, Biber P, Ray A (2016) The use of marine aquaculture solid waste for nursery of salt marsh plants Spartina alterniflora and *Juncus roemerianus*. Aquac Rep 3:108–114

Khandaker M, Kotzen B (2018) The potential for combining living wall and vertical farming systems with aquaponics with special emphasis on substrates. Aquacult Res 23 January 2018. https://doi.org/10.1111/are.13601

Kotzen B, Appelbaum S (2010) An investigation of aquaponics using brackish water resources in the Negev Desert. J Appl Aquacult 22(4):297–320. ISSN 1045-4438 (print), 1545-0805 (online). https://doi.org/10.1080/10454438.2010.527571

König B, Junge R, Bittsanszky A, Villarroel M, Komives T (2016) On the sustainability of aquaponics. Ecocycles 2(1):26–32

Kuhn DD, Boardman GD (2008) Use of microbial flocs generated from tilapia effluent as a nutritional supplement for shrimp, Litopenaeus vannamei, in recirculating aquaculture systems. J World Aquacult Soc 39:72–82

Lakhiar IA, Gao J, Naz ST, Chandio FA, Buttar NA (2018) Modern plant cultivation technologies in agriculture under controlled environment: a review on aeroponics. J Plant Interact 13 (1):338–352. https://doi.org/10.1080/17429145.2018.1472308

Langton RW, Haines KC, Lyon RE (1977) Ammonia nitrogen produced by the bivalve mollusc *Tapes japonica* and its recovery by the red seaweed *Hypnea musciformis* in a tropical mariculture system. Helgol Wiss Meeresunters 30:217–229

Lenz GL, Durigon EG, Lapa KG, Emerenciano MGC (2017) Produção de alface (*Lactuca sativa*) em efluentes de um cultivo de tilápias mantidas em sistema BFT em baixa salinidade. Bol Inst Pesca 43:614–630

Martínez-Córdova LR, Emerenciano M, Miranda-Baeza A, Martínez-Porchas M (2015) Microbial-based systems for aquaculture of fish and shrimp: an updated review. Rev Aquac 7(2):131–148

Martínez-Córdova LR, Martínez-Porchas M, Emerenciano MG, Miranda-Baeza A, Gollas-Galván T (2017) From microbes to fish the next revolution in food production. Crit Rev Biotechnol 37:287–295

Megahed M (2010) The effect of microbial biofloc on water quality, survival and growth of the green tiger shrimp (*Penaeus semisulcatus*) fed with different crude protein levels. J Arab Aquacult Soc 5:119–142

Moheimani NR (2016) *Tetraselmis suecica* culture for CO_2 bioremediation of untreated flue gas from a coal-fired power station. J Appl Phycol 28(4):2139–2146

Moheimani NR, Borowitzka MA (2007) Limits to productivity of the alga Pleurochrysis carterae (Haptophyta) grown in outdoor raceway ponds. Biotechnol Bioeng 96(1):27–36

Moheimani NR, Parlevliet D (2013) Sustainable solar energy conversion to chemical and electrical energy. Renew Sust Energ Rev 27:494–504

Moheimani NR, Isdepsky A, Lisec J, Raes E, Borowitzka MA (2011) Coccolithophorid algae culture in closed photobioreactors. Biotechnol Bioeng 108(9):2078–2087

Moheimani NR, Webb JP, Borowitzka MA (2012) Bioremediation and other potential applications of coccolithophorid algae: a review. Algal Res 1(2):120–133

Moheimani NR, Parlevliet D, McHenry MP, Bahri PA, de Boer K (2015) Past, present and future of microalgae cultivation developments. In: Biomass and biofuels from microalgae. Springer, Cham, pp 1–18

Moheimani NR, Vadiveloo A, Ayre JM, Pluske JR (2018) Nutritional profile and in vitro digestibility of microalgae grown in anaerobically digested piggery effluent. Algal Res 35:362–369

Munoz R, Guieysse B (2006) Algal–bacterial processes for the treatment of hazardous contaminants: a review. Water Res 40(15):2799–2815

NASA Spinoff, Experiments Advance Gardening at Home and in Space, https://spinoff.nasa.gov/Spinoff2008/ch_3.html

National Geographic, About the common earthworm, https://www.nationalgeographic.com/animals/invertebrates/c/common-earthworm

Neori A, Shpigel M, Ben-Ezra D (2000) A sustainable integrated system for culture of fish, seaweed, and abalone. Aquaculture 186:279–291

Nwoba EG, Ayre JM, Moheimani NR, Ubi BE, Ogbonna JC (2016) Growth comparison of microalgae in tubular photobioreactor and open pond for treating anaerobic digestion piggery effluent. Algal Res 17:268–276

Nwoba EG, Moheimani NR, Ubi BE, Ogbonna JC, Vadiveloo A, Pluske JR, Huisman JM (2017) Macroalgae culture to treat anaerobic digestion piggery effluent (ADPE). Bioresour Technol 227:15–23

Oswald WJ (1988) Role of microalgae in liquid waste treatment and reclamation. In: Lembi CA, Robert Waaland J (eds) Algae and human affairs. Sponsored by the Phycological Society of America, Inc

Oswald WJ, Gotaas HB (1957) Photosynthesis in sewage treatment. Trans Am Soc Civ Eng 122 (1):73–105

Park JBK, Craggs RJ (2010) Wastewater treatment and algal production in high rate algal ponds with carbon dioxide addition. Water Sci Technol 61(3):633–639

Parkin GF, Owen WF (1986) Fundamentals of anaerobic digestion of wastewater sludges. J Environ Eng 112(5):867–920

Pinheiro I, Arantes R, Santo CME, Seiffert WQ (2017) Production of the halophyte Sarcocornia ambigua and Pacific white shrimp in an aquaponic system with biofloc technology. Ecol Eng 100:261–267

Pinho SM (2018) Tilapia nursery in aquaponics systems using bioflocs technology. A thesis presented at aquaculture Centre of Sao Paulo State University. Degree of Master of Science, Jaboticabal, Sao Paulo, Brazil, February 2018

Pinho SM, Molinari D, De Mello GL, Fitzsimmons KM, Emerenciano MGC (2017) Effluent from a biofloc technology (BFT) tilapia culture on the aquaponics production of differente lettuce varieties. Ecol Eng 103:146–153

Poleo G, Aranbarrio JV, Mendoza L, Romero O (2011) Cultivo de cachama blanca en altas densidades y en dos sistemas cerrados. Pesq Agrop Brasileira 46(4):429–437

Poli MA, Schveitzer R, Nuñerr APO (2015) The use of biofloc technology in a South American catfish (*Rhamdia quelen*) hatchery: effect of suspended solids in the performance of larvae. Aquac Eng 66:17–21

Rahman SSA (2010) Effluent water characterization of intensive Tilapia culture units and its application in an integrated lettuce aquaponic production facility. A thesis submitted to the graduate faculty. Auburn University, Degree of Master of Science, Auburn Alabama, December 13 2010

Rakocy JE (2012) Aquaponics – integrating fish and plant culture. In: Tidwell JH (ed) Aquaculture Production Systems, 1st edn. Wiley-Blackwell, Oxford, pp 343–386

Ray AJ, Lewis BL, Browdy CL, Leffler JW (2010) Suspended solids removal to improve shrimp (Litopenaeus vannamei) production and an evaluation of a plant-based feed in minimal-exchange, superintensive culture systems. Aquaculture 299:89–98

Richmond A, Becker EW (1986) Technological aspects of mass cultivation, a general outline. In: CRC handbook of microalgal mass culture, CRC Press, Boca Raton, pp 245–263

Rocha AF, Biazzetti Filho ML, Stech MR, Silva RP (2017) Lettuce production in aquaponic and biofloc systems with silver catfish *Rhamdia quelen*. Bol Inst Pesca 44:64–73

Schoumans OF, Chardon WJ, Bechmann ME, Gascuel-Odoux C, Hofman G, Kronvang B et al (2014) Mitigation options to reduce phosphorus losses from the agricultural sector and improve surface water quality: a review. Sci Total Environ 468:1255–1266

Sgnaulin T, Mello GL, Thomas MC, Esquivel-Garcia JR, Oca GARM, Emerenciano MGC (2018) Biofloc technology (BFT): an alternative aquaculture system for Piracanjuba *Brycon orbignyanus*? Aquaculture 485:119–123

Sharpley AN, Kleinman PJ, Heathwaite AL, Gburek WJ, Folmar GJ, Schmidt JP (2008) Phosphorus loss from an agricultural watershed as a function of storm size. J Environ Qual 37 (2):362–368

Shifrin NS, Chisholm SW (1981) Phytoplankton lipids: interspecific differences and effects of nitrate, silicate and light-dark cycles. J Phycol 17(4):374–384

Smith VH (1983) Low nitrogen to phosphorus ratios favor dominance by blue-green algae in lake phytoplankton. Science 221(4611):669–671

Stoknes K, Scholwin F, Krzesiński W, Wojciechowska E, Jasińska A (2016) Efficiency of a novel "food to waste to food" system including anaerobic digestion of food waste and cultivation of vegetables on digestate in a bubble-insulated greenhouse. Waste Manag 56:466–476. https://doi.org/10.1016/j.wasman.2016.06.027

Stokstad E (2010) Down on the shrimp farm. Science 328:1504–1505

Storey, A., 2015, Vertical farming costs and the math behind them, posted 6 October 2015 https://blog.brightagrotech.com/vertical-farming-costs-and-the-math-behind-them/

Tibbitts TW, Cao W, and Wheeler RM (1994) Growth of potatoes for CELSS. NASA Contractor Report 177646

Touliatos D, Dodd IC, McAinsh M (2016) Vertical farming increases lettuce yield per unit areacompared to conventional horizontal hydroponics. Food and Energy Security 5(3):184–191

Turcios AE, Papenbrock J (2014) Sustainable treatment of aquaculture effluents – what can we learn for the past for the future? Sustainability 6:836–856

Virsile A, Olle M, Duchovskis P (2017) LED lighting in horticulture. In: Gupta D (ed) Light emitting diodes for agriculture – smart lighting. Springer, Singapore

Waller U, Buhmann AK, Ernst A, Hanke V, Kulakowski A, Wecker B, Orellana J, Papenbrock J (2015) Integrated multi-trophic aquaculture in a zero-exchange recirculation aquaculture system for marine fish and hydroponic halophyte production. Aquac Int 23:1473–1489

Weathers PJ, Zobel RW (1992) 1992, Aeroponics for the culture of organisms. Tissues Cells Biotechnol Adv 10(1):93–115

Wijihastuti RS, Moheimani NR, Bahri PA, Cosgrove JJ, Watanabe MM (2017) Growth and photosynthetic activity of *Botryococcus braunii* biofilms. J Appl Phycol 29(3):1123–1134

Zhao Z, Xu Q, Luo L, Wang CA, Li J, Wang L (2014) Effect of feed C/N ratio promoted bioflocs on water quality and production performance of bottom and filter feeder carp in minimum-water exchanged pond polyculture system. Aquaculture 434:442–448

Part III
Perspective for Sustainable Development

Chapter 13
Fish Diets in Aquaponics

Lidia Robaina, Juhani Pirhonen, Elena Mente, Javier Sánchez, and Neill Goosen

Abstract Fish and feed waste provide most of the nutrients required by the plants in aquaponics if the optimum ratio between daily fish feed inputs and the plant growing area is sustained. Thus, the fish feed needs to fulfil both the fish's and plant's nutritional requirements in an aquaponic system. A controlled fish waste production strategy where the nitrogen, phosphorus and mineral contents of fish diets are manipulated and used provides a way of influencing the rates of accumulation of nutrients, thereby reducing the need for the additional supplementation of nutrients. To optimize the performance and cost-effectiveness of aquaponic production, fish diets and feeding schedules should be designed carefully to provide nutrients at the right level and time to complement fish, bacteria and plants. To achieve this, a species-specific tailor-made aquaponic feed may be optimized to suit the aquaponic system as a whole. The optimal point would be determined based on overall system performance parameters, including economic and environmental sustainability measures. This chapter thus focuses on fish diets and feed and reviews the state of the art

L. Robaina (✉)
Aquaculture Research Group (GIA), Ecoaqua Institute, University of Las Palmas de Gran Canaria, Telde, Gran Canaria, Spain
e-mail: lidia.robaina@ulpgc.es

J. Pirhonen
Department of Biological and Environmental Science, University of Jyväskylä, Jyväskylä, Finland
e-mail: juhani.pirhonen@jyu.fi

E. Mente
Department of Ichthyology and Aquatic Environment, University of Thessaly, Volos, Greece
e-mail: emente@uth.gr

J. Sánchez
Department of Physiology, Faculty of Biology, Regional Campus of International Excellence "Campus Mare Nostrum", University of Murcia, Murcia, Spain
e-mail: javisan@um.es

N. Goosen
Department of Process Engineering, Stellenbosch University, Stellenbosch, South Africa
e-mail: njgoosen@sun.ac.za

S. Goddek et al. (eds.), *Aquaponics Food Production Systems*,
https://doi.org/10.1007/978-3-030-15943-6_13

in fish diets, ingredients and additives, as well as the nutritional/sustainable challenges that need to be considered when producing specific aquaponic feeds.

Keywords Aquaponic diets · Sustainability · Feed by-products · Nutrient flow · Nutritional requirements · Feeding times

13.1 Introduction

Aquatic food is recognized to be beneficial to human nutrition and health and will play an essential role in future sustainable healthy diets (Beveridge et al. 2013). In order to achieve this, the global aquaculture sector must contribute to increasing the quantity and quality of fish supplies between now and 2030 (Thilsted et al. 2016). This growth should be promoted not only by increasing the production and/or number of species but also by systems diversification. However, fish from aquaculture has only recently been included in the food security and nutrition (FSN) debate and the future strategies and policies, demonstrating the important role of this production to prevent malnutrition in the future (Bénét et al. 2015), as fish provide a good source of protein and unsaturated fats, as well as minerals and vitamins. It is important to note that many African nations are promoting aquaculture as the answer to some of their current and future food production challenges. Even in Europe, fish supply is currently not self-sufficient (with an unbalanced domestic supply/demand), being increasingly dependent on imports. Therefore, ensuring the successful and sustainable development of global aquaculture is an imperative agenda for the global and European economy (Kobayashi et al. 2015). Sustainability is generally required to show three key aspects: environmental acceptability, social equitability and economic viability. Aquaponic systems provide an opportunity to be sustainable, by combining both animal and plant production systems in a cost-efficient, environmentally friendly and socially beneficial ways. For Staples and Funge-Smith (2009), sustainable development is the balance between ecological well-being and human well-being, and in the case of aquaculture, an ecosystem approach has been only recently understood as a priority area for research.

Aquaculture has been the fastest growing food production sector during the last 40 years (Tveterås et al. 2012), being one of the most promising farming activities to meet near-future world food needs (Kobayashi et al. 2015). Total production statistics from aquaculture (FAO 2015) reveal an annual increment in global production of 6%, which is expected to provide up to 63% of global fish consumption by 2030 (FAO 2014), for an estimated population of nine billion people in 2050. In the case of Europe, the predicted increase is seen not only within the marine sector but also in terrestrially produced products. Some of the expected challenges to the growth of aquaculture during the coming years are the reduction in the use of antibiotics and other pathological treatments, the development of efficient aquaculture systems and equipment, together with species diversification and increased sustainability in the area of feed production and feed use. The shift from fishmeal (FM) in feed to other protein sources is also an important challenge, as well as the 'fish-in-fish-out' ratios. There is a long history, reaching back to the 1960s, of promoting the growth of the aquaculture sector towards proper sustainability including the encouragement to adapt and create new and more sustainable feed formulas, reduce feed spilling and reduce the food conversion ratio (FCR). Although aquaculture is recognized as the most efficient animal production sector, when compared with terrestrial animal production, there is still room for improvement in terms of resource efficiency, diversification of species or methods of production, and moreover a clear need for an ecosystem approach taking full advantage of the biological potential of the organisms and providing adequate consideration of environmental and societal factors (Kaushik 2017). This growth in aquaculture production will need to be supported by an increase in the expected total feed production. Approximately three million additional tons of feed will need to be produced each year to support the expected aquaculture growth by 2030. Moreover, replacing fishmeal and fish oil (FO) with plant and terrestrial substitutes is needed which requires essential research into formula feed for animal farming.

The animal and aquafeed industries are part of a global production sector, which is also the focus for future development strategies. Alltech's yearly survey (Alltech 2017) reveals that total animal feed production broke through 1 billion metric tons, with a 3.7% increase in production from 2015 despite a 7% decrease in the number of feed mills. China and the USA dominated production in 2016, accounting for 35% of the world's total feed production. The survey indicates that the top 10 producing countries have more than half of the world's feed mills (56%) and account for 60% of total feed production. This concentration in production means that many of the key ingredients traditionally used in formulations for commercial aquaculture feeds are internationally traded commodities, which subjects aquafeed production to any global market volatility. Fishmeal for example is expected to double in price by 2030, whilst fish oil is likely to increase by over 70% (Msangi et al. 2013). This illustrates the importance of reducing the amount of these ingredients in fish feed whilst increasing the interest and focus on new or alternative sources (García-Romero et al. 2014a, b; Robaina et al. 1998, 1999; Terova et al. 2013; Torrecillas et al. 2017).

Whilst new offshore platforms have been developed for aquaculture production, there is also a significant focus on marine and freshwater recirculating aquaculture systems (RAS), as these systems use less water per kg fish feed used, which increases fish production whilst reducing environmental impacts of aquaculture including reductions in water usage (Ebeling and Timmons 2012; Kingler and Naylor 2012). RAS can be integrated with plant production in aquaponic systems, which readily fit into local and regional food system models (see Chap. 15) that can be practised in or near large population centres (Love et al. 2015a). Water, energy and fish feed are the three largest physical inputs for aquaponic systems (Love et al. 2014, 2015b). Approximately 5% of feed is not consumed by the farmed fish, whilst the remaining 95% is ingested and digested (Khakyzadeh et al. 2015). Of this share, 30–40% is retained and converted into new biomass, whilst the

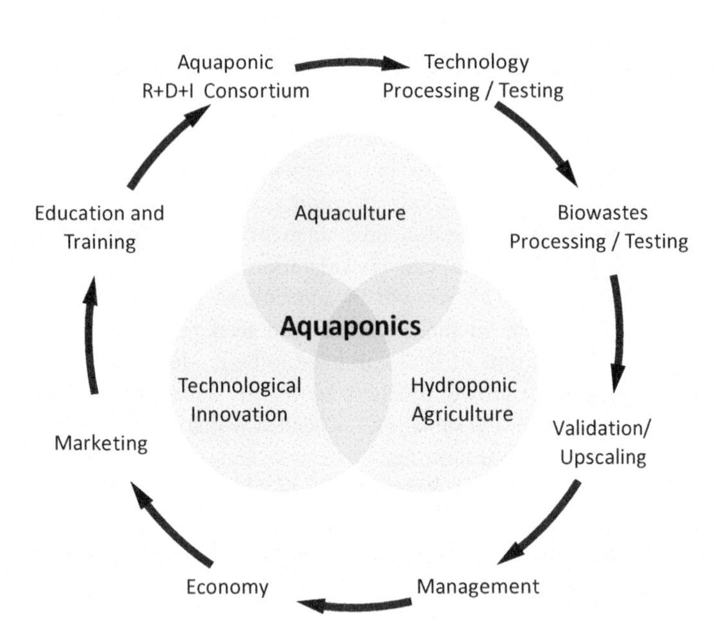

Fig. 13.1 Schematic representation of a multidisciplinary approach to locally valorize bio-by-products for aquaponic diets. (Based on 'R+D+I towards aquaponic development in the ultraperipheric islands and the circular economy'; ISLANDAP project, Interreg MAC/1.1a/2072014-2019)

remaining 60–70% is released in the form of faeces, urine and ammonia (FAO 2014). On average, 1 kg of feed (30% crude protein) globally releases about 27.6 g of N, and 1 kg of fish biomass releases about 577 g of BOD (biological oxygen demand), 90.4 g of N and 10.5 g of P (Tyson et al. 2011).

Aquaponics is currently a small but rapidly growing sector which is clearly suited to take advantage of the following political and socio-economic challenges, where 1) aquatic produce meets the need for food security and nutrition, 2) fish self-sufficient regions are established around the world, 3) aquaculture is a key sector but global ingredients and global feed production comes under focus, 4) innovation in agriculture promotes biodiversity in more sustainable ways and as part of the circular economy and 5) there is a greater take up of locally produced foods. These aspects tie in with the recommendations from the International Union for the Conservation of Nature (Le Gouvello et al. 2017), regarded the sustainability of the aquaculture and fish feed, which has recommended that efforts should be made to localize aquaculture production and the circular approach, and for putting in place a quality control programme for new products and by-products, as well as processing local fish feed within regions. So far, aquaponics as 'small-scale aquaculture farms' could provide examples for the implementation of the bioeconomy and local-scale production, thus promoting ways of using products and by-products from organic matter not suitable for use for other purposes, e.g. farmed insects and worms, macro- and microalgae, fish and by-product hydrolysates, new agro-ecology-produced plants and locally produced bioactives and micronutrients, whilst reducing the environmental footprint with quality food (fish and plants) production and moving towards zero waste generation. Moreover, aquaponics provides a good example for promoting a multidisciplinary way of learning about sustainable production and bioresource valorization, e.g. the 'Islandap Project' (INTERREG V-A MAC 2014-2020) (Fig. 13.1).

The following sections of this chapter review the state of the art of fish diets, ingredients and additives, as well as nutritional/sustainable challenges to consider when producing specific aquaponic feeds.

13.2 Sustainable Development of Fish Nutrition

The sustainable development of fish nutrition in aquaculture will need to correspond with the challenges that aquaponics delivers with respect to the growing need for producing high-quality food. Manipulating the nitrogen, phosphorus and the mineral content of fish diets used in aquaponics is one way of influencing the rates of the accumulation of nutrients, thereby reducing the need for the artificial and external supplementation of nutrients. According to Rakocy et al. (2004), fish and feed waste provide most of the nutrients required by plants if the optimum ratio between daily fish feed input and plant growing areas is sustained. Solid fish waste called 'sludge' in aquaponic systems results in losing approximately half of the available input nutrients, especially phosphorus, that theoretically could be used for plant biomass production but information is still limited (Delaide et al. 2017; Goddek et al. 2018). Whilst the goal of sustainability in fish nutrition in aquaculture will in the future be achieved by using tailor-made diets, fish feed in aquaponics needs to fulfil the nutritional requirements both for fish and for plants. Increases in sustainability will in part derive from less dependence on fishmeal (FM) and fish oil (FO) and novel, high-energy, low-carbon footprint raw natural ingredients. To safeguard biodiversity and the sustainable use of natural resources, the use of wild fisheries-based FM and FO needs to be limited in aquafeeds (Tacon and Metian 2015). However, fish performance, health and final product quality may be altered when substituting dietary FM with alternative ingredients. Thus, fish nutrition research is focused on the efficient use and transformation of the dietary components to provide the necessary essential nutrients that will maximize growth performance and achieve sustainable and resilient aquaculture. Replacing FM, which is an excellent but costly protein source in fish diets, is not straightforward due to its unique amino acid profile, high nutrient digestibility, high palatability, adequate amounts of micronutrients, as well as having a general lack of anti-nutritional factors (Gatlin et al. 2007).

Many studies have shown that FM can be successfully replaced by soybean meal in aquafeeds, but soybean meal has anti-nutritional factors such as trypsin inhibitors, soybean agglutinin and saponin, which limit its use and high replacement percentages in farming carnivorous fish. High FM replacement by plant meals in fish diets can also reduce nutrient bioavailability in fish, which results in nutrient alterations in the final quality of the product (Gatlin et al. 2007). It can also cause undesirable disturbances to the aquatic environment (Hardy 2010) and reduce fish growth due to the reduced levels of essential amino acids (especially methionine and lysine), and reduced palatability (Krogdahl et al. 2010). Gerile and Pirhonen (2017) noted that a 100% FM replacement with corn gluten meal significantly reduced growth rate of rainbow trout but FM replacement did not affect oxygen consumption or swimming capacity.

High levels of plant material can also affect the physical quality of the pellets, and may complicate the manufacturing process during extrusion. Most of the alternative plant-derived nutrient sources for fish feeds contain a wide variety of anti-nutritional factors that interfere with fish protein metabolism by impairing digestion and utilization, therefore leading to increased N release in the environment which can affect fish health and welfare. In addition, diets including high levels of phytic acid altered phosphorus and protein digestion that lead to high N and P release into the surrounding environment. Feed intake and palatability, nutrient digestibility and retention may vary according to fish species' tolerance and levels and can change the quantity and composition of the fish waste. Taking into consideration these results, fish diet formulations in aquaponics should investigate 'the tolerance' dietary levels of anti-nutritional factors (i.e. phytate) for different feed ingredients and for each fish species used in aquaponics and also the effects of the addition of minerals such as Zn and phosphate in the diets. It also should be noted that even if plant material is regarded as an ecologically sound option to replace FM in aquafeeds, plants need irrigation, and thus may induce ecological impacts in the form of their water and ecological footprints (Pahlow et al. 2015) from nutrient run-off from the fields.

Terrestrial animal by-products such as non-ruminant processed animal proteins (PAPs) derived from monogastric farmed animals (e.g. poultry, pork) that are fit for human consumption at the point of slaughter (Category 3 materials, EC regulation 142/2011; EC regulation 56/2013) could also replace FM and support the circular economy. They have higher protein content, more favourable amino acid profiles and fewer carbohydrates compared to plant feed ingredients whilst also lacking anti-nutritional factors (Hertrampf and Piedad-Pascual 2000). It has been shown that meat and bone meals may serve as a good phosphorus source when it is included in the diet of Nile tilapia (Ashraf et al. 2013), although it has been strictly banned in the feed of ruminant animals due to the danger of initiating bovine spongiform encephalopathy (mad cow disease). Certain insect species, such as black soldier fly (*Hermetia illucens*), could be used as an alternative protein source for sustainable fish feed diets. The major environmental advantages of insect farming are that (a) less land and water are required, (b) that greenhouse gas emissions are lower and that (c) insects have high feed conversion efficiencies (Henry et al. 2015). However, there is a continuing need for further research to provide evidence on quality and safety issues and screening for risks to fish, plants, people and the environment.

It is important to note that fish cannot synthesize several essential nutrients required for their metabolism and growth and depend on the feed for this supply. However, there are certain animal groups that can use nutrient-deficient diets, as they bear symbiotic microorganisms that can provide these compounds (Douglas 2010), and thus, fish can obtain maximal benefit when the microbial supply of their essential nutrients is scaled to demand. Undersupply limits fish growth, whilst oversupply can be harmful due to the need for the fish to neutralize toxicity caused by non-essential compounds. The extent to which the microbial function varies with the demands of different fish species and what are the underlying mechanisms are largely unknown. Importantly, an aquatic animal's gut microbiota can in theory play a critical role in providing the necessary nutrients and obtaining sustainability in fish farming (Kormas et al. 2014; Mente et al. 2016). Further research in this field will help facilitate the selection of ingredients to be used in fish feeds that promote gut microbiota diversity to improve fish growth and health.

Research into the utilization of alternative plant and animal protein sources and low trophic fish feed ingredients is ongoing. The substitution of marine sourced raw ingredients in fish feed, which could be used directly for human food purposes should decrease fishing pressure and contribute to preserving biodiversity. Low trophic-level organisms, such as black soldier fly, which may serve as aquafeed ingredients may be grown on by-products and waste of other agricultural industrial practices given different nutritional quality meals, thereby adding additional environmental benefits. However, efforts to succeed with the circular economy and the recycling of organic and inorganic nutrients should be handled with care since undesirable compounds in raw materials and seafood products could increase the risk to animal health, welfare, growth performance and safety of the final product for consumers. Research and continuous monitoring and reporting on contaminants of farmed aquatic animals in relation to the maximum limits in feed ingredients and diets are essential to inform revisions in and introductions of new regulations.

13.3 Feed Ingredients and Additives

13.3.1 Protein and Lipid Sources for Aquafeeds

Since the end of the twentieth century, there have been significant changes in the composition of aquafeeds but also advances in manufacturing. These transformations have originated from the need to improve the economic profitability of aquaculture as well as to mitigate its environmental impacts. However, the driving forces behind these changes is the need to decrease the amount of fishmeal (FM) and fish oil (FO) in the feeds, which have traditionally constituted the largest proportion of the feeds, especially for carnivorous fish and shrimp. Partly because of overfishing but especially due to the continuous increase in global aquaculture volume, there is

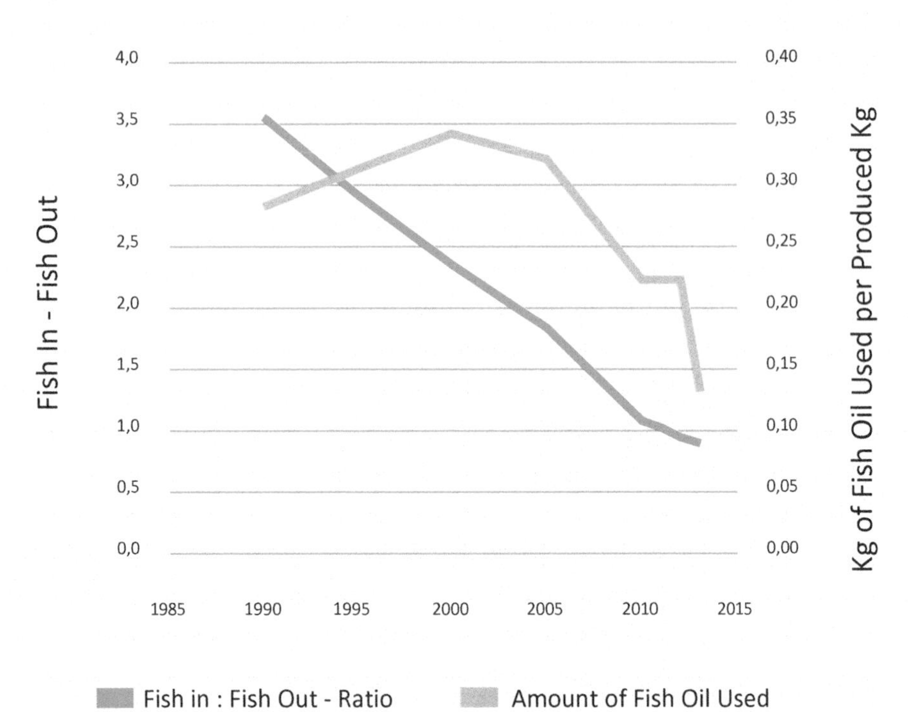

Fig. 13.2 Fish-in-fish-out ratio (blue line, left y-axis) and the amount of fish oil used (yellow line, right y-axis) for rainbow trout feed in Finland between 1990 and 2013. (Data from www.raisioagro.com)

an increasing need for alternative proteins and oils to replace FM and FO in aquafeeds.

The composition of fish feeds has changed considerably as the proportion of FM in diets has decreased from >60% in the 1990s to <20% in modern diets for carnivorous fish such as Atlantic salmon (*Salmo salar*), and FO content has decreased from 24% to 10% (Ytrestøyl et al. 2015). As a consequence the so-called fish-in-fish-out (FIFO) ratio has decreased below 1 for salmon and rainbow trout meaning that the amount of fish needed in the feed to produce 1 kg of fish meat is less than 1 kg (Fig. 13.2). Thus, carnivorous fish culture in the twenty-first century is a net producer of fish. On the other hand, feeds for lower trophic omnivorous fish species (e.g. carp and tilapia) may contain less than 5% FM (Tacon et al. 2011). Farming such low trophic fish species is ecologically more sustainable than for higher trophic species, and FIFO for tilapia was 0.15 and for cyprinids (carp species) only 0.02 in 2015 (IFFO). It should be noted that total FM replacement in the diets of tilapia (Koch et al. 2016) and salmon (Davidson et al. 2018) is not possible without significantly affecting production parameters.

Today, the major supply of proteins and lipids in fish feed comes from plants, but also commonly from other sectors, including meals and fats from meat and poultry by-products and blood meal (Tacon and Metian 2008). Additionally, waste and by-products from fish processing (offal and waste trimmings) are commonly used to produce FM and FO. However, due to EU regulations (EC 2009), the use of FM of a species is not allowed as feed for the same species, e.g. salmon cannot be fed FM containing salmon trimmings.

FM and FO replacements with other ingredients can affect the quality of the product that is sold to customers. Fish has the reputation of being a healthy food, especially due to its high content of poly and highly unsaturated fatty acids. Most importantly, seafood is the only source of EPA (eicosapentaenoic acid) and DHA (docosahexaenoic acid), both of which are omega-3 fatty acids, and essential nutrients for many functions in the human body. If FM and FO are replaced with products from terrestrial origin, this will directly affect the quality of the fish flesh, most of all its fatty acid composition, as the proportion of omega-3 fatty acids (especially EPA and DHA) will decrease whilst the amount of omega-6 fatty acids will increase along with the increase of plant material that is replacing FM and FO (Lazzarotto et al. 2018). As such, the health benefits of fish consumption are partly lost, and the product that ends up on the plate is not necessarily what consumers expected to buy. However, in order to overcome the problem of decreased omega-3 fatty acids in the final product resulting from lower fish ingredients in aquafeeds, fish farmers could employ so-called finishing diets with high FO content during the final stages of cultivation (Suomela et al. 2017).

A new interesting option for replacing FO in fish feeds is the possibility of genetic engineering, i.e. genetically modified plants which can produce EPA and DHA, e.g. oil from genetically modified *Camelina sativa* (common name of camelina, gold-of-pleasure or false flax which is known to have high levels of omega-3 fatty acids) was successfully used to grow salmon, ending up with very high concentration on EPA and DHA in the fish (Betancor et al. 2017). The use of genetically modified organisms in human food production, however, is subject to regulatory approval and may not be an option in the short term.

Another new possibility to replace FM in aquafeeds is proteins made of insects (Makkar et al. 2014). This new option has become possible within the EU only recently when the EU changed legislation, allowing insect meals in aquafeeds (EU 2017). The species permitted to be used are black soldier fly (*Hermetia illucens*), common housefly (*Musca domestica*), yellow mealworm (*Tenebrio molitor*), lesser mealworm (*Alphitobius diaperinus*), house cricket (*Acheta domesticus*), banded cricket (*Gryllodes sigillatus*) and field cricket (*Gryllus assimilis*). Insects must be reared on certain permitted substrates. Growth experiments done with different fish species show that replacing FM with meal made of black soldier fly larvae does not necessarily compromise growth and other production parameters (Van Huis and Oonincx 2017). On the other hand, meals made of yellow mealworm could replace FM only partially to avoid decrease in growth (Van Huis and Oonincx 2017). However, FM replacement with insect meal can cause a drop in omega-3 fatty acids, as they are void of EPA and DHA (Makkar and Ankers 2014).

In contrast to insects, microalgae typically have nutritionally favourable amino acid and fatty acid (including EPA and DHA) profiles but there is also a wide variation between species in this respect. Partial replacement of FM and FO in aquafeeds with certain microalgae have given promising results (Camacho-Rodríguez et al. 2017; Shah et al. 2018) and in the future the use of microalgae in aquafeeds can be expected to increase (White 2017) even though their use may be limited by price.

This short summary of potential feed ingredients indicates that there is wide range of possibilities to at least partially replace FM and FO in fish feeds. In general, the amino acid profile of FM is optimal for most fish species and FO contains DHA and EPA which are practically impossible to provide from terrestrial oils, albeit genetic engineering may change the situation in the future. However, GMO products need first to be accepted in legislation and then by customers.

13.3.2 The Use of Specialist Feed Additives Tailored for Aquaponics

Tailoring aquafeeds which are specific to aquaponic systems is more challenging than conventional aquaculture feed development, as the nature of aquaponic systems requires that the aquafeeds not only supply nutrition to the cultured animals but also to the cultured plants and the microbial communities inhabiting the system. Current aquaponic practice utilizes aquafeeds formulated to provide optimal nutrition to the cultured aquatic animals; however, as the major nutrient input into aquaponic systems (Roosta and Hamidpour 2011; Tyson et al. 2011; Junge et al. 2017), the feeds also need to take into account the nutrient requirements of the plant production component. This is especially important for commercial-scale aquaponic systems, where the productivity of the plant production system has a major impact on overall system profitability (Adler et al. 2000; Palm et al. 2014; Love et al. 2015a) and where improved production performance of the plant component can significantly improve overall system profitability.

Thus, the overall aim of developing tailored aquaponic feeds would be to design a feed which strikes a balance between providing additional plant nutrients, whilst maintaining acceptable aquaponic system operation (i.e. sufficient water quality for animal production, biofilter and anaerobic digester performance, and nutrient absorption by plants). In order to achieve this, the final tailored aquaponic feed may not be optimal for either aquatic animal or plant production individually, but would rather be optimal for the aquaponic system as a whole. The optimal point would be determined based on overall system performance parameters, e.g. economic and/or environmental sustainability measures.

One of the major challenges in increasing production output from coupled aquaponic systems is the relatively low concentrations of both macro- and micro plant nutrients (mostly in the inorganic form) in the recirculating water, compared to conventional hydroponic systems. These low levels of nutrients can result in nutrient deficiencies in the plants and suboptimal plant production rates (Graber and Junge 2009; Kloas et al. 2015; Goddek et al. 2015; Bittsanszky et al.

2016; Delaide et al. 2017). A further challenge is the significant amounts of sodium chloride in conventional fishmeal-based aquafeeds and the potential accumulation of sodium in aquaponic systems (Treadwell et al. 2010). Different approaches can be developed to address these challenges such as technological solutions, e.g. decoupled aquaponic systems (Goddek et al. 2016) (also see Chap. 8), direct nutrient supplementation in the plant production system via foliar spray or addition to the recirculating water (Rakocy et al. 2006; Roosta and Hamidpour 2011) , or the culture of better salt-tolerant plant (see Chap. 12). A new approach is the development of tailored aquafeeds specifically for use in aquaponics.

In order to address plant nutrient shortages in aquaponics, tailored aquaponic feeds need to increase the amount of plant-available nutrients, either by increasing the concentrations of specific nutrients after excretion by the cultured animals, or by rendering the nutrients more bio-available after excretion and biotransformation, for rapid uptake by the plants. Achieving this increased nutrient excretion is, however, not as simple as supplementing increased amounts of the desired nutrients to the aquaculture diets, as there are many (often conflicting) factors that need to be considered in an integrated aquaponic system. For example, although optimal plant production will require increased concentrations of specific nutrients, certain minerals, e.g. certain forms of iron and selenium, can be toxic to fish even at low concentrations and would therefore have maximum allowable levels in the circulating water (Endut et al. 2011; Tacon 1987). Apart from total nutrient levels, the ratio between nutrients (e.g. the P:N ratio) is also important for plant production (Buzby and Lin 2014), and imbalances in the ratios between nutrients can lead to accumulation of certain nutrients in aquaponic systems (Kloas et al. 2015). Furthermore, even if an aquaponic feed increases plant nutrient levels, the overall system water quality and pH still needs to be maintained within acceptable limits to ensure acceptable animal production, efficient nutrient absorption by plant roots, optimal operation of biofilters and anaerobic digesters (Goddek et al. 2015b; Rakocy et al. 2006) and to avoid precipitation of certain important nutrients like phosphates, as this will render them unavailable to plants (Tyson et al. 2011). To achieve this overall balance is no mean feat, as there are complex interactions between the different forms of nitrogen in the system (NH_3, NH_4^+, NO_2^-, NO_3^-), the system pH and the assortment of metals and other ions present in the system (Tyson et al. 2011; Goddek et al. 2015; Bittsanszky et al. 2016).

Common Nutrient Shortages in Aquaponic Systems

Plants require a range of macro- and micronutrients for growth and development. Aquaponic systems are commonly deficient in the plant macronutrients potassium (K), phosphorus (P), iron (Fe), manganese (Mn) and sulphur (S) (Graber and Junge 2009; Roosta and Hamidpour 2011). Nitrogen (N) is present in different forms in aquaponic systems, and is excreted as part of the protein metabolism of the cultured aquatic animals (Rakocy et al. 2006; Roosta and Hamidpour 2011; Tyson et al. 2011) after which it enters the nitrogen cycle in the integrated aquaponic environment. (Nitrogen is discussed in detail in Chap. 9 and is therefore excluded from the present discussion.)

The use of selected specialist aquaculture feed additives can contribute to the development of tailored aquafeeds specifically for aquaponics, by providing additional nutrients to the cultured aquatic animals and/or plants, or by adjusting the ratio of nutrients. Aquaculture feed additives are diverse, with a wide range of functions and mechanisms of working. Functions can be nutritive and non-nutritive, and the additives can be targeted towards action in the feeds or towards the physiological processes of the cultured aquatic animals (Encarnação 2016). For the purposes of this chapter, emphasis is on three specific types of additives which could assist the tailoring of aquaponic diets: (1) mineral supplements added directly to the feeds, (2) minerals that are added co-incidentally as part of additives that serve a non-mineral purpose and (3) additives which render minerals, which are already present in the feeds, more available to the cultured aquatic animals and/or plants in aquaponic systems.

1. *Direct mineral supplementation in aquaponic feeds*

Supplementing minerals directly in aquaculture diets used in aquaponic systems is one potential method to either increase the amount of minerals excreted by the cultured animals or to add specific minerals required by the plants in aquaponic systems. Minerals are routinely added in the form of mineral premixes to aquaculture diets, to supply the cultured aquatic animals with the essential elements required for growth and development (Ng et al. 2001; NRC 2011). Any minerals not absorbed by the fish during digestion are excreted, and if these are in the soluble (mostly ionic) form in the aquaponic system, these are available for plant uptake (Tyson et al. 2011; Goddek et al. 2015). It is unclear how feasible such an approach would be, as there is scant information about the efficacy of adding mineral supplements to aquafeeds for the purpose of enhancing aquaponic plant production. In general, mineral requirements and metabolism in aquaculture are poorly understood compared to terrestrial animal production, and the feasibility of this approach is therefore not well described. Potential advantages to this approach would be that it could prove to be a fairly simple intervention to improve overall system performance, it could allow supplementation of a wide range of nutrients, and it is likely to be relatively low cost. However, substantial research is still required to avoid any major potential pitfalls that may arise. One of these centres on the fact that the supplemented minerals destined for the plants first need to pass through the digestive tract of the cultured aquatic animals and these could either be absorbed fully or partially during this passage. This could lead to unwanted accumulation of minerals in the aquatic animals, or interference in normal intestinal nutrient and/or mineral absorption and physiological processes (Oliva-Teles 2012). Significant interactions can occur between dietary minerals in aquaculture diets (Davis and Gatlin 1996), and these need to be determined before direct mineral supplementation in aquaponic diets can be employed. Other potential effects may include altered physical structure and chemosensory characteristics of the feeds, which in turn could affect feed palatability. Clearly, there is still substantial research required before this method of tailoring aquaponic feeds can be adopted.

2. *Co-incidental addition of minerals by way of feed additives*

Certain classes of feed additives are added to aquafeeds in the form of ionic compounds and where only one of the ions contributes towards the intended activity. The other ion is viewed as a co-incidental and unavoidable addition to the aquafeed and is often not considered in any aquaculture research. One specific example of such a class of often-used feed additives are the organic acid salts, where the intended active ingredient in the aquafeed is the anion of an organic acid (e.g. formate, acetate, butyrate or lactate) and the accompanying cation is often ignored in the cultured animals' nutrition. Thus, if the accompanying cation is chosen purposefully to be an important macro- or micro plant nutrient, there is the potential that it could be excreted by the cultured animals into the system water and be available for uptake by the plants.

Short-chain organic acids and their salts have become well known and often used in feed additives in both terrestrial animal nutrition and in aquaculture, where the compounds are employed as performance enhancers and agents to improve disease resistance. These compounds can have different mechanisms of functioning, including acting as antimicrobials, antibiotics or growth promoters, enhancing nutrient digestibility and utilization and acting as directly metabolizable energy source (Partanen and Mroz 1999; Lückstädt 2008; Ng and Koh 2017). Either the native organic acids or their salts can be utilized in aquaculture diets, but the salt forms of the compounds are often preferred by manufacturers as they are less corrosive to feed manufacturing equipment, are less pungent and are available in solid (powder) form, which simplifies addition to formulated feed during manufacturing (Encarnação 2016; Ng and Koh 2017). For a comprehensive review on the use of organic acids and their salts in aquaculture, readers are referred to the work of Ng and Koh (2017).

Employing organic acid salts in aquaponics has the potential to have dual benefits in the system, where the anion could enhance the performance and disease resistance of the cultured aquatic animals, whilst the cation (e.g. potassium) could increase the amount of essential plant nutrients excreted. The potential advantage of this approach is that dietary inclusion levels of organic acid salts can be relatively high for a feed additive, and research regularly reports total organic acid salt inclusion of up to 2% by weight (Encarnação 2016), although commercial manufacturers tend to recommend lower levels of approximately 0.15–0.5% (Ng and Koh 2017). The cation of organic acid salts could constitute a significant proportion of the overall weight of the salt, and as these are fed daily to the cultured animals, they could contribute a significant amount of nutrients to the plants in an aquaponic system over the course of a growing season. No published research is currently available that reports findings for this line of enquiry and as with direct mineral supplementation to aquaponic feeds, this approach needs to be validated through future research to determine the fate of the cations added as part of the organic acid salts (whether they are excreted or absorbed by the aquatic animals), and whether there are any interactions with minerals or nutrients. It remains, however, an exciting future avenue of investigation.

3. *Feed additives that render nutrients more bio-available to plants*

Increasing amounts of plant ingredients are used in formulated aquafeeds, yet minerals from plant raw materials are less bio-available to cultured aquatic animals, mainly due to the presence of anti-nutritional factors in the plant-based dietary ingredients (Naylor et al. 2009; Kumar et al. 2012; Prabhu et al. 2016). This means that a higher proportion of minerals are excreted in the faeces in bound form, requiring 'liberation' before being available for plant uptake. One typical example is organic phosphorus occurring as phytate, which can bind to other minerals to form insoluble compounds, where microbial action in the environment is required before the phosphorus is released as plant-available, soluble phosphate (Kumar et al. 2012).

The use of exogenous enzymes in tailored aquaponic diets could potentially contribute towards releasing increased amounts of nutrients from high-plant content aquafeeds for both animal and plant nutrition in aquaponic systems. The most often employed enzymes in aquafeeds are proteases, carbohydrates and phytases, both to improve nutrient digestion and to degrade anti-nutritional compounds like phytate (Encarnação 2016), which can result in additional nutrients being released from aquafeeds. Although it is known that exogenous enzyme supplementation leads to improved nutrient utilization in the cultured animals, it is unclear whether additional nutrients would be excreted in plant-available form, therefore avoiding a separate remineralization step in aquaponic systems (see Chap. 10). Additionally, interactions between exogenous enzymes and nutrients in different parts of the digestive tract of fish are possible (Kumar et al. 2012), which will have further implications for the amounts of nutrients excreted for plant growth. Further research is therefore also required to determine the utility of exogenous enzymes specifically for use in aquaponic feeds.

13.4 Physiological Rhythms: Matching Fish and Plant Nutrition

The design of feeds for fish is crucial in aquaponics because fish feed is the single or at least the main input of nutrients for both animals (macronutrients) and plants (minerals) (Fig. 13.3).

Nitrogen is introduced to the aquaponic system through protein in fish feed which is metabolized by fish and excreted in the form of ammonia. The integration of recirculating aquaculture with hydroponics can reduce the discharge of unwanted nutrients to the environment as well as generate profits. In an early economic study, phosphorus removal in an integrated trout and lettuce/basil aquaponic system proved to be cost-saving (Adler et al. 2000). Integrating fish feeding rates is also paramount to fulfil the nutritional requirements of plants. Actually, farmers need to know the amount of feed used in the aquaculture unit to calculate how much nutrient needs to be supplemented to promote plant growth in the hydroponic unit. For instance, in a

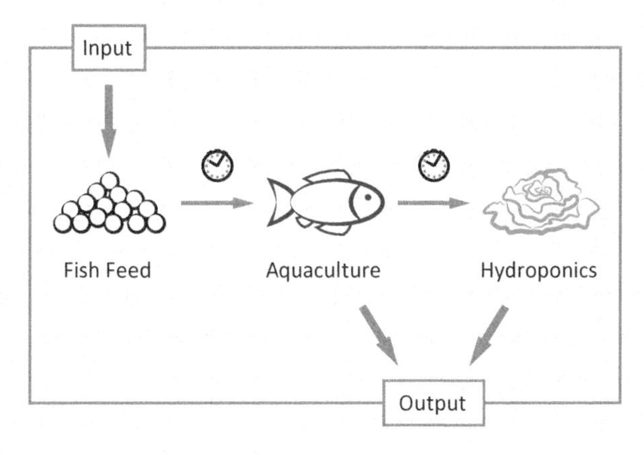

Fig. 13.3 Nutrient flow in an aquaponic system. Note that fish feed, through wastewater from the aquaculture system, provides the minerals required for plants to grow in the hydroponic system. Meal timing should be designed to match feeding/excretion rhythms in fish and nutrient uptake rhythms in plants

tilapia-strawberry aquaponic system, the total amount of feed required to produce ions (e.g. NO_3^-, Ca_2^+, $H_2PO_4^-$ and K^+) for plants was calculated at different fish densities, with better result for small fish density 2 kg fish/m^3 to reduce the cost of hydroponic solution supplementation (Villarroel et al. 2011).

It is well known that plants have daily rhythms and circadian rhythmicity in leaf movements was first described in plants by de Mairan in the early eighteenth century (McClung 2006). Circadian rhythms in plants control everything from the time of flowering to plant nutrition and thus these rhythms need to be taken into account especially when using artificial horticultural lighting. Fish are also tied into daily rhythms in most physiological functions, including feeding and nutrient uptake. It should not be surprising that fish exhibit feeding rhythms because food availability and the occurrence of predators are hardly constant but restricted to a particular time of day/night (López-Olmeda and Sánchez-Vázquez 2010). Thus, fish should be fed at the right time according to their appetite rhythms: meals scheduled during daytime for diurnal fish species, and at night for nocturnal fish. It is well known that fish show daily patterns of deamination of proteins and nitrogenous wastes related to their nutritional status and feeding rhythms (Kaushik 1980). Feeding time affects nitrogen excretion, as Gelineau et al. (1998) reported ammonia production and protein catabolism were lower in fish fed at dawn (in phase with their feeding rhythm) than in those fed at midnight (out of phase). Most interestingly, urea excretion shows circadian rhythmicity that persist in starved fish under constant conditions (Kajimura et al. 2002), revealing its endogenous origin. Furthermore, urea permeability (determined as body urea contents after immersion in a urea solution) coincided with the acrophase, i.e. the peak of the daily excretion rhythm, indicating that urea does not permeate cells by simple diffusion but there is a circadian control. Plants also show daily rhythms in nitrogen uptake, as early described by Pearson and Steer (1977), who found a daily pattern of nitrate uptake and nitrate reductase in peppers kept in a constant environment. The nitrate concentration in the leaves of spinach also increased during the night as the uptake rate of nitrate by the roots increase at that time (Steingrover et al. 1986). In aquaponics, the evidence thus points to the need for matching excretion rhythms in fish and nutrient uptake rhythms in plants. To optimize the performance and cost-effectiveness of aquaponic systems, fish diets and feeding schedules should be designed carefully to provide nutrients at the right level and the right time to complement both fish and plants.

References

Adler PR, Harper JK, Wade EM, Takeda F, Summerfelt ST (2000) Economic analysis of an aquaponic system for the integrated production of rainbow trout and plants. Int J Recirculat Aquacult 1:15–34

Alltech (2017) Alltech yearly survey 2017. https://go.alltech.com/alltech-feed-survey-2017

Ashraf S, Rania SM, Ehab REH (2013) Meat and bone meal as a potential source of phosphorus in plant-protein-based diets for Nile tilapia (*Oreochromis niloticus*). Aquacult Intl 21:375–385

Béné C, Barange M, Subasinghe R, Pinstrup-Andersen P, Merino G, Hemre G-I, Williams M (2015) Feeding 9 billion by 2050 – putting fish back on the menú. Food Sec 7:261–274. https://doi.org/10.1007/s12571-015-0427-z

Betancor MB, Li K, Sprague M, Bardal T, Sayanova O, Usher S, Han L, Måsøval K, Torrissen O, Napier JA, Tocher DR, Olsen RO (2017) An oil containing EPA and DHA from transgenic *Camelina sativa* to replace marine fish oil in feeds for Atlantic salmon (*Salmo salar* L.): Effects on intestinal transcriptome, histology, tissue fatty acid profiles and plasma biochemistry. PLoS ONE 12(4):e0175415. https://doi.org/10.1371/journal.pone.0175415

Beveridge MCM, Thilsted SH, Phillips MJ, Metian M, Troell M, Hall SJ (2013) Meeting the food and nutrition needs of the poor. J Fish Biol:1067. https://doi.org/10.1111/jfb.12187

Bittsanszky A, Uzinger N, Gyulai G, Mathis A, Junge R, Villarroel M, Kotzen B, Komives T (2016) Nutrient supply of plants in aquaponic systems. Ecocycles 2:17–20

Buzby KM, Lin L-S (2014) Scaling aquaponic systems: Balancing plant uptake with fish output. Aquacult Eng 63:39–44. https://doi.org/10.1016/j.aquaeng.2014.09.002.

Camacho-Rodríguez J, Macías-Sánchez MD, Cerón-García MC, Alarcón FJ, Molina-Grima E (2017) Microalgae as a potential ingredient for partial fish meal replacement in aquafeeds: nutrient stability under different storage conditions. J Appl Phycol. https://doi.org/10.1007/s10811-017-1281-5

Davidson J, Kenney PB, Barrows FT, Good C, Summerfelt ST (2018) Fillet quality and processing attributes of post-smolt Atlantic salmon, *Salmo salar*, fed a fishmeal-free diet and a fishmeal-based diet in recirculation aquaculture systems. J World Aquacult Soc 49:183–196. https://doi.org/10.1111/jwas.12452

Davis DA, Gatlin DM (1996) Dietary mineral requirement of fish and marine crustaceans. Rev Fish Sci 4:75–99. https://doi.org/10.1080/10641269609388579

Delaide B, Delhaye G, Dermience M, Gott J, Soyeurt H, Jijakli MH (2017) Plant and fish production performance, nutrient mass balances, energy and water use of the PAFF Box, a small-scale aquaponic system. Aquac Eng 78:130–139

Douglas AE (2010) The symbiotic habit. Princeton University Press, Princeton

Ebeling JM, Timmons MB (2012) Recirculating aquaculture systems. In: Tidwell JH (ed) Aquaculture production systems. Wiley, Hoboken

EC (2009) Regulation (EC) No 1069/2009 of the European parliament and of the council of 21 October 2009 laying down health rules as regards animal by-products and derived products not intended for human consumption and repealing regulation (EC) No 1774/2002 (animal by-products regulation). Off J Eur Union L 300/1. https://doi.org/10.3000/17252555.L_2009.300.eng

Encarnação P (2016) Chapter 5: Functional feed additives in aquaculture feeds. In: Nates SF (ed) Aquafeed formulation. Academic, San Diego, pp 217–237

Endut A, Jusoh A, Ali N, Wan Nik WB (2011) Nutrient removal from aquaculture wastewater by vegetable production in aquaponics recirculation system. Desalinat Water Treat 32:422–430. https://doi.org/10.5004/dwt.2011.2761

EU (2017) Commission Regulation (EU) 2017/893 of 24 May 2017 amending Annexes I and IV to Regulation (EC) No 999/2001 of the European Parliament and of the Council and Annexes X, XIV and XV to Commission Regulation (EU) No 142/2011 as regards the provisions on processed animal protein. Off J Eur Union L138/92. http://data.europa.eu/eli/reg/2017/893/oj

FAO (2014) The State of World Fisheries and Aquaculture. Food and Agricultural Organization, Rome, Italy

FAOSTAT (2015) Fish and fishery products – world apparent consumption statistics based on food balance sheets (1961–)

García-Romero J, Gines R, Izquierdo M, Robaina L (2014a) Marine and freshwater crab meals in diet for red porgy (*Pagrus pagrus*): effect on fillet fatty acid profile and flesh quality. Aquaculture 420–421:231–239. https://doi.org/10.1016/j.aquaculture.2013.10.035

García-Romero J, Gines R, Vargas R, Izquierdo M, Robaina L (2014b) Marine and freshwater crab meals in diet for red porgy (*Pagrus pagrus*): Digestibility, ammonia-N excretion, phosphorus and calcium retention. *Aquaculture* 428–429:158–165. https://doi.org/10.1016/j.aquaculture.2014.02.035

Gatlin DM, Barrows FT, Brown P, Dabrowski K, Gaylord TG, Hardy RW, Herman E, Hu GS, Krogdahl A, Nelson R, Rust M, Sealey W, Skonberg D, Souza EJ, Stone D, Wilson R, Wurtele E (2007) Expanding the utilization of sustainable plant products in aquafeeds; A review. Aquacult Res 38:551–579

Gelineau A, Medale F, Boujard T (1998) Effect of feeding time on postprandial nitrogen excretion and energy expenditure in rainbow trout. J Fish Biol 52:655–664

Gerile S, Pirhonen J (2017) Replacement of fishmeal with corn gluten meal in feeds for juvenile rainbow trout (*Oncorhynchus mykiss*) does not affect oxygen consumption during forced swimming. Aquaculture 479:616–618. https://doi.org/10.1016/j.aquaculture.2017.07.002

Goddek S, Delaide B, Mankasingh U, Ragnarsdottir KV, Jijakli H, Thorarinsdottir R (2015) Challenges of sustainable and commercial aquaponics. Sustainability 7:4199–4224. https://doi.org/10.3390/su7044199.

Goddek S, Espinal CA, Delaide B, Jijakli MH, Schmautz Z, Wuertz S, Keesman KJ (2016) Navigating towards decoupled Aquaponic systems: a system dynamics design approach. Water 8:303. https://doi.org/10.13140/RG.2.1.3930.0246

Goddek S, Delaide BPL, Joyce A, Wuertz S, Jijakli HM, Grosse A, Eding EH, Bläser I, Reuterg M, Keizer LCP, Morgenstern R, Körner O, Verreth J, Keesman KJ (2018) Nutrient mineralization and organic matter reduction performance of RAS-based sludge in sequential UASB-EGSB reactors. Aquacult Eng 83:10–19. https://doi.org/10.1016/j.aquaeng.2018.07.003

Graber A, Junge R (2009) Aquaponic systems: nutrient recycling from fish wastewater by vegetable production. Desalination 246:147–156

Hardy RW (2010) Utilization of plant proteins in fish diets: effects of global demand and supplies of fishmeal. Review article. Aquacult Res 41:770–776

Henry M, Gasco L, Piccolo G, Fountoulaki E (2015) Review on the use of insects in the diet of farmed fish: past and future. Anim Feed Sci Technol 203:1–22

Hertrampf JW, Piedad-Pascual F (2000) Handbook on ingredients for aquaculture feeds. Kluwer Academic Publishers, Dordrecht. 624 pp

IFFO, The Marine Ingredients Association. http://www.iffo.net

Junge R, König B, Villarroel M, Komives T, Jijakli H (2017) Strategic points in aquaponics. Water 9(3):182. https://doi.org/10.3390/w9030182

Kajimura M, Iwata K, Numata H (2002) Diurnal nitrogen excretion rhythm of the functionally ureogenic gobiid fish *Mugilogobius abei*. Comp Biochem Physiol B 131:227–239

Kaushik SJ (1980) Influence of the nutritional status on the daily pattern of nitrogen excretion in the carp (*Cyprinus carpio* L.) and the rainbow trout (*Salmo gairdneri* R.). Reprod Nut Develop 20:1751–1765

Kaushik S (2017) Aquaculture deals with the production of all kinds of aquatic organisms through human intervention, 2017 meeting. International Council of Academies of Engineering and Technological Sciences (CAETS – http://www.caets.org). Madrid, 14–15 November 2017

Khakyzadeh V, Luque R, Zolfigol MA, Vahidian HR, Salehzadeh H, Moradi V, Soleymani AR, Moosavi-Zare AR, Xu K (2015) Waste to wealth: a sustainable aquaponic system based on residual nitrogen photoconversion. Royal Society of Chemistry 5:3917–3921. https://doi.org/10.1039/C4RA15242E

Kingler D, Naylor R (2012) Searching for Solutions in Aquaculture: Charting a Sustainable Course. Anual Rev Environ Resour 37:247–276

Kloas W, Groß R, Baganz D, Graupner J, Monsees H, Schmidt U, Staaks G, Suhl J, Tschirner M, Wittstock B, Wuertz S, Zikova A, Rennert B (2015) A new concept for aquaponic systems to improve sustainability, increase productivity, and reduce environmental impacts. Aquacult Environ Interact 7:179–192. https://doi.org/10.3354/aei00146

Koch JF, Rawlesb SD, Webster CD, Cummins V, Kobayashic Y, Thompson KR, Gannam AL, Twibell RG, Hyded NM (2016) Optimizing fish meal-free commercial diets for Nile tilapia, *Oreochromis niloticus*. Aquaculture 452:357–366. https://doi.org/10.1016/j.aquaculture.2015.11.017

Kormas KA, Meziti A, Mente E, Fretzos A (2014) Dietary differences are reflected on the gut prokaryotic community structure of wild and commercially reared sea bream (*Sparus aurata*). Microbiology open. https://doi.org/10.1002/mbo3.202

Kobayashi M, Msangi S, Batka M, Vannuccini S, Dey MM, Anderson JL (2015) Fish to 2030: the role and opportunity for aquaculture. Aquacult Econ Manage 19:282–300. https://doi.org/10.1080/13657305.2015.994240

Krogdahl A, Penn M, Thorsen J, Refstie S, Bakke AM (2010) Important anti-nutrients in plant feedstuffs for aquaculture: An update on recent findings regarding responses in salmonids. Aquacult Res 41:333–344

Kumar V, Sinha AK, Makkar HPS, De Boeck G, Becker K (2012) Phytate and phytase in fish nutrition. J Anim Physiol Anim Nutr 96:335–364. https://doi.org/10.1111/j.1439-0396.2011.01169.x

Lazzarotto, V., Médale, F., Larroquet, L. & Corraze, G. (2018). Long-term dietary replacement of fishmeal and fish oil in diets for rainbow trout (*Oncorhynchus mykiss*): Effects on growth, whole body fatty acids and intestinal and hepatic gene expression. PLoS One 13(1) https://doi.org/10.1371/journal.pone.0190730

Le Gouvello, Raphaëla et François Simard (eds) (2017). Durabilité des aliments pour le poisson en aquaculture: Réflexions et recommandations sur les aspects technologiques, économique sociaux et environnementaux. Gland, Suisse: UICN, et Paris, France : Comité français de l'UICN. 296 pp

López-Olmeda JF, Sánchez-Vázquez FJ (2010) Feeding rhythms in fish: from behavioural to molecular approach. In: Kulczykowska E, Popek W. Kapoor BG (eds) Biological clock in fish. CRC Press, Enfield, pp 155–184

Love DC, Fry JP, Genello L, Hill ES, Frederick A, Li X, Semmens K (2014) An International survey of aquaponics practitioners. PLoS One 9(7):e102662. https://doi.org/10.1371/journal.pone.0102662

Love DC, Fry JP, Li X, Hill ES, Genello L, Semmens K, Thompson RE (2015a) Commercial aquaponics production and profitability: findings from an international survey. Aquaculture 435:67–74. https://doi.org/10.1016/j.aquaculture.2014.09.023

Love DC, Uhl MS, Genello L (2015b) Energy and water use of a small-scale raft aquaponics system in Baltimore, Maryland, United States. Aquacult Eng 68:19–27

Lückstädt C (2008) The use of acidifiers in fish nutrition. CAB Rev Perspect Agricult Veterinary Sci Nutr Natur Resour 3:1–8

Makkar HPS, Ankers P (2014) Towards sustainable animal diets: a survey-based study. Anim Feed Sci Technol 198:309–322. https://doi.org/10.1016/j.anifeedsci.2014.09.018

Makkar HPS, Tran G, Heuzé V, Ankers P (2014) State-of-the-art on use of insects as animal feed. Anim Feed Sci Techn 197:1–33. https://doi.org/10.1016/j.anifeedsci.2014.07.008

McClung CR (2006) Plant circadian rhythms. Plant Cell 18:792–803

Mente E, Gannon AT, Nikouli E, Hammer H, Kormas KA (2016) Gut microbial communities associated with the molting stages of the giant freshwater prawn *Macrobrachium rosenbergii*. Aquaculture 463:181–188
Msangi S, Kobayashi M, Batka M, Vannuccini S, Dey MM, Anderson JL (2013) Fish to 2030: prospects for fisheries and aquaculture. World Bank Report Number 83177-GLB. http://documents.worldbank.org/curated/en/458631468152376668/
Naylor RL, Hardy RW, Bureau DP, Chiu A, Elliott M, Farrell AP, Forster I, Gatlin DM, Goldburg RJ, Hua K, Nichols PD (2009) Feeding aquaculture in an era of finite resources. Proc Natl Acad Sci USA 106:15103–15110
Ng W-K, Koh C-B (2017) The utilization and mode of action of organic acids in the feeds of cultured aquatic animals. Rev Aquacult 9:342–368. https://doi.org/10.1111/raq.12141
Ng W-K, Ang L-P, Liew F-L (2001) An evaluation of mineral supplementation of fish meal-based diets for African catfish. Aquacult Int 9:277–282
NRC. N.R.C (2011) Nutrient requirements of fish and shrimp. The National Academies Press, Washington, District Columbia
Oliva-Teles A (2012) Nutrition and health of aquaculture fish. J Fish Dis 35:83–108. https://doi.org/10.1111/j.1365-2761.2011.01333.x
Pahlow M, Oel PR, Mekonnen MM, Hoekstra AY (2015) Increasing pressure on freshwater resources due to terrestrial feed ingredients for aquaculture production. Sci Total Environ 536:847–857. https://doi.org/10.1016/j.scitotenv.2015.07.124
Palm HW, Seidemann R, Wehofsky S, Knaus U (2014) Significant factors affecting the economic sustainability of closed aquaponic system. Part I: system design, chemo-physical parameters and general aspects. AACL Bioflux 7:20–32
Partanen KH, Mroz Z (1999) Organic acids for performance enhancement in pig diets. Nutr Res Rev 12:117–145
Pearson CJ, Steer BT (1977) Daily changes in nitrate uptake and metabolism in *Capsicum annuum*. Planta 137(2):107–112. https://doi.org/10.1007/BF00387546
Prabhu PAJ, Schrama JW, Kaushik SJ (2016) Mineral requirements of fish: a systematic review. Rev Aquacult 8:172–219
Rakocy JE, Shultz RC, Bailey DS, Thoman ES (2004) Aquaponic production of tilapia and basil: comparing a batch and staggered cropping system. Acta Hortic 648:63–69. https://doi.org/10.17660/ActaHortic.2004.648.8
Rakocy JE, Masser MP, Losordo TM (2006) Recirculating aquaculture tank production systems: aquaponics- integrating fish and plant culture. Southern Regional Aquaculture Center, pp 1–16
Robaina L, Izquierdo MS, Moyano FJ, Socorro J, Vergara JM, Montero D (1998) Increase of the dietary n-3/n-6 fatty acid ratio and addition of phosphorus improves liver histological alterations induced by feeding diets containing soybean meal to gilthead seabream, *Sparus aurata*. Aquaculture 161:281–293
Robaina L, Corraze G, Aguirre P, Blanc D, Melcion JP, Kaushik S (1999) Digestibility, postprandial ammonia excretion and selected plasma metabolites in European sea bass (*Dicentrarchus labrax*) fed pelleted or extruded diets with or without wheat gluten. Aquaculture 179:45–56
Roosta HR, Hamidpour M (2011) Effects of foliar application of some macro- and micro-nutrients on tomato plants in aquaponic and hydroponic systems. Scientia Horticulturae 129:396–402. https://doi.org/10.1016/j.scienta.2011.04.006
Shah MR, Giovanni Antonio Lutzu GA, Alam A, Sarker P, Chowdhury MAK, Parsaeimehr A, Liang Y, Daroch M (2018) Microalgae in aquafeeds for a sustainable aquaculture industry. J Appl Phycol 30:197–213. https://doi.org/10.1007/s10811-017-1234-z
Staples D, Funge-Smith S (2009) Ecosystem approach to fisheries and aquaculture: Implementing the FAO Code of Conduct for Responsible Fisheries. FAO Regional Office for Asia and the Pacific, Bangkok, Thailand. RAP Publication 2009/11, 48 pp
Steingrover E, Ratering P, Siesling J (1986) Daily changes in uptake, reduction and storage of nitrate in spinach grown at low light intensity. Physiol Plantarum 66:555–556
Suomela JP, Tarvainen M, Kallio H, Airaksinen S (2017) Fish oil finishing diet maintains optimal n-3 long-chain fatty acid content in European whitefish (*Coregonus lavaretus*). Lipids 52:849–855. https://doi.org/10.1007/s11745-017-4290-x

Tacon AGJ (1987) The nutrition and feeding of farmed fish and shrimp - a training manual. Food and Agricultural Organisation of the United Nations, Rome, Italy

Tacon AGJ, Metian M (2008) Global overview on the use of fish meal and fish oil in industrially compounded aquafeeds: trends and future prospects. Aquaculture 285:146–158. https://doi.org/10.1016/j.aquaculture.2008.08.015

Tacon AGJ, Metian M (2015) Feed matters: Satisfying the feed demand of aquaculture. Rev Fish Sci Aquacult 23(1):1–10. https://doi.org/10.1080/23308249.2014.987209

Tacon AGJ, Hasan MR, Metian M (2011) Demand and supply of feed ingredients for farmed fish and crustaceans: trends and prospects. FAO Fisheries and Aquaculture Technical Paper no. 564. FAO, Rome, 87 pp

Terova G, Robaina LE, Izquierdo MS, Cattaneo AG, Molinari S, Bernardini G, Saroglia M (2013) PepT1 mRNA expression levels in sea bream (*Sparus aurata*) fed different plant protein sources. Springer Plus 2:17. https://doi.org/10.1186/2193-1801-2-17

Thilsted SH, Thorne-Lyman A, Webb P, Bogard JR, Subasinghe R, Phillips MJ, Allison EH (2016) Sustaining healthy diets: The role of capture fisheries and aquaculture for improving nutrition in the post-2015 era. Food Policy 61:126–131. https://doi.org/10.1016/j.foodpol.2016.02.005

Torrecillas S, Robaina L, Caballero MJ, Montero D, Calandra G, Mompel D, Karalazos V, Sadasivam K, Izquierdo M (2017) Combined replacement of fishmeal and fish oil in European sea bass (*Dicentrarchus labrax*): production performance, tissue composition and liver morphology. Aquaculture 474. https://doi.org/10.1016/j.aquaculture.2017.03.031

Treadwell D, Taber S, Tyson R, Simonne E (2010) HS1163: Foliar-applied micronutrients in aquaponics: s guide to use and sourcing, IFAS Extension. University of Florida

Tveterås S, Asche F, Bellemare MF, Smith MD, Guttormsen AG, Lem A, Lien K, Vannuccini S (2012) Fish is food – the FAO's fish price index. PLoS ONE 7:e36731

Tyson RV, Treadwell DD, Simonne EH (2011) Opportunities and challenges to sustainability in aquaponic systems. HortTechnology 21:6–13

Van Huis A, Oonincx DGAB (2017) The environmental sustainability of insects as food and feed. A review. Agron Sustain Dev 37:43. https://doi.org/10.1007/s13593-017-0452-8

Villarroel M, Alvariño JMR, Duran JM (2011) Aquaponics: integrating fish feeding rates and ion waste production for strawberry hydroponics. Spanish J Agricult Res 9:537–545

White C (2017) Algae-based aquafeed firms breaking down barriers for fish-free feeds. https://www.seafoodsource.com/news/aquaculture/algae-based-aquafeed-firms-breaking-down-barriers-for-fish-free-feeds

Ytrestøy T, Aas TS, Åsgård T (2015) Utilisation of feed resources in production of Atlantic salmon (*Salmo salar*) in Norway. Aquaculture 448:365–374. https://doi.org/10.1016/j.aquaculture.2015.06.023

Chapter 14
Plant Pathogens and Control Strategies in Aquaponics

Gilles Stouvenakers, Peter Dapprich, Sebastien Massart, and M. Haïssam Jijakli

Abstract Among the diversity of plant diseases occurring in aquaponics, soil-borne pathogens, such as *Fusarium* spp., *Phytophthora* spp. and *Pythium* spp., are the most problematic due to their preference for humid/aquatic environment conditions. *Phytophthora* spp. and *Pythium* spp. which belong to the Oomycetes pseudo-fungi require special attention because of their mobile form of dispersion, the so-called zoospores that can move freely and actively in liquid water. In coupled aquaponics, curative methods are still limited because of the possible toxicity of pesticides and chemical agents for fish and beneficial bacteria (e.g. nitrifying bacteria of the biofilter). Furthermore, the development of biocontrol agents for aquaponic use is still at its beginning. Consequently, ways to control the initial infection and the progression of a disease are mainly based on preventive actions and water physical treatments. However, suppressive action (suppression) could happen in aquaponic environment considering recent papers and the suppressive activity already highlighted in hydroponics. In addition, aquaponic water contains organic matter that could promote establishment and growth of heterotrophic bacteria in the system or even improve plant growth and viability directly. With regards to organic hydroponics (i.e. use of organic fertilisation and organic plant media), these bacteria could act as antagonist agents or as plant defence elicitors to protect plants from diseases. In the future, research on the disease suppressive ability of the aquaponic biotope must be increased, as well as isolation, characterisation and formulation of microbial plant pathogen antagonists. Finally, a good knowledge in the rapid identification of pathogens, combined with control methods and diseases monitoring, as recommended in integrated plant pest management, is the key to an efficient control of plant diseases in aquaponics.

G. Stouvenakers (✉) · S. Massart · M. H. Jijakli
Integrated and Urban Plant Pathology Laboratory, Université de Liège, Gembloux Agro-Bio Tech, Gembloux, Belgium
e-mail: g.stouvenakers@uliege.be; sebastien.massart@uliege.be; mh.jijakli@uliege.be

P. Dapprich
Department of Agriculture, Fachhochschule Südwestfalen University of Applied Sciences, Soest, Germany
e-mail: dapprich.peter@fh-swf.de

S. Goddek et al. (eds.), *Aquaponics Food Production Systems*,
https://doi.org/10.1007/978-3-030-15943-6_14

Keywords Pathogens · Fungi · Aquaponics · Biotope · Suppressive · Plant pest management · Biocontrol agents

14.1 Introduction

Nowadays, aquaponic systems are the core of numerous research efforts which aim at better understanding these systems and at responding to new challenges of food production sustainability (Goddek et al. 2015; Villarroel et al. 2016). The cumulated number of publications mentioning "aquaponics" or derived terms in the title went from 12 in early 2008 to 215 in 2018 (January 2018 Scopus database research results). In spite of this increasing number of papers and the large area of study topics they are covering, one critical point is still missing, namely plant pest management (Stouvenakers et al. 2017). According to a survey on EU Aquaponic Hub members, only 40% of practitioners have some notions about pests and plant pest control (Villarroel et al. 2016).

In aquaponics, the diseases might be similar to those found in hydroponic systems under greenhouse structures. Among the most problematic pathogens, in term of spread, are hydrophilic fungi or fungus-like protists which are responsible for root or collar diseases. To consider plant pathogen control in aquaponics, firstly, it is important to differentiate between coupled and decoupled systems. Decoupled systems allow disconnection between water from the fish and crop compartment (see Chap. 8). This separation allows the optimisation and a better control of different parameters (e.g. temperature, mineral or organic composition and pH) in each compartment (Goddek et al. 2016; Monsees et al. 2017). Furthermore, if the water from the crop unit does not come back to the fish part, the application of phytosanitary treatments (e.g. pesticides, biopesticides and chemical disinfection agents) could be allowed here. Coupled systems are built in one loop where water recirculates in all parts of the system (see Chaps. 5 and 7). However, in coupled systems, plant pest control is more difficult due to the both presence of fish and beneficial microorganisms which transform fish sludge into plant nutrients. Their existence limits or excludes the application of already available disinfecting agents and chemical treatments. Furthermore no pesticides or biopesticides have been specifically developed for aquaponics (Rakocy et al. 2006; Rakocy 2012; Somerville et al. 2014; Bittsanszky et al. 2015; Nemethy et al. 2016; Sirakov et al. 2016). Control measures are consequently mainly based on non-curative physical practices (see Sect. 14.3.1) (Nemethy et al. 2016; Stouvenakers et al. 2017).

On the other hand, recent studies highlight that aquaponic plant production offers similar yields when compared to hydroponics although concentrations of mineral plant nutrients are lower in aquaponic water. Furthermore, when aquaponic water is complemented with some minerals to reach hydroponic concentrations of mineral nutritive elements, even better yields can be observed (Pantanella et al. 2010; Pantanella et al. 2015; Delaide et al. 2016; Saha et al. 2016; Anderson et al. 2017;

Wielgosz et al. 2017; Goddek and Vermeulen 2018). Moreover, some informal observations from practitioners in aquaponics and two recent scientific studies (Gravel et al. 2015; Sirakov et al. 2016) report the possible presence of beneficial compounds and/or microorganisms in the water that could play a role in biostimulation and/or have antagonistic (i.e. inhibitory) activity against plant pathogens. Biostimulation is defined as the enhancement of plant quality traits and plant tolerance against abiotic stress using any microorganism or substance.

With regard to these aspects, this chapter has two main objectives. The first is to give a review of microorganisms involved in aquaponic systems with a special focus on plant pathogenic and plant beneficial microorganisms. Factors influencing these microorganisms will be also considered (e.g. organic matter). The second is to review available methods and future possibilities in plant diseases control.

14.2 Microorganisms in Aquaponics

Microorganisms are present in the entire aquaponics system and play a key role in the system. They are consequently found in the fish, the filtration (mechanical and biological) and the crop parts. Commonly, the characterisation of microbiota (i.e. microorganisms of a particular environment) is carried out on circulating water, periphyton, plants (rhizosphere, phyllosphere and fruit surface), biofilter, fish feed, fish gut and fish faeces. Up until now, in aquaponics, most of microbial research has focused on nitrifying bacteria (Schmautz et al. 2017). Thus, the trend at present is to characterise microorganisms in all compartments of the system using modern sequencing technologies. Schmautz et al. (2017) identified the microbial composition in different parts of the system, whereas Munguia-Fragozo et al. (2015) give perspectives on how to characterize aquaponics microbiota from a taxonomical and functional point of view by using cutting-edge technologies. In the following sub-sections, focus will be only brought on microorganisms interacting with plants in aquaponic systems organised into plant beneficial and plant pathogenic microorganisms.

14.2.1 Plant Pathogens

Plant pathogens occurring in aquaponic systems are theoretically those commonly found in soilless systems. A specificity of aquaponic and hydroponic plant culture is the continuous presence of water in the system. This humid/aquatic environment suits almost every plant pathogenic fungus or bacteria. For root pathogens some are particularly well adapted to these conditions like pseudo-fungi belonging to the taxa of Oomycetes (e.g. root rot diseases caused by *Pythium* spp. and *Phytophthora* spp.) which are able to produce a motile form of dissemination called zoospores. These zoospores are able to move actively in liquid water and thus are able to spread over the entire system extremely quickly. Once a plant is infected, the disease can rapidly spread out the system, especially because of the water's recirculation (Jarvis 1992; Hong and Moorman 2005; Sutton et al. 2006; Postma et al. 2008; Vallance et al. 2010; Rakocy 2012; Rosberg 2014; Somerville et al. 2014). Though Oomycetes are among the most prevalent pathogens detected during root diseases, they often form a complex with other pathogens. Some *Fusarium* species (with existence of species well adapted to aquatic environment) or species from the genera *Colletotrichum*, *Rhizoctonia* and *Thielaviopsis* can be found as part of these complexes and can also cause significant damage on their own (Paulitz and Bélanger 2001; Hong and Moorman 2005; Postma et al. 2008; Vallance et al. 2010). Other fungal genera like *Verticillium* and *Didymella*, but also bacteria, such as *Ralstonia*, *Xanthomonas*, *Clavibacter*, *Erwinia* and *Pseudomonas*, as well as viruses (e.g. tomato mosaic, cucumber mosaic, melon necrotic spot virus, lettuce infectious virus and tobacco necrosis), can be detected in hydroponics or irrigation water and cause vessel, stem, leaf or fruit damage (Jarvis 1992; Hong and Moorman 2005). However note that not all microorganisms detected are damaging or lead to symptoms in the crop. Even species of the same genus can be either harmful or beneficial (e.g. *Fusarium*, *Phoma*, *Pseudomonas*). Disease agents discussed above are mainly pathogens linked to water recirculation but can be identified in greenhouses also. Section 14.2.2 shows the results of the first international survey on plant diseases occurring specifically in aquaponics, while Jarvis (1992) and Albajes et al. (2002) give a broader view of occurring pathogens in greenhouse structures.

In hydroponics or in aquaponic systems, plants generally grow under greenhouse conditions optimized for plant production, especially for large-scale production where all the environmental parameters are computer managed (Albajes et al. 2002; Vallance et al. 2010; Somerville et al. 2014; Parvatha Reddy 2016). However, optimal conditions for plant production can also be exploited by plant pathogens. In fact, these structures generate warm, humid, windless and rain-free conditions that can encourage plant diseases if they are not correctly managed (ibid.). To counteract this, compromises must be made between optimal plant conditions and disease prevention (ibid.). In the microclimate of the greenhouse, an inappropriate management of the vapour-pressure deficit can lead to the formation of a film or a drop of water on the plants surface. This often promotes plant pathogen development. Moreover, to maximise the yield in commercial hydroponics, some other parameters (e.g. high plant density, high fertilisation, to extend the period production) can enhance the susceptibility of plants to develop diseases (ibid.).

The question now is to know by which route the initial inoculum (i.e. the first step in an epidemiological cycle) is brought into the system. The different steps in plant disease epidemiological cycle (EpC) are represented in Fig. 14.1. In aquaponics, as in greenhouse hydroponic culture, it can be considered that entry of pathogens could be linked to water supply, introduction of infected plants or seeds, the growth material (e.g. reuse of the media), air exchange (dust and particles carriage), insects (vectors of diseases and particles carriage) and staff (tools and clothing) (Paulitz and Bélanger 2001; Albajes et al. 2002; Hong and Moorman 2005; Sutton et al. 2006; Parvatha Reddy 2016).

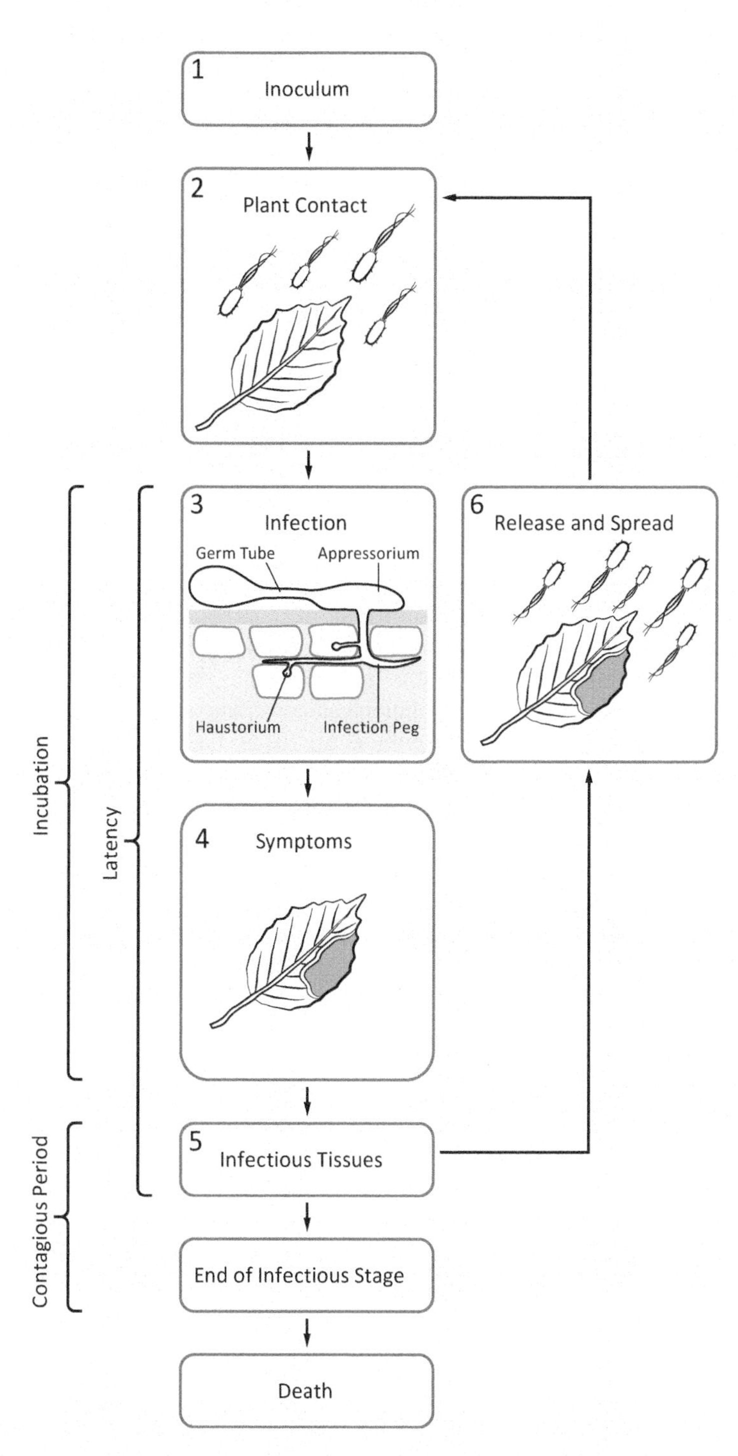

Fig. 14.1 Basic steps (*1* to *6*) in plant disease epidemiological cycle (EpC) according to Lepoivre (2003). (*1*) Arrival of the pathogen inoculum, (*2*) contact with the host plant, (*3*) tissues penetration and infection process by the pathogen, (*4*) symptoms development, (*5*) plant tissues that become infectious, (*6*) release and spread of infectious form of dispersion

Once the inoculum is in contact with the plant (step 2 in the EpC), several cases of infection (step 3 in the EpC) are possible (Lepoivre 2003):

- The pathogen-plant relationship is incompatible (non-host relation) and disease does not develop.
- There is a host relation but the plant does not show symptoms (the plant is tolerant).
- The pathogen and the plant are compatible but defence response is strong enough to inhibit the progression of the disease (the plant is resistant: interaction between host resistance gene and pathogen avirulence gene).
- The plant is sensitive (host relation without gene for gene recognition), and the pathogen infects the plant, but symptoms are not highly severe (step 4 in the EpC).
- And lastly, the plant is sensitive and disease symptoms are visible and severe (step 4 in the EpC).

Regardless of the degree of resistance, some environmental conditions or factors can influence the susceptibility of a plant to be infected, either by a weakening of the plant or by promoting the growth of the plant pathogen (Colhoun 1973; Jarvis 1992; Cherif et al. 1997; Alhussaen 2006; Somerville et al. 2014). The main environmental factors influencing plant pathogens and disease development are temperature, relative humidity (RH) and light (ibid.). In hydroponics, temperature and oxygen concentrations within the nutrient solution can constitute additional factors (Cherif et al. 1997; Alhussaen 2006; Somerville et al. 2014). Each pathogen has its own preference of environmental conditions which can vary during its epidemiologic cycle. But in a general way, high humidity and temperature are favourable to the accomplishment of key steps in the pathogen's epidemic cycle such as spore production or spore germination (Fig. 14.1, step 5 in the EpC) (Colhoun 1973; Jarvis 1992; Cherif et al. 1997; Alhussaen 2006; Somerville et al. 2014). Colhoun (1973) sums up the effects of the various factors promoting plant diseases in soil, whereas Table 14.1 shows the more specific or adding factors that may encourage plant pathogen development linked to aquaponic greenhouse conditions.

In the epidemiological cycle, once the infective stage is reached (step 5 in the EpC), the pathogens can spread in several ways (Fig. 14.1, step 6 in the EpC) and infect other plants. As explained before, root pathogens belonging to Oomycetes taxa can actively spread in the recirculating water by zoospores release (Alhussaen 2006; Sutton et al. 2006). For other fungi, bacteria and viruses responsible for root or aerial diseases, the dispersion of the causal agent can occur by propagation of infected material, mechanical wounds, infected tools, vectors (e.g. insects) and particles (e.g. spores and propagules) ejection or carriage allowed by drought, draughts or water splashes (Albajes et al. 2002; Lepoivre 2003).

14.2.2 Survey on Aquaponic Plant Diseases

During January 2018, the first international survey on plant diseases was made among aquaponics practitioner members of the COST FA1305, the American Aquaponics Association and the EU Aquaponics Hub. Twenty-eight answers were

Table 14.1 Adding factors encouraging plant pathogen development under aquaponic greenhouse structure compared to classical greenhouse culture

Promoting factor	Profiting to	Causes	References
Nutrient film technique (NFT), Deep flow technique (DFT)	*Pythium* spp., *Fusarium* spp.	Easy spread by water recirculation; possibility of post contamination after a disinfection step; poor content in oxygen in the nutrient solution	Koohakan et al. (2004) and Vallance et al. (2010)
Inorganic media (e.g. rockwool)	Higher content in bacteria (no information about their possible pathogenicity)	Unavailable organic compounds in the media	Khalil and Alsanius (2001), Koohakan et al. (2004), Vallance et al. (2010)
Organic media (e.g. coconut fibre and peat)	Higher content in fungi; higher content in *Fusarium* spp. for coconut fibre	Available organic compounds in the media	Koohakan et al. (2004), Khalil et al. (2009), and Vallance et al. (2010)
Media with high water content and low content in oxygen (e.g. rockwool)	*Pythium* spp.	Zoospores mobility; plant stress	Van Der Gaag and Wever (2005), Vallance et al. (2010), and Khalil and Alsanius (2011)
Media allowing little water movement (e.g. rockwool)	*Pythium* spp.	Better condition for zoospores dispersal and chemotaxis movement; no loss of zoospore flagella	Sutton et al. (2006)
High temperature and low concentration of DO in the nutrient solution	*Pythium* spp.	Plant stressed and optimal condition for *Pythium* growth	Cherif et al. (1997), Sutton et al. (2006), Vallance et al. (2010), and Rosberg (2014)
High host plant density and resulting microclimate	Pathogens growth; diseases spread	Warm and humid environment	Albajes et al. (2002) and Somerville et al. (2014)
Deficiencies, excess or imbalance of macro/micronutrients	Fungi, viruses and bacteria	Plant physiological modifications (e.g. action on defence response, transpiration, integrity of cell walls); plant morphological modifications (e.g. higher susceptibility to pathogens, attraction of pests); nutrient resources in host tissues for pathogens; direct action on the pathogen development cycle	Colhoun (1973), Snoeijers and Alejandro (2000), Mitchell et al. (2003), Dordas (2008), Veresoglou et al. (2013), Somerville et al. (2014), and Geary et al. (2015)

received describing 32 aquaponic systems from around the world (EU, 21; North America, 5; South America, 1; Africa, 4; Asia, 1). The first finding was the small response rate. Among the possible explanations for the reluctance to reply to the questionnaire was that practitioners did not feel able to communicate about plant pathogens because of a lack of knowledge on this topic. This had already been observed in the surveys of Love et al. (2015) and Villarroel et al. (2016).

Key information obtained from the survey are:

- 84.4% of practitioners observe disease in their system.
- 78.1% cannot identify the causal agent of a disease.
- 34.4% do not apply disease control measures.
- 34.4% use physical or chemical water treatment.
- 6.2% use pesticides or biopesticides in coupled aquaponic system against plant pathogens.

These results support the previous arguments saying that aquaponic plants do get diseases. Yet, practitioners suffer from a lack of knowledge about plant pathogens and disease control measures actually used are essentially based on non-curative actions (90.5% of cases).

In the survey, a listing of plant pathogens occurring in their aquaponic system was provided. Table 14.2 shows the results of this identification. To remedy the lack of practitioner's expertise about plant disease diagnostics, a second survey version was

Table 14.2 Results of the first identifications of plant pathogens in aquaponics from the 2018 international survey analysis and from existing literature

Plant host	Plant pathogen	References or survey results
Allium schoenoprasu	*Pythium* sp.[(b)]	Survey
Beta vulgaris (swiss chard)	*Erysiphe betae*[(a)]	Survey
Cucumis sativus	*Podosphaera xanthii*[(a)]	Survey
Fragaria spp.	*Botrytis cinerea*[(a)]	Survey
Lactuca sativa	*Botrytis cinerea*[(a)]	Survey
	Bremia lactucae[(a)]	Survey
	Fusarium sp.[(b)]	Survey
	Pythium dissotocum[(b)]	Rakocy (2012)
	Pythium myriotylum[(b)]	Rakocy (2012)
	Sclerotinia sp.[(a)]	Survey
Mentha spp.	*Pythium* sp.[(b)]	Survey
Nasturtium officinale	*Aspergillus sp.*[(a)]	Survey
Ocimum basilicum	*Alternaria* sp.[(a)]	Survey
	Botrytis cinerea[(a)]	Survey
	Pythium sp.[(b)]	Survey
	Sclerotinia sp.[(a)]	Survey
Pisum sativum	*Erysiphe pisi*[(a)]	Survey
Solanum lycopersicum	*Pseudomononas solanacearum*[(a)]	McMurty et al. (1990)
	Phytophthora infestans[(a)]	Survey

Plant pathogens identified by symptoms in the aerial plant part are annotated by (a) and in root part by (b) in exponent

sent with the aim to identify symptoms without disease name linkage (Table 14.3). Table 14.2 mainly identifies diseases with specific symptoms, i.e. symptoms that can be directly linked to a plant pathogen. It is the case of *Botrytis cinerea* and its typical grey mould, powdery mildew (*Erysiphe* and *Podosphaera* genera in the table) and its white powdery mycelium/conidia, and lastly *Sclerotinia* spp. and its sclerotia production. The presence of 3 plant pathologists in the survey respondents expands the list, with the identification of some root pathogens (e.g. *Pythium* spp.). General symptoms that are not specific enough to be directly related to a pathogen without further verification (see diagnosis in Sect. 14.3) are consequently found in Table 14.3. But it is important to highlight that most of the symptoms observed in this table could also be the consequence of abiotic stresses. Foliar chlorosis is one of the most explicit examples because it can be related to a large number of pathogens (e.g. for lettuces: *Pythium* spp., *Bremia lactucae*, *Sclerotinia* spp., beet western yellows virus), to environmental conditions (e.g. temperature excess) and to mineral deficiencies (nitrogen, magnesium, potassium, calcium, sulfur, iron, copper, boron, zinc, molybdenum) (Lepoivre 2003; Resh 2013).

Table 14.3 Review of occurring symptoms in aquaponics from the 2018 international survey analysis

Symptoms	Plants species
Foliar chlorosis	*Allium schoenoprasum* [1], *Amaranthus viridis* [1], *Coriandrum sativum* [1], *Cucumis sativus* [1], Ocimum *basilicum* [6], *Lactuca sativa* [4], *Mentha* spp. [2], *Petroselinum crispum* [1], *Spinacia oleracea* [2], *Solanum lycopersicum* [1], *Fragaria* spp. [1]
Foliar necrosis	*Mentha* spp. [2], *Ocimum basilicum* [1],
Stem necrosis	*Solanum lycopersicum* [1],
Collar necrosis	*Ocimum basilicum* [1]
Foliar Mosaic	*Cucumis sativus* [1], *Mentha* spp. [1], *Ocimum basilicum* [1],
Foliar wilting	*Brassica oleracea* Acephala group [1], *Lactuca sativa* [1], *Mentha* spp. [1], *Cucumis sativus* [1], *Ocimum basilicum* [1], *Solanum lycopersicum* [1]
Foliar, stem and collar mould	*Allium schoenoprasum* [1], *Capsicum annuum* [1], *Cucumis sativus* [1], *Lactuca sativa* [2], *Mentha* spp. [2], *Ocimum basilicum* [4], *Solanum lycopersicum* [1]
Foliar spots	*Capsicum annuum* [1], *Cucumis sativus* [1], *Lactuca sativa* [2], *Mentha* spp. [1], *Ocimum basilicum* [5]
Damping off	*Spinacia oleracea* [1], *Ocimum basilicum* [1], *Solanum lycopersicum* [1], seedlings in general [5]
Crinkle	*Beta vulgaris* (swiss chard) [1], *Capsicum annuum* [1], *Lactuca sativa* [1], *Ocimum basilicum* [1]
Browning or decaying root	*Allium schoenoprasum* [1], *Amaranthus viridis* [1], *Beta vulgaris* (swiss chard) [1], *Coriandrum sativum* [1], *Lactuca sativa* [1], *Mentha* spp. [2], *Ocimum basilicum* [2], *Petroselinum crispum* [2], *Solanum lycopersicum* [1], *Spinacia oleracea* [1]

Numbers in exponent represent the occurrence of the symptom for a specific plant on a total of 32 aquaponic systems reviewed

14.2.3 Beneficial Microorganisms in Aquaponics: The Possibilities

As explained in the introduction, several publications focused on bacteria involved in the nitrogen cycle, while others already emphasise the potential presence of beneficial microorganisms interacting with plant pathogens and/or plants (Rakocy 2012; Gravel et al. 2015; Sirakov et al. 2016). This section reviews the potential of plant beneficial microorganisms involved in aquaponics and their modes of action.

Sirakov et al. (2016) screened antagonistic bacteria against *Pythium ultimum* isolated from an aquaponic system. Among the 964 tested isolates, 86 showed a strong inhibitory effect on *Pythium ultimum* in vitro. Further research must be achieved to taxonomically identify these bacteria and evaluate their potential in in vivo conditions. The authors assume that many of these isolates belong to the genus *Pseudomonas*. Schmautz et al. (2017) came to the same conclusion by identifying *Pseudomonas* spp. in the rhizosphere of lettuce. Antagonistic species of the genus *Pseudomonas* were able to control plant pathogens in natural environments (e.g. in suppressive soils) while this action is also affected by environmental conditions. They can protect plants against pathogens either in an active or a passive way by eliciting a plant defence response, playing a role in plant growth promotion, compete with pathogens for space and nutrients (e.g. iron competition by release of iron-chelating siderophores), and/or finally by production of antibiotics or antifungal metabolites such as biosurfactants (Arras and Arru 1997; Ganeshan and Kumar 2005; Haas and Défago 2005; Beneduzi et al. 2012; Narayanasamy 2013). Although no identification of microorganisms was done by Gravel et al. (2015)), they report that fish effluents have the capacity to stimulate plant growth, decrease the mycelial growth of *Pythium ultimum* and *Pythium oxysporum* in vitro and reduce the colonization of tomato root by these fungi.

Information about the possible natural plant protection capacity of aquaponic microbiota is scarce, but the potential of this protective action can be envisaged with regard to different elements already known in hydroponics or in recirculated aquaculture. A first study was conducted in 1995 on suppressive action or suppressiveness promoted by microorganisms in soilless culture (McPherson et al. 1995). Suppressiveness in hydroponics, here defined by Postma et al. (2008)), has been "referred to the cases where (i) the pathogen does not establish or persist; or (ii) establishes but causes little or no damage". The suppressive action of a milieu can be related to the abiotic environment (e.g. pH and organic matter). However, in most situations, it is considered to be related directly or indirectly to microorganisms' activity or their metabolites (James and Becker 2007). In soilless culture, the suppressive capacity shown by water solution or the soilless media is reviewed by Postma et al. (2008) and Vallance et al. (2010). In these reviews, microorganisms responsible for this suppressive action are not clearly identified. In contrast, plant pathogens like *Phytophthora cryptogea*, *Pythium* spp., *Pythium aphanidermatum* and *Fusarium oxysporum* f.sp. *radicis-lycopersici* controlled or suppressed by the natural microbiota are exhaustively described. In the various articles reviewed by

Postma et al. (2008) and Vallance et al. (2010), microbial involvement in the suppressive effect is generally verified via a destruction of the microbiota of the soilless substrate by sterilisation first and eventually followed by a re-inoculation. When compared with an open system without recirculation, suppressive activity in soilless systems could be explained by the water recirculation (McPherson et al. 1995; Tu et al. 1999, cited by Postma et al. 2008) which could allow a better development and spread of beneficial microorganisms (Vallance et al. 2010).

Since 2010, suppressiveness of hydroponic systems has been generally accepted and research topics have been more driven on isolation and characterization of antagonistic strains in soilless culture with *Pseudomonas* species as main organisms studied. If it was demonstrated that soilless culture systems can offer suppressive capacity, there is no similar demonstration of such activity in aquaponics systems. However, there is no empiric indication that it should not be the case. This optimism arises from the discoveries of Gravel et al. (2015) and Sirakov et al. (2016) described in the second paragraph of this section. Moreover, it has been shown in hydroponics (Haarhoff and Cleasby 1991 cited by Calvo-bado et al. 2003; Van Os et al. 1999) but also in water treatment for human consumption (reviewed by Verma et al. 2017) that slow filtration (described in Sect. 14.3.1) and more precisely slow sand filtration can also act against plant pathogens by a microbial suppressive action in addition to other physical factors. In hydroponics, slow filtration has been demonstrated to be effective against the plant pathogens reviewed in Table 14.4. It is assumed that the microbial suppressive activity in the filters is most probably due to species of *Bacillus* and/or *Pseudomonas* (Brand 2001; Déniel et al. 2004; Renault et al. 2007; Renault et al. 2012). The results of Déniel et al. (2004) suggest that in hydroponics, the mode of action of *Pseudomonas* and *Bacillus* relies on competition for nutrients and antibiosis, respectively. However, additional modes of action could be present for these two genera as already explained for *Pseudomonas* spp. *Bacillus* species can, depending on the environment, act either indirectly by plant biostimulation or elicitation of plant defences or directly by antagonism via production of antifungal and/or antibacterial substances. Cell wall-degrading enzymes, bacteriocins, and antibiotics, lipopeptides (i.e. biosurfactants), are identified as key molecules for the latter action (Pérez-García et al. 2011; Beneduzi et al. 2012;

Table 14.4 Review of plant pathogens effectively removed by slow filtration in hydroponics

Plant pathogens	References
Xanthomonas campestris pv. *Pelargonii*	Brand (2001)
Fusarium oxysporum	Wohanka (1995), Ehret et al. (1999) cited by Ehret et al. (2001), van Os et al. (2001), Déniel et al. (2004), and Furtner et al. (2007)
Pythium spp.	Déniel et al. (2004)
Pythium aphanidermatum	Ehret et al. (1999) cited by Ehret et al. (2001), and Furtner et al. (2007)
Phytophthora cinnamomi	Van Os et al. (1999), 4 references cited by Ehret et al. (2001)
Phytophthora cryptogea	Calvo-bado et al. (2003)
Phytophthora cactorum	Evenhuis et al. (2014)

Narayanasamy 2013). All things considered, the functioning of a slow filter is not so different from the functioning of some biofilters used in aquaponics. Furthermore, some heterotrophic bacteria like *Pseudomonas* spp. were already identified in aquaponics biofilters (Schmautz et al. 2017). This is in accordance with the results of other researchers who frequently detected *Bacillus* and/or *Pseudomonas* in RAS (recirculated aquaculture system) biofilters (Tal et al. 2003; Sugita et al. 2005; Schreier et al. 2010; Munguia-Fragozo et al. 2015; Rurangwa and Verdegem 2015). Nevertheless, up until now, no study about the possible suppressiveness in aquaponic biofilters has been carried out.

14.3 Protecting Plants from Pathogens in Aquaponics

At the moment aquaponic practitioners operating a coupled system are relatively helpless against plant diseases when they occur, especially in the case of root pathogens. No pesticide nor biopesticide is specifically developed for aquaponic use (Rakocy 2007; Rakocy 2012; Somerville et al. 2014; Bittsanszky et al. 2015; Sirakov et al. 2016). In brief, curative methods are still lacking. Only Somerville et al. (2014) list the inorganic compounds that may be used against fungi in aquaponics. In any case, an appropriate diagnostic of the pathogen(s) causing the disease is mandatory in order to identify the target(s) for curative measures. This diagnosis requires good expertise in terms of observation capacity, plant pathogen cycle understanding and analysis of the situation. However, in case of generalist (not specific) symptoms and depending on the degree of accuracy needed, it is often necessary to use laboratory techniques to validate the hypothesis with respect to the causal agent (Lepoivre 2003). Postma et al. (2008) reviewed the different methods to detect plant pathogens in hydroponics, and four groups were identified:

1. Direct macroscopic and microscopic observation of the pathogen
2. Isolation of the pathogen
3. Use of serological methods
4. Use of molecular methods

14.3.1 Non-biological Methods of Protection

Good agricultural practices (GAP) for plant pathogens control are the various actions aiming to limit crop diseases for both yield and quality of produce (FAO 2008). GAP transposable to aquaponics are essentially non-curative physical or cultivation practices that can be divided in preventive measures and water treatment.

Preventive Measures

Preventive measures have two distinct purposes. The first is to avoid the entry of the pathogen inoculum into the system and the second is to limit (i) plant infection,

(ii) development and (iii) spread of the pathogen during the growing period. Preventive measures aiming to avoid the entry of the initial inoculum in the greenhouse are, for example, a fallow period, a specific room for sanitation, room sanitation (e.g. plant debris removal and surface disinfection), specific clothes, certified seeds, a specific room for plant germination and physical barriers (against insect vectors) (Stanghellini and Rasmussen 1994; Jarvis 1992; Albajes et al. 2002; Somerville et al. 2014; Parvatha Reddy 2016). Among the most important practices used for the second type of preventive measures are, the use of resistant plant varieties, tools disinfection, avoidance of plant abiotic stresses, good plant spacing, avoidance of algae development and environmental conditions management. The last measure, i.e. environmental conditions management, means to control all greenhouse parameters in order to avoid or limit diseases by intervening in their biological cycle (ibid.). Generally, in large-scale greenhouse structures, computer software and algorithms are used to calculate the optimal parameters allowing both plant production and disease control. The parameters measured, among others, are temperature (of the air and the nutrient solution), humidity, vapour pressure deficit, wind speed, dew probability, leaf wetness and ventilation (ibid.). The practitioner acts on these parameters by manipulating the heating, the ventilation, the shading, the supplement of lights, the cooling and the fogging (ibid.).

Water Treatment

Physical water treatments can be employed to control potential water pathogens. Filtration (pore size less than 10 μm), heat and UV treatments are among the most effective to eliminate pathogens without harmful effects on fish and plant health (Ehret et al. 2001; Hong and Moorman 2005; Postma et al. 2008; Van Os 2009; Timmons and Ebeling 2010). These techniques allow the control of disease outbreaks by decreasing the inoculum, the quantity of pathogens and their proliferation stages in the irrigation system (ibid.). Physical disinfection decreases water pathogens to a certain level depending on the aggressiveness of the treatment. Generally, the target of heat and UV disinfection is the reduction of the initial microorganisms population by 90–99.9% (ibid.). The filtration technique most used is slow filtration because of its reliability and its low cost. The substrates of filtration generally used are sand, rockwool or pozzolana (ibid.). Filtration efficiency is essentially dependent on pore size and flow. To be effective as disinfection treatment, the filtration needs to be achieved with a pore size less than 10 μm and a flow rate of 100 $l/m^2/h$, even if less binding parameters show satisfactory performances (ibid.). Slow filtration does not eliminate all of the pathogens; more than 90% of the total aerobic bacteria remain in the effluent (ibid.). Nevertheless, it allows a suppression of plant debris, algae, small particles and some soil-borne diseases such as *Pythium* and *Phytophthora* (the efficiency is genus dependent). Slow filters do not act only by physical action but also show a microbial suppressive activity, thanks to antagonistic microorganisms, as discussed in Sect. 14.2.3 (Hong and Moorman 2005; Postma et al. 2008; Van Os 2009; Vallance et al. 2010). Heat treatment is very effective against plant pathogens. However it requires temperatures reaching 95 ° C during at least 10 seconds to suppress all kind of pathogens, viruses included. This practice consumes a lot of energy and imposes water cooling (heat exchanger and transitional tank) before reinjection of the treated water back into the irrigation loop. In addition, it has the disadvantage of killing all microorganisms including the beneficial ones (Hong and Moorman 2005; Postma et al. 2008; Van Os 2009). The last technique and probably the most applied is UV disinfection. 20.8% of EU Aquaponics Hub practitioners use it (Villarroel et al. 2016). UV radiation has a wavelength of 200 to 280 nm. It has a detrimental effect on microorganisms by direct damage of the DNA. Depending on the pathogen and the water turbulence, the energy dose varies between 100 and 250 mJ/cm^2 to be effective (Postma et al. 2008; Van Os 2009).

Physical water treatments eliminate the most of the pathogens from the incoming water but they cannot eradicate the disease when it is already present in the system. Physical water treatment does not cover all the water (especially the standing water zone near the roots), nor the infected plant tissue. For example, UV treatments often fail to suppress *Pythium* root rot (Sutton et al. 2006). However, if physical water treatment allows a reduction of plant pathogens, theoretically, they also have an effect on nonpathogenic microorganisms potentially acting on disease suppression. In reality, heat and UV treatments create a microbiological vacuum, whereas slow filtration produces a shift in effluent microbiota composition resulting in a higher disease suppression capacity (Postma et al. 2008; Vallance et al. 2010). Despite the fact that UV and heat treatment in hydroponics eliminate more than 90% of microorganisms in the recirculating water, no diminution of the disease suppressiveness was observed. This was probably due to a too low quantity of water treated and a re-contamination of the water after contact with the irrigation system, roots and plant media (ibid.).

Aquaponic water treatment by means of chemicals is limited in continuous application. Ozonation is a technique used in recirculated aquaculture and in hydroponics. Ozone treatment has the advantage to eliminate all pathogens including viruses in certain conditions and to be rapidly decomposed to oxygen (Hong and Moorman 2005; Van Os 2009; Timmons and Ebeling 2010; Gonçalves and Gagnon 2011). However it has several disadvantages. Introducing ozone in raw water can produce by-products oxidants and significant amount of residual oxidants (e.g. brominated compound and haloxy anions that are toxic for fish) that need to be removed, by UV radiation, for example, prior to return to the fish part (reviewed by Gonçalves and Gagnon 2011). Furthermore, ozone treatment is expensive, is irritant for mucous membranes in case of human exposure, needs contact periods of 1 to 30 minutes at a concentration range of 0.1–2.0 mg/L, needs a temporal sump to reduce completely from O_3 to O_2 and can oxidize elements present in the nutrient solution, such as iron chelates, and thus makes them unavailable for plants (Hong and Moorman 2005; Van Os 2009; Timmons and Ebeling 2010; Gonçalves and Gagnon 2011).

14.3.2 Biological Methods of Protection

In hydroponics, numerous scientific papers review the use of antagonistic microorganisms (i.e. able to inhibit other organisms) to control plant pathogens but until now no research has been carried out for their use in aquaponics. The mode of action of these antagonistic microorganisms is according to Campbell (1989)), Whipps (2001) and Narayanasamy (2013) grouped in:

1. Competition for nutrients and niches
2. Parasitism
3. Antibiosis
4. Induction of diseases resistance in plants

The experiments introducing microorganisms in aquaponic systems have been focused on the increase of nitrification by addition of nitrifying bacteria (Zou et al. 2016) or the use of plant growth promoters (PGPR) such as *Azospirillum brasilense* and *Bacillus* spp. to increase plant performance (Mangmang et al. 2014; Mangmang et al. 2015a; Mangmang et al. 2015b; Mangmang et al. 2015c; da Silva Cerozi and Fitzsimmons 2016; Bartelme et al. 2018). There is now an urgent need to work on biocontrol agents (BCA) against plant pathogens in aquaponics with regard to the restricted use of synthetic curative treatments, the high value of the culture and the increase of aquaponic systems in the world. BCA are defined, in this context, as viruses, bacteria and fungi exerting antagonistic effects on plant pathogens (Campbell 1989; Narayanasamy 2013).

Generally, the introduction of a BCA is considered to be easier in soilless systems. In fact, the hydroponic root environment is more accessible than in soil and the microbiota of the substrate is also unbalanced due to a biological vacuum. Furthermore, environmental conditions of the greenhouse can be manipulated to achieve BCA growth needs. Theoretically all these characteristics allow a better introduction, establishment and interaction of the BCA with plants in hydroponics than in soil (Paulitz and Bélanger 2001; Postma et al. 2009; Vallance et al. 2010). However, in practice, the effectiveness of BCA inoculation to control root pathogens can be highly variable in soilless systems (Postma et al. 2008; Vallance et al. 2010; Montagne et al. 2017). One explanation for this is that BCA selection is based on in vitro tests which are not representing real conditions and subsequently a weak adaptation of these microorganisms to the aquatic environment used in hydroponics or aquaponics (Postma et al. 2008; Vallance et al. 2010). To control plant pathogens and more especially those responsible for root rots, a selection and identification of microorganisms involved in aquatic systems which show suppressive activity against plant pathogens is needed. In soilless culture, several antagonistic microorganisms can be picked due to their biological cycle being similar to root pathogens or their ability to grow in aqueous conditions. Such is the case of nonpathogenic

Pythium and *Fusarium* species and bacteria, where *Pseudomonas*, *Bacillus* and *Lysobacter* are the genera most represented in the literature (Paulitz and Bélanger 2001; Khan et al. 2003; Chatterton et al. 2004; Folman et al. 2004; Sutton et al. 2006; Liu et al. 2007; Postma et al. 2008; Postma et al. 2009; Vallance et al. 2010; Sopher and Sutton 2011; Hultberg et al. 2011; Lee and Lee 2015; Martin and Loper 1999; Moruzzi et al. 2017; Thongkamngam and Jaenaksorn 2017). The direct addition of some microbial metabolites such as biosurfactants has also been studied (Stanghellini and Miller 1997; Nielsen et al. 2006; Nielsen et al. 2006). Although some microorganisms are efficient at controlling root pathogens, there are other problems that need to be overcome in order to produce a biopesticide. The main challenges are to determine the means of inoculation, the inoculum density, the product formulation (Montagne et al. 2017), the method for the production of sufficient quantity at low cost and the storage of the formulated product. Ecotoxicological studies on fish and living beneficial microorganisms in the system are also an important point. Another possibility that could be exploited is the use of a complex of antagonistic agents, as observed in suppressive soil techniques (Spadaro and Gullino 2005; Vallance et al. 2010). In fact, microorganisms can work in synergy or with complementary modes of action (ibid.). The addition of amendments could also enhance the BCA potential by acting as prebiotics (see Sect. 14.4).

14.4 The Role of Organic Matter in Biocontrol Activity in Aquaponic Systems

In Sect. 14.2.3, the suppressiveness of aquaponic systems was suggested. As stated before, the main hypothesis is related to the water recirculation as it is for hydroponic systems. However, a second hypothesis exists and this is linked to the presence of organic matter in the system. Organic matter that could drive a more balanced microbial ecosystem including antagonistic agents which is less suitable for plant pathogens (Rakocy 2012).

In aquaponics, organic matter comes from water supply, uneaten feeds, fish faeces, organic plant substrate, microbial activity, root exudates and plant residues (Waechter-Kristensen et al. 1997; Naylor et al. 1999; Waechter-Kristensen et al. 1999). In such a system, heterotrophic bacteria are organisms able to use organic matter as a carbon and energy source, generally in the form of carbohydrates, amino acids, peptides or lipids (Sharrer et al. 2005; Willey et al. 2008; Whipps 2001). In recirculated aquaculture (RAS), they are mainly localised in the biofilter and consume organic particles trapped in it (Leonard et al. 2000; Leonard et al. 2002). However, another source of organic carbon for heterotrophic bacteria is humic substances present as dissolved organic matter and responsible for the yellow-brownish coloration of the water (Takeda and Kiyono 1990 cited by Leonard et al. 2002; Hirayama et al. 1988). In the soil as well as in hydroponics, humic acids are known to stimulate plant growth and sustain the plant under abiotic stress conditions (Bohme 1999; du Jardin 2015). Proteins in the water can be used by plants as an alternative nitrogen source thus enhancing their growth and pathogen resistance (Adamczyk et al. 2010). In the recirculated water, the abundance of free-living heterotrophic bacteria is correlated with the amount of biologically available organic carbon and carbon-nitrogen ratio (C/N) (Leonard et al. 2000; Leonard et al. 2002; Michaud et al. 2006; Attramadal et al. 2012). In the biofilter, an increase in the C/N ratio increases the abundance of heterotrophic bacteria at the expense of the number of autotrophic bacteria responsible for the nitrification process (Michaud et al. 2006; Michaud et al. 2014). As implied, heterotrophic microorganisms can have a negative impact on the system because they compete with autotrophic bacteria (e.g. nitrifying bacteria) for space and oxygen. Some of them are plant or fish pathogens, or responsible for off-flavour in fish (Chang-Ho 1970; Funck-Jensen and Hockenhull 1983; Jones et al. 1991; Leonard et al. 2002; Nogueira et al. 2002; Michaud et al. 2006; Mukerji 2006; Whipps 2001; Rurangwa and Verdegem 2015). However, heterotrophic microorganisms involved in the system can also be positive (Whipps 2001; Mukerji 2006). Several studies using organic fertilizers or organic soilless media, in hydroponics, have shown interesting effects where the resident microbiota were able to control plant diseases (Montagne et al. 2015). All organic substrates have their own physico-chemical properties. Consequently, the characteristics of the media will influence microbial richness and functions. The choice of a specific plant media could therefore influence the microbial development so as to have a suppressive effect on pathogens (Montagne et al. 2015; Grunert et al. 2016; Montagne et al. 2017). Another possibility of pathogen suppression related to organic carbon is the use of organic amendments in hydroponics (Maher et al. 2008; Vallance et al. 2010). By adding composts in soilless media like it is common use in soil, suppressive effects are expected (Maher et al. 2008). Enhancing or maintaining a specific microorganism such as *Pseudomonas* population by adding some formulated carbon sources (e.g. nitrapyrin-based product) as reported by Pagliaccia et al. (2007) and Pagliaccia et al. (2008) is another possibility. The emergence of organic soilless culture also highlights the involvement of beneficial microorganisms against plant pathogens supported by the use of organic fertilizers. Fujiwara et al. (2013), Chinta et al. (2014), and Chinta et al. (2015) reported that fertilization with corn steep liquor helps to control *Fusarium oxysporum* f.sp. *lactucae* and *Botrytis cinerea* on lettuces and *Fusarium oxysporum* f.sp. *radicis-lycopersici* on tomato plants. And even if hardly advised for aquaponic use, 1 g/L of fish-based soluble fertilizer (Shinohara et al. 2011) suppresses bacterial wilt on tomato caused by *Ralstonia solanacearum* in hydroponics (Fujiwara et al. 2012).

Finally, though information about the impact of organic matter on plant protection in aquaponics is scarce, the various elements mentioned above show their potential capacity to promote a system-specific and plant pathogen-suppressive microbiota.

14.5 Conclusions and Future Considerations

This chapter aimed to give a first report of plant pathogens occurring in aquaponics, reviewing actual methods and future possibilities to control them. Each strategy has advantages and disadvantages and must be thoroughly designed to fit each case. However, at this time, curative methods in coupled aquaponic systems are still limited and new perspectives of control must be found. Fortunately, suppressiveness in terms of aquaponic systems could be considered, as already observed in hydroponics (e.g. in plant media, water, and slow filters). In addition, the presence of organic matter in the system is an encouraging factor when compared to soilless culture systems making use of organic fertilisers, organic plant media or organic amendments.

For the future, it seems important to investigate this suppressive action followed by identification and characterization of the responsible microbes or microbe consortia. Based on the results, several strategies could be envisaged to enhance the capacity of plants to resist pathogens. The *first* is biological control by conservation, which means favouring beneficial microorganisms by manipulating and managing water composition (e.g. C/N ratio, nutrients and gases) and parameters (e.g. pH and temperature). But identification of these influencing factors needs to be realized first. This management of autotrophic and heterotrophic bacteria is also of key importance to sustain good nitrification and keep healthy fish. The *second* strategy is augmentative biological control by additional release of beneficial microorganisms already present in the system in large numbers (inundative method) or in small numbers but repeated in time (inoculation method). But prior identification and multiplication of an aquaponic BCA should be achieved. The *third* strategy is importation, i.e. introducing a new microorganism normally not present into the system. In this case, selection of a microorganism adapted and safe for aquaponic environment is needed. For the two last strategies, the site of inoculation in the system must be considered depending on the aim desired. Sites where microbial activity could be enhanced are the recirculated water, the rhizosphere (plant media included), the biofilter (such as in slow sand filters where BCA addition is already tested) and the phyllosphere (i.e. aerial plant part). Whatever the strategy, the ultimate goal should be to lead the microbial communities to provide a stable, ecologically balanced microbial environment allowing good production of both plant and fish.

To conclude, following the requirements of integrated plant pest management (IPM) is a necessity to correctly manage the system and avoid development and spread of plant diseases (Bittsanszky et al. 2015; Nemethy et al. 2016). The principle of IPM is to apply chemical pesticides or other agents as a last resort when economic injury level is reached. Consequently, control of pathogens will need to be firstly based on physical and biological methods (described above), their combination and an efficient detection and monitoring of the disease (European Parliament 2009).

References

Adamczyk B, Smolander A, Kitunen V, Godlewski M (2010) Proteins as nitrogen source for plants: a short story about exudation of proteases by plant roots. Plant Signal Behav 5:817–819. https://doi.org/10.4161/psb.5.7.11699

Albajes R, Lodovica Gullino M, Van Lenteren JC, Elad Y (2002) Integrated pest and disease management in greenhouse crops. Kluwer Academic Publishers. https://doi.org/10.1017/CBO9781107415324.004

Alhussaen K (2006) Pythium and phytophthora associated with root disease of hydroponic lettuce. University of Technologie Sydney Faculty of Science. https://opus.lib.uts.edu.au/handle/10453/36864

Anderson TS, de Villiers D, Timmons MB (2017) Growth and tissue elemental composition response of Spinach (Spinacia oleracea) to hydroponic and aquaponic water quality conditions. Horticulturae 3:32. https://doi.org/10.3390/horticulturae3020032

Arras G, Arru S (1997) Mechanism of action of some microbial antagonists against fungal pathogens. Ann Microbiol Enzimol 47:97–120

Attramadal KJK, Salvesen I, Xue R, Øie G, Størseth TR, Vadstein O, Olsen Y (2012) Recirculation as a possible microbial control strategy in the production of marine larvae. Aquac Eng 46:27–39. https://doi.org/10.1016/j.aquaeng.2011.10.003

Bartelme RP, Oyserman BO, Blom JE, Sepulveda-Villet OJ, Newton RJ (2018) Stripping away the soil: plant growth promoting microbiology opportunities in aquaponics. Front Microbiol 9:8. https://doi.org/10.3389/fmicb.2018.00008

Beneduzi A, Ambrosini A, Passaglia LMP (2012) Plant growth-promoting rhizobacteria (PGPR): their potential as antagonists and biocontrol agents. Genet Mol Biol 35(4):1044–1051

Bittsanszky A, Gyulai G, Junge R, Schmautz Z, Komives T, Has CAR, Otto H (2015) Plant protection in ecocycle-based agricultural systems : aquaponics as an example. In: International Plant Protection Congress (IPPC), Berlin, Germany, pp 2–3. https://doi.org/10.13140/RG.2.1.4458.0321

Bohme M (1999) Effects of Lactate, Humate and Bacillus subtilis on the growth of tomato plants in hydroponic systems. In: International symposium on growing media and hydroponics. Acta Horticulturae, pp 231–239

Brand T (2001) Importance and characterization of the biological component in slow filters. Acta Hortic 554:313–321

Calvo-bado LA, Pettitt TR, Parsons N, Petch GM, Morgan JAW, Whipps JM (2003) Spatial and Temporal Analysis of the Microbial Community in Slow Sand Filters Used for Treating Horticultural Irrigation Water. Appl Enviromental Microbiol 69:2116–2125. https://doi.org/10.1128/AEM.69.4.2116

Campbell R (1989) Biological control of microbial plant pathogens. Cambridge University Press, Cambridge

Chang-Ho Y (1970) The effect of pea root exudate on the germination of Pythium aphanidermatum zoospore cysts. Can J Bot 48:1501–1514

Chatterton S, Sutton JC, Boland GJ (2004) Timing Pseudomonas chlororaphis applications to control Pythium aphanidermatum, Pythium dissotocum, and root rot in hydroponic peppers. Biol Control 30:360–373. https://doi.org/10.1016/j.biocontrol.2003.11.001

Cherif M, Tirilly Y, Belanger RR (1997) Effect of oxygen concentration on plant growth, lipid peroxidation, and receptivity of tomato roots to Pythium under hydroponic conditions. Eur J Plant Pathol 103:255–264

Chinta YD, Kano K, Widiastuti A, Fukahori M, Kawasaki S, Eguchi Y, Misu H, Odani H, Zhou S, Narisawa K, Fujiwara K, Shinohara M, Sato T (2014) Effect of corn steep liquor on lettuce root rot (Fusarium oxysporum f.sp. lactucae) in hydroponic cultures. J Sci Food Agric 94:2317–2323. https://doi.org/10.1002/jsfa.6561

Chinta YD, Eguchi Y, Widiastuti A, Shinohara M, Sato T (2015) Organic hydroponics induces systemic resistance against the air-borne pathogen, *Botrytis cinerea* (gray mould). J Plant Interact 10:243–251. https://doi.org/10.1080/17429145.2015.1068959

Colhoun J (1973) Effects of environmental factors on plant disease. Annu Rev Phytopathol 11:343–364

da Silva Cerozi B, Fitzsimmons K (2016) Use of Bacillus spp. to enhance phosphorus availability and serve as a plant growth promoter in aquaponics systems. Sci Hortic (Amsterdam) 211:277–282. https://doi.org/10.1016/j.scienta.2016.09.005

Delaide B, Goddek S, Gott J, Soyeurt H, Jijakli HM (2016) Lettuce (Lactuca sativa L. var. Sucrine) growth performance in complemented aquaponic solution outperforms hydroponics. Water 8:1–11. https://doi.org/10.3390/w8100467

Déniel F, Rey P, Chérif M, Guillou A, Tirilly Y (2004) Indigenous bacteria with antagonistic and plant-growth-promoting activities improve slow-filtration efficiency in soilless cultivation. Can J Microbiol 50:499–508. https://doi.org/10.1139/w04-034

Dordas C (2008) Role of nutrients in controlling plant diseases in sustainable agriculture: a review. Agron Sustain Dev 28:33–46. https://doi.org/10.1051/agro:2007051

du Jardin P (2015) Plant biostimulants: definition, concept, main categories and regulation. Sci Hortic (Amsterdam) 196:3–14. https://doi.org/10.1016/j.scienta.2015.09.021

Ehret DL, Bogdanoff C, Utkhede R, Lévesque A, Menzies JG, Ng K, Portree J (1999) Disease control with slow filtration for greenhouse crops grown in recirculation. Pacific Agri-Food Res Cent Cent Tech Rep 155:37

Ehret DL, Alsanius B, Wohanka W, Menzies JG, Utkhede R (2001) Disinfestation of recirculating nutrient solutions in greenhouse horticulture. Agronomie 21:323–339. https://doi.org/10.1051/agro:2001127

European Parliament (2009) Directive 2009/128/EC of the European Parliament and the Council of 21 October 2009 establishing a framework for Community action to achieve the sustainable use of pesticides. October 309, pp 71–86. https://doi.org/10.3000/17252555.L_2009.309

Evenhuis B, Nijhuis E, Lamers J, Verhoeven J, Postma J (2014) Alternative methods to control phytophthora cactorum in strawberry cultivated in soilless growing media. Acta Hortic 1044:337–342. https://doi.org/10.17660/ActaHortic.2014.1044.44

FAO (2008) Good agricultural practices [WWW Document]. http://www.fao.org/prods/gap/. Accessed 27 Feb 2018

Folman LB, De Klein MJEM, Postma J, Van Veen JA (2004) Production of antifungal compounds by Lysobacter enzymogenes isolate 3.1T8 under different conditions in relation to its efficacy as a biocontrol agent of Pythium aphanidermatum in cucumber. Biol. Control 31:145–154. https://doi.org/10.1016/j.biocontrol.2004.03.008

Fujiwara K, Aoyama C, Takano M, Shinohara M (2012) Suppression of Ralstonia solanacearum bacterial wilt disease by an organic hydroponic system. J Gen Plant Pathol. https://doi.org/10.1007/s10327-012-0371-0

Fujiwara K, Iida Y, Iwai T, Aoyama C, Inukai R, Ando A, Ogawa J, Ohnishi J, Terami F, Takano M, Shinohara M (2013) The rhizosphere microbial community in a multiple parallel mineralization system suppresses the pathogenic fungus Fusarium oxysporum. Microbiologyopen 2:997–1009. https://doi.org/10.1002/mbo3.140

Funck-Jensen D, Hockenhull J (1983) Is damping-off, caused by Pythium, less of a problem in hydroponics than in traditional growing systems? Acta Hortic 133:137–145

Furtner B, Bergstrand K, Brand T (2007) Abiotic and biotic factors in slow filters integrated to closed hydroponic systems. Eur J Hortic Sci 72:104–112

Ganeshan G, Kumar AM (2005) Pseudomonas fluorescens, a potential bacterial antagonist to control plant diseases. J Plant Interact 1:123–134. https://doi.org/10.1080/17429140600907043

Geary B, Clark J, Hopkins BG, Jolley VD (2015) Deficient, adequate and excess nitrogen levels established in hydroponics for biotic and abiotic stress-interaction studies in potato. J Plant Nutr 38:41–50. https://doi.org/10.1080/01904167.2014.912323

Goddek S, Vermeulen T (2018) Comparison of Lactuca sativa growth performance in rainwater and RAS-water-based hydroponic nutrient solutions. Aquac Int 10. https://doi.org/10.1007/s10499-018-0293-8

Goddek S, Delaide B, Mankasingh U, Vala Ragnarsdottir K, Jijakli H, Thorarinsdottir R (2015) Challenges of sustainable and commercial aquaponics. Sustainability 7:4199–4224. https://doi.org/10.3390/su7044199

Goddek S, Espinal CA, Delaide B, Jijakli MH, Schmautz Z, Wuertz S, Keesman KJ (2016) Navigating towards decoupled aquaponic systems: a system dynamics design approach. Water (Switzerland) 8:1–29. https://doi.org/10.3390/W8070303

Gonçalves AA, Gagnon GA (2011) Ozone application in recirculating aquaculture system: an overview. Ozone Sci Eng 33:345–367. https://doi.org/10.1080/01919512.2011.604595

Gravel V, Dorais M, Dey D, Vandenberg G (2015) Fish effluents promote root growth and suppress fungal diseases in tomato transplants. Can J Plant Sci 95:427–436

Grunert O, Hernandez-Sanabria E, Vilchez-Vargas R, Jauregui R, Pieper DH, Perneel M, Van Labeke M-C, Reheul D, Boon N (2016) Mineral and organic growing media have distinct community structure, stability and functionality in soilless culture systems. Sci. Rep. 6:18837. https://doi.org/10.1038/srep18837

Haarhoff J, Cleasby JL (1991) Biological and physical mechanisms in slow sand filtration. In: Logsdon GS (ed) Slow sand filtration, vol 1. American Society of Civil Engineers, New York, pp 19–68

Haas D, Défago G (2005) Biological control of soil-borne pathogens by fluorescent pseudomonads. Nat Rev Microbiol 3:307–319. https://doi.org/10.1038/nrmicro1129

Hirayama K, Mizuma H, Mizue Y (1988) The accumulation of dissolved organic substances in closed recirculation culture systems. Aquac Eng 7:73–87. https://doi.org/10.1016/0144-8609(88)90006-4

Hong CX, Moorman GW (2005) Plant pathogens in irrigation water: challenges and opportunities. CRC Crit Rev Plant Sci 24:189–208. https://doi.org/10.1080/07352680591005838

Hultberg M, Holmkvist A, Alsanius B (2011) Strategies for administration of biosurfactant-producing pseudomonads for biocontrol in closed hydroponic systems. Crop Prot 30:995–999. https://doi.org/10.1016/j.cropro.2011.04.012

James, Becker JO (2007) Identifying microorganisms involved in specific pathogen suppression in soil. Annu Rev Phytopathol 45:153–172. https://doi.org/10.1146/annurev.phyto.45.062806.094354

Jarvis WR (1992) Managing disease in greenhouse crops. The American Phytopathological Society, St. Paul

Jones SW, Donaldson SP, Deacon JW (1991) Behaviour of zoospores and zoospore cysts in relation to root infection by *Pythium aphanidermatum*. New Phytol 117:289–301. https://doi.org/10.1111/j.1469-8137.1991.tb04910.x

Khalil S, Alsanius BW (2001) Dynamics of the indigenous microflora inhabiting the root zone and the nutrient solution of tomato in a commercial closed greenhouse system. Gartenbauwissenschaft 66:188–198

Khalil S, Alsanius WB (2011) Effect of growing medium water content on the biological control of root pathogens in a closed soilless system. J Hortic Sci Biotechnol 86:298–304. https://doi.org/10.1080/14620316.2011.11512764

Khalil S, Hultberg M, Alsanius BW (2009) Effects of growing medium on the interactions between biocontrol agents and tomato root pathogens in a closed hydroponic system. J Hortic Sci Biotechnol 84:489–494. https://doi.org/10.1080/14620316.2009.11512553

Khan A, Sutton JC, Grodzinski B (2003) Effects of Pseudomonas chlororaphis on Pythium aphanidermatum and root rot in peppers grown in small-scale hydroponic troughs. Biocontrol Sci Technol 13:615–630. https://doi.org/10.1080/0958315031000151783

Koohakan P, Ikeda H, Jeanaksorn T, Tojo M, Kusakari S-I, Okada K, Sato S (2004) Evaluation of the indigenous microorganisms in soilless culture: occurrence and quantitative characteristics in the different growing systems. Sci Hortic (Amsterdam) 101:179–188. https://doi.org/10.1016/j.scienta.2003.09.012

Lee S, Lee J (2015) Beneficial bacteria and fungi in hydroponic systems: types and characteristics of hydroponic food production methods. Sci Hortic (Amsterdam) 195:206–215. https://doi.org/10.1016/j.scienta.2015.09.011

Leonard N, Blancheton JP, Guiraud JP (2000) Populations of heterotrophic bacteria in an experimental recirculating aquaculture system. Aquac Eng 22:109–120

Leonard N, Guiraud JP, Gasset E, Cailleres JP, Blancheton JP (2002) Bacteria and nutrients – nitrogen and carbon – in a recirculating system for sea bass production. Aquac Eng 26:111–127

Lepoivre P (2003) Phytopathologie, 1st editio. ed. Les Presses Agronomiques de Gembloux Bruxelles: De Boeck

Liu W, Sutton JC, Grodzinski B, Kloepper JW, Reddy MS (2007) Biological control of pythium root rot of chrysanthemum in small-scale hydroponic units. Phytoparasitica 35:159–178. https://doi.org/10.1007/BF02981111

Love DC, Fry JP, Li X, Hill ES, Genello L, Semmens K, Thompson RE (2015) Commercial aquaponics production and profitability: Findings from an international survey. Aquaculture 435:67–74. https://doi.org/10.1016/j.aquaculture.2014.09.023

Maher M, Prasad M, Raviv M (2008) Ch 11 – organic soilless media components. In: Soilless culture: theory and practice. Elsevier B.V., Amsterdam, pp 459–504. https://doi.org/10.1016/B978-044452975-6.50013-7

Mangmang JS, Deaker R, Rogers G (2014) Response of lettuce seedlings fertilized with fish effluent to Azospirillum brasilense inoculation. Biol Agric Hortic 31:61–71. https://doi.org/10.1080/01448765.2014.972982

Mangmang JS, Deaker R, Rogers G (2015a) Inoculation effect of Azospirillum brasilense on basil grown under aquaponics production system. Org Agric 6:65–74. https://doi.org/10.1007/s13165-015-0115-5

Mangmang JS, Deaker R, Rogers G (2015b) Maximizing fish effluent utilization for vegetable seedling production by Azospirillum Brasilense. Procedia Environ Sci 29:179. https://doi.org/10.1016/j.proenv.2015.07.248

Mangmang JS, Deaker R, Rogers G (2015c) Response of Cucumber Seedlings Fertilized with Fish Effluent to Azospirillum brasilense. Int J Veg Sci 5260:150409121518007. https://doi.org/10.1080/19315260.2014.967433

Martin FN, Loper JE (1999) Soilborne plant diseases caused by Pythium spp.: ecology, epidemiology, and prospects for biological control soilborne plant diseases caused by Pythium spp.: ecology, epidemiology, and prospects for biological control. CRC Crit Rev Plant Sci 18:11–181

McMurty MR, Nelson PV, Sanders DC, Hodges L (1990) Sand culture of vegetables using recirculated aquacultural effluents. Appl Agric Res 5:280–284

McPherson GM, Harriman MR, Pattison D (1995) The potential for spread of root diseases in recirculating hydroponic systems and their control with disinfection. Med Fac Landbouww Univ Gent 60:371–379

Michaud L, Blancheton JP, Bruni V, Piedrahita R (2006) Effect of particulate organic carbon on heterotrophic bacterial populations and nitrification efficiency in biological filters. Aquac Eng 34:224–233. https://doi.org/10.1016/j.aquaeng.2005.07.005

Michaud L, Giudice AL, Interdonato F, Triplet S, Ying L, Blancheton JP (2014) C/N ratio-induced structural shift of bacterial communities inside lab-scale aquaculture biofilters. Aquac Eng 58:77–87. https://doi.org/10.1016/j.aquaeng.2013.11.002

Mitchell CE, Reich PB, Tilman D, Groth JV (2003) Effects of elevated CO_2, nitrogen deposition, and decreased species diversity on foliar fungal plant disease. Glob Chang Biol 9, 438–451. https://doi.org/10.1046/j.1365-2486.2003.00602.x

Monsees H, Kloas W, Wuertz S (2017) Decoupled systems on trial: Eliminating bottlenecks to improve aquaponic processes. PLoS One 12:1–18. https://doi.org/10.1371/journal.pone.0183056

Montagne V, Charpentier S, Cannavo P, Capiaux H, Grosbellet C, Lebeau T (2015) Structure and activity of spontaneous fungal communities in organic substrates used for soilless crops. Sci Hortic (Amsterdam) 192:148–157. https://doi.org/10.1016/j.scienta.2015.06.011

Montagne V, Capiaux H, Barret M, Cannavo P, Charpentier S, Grosbellet C, Lebeau T (2017) Bacterial and fungal communities vary with the type of organic substrate: implications for biocontrol of soilless crops. Environ Chem Lett 15:537–545. https://doi.org/10.1007/s10311-017-0628-0

Moruzzi S, Firrao G, Polano C, Borselli S, Loschi A, Ermacora P, Loi N, Martini M (2017) Genomic-assisted characterisation of Pseudomonas sp. strain Pf4, a potential biocontrol agent in hydroponics. Biocontrol Sci Technol 27:969–991. https://doi.org/10.1080/09583157.2017.1368454

Mukerji KG (2006) Microbial Activity in the Rhizosphere. Springer/GmbH & Co, Dordrecht/Berlin and Heidelberg. https://doi.org/10.1017/CBO9781107415324.004

Munguia-Fragozo P, Alatorre-Jacome O, Rico-Garcia E, Torres-Pacheco I, Cruz-Hernandez A, Ocampo-Velazquez RV, Garcia-Trejo JF, Guevara-Gonzalez RG (2015) Perspective for Aquaponic Systems: (Omic) Technologies for Microbial Community Analysis. Biomed Res Int 2015:10. https://doi.org/10.1155/2015/480386

Narayanasamy P (2013) Biological management of diseases of crops: volume 1: characteristics of biological control agents, Biological management of diseases of crops. Springer, Dordrecht. https://doi.org/10.1007/978-94-007-6380-7_1

Naylor SJ, Moccia RD, Durant GM (1999) The chemical composition of settleable solid fish waste (manure) from commercial rainbow trout farms in Ontario. Canada N Am J Aquac 61:21–26. https://doi.org/10.1577/1548-8454(1999)061<0021:TCCOSS>2.0.CO;2

Nemethy S, Bittsanszky A, Schmautz Z, Junge R, Komives T (2016) Protecting plants from pests and diseases in aquaponic systems. In: Ecological footprint in Central Europe. The University College of Tourism and Ecology Press, Sucha Beskidzka, pp 1–8

Nielsen CJ, Ferrin DM, Stanghellini ME (2006) Efficacy of biosurfactants in the management of Phytophthora capsici on pepper in recirculating hydroponic systems. Can J Plant Pathol 28:450–460. https://doi.org/10.1080/07060660609507319

Nogueira R, Melo LF, Purkhold U, Wuertz S, Wagner M (2002) Nitrifying and heterotrophic population dynamics in biofilm reactors: effects of hydraulic retention time and the presence of organic carbon. Water Res 36:469–481

Pagliaccia D, Ferrin D, Stanghellini ME (2007) Chemo-biological suppression of root-infecting zoosporic pathogens in recirculating hydroponic systems. Plant Soil 299:163–179. https://doi.org/10.1007/s11104-007-9373-7

Pagliaccia D, Merhaut D, Colao MC, Ruzzi M, Saccardo F, Stanghellini ME (2008) Selective enhancement of the fluorescent pseudomonad population after amending the recirculating nutrient solution of hydroponically grown plants with a nitrogen stabilizer. Microb Ecol 56:538–554. https://doi.org/10.1007/s00248-008-9373-z

Pantanella E, Cardarelli M, Colla G, Rea E, Marcucci A (2010) Aquaponics vs. hydroponics: production and quality of lettuce crop, pp 887–893. https://doi.org/10.17660/ActaHortic.2012.927.109

Pantanella E, Cardarelli M, Di Mattia E, Colla G (2015) Aquaponics and food safety: effects of UV sterilization on total coliforms and lettuce production. In: Conference and exhibition on soilless culture, pp 71–76

Parvatha Reddy P (2016) Sustainable crop protection under protected cultivation. Springer. https://doi.org/DOI, Springer. https://doi.org/10.1007/978-981-287-952-3_7

Paulitz TC, Bélanger RR (2001) Biological control in greenhouse systems. Annu Rev Phytopathol 39:103–133

Pérez-García A, Romero D, de Vicente A (2011) Plant protection and growth stimulation by microorganisms: biotechnological applications of Bacilli in agriculture. Curr Opin Biotechnol 22(2):187–193. https://doi.org/10.1016/j.copbio.2010.12.003

Postma J, van Os E, Bonants PJM (2008) Ch 10 – pathogen detection and management strategies in soilless plant growing system. In: Soilless culture: theory and practice. Elsevier B.V., Amsterdam, pp. 425–457. https://doi.org/10.1016/B978-0-444-52975-6.50012-5

Postma J, Stevens LH, Wiegers GL, Davelaar E, Nijhuis EH (2009) Biological control of Pythium aphanidermatum in cucumber with a combined application of Lysobacter enzymogenes strain 3.1T8 and chitosan. Biol Control 48:301–309. https://doi.org/10.1016/j.biocontrol.2008.11.006

Rakocy J (2007) Ten guidelines for aquaponic systems. Aquaponics J 46:14–17

Rakocy JE (2012) Aquaponics – integrating fish and plant culture. In: Tidwell JH (ed) Aquaculture production systems. Wiley, New York, pp 343–386

Rakocy JE, Maseer PM, Losordo TM (2006) Recirculating aquaculture tank production systems: Aquaponics – integrated fish and plant culture. South Reg Aquac Cent Publication No. 454, 16pp. http://srac.tamu.edu/getfile.cfm?pubid=105

Renault D, Déniel F, Benizri E, Sohier D, Barbier G, Rey P (2007) Characterization of Bacillus and Pseudomonas strains with suppressive traits isolated from tomato hydroponic-slow filtration unit. Can J Microbiol 53:784–797. https://doi.org/10.1139/W07-046

Renault D, Vallance J, Déniel F, Wery N, Godon JJ, Barbier G, Rey P (2012) Diversity of bacterial communities that colonize the filter units used for controlling plant pathogens in soilless cultures. Microb Ecol 63:170–187. https://doi.org/10.1007/s00248-011-9961-1

Resh HM (2013) Hydroponic food production : a definitive guidebook for the advanced home gardener and the commercial hydroponic grower, 7th edn. CRC Press, Boca Raton

Rosberg AK (2014) Dynamics of root microorganisms in closed hydroponic cropping systems. Department of Biosystems and Technology, Swedish University of Agricultural Science, Alnarp

Rurangwa E, Verdegem MCJ (2015) Microorganisms in recirculating aquaculture systems and their management. Rev Aquac 7:117–130. https://doi.org/10.1111/raq.12057

Saha S, Monroe A, Day MR (2016) Growth, yield, plant quality and nutrition of basil (*Ocimum basilicum* L.) under soilless agricultural systems. Ann Agric Sci 61:181–186. https://doi.org/10.1016/j.aoas.2016.10.001

Schmautz Z, Graber A, Jaenicke S, Goesmann A, Junge R, Smits THM (2017) Microbial diversity in different compartments of an aquaponics system. Arch Microbiol 1–8. https://doi.org/10.1007/s00203-016-1334-1

Schreier HJ, Mirzoyan N, Saito K (2010) Microbial diversity of biological filters in recirculating aquaculture systems. Curr Opin Biotechnol 21:318–325. https://doi.org/10.1016/j.copbio.2010.03.011

Sharrer MJ, Summerfelt ST, Bullock GL, Gleason LE, Taeuber J (2005) Inactivation of bacteria using ultraviolet irradiation in a recirculating salmonid culture system. Aquac Eng 33:135–149. https://doi.org/10.1016/j.aquaeng.2004.12.001

Shinohara M, Aoyama C, Fujiwara K, Watanabe A, Ohmori H, Uehara Y, Takano M (2011) Microbial mineralization of organic nitrogen into nitrate to allow the use of organic fertilizer in hydroponics. Soil Sci Plant Nutr 57:190–203. https://doi.org/10.1080/00380768.2011.554223

Sirakov I, Lutz M, Graber A, Mathis A, Staykov Y (2016) Potential for combined biocontrol activity against fungal fish and plant pathogens by bacterial isolates from a model aquaponic system. Water 8:1–7. https://doi.org/10.3390/w8110518

Snoeijers SS, Alejandro P (2000) The effect of nitrogen on disease development and gene expression in bacterial and fungal plant pathogens. Eur J Plant Pathol 106:493–506

Somerville C, Cohen M, Pantanella E, Stankus A, Lovatelli A (2014) Small-scale aquaponic food production – integrated fish and plant farming. FAO, Rome

Sopher CR, Sutton JC (2011) Quantitative relationships of Pseudomonas chlororaphis 63-28 to Pythium root rot and growth in hydroponic peppers. Trop Plant Pathol 36:214–224. https://doi.org/10.1590/S1982-56762011000400002

Spadaro D, Gullino ML (2005) Improving the efficacy of biocontrol agents against soilborne pathogens. Crop Prot 24:601–613. https://doi.org/10.1016/j.cropro.2004.11.003

Stanghellini ME, Miller RM (1997) Their identity and potencial efficacy in the biological control of zoosporic plant pathogens. Plant Dis 81:4–12

Stanghellini ME, Rasmussen SL (1994) Hydroponics: a solution for zoosporic pathogens. Plant Dis 78:1129–1138

Stouvenakers G, Sébastien M, Haissam JM (2017) Biocontrol properties of recirculating aquaculture water against hydroponic root pathogens. Oral presentation at Aquaculture Europe 2017 meeting, Dubrovnik, Croatia

Sugita H, Nakamura H, Shimada T (2005) Microbial communities associated with filter materials in recirculating aquaculture systems of freshwater fish. Aquaculture 243:403–409. https://doi.org/10.1016/j.aquaculture.2004.09.028

Sutton JC, Sopher CR, Owen-Going TN, Liu W, Grodzinski B, Hall JC, Benchimol RL (2006) Etiology and epidemiology of Pythium root rot in hydroponic crops: current knowledge and perspectives. Summa Phytopathol 32:307–321. https://doi.org/10.1590/S0100-54052006000400001

Takeda S, Kiyono M (1990) The characterisation of yellow substances accumulated in a closed recirculating system for fish culture. In: Proceedings of the second Asian fisheries forum, pp 129–132

Tal Y, Watts JEM, Schreier SB, Sowers KR, Schreier HJ, Schreier HJ (2003) Characterization of the microbial community and nitrogen transformation processes associated with moving bed bioreactors in a closed recirculated mariculture system. Aquaculture 215:187–202

Thongkamngam T, Jaenaksorn T (2017) Fusarium oxysporum (F221-B) as biocontrol agent against plant pathogenic fungi in vitro and in hydroponics. Plant Prot Sci 53:85–95. https://doi.org/10.17221/59/2016-PPS

Timmons MB, Ebeling JM (2010) Recirculating aquaculture, 2nd edn. NRAC Publication, Ithaca

Tu JC, Papadopoulos AP, Hao X, Zheng J (1999) The relationship of a pythium root rot and rhizosphere microorganisms in a closed circulating and an open system in stone wool culture of tomato. Acta Hort (ISHS) 481:577–583

Vallance J, Déniel F, Le Floch G, Guérin-Dubrana L, Blancard D, Rey P (2010) Pathogenic and beneficial microorganisms in soilless cultures. Agron Sustain Dev 31:191–203. https://doi.org/10.1051/agro/2010018

Van Der Gaag DJ, Wever G (2005) Conduciveness of different soilless growing media to Pythium root and crown rot of cucumber under near-commercial conditions. Eur J Plant Pathol 112:31–41. https://doi.org/10.1007/s10658-005-1049-7

Van Os EA (2009) Comparison of some chemical and non-chemical treatments to disinfect a recirculating nutrient solution. Acta Hortic 843:229–234

Van Os EA, Amsing JJ, Van Kuik AJ, Willers H (1999) Slow sand filtration: a potential method for the elimination of pathogens and nematodes in recirculating nutrient solutions from glasshouse-grown crops. Acta Hortic 481:519–526

van Os EA, Bruins M, Wohanka W, Seidel R (2001) Slow filtration: a technique to minimise the risks of spreading root-infecting pathogens in closed hydroponic systems. In: International symposium on protected cultivation in mild winter climates: current trends for sustainable techniques, pp 495–502

Veresoglou SD, Barto EK, Menexes G, Rillig MC (2013) Fertilization affects severity of disease caused by fungal plant pathogens. Plant Pathol 62:961–969. https://doi.org/10.1111/ppa.12014

Verma S, Daverey A, Sharma A (2017) Slow sand filtration for water and wastewater treatment – a review. Environ Technol Rev 6:47–58. https://doi.org/10.1080/21622515.2016.1278278

Villarroel M, Junge R, Komives T, König B, Plaza I, Bittsánszky A, Joly A (2016) Survey of aquaponics in Europe. Water (Switzerland) 8:3–9. https://doi.org/10.3390/w8100468

Waechter-Kristensen B, Sundin P, Gertsson UE, Hultberg M, Khalil S, Jensen P, Berkelmann-Loehnertz B, Wohanka W (1997) Management of microbial factors in the rhizosphere and nutrient solution of hydroponically grown tomato. Acta Hortic. https://doi.org/10.17660/ActaHortic.1997.450.40

Waechter-Kristensen B, Caspersen S, Adalsteinsson S, Sundin P, Jensén P (1999) Organic compounds and micro-organisms in closed, hydroponic culture: occurrence and effects on plant growth and mineral nutrition. Acta Hortic 481:197–204

Whipps JM (2001) Microbial interactions and biocontrol in the rhizosphere. J Exp Bot 52:487–511. https://doi.org/10.1093/jexbot/52.suppl_1.487

Wielgosz ZJ, Anderson TS, Timmons MB (2017) Microbial effects on the production of aquaponically Grown Lettuce. Horticulturae 3:46. https://doi.org/10.3390/horticulturae3030046

Willey JM, Sherwood LM, Woolverton CJ (2008) Prescott, Harley, & Klein's micobiology, 7th edn. McGrawHill Higher Education, New York

Wohanka W (1995) Disinfection of recirculating nutrient solutions by slow sand filtration. Acta Hortic. https://doi.org/10.17660/ActaHortic.1995.382.28

Zou Y, Hu Z, Zhang J, Xie H, Liang S, Wang J, Yan R (2016) Attempts to improve nitrogen utilization efficiency of aquaponics through nitrifies addition and filler gradation. Environ Sci Pollut Res 23:6671–6679. https://doi.org/10.1007/s11356-015-5898-0

Chapter 15
Smarthoods: Aquaponics Integrated Microgrids

Florijn de Graaf and Simon Goddek

Abstract With the pressure to transition towards a fully renewable energy system increasing, a new type of power system architecture is emerging: the microgrid. A microgrid integrates a multitude of decentralised renewable energy technologies using smart energy management systems, in order to efficiently balance the local production and consumption of renewable energy, resulting in a high degree of flexibility and resilience. Generally, the performance of a microgrid increases with the number of technologies present, although it remains difficult to create a fully autonomous microgrid within economic reason (de Graaf F, New strategies for smart integrated decentralised energy systems, 2018). In order to improve the self-sufficiency and flexibility of these microgrids, this research proposes integrating a neighbourhood microgrid with an urban agriculture facility that houses a decoupled multi-loop aquaponics facility. This new concept is called *Smarthood*, where all Food–Water–Energy flows are circularly connected. In doing so, the performance of the microgrid greatly improves, due to the high flexibility present within the thermal mass, pumps and lighting systems. As a result, it is possible to achieve 95.38% power and 100% heat self-sufficiency. This result is promising, as it could pave the way towards realising these fully circular, decentralised Food–Water–Energy systems.

Keywords Smart grids · Urban agriculture · Microgrids · Blockchains · Self-sufficiency · Energy system modelling

F. de Graaf (✉)
Spectral, Amsterdam, The Netherlands
e-mail: florijn@spectral.energy

S. Goddek
Mathematical and Statistical Methods (Biometris), Wageningen University, Wageningen, The Netherlands
e-mail: simon.goddek@wur.nl; simon@goddek.nl

S. Goddek et al. (eds.), *Aquaponics Food Production Systems*,
https://doi.org/10.1007/978-3-030-15943-6_15

15.1 Introduction

Switching towards a fully sustainable energy system will partly require switching from a centralised generation and distribution system, towards a decentralised system, due to the rise of decentralised energy generation technologies using wind and rooftop solar radiation. In addition, integrating the heat and transport sectors into the electricity system will lead to a very significant increase in peak demand. These developments require massive and costly adaptations to the energy infrastructure, while the utilisation of existing production assets is expected to drop from 55% to 35% by 2035 (Strbac et al. 2015). This poses a major challenge, but also an opportunity: if the energy flows can be balanced locally in *microgrids*, the demand for expensive infrastructure upgrade can be minimised, while providing extra stability to the main grid. For these reasons, 'microgrids have been identified as a key component of the Smart Grid for improving power reliability and quality, increasing system energy efficiency' (Strbac et al. 2015).

Microgrids can provide much-needed resilience and flexibility, and are therefore likely to play an important role in the energy system of the future. It is estimated that by 2050, over half of EU households will be generating their own electricity (Pudjianto et al. 2007). Unlocking flexible resources within microgrids is therefore needed in order to balance the intermittent renewable energy generation.

Urban agriculture systems, such as aquaponics (dos Santos 2016), can provide this much-needed energy flexibility (Goddek and Körner 2019; Yogev et al. 2016). Plants can grow within a wide range of external conditions, since they are used to doing so in nature. The same applies to fish in an aquaculture system, which can thrive in a broad temperature range. These flexible operating conditions allow for a buffering effect on energy input requirements, which create a large degree of flexibility within the system. The high thermal mass embodied by the aquaculture system allows for vast amounts of heat to be stored within the system. The lights can be turned on and off depending on the abundance of electricity, allowing for excess electricity generation to essentially be curtailed by turning it into valuable biomass. Pumps can be operated in synchronicity with peak power generation times (e.g. noon) to limit net peak power (peak shaving). Optimal distillation units (Chap. 8) also have a very flexible heat demand and can be turned off as soon as there is an oversupply of heat or electricity (i.e. the heat pump would then convert electric energy into thermal energy). All these aspects make aquaponic systems well-suited to provide flexibility to a microgrid.

Next to providing flexibility in consumption, a multi-loop aquaponics system can be further integrated to also provide flexibility in production. Biogas is produced as a byproduct from the UASB in the aquaponic facility. This biogas can be combusted in order to produce both heat and power, by incorporating a micro-CHP in the microgrid. Integrating aquaponic systems within microgrids can therefore enhance energy flexibility both on the demand and supply sides.

15.2 The Smarthoods Concept

To unlock the full potential of the Food–Water–Energy nexus with respect to decentralised microgrids, a fully integrated approach focuses not only on energy (microgrid) and food (aquaponics) but also on utilising the local water cycle. The integration of various water systems (such as rainwater collection, storage and wastewater treatment) within aquaponic-integrated microgrids yields the biggest potential for efficiency, resilience and circularity. The concept of a fully integrated and decentralised Food–Water–Energy microgrid will from now on be referred to as a *Smarthood* (smart neighbourhood) and is depicted in Fig. 15.2.

The benefit of implementing aquaponics into the Smarthoods concept is its potential to contribute to optimise integrated nutrient, energy and water flows (Fig. 15.1). This integration potential goes well beyond the already-mentioned

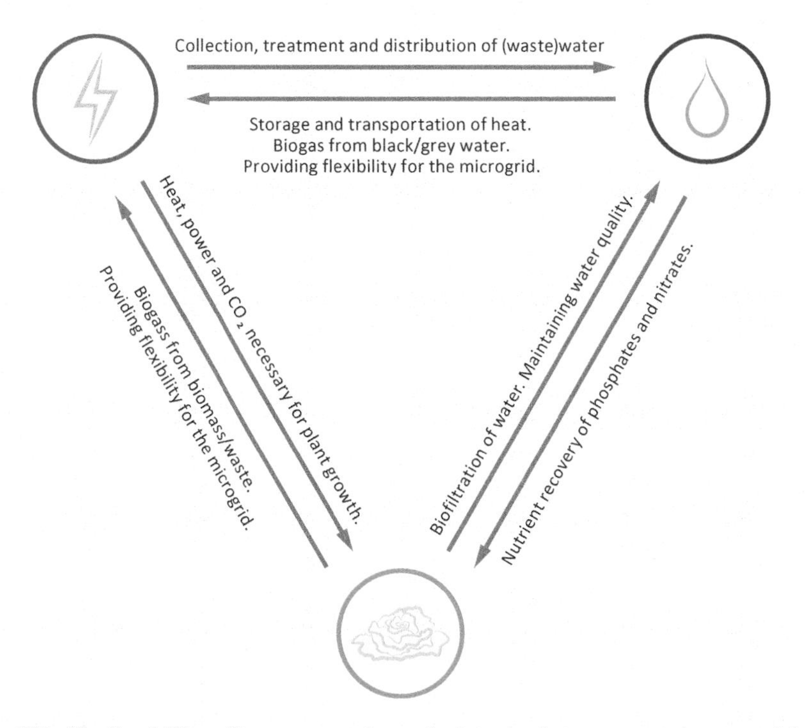

Fig. 15.1 The Food–Water–Energy nexus shows the interplay between energy, water and food production (based on IRENA 2015)

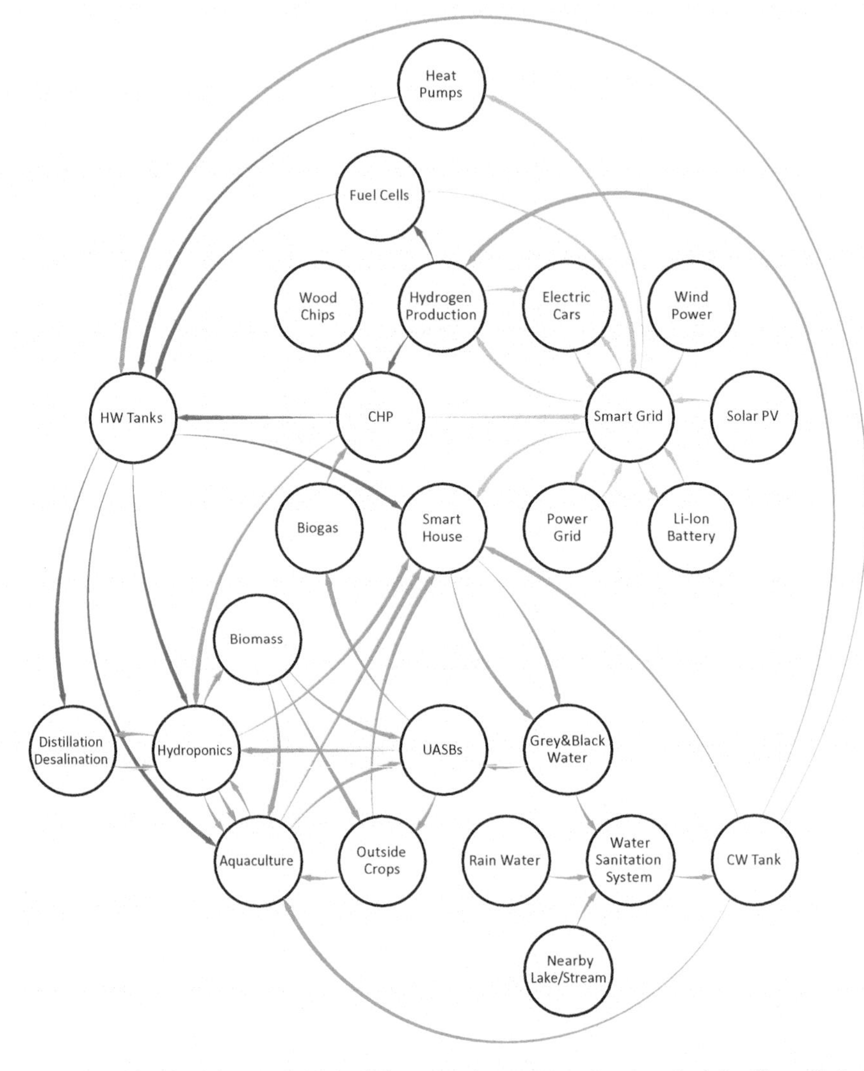

Fig. 15.2 The integration of decoupled aquaponics systems (as described in Chap. 8) in a decentralised local environment as designed for the Smarthoods concept. The green arrows show to what extent an aquaponics system can interact with the overall system. The red arrows represent heat flows, the blue arrows water flows and the yellow arrows power flows

crossovers between the energy and food systems. For instance, occurring biodegradable waste streams can be treated in anaerobic reactors (e.g. UASBs) and generate both biogas and bio-fertiliser (Goddek et al. 2018). Even the demineralized waste sludge can be utilised as liquid manure on conventional cropland.

Fig. 15.3 Illustrated photo of *De Ceuvel* (source: Metabolic – www.metabolic.nl)

Example 15.1

An early example of an urban integrated aquaponic microgrid development is *De Ceuvel*, a previously abandoned shipyard in Amsterdam-North that has been converted into a self-sufficient office space and recreational hub. *De Ceuvel* serves as a testbed for new technologies and policies aimed at creating a circular economy. It features an all-electric microgrid including solar PV, heat pumps and peer-to-peer energy trading over the blockchain using their own energy token: the *Jouliette*.[1] A small aquaponic facility produces herbs and vegetables for the on-site restaurant. The same restaurant utilises biogas extracted from locally produced organic waste for their cooking activities as well as space heating. In addition, there is a lab present that is used for testing the water quality and extracting phosphates and nitrates.

Although *De Ceuvel* is currently not actively using the aquaponics facility to increase the flexibility of its microgrid, sensors are being installed to monitor the energy and nutrient flows in order to assess its performance. This data will be used to aid in the development of newer and smarter urban integrated aquaponics microgrids, such as the Smarthoods concept proposed in this chapter. Early use cases found in urban living labs like De Ceuvel are essential to the successful development of the Smarthoods concept (Fig. 15.3).

[1]https://www.jouliette.net

Although a holistic approach to urban FWE systems such as the *Smarthoods* concepts yields many benefits, the integration of aquaponics systems within microgrids remains very case-dependent. Aquaponic food production systems are characterised by a higher yield and a lower water, nutrient and energy footprint than conventional agricultural systems; however, they are also more costly to build. They are therefore best suited in locations that require high yields due to, for instance, space limitations. In dense urban areas, there may not always be sufficient space to build an aquaponics facility, whereas for rural areas the cost of land may be too low to warrant building a state-of-the-art aquaponics facility; a standard agricultural facility with lower financing costs and yield will be more suited in such cases. The most optimal use case for an integrated aquaponic facility is one where sufficient space is available, and a high yield per area is required to offset the cost of land use. Suburban neighbourhoods and other urban areas (e.g. an abandoned warehouse) are therefore most likely to see the first implementation of microgrids integrated with an aquaponic facility (see Example 15.1).

15.3 Goal

The goal of this research is to quantify the degree of self-sufficiency and flexibility for a microgrid integrated with a decoupled multi-loop aquaponics system.

15.4 Method

A neighbourhood of 50 households was assumed a 'Smarthood', with a decoupled multi-loop aquaponics facility present that is capable of providing fish and vegetables for all the 100 inhabitants of the Smarthood.

For the detailed modelling of the Smarthood, a hypothetical reference case of a suburban neighbourhood in Amsterdam was used, consisting of 50 households (houses) with an average household occupancy of 2 persons per household (100 persons total). In addition, one urban aquaponic facility consists of a greenhouse, aquaculture system, a UASB and a distillation unit. The dimensioning of the different components is motivated using data for a typical Dutch household and greenhouse (see Table 15.1).

15.4.1 The Energy System Model

An Energy System Model (ESM) was made that can simulate the energy flows of a wide range of components, whose main specifications are shown in Table 15.2. The ESM is capable of calculating energy flows for each component for each hour of the year.

Table 15.1 Food and energy requirements per persona/household in the Netherlands

	Average (per capita/year)	Total (100 persons)	Source
Food			
Vegetable consumption (the Netherlands)	33 kg[a] (whereas 73 kg are recommended)	7300 kg	EFSA (2018)
Required greenhouse area	Approx. 4 m^2	400 m^2	Estimated based on min. consumption recommendation
Fish consumption	20 kg	2000 kg	FAO (2015)
Required aquaculture volume[b]	0.2 m^3	20 m^3	Estimated
Energy			
Household electricity consumption (Netherlands)	3000 kWh_e/house/year	150 MWh_e/year	CBS (2018)
Household heat consumption (Netherlands)	6500 kWh_{th}/house/year	325 MWh_{th}/year	CBS (2018)
RAS electricity consumption	0.05–0.15 kW_e/m^3	1–3 kW_e 8,76–28,26 MWh_e/year	(Espinal, pers. communication)

[a]The average Dutch person eats 50 kg of vegetables per year. However, only 33 kg of vegetables that can be grown in hydroponics systems, which are fruiting vegetables 31.87 g/day, brassica vegetables 22.11 g/day, leaf vegetables 12.57 g/day, legume vegetables 19.74 g/day, stem vegetables 4.29 g/day
[b]Considering a max. fish density of 80 kg/m^3

Table 15.2 Production components

Component	Size	Specifications
Solar PV	40 $kW_{p,e}$	Eta: 0.15
Urban wind turbine	20 $kW_{p,e}$	Eta: 0.33
Heat pump	10 $kW_{p,e}$	COP: 4.0
CHP	20 $kW_{p,e}$	Eta_{el}: 0.24, Eta_{th} = 0.61
Fuel cell	10 $kW_{p,e}$	Eta: 0.55
Electrolyser	20 $kW_{p,e}$	Eta: 0.45
Battery	200 kWh	Eta: 0.90
Hot water tank	930 kWh	40–60°C
Hydrogen tank	1000 kWh	30 kg of H_2 storage

The energy system was modelled in MATLAB using energy profile data for Amsterdam obtained through DesignBuilder. The numerical time-series model incorporates a wide selection of energy technologies, listed in Table 15.2 with their relevant specifications (Fig. 15.4).

The Energy System Model (ESM) uses simple conditional statements for the decision-making process, i.e. it is a rule-based control system. In the current version of this model, the control is centralised, with the objective of self-consumption

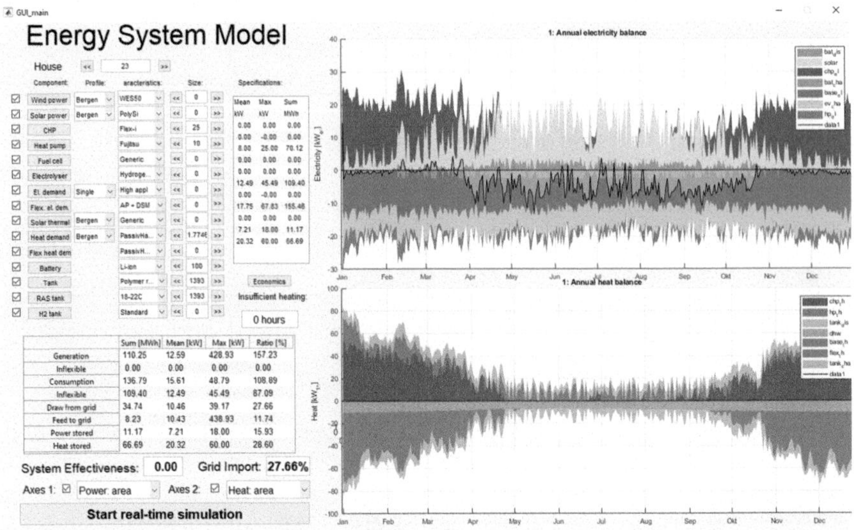

Fig. 15.4 The aquaponics microgrid model (F. de Graaf 2018), showing the energy balances for power (upper diagram) and heat (lower diagram) for the reference case (Amsterdam)

maximisation for the system as a whole (in a future version, the control architecture will be decentralised, see Sect. 15.5). The conditional statements to achieve this can be stated as follows:

1. Keep the heat storage to a minimum.
2. Forecast the predicted inflexible electricity production and consumption.
3. (a) If the battery will be full, turn on flexible consumption.
   (b) If the battery will be empty, turn on flexible generation.

By keeping the heat storage to a minimum, the buffer for flexible energy balancing is maximised. If there is an overproduction of inflexible electricity (i.e. electricity production that cannot be flexibly scheduled or controlled, such as solar or wind), the heat pump can be turned on to create a buffer provided by hot water storage and the thermal mass of the aquaponic RAS system. Conversely, if there is an underproduction of electricity, flexible generation such as the CHP and the fuel cell can be turned on, thereby utilising the thermal storage capacity.

For both heat and power, the energy balance is equivalent to

$$P_{gen,flex} + P_{gen,inflex} + P_{grid} = P_{cons,inflex} + P_{cons,flex} + P_{storage} \tag{15.1}$$

Flexible generations include the heat pump, Combined Heat and Power (CHP) unit, fuel cell, battery and smart/flexible devices (e.g. aquaponic pumps). Wind, solar photovoltaics (PV) and solar collectors are classified as inflexible generation. Non-flexible devices make up the bulk of electricity consumption, especially in winter (due to the need for instant lighting) (Fig. 15.5).

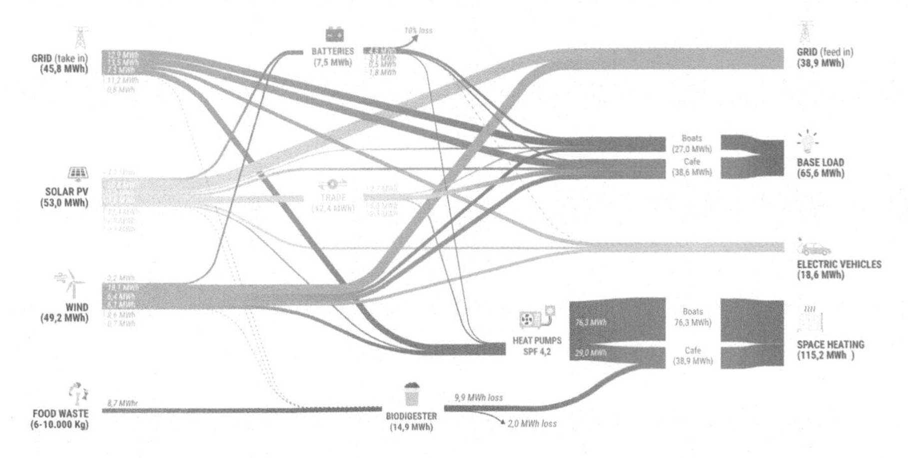

Fig. 15.5 Example of the energy flows (Sankey diagram) of a possible integrated microgrid configuration at De Ceuvel (de Graaf 2018), including a biodigester for the production of biogas. This particular configuration does not include the Combined Heat and Power unit that is present within the Smarthood concept, nor does it take into account a large aquaponics facility

15.5 Results

The total electrical and thermal consumption of both the houses and the aquaponic greenhouse facility (modelled from the data in Tables 15.1 and 15.2) is shown in Table 15.3. The aquaponic greenhouse facility is responsible for 38.3% of power consumption and 51.4% of heat consumption. The power demand for an aquaponics facility integrated in a residential microgrid is therefore slightly over one-third of the total local energy demand, given that all of the residential energy and vegetable/fish production is done locally. The heat demand comprises roughly 50% of the total heat demand, which can be attributed for a large part to the distillation unit running on high-temperature water.

As can be seen in Figs. 15.4 and 15.6, the Smarthoods energy system is capable of balancing production and demand most of the time. The total share of imported electricity from the grid is 4.62% for the reference case. At times, a slight imbalance of power can be observed, which can be attributed to suboptimal control for the current version of the model for the most part. The CHP, for instance, switches from an on- to off-state multiple times over the course of several hours, resulting in an overproduction of electricity. Such behaviour will not occur for a more optimised control system, since the CHP can be ramped down in coordination with the heat pump in order to deliver the precise amount of electricity and heat needed.

15.5.1 Flexibility

The system is highly flexible as a result of the CHP and the aquaponics facility with its flexible lighting and pumps, and high thermal buffering capacity, as well as the

Table 15.3 Electrical and thermal load for different aspects of the microgrid

	Residential	Aquaponic facility
Electrical average demand	17.2 kW	10.2 kW
Electrical peak demand	47.6 kW_p	15.2 kW_p
Electrical total demand	**143.2 MWh/year**	**89.2 MWh/year**
Thermal average demand	37.1 kW	39.3 kW
Thermal peak demand	148.4 kW	121.2 kW
Thermal total demand	**325.0 MWh_{th}/year**	**344.2 MWh_{th}/year**

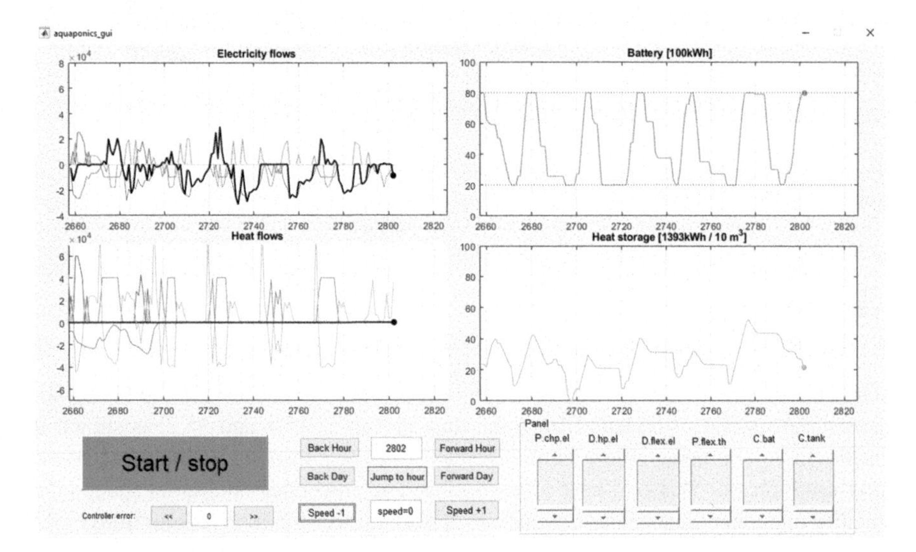

Fig. 15.6 Time-series graphical diagrams for the power (top-left) and heat (bottom-left) energy balances (in W) of the Smarthood system. Storage capacity (in kWh) is indicated on the right side for power (top-right) and heat (bottom-right). The x-axis represents number of hours since the start of the year. The black line represents the imbalance of energy

battery, and the hydrogen system. The aquaponic system, especially, greatly increases the overall flexibility of the system, as it can function for a wide range of energy input, as can be derived from Table 15.4. As a result of this flexibility, the system manages to achieve near total (95.38%) power self-sufficiency and 100% heat self-sufficiency.

15.6 Discussion

Self-Sufficiency The energy system proposed for the Smarthood concept is capable of achieving near full grid-independence through the use of the flexibility provided by the various system components. The aquaponic system, especially, has a positive

Table 15.4 Flexible demand of the aquaponic system

Component	Order of magnitude	Flexibility
Pumps	$0.05–0.15\ kW_e\ M^{-3}$ $1–3\ kW_e$ 8,76–28,26 MWh_e/year	Not all pumps have to run continuously. Main processes (oxygen control, ammonia control, CO_2 control, tank exchanges, suspended solids control) must run continuously. Smaller processes such as pH buffer dosing, backwash routines, water exchanges or back-up oxygenation do not have to run continuously
Lighting	80–150 W/m^2 With a capacity factor of 10–20% this leads to 28–105 MWh_e/year kW_e	Plants need ~4–6 h of darkness, the rest of the day they can be lit artificially. This leaves approx. 0 (summer) to 12 (winter) hours of flexible additional lighting
Space heating (underfloor) and aquaculture tank heating	444 kW_{th}/m^2/year 177,8 MWh_{th}/year	Due to the high thermal mass of the concrete floor and the large water volume in the RAS tank, the heat load is extremely flexible
Distillation unit	50 kW_{th} MWh_e/year 166,4 MWh_{th}/year	The distillation unit operates on hot water (70–90 °C) and can be operated with a significant degree of flexibility (MemSys 2017)

effect on the overall flexibility of the system. With 95.38% power self-sufficiency, this system performs better than any other economically feasible system assessed in previous research (de Graaf 2018).

Control Architecture Facilitating a decentralised local energy economy, such as the one proposed in the *Smarthoods* concept, requires a platform that keeps track of all the peer-to-peer transactions occurring within the neighbourhood. The corresponding peer-to-peer network can be classified as a multi-agent system (MAS) approach, in which multiple nodes (e.g. households or utility buildings) function as independent agents with their own objective (e.g. minimise cost or maximise energy saving) and corresponding decision-making process. Such a decentralised, multi-agent decision-making approach is necessary due to the complexity of the system. There is simply too much information and too many variables for the computation of a hierarchical, top-down and centralised control architecture.

Blockchain A blockchain-based multi-agent system control architecture could potentially provide the necessary framework to accommodate a decentralised peer-to-peer network. A vast number of distributed nodes ensure stability and security for the network, and an alternative to mining can be used: minting. With minting, tokens/coins are generated based on the data provided by a real-world device such as a smart energy metre. Provided that these sources of information can be trusted, i.e. that these devices can be tamper-proofed, a secure and independent ledger can be created in which various stakeholders can exchange goods (e.g. electricity) and services (e.g. demand-side management). Using smart contracts, complex services such as flexibility trading can be programmed into the control architecture of the system.

Internet of Things The constituent components within the Smarthood system, such as heat pumps, greenhouse lighting or the UASB, can all be controlled using Internet-connected sensors and actuators, known as Internet of Things. An IoT sensor network allows for the extensive acquisition of data, ranging from fish tank nutrient concentration to, for instance, battery load cycles, all on a per-time-step basis. This data can be used to verify the numerical model and optimise the dynamical control of the system.

Artificial Intelligence Optimising the control of the Smarthood system can be done by analysing the data using artificial intelligence algorithms, such as genetic programming (evolutionary algorithms) or machine reinforcement learning. With machine reinforcement learning, for instance, a set of actions and their influence on the environment are passed to the algorithm as input arguments, along with the current state of the system and a cumulative objective/cost function. An incrementally improving, heuristic decision-making process can be implemented at each household that will dynamically adapt to situations in order to find a near-optimal decision-making programme that will manage the energy flows within the house and the *Smarthood*. Each house can run such an algorithm, and as a result, a multi-nodal control system architecture, known as a multi-agent system (MAS), can be created that is relatively computationally inexpensive (compared to centralised control)—and close to optimal.

Legal Barriers The highly innovative nature of various aspects of the Smarthood concept, such as the polygeneration microgrid, the multi-loop aquaponic system and the unconventional urban planning requirements, brings along a unique set of challenges to overcome. For many of these challenges, the current regulatory framework is insufficient to accommodate the developments proposed in the Smarthoods concept.

Microgrids, for instance, work best when there is a local marketplace in which various *prosumers* (consumers that simultaneously produce power) can engage in frictionless peer-to-peer energy trading in a free market. Market forces will then work to create a local energy market in which a fluctuating energy price will result from local supply and demand. This price fluctuation will consequently incentivise smart energy solutions such as energy storage, demand-side management or flexible energy generation. In most EU countries, a free local market is currently impossible due to regulations; taxes have to be paid for every kWh that passes through the electricity metre, the electricity price for consumers is fixed and prosumers are not allowed to participate in the energy market without the intervention of a third party called the *aggregator*. With the expected increase in the development of microgrid projects, regulators will have to find ways to facilitate local energy markets in order to unlock the full potential of highly integrated microgrids (see Example 15.2).

Example 15.2
A recent advancement within the regulatory framework in the Netherlands is the introduction of the *experimenteerregeling,* an experimental law that allows a small number of carefully selected projects (such as de Ceuvel, example Z.1) to allow energy cooperatives to become their own distribution system operator, as if they were behind a single metre connection. This law is indicative of the awareness amongst Dutch regulatory bodies of previously mentioned legal barriers, and will therefore most likely lead to the current electricity law to be revised in the near future in order to better accommodate microgrid developments.

There are also some legal barriers in most EU countries with respect to reusing treated black water for fish and plant production, as it has to be ensured that human pathogens are fully eliminated. More information on the legal framework of aquaponics can be found in Chap. 20.

15.7 Conclusions

The goal of this research was to quantify the degree of flexibility and self-sufficiency that an aquaponics integrated microgrid can provide. In order to attain this answer, a neighbourhood of 50 households was assumed a 'Smarthood', with a decoupled multi-loop aquaponics facility present that is capable of providing fish and vegetables for all the 100 inhabitants of the Smarthood.

The results are promising: thanks to the high degree of flexibility inherent in the aquaponic system as a result of high thermal mass, flexible pumps and adaptive lighting, the overall degree of self-sufficiency is 95.38%, making it nearly completely self-sufficient and grid independent. With the aquaponics system being responsible for 38.3% of power consumption and 51.4% of heat consumption, the impact of the aquaponics facility on the total system's energy balance is very high.

Earlier research (de Graaf 2018) has indicated that it is very difficult to achieve self-consumption levels over 60% without relying on an external biomass source to drive a CHP. Even with this source included, the maximum techno-economically feasible self-consumption did not exceed 89%. In the Smarthood, biomass inputs for the CHP are partially derived from the aquaponic system itself, and the recycling of grey and black water. A higher self-consumption combined with a lower dependence on external biomass inputs, and a resulting self-consumption of 95%, makes the proposed aquaponic-integrated microgrid perform better from a self-sufficiency point of view than any other renewable microgrid known to the authors.

The authors of this chapter therefore strongly believe that with enough experimentation, integrating aquaponic greenhouse systems within microgrids yields great potential for creating highly self-sufficient Food–Water–Energy systems at a local level.

References

CBS (2018) Energieverbruik particuliere woningen; woningtype en regio's [WWW Document]. URL http://statline.cbs.nl/Statweb/publication/?VW=T&DM=SLNL&PA=81528NED&D1=a&D2=a&D3=0&D4=a&HD=161202-1617&HDR=T,G3&STB=G2,G1. Accessed 19 Oct 2018

de Graaf F (2018) New strategies for smart integrated decentralised energy systems

dos Santos MJPL (2016) Smart cities and urban areas—aquaponics as innovative urban agriculture. Urban For Urban Green 20:402–406. https://doi.org/10.1016/j.ufug.2016.10.004

EFSA (2018) The EFSA comprehensive European food consumption database | European [WWW Document]. URL https://www.efsa.europa.eu/en/food-consumption/comprehensive-database. Accessed 19 Oct 2018

FAO (2015) Statistical pocketbook - world food and agriculture

Goddek S, Körner O (2019) A fully integrated simulation model of multi-loop aquaponics: a case study for system sizing in different environments. Agric Syst 171:143–154

Goddek S, Delaide BPL, Joyce A, Wuertz S, Jijakli MH, Gross A, Eding EH, Bläser I, Reuter M, Keizer LCP, Morgenstern R, Körner O, Verreth J, Keesman KJ (2018) Nutrient mineralization and organic matter reduction performance of RAS-based sludge in sequential UASB-EGSB reactors. Aquac Eng 83:10–19. https://doi.org/10.1016/J.AQUAENG.2018.07.003

IRENA (2015) Renewable energy capacity statistics 2015, Masdar: IRENA

MemSys (2017) Membrane distillation system - MDS 500

Pudjianto D, Ramsay C, Strbac G (2007) Virtual power plant and system integration of distributed energy resources. IET Renew Power Gener 1:10. https://doi.org/10.1049/iet-rpg:20060023

Strbac G, Hatziargyriou N, Lopes JP, Moreira C, Dimeas A, Papadaskalopoulos D (2015) Microgrids: enhancing the resilience of the European megagrid. IEEE Power Energ Mag 13:35–43. https://doi.org/10.1109/MPE.2015.2397336

Yogev U, Barnes A, Gross A (2016) Nutrients and energy balance analysis for a conceptual model of a three loops off grid. Aquaponics Water 8:589. https://doi.org/10.3390/W8120589

Chapter 16
Aquaponics for the Anthropocene: Towards a 'Sustainability First' Agenda

James Gott, Rolf Morgenstern, and Maja Turnšek

Abstract 'The Anthropocene' has emerged as a unique moment in earth history where humanity recognises its devastating capacity to destabilise the planetary processes upon which it depends. Modern agriculture plays a central role in this problematic. Food production innovations are needed that exceed traditional paradigms of the Green Revolution whilst at the same time are able to acknowledge the complexity arising from the sustainability and food security issues that mark our times. Aquaponics is one technological innovation that promises to contribute much towards these imperatives. But this emergent field is in an early stage that is characterised by limited resources, market uncertainty, institutional resistance and high risks of failure—a developmental environment where hype prevails over demonstrated outcomes. Given this situation, the aquaponics research community potentially holds an important place in the development path of this technology. But the field needs to craft a coherent and viable vision for this technology that can move beyond misplaced techno-optimist accounts. Turning to sustainability science and STS research, we discuss the urgent need to develop what we call a 'critical sustainability knowledge' for aquaponics, giving pointers for possible ways forward, which include (1) expanding aquaponic research into an interdisciplinary research domain, (2) opening research up to participatory approaches in real-world contexts and (3) pursuing a solution-oriented approach for sustainability and food security outcomes.

Keywords Anthropocene · Green Revolution · Techno-optimism · STS research (Science, Technology and Society research)

J. Gott (✉)
Geography and Environment, University of Southampton, Southampton, UK
e-mail: j.gott@soton.ac.uk

R. Morgenstern
Department of Agriculture, University of Applied Sciences of South Westphalia, Soest, Germany
e-mail: morgenstern.rolf@fh-swf.de

M. Turnšek
Faculty of Tourism, University of Maribor, Brežice, Slovenia
e-mail: maja.turnsek@um.si

S. Goddek et al. (eds.), *Aquaponics Food Production Systems*,
https://doi.org/10.1007/978-3-030-15943-6_16

16.1 Introduction

Key drivers stated for aquaponic research are the global environmental, social and economic challenges identified by supranational authorities like the Food and Agriculture Organization (FAO) of the United Nations (UN) (DESA 2015) whose calls for sustainable and stable food production advance the 'need for new and improved solutions for food production and consumption' (1) (Junge et al. 2017; Konig et al. 2016). There is growing recognition that current agricultural modes of production cause wasteful overconsumption of environmental resources, rely on increasingly scarce and expensive fossil fuel, exacerbate environmental contamination and ultimately contribute to climate change (Pearson 2007). In our time of 'peak-everything' (Cohen 2012), 'business as usual' for our food system appears at odds with a sustainable and just future of food provision (Fischer et al. 2007). A food system revolution is urgently needed (Kiers et al. 2008; Foley et al. 2011), and as the opening chapters (Chaps. 1 and 2) of this book attest, aquaponics technology shows much promise. The enclosed systems of aquaponics offer an especially alluring convergence of potential resolutions that could contribute towards a more sustainable future (Kőmíves and Ranka 2015). But, we ask, what kind of sustainable future might aquaponics research and aquaponics technology contribute towards? In this chapter, we take a step back to consider the ambitions of our research and the functions of our technology.

In this chapter we situate current aquaponic research within the larger-scale shifts of outlook occurring across the sciences and beyond due to the problematic that has become known as 'the Anthropocene' (Crutzen and Stoermer 2000b). Expanding well beyond the confines of its original geological formulation (Lorimer 2017), the Anthropocene concept has become no less than 'the master narrative of our times' (Hamilton et al. 2015). It represents an urgent realisation that demands deep questions be asked about the way society organises and relates to the world, including the *modus operandi* of our research (Castree 2015). However, until now, the concept has been largely sidelined in aquaponic literature. This chapter introduces the Anthropocene as an obligatory frame of reference that must be acknowledged for any concerted effort towards future food security and sustainability.

We discuss how the Anthropocene unsettles some key tenets that have underpinned the traditional agriscience of the Green Revolution (Stengers 2018) and how this brings challenges and opportunities for aquaponic research. Aquaponics is an innovation that promises to contribute much towards the imperatives of sustainability and food security. But this emergent field is in an early stage that is characterised by limited resources, market uncertainty, institutional resistance with high risks of failure and few success stories—an innovation environment where hype prevails over demonstrated outcomes (König et al. 2018). We suggest this situation is characterised by a misplaced techno-optimism that is unconducive to the deeper shifts towards sustainability that are needed of our food system.

Given this, we feel the aquaponics research community has an important role to play in the future development of this technology. We suggest a refocusing of aquaponics research around the key demands of our food system—sustainability and food security. Such a task entails we more thoroughly consider the nature of sustainability, and so we draw on the insights from the fields of sustainability science and STS. Addressing sustainability in the Anthropocene obligates the need to attend more holistically the interacting biophysical, social, economic, legal and ethical dimensions that encroach on aquaponic systems (Geels 2011). This is no small task that places great demands on the way we produce and use knowledge. For this reason we discuss the need to develop what we call a 'critical sustainability knowledge' for aquaponics, giving pointers for possible ways forward, which include (1) expanding aquaponic research into an interdisciplinary research domain, (2) opening research up to participatory approaches in real-world contexts and (3) pursuing a solution-oriented approach for sustainability and food security outcomes.

16.2 The Anthropocene and Agriscience

'Today, humankind has begun to match and even exceed some of the great forces of nature [...] [T]he Earth System is now in a no analogue situation, best referred to as a new era in the geological history, the Anthropocene' (Oldfield et al. 2004: 81).

The scientific proposal that the Earth has entered a new epoch—'the Anthropocene'—as a result of human activities was put forward at the turn of the new millennium by the chemist and Nobel Laureate Paul Crutzen and biologist Eugene Stoermer (Crutzen and Stoermer 2000a). Increasing quantitative evidence suggests that anthropogenic material flows stemming from fossil fuel combustion, agricultural production and mineral extraction now rival in scale those natural flows supposedly occurring outside of human activity (Steffen et al. 2015a). This is a moment marked by unprecedented and unpredictable climatic, environmental and ecological events (Williams and Jackson 2007). The benign era of the Holocene has passed, so the proposal claims; we have now entered a much more unpredictable and dangerous time where humanity recognises its devastating capacity to destabilise planetary processes upon which it depends (Rockström et al. 2009, Steffen et al. 2015b; See chapter 1). The Anthropocene is therefore a moment of realisation, where the extent of human activities must be reconciled within the boundaries of biophysical processes that define the safe operating space of a stable and resilient Earth system (Steffen et al. 2015b).

A profound intertwining of the fates of nature and humankind has emerged (Zalasiewicz et al. 2010). The growing awareness of environmental and human calamity—and our belated, tangled role within it—puts to test our faith in the key modernist assumption, namely, the dualisms separating humans from nature (Hamilton et al. 2015). This is a shocking and unprecedented moment because modernist epistemologies have proven exceedingly powerful, contributing significantly towards the organisation of society to the present day (Latour 1993). Conceptions of unique and stable human agency, the presumption of progressive norms such as liberty or universal dignity, and the existence of an objective world separate from human doings are all put to test (Latour 2015; Hamilton et al. 2015).

This insight, without doubt, applies to the food system of which we all inherit. The Green Revolution[1] was underpinned with modern aspirations, being founded on ideas such as linear notions of progress, the power of human reason and faith in the inevitable technological resolution of human problems (Cota 2011). These conceptions, which have traditionally secured the role of science in society, begin to appear increasingly unreliable with the advent of the Anthropocene (Savransky 2013; Stengers 2015). The inconvenient truth is that the technoscientific interventions, which have been implemented as modern agrarian solutions onto our world over the last century, have carried with them serious and unexpected outcomes. What's more, these escalating biophysical disruptions (e.g. greenhouse gas emissions and nitrogen and phosphorous cycle perturbations) that have only recently become perceived must be added to a much broader series of environmental, biological and social repercussions brought about by particular aspects of our modernised food system.

The Anthropocene problematic leaves little doubt that our contemporary food system faces enormous challenges (Kiers et al. 2008; Baulcombe et al. 2009; Pelletier and Tyedmers 2010). Prominent studies point to agriculture as the single largest contributor to the rising environmental risks posed in the Anthropocene (Struik and Kuyper 2014; Foley et al. 2011). Agriculture is the single largest user of freshwater in the world (Postel 2003); the world's largest contributor to altering the global nitrogen and phosphorus cycles and a significant source (19–29%) of greenhouse gas emissions (Vermeulen et al. 2012; Noordwijk 2014). Put simply, 'agriculture is a primary driver of global change' (Rockström et al. 2017:6). And yet, it is from within the new epoch of the Anthropocene that the challenge of feeding humanity must be resolved. The number of hungry people in the world persists at approximately 900 million (FAO, Ifad and WFP. 2013). Even then, in order to feed the world by 2050, best estimates suggest that production must roughly double to keep pace with projected demands from population growth, dietary changes (particularly meat consumption) and increasing bioenergy use (Kiers et al. 2008; Baulcombe et al. 2009; Pelletier and Tyedmers 2010; Kearney 2010). Complicating matters even further is the need not simply to produce more, but also to manage the entire food system more efficiently. In a world where 2 billion suffer from micronutrient deficiencies, whilst 1.4 billion adults are over-nourished, the need for better distribution, access and nutrition is glaring, as is the drastic need to reduce the deplorable levels of waste (conservative estimates suggest 30%) in the farm-to-fork supply chain (Parfitt et al. 2010; Lundqvist et al. 2008; Stuart 2009).

---

[1]The Green Revolution refers to a set of research and technology transfer initiatives occurring from the 1930s and the late 1960s that increased agricultural production worldwide, particularly in the developing world. As Farmer (1986) describes, these initiatives resulted in the adoption of new technologies, including: 'New, high-yielding varieties of cereals... in association with chemical fertilizers and agro-chemicals, and with controlled water-supply... and new methods of cultivation, including mechanization. All of these together were seen as a "package of practices" to supersede "traditional" technology and to be adopted as a whole'.

The Anthropocene problematic presents serious questions about modern industrial agriculture, which in many guises is now deemed inefficient, destructive and inadequate for our new global situation. But the fallout of this situation is more considerable still, for the Anthropocene strikes a challenge at the very agricultural paradigm currently dominating food provision (Rockström et al. 2017). For this reason the challenge extends well beyond 'the farm' and incorporates a much wider set of structures, practices and beliefs that continue to enact and propel the modern agricultural paradigm into our newly demanding epoch. With this comes the urgent need to reconsider the methods and practices, ambitions and goals that define our current agriscience research. Are they fit for the challenges of our new epoch, or do they merely reproduce inadequate visions of modernist food provision?

16.3 Getting Beyond the Green Revolution

The Anthropocene marks a step change in the relation between humans and our planet. It demands a rethink of the current modes of production that currently propel us on unsustainable trajectories. Until now, such reflexive commitments have not been required of agriscience research and development. It is worth remembering that the Green Revolution, in both its ambitions and methods, was for some time uncontroversial; agriculture was to be intensified and productivity per unit of land or labour increased (Struik 2006). Without doubt, this project, whose technological innovations were vigorously promoted by governments, companies and foundations around the world (Evenson and Gollin 2003), was phenomenally successful across vast scales. More calories produced with less average labour time in the commodity system was the equation that allowed the cheapest food in world history to be produced (Moore 2015). In order to simplify, standardise and mechanise agriculture towards increases in productivity per worker, plant and animal, a series of biophysical barriers had to be overridden. The Green Revolution achieved this largely through non-renewable inputs.

In the Anthropocene, this agricultural paradigm that marked the Green Revolution runs up against (geological) history. Growing awareness is that this 'artificialised' agricultural model, which substitutes each time more ecological processes with finite chemical inputs, irrigation and fossil fuel (Caron et al. 2014), literally undermines the foundations of future food provision. The biophysical contradictions of late-capitalist industrial agriculture have become increasingly conspicuous (Weis 2010). Moreover, the dramatic environmental, economic and social consequences of contemporary models of high-intensity artificialised agriculture have become an escalating concern for a globalised food system manifesting accelerating contradictions (Kearney 2010; Parfitt et al. 2010).

During the post-war period (mid-40s–70s), secure economic growth was founded on the accelerated extraction of fossil fuel, and as Cota (Cota 2011) notes, agriscience development during this time progressed more in tune with the geochemical sciences than the life sciences. Agricultural production designed around the cheapest maximum yields had been simplified and unified into monocrops, made to depend on mechanisation and agrochemical products. Although highly effective when first implemented, the efficiency of these commercial inputs has witnessed diminishing returns (Moore 2015). Following the oil crises of the 70s, the productivist ideals of the Green Revolution fell more upon the life sciences, particularly in the guise of agri-biotech, which has grown into a multibillion-dollar industry.

Feeding the globe's exploding population has been the key concern in a decade-long productivist narrative that has served to secure the prominent position of agricultural biotech in our current food system (Hunter et al. 2017). The great shock is that this highly advanced sector has done little to improve intrinsic yields. World agricultural productivity growth slowed from 3% a year in the 1960s to 1.1% in the 1990s (Dobbs et al. 2011). Recently, the yields of key crops have in some places approached plateaux in production (Grassini et al. 2013). Mainstream agroscientists have voiced concern that the maximum yield potential of current varieties is fast approaching (Gurian-Sherman 2009). On top of this, climate change is estimated to have already reduced global yields of maize and wheat by 3.8% and 5.5%, respectively (Lobell et al. 2011), and some warn of sharp declines in crop productivity when temperatures exceed critical physiological thresholds (Battisti and Naylor 2009).

The waning efficiency gains of artificial inputs added to the biological limits of traditional varieties is a situation that, for some, further underscores the need to accelerate the development of genetically engineered varieties (Prado et al. 2014). Even then, the greatest proponents of GM—the biotech firms themselves—are aware that GM interventions rarely work to increase yield, but rather to *maintain* it through pesticide and herbicide resistance (Gurian-Sherman 2009). As such, agricultural production has become locked into a cycle that requires the constant replacement of new crop varieties and product packages to overcome the growing negative environmental and biological impingements upon yield [2]. Melinda Cooper's (2008: 19) influential analysis of agro-biotechnology has traced how neoliberal modes of production become relocated ever more within the genetic, molecular and cellular levels. As such, the commercialisation of agrarian systems increasingly extends towards the capture of germplasm and DNA, towards 'life itself' (Rose 2009). Cooper's (2008) diagnosis is that we are living in an era of capitalist delirium characterised by its attempt to overcome biophysical limits of our earth through the speculative biotechnological reinvention of the future. In this respect, some have argued that rather than overcoming weaknesses of the conventional paradigm, the narrow focus of GM interventions seems only to intensify its central characteristics (Altieri 2007).

Amidst the deceleration of yield increases, the estimated targets of 60–100% increases in production needed by 2050 (Tilman et al. 2011; Alexandratos and Bruinsma 2012) appear increasingly daunting. As compelling and clear as these targets may be, concerns have been raised that productivist narratives have eclipsed other pressing concerns, namely, the environmental sustainability of production (Hunter et al. 2017) and food security (Lawrence et al. 2013). The current agricultural paradigm has held production first and sustainability as a secondary task of mitigation (Struik et al. 2014).

Thirty years of frustrated sustainability talk within the productivist paradigm are testament to the severe difficulties for researchers and policymakers alike to bridge the gap between sustainability theory and practice (Krueger and Gibbs 2007). 'Sustainability' as a concept had initially had revolutionary potential. Key texts such as the Club of Rome's *The Limits of Growth* (Meadows et al. 1972), for instance, contained an imminent critique of global development narratives. But researchers have pointed out the way that 'sustainability' throughout the 80s and 90s became assimilated into neoliberal growth discourse (Keil 2007). We now have a situation where, on the one hand, global sustainability is almost unanimously understood as a prerequisite to attain human development across all scales—from local, to city, nation and the world (Folke et al. 2005)—whilst on the other, despite substantial efforts in many levels of society towards the creation of a sustainable future, key global-scale indicators show that humanity is actually moving away from sustainability rather than towards it (Fischer et al. 2007). This is in spite of the increasing regularity of high-profile reports that evermore underscore the grave risks of existing trends to the long-term viability of ecological, social and economic systems (Steffen et al. 2006; Stocker 2014; Assessment 2003; Stern 2008). This situation—the widening gap between our current trajectory and all meaningful sustainability targets—has been discussed as the so-called 'paradox of sustainability' (Krueger and Gibbs 2007). Prevailing discourse on food security and sustainability continues to galvanise growth-oriented developmental imperatives (Hunter et al. 2017).

Agriscience research and development proliferated in accordance with the dominant politico-economic structures that defined planetary development over the last 30 years (Marzec 2014). Although the negative effects of the so-called 'Chicago School' of development have by now been well documented (Harvey 2007), biotechnological innovation remains rooted within neoliberal discourse (Cooper 2008). These narratives consistently present global markets, biotech innovation and multinational corporate initiatives as the structural preconditions for food security and sustainability. The empirical credibility of such claims has long been challenged (Sen 2001), but seem especially relevant amidst the accumulating history of chronic distributional failures and food crises that mark our times. It is worth repeating Nally's (2011; 49) point: 'The spectre of hunger in a world of plenty seems set to continue into the 21st century. . . this is not the failure of the modern food regime, but the logical expression of its central paradoxes'. The situation is one where malnutrition is seen no longer as a failure of an otherwise efficiently functioning system, but rather as an endemic feature within the systemic production of scarcity (Nally 2011). In the face of such persisting inconsistencies, commentators note that neoliberal appeals to human prosperity, food security and green growth appear out of touch and often ideologically driven (Krueger and Gibbs 2007).

The Anthropocene is a time where ecological, economic and social disaster walk hand in hand as modern economies and institutions geared towards unlimited growth crash against the finite biophysical systems of the earth (Altvater et al. 2016; Moore

2015). Cohen (2013) describes the Anthropocene as an 'eco-eco' disaster, paying heed to the rotten relationship in which eco*nomic* debt becomes compounded against the eco*logical* debt of species extinction. Now more than ever, faith in the modernising powers of neoliberal food interventions proclaiming just and sustainable futures wears thin (Stengers 2018), yet the resemblance noted by some commentators (Gibson-Graham 2014), between our food system and the unhinged financial systems of our neoliberal economies charts an alarming trend. It's worth noting this resemblance runs deeper than the mere production of debt (one being calorific and genetic, the other economic). The truth is our food system hinges on a cash nexus that links trade tariffs, agricultural subsidies, enforcement of intellectual property rights and the privatisation of public provisioning systems. Viewed from above, these procedures constitute a pseudo-corporate management of the food system, which according to Nally (2011: 37) should be seen as a properly *biopolitical* process designed for managing life, "including the lives of the hungry poor who are 'let die' as commercial interests supplant human needs". Petrochemicals and micronutrients, it seems, are not the only things being consumed in the Anthropocene; futures are (Collings 2014; Cardinale et al. 2012).

What once might have been considered necessary side effects of the modernising imperative of the Green Revolution, the so-called 'externalities' of our current food system, are increasingly exposed as a kind of 'deceptive efficiency' bent towards rapid production and profit and very little else (Weis 2010). The disturbing realisation is that the food system we inherit from the Green Revolution creates value only when a great number of costs (physical, biological, human, moral) are allowed to be overlooked (Tegtmeier and Duffy 2004). A growing number of voices remind us that costs of production go beyond the environment into matters such as the exclusion of deprived farmers, the promotion of destructive diets (Pelletier and Tyedmers 2010) and more generally the evacuation of social justice and political stability from matters of food provision (Power 1999). The relation between agrarian technological intervention, food security and sustainability emerges as far wider and complex issue than could be acknowledged by narratives of the Green Revolution.

Situating the contemporary food system within dominant recent historical processes, the above discussion has paid particular attention to destructive links between modern agriculture and the economic logics of late capitalism. It is important, however, to remember that numerous commentators have cautioned against oversimplified or deterministic accounts regarding the relationship between capitalist relations of production and the Anthropocene problematic (Stengers 2015; Haraway 2015; Altvater et al. 2016). Such a discussion is made possible by close to four decades of critical investigations by feminists, science and technology scholars, historians, geographers, anthropologists and activists, which have endeavoured in tracing the links between hegemonic forms of science and the social/environmental destruction caused by industrial capitalism (Kloppenburg 1991). This 'deconstructive' research ethic developed important understandings of the way modern agriscience progressed down trajectories that involve the neglect of particular physical, biological, political and social contexts and histories (Kloppenburg 1991). In many instances, the modernising narratives of 'development' like those put to work in the Green Revolution became seen—by anthropologists, historians and indigenous communities alike—as a kind of modified successor to pre-war colonial discourse (Scott 2008; Martinez-Torres and Rosset 2010). In anthropological terms, what these studies taught us was that although modern agriculture was rooted in developmental narratives of universal prosperity, in reality, 'progress' was achieved through the displacement or indeed destruction of a great diversity of agricultural perspectives, practices, ecologies and landscapes. It is for this reason Cota (2011: 6) reminds us of the importance of the critical work that explicitly positioned the biopolitical paradigm of industrial agriculture 'not first and foremost as an economic kind of imperialism, but more profoundly as an epistemic and culturally specific kind of imperialism'.

This is a key point. The Green Revolution was not merely a technical, nor economic intervention, but involved the spread of a more profound reconfiguration of the epistemological registers of food provision itself. It was a process that deeply influenced the way agricultural knowledge was produced, propagated and implemented. As Cota (2011: 6) explains: 'the use of physicalist and probabilistic discourse, a purely instrumental conception of nature and work, the implementation of statistical calculations disconnected from local conditions, [as well as] the reliance on models without recognizing historic specificities' were all ways of enacting the biopolitical agenda of the Green Revolution. This list of commitments describes the fundamentals at the sharp end of the Green Revolution, but as we have seen, such commitments alone have proven insufficient for the task of creating a just and sustainable food system. It becomes apparent that any research agenda fit for the Anthropocene must learn to move beyond the modern food paradigm by forging a different research ethic with different commitments.

16.4 Paradigm Shift for a New Food System

To claim that Agriculture is 'at a crossroads' (Kiers et al. 2008) does not quite do justice to the magnitude of the situation. The gaping 'sustainability gap' (Fischer et al. 2007) amidst unanimous calls for sustainability are increasingly being met with common response amongst researchers: pleas for revolutionary measures and paradigm shifts. Foley et al. (2011: 5) put it quite directly: 'The challenges facing agriculture today are unlike anything we have experienced before, and they require revolutionary approaches to solving food production and sustainability problems. In short, new agricultural systems must deliver more human value, to those who need it most, with the least environmental harm'. Somehow, world agriculture's current role as the single largest driver of global environmental change must shift into a 'critical agent of a world transition' towards global sustainability within the biophysical safe operating space of the Earth (Rockström et al. 2017).

The Anthropocene lays steep demands: Agriculture must be intensified; it must meet the needs of a growing population, but at the same time it is mandatory that the pressures exerted by our food production systems stay within the carrying capacity of Planet Earth. It is increasingly understood that future food security depends on the development of technologies that increase the efficiency of resource use whilst simultaneously preventing the externalisation of costs (Garnett et al. 2013). The search for alternatives to our current agricultural paradigm has brought to the fore ideas such as agroecology (Reynolds et al. 2014) and 'sustainable intensification', with the acknowledgement that real progress must be made towards 'ecological intensification', that is, increasing agricultural output by capitalising on the ecological processes in agroecosystems (Struik and Kuyper 2014).

There has been well-documented debate on what constitutes 'sustainable intensification' (SI) of agriculture as well as the role it might play in addressing global food security (Struik and Kuyper 2014; Kuyper and Struik 2014; Godfray and Garnett 2014). Critics have cautioned against the top-down, global analyses that are often framed in narrow, production-oriented perspectives, calling for a stronger engagement with the wider literature on sustainability, food security and food sovereignty (Loos et al. 2014). Such readings revisit the need for developing regionally grounded, bottom-up approaches, with a growing consensus claiming that an SI agenda fit for the Anthropocene does not entail 'business-as-usual' food production with marginal improvements in sustainability but rather a radical rethinking of food systems not only to reduce environmental impacts but also to enhance animal welfare, human nutrition and support rural/urban economies with sustainable development (Godfray and Garnett 2014).

While traditional 'sustainable intensification' (SI) has been criticised by some as too narrowly focused on production, or even as a contradiction in terms altogether (Petersen and Snapp 2015), others make it clear that the approach must be broadly conceived, with the acknowledgement that there is no single universal pathway to sustainable intensification (Garnett and Godfray 2012). Important here is the growing appreciation of 'multifunctionality' in agriculture (Potter 2004). If, during the twentieth century, 'Malthusian' demographics discourse had secured the narrow goal of agricultural development on increasing production, the growing rediscovery of the multiple dimensions of farming currently taking place is altering the perception of the relationship between agriculture and society.

'Multifunctionality' as an idea was initially contested in the context of the controversial GATT and WTO agricultural and trade policy negotiations (Caron et al. 2008), but has since gained wide acceptance, leading to a more integrative view of our food system (Potter 2004). In this view, progress in seeing agriculture as an important type of 'land use' competing with other land functions (Bringezu et al. 2014) interrelates with a number of other perspectives. These have been conceptualised through several important categories: (1) as a source of employment and livelihood for a rural and future urban population (McMichael 1994); (2) as a key part of cultural heritage and identity (van der Ploeg and Ventura 2014); (3) as the basis of complex value chain interactions in 'food systems' (Perrot et al. 2011); (4) as a sector in regional, national and global economies (Fuglie 2010); (5) as modifier and storehouse of genetic resources (Jackson et al. 2010); (6) as a threat to environmental integrity that exerts destructive pressures on biodiversity (Brussaard et al. 2010; Smil 2011); and (7) as a source of greenhouse gas emissions (Noordwijk

2014). This list is by no means comprehensive, but what is important is that each of these interacting dimensions is understood to impact sustainability and food security in one way or another and must be apprehended by serious attempts towards SI.

Sustainability outcomes are increasingly seen as a complex interplay between local and global concerns (Reynolds et al. 2014). Biophysical, ecological and human needs intermix within the complexities and idiosyncrasies of 'place' (Withers 2009). The 'one size fits all' solutions, characteristics of the Green Revolution, fail to acknowledge these unique sustainability potentials and demands. The result is that changes in food production and consumption must be perceived through a multiplicity of scales *and* styles. To this end, Reynolds et al. (2014) suggest an approach to sustainability that takes advantage of the insights of agroecological principles. They forward a 'custom-fit' food production focus 'explicitly tailored to the environmental and cultural individuality of place and respectful of local resource and waste assimilative limits, thus promoting biological and cultural diversity as well as steady-state economics'.

If the issues at stake are inherently *multidimensional*, others have also underlined that they are *contested*. Trade-offs between the plethora of biophysical and human concerns are inevitable and often exceedingly complex. Sustainability thresholds are diverse, often normative, and can seldom all be realised in full simultaneously (Struik and Kuyper 2014). It has been emphasised that new directions towards sustainability and food security require simultaneous change at the level of formal and informal social rules and incentive systems (i.e. institutions) that orient human interaction and behaviour, and hence that 'institutional innovation' is held to be a key entry point in addressing challenges (Hall et al. 2001). Insomuch as the complexity of sustainable intensification derives from human framings (which entail and flow from contexts, identities, intentions, priorities and even contradictions), they are, as Kuyper and Struik (2014: 72) put it, 'beyond the command of science'. Attempting to reconcile the many dimensions of food production towards sustainable ends and within the bounds of our finite planet involves a great deal of uncertainty, irreducibility and contestation (Funtowicz and Ravetz 1995); it requires an awareness and acknowledgement that such issues are shot through with *political* implication.

Food systems and sustainability research have come a long way in expanding the narrow focus of the Green Revolution, bringing greater clarity to the steep challenges we face in the pursuit of a more environmentally and socially sustainable food system. Thanks to a broad range of work, it is now apparent that food production lies at the heart of a nexus of interconnected and multi-scalar processes, on which humanity relies upon to meet a host of multidimensional—often contradictory—needs (physical, biological, economic, cultural). As Rockström et al. (2017: 7) have stated: 'World agriculture must now meet social needs and fulfil sustainability criteria that enables food and all other agricultural ecosystem services (i.e., climate stabilization, flood control, support of mental health, nutrition, etc.) to be generated within a safe operating space of a stable and resilient Earth system'. It is precisely within these recalibrated agricultural goals that aquaponics technology must be developed.

16.5 Aquaponic Potential or Misplaced Hope?

Contemporary aquaponic research has shown keen awareness of particular concerns raised in the Anthropocene problematic. Justifications for aquaponic research have tended to foreground the challenge of food security on a globe with an increasing human population and ever strained resource base. For instance, König et al. (2016) precisely situate aquaponics within the planetary concerns of Anthropocene discourse when they state: 'Assuring food security in the twenty-first century within sustainable planetary boundaries requires a multi-faceted agro-ecological intensification of food production and the decoupling from unsustainable resource use'. Towards these important sustainability goals, it is claimed that aquaponic technology shows much promise (Goddek et al. 2015). The innovative enclosed systems of aquaponics offer an especially alluring convergence of potential resolutions that could contribute towards a more sustainable future.

Proponents of aquaponics often stress the ecological principles at the heart of this emerging technology. Aquaponic systems harness the positive potential of a more or less simple ecosystem, in order to reduce the use of finite inputs whilst simultaneously reducing waste by-products and other externalities. On these grounds, aquaponic technology can be viewed as a primary example of 'sustainable intensification' (Garnett et al. 2013) or, more precisely, as a form of 'ecological intensification' since its founding principles are based on the management of service-providing organisms towards quantifiable and direct contributions to agricultural production (Bommarco et al. 2013). From this agroecological principle flow a great number of potential sustainability benefits. Chapters 1 and 2 of this book do an exemplary job of highlighting these, detailing the challenges faced by our food system and situating aquaponics science as the potential locus for a range of sustainability and food security interventions. There is no need to repeat these points again, but it is worth noting this perceived convergence of potential resolutions is what drives research and strengthens the 'conviction that this technology has the potential to play a significant role in food production in the future' (7) (Junge et al. 2017).

However, despite the considerable claims made by its proponents, the future of aquaponics is less than certain. Just what kind of role aquaponics might play in transitions to sustainable food provision is still largely up for debate—crucially, we must stress, *the publication of sustainability and food security outcomes of aquaponic systems remain conspicuous by their absence across Europe* (König et al. 2018). On paper, the 'charismatic' attributes of aquaponics ensure that it can easily be presented as a 'silver bullet' type of innovation that gets to the heart of our food system's deepest sustainability and food security issues (Brooks et al. 2009). Such images have been able to garner considerable attention for aquaponics far beyond the confines of academic research—consider, for instance, the significant production of online aquaponic 'hype' in comparison to similar fields, usefully pointed out by Junge et al. (2017). It is here we may take time to point out the relationship between the perceived potential of aquaponics and 'techno-optimism'.

The introduction of every new technology is accompanied by myths that spur further interest in that technology (Schoenbach 2001). Myths are circulated amongst early adopters and are picked up by the general media often long before the scientific community has time to thoroughly analyse and answer to their claims. Myths, as Schoenbach states (2001, 362), are widely believed because they 'comprise a clear-cut and convincing explanation of the world'. These powerful explanations are able to energise and align individual, community and also institutional action towards particular ends. The 'beauty' of aquaponics, if we can call it that, is that the concept can often render down the complexity of sustainability and food security issues into clear, understandable and scalable systems metaphors. The ubiquitous image of the aquaponic cycle—water flowing between fish, plants and bacteria—that elegantly resolves food system challenges is exemplary here. However, myths on technology, whether optimistic or pessimistic, share a techno-deterministic vision of the relation between technology and society (Schoenbach 2001). Within the techno-deterministic vision of technology, it is the technology that causes important changes in society: if we manage to change the technology, we thus manage to change the world. Regardless whether the change is for the better (techno-optimism) or the worse(technophobia), the technology by itself creates an effect.

Techno-determinist views have been thoroughly critiqued on sociological, philosophical (Bradley 2011), Marxist (Hornborg 2013), material-semiotic (Latour 1996) and feminist (Haraway 1997) grounds. These more nuanced approaches to technological development would claim that technology by itself does not bring change to society; it is neither inherently good nor bad but is always embedded within society's structures, and it is those structures that enable the use and effect of the technology in question. To one degree or another, technology is an emergent entity, the effects of which we cannot know in advance (de Laet and Mol 2000). This might seem like an obvious point, but techno-determinism remains a strong, if often latent, feature within our contemporary epistemological landscape. Our innovation-driven, technological societies are maintained by discursive regimes that hold on to the promise of societal renewal through technological advancement (Lave et al. 2010). Such beliefs have been shown to have an important normative role within expert communities whether they be scientists, entrepreneurs or policymakers (Franklin 1995; Soini and Birkeland 2014).

The rise of aquaponics across Europe is intertwined with specific interests of various actors. We can identify at least five societal processes that led to the development of aquaponics: (a) interest of public authorities in funding high-tech solutions for problems of sustainability; (b) venture capital financing, motivated by the successes in IT startups, looking for 'the next big thing' that will perhaps discover the new 'unicorn' (startup companies valued at over $1 billion); (c) mass media event-focused interest in snapshot reporting on positive stories of new aquaponics startups, fuelled by the public relations activities of these startups, with rare media follow-up reporting on the companies that went bust; (d) internet-supported growth of enthusiastic, do-it-yourself aquaponics communities, sharing both sustainability values and love for tinkering with new technology; (e) interests of urban developers to find economically viable solutions for vacant urban spaces and greening of urban space; and (f) research communities focused on developing technological solutions to impending sustainability and food security problems. To a greater or lesser degree, the spectre of techno-optimistic hope permeates the development of aquaponics.

Although the claims of techno-optimist positions are inspiring and able to precipitate the investment of money, time and resources from diverse actors, the potential for such standpoints to generate justice and sustainability has been questioned on scales from local (Leonard 2013) and regional issues (Hultman 2013) to global imperatives (Hamilton 2013). And it is at this point, we might consider the ambitions of our own field. A good starting point would be the 'COST action FA1305', which has been an important facilitator of Europe's aquaponic research output over recent years, with a number of publications acknowledging the positive impact of the action in enabling research (Miličić et al. 2017; Delaide et al. 2017; Villarroel et al. 2016). Like all COST actions, this EU-funded transnational networking instrument has acted as a hub for aquaponic research in Europe, galvanising and broadening the traditional networks amongst researchers by bringing together experts from science, experimental facilities and entrepreneurs. The original mission statement of COST action FA1305 reads as follows:

> *Aquaponics has a key role to play in food provision and tackling global challenges such as water scarcity, food security, urbanization, and reductions in energy use and food miles.* The EU acknowledges these challenges through its Common Agriculture Policy and policies on Water Protection, Climate Change, and Social Integration. A European approach is required in the globally emerging aquaponics research field building on the foundations of Europe's status as a global centre of excellence and technological innovation in the domains of aquaculture and hydroponic horticulture. The EU Aquaponics Hub aims to the development of aquaponics in the EU, by leading the research agenda through the creation of a networking hub of expert research and industry scientists, engineers, economists, aquaculturists and horticulturalists, and contributing to the training of young aquaponic scientists. The EU Aquaponics Hub focuses on three primary systems in three settings; (1) "cities and urban areas" – urban agriculture aquaponics, (2) "developing country systems" – devising systems and technologies for food security for local people and (3) "industrial scale aquaponics" – *providing competitive systems delivering cost effective, healthy and sustainable local food in the EU.* (http://www.cost.eu/COST_Actions/fa/FA1305, 12.10.2017, emphasis added).

As the mission statement suggests, from the outset of COST action FA1305, high levels of optimism were placed on the role of aquaponics in tackling sustainability and food security challenges. The creation of the COST EU Aquaponics Hub was to 'provide a necessary forum for 'kick-starting' aquaponics as a serious and potentially viable industry for sustainable food production in the EU and the world' (COST 2013). Indeed, from the authors' own participation within COST FA1305, our lasting experience was without doubt one of being part of a vibrant, enthused and highly skilled research community that were more or less united in their ambition to make aquaponics work towards a more sustainable future. Four years down the line since the Aquaponic Hub's mission statement was issued, however, the sustainability and food security potential of aquaponics remains just that—potential. At present it is uncertain what precise role aquaponics can play in Europe's future food system (König et al. 2018).

The commonly observed narrative that aquaponics provides a sustainable solution to the global challenges agriculture faces unveils a fundamental misconception of what it is actually capable to achieve. The plant side of aquaponics is horticulture, not agriculture, producing vegetables and leafy greens with high water content and low nutritional value compared to the staple foods agriculture on farmland produces. A quick comparison of current agricultural area, horticultural area and protected horticultural area, 184.332 km^2, 2.290 km^2 (1,3%) and 9,84 km^2 (0,0053%), in Germany, reveals the flaw in the narrative. Even if considering a much higher productivity in aquaponics through the utilisation of controlled environment systems, aquaponics is not even close to having the potential to make a real impact on agricultural practice. This becomes even more obvious when the ambition to be a 'food system of the future' ends in the quest for high-value crops (e.g. microgreens) that can be marketed as gourmet gastronomy.

It is well known that the development of sustainable technology is characterised by uncertainties, high risks and large investments with late returns (Alkemade and Suurs 2012). Aquaponics, in this regard, is no exception; only handful commercially operating systems exist across Europe (Villarroel et al. 2016). There appears considerable resistance to the development of aquaponic technology. Commercial projects have to contend with comparatively high technological and management complexity, significant marketing risks, as well as an uncertain regulatory situation that until now persists (Joly et al. 2015). Although it is difficult to pin down the rate of startup failure, the short history of commercial aquaponics across Europe might well be summed up as 'Small successes and big failures' (Haenen 2017). It is worth pointing out also that the pioneers already involved in aquaponics at the moment across Europe are unclear if their technology is bringing about any improvements in sustainability (Villarroel et al. 2016). Recent analysis from König et al. (2018) has shown how the challenges to aquaponics development derive from a host of structural concerns, as well as the technology's inherent complexity. Combined, these factors result in a high-risk environment for entrepreneurs and investors, which has produced a situation whereby startup facilities across Europe are forced to focus on production, marketing and market formation over the delivery of sustainability credentials (König et al. 2018). Aside from the claims of great potential, the sombre reality is that it remains to be seen just what impact aquaponics can have on the entrenched food production and consumption regimes operating in contemporary times. The place for aquaponic technology in the transition towards more sustainable food systems, it seems, has no guarantee.

Beyond the speculation of techno-optimism, aquaponics has emerged as a highly complex food production technology that holds potential but is faced with steep challenges. In general, there exists a lack of knowledge about how to direct research activities to develop such technologies in a way that preserves their promise of sustainability and potential solutions to pressing food system concerns (Elzen et al. 2017). A recent survey conducted by Villarroel et al. (2016) found that from

68 responding aquaponic actors spread across 21 European countries, 75% were involved in research activities and 30.8% in production, with only 11.8% of those surveyed actually selling fish or plants in the past 12 months. It is clear that the field of aquaponics in Europe is still mainly shaped by actors from research. In this developmental environment, we believe the next phase of aquaponic research will be crucial to developing the future sustainability and food security potential of this technology.

Interviews (König et al. 2018) and the quantitative surveys (Villarroel et al. 2016) of the European aquaponic field have indicated there is mixed opinion regarding the vision, motivations and expectations about the future of aquaponics. In light of this, Konig et al. (2018) have raised concerns that a diversity of visions for aquaponic technology might hinder the coordination between actors and ultimately disrupt the development of 'a realistic corridor of acceptable development paths' for the technology (König et al. 2018). From an innovation systems perspective, emergent innovations that display an unorganised diversity of visions can suffer from 'directionality failure' (Weber and Rohracher 2012) and ultimately fall short of their perceived potentials. Such perspectives run in line with positions from sustainability science that stress the importance of 'visions' for creating and pursuing desirable futures (Brewer 2007). In light of this, we offer up one such vision for the field of aquaponics. We argue that aquaponics research must refocus on a radical sustainability and food security agenda that is fit for the impending challenges faced in the Anthropocene.

16.6 Towards a 'Sustainability First' Paradigm

As we saw earlier, it has been stressed that the goal to move towards sustainable intensification grows from the acknowledgment of the limits of the conventional agricultural development paradigm and its systems of innovation. Acknowledging the need for food system innovations that exceed the traditional paradigm and that can account for the complexity arising from sustainability and food security issues, Fischer et al. (2007) have called for no less than 'a new model of sustainability' altogether. Similarly, in their recent plea for global efforts towards sustainable intensification, Rockström et al. (2017) have pointed out that a paradigm shift in our food system entails challenging the dominant research and development patterns that maintain the 'productivity first' focus whilst subordinating sustainability agendas to a secondary, 'mitigating' role. Instead, they call for a reversal of this paradigm so that 'sustainable principles become the *entry point* for generating productivity enhancements'. Following this, we suggest a *sustainability first* vision for aquaponics as one possible orientation that can both offer coherence to the field and guide its development towards the proclaimed goals of sustainability and food security.

As with most calls for sustainability, our *sustainability first* proposal might sound rather obvious and unchallenging at first glance, if not completely redundant—surely, we could say, aquaponics is all about sustainability. But history would remind us that making sustainability claims is an agreeable task, whereas securing sustainability outcomes is far less certain (Keil 2007). As we have argued, the 'sustainability' of aquaponics currently exists as potential. Just how this potential translates into sustainability outcomes must be a concern for our research community.

Our 'sustainability first' proposal is far from straightforward. First and foremost, this proposal demands that, if our field is to justify itself on the grounds of sustainability, we must get to grips with the nature of sustainability itself. In this regard, we feel there is much to be learned from the growing arena of sustainability science as well as Science and Technology Studies (STS). We will find that maintaining a sustainability focus within aquaponic research represents a potentially huge shift in the direction, composition and ambition of our research community. Such a task is necessary if we are to direct the field towards coherent and realistic goals that remain focussed on sustainability and food security outcomes that are relevant for the Anthropocene.

Taking sustainability seriously is a massive challenge. This is because, at its core, sustainability is fundamentally an *ethical* concept raising questions about the value of nature, social justice, responsibilities to future generations, etc. and encompasses the multidimensional character of human-environment problems (Norton 2005). As we discussed earlier, the sustainability thresholds that might be drawn up concerning agricultural practices are diverse and often cannot be reconciled in entirety, obligating the need for 'trade-offs' (Funtowicz and Ravetz 1995). Choices have to be made in the face of these trade-offs and most often the criteria upon which such choices are based depend not only upon scientific, technical or practical concerns but also on norms and moral values. It goes without saying, there is little consensus on how to make these choices nor is there greater consensus on the norms and moral values themselves. Regardless of this fact, inquiries into values are largely absent from the mainstream sustainability science agenda, yet as Miller et al. (2014) assert, 'unless the values [of sustainability] are understood and articulated, the unavoidable political dimensions of sustainability will remain hidden behind scientific assertions'. Such situations prevent the coming together of and democratic deliberation between communities—a certain task for achieving more sustainable pathways.

Taking note of the prominent place of values in collective action towards sustainability and food security, scholars from the field of science and technology studies have highlighted that rather than be treated as an important externality to research processes (often dealt with separately or after the fact), values must be moved upstream in research agendas (Jasanoff 2007). When values become a central part of sustainability research, along comes the acknowledgement that decisions can no longer be based on technical criteria alone. This has potentially huge impacts on the research process, because traditionally what might have been regarded as the sole remit of 'expert knowledge' must now be opened up to other knowledge streams (for instance, 'lay', indigenous and practitioner knowledge) with all the epistemological difficulty this entails (Lawrence 2015). In response to these problems, sustainability science has emerged as a field that aims to transcend disciplinary boundaries and seeks to involve non-scientists in solution-oriented, context-determined, research processes that are focused on outcome generation (Miller et al. 2014).

A key question in these discussions is knowledge. Sustainability problems are often caused by the complex interplay of diverse social–ecological factors, and the knowledge needed for effectively governing these challenges has become progressively more dispersed and specialised (Ansell and Gash 2008). The knowledge required for understanding how sustainability concerns hang together is too complex to be organised by a single body and results in the need to integrate different types of knowledge in new ways. This is certainly the case for our own field: like other modes of sustainable intensification (Caron et al. 2014), aquaponic systems are characterised by inherent complexity (Junge et al. 2017) which places great emphasis on new forms of knowledge production (FAO 2013). Complexity of aquaponic systems derives not only from their 'integrated' character but stems also from the wider economic, institutional and political structures that impact the delivery of aquaponics and its sustainability potential (König et al. 2016). Developing solutions towards sustainable aquaponic food systems may well involve contending with diverse realms of understanding from engineering, horticultural, aquacultural, microbiological, ecological, economic and public health research, to the practical and experiential knowledge concerns of practitioners, retailers and consumers. What this amounts to is not just a grouping together of ideas and positions, but entails developing entirely novel modes of knowledge production and an appreciation to bridge 'knowledge gaps' (Caron et al. 2014). Abson et al. (2017) have identified three key requirements of new forms of knowledge production that can foster sustainability transformations: (i) the explicit inclusion of values, norms and context characteristics into the research process to produce 'socially robust' knowledge; (ii) mutual learning processes between science and society, involving a rethink of the role of science in society; and (iii) a problem- and solution-oriented research agenda. Drawing upon these three insights can help our field develop what we call a 'critical sustainability knowledge' for aquaponics. Below we discuss three areas our research community can address that we consider crucial to unlocking the sustainability potential of aquaponics: partiality, context and concern. Developing an understanding of each of these points will help our field pursue a solution-oriented approach for aquaponic sustainability and food security outcomes.

16.7 'Critical Sustainability Knowledge' for Aquaponics

16.7.1 Partiality

Despite contemporary accounts of sustainability that underline its complex, multidimensional and contested character, in practice, much of the science that engages with sustainability issues remains fixed to traditional, disciplinary perspectives and actions (Miller et al. 2014). Disciplinary knowledge, it must be said, has obvious value and has delivered huge advances in understanding since antiquity. Nevertheless, the appreciation and application of sustainability issues through traditional disciplinary channels has been characterised by the historic failure to facilitate the deeper societal change needed for issues such as the one we contend with here—the sustainable transformation of the food system paradigm (Fischer et al. 2007).

The articulation of sustainability problems through traditional disciplinary channels often leads to 'atomised' conceptualisations that view biophysical, social and economic dimensions of sustainability as compartmentalised entities and assume these can be tackled in isolation (e.g. Loos et al. 2014). Instead of viewing sustainability issues as a convergence of interacting components that must be addressed together, disciplinary perspectives often promote 'techno-fixes' to address what are often complex multidimensional problems (e.g. Campeanu and Fazey 2014). A common feature of such framings is that they often imply that sustainability problems can be resolved without consideration of the structures, goals and values that underpin complex problems at deeper levels, typically giving little consideration to the ambiguities of human action, institutional dynamics and more nuanced conceptions of power.

The practice of breaking a problem down into discrete components, analysing these in isolation and then reconstructing a system from interpretations of the parts has been a hugely powerful methodological insight that traces its history back to the dawn of modernity with the arrival of Cartesian reductionism (Merchant 1981). Being a key tenet of the production of objective knowledge, this practice forms the bedrock of most disciplinary effort in the natural sciences. The importance of objective knowledge, of course, is in that it provides the research community with 'facts'; precise and reproducible insights about generally dispersed phenomena. The production of facts was the engine room of innovation that propelled the Green Revolution. Science fuelled 'expert knowledge' and provided penetrating information about dynamics in our food production systems that remained invariant through change in time, space or social location. Building a catalogue of this kind of knowledge, and deploying it as what Latour (1986) calls 'immutable mobiles', formed the basis of the universal systems of monocropping, fertilisation and pest control that characterise the modern food system (Latour 1986).

But this form of knowledge production has weaknesses. As any scientist knows, in order to gain significant insights, this method must be strictly applied. It has been shown that this knowledge production is 'biased toward those elements of nature which yield to its method and toward the selection of problems most tractable to solutions with the knowledge thereby produced' (Kloppenburg 1991). A clear example of this would be our imbalanced food security research agenda that heavily privileges production over conservation, sustainability or food sovereignty issues (Hunter et al. 2017). Most high-profile work on food security concentrates on production (Foley et al. 2011), emphasising material flows and budgets over deeper issues such as the structures, rules and values that shape food systems. The simple fact is that because we know more about material interventions it is easier to design, model and experiment on these aspects of the food system. As Abson et al. (2017: 2) point out: 'Much scientific lead sustainability applications assume some of the most challenging drivers of unsustainability can be viewed as "fixed system properties" that can be addressed in isolation'. In pursuing the paths along which experimental success is most often realised, 'atomised' disciplinary approaches neglect those areas where other approaches might prove rewarding. Such epistemological 'blind spots' mean that sustainability interventions are often geared towards highly tangible aspects that may be simple to envisage and implement, yet have weak potential for 'leveraging' sustainable transition or deeper system change (Abson et al. 2017). Getting to grips with the limits and partialities of our disciplinary knowledge is one aspect that we stress when we claim the need to develop a 'critical sustainability knowledge' for aquaponics.

Viewed from disciplinary perspectives the sustainability credentials of aquaponic systems can be more or less simple to define (for instance, water consumption, efficiency of nutrient recycling, comparative yields, consumption of non-renewable inputs, etc.). Indeed, the more narrowly we define the sustainability criteria, the more straightforward it is to test such parameters, and the easier it is to stamp the claim of sustainability on our systems. The problem is that we can engineer our way to a form of sustainability that only few might regard as sustainable. To paraphrase Kläy et al. (2015), when we transform our original concern of how to realise a sustainable food system into a 'matter of facts' (Latour 2004) and limit our research effort to the analysis of these facts, we subtly but profoundly change the problem and direction of research. Such an issue was identified by Churchman (1979:4–5) who found that because science addresses mainly the identification and the solution of problems, and not the systemic and related ethical aspects, there is always the risk that the solutions offered up may even increase the unsustainability of development—what he called the 'environmental fallacy' (Churchman 1979).

We might raise related concerns for our own field. Early research in aquaponics attempted to answer questions concerning the environmental potential of the technology, for instance, regarding water discharge, resource inputs and nutrient recycling, with research designed around small-scale aquaponic systems. Although admittedly narrow in its focus, this research generally held sustainability concerns in focus. Recently, however, we have detected a change in research focus. This is raised in Chap. 1 of this book, whose authors share our own view, observing that research 'in recent years has increasingly shifted towards economic feasibility in order to make aquaponics more productive for large-scale farming applications'. Discussions, we have found, are increasingly concerned with avenues of efficiency and profitability that often fix the potential of aquaponics against its perceived competition with other large-scale production methods (hydroponics and RAS). The argument appears to be that only when issues of system productivity are solved, through efficiency measures and technical solutions such as optimising growth conditions of plants and fish, aquaponics becomes economically competitive with other industrial food production technologies and is legitimated as a food production method.

We would certainly agree that economic viability is an important constituent of the long-term resilience and sustainability potential of aquaponics. However, we would caution against too narrowly defining our research ethic—and indeed, the future vision of aquaponics—based on principles of production and profit alone. We worry that when aquaponic research is limited to efficiency, productivity and market competitiveness, the old logics of the Green Revolution are repeated and our claims to food security and sustainability become shallow. As we saw earlier, productionism has been understood as a process in which a logic of production overdetermines other activities of value within agricultural systems (Lilley and Papadopoulos 2014). Since sustainability inherently involves a complex diversity of values, these narrow avenues of research, we fear, risk the articulation of aquaponics within a curtailed vision of sustainability. Asking the question 'under what circumstances can aquaponics outcompete traditional large-scale food production methods?' is not the same as asking 'to what extent can aquaponics meet the sustainability and food security demands of the Anthropocene?'.

16.7.2 Context

Knowledge production through traditional disciplinary pathways involves a loss of context that can narrow our response to complex sustainability issues. The multidimensional nature of food security implies that 'a single globally valid pathway to sustainable intensification does not exist' (Struik and Kuyper 2014). The physical, ecological and human demands placed on our food systems are context-bound and, as such, so are the sustainability and food security pressures which flow from these needs. Intensification requires contextualisation (Tittonell and Giller 2013). Sustainability and food security are outcomes of 'situated' practices, and cannot be extracted from the idiosyncrasies of context and 'place' that are increasingly seen as important factors in the outcomes of such (Altieri 1998; Hinrichs 2003; Reynolds et al. 2014). Added to this, the Anthropocene throws up an added task: localised forms of knowledge must be coupled with 'global' knowledge to produce sustainable solutions. The Anthropocene problematic places a strong need upon us to recognise the interconnectedness of the world food system and our globalised place within it: The particular way sustainable intensification is achieved in one part of the planet is likely to have ramifications elsewhere (Garnett et al. 2013). Developing a 'critical sustainability knowledge' means opening up to the diverse potentials and restraints that flow from contextualised sustainability concerns.

One of the main ruptures proposed by ecological intensification is the movement away from the chemical regulation that marked the driving force of agricultural development during the industrial revolution and towards biological regulation. Such a move reinforces the importance of local contexts and specificities. Although dealing most often with traditional, small-holder farming practices, agroecological methods have shown how context can be attended to, understood, protected and celebrated in its own right (Gliessman 2014). Studies of 'real' ecosystems in all their contextual complexity may lead to a 'feeling for the ecosystem'—critical to the pursuit of understanding and managing food production processes (Carpenter 1996).

The relevance of agroecological ideas need not be restricted to 'the farm'; the nature of closed-loop aquaponics systems demands a 'balancing' of co-dependent ecological agents (fish, plants, microbiome) within the limits and affordances of each particular system. Although the microbiome of aquaponics systems has only just begun to be analysed (Schmautz et al. 2017), complexity and dynamism is expected to exceed Recirculating Aquaculture Systems, whose microbiology is known to be affected by feed type and feeding regime, management routines, fish-associated microflora, make-up water parameters and selection pressure in the biofilters (Blancheton et al. 2013). What might be regarded as 'simple' in comparison to other farming methods, the ecosystem of aquaponics systems is nevertheless dynamic and requires care. Developing an 'ecology of place', where context is intentionality and carefully engaged with, can serve as a creative force in research, including scientific understanding (Thrift 1999; Beatley and Manning 1997).

The biophysical and ecological dynamics of aquaponic systems are central to the whole conception of aquaponics, but sustainability and food security potentials do not derive solely from these parameters. As König et al. (2016) point out, for aquaponic systems: 'different settings potentially affects the delivery of all aspects of sustainability: economic, environmental and social' (König et al. 2016). The huge configurational potential of aquaponics—from miniature to hectares, extensive to intensive, basic to high-tech systems—is quite atypical across food production technologies (Rakocy et al. 2006). The integrative character and physical plasticity of aquaponic systems means that the technology can be deployed in a wide variety of applications. This, we feel, is precisely the strength of aquaponic technology. Given the diverse and heterogeneous nature of sustainability and food security concerns in the Anthropocene, the great adaptability, or even 'hackability' (Delfanti 2013), of aquaponics offers much potential for developing 'custom-fit' food production (Reynolds et al. 2014) that is explicitly tailored to the environmental, cultural and nutritional demands of place. Aquaponic systems promise avenues of food production that might be targeted towards local resource and waste assimilative limits, material and technological availability, market and labour demands. It is for this reason that the pursuit of sustainability outcomes may well involve different technological developmental paths dependent upon locale (Coudel et al. 2013). This is a point that is beginning to receive increasing acknowledgement, with some commentators claiming that the urgency of global sustainability and food security issues in the Anthropocene demand an open and multidimensional approach to technological innovation. For instance, Foley et al. (2011:5) state: 'The search for agricultural solutions should remain technology neutral. There are multiple paths to improving the production, food security and environmental performance of agriculture, and we should not be locked into a single approach a priori, whether it be conventional agriculture, genetic modification or organic farming' (5) (Foley et al. 2011). We would highlight this point for aquaponics, as König et al. (2018: 241) have already done: 'there are several sustainability problems which aquaponics could address, but which may be impossible to deliver in one system setup. Therefore, future pathways will always need to involve a diversity of approaches'.

But the adaptability of aquaponics might be seen as a double-edged sword. Inspiration for specific 'tailor-made' sustainability solutions brings with it the difficulty of generalising aquaponic knowledge for larger-scale and repeatable purposes. Successful aquaponics systems respond to local specificities in climate, market, knowledge, resources, etc. (Villarroel et al. 2016; Love et al. 2015; Laidlaw and Magee 2016), but this means that changes at scale cannot easily proceed from the fractal replication of non-reproducible local success stories. Taking similar issues as these into account, other branches of ecological intensification research have suggested that the expression 'scaling up' must be questioned (Caron et al. 2014). Instead, ecological intensification is beginning to be viewed as a transition of multi-scalar processes, all of which follow biological, ecological, managerial and political 'own rules', and generate unique trade-off needs (Gunderson 2001).

Understanding and intervening in complex systems like this presents huge challenges to our research, which is geared towards the production of 'expert knowledge', often crafted in the lab and insulated from wider structures. The complex problem of food security is fraught with uncertainties that cannot be adequately resolved by resorting to the puzzle-solving exercises of Kuhnian 'normal science' (Funtowicz and Ravetz 1995). The necessity to account for 'specificity' and 'generality' in complex sustainably issues produces great methodological, organisational and institutional difficulties. The feeling is that to meet contextualised sustainability and food security goals, 'universal' knowledge must be connected to 'place-based' knowledge (Funtowicz and Ravetz 1995). For Caron et al. (2014), this means that 'scientists learn to continually go back and forth...' between these two dimensions, '. . .both to formulate their research question and capitalize their results. . . Confrontation and hybridization between heterogeneous sources of knowledge is thus essential' (Caron et al. 2014). Research must be opened up to wider circles of stakeholders and their knowledge streams.

Given the huge challenge on all accounts that such a scheme entails, a tempting resolution might be found in the development of more advanced 'environment-controlled' aquaponic farming techniques. Such systems work by cutting out external influences in production, maximising efficiency by minimising the influence of suboptimal, location-specific variables (Davis 1985). But we question this approach on a number of accounts. Given that the impulse of such systems lies in buffering food production from 'localised inconsistencies', there is always a risk that the localised sustainability and food security needs might also be externalised from system design and management. Cutting out localised anomalies in the search of the 'perfect system' must certainly offer tantalising efficiency potentials on paper, but we fear this type of problem-solving bypasses the specificity-generality problematic of sustainability issues in the Anthropocene without confronting them. Rather than a remedy, the result may well be an extension of the dislocated, 'one size fits all' approach to food production that marked the Green Revolution.

Current aquaponics research that follows either of the informal schools of 'decoupling' or 'closing the cycle' might well be an example of such framings. By pushing the productivity limits of either production side—aquaculture or hydroculture—inherent operational compromises of the ecological aquaponic principle become more apparent and become viewed as barriers to productivity that must be overcome. Framing the aquaponic problem like this results in solutions that involve more technology: patented one-way valves, condensation traps, high-tech oxygenators, LED lighting, additional nutrient dispensers, nutrient concentrators and so on. These directions repeat the knowledge dynamic of modern industrial agriculture that overly concentrated the expertise and power of food production systems into the hands of applied scientists engaged in the development of inputs, equipment and remote system management. We are unsure of how such technocratic measures might fit within a research ethic that places sustainability first. This is not an argument against high-tech, closed environment systems; we simply hope to emphasise that within a *sustainability first* paradigm, our food production technologies must be justified on the grounds of generating context-specific sustainability and food security outcomes.

Understanding that sustainability cannot be removed from the complexities of context or the potentials of place is to acknowledge that 'expert knowledge' alone cannot be held as guarantor of sustainable outcomes. This strikes a challenge to modes of centralised knowledge production based on experiments under controlled conditions and the way science might contribute to the innovation processes (Bäckstrand 2003). Crucial here is the design of methodological systems that ensure both the robustness and genericity of scientific knowledge is maintained along with its relevance to local conditions. Moving to conceptions like this requires a huge shift in our current knowledge production schemes and not only implies better integration of agronomic with human and political sciences but suggests a path of knowledge co-production that goes well beyond 'interdisciplinarity' (Lawrence 2015).

Here it is important to stress Bäckstrand's (2003: 24) point that the incorporation of lay and practical knowledge in scientific processes 'does not rest on the assumption that lay knowledge is necessarily "truer", "better" or "greener"'. Rather, as Leach et al. (2012: 4) point out, it stems from the idea that 'nurturing more diverse approaches and forms of innovation (social as well as technological) allows us to respond to uncertainty and surprise arising from complex, interacting biophysical and socioeconomic shocks and stresses'. Faced with the uncertainty of future environmental outcomes in the Anthropocene, a multiplicity of perspectives can prevent the narrowing of alternatives. In this regard, the potential wealth of experimentation occurring in 'backyard' and community projects across Europe represents an untapped resource which has until now received little attention from research circles. 'The small-scale sector…' Konig et al. (2018: 241) observe, '…shows optimism and a surprising degree of self-organization over the internet. There might be room for creating additional social innovations'. Given the multidimensional nature of issues in the Anthropocene, grassroots innovations, like the backyard aquaponics sector, draw from local knowledge and experience and work towards social and organisational forms of innovation that are, in the eyes of Leach et al. (2012: 4), 'at least as crucial as advanced science and technology'. Linking with community aquaponics groups potentially offers access to vibrant local food groups, local government and local consumers who are often enthusiastic about the prospects of collaborating with researchers. It is worth noting that in an increasingly competitive funding climate, local communities offer a well of resources—intellectual, physical and monetary—that often get overlooked but which can supplement more traditional research funding streams (Reynolds et al. 2014).

As we know, currently, large-scale commercial projects face high marketing risks, strict financing deadlines, as well as high technological and management complexity that makes collaboration with outside research organisations difficult. Because of this, we would agree with König et al. (2018) who find advantages for experimentation with smaller systems that have reduced complexity and are tied down by fewer legal regulations. The field must push to integrate these organisations within participatory, citizen-science research frameworks, allowing academic research to more thoroughly mesh with forms of aquaponics working in the world. In the absence of formalised sustainability measures and protocols, aquaponic enterprises risk legitimation issues when their produce is marketed on claims of sustainability. One clear possibility of participatory research collaborations would be the joint production of much needed 'situation-specific sustainability goals' for facilities that could form the 'basis for system design' and bring 'a clear marketing strategy' (König et al. 2018). Working towards outcomes like these might also improve the transparency, legitimacy and relevance of our research endeavours (Bäckstrand 2003).

The European research funding climate has begun to acknowledge the need to shift research orientation by including the requirement in recent project funding calls of implementing the so-called 'living labs' into research projects (Robles et al. 2015). Starting in June 2018, the Horizon 2020 project proGIreg (H2020-SCC-2016-2017) is going to include a living lab for the exemplary implementation of the so-called nature-based systems (NBS), one of which will be a community designed, community-built and community-operated aquaponic system in a passive solar greenhouse. The project, with 36 partners in 6 countries, aims to find innovative ways to productively utilise green infrastructure of urban and peri-urban environments, building upon the co-production concepts developed in its currently running sibling project, CoProGrün.

The researchers' working packages regarding the aquaponic part of the project are going to be threefold. One part will be about raising the so-called technology readiness level (TRL) of aquaponics, a research task without explicit collaboration with laypersons and the community. Resource utilisation of current aquaponic concepts and resource optimisation potential of additional technical measures are the core objectives of this task. While at first glance this task seems to follow the above-criticised paradigm of productivity and yield increase, evaluation criteria for different measures will include more multifaceted aspects such as ease of implementation, understandability, appropriateness and transferability. A second focus will be support of the community planning, building and operational processes, which seeks to integrate objective knowledge and practitioner knowledge generation. A meta-objective of this process will be the observation and the moderation of the relevant community collaboration and communication processes. In this approach, moderation is actively expected to alter observation, illustrating a deviation from the traditional research routines of fact building and repeatability. A third package encompasses research on political, administrative, technical and financial obstacles. The intention here is to involve a wider collection of stakeholders, from politicians and decision-makers to planners, operators and neighbours, with research structures developed to bring together each of these specific perspectives. Hopefully, this more holistic method opens a path to the 'sustainability first' approach proposed in this chapter.

16.7.3 Concern

Recognising aquaponics as a multifunctional form of food production faces large challenges. As has been discussed, grasping the notion of 'multifunctional agriculture' is more than just a critical debate on what constitutes 'post-productionism' (Wilson 2001); this is because it seeks to move understandings of our food system to positions that better encapsulates the diversity, nonlinearity and spatial heterogeneity that are acknowledged as key ingredients to a sustainable and just food system. It is important to remember that the very notion of 'multifunctionality' in agriculture arose during the 1990s as 'a consequence of the undesired and largely unforeseen environmental and societal consequences and the limited cost-effectiveness of the European Common Agricultural Policy (CAP), which mainly sought to boost agrarian outputs and the productivity of agriculture' (270) (Cairol et al. 2009). Understanding that our political climates and institutional structures have been unconducive to sustainable change is a point we must not forget. As others have pointed out in adjacent agronomic fields, understanding and unlocking the richness of food production contributions to human welfare and environmental health will necessarily involve a *critical* dimension (Jahn 2013). This insight, we feel, must feature more strongly in aquaponics research.

We chose the word 'concern' here carefully. The word concern carries different connotations to 'critique'. Concern carries notions of anxiety, worry and trouble. Anxiety comes when something disrupts what could be a more healthy or happy or secure existence. It reminds us that to do research in the Anthropocene is to acknowledge our drastically unsettling place in the world. That our 'solutions' always carry the possibility of trouble, whether this be ethical, political or environmental. But concern has more than just negative connotations. To concern also means to 'be about', to 'relate to' and also 'to care'. It reminds us to question what our research is about. How our disciplinary concerns relate to other disciplines as well as wider issues. Crucially, sustainability and food security outcomes require us to care about the concerns of others.

Considerations such as these make up a third aspect of what we mean when we call for a *'Critical sustainability knowledge'* for aquaponics. As a research community, it is crucial that we develop an understanding of the structural factors which impinge upon and restrict the effective social, political and technological innovation of aquaponics. Technical change relies upon infrastructure, financing capacities, market organisations as well as labour and land rights conditions (Röling 2009). When the role of this wider framing is assumed only as an 'enabling environment', often the result is that such considerations are left outside of the research effort. This is a point which serves to easily justify the failure of technology-based, top-down development drives (Caron 2000). In this regard, the techno-optimistic discourse of contemporary aquaponics, in its failure to apprehend wider structural resistance to the development of sustainable innovation, would serve as a case example.

As an important potential form of sustainable intensification, aquaponics needs to be recognised as being embedded in and linked to different social, economic and organisational forms at various scales potentially from household, value chain, food system and beyond including also other political levels. Thankfully, moves towards attending to the wider structural difficulties that aquaponic technology faces have recently been made, with König et al. (2018) offering a view of aquaponics through an 'emerging technological innovation system' lens. König et al. (2018) have shown how the challenges to aquaponics development derive from: (1) system complexity, (2) the institutional setting and (3) the sustainability paradigm it attempts to impact. The aquaponic research field needs to respond to this diagnosis.

The slow uptake and high chance of failure that aquaponics technology currently exhibits is an expression of the wider societal resistance that makes sustainable innovation such a challenge, as well as our inability to effectively organise against such forces. As König et al. (2018) note, the high-risk environment that currently exists for aquaponic entrepreneurs and investors forces startup facilities across Europe to focus on production, marketing and market formation, over the delivery of sustainability credentials. Along these lines, Alkemade and Suurs (2012) remind us, 'market forces alone cannot be relied upon to realize desired sustainability transitions'; rather, they point out, insight into the dynamics of innovation processes is needed if technological change can be guided along more sustainable trajectories (Alkemade and Suurs 2012).

The difficulties aquaponic businesses face in Europe suggest the field currently lacks the necessary market conditions, with 'consumer acceptance'—an important factor enabling the success of novel food system technologies—acknowledged as a possible problem area. From this diagnosis, there has been raised the problem of 'consumer education' (Miličić et al. 2017). Along with this, we would stress that collective education is a key concern for questions of food system sustainability. But accounts like these come with risks. It is easy to fall back on traditional modernist conceptions regarding the role of science in society, assuming that 'if only the public understood the facts' about our technology they would choose aquaponics over other food production methods. Accounts like these assume too much, both about the needs of 'consumers', as well as the value and universal applicability of expert knowledge and technological innovation. There is a need to seek finer-grain and more nuanced accounts of the struggle for sustainable futures that move beyond the dynamic of consumption (Gunderson 2014) and have greater sensitivity to the diverse barriers communities face in accessing food security and implementing sustainable action (Carolan 2016; Wall 2007).

Gaining insight into innovation processes puts great emphasis upon our knowledge-generating institutions. As we have discussed above, sustainability issues demand that science opens up to public and private participatory approaches entailing knowledge co-production. But in terms of this point, it's worth noting that huge challenges lay in store. As Jasanoff (2007: 33) puts it: 'Even when scientists recognize the limits of their own inquiries, as they often do, the policy world, implicitly encouraged by scientists, asks for more research'. The widely held assumption that *more* objective knowledge is the key to bolstering action towards sustainability runs contrary to the findings of sustainability science. Sustainability outcomes are actually more closely tied deliberative knowledge processes: building greater awareness of the ways in which experts and practitioners frame sustainability issues; the values that are included as well as excluded; as well as effective ways of facilitating communication of diverse knowledge and dealing with conflict if and when it arises (Smith and Stirling 2007; Healey 2006; Miller and Neff 2013; Wiek et al. 2012). As Miller et al. (2014) point out, the continuing dependence upon objective knowledge to adjudicate sustainability issues represents the persistence of the modernist belief in rationality and progress that underwrites almost all knowledge-generating institutions (Horkheimer and Adorno 2002; Marcuse 2013).

It is here where developing a critical sustainability knowledge for aquaponics shifts our attention to our own research environments. Our increasingly 'neoliberalised' research institutions exhibit a worrying trend: the rollback of public funding for universities, the increasing pressure to get short-term results, the separation of research and teaching missions, the dissolution of the scientific author, the contraction of research agendas to focus on the needs of commercial actors, an increasing reliance on market take-up to adjudicate intellectual disputes and the intense fortification of intellectual property in the drive to commercialise knowledge, all of which have been shown to impact on the production and dissemination of our research, and indeed all are factors that impact the nature of our science (Lave et al. 2010). One question that must be confronted is whether our current research environments are fit for the examination of complex sustainability and long-term food security targets that must be part of aquaponic research. This is the key point we would like to stress—if sustainability is an outcome of multidimensional collective deliberation and action, our own research endeavours, thoroughly part of the process, must be viewed as something that can be innovated towards sustainability outcomes also. The above-mentioned Horizon 2020 project proGIreg may be an example of some ambitious first steps towards crafting new research environments, but we must work hard to keep the research process itself from slipping out of view. Questions might be raised about how these potentially revolutionary measures of 'living labs' might be implemented from within traditional funding logics. For instance, calls for participatory approaches foreground the conceptual importance of open-ended outcomes, while at the same time requiring the intended spending of such living labs to be predefined. Finding productive ways out of traditional institutional barriers is an ever-present concern.

Our modern research environments can no longer be regarded as having a privileged isolation from the wider issues of society. More than ever our innovation-driven biosciences are implicated in the agrarian concerns of the Anthropocene (Braun and Whatmore 2010). The field of Science and Technology Studies teaches us that technoscientific innovations come with serious ethico-political implication. A 30-year-long discussion in this field has moved well beyond the idea that technologies are simply 'used' or 'misused' by different socio-political interests after the hardware has been 'stabilised' or legitimated through objective experimentation in neutral lab spaces (Latour 1987; Pickering 1992). The 'constructivist' insight in STS analyses goes beyond the identification of politics inside labs (Law and Williams 1982; Latour and Woolgar 1986 [1979]) to show that the technologies we produce are not 'neutral' objects but are in fact infused with 'world-making' capacities and political consequence.

The aquaponics systems we help to innovate are filled with future making capacity, but the consequences of technological innovation are seldom a focus of study. To paraphrase Winner (1993), what the introduction of new artefacts means for people's sense of self, for the texture of human/nonhuman communities, for qualities of everyday living within the dynamic of sustainability and for the broader distribution of power in society, these have not traditionally been matters of explicit concern. When classic studies (Winner 1986) ask the question 'Do artefacts have politics?', this is not only a call to produce more accurate examinations of technology by including politics in accounts of the networks of users and stakeholders, though this is certainly needed; it also concerns us researchers, our modes of thought and ethos that affect the politics (or not) we attribute to our objects (de la Bellacasa 2011; Arboleda 2016). Feminist scholars have highlighted how power relations are inscribed into the very fabric of modern scientific knowledge and its technologies. Against alienated and abstract forms of knowledge, they have innovated key theoretical and methodological approaches that seek to bring together objective and subjective views of the world and to theorise about technology from the starting point of practice (Haraway 1997; Harding 2004). Aware of these points, Jasanoff (2007) calls for the development of what she calls 'technologies of humility': 'Humility instructs us to think harder about how to reframe problems so that their ethical dimensions are brought to light, which new facts to seek and when to resist asking science for clarification. Humility directs us to alleviate known causes of people's vulnerability to harm, to pay attention to the distribution of risks and benefits, and to reflect on the social factors that promote or discourage learning'.

An important first step for our field to take towards understanding better the political potentials of our technology would be to encourage the expansion of the field out into critical research areas that are currently underrepresented. Across the Atlantic in the US and Canada similar moves like this have already been made, where an interdisciplinary approach has progressively developed into the critical field of political ecology (Allen 1993). Such projects not only aim to combine agriculture and land use patterns with technology and ecology, but furthermore, also emphasise the integration of socioeconomic and political factors (Caron et al. 2014). The aquaponics research community in America has begun to acknowledge the expanding resources of food sovereignty research, exploring how urban communities can be re-engaged with the principles of sustainability, whilst taking more control over their food production and distribution (Laidlaw and Magee 2016). Food sovereignty has become a huge topic that precisely seeks to intervene into food systems that are overdetermined by disempowering capitalist relations. From food sovereignty perspectives, the corporate control of the food system and the commodification of food are seen as predominant threats to food security and the natural environment (Nally 2011). We would follow Laidlaw and Magee's (2016) view that community-based aquaponics enterprises 'represent a new model for how to blend local agency with scientific innovation to deliver food sovereignty in cities'.

Developing a *'critical sustainability knowledge'* for aquaponics means resisting the view that society and its institutions are simply neutral domains that facilitate the linear progression towards sustainable innovation. Many branches of the social sciences have contributed towards an image of society that is infused with asymmetric power relations, a site of contestation and struggle. One such struggle concerns the very meaning and nature of sustainability. Critical viewpoints from wider fields would underline that aquaponics is a technology ripe with both political potential and limitation. If we are serious about the sustainability and food security credentials of aquaponics, it becomes crucial that we examine more thoroughly how our expectations of this technology relate to on-the-ground experience, and in turn, find ways of integrating this back into research processes. We follow Leach et al. (2012) here who insist on the need for finer-grained considerations regarding the performance of sustainable innovations. Apart from the claims, just who or what stands to benefit from such interventions must take up a central place in the aquaponic innovation process. Lastly, as the authors of Chap. 1 have made clear, the search for a lasting paradigm shift will require the ability to place our research into policy circuits that make legislative environments more conducive to aquaponics development and enable larger-scale change. Influencing policy requires an understanding of the power dynamics and political systems that both enable and undermine the shift to sustainable solutions.

16.8 Conclusion: Aquaponic Research into the Anthropocene

The social–biophysical pressures *of* and *on* our food system converge in the Anthropocene towards what becomes seen as an unprecedented task for the global community, requiring 'nothing less than a planetary food revolution' (Rockström et al. 2017). The Anthropocene requires food production innovations that exceed traditional paradigms, whilst at the same time are able to acknowledge the complexity arising from the sustainability and food security issues that mark our times. Aquaponics is one technological innovation that promises to contribute much towards these imperatives. But this emergent field is in an early stage that is characterised by limited resources, market uncertainty, institutional resistance and high risks of failure—an innovation environment where hype prevails over demonstrated outcomes. The aquaponics research community potentially holds an important place in the development path of this technology. As an aquaponics research community, we need to craft viable visions for the future.

We propose one such vision when we call for a 'sustainability first' research programme. Our vision follows Rockström et al.'s (2017) diagnosis that paradigm change requires shifting the research ethic away from traditional productivist avenues so that sustainability becomes the central locus of the innovation process. This task is massive because the multidimensional and context-bound nature of sustainability and food security issues is such that they cannot be resolved solely through technical means. The ethical- and value-laden dimensions of sustainability require a commitment to confront the complexities, uncertainty, ignorance and contestation that ensue such issues. All this places great demands on the knowledge we produce; not only how we distribute and exchange it, but also its very nature.

We propose the aquaponic field needs to pursue a 'critical sustainability knowledge'. When König et al. (2018) ask what sustainability experimentation settings would be needed to enable science, business, policy and consumers to 'answer sustainability questions without repeating the development path of either [RAS or hydroponics]', the point is clear—we need to learn from the failures of the past. The current neoliberal climate is one that consistently opens 'sustainability' discussion up to (mis)appropriation as 'agribusiness mobilises its resources in an attempt to dominate discourse and to make its meaning of "alternative agriculture" the universal meaning' (Kloppenburg 1991). We need to build a critical sustainability knowledge that is wise to the limits of technocratic routes to sustainability, which is sensitive to the political potential of our technologies as well as the structural forms of resistance that limit their development.

A critical sustainability knowledge builds awareness of the limits of its own knowledge pathways and opens up to those other knowledge streams that are often pushed aside in attempts to expand scientific understanding and technological capacity. This is a call for interdisciplinarity and the depth it brings, but it goes further than this. Sustainability and food security outcomes have little impact if they can only be generated in the lab. Research must be contextualised: we need 'to produce and embed scientific knowledge into local innovation systems' (51) (Caron et al. 2014). Building co-productive links with aquaponics communities already existing in society means forging the social and institutional structures that can enable our communities to continually learn and adapt to new knowledge, values, technologies and environmental change. Together, we need to deliberate on the visions and the values of our communities and explore the potential sociotechnical pathways that might realise such visions. Central to this, we need systems of organising and testing the sustainability and food security claims that are made of this technology (Pearson et al. 2010; Nugent 1999) so that greater transparency and legitimation might be brought to the entire field: entrepreneurs, enterprises, researchers and activists alike.

If all this seems like a tall order, that's because it is. The Anthropocene calls for a huge rethink in the way society is being organised, and our food system is central to this. There is a chance, we believe, that aquaponics has a part to play in this. But if our hopes are not to get lost in the hype bubble of hollow sustainability chatter that marks our neoliberal times, we have to demonstrate that aquaponics offers something different. As a final remark, we revisit de la Bellacasa's (2015) point that: 'agricultural intensification is not only a quantitative orientation (yield increase), but entails a "way of life"'. If this is the case, then the pursuit of *sustainable intensification* demands that we find a new way of living. We need sustainability solutions that acknowledge this fact and research communities that are responsive to it.

[1] For instance, consider the following statement issued by Monsanto: 'The main uses of GM crops are to make them insecticide- and herbicide tolerant. They don't inherently increase the yield. They protect the yield'. Quoted in E. Ritch, 'Monsanto Strikes Back at Germany, UCS', Cleantech.com (April 17, 2009). Accessed on July 18, 2009.

[2] Especially important here are the effects of climate change, as well as the 'superweed' phenomenon of increasingly resistant pests that significantly diminish yields.

[3] Productivist discourse invariably ignores Amartya Sen's (1981, 154; Roberts 2008, 263; WFP 2009, 17) classic point that the volume and availability of food alone is not a sufficient explanation for the persistence of world hunger. It is well established that enough food exists to feed in excess of the world's current population (OECD 2009, 21)

[4] Although the calculations are complex and contested, one common estimate is that industrial agriculture requires an average 10 calories of fossil fuels to produce a single calorie of food (Manning 2004), which might rise to 40 calories in beef (Pimentel 1997).

[5] Externalities of our current food system are often ignored or heavily subsidised away. Moore (2015: 187) describes the situation as 'a kind of "ecosystem services" in reverse': 'Today, a billion pounds of pesticides and herbicides are used each year in American agriculture. The long recognized health impacts have been widely studied. Although the translation of such "externalities" into the register of accumulation is imprecise, their scale is impressive, totalling nearly $17 billion in unpaid costs for American agriculture in the early twenty-first century'. On externalities see: Tegtmeier and Duffy (2004).

References

Abson DJ, Fischer J, Leventon J, Newig J, Schomerus T, Vilsmaier U, von Wehrden H, Abernethy P, Ives CD, Jager NW (2017) Leverage points for sustainability transformation. Ambio 46:30–39

Alexandratos N, Bruinsma J (2012) World agriculture towards 2030/2050: the 2012 revision, ESA working paper. FAO, Rome

Alkemade F, Suurs RA (2012) Patterns of expectations for emerging sustainable technologies. Technol Forecast Soc Chang 79:448–456

Allen P (1993) Food for the future: conditions and contradictions of sustainability. Wiley, Chichester

Altieri MA (1998) Ecological impacts of industrial agriculture and the possibilities for truly sustainable farming. Mon Rev 50:60

Altieri MA (2007) Fatal harvest: old and new dimensions of the ecological tragedy of modern agriculture. In: Sustainable resource management. Edward Elgar, Londres, pp 189–213

Altvater E, Crist E, Haraway D, Hartley D, Parenti C, Mcbrien J (2016) Anthropocene or capitalocene?: nature, history, and the crisis of capitalism. PM Press, Oakland

Ansell C, Gash A (2008) Collaborative governance in theory and practice. J Public Adm Res Theory 18:543–571

Arboleda M (2016) Revitalizing science and technology studies: a Marxian critique of more-than-human geographies. Environt Plan D Soc Space 35:360–378. https://doi.org/10.1177/0263775816664099

Assessment ME (2003) Ecosystems and human well-being: biodiversity synthesis; a report of the millennium ecosystem assessment. Island Press, Washington, DC

Bäckstrand K (2003) Civic science for sustainability: reframing the role of experts, policy-makers and citizens in environmental governance. Glob Environ Politics 3:24–41

Battisti DS, Naylor RL (2009) Historical warnings of future food insecurity with unprecedented seasonal heat. Science 323:240–244

Baulcombe D, Crute I, Davies B, Dunwell J, Gale M, Jones J, Pretty J, Sutherland W, Toulmin C (2009) Reaping the benefits: science and the sustainable intensification of global agriculture. The Royal Society, London

Beatley T, Manning K (1997) The ecology of place: planning for environment, economy, and community. Island Press, Washington, DC

Blancheton J, Attramadal K, Michaud L, D'Orbcastel ER, Vadstein O (2013) Insight into bacterial population in aquaculture systems and its implication. Aquac Eng 53:30–39

Bommarco R, Kleijn D, Potts SG (2013) Ecological intensification: harnessing ecosystem services for food security. Trends Ecol Evol 28:230–238

Bradley A (2011) Originary technicity: the theory of technology from Marx to Derrida. Palgrave Macmillan Basingstoke, Hants

Braun B, Whatmore SJ (2010) Political matter: technoscience, democracy, and public life. University of Minnesota Press, Minneapolis

Brewer GD (2007) Inventing the future: scenarios, imagination, mastery and control. Sustain Sci 2:159–177

Bringezu S, Schütz H, Pengue W, O'Brien M, Garcia F, Sims R, Howarth RW, Kauppi L, Swilling M, Herrick J (2014) Assessing global land use: balancing consumption with sustainable supply. United Nations Environment Programme, Nairobi

Brooks S, Leach M, Millstone E, Lucas H (2009) Silver bullets, grand challenges and the new philanthropy. STEPS Centre, Brighton

Brussaard L, Caron P, Campbell B, Lipper L, Mainka S, Rabbinge R, Babin D, Pulleman M (2010) Reconciling biodiversity conservation and food security: scientific challenges for a new agriculture. Curr Opin Environ Sustain 2:34–42

Câmpeanu CN, Fazey I (2014) Adaptation and pathways of change and response: a case study from Eastern Europe. Glob Environ Chang 28:351–367

Cairol D, Coudel E, Knickel K, Caron P, Kröger M (2009) Multifunctionality of agriculture and rural areas as reflected in policies: the importance and relevance of the territorial view. J Environ Policy Plan 11:269–289

Cardinale BJ, Duffy JE, Gonzalez A, Hooper DU, Perrings C, Venail P, Narwani A, Mace GM, Tilman D, Wardle DA, Kinzig AP, Daily GC, Loreau M, Grace JB, Larigauderie A, Srivastava DS, Naeem S (2012) Biodiversity loss and its impact on humanity. Nature 486:59–67

Carolan M (2016) Adventurous food futures: knowing about alternatives is not enough, we need to feel them. Agric Hum Values 33:141–152

Caron P (2000) Decentralisation and multi-levels changes: challenges for agricultural research to support co-ordination between resource poor stakeholders and local governments

Caron P, Biénabe E, Hainzelin E (2014) Making transition towards ecological intensification of agriculture a reality: the gaps in and the role of scientific knowledge. Curr Opin Environ Sustain 8:44–52

Caron P, Reig E, Roep D, Hediger W, Cotty T, Barthelemy D, Hadynska A, Hadynski J, Oostindie H, Sabourin E (2008) Multifunctionality: refocusing a spreading, loose and fashionable concept for looking at sustainability? Int J Agric Resour Gov Ecol 7:301–318

Carpenter SR (1996) Microcosm experiments have limited relevance for community and ecosystem ecology. Ecology 77:677–680

Castree N (2015) Changing the Anthropo (s) cene geographers, global environmental change and the politics of knowledge. Dialogues Hum Geogr 5:301–316

Churchman CW (1979) The systems approach and its enemies. Basic Books, New York

Cohen T (2012) Anecographics: climate change and "Late" deconstruction. In: Impasses of the post-global: theory in the era of climate change, vol 2. Open Humanities Press, London, pp 32–57

Cohen T (2013) Telemorphosis: theory in the era of climate change, vol 1. Open Humanities Press

Collings DA (2014) Stolen future, broken present: the human significance of climate change. Open Humanities Press, London

Cooper M (2008) Life as surplus: biotechnology and capitalism in the neoliberal era. University of Washington Press, Seattle

COST (2013) Memorandum of understanding for the implementation of a European Concerted Research Action designated as COST Action FA1305: The EU Aquaponics Hub: realising sustainable integrated fish and vegetable production for the EU. European Cooperation in the field of Scientific and Technical Research – COST, Brussels

Cota GM (2011) Introduction: the posthuman life of agriculture. In: Cota GM (ed) Another technoscience is possible: agricultural lessons for the Posthumanities. Open Humanities Press, London

Coudel E, Devautour H, Soulard C-T, Faure G, Hubert B (2013) Renewing innovation systems in agriculture and food: how to go towards more sustainability? Wageningen Academic Publishers, Wageningen

Crutzen PJ, Stoermer EF (2000a) The "Anthropocene". Global Change Newsl 41:17–18. International Geosphere–Biosphere Programme (IGBP)

Crutzen PJ, Stoermer EF (2000b) Global change newsletter. Anthropocene 41:17–18

Davis N (1985) Controlled-environment agriculture-past, present and future. Food Technol 39:124–126

de La Bellacasa MP (2011) Matters of care in technoscience: assembling neglected things. Soc Stud Sci 41:85–106

de La Bellacasa MP (2015) Making time for soil: Technoscientific futurity and the pace of care. Soc Stud Sci 45:691–716

de Laet M, Mol A (2000) The Zimbabwe bush pump: mechanics of a fluid technology. Soc Stud Sci 30:225–263

Delaide B, Delhaye G, Dermience M, Gott J, Soyeurt H, Jijakli MH (2017) Plant and fish production performance, nutrient mass balances, energy and water use of the PAFF box, a small-scale aquaponic system. Aquac Eng 78:130

Delfanti A (2013) Biohackers. The politics of open science. Pluto Press, London

Desa U (2015) World population prospects: the 2015 revision, key findings and advance tables. Working Paper No Rep, No. ESA/P/WP. 241

Dobbs R, Oppenheim J, Thompson F, Brinkman M, Zornes M (2011) Resource revolution: meeting the world's energy, materials, food, and water needs. McKinsey & Company, London

Elzen,B, Janssen A, Bos A (2017) Portfolio of promises: designing and testing a new tool to stimulate transition towards sustainable agriculture. AgroEcological transitions. Wageningen University & Research

Evenson RE, Gollin D (2003) Assessing the impact of the green revolution, 1960 to 2000. Science 300:758–762

FAO (2013) Annotated bibliography on ecological intensification. Report of the Liberation EU Collaborative funded project (LInking farmland Biodiversity to Ecosystem seRvices for effective ecofunctional intensification)

Farmer B (1986) Perspectives on the 'Green Revolution'in South Asia. Mod Asian Stud 20 (1):175–199

Fischer J, Manning AD, Steffen W, Rose DB, Daniell K, Felton A, Garnett S, Gilna B, Heinsohn R, Lindenmayer DB (2007) Mind the sustainability gap. Trends Ecol Evol 22:621–624

Folke C, Hahn T, Olsson P, Norberg J (2005) Adaptive governance of social-ecological systems. Annu Rev Environ Resour 30:441–473

Foley JA, Ramankutty N, Brauman KA, Cassidy ES, Gerber JS, Johnston M, Mueller ND, O'Connell C, Ray DK, West PC (2011) Solutions for a cultivated planet. Nature 478:337–342

Franklin S (1995) Science as culture, cultures of science. Annu Rev Anthropol 24:163–184

Fuglie KO (2010) Total factor productivity in the global agricultural economy: evidence from FAO data. In: The shifting patterns of agricultural production and productivity worldwide. Midwest Agribusiness Trade Research and Information Center, Ames, pp 63–95

Funtowicz SO, Ravetz JR (1995) Science for the post normal age. In: Perspectives on ecological integrity. Springer, Dordrecht

Garnett T, Appleby MC, Balmford A, Bateman IJ, Benton TG, Bloomer P, Burlingame B, Dawkins M, Dolan L, Fraser D (2013) Sustainable intensification in agriculture: premises and policies. Science 341:33–34

Garnett T, Godfray C (2012) Sustainable intensification in agriculture. Navigating a course through competing food system priorities. In: Food climate research network and the Oxford Martin programme on the future of food, vol 51. University of Oxford, Oxford

Geels FW (2011) The multi-level perspective on sustainability transitions: responses to seven criticisms. Environ Innov Soc Trans 1:24–40

Gibson-Graham J (2014) Rethinking the economy with thick description and weak theory. Curr Anthropol 55:S147–S153

Gliessman SR (2014) Agroecology: the ecology of sustainable food systems. CRC press, Bosa Roca

Goddek S, Delaide B, Mankasingh U, Ragnarsdottir K, Jijakli H, Thorarinsdottir R (2015) Challenges of sustainable and commercial aquaponics. Sustainability 7:4199–4224

Godfray HCJ, Garnett T (2014) Food security and sustainable intensification. Philos Trans R Soc B 369:20120273

Grassini P, Eskridge KM, Cassman KG (2013) Distinguishing between yield advances and yield plateaus in historical crop production trends. Nat Commun 4:2918

Gunderson LH (2001) Panarchy: understanding transformations in human and natural systems. Island Press, Washington, DC

Gunderson R (2014) Problems with the defetishization thesis: ethical consumerism, alternative food systems, and commodity fetishism. Agric Hum Values 31:109–117

Gurian-Sherman D (2009) Failure to yield: evaluating the performance of genetically engineered crops, Union of Concerned Scientists. UCS Publications, Cambridge, MA

H2020-SCC-2016-2017 (smart and sustainable cities), Topic: SCC-02-2016-2017, Proposal number: 776528-1, Proposal acronym: proGIreg

Haenen I (2017) Small successes and big failures: lessons from the aquaponics facility at Uit Je Eigen Stad (UJES). 'Aquaponics.biz': A COST FA1305 Conference on Aquaponics SMEs, Rotterdam

Hall A, Bockett G, Taylor S, Sivamohan M, Clark N (2001) Why research partnerships really matter: innovation theory, institutional arrangements and implications for developing new technology for the poor. World Dev 29:783–797

Hamilton C (2013) Earthmasters: the dawn of the age of climate engineering. Yale University Press, New Haven

Hamilton C, Gemenne F, Bonneuil C (2015) The anthropocene and the global environmental crisis: rethinking modernity in a new epoch. Routledge, Oxon

Haraway D (2015) Anthropocene, capitalocene, plantationocene, chthulucene: making kin. Environ Humanit 6:159–165

Haraway DJ (1997) Modest-Witness@Second-Millennium.FemaleMan-Meets-OncoMouse : feminism and technoscience. Routledge, New York/London

Harding S (2004) Introduction: standpoint theory as a site of political, philosophic, and scientific debate. Routledge, New York

Harvey D (2007) A brief history of neoliberalism. Oxford University Press, New York

Healey P (2006) Urban complexity and spatial strategies: towards a relational planning for our times. Routledge, New York

Hinrichs CC (2003) The practice and politics of food system localization. J Rural Stud 19:33–45

Horkheimer M, Adorno TW (2002) Dialectic of enlightenment. Stanford University Press, Palo Alto

Hornborg A (2013) Technology as fetish: Marx, Latour, and the cultural foundations of capitalism. Theory Cult Soc 31:119–140

Hultman M (2013) The making of an environmental hero: a history of ecomodern masculinity, fuel cells and Arnold Schwarzenegger. Environ Humanit 2:79–99

Hunter MC, Smith RG, Schipanski ME, Atwood LW, Mortensen DA (2017) Agriculture in 2050: recalibrating targets for sustainable intensification. Bioscience 67:386–391

Jackson L, van Noordwijk M, Bengtsson J, Foster W, Lipper L, Pulleman M, Said M, Snaddon J, Vodouhe R (2010) Biodiversity and agricultural sustainagility: from assessment to adaptive management. Curr Opin Environ Sustain 2:80–87

Jahn T (2013) Sustainability science requires a critical orientation. GAIA 22:29–33

Jasanoff S (2007) Technologies of humility. Nature 450:33

Joly A, Junge R, Bardocz T (2015) Aquaponics business in Europe: some legal obstacles and solutions. Ecocycles 1:3–5

Junge R, König B, Villarroel M, Komives T, Jijakli MH (2017) Strategic points in aquaponics. Water 9:182

Kearney J (2010) Food consumption trends and drivers. Philos. Trans. R. Soc., B 365:2793–2807

Keil R (2007) Sustaining modernity, modernizing nature. In: The sustainable development paradox: urban political ecology in the US and Europe. Guilford Press, New York, pp 41–65

Kiers ET, Leakey RR, Izac A-M, Heinemann JA, Rosenthal E, Nathan D, Jiggins J (2008) Agriculture at a crossroads. Science 320:320–321

Kloppenburg J (1991) Social theory and the de/reconstruction of agricultural science: local knowledge for an alternative agriculture. Rural Sociol 56:519–548

Kläy A, Zimmermann AB, Schneider F (2015) Rethinking science for sustainable development: reflexive interaction for a paradigm transformation. Futures 65:72–85

Kőmíves T, Ranka J (2015) On the aquaponic corner section of our journal. Ecocycles 1:1–2

König B, Janker J, Reinhardt T, Villarroel M, Junge R (2018) Analysis of aquaponics as an emerging technological innovation system. J Clean Prod 180:232–243

Konig B, Junge R, Bittsanszky A, Villarroel M, Komives T (2016) On the sustainability of aquaponics. Ecocycles 2:26–32

Krueger R, Gibbs D (2007) The sustainable development paradox: urban political economy in the United States and Europe. Guilford Press, New York

Kuyper TW, Struik PC (2014) Epilogue: global food security, rhetoric, and the sustainable intensification debate. Curr Opin Environ Sustain 8:71–79

Laidlaw J, Magee L (2016) Towards urban food sovereignty: the trials and tribulations of community-based aquaponics enterprises in Milwaukee and Melbourne. Local Environ 21:573–590

Leach M, Rockström J, Raskin P, Scoones I, Stirling A, Smith A, Thompson J, Millstone E, Ely A, Arond E (2012) Transforming innovation for sustainability. Ecol Soc 17

Latour B (1986) Visualization and cognition. In: Knowledge and society, vol 6. Open University Press, Milton Keynes, pp 1–40

Latour B (1987) Science in action: how to follow scientists and engineers through society. Harvard University Press, Cambridge, MA

Latour B (1993) We have never been modern. Harvester Wheatsheaf, New York/London

Latour B (1996) Aramis, or, the love of technology. Harvard University Press, Cambridge, MA
Latour B (2004) Why has critique run out of steam? From matters of fact to matters of concern. Crit Inq 30:225–248
Latour B (2015) Telling friends from foes in the time of the Anthropocene. In: Hamilton C, Bonneuil C, Gemenne F (eds) The Anthropocene and the global environmental crisis: rethinking modernity in a new epoch. Routledge, Abingdon/Oxon
Latour B, Woolgar S (1986[1979]) Laboratory life: the construction of scientific facts. Princeton University Press, Princeton
Lave R, Mirowski P, Randalls S (2010) Introduction: STS and neoliberal science, vol 32. Sage, London, p 463
Law J, Williams RJ (1982) Putting facts together: a study of scientific persuasion. Soc Stud Sci 12:535–558
Lawrence RJ (2015) Advances in transdisciplinarity: epistemologies, methodologies and processes. Futures 65:1–9
Lawrence G, Richards C, Lyons K (2013) Food security in Australia in an era of neoliberalism, productivism and climate change. J Rural Stud 29:30–39
Leonard L (2013) Ecomodern discourse and localized narratives: waste policy, community mobilization and governmentality in Ireland. In: Organising waste in the City: international perspectives on narratives and practices. Policy Press, Chicago, pp 181–200
Lilley S, Papadopoulos D (2014) Material returns: cultures of valuation, biofinancialisation and the autonomy of politics. Sociology 48:972–988
Lobell DB, Schlenker W, Costa-Roberts J (2011) Climate trends and global crop production since 1980. Science 333:616–620
Loos J, Abson DJ, Chappell MJ, Hanspach J, Mikulcak F, Tichit M, Fischer J (2014) Putting meaning back into "sustainable intensification". Front Ecol Environ 12:356–361
Lorimer J (2017) The Anthropo-scene: a guide for the perplexed. Soc Stud Sci 47:117–142
Love DC, Fry JP, Li X, Hill ES, Genello L, Semmens K, Thompson RE (2015) Commercial aquaponics production and profitability: findings from an international survey. Aquaculture 435:67–74
Lundqvist J, de Fraiture C, Molden D (2008) Saving water: from field to fork: curbing losses and wastage in the food chain. Stockholm International Water Institute, Stockholm
LAWRENCE, R. J. 2015. Advances in transdisciplinarity: Epistemologies, methodologies and processes.Futures, 65, 1-9
Marcuse H (2013) One-dimensional man: studies in the ideology of advanced industrial society. Routledge, London
Martinez-Torres ME, Rosset PM (2010) La Vía Campesina: the birth and evolution of a transnational social movement. J Peasant Stud 37:149–175
Marzec RP (2014) Neoliberalism, Environmentality, and the specter of Sajinda khan. In: Di Leo JR, Mehan U (eds) Capital at the Brink: overcoming the destructive legacies of neoliberalism. Open Humanities Press, London
Mcmichael P (1994) The global restructuring of agro-food systems. Cornell University Press, Ithaca
Meadows DH, Meadows DL, Randers J, Behrens WW (1972) The limits to growth, vol 102. Universe Books, New York, p 27
Merchant C (1981) The death of nature: women, ecology, and scientific revolution. HarperOne, San Francisco
Miličić V, Thorarinsdottir R, Santos MD, Hančič MT (2017) Commercial aquaponics approaching the european market: to consumers' perceptions of aquaponics products in europe. Water 9:80
Miller TR, Neff MW (2013) De-facto science policy in the making: how scientists shape science policy and why it matters (or, why STS and STP scholars should socialize). Minerva 51:295–315
Miller TR, Wiek A, Sarewitz D, Robinson J, Olsson L, Kriebel D, Loorbach D (2014) The future of sustainability science: a solutions-oriented research agenda. Sustain Sci 9:239–246

Moore JW (2015) Capitalism in the web of life: ecology and the accumulation of capital. Verso Books, New York
Nally D (2011) The biopolitics of food provisioning. Trans Inst Br Geogr 36:37–53
Noordwijk V (2014) Climate change: agricultural mitigation. In: van Alfen NK (ed) Encyclopedia of agriculture and food systems. Elsevier, San Diego
Norton BG (2005) Sustainability: a philosophy of adaptive ecosystem management. University of Chicago Press, Chicago
Nugent RA (1999) Measuring the sustainability of urban agriculture. For hunger-proof cities. Sustainable urban food systems. IDRC, Ottawa, pp 95–99
Oldfield F, Richardson K, Schellnhuber H, Turner B II, Wasson R, Planet A (2004) Global change and the earth system. Springer, New York
Parfitt J, Barthel M, Macnaughton S (2010) Food waste within food supply chains: quantification and potential for change to 2050. Philosophical Transactions of the Royal Society B 365:3065–3081
Pearson CJ (2007) Regenerative, semiclosed systems: a priority for twenty-first-century agriculture. Bioscience 57:409
Pearson LJ, Pearson L, Pearson CJ (2010) Sustainable urban agriculture: stocktake and opportunities. Int J Agric Sustain 8:7–19
Pelletier N, Tyedmers P (2010) Forecasting potential global environmental costs of livestock production 2000–2050. Proc Natl Acad Sci 107:18371–18374
Perrot N, Trelea I-C, Baudrit C, Trystram G, Bourgine P (2011) Modelling and analysis of complex food systems: state of the art and new trends. Trends Food Sci Technol 22:304–314
Petersen B, Snapp S (2015) What is sustainable intensification? Views from experts. Land Use Policy 46:1–10
Pickering A (1992) Science as practice and culture. University of Chicago Press, Chicago
Postel SL (2003) Securing water for people, crops, and ecosystems: new mindset and new priorities. Nat Res Forum:89–98. Wiley Online Library
Potter C (2004) Multifunctionality as an agricultural and rural policy concept. Sustaining agriculture and the rural environment: governance, policy and multifunctionality. Edward Elgar, Northampton, pp 15–35
Power EM (1999) Combining social justice and sustainability for food security. In: For hunger-proof cities: sustainable urban food systems. International Development Research Centre, Ottawa, pp 30–37
Prado JR, Segers G, Voelker T, Carson D, Dobert R, Phillips J, Cook K, Cornejo C, Monken J, Grapes L (2014) Genetically engineered crops: from idea to product. Annu Rev Plant Biol 65:769
Rakocy JE, Masser MP, Losordo TM (2006) Recirculating aquaculture tank production systems: aquaponics—integrating fish and plant culture. SRAC Publication 454:1–16
Reynolds HL, Smith AA, Farmer JR (2014) Think globally, research locally: paradigms and place in agroecological research. Am J Bot 101:1631–1639
Robles AG (2015) Introducing ENoLL and its living lab community. European Network of Living Labs, Brussels
Rockstrom J, Steffen W, Noone K, Persson A, Chapin FS, Lambin E, Lenton TM, Scheffer M, Folke C, Schellnhuber HJ, Nykvist B, De Wit CA, Hughes T, van der Leeuw S, Rodhe H, Sorlin S, Snyder PK, Costanza R, Svedin U, Falkenmark M, Karlberg L, Corell RW, Fabry VJ, Hansen J, Walker B, Liverman D, Richardson K, Crutzen P, Foley J (2009) Planetary boundaries: exploring the safe operating space for humanity. Ecol Soc 14:32
Rockström J, Williams J, Daily G, Noble A, Matthews N, Gordon L, Wetterstrand H, Declerck F, Shah M, Steduto P, De Fraiture C, Hatibu N, Unver O, Bird J, Sibanda L, Smith J (2017) Sustainable intensification of agriculture for human prosperity and global sustainability. Ambio 46:4–17
Röling N (2009) Pathways for impact: scientists' different perspectives on agricultural innovation. Int J Agric Sustain 7:83–94

Rose N (2009) The politics of life itself: biomedicine, power, and subjectivity in the twenty-first century. Princeton University Press, Princeton
Savransky M (2013) An ecology of times: modern knowledge, non-modern temporalities. In: Movements in time: revolution, social justice and times of change, vol 265. Cambridge Scholars Publishing, Newcastle upon Tyne
Schmautz Z, Graber A, Jaenicke S, Goesmann A, Junge R, Smits TH (2017) Microbial diversity in different compartments of an aquaponics system. Arch Microbiol 199:613–620
Scott JC (2008) Weapons of the weak: everyday forms of peasant resistance. Yale University Press, New Haven
Schoenbach K (2001) Myths of media and audiences: inaugural lecture as professor of general communication science, University of Amsterdam. Eur J Commun 16:361–376
Sen A (2001) Development as freedom. Oxford Paperbacks, Oxford
Smil V (2011) Harvesting the biosphere: the human impact. Popul Dev Rev 37:613–636
Smith A, Stirling A (2007) Moving outside or inside? Objectification and reflexivity in the governance of socio-technical systems. J Environ Policy Plan 9:351–373
Soini K, Birkeland I (2014) Exploring the scientific discourse on cultural sustainability. Geoforum 51:213–223
Steffen W, Broadgate W, Deutsch L, Gaffney O, Ludwig C (2015a) The trajectory of the Anthropocene: the great acceleration. Anthropocene Rev 2:81–98
Steffen W, Richardson K, Rockström J, Cornell SE, Fetzer I, Bennett EM, Biggs R, Carpenter SR, De Vries W, De Wit CA (2015b) Planetary boundaries: guiding human development on a changing planet. Science 347:1259855
Steffen W, Sanderson RA, Tyson PD, Jäger J, Matson PA, Moore B III, Oldfield F, Richardson K, Schellnhuber HJ, Turner BL (2006) Global change and the earth system: a planet under pressure. Springer, New York
Stengers I (2015) In catastrophic times: resisting the coming barbarism. Open Humanities Press, London
Stengers I (2018) Another science is possible: a manifesto for slow science. Polity Press, Cambridge
Stern N (2008) The economics of climate change. Am Econ Rev 98:1–37
Stocker T (2014) Climate change 2013: the physical science basis: working group I contribution to the fifth assessment report of the intergovernmental panel on climate change. Cambridge University Press, Cambridge
Struik P (2006) Trends in agricultural science with special reference to research and development in the potato sector. Potato Res 49:5
Struik P, Kuyper T, Brussaard L, Leeuwis C (2014) Deconstructing and unpacking scientific controversies in intensification and sustainability: why the tensions in concepts and values? Curr Opin Environ Sustain 8:80–88
Struik PC, Kuyper TW (2014) Editorial overview: sustainable intensification to feed the world: concepts, technologies and trade-offs. Curr Opin Environ Sustain 8:vi–viii
Stuart T (2009) Waste: uncovering the global food scandal. WW Norton & Company, New York
Tegtmeier EM, Duffy MD (2004) External costs of agricultural production in the United States. Int J Agric Sustain 2:1–20
Thrift N (1999) Steps to an ecology of place. Human geography today. Polity Press, Cambridge, pp 295–322
Tilman D, Balzer C, Hill J, Befort BL (2011) Global food demand and the sustainable intensification of agriculture. Proc Natl Acad Sci 108:20260–20264
Tittonell P, Giller KE (2013) When yield gaps are poverty traps: the paradigm of ecological intensification in African smallholder agriculture. Field Crop Res 143:76–90
van der Ploeg JD, Ventura F (2014) Heterogeneity reconsidered. Curr Opin Environ Sustain 8:23–28
Vermeulen SJ, Campbell BM, Ingram JS (2012) Climate change and food systems. Annu Rev Environ Resour 37:195
Villarroel M, Junge R, Komives T, König B, Plaza I, Bittsánszky A, Joly A (2016) Survey of aquaponics in europe. Water 8:468
Wall D (2007) Realist utopias? Green alternatives to capitalism. Environ Politics 16:518–522

Weber KM, Rohracher H (2012) Legitimizing research, technology and innovation policies for transformative change: combining insights from innovation systems and multi-level perspective in a comprehensive 'failures' framework. Res Policy 41:1037–1047

Weis T (2010) The accelerating biophysical contradictions of industrial capitalist agriculture. J Agrar Chang 10:315–341

Wiek A, Ness B, Schweizer-Ries P, Brand FS, Farioli F (2012) From complex systems analysis to transformational change: a comparative appraisal of sustainability science projects. Sustain Sci 7:5–24

Williams JW, Jackson ST (2007) Novel climates, no-analog communities, and ecological surprises. Front Ecol Environ 5:475–482

Wilson GA (2001) From productivism to post-productivism... and back again? Exploring the (un) changed natural and mental landscapes of European agriculture. Trans Inst Br Geogr 26:77–102

Winner L (1986) The whale and the reactor: a search for limits in an age of high technology. University of Chicago Press, Chicago

Withers CW (2009) Place and the "Spatial Turn" in geography and in history. J Hist Ideas 70:637–658

Zalasiewicz J, Williams M, Steffen W, Crutzen P (2010) The new world of the Anthropocene. Environ Sci Technol 44:2228–2231

Part IV
Management and Marketing

Chapter 17
Insight into Risks in Aquatic Animal Health in Aquaponics

Hijran Yavuzcan Yildiz, Vladimir Radosavljevic, Giuliana Parisi, and Aleksandar Cvetkovikj

Abstract Increased public interest in aquaponics necessitates a greater need to monitor fish health to minimize risk of infectious and non-infectious disease outbreaks which result from problematic biosecurity. Fish losses due to health and disease, as well as reporting of poor management practices and quality in produce, which could in a worst-case scenario affect human health, can lead to serious economic and reputational vulnerability for the aquaponics industry. The complexity of aquaponic systems prevents using many antimicrobial/antiparasitic agents or disinfectants to eradicate diseases or parasites. In this chapter, we provide an overview of potential hazards in terms of risks related to aquatic animal health and describe preventive approaches specific to aquaponic systems.

Keywords Aquaponics · Fish health · Biosecurity · Treatment strategies

H. Yavuzcan Yildiz (✉)
Department of Fisheries and Aquaculture, Ankara University, Ankara, Turkey
e-mail: yavuzcan@ankara.edu.tr

V. Radosavljevic
Department of Fish Diseases, National Reference Laboratory for Fish Diseases, Institute of Veterinary Medicine of Serbia, Belgrade, Serbia
e-mail: vladimiradosavljevic@yahoo.co.uk

G. Parisi
Department of Agriculture, Food, Environment and Forestry (DAGRI), Animal Science Section, University of Florence, Florence, Italy
e-mail: giuliana.parisi@unifi.it

A. Cvetkovikj
Department of Parasitology and Parasitic diseases, Faculty of Veterinary Medicine, Ss. Cyril and Methodius University in Skopje, Skopje, Republic of Macedonia

Department of Fisheries, Institute of Animal Science, Ss. Cyril and Methodius University in Skopje, Skopje, Republic of Macedonia
e-mail: acvetkovic@fvm.ukim.edu.mk; acvetkovikj@gmail.com

S. Goddek et al. (eds.), *Aquaponics Food Production Systems*,
https://doi.org/10.1007/978-3-030-15943-6_17

17.1 Introduction

The European Food Safety Authority reported a variety of drivers and potential issues associated with new trends in food production, and aquaponics was identified as a new food production process/practice (Afonso et al. 2017). As a new food production process, aquaponics can be defined as 'the combination of animal aquaculture and plant culture, through a microbial link and in a symbiotic relationship'. In aquaponics, the basic approach is to get benefit from the complementary functions of the organisms and nutrient recovery. The aquaculture part of the system applies principles that are similar to recirculating aquaculture systems (RAS). Aquaponics has gained momentum due to its superior features compared to traditional production systems. Thus, aquaponics seems capable of maintaining ecosystems and strengthening capacity for adaptation to climate change, extreme weather, drought, flooding and other disasters. These attributes are within reach, but as in other agri-/aquacultural production, aquaponics is not free of risks. Given the complexity of aquaponics as an environment for co-production of aquatic animals with plants, the hazards and risks may be more complicated.

The focus in this chapter is on categories of risk (i.e. animal health/disease) rather than specific risks (e.g. flectobacillosis disease). In traditional aquaculture, some of the more common types of production risks are diseases resulting from pathogens, unsuitable water quality and system failure. Snieszko (1974) reported that infectious diseases of fish occur when susceptible fish are exposed to virulent pathogens under certain environmental conditions. Thus, the interaction of pathogens, water quality and fish resistance is linked to occurrence of disease. Previous research using risk methods has studied the routes of introduction of aquatic animal pathogens in order to secure safe trade (e.g. import risk analyses) and support biosecurity (Peeler and Taylor 2011). Considering the similarity of aquaponics to RAS, it is expected that the health problems of aquatic animals in aquaponics may be identical to aquatic animals in RAS. Specifically, fluctuations in water quality may increase susceptibility of fish to pathogens (i.e. disease-causing organisms such as virus, bacteria, parasite, fungi) in RAS and cause disease outbreaks. Microorganisms in closed systems such as RAS or aquaponics are of significance in terms of maintaining fish health. Thus, Xue et al. (2017) reported the potential correlation between fish diseases and environmental bacterial populations in RAS. High pathogen density and limited medication possibilities make the system prone to disease problems. Disease or impaired health can cause catastrophic losses with decreased survival or poor feed conversion ratios. Regardless of which potential risk becomes problematic, each has the same impact: an overall decline in the production of a marketable quality product that then results in financial loss (McIntosh 2008). Diseases can be prevented only when the risks are recognized and managed before disease occurs (Nowak 2004). The severity of risks differs and will likely change depending on when each is encountered during the production cycle.

17.2 Aquaponics and Risk: A Development Perspective for Fish Health

Fish pathogens are prevalent in the aquatic environment, and fish are generally able to resist them unless overloaded by the allostatic load (Yavuzcan Yıldız and Seçer 2017). Allostasis refers to the 'stability through change' proposed by Sterling and Eyer (1988). Put simply this is the effort of fish to maintain homeostasis through changes in physiology. Allostatic load of fish in aquaponics may be a challenging factor as aquaponics is a complex system mainly in terms of the water quality and the microbial community in the system. Hence, the diseases of fish are generally species- and system-specific. Specific aquaponic diseases have not been described yet. From aquaculture, it is known that fish diseases are difficult to detect and are usually the end result of the interaction between various factors involving the environment, nutritional status of the fish, the immune robustness of the fish, existence of an infectious agent and/or poor husbandry and management practices. In order to sustain aquaponic systems, an aquatic health management approach needs to be developed considering the species cultured, the complexity of environments in aquaponics and the type of the aquaponic system management. Profitability in aquaponic production can be affected by even small percentage decreases in production, as seen in aquaculture (Subasinghe 2005).

Aquaponics is a sustainable, innovative approach for future food production systems, but this integrated system for production currently shows difficulties in moving from the experimental stage or small-scale modules to large-scale production. It could be hypothesized that the lack of economic success of this highly sustainable production system is due to major bottlenecks not scientifically addressed yet. Without a doubt, the cost-effectiveness and technical capabilities of aquaponic systems need further research to realize a scaling up of production (Junge et al. 2017). Research activity and innovations applied since the 1980s have transformed aquaponic technology into a viable system of food production, and although small-scale plants and research-structured plants are already viable, commercial-scale aquaponics are not often economically viable. The claimed advantages attributed and recognized for aquaponic systems are the following: significant reduction in the usage of water (compared to traditional soil methods of growing plants), bigger and healthier vegetables than when grown in soil, production of plants does not require artificial fertilizer and aquaponic products are free of antibiotics, pesticides and herbicides.

17.2.1 Risk Analysis Overview

Risk is defined as 'uncertainty about and severity of the consequences of an activity' (Aven 2016), and the risk picture reflects (i) probabilities/frequencies of hazards/ threats, (ii) expected losses given the occurrence of such a hazard/threat and (iii)

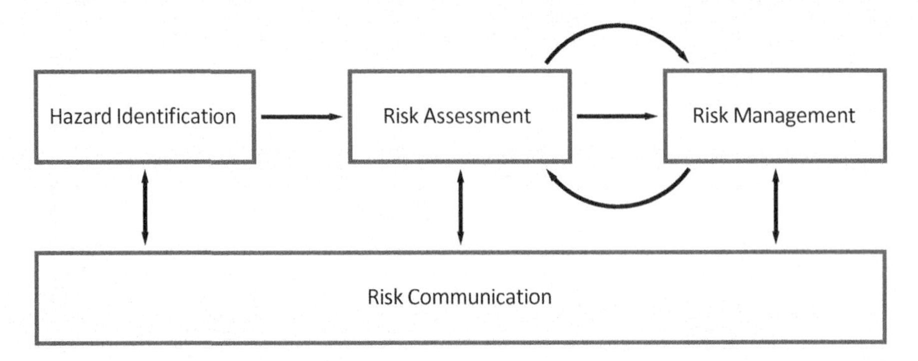

Fig. 17.1 Risk analysis (OIE 2017)

factors that could create large deviations between expected outcomes and the actual outcomes (uncertainties, vulnerabilities). Risk analysis offers tools to judge risk and assist in decision-making (Ahl et al. 1993; MacDiarmid 1997). Risk analysis is based on systematic use of the available information for decision-making, using the components of hazard identification, risk assessment, risk management and risk communication as indicated by World Organisation of Animal Health (OIE) (Fig. 17.1). This framework is commonly used for pathogen risk analysis (Peeler et al. 2007).

Risk analysis in food production, including aquaponics, can be applied to many cases, such as food security, invasive species, production profitability, trade and investment, and for consumer preference for safe, high-quality products (Bondad-Reantaso et al. 2005; Copp et al. 2016). The benefits of applying risk analysis in aquaculture became more clearly linked to this sector's sustainability, profitability and efficiency, and this approach can also be effective for the aquaponics sector. Therefore, disease introduction and potential transmission of pathogens can be evaluated in the context of risk to aquatic animal health (Peeler et al. 2007). Various international agreements, conventions and protocols cover human, animal and plant health, aquaculture, wild fisheries and the general environment in the field of risk. The most comprehensive and broad agreements and protocols are the World Trade Organization's (WTO) Sanitary and Phytosanitary Agreement, United Nations Environmental Program's (UNEP) Convention on Biological Diversity and the supplementary agreement Cartagena Protocol on Biosafety and the Codex Alimentarius (Mackenzie et al. 2003; Rivera-Torres 2003).

A key challenge regarding the field of risk relates to our depth of knowledge. Risk decisions are related situations characterized by large uncertainties (Aven 2016). Specifically, animal health risk analysis depends on knowledge gained from studies of epidemiology and statistics. Oidtmann et al. (2013) point out that the main constraint in developing risk-based surveillance (RBS) designs in the aquatic context is the lack of published data to advance the design of RBS. Thus, to increase robust

Table 17.1 Composite research needs for aquatic animal health in aquaponics

Research area	Research need
Basic research	Understanding the aquatic animal health and welfare concept in aquaponics in terms of the species of aquatic organisms and the system used
	Understanding the stress/stressor concept for aquatic organisms in aquaponics by the species and the system used
	Understanding the allostatic load for aquatic organisms and the emergence of diseases
	Understanding the welfare concept in aquaponics
	Characterizing the critical water quality parameters against aquatic animal health
	Understanding the sensitivity of aquatic organisms to the aquaponic environment
	Revealing the microbial profile for the different systems of the aquaponics
Health indicators	Developing and validating health indicators for aquatic animals raised in the aquaponic systems
Database development	Field data on the health/disease of aquatic animals in the aquaponics
	Field data on the microbial profile including pathogens

knowledge of risks in aquaponics, studies that both increase scientific data and reduce specific weaknesses and uncertain fields in aquaponics operations are needed. Some research areas that require more data for risk analysis in aquaponic systems are presented below (Table 17.1).

In terms of risk analysis for aquatic animal diseases or health in aquaponic systems, the OIE Aquatic Animal Health Code (the Aquatic Code) can be considered because the Aquatic Code sets out standards for the improvement of aquatic animal health and welfare of farmed fish worldwide and for safe international trade of aquatic animals and their products. This Code also includes use of antimicrobial agents in aquatic animals (OIE 2017).

17.3 Hazard Identification

In risk analysis, a hazard is generally specified by describing what might go wrong and how this might happen (Ahl et al. 1993). A hazard refers not only to the magnitude of an adverse effect but also to the likelihood of the adverse effect occurring (Müller-Graf et al. 2012). Hazard identification is important for revealing the factors that may favour the establishment of a disease and/or potential pathogen threat, or otherwise detrimental for fish welfare. Biological pathogens are recognised as hazard in aquaculture by Bondad-Reantaso et al. (2008). A broad range of factors can be taken into consideration as long as they are associated with disease occurrence, i.e. they are hazards.

Table 17.2 List of potential hazards for aquatic animal health in aquaponics

	Hazard identification	Hazard specification
Abiotic	pH	Too high/too low/rapid change
	Water temperature	Too high/too low/rapid change
	Suspended solids	Too high
	Dissolved oxygen content	Too low
	Carbon dioxide content	Too high
	Ammonia content	Too high, pH dependent
	Nitrite content	Too high
	Nitrate content	Extremely high
	Metal content	Too high, pH dependent
Biotic	Stocking density	Too high/too low
	Biofouling	
Feeding	Nutrients by the fish species	Surplus/deficiency
	Feeding frequency	Inadequate/improper feeding
	Dietary toxins	
	Feed additives	Unsuitable growth promoters
Management	Aquaponic system design	Poor system design
	Fish species	Unsuitable for aquaponics
	Operational issues (water circulation, biofilter, mechanical)	
	Chemotherapeutants use	Threat for the microbial balance
	Staff hygiene	
	Biosecurity	
Welfare	Stressors	Too high
	Allostatic load	High
	Rearing conditions	Suboptimal
Diseases	Nutritional diseases	
	Environmental diseases	
	Infectious diseases	

The sustainability of aquaponics is linked with a variety of factors, including system design, fish feed and faeces features, fish welfare and elimination of pathogens from the system (Palm et al. 2014a, b). Goddek (2016) reported that aquaponic systems are characterized by a wide range of microflora as fish and biofiltration exist in the same water mass. Since a great variety of microflora exists in aquaponic practices, the occurrence of pathogens and risks for human health should also be considered in order to guarantee food safety. In terms of sustainability of aquaponic systems, pathogen elimination to prevent losses due to diseases may be a challenging factor when aquatic animal production is intensified.

The use of chemotherapeutants in aquaculture to fight pathogens presents a number of potential hazards and risks to production systems, the environment and human health (Bondad-Reantaso and Subasinghe 2008) (Table 17.2).

To eliminate hazards, the fish rearing and plant cultivation phases should be considered separately. The biggest risks in fish rearing are related to water quality, fish density, feeding quality and quantity and disease (Yavuzcan Yildiz et al. 2017). Depending on the species of fish reared, the level of risk can increase if the species is not appropriate for the conditions of the particular system. For example, potassium is often supplemented in aquaponic systems to promote plant growth, but results in reduced performance in hybrid striped bass. Normally, freshwater and high-density culture-tolerant species are utilized in aquaponics. The most common species of fish in commercial systems are tilapia and ornamental fish. Channel catfish, largemouth bass, crappies, rainbow trout, pacu, common carp, koi carp, goldfish, Asian sea bass (or barramundi) and Murray cod are among the species that have been trialled (Rakocy et al. 2006). Tilapia, a warm-water species, highly tolerant of fluctuating water parameters (pH, temperature, oxygen and dissolved solids), is the species largely reared in most commercial aquaponic systems in North America and elsewhere. The results of a recent online survey, based on answers from 257 respondents, showed that tilapia is reared in 69% of aquaponic plants (Love et al. 2015). Tilapia presents an economic interest in some markets but not in others. In the same survey (Love et al. 2015), other species utilized were ornamental fish (43%), catfish (25%), other aquatic animals (18%), perch (16%), bluegill (15%), trout (10%) and bass (7%). One of the major weaknesses in aquaponic systems is the management of water quality to meet the requirements of the tank-reared fish, while cultivated crops are treated as the second step of the process. Fish require water with appropriate parameters for oxygen, carbon dioxide, ammonia, nitrate, nitrite, pH, chlorine and others. A high level of suspended solids can affect the health status of fish (Yavuzcan Yildiz et al. 2017), provoking damages to gill structure, such as the epithelium lifting, hyperplasia in the pillar system and reduction of epithelial volume (Au et al. 2004). Fish stocking density and feeding (feeding rate and volume, feed composition and characteristics) affect the digestion processes and metabolic activities of fish and, accordingly, the catabolites, total dissolved solids (TDS) and waste by-products (faeces and uneaten feed) in the rearing water. The basic principle on which the aquaponic system is based is the utilization of catabolites in water for plant growth. Aquaponic systems require 16 essential nutrients and all these macro- and micronutrients must be balanced for optimal plant growth. An excess of one nutrient can negatively affect the bioavailability of others (Rakocy et al. 2006). Therefore, the continuous monitoring of water parameters is essential to maintain water quality appropriate for fish and crop growth and to maximize the benefits of the process. Reduced water exchange and low crop growth rate can create toxic nutrient concentrations in water for fish and crops. On the other hand, the addition of some micronutrients (Fe^{+2}, Mn^{+2}, Cu^{+2}, B^{+3} and Mo^{+6}), normally scarce in water where fish are reared, is essential to adequately sustain crop production. In comparison to hydroponic culture, crops in aquaponic systems require lower levels of total dissolved solid (TDS, 200–400 ppm) or EC (0.3–0.6 mmhos/cm) and require, like fish, a high level of dissolved oxygen in water (Rakocy et al. 2006) for root respiration.

17.4 Fish Health Management

17.4.1 Fish Diseases and Prevention

While fish diseases caused by bacteria, viruses, parasites or fungi can have a significant negative impact on aquaculture (Kabata 1985), the appearance of a disease in aquaponic systems can be even more devastating. Maintenance of fish health in aquaponic systems is more difficult than in RAS, and, in fact, control of fish diseases is one of the main challenges for successful aquaponics (Sirakov et al. 2016). Diseases which affect fish can be divided into two categories: infectious and non-infectious fish diseases. Infectious diseases are caused by different microbial pathogens transmitted either from the environment or from other fish. Pathogens can be transmitted between the fish (horizontal transmission) or vertically, by (externally or internally) infected eggs or infected milt. More than half of the infectious disease outbreaks in aquaculture (54.9%) are caused by bacteria, followed by viruses, parasites and fungi (McLoughlin and Graham 2007). Often, although clinical signs or lesions are not present, fish can carry pathogens in a subclinical or carrier state (Winton 2002). Fish diseases can be caused by ubiquitous bacteria, present in any water containing organic enrichment. Under certain conditions, bacteria quickly become opportunistic pathogens. The presence of low numbers of parasites on the gills or skin usually does not lead to significant health problems. The capability of a pathogen to cause clinical disease depends on the interrelationship of six major components related to fish and the environment in which they live (physiological status, host, husbandry, environment, nutrition and pathogen). If any of the components is weak, it will affect the health status of the fish (Plumb and Hanson 2011). Non-infectious diseases are usually related to environmental factors, inadequate nutrition or genetic defects (Parker 2012). Successful fish health management is accomplished through disease prevention, reduction of infectious disease incidence and reduction of disease severity when it occurs. Avoidance of contact between the susceptible fish and a pathogen should be a critical goal, in order to prevent outbreak of infectious disease.

Three main measures to achieve this goal are:

- Use of pathogen-free water supply.
- Use of certified pathogen-free stocks.
- Strict attention to sanitation (Winton 2002).

Implementation of these measures will decrease fish exposure to pathogenic agents. However, it is practically impossible to define all agents which could cause disease in the aquatic environment and to completely prevent host exposure to pathogens. Certain factors, such as overcrowding, increase fish susceptibility to infection and pathogen transmission. For that reason, many pathogens which do not cause disease in wild fish can cause disease outbreaks with high mortality rates in high-density fish production systems. To avoid this, the infection level of fish in aquaponics must be continually monitored. Maintaining biosecurity in aquaponics is important not only from an economic point of view but also for fish welfare. Appearance of any fish pathogen in constrained tank space and under high population density will inevitably pose a threat to fish health, both to the individuals that are affected by the pathogen and those still unaffected.

The goal of biosecurity is the implementation of practices and procedures which will reduce the risks of:

- Introduction of pathogens into the facility.
- Spread of pathogens throughout the facility.
- Presence of conditions which can increase susceptibility to infection and disease (Bebak-Williams et al. 2007).

The achievement of this goal involves management protocols to prevent specific pathogens from entering the production system. Quarantine is an important biosecurity component for prevention of contact with infectious agents and is used when fish are moved from one area to another. All newly acquired fish are quarantined before they are introduced into established populations. Fish under quarantine are isolated for a specific period of time before release into contact with a resident population, preferably in a separate area with dedicated equipment (Plumb and Hanson 2011). New fish remain in quarantine until shown to be disease-free. It is advisable in some cases to quarantine new fish in an isolation tank for 45 days before adding them to the main system (Somerville et al. 2014). During quarantine, fish are monitored for signs of disease and sampled for presence of infectious agents. Prophylactic treatments may be initiated during the quarantine period in order to remove initial loads of external parasites.

For disease prevention, certain measures are recommended to reduce risk factors:

- Administer commercial vaccines against various fish viral and bacterial pathogens. Most common routes of application are by injection, by immersion or via food.
- Breed strains of fish which are more resistant to certain fish pathogens. Although Evenhuis et al. (2015) report that fish strains with increased simultaneous resistance to two bacterial diseases (columnaris and bacterial cold water disease) are available, there is evidence that increased susceptibility to other pathogens may occur (Das and Sahoo 2014; Henryon et al. 2005).
- Take preventive and corrective measures to prevent stress in fish. Since multiple stressors are present in every step of aquaponic production, avoidance and management of stress through monitoring and prevention minimize its influence on fish health.

- Avoid high stocking density, which causes stress and may increase the incidence of disease even if other environmental factors are acceptable. Also, high stocking density increases the possibility of skin lesions, which are sites of various pathogen entries into the organism.
- Regularly remove contaminants from water (uneaten food, faeces and other particulate organics). Dead or dying fish should be removed promptly as they can serve as potential disease sources to the remaining stock and a breeding ground for others, as well as fouling the water when decomposing (Sitjà-Bobadilla and Oidtmann 2017).
- Disinfect all equipment used for tank cleaning and fish manipulation. After adequate disinfection, all equipment should be rinsed with clear water. Use of footbaths and hand washing with disinfecting soap at the entrance and within the buildings are recommended. These steps directly decrease the potential for the spread of pathogens (Sitjà-Bobadilla and Oidtmann 2017). Certain chemicals used as disinfectants (such as benzalkonium chloride, chloramine B and T, iodophors) are effective for disease prevention.
- Administer dietary additives and immunostimulants for improvement of health and to reduce the impacts of disease. Such diets contain various ingredients important for improvement of health and disease resistance (Anderson 1992; Tacchi et al. 2011). There exists a wide range of products and molecules, including natural plant products, immunostimulants, vitamins, microorganisms, organic acids, essential oils, prebiotics, probiotics, synbiotics, nucleotides, vitamins, etc. (Austin and Austin 2016; Koshio 2016; Martin and Król 2017).
- Segregate fish by age and species for disease prevention, since susceptibility to certain pathogens varies with age, and certain pathogens are specific to some fish species. Generally, young fish are more susceptible to pathogens than older fish (Plumb and Hanson 2011).

Maintaining the health of fish in aquaponics requires adequate health management and continuous attention. Optimal fish health is best achieved through biosecurity measures, adequate production technology and husbandry management practices which enable optimal conditions. As mentioned, avoidance through optimal rearing conditions and biosecurity procedures are the best way to avoid fish diseases. Invariably, however, a pathogen may appear in the system. The first and most important action is to identify the pathogen correctly.

17.4.2 *Disease Diagnosis (Identification of Diseased Fish)*

Early recognition of diseased fish is important in maintaining health of the aquaculture unit in the aquaponic system. Accurate diagnosis and prompt response will stop the spread of disease to other fish, thus minimizing losses.

Examination of live fish starts by observing their behaviour. Constant and careful daily observation enables early recognition of diseased fish. As a rule, fish should be observed for behavioural changes before, during and after feeding.

Healthy fish exhibit fast, energetic swimming movements and a strong appetite. They swim in normal, species-specific patterns and have intact skin without discolorations (Somerville et al. 2014). Diseased fish exhibit various behavioural changes with or without visible change in physical appearance. The most obvious indicator of deteriorating fish health is the reduction (cessation) of feeding activity, usually as a result of an environmental stress and/or an infectious/parasitic disease. The most obvious sign of disease is the presence of dead or dying animals (Parker 2012; Plumb and Hanson 2011).

Behavioural changes in diseased fish may include abnormal swimming (swimming near the surface, along the tank sides, crowding at the water inlet, whirling, twisting, darting, swimming upside down), flashing, scratching on the bottom or sides of the tank, unusually slow movement, loss of equilibrium, weakness, hanging listlessly below the surface, lying on the bottom and gasping at the water surface (sign of low oxygen level) or not reacting to external stimuli. In addition to behavioural changes, diseased fish exhibit physical signs that can be seen by the unaided eye. These gross signs can be external, internal or both and may include loss of body mass; distended abdomen or dropsy; spinal deformation; darkening or lightening of the skin; increased mucus production; discoloured areas on the body; skin erosions, ulcers or sores; fin damage; scale loss; cysts; tumours; swelling on the body or gills; haemorrhages, especially on the head and isthmus, in the eyes and at the base of fins; and bulging eyes (pop-eye, exophthalmia) or endophthalmia (sunken eyes). The internal signs are changes in the size, colour and texture of the organs or tissues, accumulation of fluids in the body cavities and presence of pathological formations such as tumours, cysts, haematomas and necrotic lesions (Noga 2010; Parker 2012; Plumb and Hanson 2011; Winton 2002).

Upon suspicion of deteriorating fish health, the first step is to check water quality (water temperature, dissolved oxygen, pH, levels of ammonia, nitrite and nitrate) and promptly respond to any deviations from the optimal range. If the majority of fish in the tank has abnormal behaviour and shows non-specific signs of disease, there is likely a change in the environmental conditions (Parker 2012; Somerville et al. 2014). Low oxygen (hypoxia) is a frequent cause of fish mortality. Fish in water with low oxygen are lethargic, congregate near the water surface, gasp for air and have brighter pigmentation. Dying fish exhibit agonal respiration, with mouth open and opercula flared. These signs are also evident in fish carcasses. High ammonia levels cause hyperexcitability with muscular spasms, cessation of feeding and death. Chronic deviation from optimal levels results in anaemia and decreased growth and disease resistance. Nitrite-poisoned fish have behavioural changes characteristic of hypoxia with pale tan or brown gills and brown blood (Noga 2010).

When only few fish show signs of disease, it is imperative to remove them immediately in order to stop and prevent the spread of the disease agent to the other fish. In the early stages of a disease outbreak, generally only a few fish will show signs and die. In the following days, there will be a gradual increase in the daily mortality rate. The diseased fish must be carefully examined in order to determine the cause. Only a few fish diseases produce pathognomonic (specific to a given disease) behavioural and physical signs. Nevertheless, careful observation will often allow the examiner to narrow down the cause to environmental conditions or disease agents. In a serious disease outbreak, a fish veterinarian/health specialist should be contacted immediately for professional diagnosis and disease management options. In order to solve the disease problem, the diagnostician will need a detailed description of the behavioural and physical signs exhibited by the diseased fish, daily records of the water quality parameters, origin of the fish, date and size of fish at stocking, feeding rate, growth rate and daily mortality (Parker 2012; Plumb and Hanson 2011; Somerville et al. 2014).

17.5 Treatment Strategies in Aquaponics

Treatment options for diseased fish in an aquaponic system are very limited. As both fish and plants share the same water loop, medications used for disease treatments can easily harm or destroy the plants, and some may get absorbed by the plants, causing withdrawal periods or even making them unusable for consumption. The medications can also have detrimental effects on the beneficial bacteria in the system. If a medicinal treatment is absolutely necessary, it must be implemented early in the course of the disease. The diseased fish is transferred into a separate (hospital, quarantine) tank isolated from the system for treatment. When returning the fish after the treatment, it is important not to transfer the used medications into the aquaponic system. All these limitations require improvements of disease management options with minimal negative effects to the fish, the plants and the system (Goddek et al. 2015, 2016; Somerville et al. 2014; Yavuzcan Yildiz et al. 2017). One of the most used and effective, old-school treatments against the most common bacterial, fungal and parasitic infections in fish is a salt (sodium chloride) bath. Salt is beneficial for the fish, but can be detrimental to the plants in the system (Rakocy 2012), and the whole treatment procedure must be performed in a separate tank. A good option is to separate the recirculating aquaculture unit from the hydroponic unit (decoupled aquaponic systems) (see Chap. 8). Decoupling allows for fish disease and water treatment options that are not possible in coupled systems (Monsees et al. 2017) (see Chap. 7). One recent improvement for the control of fish ectoparasites and disinfection in the aquaponic systems is the use of Wofasteril (KeslaPharmaWolfen GMBH, Bitterfeld-Wolfen, Germany), a peracetic acid-containing product that leaves no residues in the system (Sirakov et al. 2016). Alternatively, hydrogen peroxide can be used, but at a much higher concentration. While these chemicals have minimal side effects, their presence is undesirable in aquaponic systems and alternative approaches, such as biological control methods, are required (Rakocy 2012).

The biological control method (biocontrol) is based on the use of other living organisms in the system, relying on natural relationships among the species (commensalism, predation, antagonism, etc.) (Sitjà-Bobadilla and Oidtmann 2017) to control fish pathogens. At present, this method is a complementary fish health management tool with high potential, especially in aquaponic systems. The most successful implementation of biocontrol in fish culture is the use of cleaner fish against sea lice (skin parasites) in salmon farms. It is best practiced in Norwegian farms where cleaning wrasse (Labridae) are co-cultured with salmon. The wrasse remove and feed on sea lice (Skiftesvik et al. 2013). Although cleaning is less common in freshwater fish, the leopard plecos (*Glyptoperichthys gibbiceps*), cohabiting with blue tilapia (*Oreochromis aureus*), successfully keeps infection with *Ichthyophthirius multifiliis* under control by feeding on the parasite cysts (Picón-Camacho et al. 2012). This biocontrol method is becoming increasingly important in aquaculture and can be considered in aquaponic systems. Additionally, it must be noted that the cleaner fish can also harbour pathogens that can be transmitted to the main cultured species. Therefore, they must also undergo preventive and quarantine procedures before introduction into the system.

Another biocontrol method, still in the exploratory application phase in fish culture, is the use of filter-feeding and filtering organisms. By reducing the pathogen loads in the water, these organisms can lower the chances of disease emergence (Sitjà-Bobadilla and Oidtmann 2017). For example, Othman et al. (2015) demonstrated the ability of freshwater mussels (*Pilsbryoconcha exilis*) to reduce the population of *Streptococcus agalactiae* in a laboratory-scale tilapia culture system. The potential of this biocontrol method in aquaponic systems is yet to be tested, and new studies are needed to explore the possibilities not only for fish disease control but also for control of plant pathogens.

The most promising and well-documented biocontrol method is the use of beneficial microorganisms as probiotics in fish feed or in the rearing water. Their usage in aquaponic systems as promoters of fish/plant growth and health is well known, and probiotics have also demonstrated effectiveness against a range of bacterial pathogens in different fish species. For example, in rainbow trout, dietary *Carnobacterium maltaromaticum* and *C. divergens* protected from *Aeromonas salmonicida* and *Yersinia ruckeri* infections (Kim and Austin 2006) and *Aeromonas sobria* GC2 incorporated into the feed successfully prevented clinical disease caused by *Lactococcus garvieae* and *Streptococcus iniae* (Brunt and Austin 2005). Dietary *Micrococcus luteus* reduced the mortalities from *Aeromonas hydrophila* infection and enhanced the growth and health of Nile tilapia (Abd El-Rhman et al. 2009). Recent research by Sirakov et al. (2016) has made good progress in simultaneous biocontrol of parasitic fungi in both fish and plants in a closed recirculating aquaponic system. In total, over 80% of the isolates (bacteria isolated from the aquaponic system) were antagonistic to both fungi (*Saprolegnia parasitica* and *Pythium ultimum*) in the *in vitro* tests. Bacteria were not classified taxonomically, and the authors assumed that they belonged to the genus *Pseudomonas* and to a group of lactic acid bacteria. These findings, although very promising, have yet to be tested in an operational aquaponic system.

As a final alternative to chemical treatment, we suggest the use of medicinal plants with antibacterial, antiviral, antifungal and antiparasitic properties. Plant extracts have various biological characteristics with minimal risk of developing resistance in the targeted organisms (Reverter et al. 2014). Many scientific reports demonstrate the effectiveness of medicinal plants against fish pathogens. For example, Nile tilapia fed with a diet containing mistletoe (*Viscum album coloratum*) increased the survivability when challenged with *Aeromonas hydrophila* (Park and Choi 2012). Indian major carp showed a significant reduction in mortality when challenged with *Aeromonas hydrophila* and fed with diets containing prickly chaff flower (*Achyranthes aspera*) and Indian ginseng (*Withania somnifera*) (Sharma et al. 2010; Vasudeva Rao et al. 2006). Medicinal plant extracts have also proven effective against ectoparasites. In goldfish, Yi et al. (2012) demonstrated the effectiveness of *Magnolia officinalis* and *Sophora alopecuroides* extracts against *Ichthyophthirius multifiliis*, and Huang et al. (2013) showed that extracts of *Caesalpinia sappan*, *Lysimachia christinae*, *Cuscuta chinensis*, *Artemisia argyi* and *Eupatorium fortunei* have 100% anthelmintic efficacy against *Dactylogyrus intermedius*. The use of medicinal plants in aquaponics is promising, but yet more research is needed to find the appropriate treatment strategy without undesirable effects. As referred by Junge et al. (2017), even though research on aquaponics has largely developed in recent years, the number of research papers published on the topic is still dramatically low compared to papers published related to aquaculture or hydroponics. Aquaponics, still considered an emerging technology, is however now characterized by having great potential for food production for the world's population that, according to the results of the UN World Population Prospects (UN 2017), numbered nearly 7.6 billion in mid-2017 and, based on the projections, it is expected to increase to 1 billion within 12 years, reaching about 8.6 billion in 2030. Nevertheless, considering the potential risks to the sustainability of aquaponics due to fish diseases, development of good ideas, and novel methods and approaches for pathogen control will be our major challenge for the future. There is a pressing need to initiate new knowledge to provide a better basis for management of fish and plant health, and to continue to develop operation and infrastructure systems for the aquaponic industry. The causes of fish losses in aquaponic systems, system-specific diseases and the interaction and alteration of microbial community, along with pathogens, are priority areas for study.

References

Abd El-Rhman AM, Khattab YAE, Shalaby AME (2009) *Micrococcus luteus* and *Pseudomonas* species as probiotics for promoting the growth performance and health of Nile tilapia, *Oreochromis niloticus*. Fish Shellfish Immunol 27:175–180. https://doi.org/10.1016/j.fsi.2009.03.020

Afonso A, Matas RG, Maggiore A, Merten C, Robinson T (2017) EFSA's activities on emerging risks in 2016. EFSA Support Publ 14:1–59. https://doi.org/10.2903/sp.efsa.2017.EN-1336

Ahl AS, Acree JA, Gipson PS, McDowell RM, Miller L, McElvaine MD (1993) Standardization of nomenclature for animal health risk analysis. Rev Sci Tech 12:1045–1053

Anderson DP (1992) Immunostimulants, adjuvants, and vaccine carriers in fish: applications to aquaculture. Annu Rev Fish Dis 2:281–307. https://doi.org/10.1016/0959-8030(92)90067-8

Au DWT, Pollino CA, Wu RSS, Shin PKS, Lau STF, Tang JYM (2004) Chronic effects of suspended solids on gill structure, osmoregulation, growth, and triiodothyronine in juvenile green grouper *Epinephelus coioides*. Mar Ecol Prog Ser 266:255–264. https://doi.org/10.3354/meps266255

Austin B, Austin DA (2016) Bacterial fish pathogens: disease of farmed and wild fish, 6th edn. Springer, Cham. https://doi.org/10.1007/978-3-319-32674-0

Aven T (2016) Risk assessment and risk management: review of recent advances on their foundation. Eur J Oper Res 253:1–13. https://doi.org/10.1016/j.ejor.2015.12.023

Bebak-Williams J, Noble A, Bowser P, Wooster G (2007) Fish health management. In: Timmons MB, Ebeling JM (eds) Recirculation aquaculture. Cayuga Aqua Ventures, Ithaca, NY, and Northeastern Regional Aquaculture Center. Publication No. 01-007, p 619–664

Bondad-Reantaso MG, Subasinghe RP (2008) Meeting the future demand for aquatic food through aquaculture: the role of aquatic animal health. In: Fisheries for global welfare and environment, 5th World Fisheries Congress 2008, pp 197–207

Bondad-Reantaso MG, Subasinghe RP, Arthur JR, Ogawa K, Chinabut S, Adlard R, Tan Z, Shariff M (2005) Disease and health management in Asian aquaculture. Vet Parasitol 132:249–272. https://doi.org/10.1016/j.vetpar.2005.07.005

Bondad-Reantaso MG, Arthur JR, Subasinghe RP (2008) Understanding and applying risk analysis in aquaculture, FAO Fisheries and Aquaculture Technical Paper No. 519

Brunt J, Austin B (2005) Use of a probiotic to control lactococcosis and streptococcosis in rainbow trout, *Oncorhynchus mykiss* (Walbaum). J Fish Dis 28:693–701. https://doi.org/10.1111/j.1365-2761.2005.00672.x

Copp GH, Russell IC, Peeler EJ, Gherardi F, Tricarico E, Macleod A, Cowx IG, Nunn AD, Occhipinti-Ambrogi A, Savini D, Mumford J, Britton JR (2016) European non-native species in aquaculture risk analysis scheme – a summary of assessment protocols and decision support tools for use of alien species in aquaculture. Fish Manag Ecol 23:1–11. https://doi.org/10.1111/fme.12074

Das S, Sahoo PK (2014) Markers for selection of disease resistance in fish: a review. Aquac Int 22:1793–1812. https://doi.org/10.1007/s10499-014-9783-5

Evenhuis JP, Leeds TD, Marancik DP, LaPatra SE, Wiens GD (2015) Rainbow trout (*Oncorhynchus mykiss*) resistance to columnaris disease is heritable and favorably correlated with bacterial cold water disease resistance. J Anim Sci 93:1546–1554. https://doi.org/10.2527/jas.2014-8566

Goddek S (2016) Three-loop aquaponics systems: chances and challenges. In: Proceedings of the international conference on Aquaponics Research Matters, Ljubljana, Slovenia, 22 March 2016

Goddek S, Delaide B, Mankasingh U, Ragnarsdottir K, Jijakli H, Thorarinsdottir R (2015) Challenges of sustainable and commercial aquaponics. Sustainability 7:4199–4224. https://doi.org/10.3390/su7044199

Goddek S, Espinal C, Delaide B, Jijakli M, Schmautz Z, Wuertz S, Keesman K (2016) Navigating towards decoupled aquaponic systems: a system dynamics design approach. Water 8:303. https://doi.org/10.3390/w8070303

Henryon M, Berg P, Olesen NJ, Kjær TE, Slierendrecht WJ, Jokumsen A, Lund I (2005) Selective breeding provides an approach to increase resistance of rainbow trout (*Oncorhynchus mykiss*) to the diseases, enteric redmouth disease, rainbow trout fry syndrome, and viral haemorrhagic septicaemia. Aquaculture 250:621–636. https://doi.org/10.1016/j.aquaculture.2004.12.022

Huang A-G, Yi Y-L, Ling F, Lu L, Zhang Q-Z, Wang G-X (2013) Screening of plant extracts for anthelmintic activity against *Dactylogyrus intermedius* (Monogenea) in goldfish (*Carassius auratus*). Parasitol Res 112:4065–4072. https://doi.org/10.1007/s00436-013-3597-7

Junge R, König B, Villarroel M, Komives T, Jijakli MH (2017) Strategic points in aquaponics. Water 9:1–9. https://doi.org/10.3390/w9030182

Kabata Z (1985) Parasites and diseases of fish cultured in the tropics. Taylor & Francis, London

Kim D-H, Austin B (2006) Innate immune responses in rainbow trout (*Oncorhynchus mykiss*, Walbaum) induced by probiotics. Fish Shellfish Immunol 21:513–524. https://doi.org/10.1016/j.fsi.2006.02.007

Koshio S (2016) Immunotherapies targeting fish mucosal immunity – current knowledge and future perspectives. Front Immunol 6:643. https://doi.org/10.3389/fimmu.2015.00643

Love DC, Fry JP, Li X, Hill ES, Genello L, Semmens K, Thompson RE (2015) Commercial aquaponics production and profitability: findings from an international survey. Aquaculture 435:67–74. https://doi.org/10.1016/j.aquaculture.2014.09.023

MacDiarmid SC (1997) Risk analysis, international trade, and animal health. In: Molak V (ed) Fundamentals of risk analysis and risk management. CRC Lewis Publishers, Boca Raton, pp 377–387

Mackenzie R, Burhenne-Guilmin F, La Viña AGM, Werksman JD, Ascencio A, Kinderlerer J, Kummer K, Tapper R (2003) An explanatory guide to the Cartagena protocol on biosafety. IUCN, Gland and Cambridge, UK, p 295

Martin SAM, Król E (2017) Nutrigenomics and immune function in fish: new insights from omics technologies. Dev Comp Immunol 75:86–98. https://doi.org/10.1016/j.dci.2017.02.024

McIntosh D (2008) Aquaculture risk management. NRAC Publication No. 107. 1 Feb 2018. http://www.mdsg.umd.edu/sites/default/files/files/107-Risk%20management.pdf

McLoughlin MF, Graham DA (2007) Alphavirus infections in salmonids - a review. J Fish Dis 30:511–531. https://doi.org/10.1111/j.1365-2761.2007.00848.x

Monsees H, Kloas W, Wuertz S (2017) Decoupled systems on trial: eliminating bottlenecks to improve aquaponic processes. PLoS One 12:e0183056. https://doi.org/10.1371/journal.pone.0183056

Müller-Graf C, Berthe F, Grudnik T, Peeler E, Afonso A (2012) Risk assessment in fish welfare, applications and limitations. Fish Physiol Biochem 38:231–241. https://doi.org/10.1007/s10695-011-9520-1

Noga E (2010) Fish disease: diagnosis and treatment, 2nd edn. Wiley-Blackwell, Ames. https://doi.org/10.1002/9781118786758

Nowak BF (2004) Assessment of health risks to southern bluefin tuna under current culture conditions. Bull Eur Assoc Fish Pathol 24:45–51

Oidtmann B, Peeler E, Lyngstad T, Brun E, Bang Jensen B, Stärk KDC (2013) Risk-based methods for fish and terrestrial animal disease surveillance. Prev Vet Med 112:13–26. https://doi.org/10.1016/j.prevetmed.2013.07.008

OIE (2017) Aquatic animal health code. Office International des Epizooties, Paris. 1 Feb 2018. http://www.oie.int/index.php?id=171&L=0&htmfile=preface.htm

Othman F, Islam MS, Sharifah EN, Shahrom-Harrison F, Hassan A (2015) Biological control of streptococcal infection in Nile tilapia *Oreochromis niloticus* (Linnaeus, 1758) using filter-feeding bivalve mussel *Pilsbryocon chaexilis* (Lea, 1838). J Appl Ichthyol 31:724–728. https://doi.org/10.1111/jai.12804

Palm HW, Seidemann R, Wehofsky S, Knaus U (2014a) Significant factors affecting the economic sustainability of closed aquaponic systems. Part I: system design, chemo-physical parameters and general aspects. AACL Bioflux 7:20–32

Palm HW, Bissa K, Knaus U (2014b) Significant factors affecting the economic sustainability of closed aquaponic systems. Part II: fish and plant growth. AACL Bioflux 7:162–175

Park K-H, Choi S-H (2012) The effect of mistletoe, *Viscum album coloratum*, extract on innate immune response of Nile tilapia (*Oreochromis niloticus*). Fish Shellfish Immunol 32:1016–1021. https://doi.org/10.1016/j.fsi.2012.02.023

Parker R (2012) Aquaculture science, 3rd edn. Delmar, Cengage Learning, Clifton Park

Peeler EJ, Taylor NG (2011) The application of epidemiology in aquatic animal health -opportunities and challenges. Vet Res 42:94. https://doi.org/10.1186/1297-9716-42-94

Peeler EJ, Murray AG, Thebault A, Brun E, Giovaninni A, Thrush MA (2007) The application of risk analysis in aquatic animal health management. Prev Vet Med 81:3–20. https://doi.org/10.1016/j.prevetmed.2007.04.012

Picón-Camacho SM, Leclercq E, Bron JE, Shinn AP (2012) The potential utility of the leopard pleco (*Glyptoperichthys gibbiceps*) as a biological control of the ciliate protozoan *Ichthyophthirius multifiliis*. Pest Manag Sci 68:557–563. https://doi.org/10.1002/ps.2293

Plumb JA, Hanson LA (2011) Health maintenance and principal microbial diseases of cultured fishes, 3rd edn. Wiley-Blackwell, Oxford. https://doi.org/10.1002/9780470958353

Rakocy J (2012) Aquaponics-integrating fish and plant culture. In: Tidwell J (ed) Aquaculture production systems. Wiley-Blackwell, Ames, pp 343–386. https://doi.org/10.1002/9781118250105

Rakocy JE, Masser MP, Losordo TM (2006) Recirculating aquaculture tank production systems: aquaponics- integrating fish and plant culture. SRAC Publication No. 454. 1 Feb 2018. http://www.aces.edu/dept/fisheries/aquaculture/documents/309884-SRAC454.pdf

Reverter M, Bontemps N, Lecchini D, Banaigs B, Sasal P (2014) Use of plant extracts in fish aquaculture as an alternative to chemotherapy: current status and future perspectives. Aquaculture 433:50–61. https://doi.org/10.1016/j.aquaculture.2014.05.048

Rivera-Torres O (2003) The biosafety protocol and the WTO, 26 B.C. Int Comp L Rev 263, http://lawdigitalcommons.bc.edu/iclr/vol26/iss2/7

Sharma A, Deo AD, Riteshkumar ST, Chanu TI, Das A (2010) Effect of *Withania somnifera* (L. Dunal) root as a feed additive on immunological parameters and disease resistance to *Aeromonas hydrophila* in Labeo rohita(Hamilton) fingerlings. Fish Shellfish Immunol 29:508–512. https://doi.org/10.1016/j.fsi.2010.05.005

Sirakov I, Lutz M, Graber A, Mathis A, Staykov Y, Smits T, Junge R (2016) Potential for combined biocontrol activity against fungal fish and plant pathogens by bacterial isolates from a model aquaponic system. Water 8:518. https://doi.org/10.3390/w8110518

Sitjà-Bobadilla A, Oidtmann B (2017) Integrated pathogen management strategies in fish farming. In: Jeney G (ed) Fish diseases: prevention and control strategies. Academic, London, pp 119–144. https://doi.org/10.1016/B978-0-12-804564-0.00005-3

Skiftesvik AB, Bjelland RM, Durif CMF, Johansen IS, Browman HI (2013) Delousing of Atlantic salmon (*Salmo salar*) by cultured vs. wild ballan wrasse (*Labrus bergylta*). Aquaculture 402–403:113–118. https://doi.org/10.1016/j.aquaculture.2013.03.032

Snieszko SF (1974) The effects of environmental stress on outbreaks of infectious diseases of fishes. J Fish Biol 6:197–208

Somerville C, Cohen M, Pantanella E, Stankus A, Lovatelli A (2014) Small-scale aquaponic food production. Integrated fish and plant farming. FAO Fisheries and Aquaculture, Rome

Sterling P, Eyer J (1988) Allostasis: a new paradigm to explain arousal pathology. In: Fischer S, Reason J (eds) Handbook of life stress, cognition and health. Wiley, Chichester, pp 629–639

Subasinghe RP (2005) Epidemiological approach to aquatic animal health management: opportunities and challenges for developing countries to increase aquatic production through aquaculture. Prev Vet Med 67:117–124. https://doi.org/10.1016/j.prevetmed.2004.11.004

Tacchi L, Bickerdike R, Douglas A, Secombes CJ, Martin SAM (2011) Transcriptomic responses to functional feeds in Atlantic salmon (*Salmo salar*). Fish Shellfish Immunol 31:704–715. https://doi.org/10.1016/j.fsi.2011.02.023

UN (2017) United Nations, Department of Economic and Social Affairs, Population Division, World Population Prospects 2017 - Data Booklet (ST/ESA/SER.A/401). 1 Feb 2018. https://esa.un.org/unpd/wpp/publications/Files/WPP2017_DataBooklet.pdf

Vasudeva Rao Y, Das BK, Jyotyrmayee P, Chakrabarti R (2006) Effect of *Achyranthes aspera* on the immunity and survival of *Labeo rohita* infected with *Aeromonas hydrophila*. Fish Shellfish Immunol 20:263–273. https://doi.org/10.1016/j.fsi.2005.04.006

Winton J (2002) Fish health management. In: Wedemayer G (ed) Fish hatchery management, 2nd edn. American Fisheries Society, Bethesda, pp 559–640

Xue S, Xu W, Wei J, Sun J (2017) Impact of environmental bacterial communities on fish health in marine recirculating aquaculture systems. Vet Microbiol 203:34–39. https://doi.org/10.1016/j.vetmic.2017.01.034

Yavuzcan Yıldız H, Seçer SF (2017) Stress and fish health: towards an understanding of allostatic load. In: Berillis P (ed) Trends in fisheries and aquatic animal health. Bentham Science Publishers, Sharjah, pp 133–154. https://doi.org/10.2174/97816810858071170101

Yavuzcan Yildiz H, Robaina L, Pirhonen J, Mente E, Domínguez D, Parisi G (2017) Fish welfare in aquaponic systems: its relation to water quality with an emphasis on feed and faeces-a review. Water 9:13. https://doi.org/10.3390/w9010013

Yi Y-L, Lu C, Hu X-G, Ling F, Wang G-X (2012) Antiprotozoal activity of medicinal plants against *Ichthyophthirius multifiliis* in goldfish (*Carassius auratus*). Parasitol Res 111:1771–1778. https://doi.org/10.1007/s00436-012-3022-7

Chapter 18
Commercial Aquaponics: A Long Road Ahead

Maja Turnšek, Rolf Morgenstern, Iris Schröter, Marcus Mergenthaler, Silke Hüttel, and Michael Leyer

Abstract Aquaponic systems are often designated as sustainable food production systems that are still facing various challenges, especially when they are considered as a commercial endeavour that needs to compete on the market. The early stages of the aquaponics industry have witnessed a number of unrealistic statements about the economic advantageousness of aquaponics. This chapter deals with these topics and discusses them critically. The latest scientific literature and current personal experiences of European commercial aquaponics farmers are taken into account on three levels: The horticulture side of production, the aquaculture side of production and the early data on the market response to aquaponics, emphasising the marketing issues and public acceptance of aquaponics. In summary, the chapter does not provide an "off-the-peg" solution to evaluate the economic performance of a particular aquaponics system. Instead it provides a broad database that enables an estimation of the efficiency of a planned system more realistically, pointing to challenges that the commercial aquaponics early adopters faced that are important lessons for future aquaponic endeavours, particularly in Europe.

Authors Maja Turnšek and Rolf Morgenstern have equally contributed to this chapter.

M. Turnšek (✉)
Faculty of Tourism, University of Maribor, Brežice, Slovenia
e-mail: maja.turnsek@um.si

R. Morgenstern · I. Schröter · M. Mergenthaler
Department of Agriculture, University of Applied Sciences of South Westphalia, Soest, Germany
e-mail: morgenstern.rolf@fh-swf.de; schroeter.iris@fh-swf.de; mergenthaler.marcus@fh-swf.de

S. Hüttel
Institute of Food and Resource Economics, Chair of Production Economics, Faculty of Agriculture, Rheinische Friedrichs-Wilhelms Universität Bonn, Bonn, Germany
e-mail: S.Huettel@ilr.uni-bonn.de

M. Leyer
Institute of Business Administration, Rostock University, Rostock, Germany
e-mail: michael.leyer@uni-rostock.de

S. Goddek et al. (eds.), *Aquaponics Food Production Systems*,
https://doi.org/10.1007/978-3-030-15943-6_18

Keywords Commercial aquaponics · Aquaponics economic myths · Marketing of aquaponics

18.1 Introduction: Beyond Myths

Although we have witnessed the first research developments in aquaponics as far back as the late 1970s (Naegel 1977; Lewis et al. 1978), there is still a long road ahead for the sound economical assessment of aquaponics. The industry is developing slowly, and thus available data is often based on model cases from research and not on commercial-based systems. After initial positive conclusions about the economic potentials of aquaponics in research-based settings of the low-investment systems in USA, primarily the system in Virgin Islands (Bailey et al. 1997) and Alberta (Savidov and Brooks 2004), commercial aquaponics encountered a high level of early enthusiasm in business contexts, often based on unrealistic expectations.

To provide a specific example, in its early market forecast, IndustryARC (2012) anticipated that aquaponics as an industry has a potential market size of around $180 million in 2013 and is expected to reach $1 billion in sales in 2020. Later they projected aquaponics to increase from $409 million in 2015 to $906.9 million by 2021 (IndustryARC 2017). The same report (IndustryARC 2012) provided a number of yet untested claims about aquaponics, for example, about the economic superiority of aquaponics in terms of output, growth time and diversification possibilities in a commercial setting. We name such claims here as "aquaponics economic myths" that have been a typical part of the early internet-fuelled hype on commercial aquaponics.

Take a look at their statement: "Aquaponics uses 90% less land and water than agriculture but has the potential to generate 3 to 4 times more food than the latter also" (IndustryARC 2012). Comments such as these are extremely vague, since it is not clear what exactly aquaponics is being compared to when the authors are referring to "agriculture". Although aquaponics does use less water than soil-based food production, since the water used in soil-based production can be lost in the soil, not being absorbed by plants compared to staying in a recirculation loop with aquaponics. The exact amount of water savings depends on the type of the system. Additionally, "3 to 4 times more food" seems highly exaggerated. Aquaponics can have yields comparable to hydroponics (e.g. Savidov and Brooks 2004; Graber and Junge, 2009). Yet the statement glosses over the fact that at least in coupled aquaponics so-called operational compromises need to be made in order to find a balance between optimum parameters for healthy plants and fish (see Chaps. 1 and 8 of this manuscript), which can lead to aquaponics having lower outputs compared to hydroponics.

Therefore, statements like the above lack a clear definition of the reference scenario and the reference unit of comparison. In an economic assessment, higher output levels can only be compared meaningfully if there is a clear reference to the input levels required to achieve this output. In the assessment of aquaponic systems, higher outputs per area might be achieved compared to conventional agriculture, yet aquaponic systems might require more energy, capital and work input. Only referring to land as an input factor assumes that other production factors are not scarce, which is hardly the case. Therefore, statements like the above neglect the "all other things being equal" principle in economic assessments. Vaclav Smil (2008) calculates and summarises energy expenditure of different agricultural activities, utilising energy as the common denominator, and this allows us to compare different agricultural methods with the aquaponic approach.

A similar myth is contained in the statement: "A major advantage pertaining to the aquaponics industry is that crop production time can be accelerated" (IndustryARC 2012). An acceleration of crop production necessarily depends on the amount of nutrients and water, oxygen and carbon dioxide in the surrounding atmosphere and light and temperature available to crops – factors that are not elements of aquaponics per se but can be added via greenhouse management practices, such as fertilisation and irrigation heating and artificial lights. These additional elements, however, increase both the costs of investment and the operational costs, often being too expensive to be economically viable (depending of course on the location, type of crops and especially the price of crops).

Another economically important advantage of aquaponics provided in the report was that "aquaponics is an adaptable process that allows for a diversification of income streams. Crops may be produced depending upon local market interest and the interest of the grower" (IndustryARC 2012). What statements like this gloss over is the fact that diversification of production always comes at a price. Crop diversification necessarily includes higher levels of knowledge and higher labour demands. The larger the variety of crops, the more difficult it is to meet optimum conditions for all the selected crops. Large-scale commercial production thus looks for constant parameters for a limited number of crops that need similar growth conditions, allowing for large outputs in order to penetrate distribution via large distribution partners such as supermarket chains, and allow for the same storage and potential processing equipment and processes. Such large-scale production is able to use economies of scale to reduce unit costs, a basic principle in economic assessment, which is not usually the case for aquaponics at smaller scales of production.

Finally, the most important statement provided in the report was that "the return of investment (ROI) for aquaponic systems ranges from 1 to 2 years depending on the farmer experience as well as the scale of farming" (IndustryARC 2012). Such statements need to be taken with extreme caution. The scarce data that is available on return on investments reports on a much longer time: According to Adler et al. (2000), it takes 7.5 years of return for an approximately $ 300.000 investment in the hypothetical scenario of a rainbow trout and lettuce system. Recently, Quagrainie et al. (2018) reported a similar period of 6.8 years for the payback of an investment in aquaponics if the products can only be sold at non-organic prices. Real data on the economics of aquaponics is extremely hard to come by, since the enterprises that have ventured into commercial aquaponics are reluctant to share their data. In cases where enterprises are performing well, they either do not share their data, since it is considered a business secret, or if they do share data, such data needs to be taken with caution since typically these companies have an interest in selling aquaponic equipment, engineering and consultancy. In addition, enterprises that have failed in achieving profitability prefer not to publicly share their failures.

These "myths" are continually circulating online amongst non-experienced aquaponic enthusiasts, fuelled by hope for both high returns and a path towards more sustainable future food production. So there is a need to go beyond the myths and look at individual enterprises and provide an in-depth analysis on the basic and the general economics of aquaponics.

Even if realistic data on aquaponics were available, it has to be considered that such analyses are based on single cases. As aquaponic systems are far from technically standardised production systems, the diversity with respect to marketing concepts is even higher. So, data on every single aquaponics system lacks generalisability and can be regarded only as a single case study. General statements are therefore not valid if the framework conditions and technical and marketing specificities are not made transparent.

Journalistic publications about aquaponics often follow a narrative that elaborates on the general challenges of global agriculture, such as shrinking agricultural areas, humus loss and desertification, and then elaborate on the advantages of aquaponic food production methods. Apart from the above-mentioned mistake that in fact controlled environment system (CES) production is compared with field production, no distinction between agriculture and horticulture is made. Whilst the term "agriculture" technically includes horticulture, agriculture in its more specific sense is the large-scale crop production on farmland. Horticulture is the cultivation of plants, usually excluding large-scale crop production on farmland, and is typically carried out in greenhouses. Following these definitions, the plant side of aquaponics is horticulture and not agriculture. Thus comparing yield and other productivity properties of aquaponics with agriculture is simply comparing apples to oranges.

To state this differently, the horticulture side of agriculture is only a very small part of it. Large-scale crop production in agriculture mostly encompasses so-called staple food production: Cereals like corn, barley and wheat, oilseed like rape and sunflower and starchy root vegetables like potatoes. The agricultural area of Germany covers 184.332 km^2 (Destatis 2015). Of that only 2.290 km^2 (1,3%) is used for horticulture. Of the horticultural area, 9.84 km^2 (0,0053%) is protected and under glass. Absolute and relative figures for other countries surely differ, but the example clearly shows that the plant side of aquaponics will only be able to substitute and thereby enhance a small fraction of our food production. Staple foods can theoretically be produced in CES under glass using hydroculture as demonstrated in NASA research (Mackowiak et al. 1989) and could surely also be cultivated in aquaponic systems, but due to the high investment needed for such production, it does not make sense to think of aquaponics replacing the production of these crops under the current global economic and resource conditions.

18.2 Hypothetical Modelling, Small-Scale Case Studies and Surveys Amongst Farmers

Early research on commercial aquaponics focused on evaluation and the development of specific, mostly research institute-led case studies. These first results were highly positive and optimistic about the future of commercial aquaponics. Bailey et al. (1997) concluded that, at least in the case of Virgin Islands, aquaponic farms can be profitable. Savidov and Brooks (2004) reported that the yields of cucumbers and tomatoes calculated on an annual basis exceeded the average values for commercial greenhouse production based on conventional hydroponics technology in Alberta. Adler et al. (2000) performed an economic analysis of a 20-year expected scenario of producing lettuce and rainbow trout and argued that the integration of the fish and plant production systems produces economic costs savings over either system alone. They concluded that an approx. $300.000 investment would have a 7.5-year payback period.

Technologically based dynamic optimization models are commonly used to represent production engineering relationships in aquaponics systems (Karimanzira et al. 2016; Körner et al. 2017). It is noticeable that so far hardly any different scales are considered, and previous studies like Tokunaga et al. (2015) and Bosma et al. (2017) are limited to small-scale aquaponics for the local production of food or are performed on data from research facilities, such as University of Virgin Islands' aquaponics systems (Bailey and Ferrarezi 2017). Furthermore, as Engle (2015) points out, the literature on the economics of aquaponics is sparse, with much of the early literature based primarily on model aquaponics. Without realistic farm data, such projections often are overly optimistic because they lack details on expenses beyond the obvious ones of fingerlings, feed and utilities and do not include the everyday risks involved in farming. In this research on the economics of aquaponics, production functions are only partially reproduced and questions of process-based optimization addressed only to some extent. Leyer and Hüttel (2017) demonstrated the potential for investment accounting as part of an initial analysis to capture various parameters of an aquaponics facility. Furthermore, Engle (2015) points to the difficulties of estimating annual costs to operate in aquaponics farms since many of these systems are quite new. She also points out that modelling is based on hypothetical situations and that more realistic farm data is needed, whereby the unexpected expenses are incurred daily, "from screens that clog, pumps that fail, or storms that cause damage".

As aquaponics started to grow both as a do-it-yourself (DIY) activity (Love et al. 2014) and as an industry (Love et al. 2015), research on real commercial farm case studies emerged. Specific case studies of aquaponics production were performed on commercial attempts, for example, in Puerto Rico (Bunyaviroch 2013) and Hawaii (Tokunaga et al. 2015), including also the case study of a small-scale aquaponics social enterprise (Laidlaw 2013) (see Chap. 24).

With the continuous rise in the number of aquaponics growers, the first in-depth analyses of the state of the art of the industry emerged, focused primarily on the USA. These studies showed a less optimistic picture of the emerging industry. Love et al. (2015) performed an international survey amongst 257 participants, who in the last 12 months sold aquaponics-related food or non-food products and services. Only 37% of these participants could be named as solely commercial producers who gained their revenue from selling only fish or plants. Thirty-six percent of the respondents combined the sales of produce with aquaponics-related material or services: Sale of supplies and equipment, consulting fees for design or construction of aquaponics facilities and fees associated with workshops, classes, public speaking or agro-tourism. Finally, approximately one third (27%) were organisations that sold only aquaponics-related materials or services and no produce. The average aquaponics production site of 143 US-based producers was 0,01 ha. By comparing this to the overall hydroponic production in Florida (29,8 ha), Love et al. (2015) concluded that the size of aquaponics producers is significantly smaller than hydroponic production and is to a large extent still more of a hobby activity than successful commercial enterprises. In terms of water volume, the aquaponics farms reported comparable sizes as typical RAS aquaculture farms in the USA. Yet nearly a quarter of respondents (24%) did not harvest any fish in the last 12 months, and the estimated overall size of fish production was 86t of fish, which is less than 1% of the farmed tilapia industry in the USA.

According to the same study, aquaponics was the primary source of income for only 30% of respondents, and only 31% of respondents reported that their operation was profitable in the last 12 months. For example, the median respondent received only $1000 to $4999 in the last 12 months, and only 10% of respondents received more than $50,000 in the last 12 months. This led Love et al. (2015) to conclude that aquaponic farms were small-scale farms, which is comparable to agriculture in general, since farms with gross revenues of less than $ 50,000 made up approximately 75% of all farms in the USA and farms with less than $50,000 typically sold only around $7800 in local food sales – making it thus necessary to combine farming income with other sources of income. It is therefore not surprising that aquaponics, like small-scale farming, relies heavily on volunteer work. Typically, there were a large number of unpaid workers, family members and volunteers working on these small units, with an average of six unpaid workers per facility.

Similarly, Engle (2015) addresses the 2012 census where 71 aquaponics farms across the USA were reported which represented 2% of all aquaculture farms. Of these, only 11% had sales of $ 50,000 or more, compared to 60% of pond-based aquaculture operations that had sales of $ 50,000 or more. Additionally, Engle (2015) points out to the difficulties of obtaining data from these farms, for example, estimating annual costs to operate in aquaponics farms, since many of these systems are quite new.

In summary, from an economic point of view, there is a research gap in so far that there are no records and analysis available which include statements about economically viable systems. Further research is needed that would take into account (a) the production possibility curves (normative), (b) the combined analysis of fish and plants including feedback between both, (c) the economic efficiency in combination with optimising the business processes and feedback (simultaneous optimization production process and economic efficiency) and (d) the consideration of different scales (scale efficiency) against the background of the environmental sustainability of this agricultural system. In addition, there are no comprehensive and reliable data that combine key factors such as production volumes, factor entitlements and cost structures, scaling and sales strategies derived from existing real investments. Further profitability analyses should consider temporal aspects and risk whilst formulating normative benchmarks that in turn can serve as the basis for investment decisions.

18.3 Hypothetical Modelling Data from Europe

In Hawaii, Baker (2010) calculated the break-even price of aquaponics lettuce and tilapia production based on a hypothetical operation. The study estimates that the break-even price of lettuce is $3.30/kg and tilapia is $11.01/kg. Although his conclusion is that this break-even can potentially be economically viable for Hawaii, such break-even prices are much too high for most European contexts, especially when marketing through retailers and conventional distribution channels. In the Philippines, Bosma (2016) concluded that aquaponics can only be financially sustainable if the producers manage to secure high-end niche markets for fish and large markets for fresh organic vegetables.

Aquaponics on tropical islands (Virgin Islands and Hawaii) and warm, frost-free zones (Australia) contrasts highly to locations further away from the equator. The advantages in warm locations are the lower costs of heating and the seasonally even availability of daylight, thus allowing for potentially low-cost systems to economically survive. A frost-free location close to the equator with little to no seasonal differences makes it cheaper and easier to set up and operate a system year-round, which allows for semi-professional family business setups in those regions. Additionally, local production in these areas is valued higher since leafy green crops are either hard to store (e.g. Australia/heat) or difficult to transport to the customers (Islands) and generally have a much higher contribution margin than in locations such as Europe and Northern America.

Aquaponics can have several advantages in an urban context. Yet, advantages are only effective if the specific urban framework conditions are taken into account and if additional communication efforts are put in place. Peri-urban agroparks are presented by Smeets (2010) as a technically and economically viable solution for urban agriculture, offering synergy potential with existing industry through residual heat and suitable logistics as well as alternative inorganic and organic materials streams, for example, CO_2, from cement production. Rooftop aquaponics utilises "empty" spaces in urban areas (Orsini et al. 2017). Rooftops are often assumed to be free of cost "because they are there". Yet every space in the city is of high value. An owner of a building will always seek revenue for the space they offer, even the utilisation of vacant rooftops. A rooftop farm carries a high economic risk and changes may have to be made to the building (vents and logistics). Rooftops are also interesting for solar energy production with less risk to the operator (see also Chap. 12).

Whilst aquaponics is often explicitly touted as a production technology suitable for urban environments and even areas with contaminated soil, the real estate cost is often completely underestimated. For example, official real estate prices in Germany can be examined via the online tool BORISplus (2018), revealing a significant gap between inner city limit prices and agricultural land prices. For example, peri-urban real estate within city limits in Dortmund, Germany, is in the 280 €/m^2–350 €/m^2 range, whereas agricultural land outside of the city limits is in the 2 €/m^2–6 €/m^2 range. In addition to that German building codes grant the privilege to farmers to erect agricultural buildings outside of the city limits. This legal and financial situation renders agricultural land in proximity to economic zones attractive for larger-scale aquaponic farms, leading to the above-mentioned concept of agroparks.

The placement of aquaponic farms raises challenges with customer perception. Citizens who have been interviewed about their preference of different urban agriculture concepts for inner city public land use showed a preference for usage that keeps the space accessible for citizens as well as a low acceptance levels for agroparks (Specht et al. 2016). The research results on the acceptance of aquaponics revealed a larger variance than the other potential utilisations, suggesting a citizen ambivalence due to a lack of information on the production method. Additional communication efforts are required as aquaponics is a highly complex and new production system unknown to most people in society including urban populations.

The potentials and risks of aquaponics in an urban context become clear from the paragraph above. Distinct strategies and contingency plans have to be developed in an urban context when planning to implement an aquaponics production facility.

Most of the data currently collected on commercial farmers is focused on locations outside of Europe. A sound economic assessment of aquaponics facilities in European latitudes and climates is difficult, because on the one hand only very few commercial plants exist in Europe and on the other hand technical equipment, scale and business models are very different in other parts of the globe, where commercial aquaponics is more widespread (Bosma et al. 2017). Whilst Goddek et al. (2015) and Thorarinsdottir (2015) provide a very good overview of European commercial plants and their challenges, they present only a few economic parameters such as (targeted) consumer prices, statements on "potentially" achievable income or break-even prices for production. Since these are only valid under the specific conditions of the investigated facilities, only limited statements can be transferred to other locations, even within Europe.

Whilst there are some specific assessments of productivity (e.g. Medina et al. 2015, Petrea et al. 2016), full market potential analysis and well-founded cost-effectiveness assessments are not known at the present time. In addition, there are initial studies on technical dynamic models using the methodology system dynamics such as Goddek et al. (2016) and Körner and Holst (2017). This illustrates how essential the availability of comprehensive data is in order to conduct sound profitability analysis.

One of the very few hypothetical modelling cases created with data from within Europe is Morgenstern's et al. (2017) model. They provided technical data from the pilot plant of the University of Applied Sciences of South Westphalia, which consisted of a commercial fish farm and a standard horticulture system. In this case, investment and full-cost calculations with comprehensive detailed technical data for systems in three different scales were modelled. Model calculations for operational costs for a start-up period of 6 years and investment costs as well as a simplified cost performance difference calculation have been conducted for three differently sized aquaponic farms rearing European catfish (*Silurus glanis*) and producing lettuce. The calculated sizes were derived from the pilot plant located at the University of Applied Sciences of South Westphalia and the aquaculture scale of the project partner. Modelled aquaculture sizes were 3 m^3, 10 m^3 and 300 m^3. A couple of general assumptions and simplifications were made for the calculations, which illustrate the above-presented critiques on the limitations of hypothetical modelling:

(a) Less than average production quality and production losses within the first 5 years have been considered. The profitability calculations are based on a matured and stable production process starting in year 6.
(b) Constant hydroculture production. The complete nutrient stream from the process water was calculated to be consumed by the hydroculture production of lettuce, regardless of seasonal differences and regardless of nutrient availability from the aquaculture.
(c) The hydroculture grow bed size has been calculated to be 60 m^2, 200 m^2 and 5.500 m^2.
(d) Heating demand for hydroculture and aquaculture has been approximated with a slightly modified methodology of KTBL (2009). The modelled location of the farm is Düsseldorf, Germany.
(e) Energy costs per kWh have been approximated for production with a combined heat and power (CHP) system with 15 ct/kWh (electricity) and 5,5 ct/kWh (heat), respectively. For simplicity, a CHP system has not been modelled.
(f) Direct marketing of the products was assumed. Fairly optimistic, but not overly optimistic, market prices have been calculated for the products. No extended marketing costs have been included in the calculation, since the marketing effort required to build a customer base and stable market have not been addressed in the project. Neglecting marketing costs assumes that market prices in direct marketing come at no costs and therefore constitute a major simplification of the calculation.
(g) No costs related to the real estate required for the farm have been included in the calculations. The rationale for this simplification is the vastly different costs for space depending on location and project context.

(h) Labour cost has been calculated at minimum wage, which is a strong assumption with regard to high levels of human capital required to run complex aquaponics systems.
(i) Mortality losses of 5% in the aquaculture system are compensated by overstocking at the start of each production cycle.

An analysis of the cost structure of the modelled production-sized aquaculture system shows that labour, fish feed and juveniles and energy are the main cost drivers, contributing roughly one third of the main costs each. At this point, it has to be emphasised that labour costs are calculated on a minimum wage basis and that costs for the occupied area of the farm have not been considered in the calculations (Fig. 18.1).

Electricity and heating costs offer potential for optimization. Pumps have a lifetime between 2 and 5 years. Inefficient pumps can be replaced with more efficient pumps in the natural machine life cycle. Cost efficiency gains for these kinds of optimizations are simple to calculate, and efficiency gains are also easy to monitor after implementation. Similar measures to reduce heating costs are relatively easy to calculate. For example, the costs and effects of additional insulation panels can be calculated, and also here the gains can be easily monitored.

Labour costs emerge as the main cost driver that shows significant optimization potential with upscaling. Larger-scale systems allow for the usage of labour-

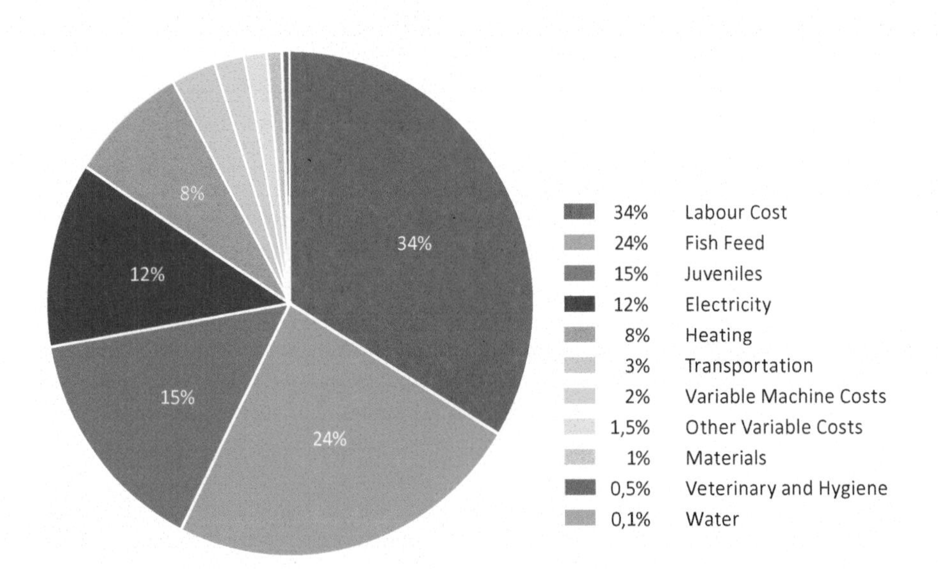

Fig. 18.1 Cost structure for aquaculture side of an aquaponics system, hypothetical model from technical data from the pilot plant of the University of Applied Sciences of South Westphalia. (Based on Morgenstern et al. 2017)

saving devices, for example, automated graders or automated feeder-filling machinery. Profitability of these kinds of optimizations has to be calculated on a per-project basis.

Likewise, a cost analysis has been performed for the hydroculture part of the modelled systems. The main cost drivers are labour, seedlings and energy costs for lighting and heating. A higher operational maturity of the production, when the initial start-up learning curve has been mastered, can make room for in-house seedling production. The integration of this production step can offer cost optimization potential. Regarding cost reduction potential of the other cost drivers, energy and labour, the above-described situation is applicable for the hydroculture part as well (Fig. 18.2).

A cost performance difference analysis has been performed for the three system sizes, showing that the microsystem and the small system are not economically viable. There is no exploitable automation and rationalisation potential because of the extremely small aquaculture sizes and the small hydroculture sizes resulting in prohibitively high labour costs. Minimum quantity surcharge and transportation fees for the fish feed and similar effects for other cost categories put an additional financial burden on these two systems.

The production size system has a positive cost performance difference when real estate costs or tenures for the required land are not taken into account (Table 18.1).

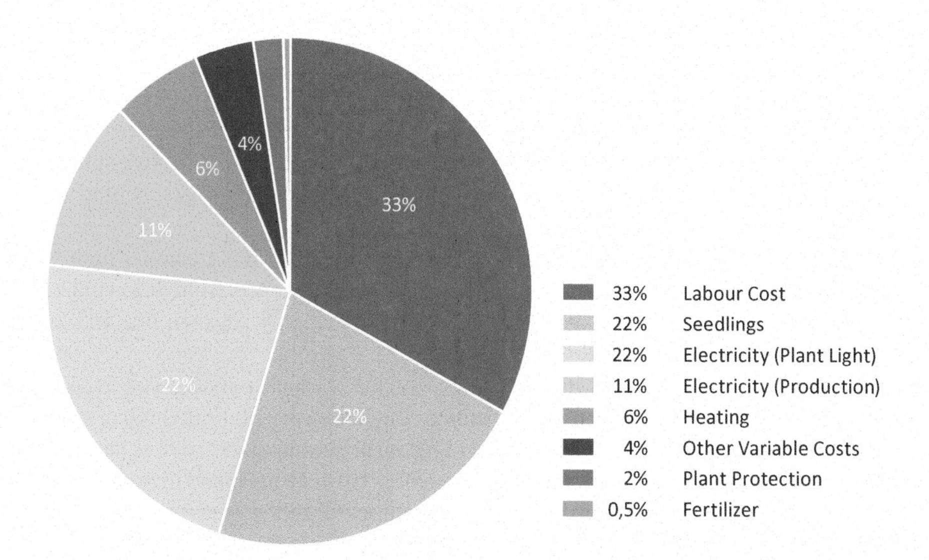

Fig. 18.2 Cost structure for hydroponics side of an aquaponics system, hypothetical model from technical data from the pilot plant of the University of Applied Sciences of South Westphalia. (Based on Morgenstern et al. 2017)

Table 18.1 Cost performance analysis of model calculation

Cost performance difference	Unit	Micro	Small	Production
Contribution margin aquaculture	€/a	−4173	−2566	114.862
Contribution margin hydroculture	€/a	691	13.827	541.087
Sum contribution margins		−3.483	11.260	655.948
Labour cost aquaculture	€/a	3.705	8.198	45.000
Labour cost hydroculture	€/a	3.148	8.395	179.443
Sum labour costs	€/a	6.853	16.593	224.443
Real estate costs tenure		n.a	n.a	n.a
Depreciation	€/a	7.573	15.229	185.269
Interest rate 2%	€/a	1.515	3.046	37.054
Cost performance difference	€/a	−19.424	−23.607	209.183

Source: Morgenstern et al. (2017)

Table 18.2 Job creation potential

	Unit	Micro	Small	Production
Sum labour cost	€/a	6.853	16.593	224.443
Sum labour time	Days/a	46	111	1.496
Number of jobs		0,21	0,5	6,8

Source: Morgenstern et al. (2017)

The analysis additionally sheds light on the job creation potential of the respective systems. The model calculation was performed under the assumption that all the required overhead tasks of the enterprise are handled by regular employees, an assumption that is rather optimistic with regard to the fact that the minimum wage has been used for the calculation.

One further assumption was made regarding the separation of jobs: The employees work on both parts of the sytem, the aquaculture and the hydroculture parts, in accordance to the work that is needed by the respective system. This requires an elevated skill set which puts another question mark behind the minimum wage calculation.

Even in the larger-sized production system, the number of jobs created is limited. The calculated number of jobs is congruent with experience from horticultural companies working with hydroponics, which usually employ between five and ten workers per hectare of greenhouse (Table 18.2).

Data on initial investments in aquaponics is on the one hand very difficult to come by and on the other hand even more difficult to compare. Some of the preliminary data collected from other sources on the initial investment needed to set up an aquaponics farm (see Table 18.3) below shows high differences between the initial investments in the systems, either real or in hypothetical modelling. Since the systems differ in the extreme amount of factors, it is highly problematic to draw any conclusions regarding the necessary initial investments. However, the initial investment in aquaponics does seem to be relatively high, which is reflective of the early stage of the industry. We estimate that an initial investment in a commercial aquaponics system in Europe starts

Table 18.3 Estimated investment costs on aquaponics, various sources

Literature source	Total investment [approx. per m^2 of growth area]	Location	Aquaculture size and type	Hydroponic size and type
Bailey et al. (1997)	$ 22,642 [$ 226/ m^2]	Virgin Islands, USA	4 tanks Tilapia No heating	100 m^2 Lettuce DWC No greenhouse
Adler et al. (2000)	$ 244,720 [$ 240/ m^2]	Shepherdstown, WV, USA	19,000 l 239 m^2 Rainbow trout No heating ($ 122,80)	cca. 120 m^2 Lettuce NFT ($ 17,150) Polyethylene greenhouse with heating and lights ($ 78,770)
Tokunaga et al. (2015)	$ 217,078 [$ 190/ m^2]	Hawai'i, USA	75.71 m^3 Tilapia	1142 m^2 Lettuce DWC
Morgenstern et al. (2017)	€ 151.468 [€ 1067/ m^2]	Model location: Düsseldorf	3 m^3 European catfish	59 m^2 grow bed area 83 m^2 greenhouse Lettuce DWC
Morgenstern et al. (2017)	€ 304.570 [€ 650/m^2]	Model location: Düsseldorf	10 m^3 European catfish	195 m^2 grow bed area 274 m^2 greenhouse Lettuce DWC
Morgenstern et al. (2017)	€ 3.705.371 [€ 302/m^2]	Model location: Düsseldorf	300 m^3 European catfish	5.568 m^2 grow bed area 6.682 m^2 greenhouse Lettuce DWC

with at least 250 EUR / m^2 of growth area but can easily require a much higher investment, depending on the outside conditions, the system size and complexity and the length of the growth season aspired to (Table 18.3).

The experimental and pioneering status of commercial aquaponics is one reason why the financing of larger commercial-scale projects can be a challenge. Most aquaponic systems have been financed through research grants or through aquaponics enthusiasts. Personal communication with German banks that are traditionally strong in financing agricultural investments and that are therefore familiar with the intricacies of crop production and animal rearing revealed that they would not finance an aquaponics project due to lack of a proven and established business model (Morgenstern et al. 2017).

18.4 Aquaponic Farms in Europe

Thorarinsdottir (2015) identified ten pilot aquaponic units in Europe, approximately half of which were at the stage of setting up still rather small-scale systems for commercial production. Villarroel et al. (2016) estimated that the number of aquaponic commercial enterprises in Europe comprised approximately 20 companies. Currently, Villarroel (2017) identifies 52 research organisations (universities, vocational schools, research institutes) and 45 commercial enterprises in Europe. Only a handful of these, however, sell aquaponic produce and could be considered as an aquaponic farm. In 2016, as a spin-off from the COST FA1305 project, the Association of Commercial Aquaponics Companies (ACAC 2018) was founded, currently involving 25 companies from all over Europe, only about a third of which focuses on food production. Others offer mostly aquaponics-related services such as engineering and consulting (Thorarinsdottir 2015).

In the remainder of this chapter, we focus on the in-depth information from the few operating aquaponic systems in Europe. Within the time of the COST action (2014–2018), a number of organisations started and tested their first marketing attempts at commercial food production with the goal of making the sales of produce as their main income. We have gathered data with three in-depth qualitative methods: (a) site visits, (b) SME presentations at the COST FA 1305 conference (date, place) and (c) semi-structured in-depth interviews.

18.4.1 Site Visits

Within the COST FA1305 project, three European pilot aquaponic systems opened their doors for site visits of COST members:

- Ponika, Slovenia, Matej Leskovec (site visit made on 23 March 2016).
- UrbanFarmers AG, system in the Netherlands, Andreas Graber (site visit made on 6 September 2016).
- Tilamur, Spain, Mariano Vidal (site visit made on 20 April 2017).

Within the site visits, we asked questions about the type and size of the system, the initial investment, the types of fish and crops produced and the reasoning behind it and the selling experiences.

18.4.2 Presentations

COST FA1305 organised a special conference devoted to presentations of aquaponics SMEs in Europe at the International Conference on Aquaponics SMEs at the University of Murcia (ES) (19 April 2017). We only analysed the presentations and answers to questions from SMEs that presented their experiences with their own commercial aquaponics system:

- Bioaqua Farm, UK: Antonio Paladino.
- NerBreen, Spain: Fernando Sustaeta.
- Ponika, Slovenia: Maja Turnsek.

- Samraekt Laugarmyri, Iceland: Ragnheidur Thorarinsdottir.
- Tilamur, Spain, Mariano Vidal.
- Uit Je Eigen Stad, Netherlands: Ivo Haenen.

18.4.3 Follow-Up In-Depth Interviews

We made additional follow-up semi-structured in-depth interviews with the aquaponics SMEs, contacting also other European aquaponic producers, some of which declined to participate.

The final interviewers were:

- ECF, Germany: Christian Echternacht (interview made on 13 February 2018)
- NerBreen, Spain: Fernando Sustaeta (interview made on 7 April 2018)
- Ponika, Slovenia: Maja Turnsek (interview made on 30 January 2018)
- Samraekt Laugarmyri, Iceland: Ragnheidur Thorarinsdottir (interview made on 19 February 18)
- Uit Je Eigen Stad, Netherlands: Ivo Haenen (interview made on 8 February 2018)

The average interview lasted approx. 60 min. and included questions on the reasons for choosing aquaponics, type and size of the system, produce and fish type, experiences with growing, distribution and price, future plans and outlook for commercial aquaponics in the future. Below we will present the results together with the literature review on each of the areas of discussion (Table 18.4).

18.5 Horticulture Side of Commercial Aquaponics in Europe

Petrea et al. (2016) conducted a comparative cost-effective analysis on different aquaponics setups, utilising five different crops: baby leaf spinach, spinach, basil, mint and tarragon in deepwater culture and light expanded clay aggregate (LECA). Whilst the study was conducted in very small systems without taking into account any upscaling opportunity or potential, several aspects of the presented results are worth discussing. The grow beds have been illuminated in different lighting regimes with fluorescent bulbs and metal halide grow lights. The cost comparisons of the electricity illuminates the significant share plant lighting has on overall electricity cost. Furthermore, the analysis sheds light on the importance of sensible crop selection. Whilst tarragon is referenced as what is often called a "high-value crop" earlier in the text, the later economic crop yield analysis shows that other crops, basil and mint, generate a higher economic value per grow bed area (Petrea et al. 2016, p. 563).

Table 18.4 General information on the interviewed aquaponics companies in Europe

Company	Location	Initial investment	Overall size of the system	Plant production size	Type of plants	Type of fish	Years of operation
ECF	Germany	1.3 mio EUR	1800 m^2	1000 m^2	Basil	Tilapia	2 years
NerBreen	Spain	2 mio EUR	6000 m^2	3000 m^2	Lettuce, strawberries, tomatoes, peppers	Tilapia	5 years the 500 m^2 pilot system, 1,5 years the large system
Ponika	Slovenia	100.000 EUR	400 m2 greenhouse + water reserve packaging and cooling area in a container	320 m^2	Fresh-cut basil, chives and mint	Carp, trout, big mouth bass (sold only carp and even that only as live fish)	2 years, after that stopped production
Samraekt Laugarmyri	Iceland	640.000 EUR		1000 m^2	TomatoLettuceHerbs	Tilapia, Arctic char	Commercial project is in planning phase
Uit je Eigen Stad	Netherlands	Embedded in larger project. Approx. 150.000 EUR–200.000 EUR	400 m^2	200 m^2	Head lettuce varieties	African catfishPink tilapia	Partially operational for 1 year. Never selling produce. Operation ceased

The comparison shows a wide span of economic value ranging from 5.70€/m^2/cycle (baby leaf spinach) up to 2110€/m^2/cycle (basil) and 23.00€/m^2/cycle (mint). Realistically, these figures should not be taken as reference points for upscaled commercial productions, but they illustrate how different the economic output of the crop production can be. Table 8 of the publication elaborates on the seasonal market price variability of the examined crops. The seasonal variations of these crops are rather moderate with slightly elevated prices in fall and winter months. Seasonal market price variations for produce and fruit with higher global market volumes like tomatoes and strawberries are usually much more pronounced, enticing producers to put effort into season extension at both ends. Artificial lighting for season extension is costly both investment-wise and regarding operational cost but might well be worthwhile, especially considering the inherent pressure to utilise process water from the aquaculture in the low light season.

No emphasis has been placed on product quality and marketability of the produced biomass within this study. Experience shows that the cultivation of certain crops is easier than the cultivation of others. Mint is generally regarded as an easy-to-grow crop, whilst the production of marketable basil is more challenging. Petrea et al. (2016) cultivated the crops in deepwater culture and ebb and flow grow beds with LECA substrate. The latter is particularly uncommon in commercial production, since it has close to zero potential for rationalisation and automation. Basil usually is produced in pot culture as opposed to a cut and come again production of mint which leaves the rootstock in the system with a faster regrowth of a marketable product. In addition to the different growth medium requirements, different crops are produced with differing temperatures, climate and light regimes. Optimal product quality can only be achieved with optimal cultivation techniques. It is important to remember that customers are used to premium quality and show little to zero tolerance for suboptimal products.

The horticulture side of commercial aquaponics faces high risks from infestations of diseases or parasites, which can be difficult to overcome because only biological controls can be used (see Chaps. 14 & 17). Large risks are involved also since most aquaponic farms require a market that will pay higher-than-average prices for the crop. Finally, aquaponics seems to be very labour-intensive since even small scale aquaponics systems are complex because of their many components and requirements (Engle 2015).

For start-ups, it might be tempting to strive for the production of a wide spectrum of novelty varieties of plant species with fruity aromas or colourful leaves. Melon sage (*Salvia elegans*) or pineapple mint (*Mentha suaveolens* 'Variegata') are examples for these kinds of varieties. According to small-scale commercial producers in Soest, Germany (non-aquaponics; personal communication summer 2016), the market demand for novelty varieties has long been recognised by retailers and is being supplied for by their large-scale producers. This segment is not a profitable niche any more but rather a market that follows yearly trends. The switch by the Berlin-based aquaponics producer ECF from a wide spectrum of the aforementioned plant species in their start-up phase to a basil monocrop that is being marketed through the German retailer REWE reflects this situation.

Christian Echternacht from ECF, Berlin, reported in the interview about the difficulty to sustainably establish a local direct marketing channel for a wider range of products of limited quantity. Drawing from their first-hand experience, the company decided to shift the plant-side production to one crop, basil in pots, and to market this crop via one single retailer in over 250 supermarkets in the city of Berlin. Interestingly the regionally labelled product (Hauptstadtbasilikum/Capital City Basil) without an organic label is placed directly next to organically labelled basil from non-regional sources and is reported to generate higher sales despite slightly higher prices.

NerBreen based in Spain with its 6000 m^2 size is currently the largest system in Europe. It is more focused on the aquaculture element and includes aquaponics as one of several means of water filtration, but the plant production is still 3000 m^2 and produces enough to create a market. They are currently undergoing the second season of production within the farm, having 5 years previous experiences with a smaller pilot plant (overall size of the pilot farm, 500 m^2). In the winter season, they now grow fresh garlic, strawberry plants without fruit (since the plants need to be maintained for 3 years) and four different types of lettuce. In the summer, they replace the garlic with cherry tomatoes and peppers but keep the strawberries and the same lettuce varieties. Since this is only the second season, it is very difficult for them to provide an average amount of produce. The last winter was very cold, and it significantly affected the growth of the lettuce. In the first season, when they were still trying to improve and gain experience, they produced about 3t of strawberries, 5t of tomatoes and 60,000 lettuce heads. Their hopes are to ramp up production, whilst at the same time, their strategy shifted from putting a focus on the quantity to quality and variety instead. In the first year of their production, they had a good season with tomatoes in terms of quantity – but the overall market was flooded with tomatoes, and the price was subsequently too low. They adjusted to this problem with a focus on more selected, niche varieties of cherry tomatoes since the price is better, and they do not to want to compete with quantity but with quality, thus trying to reach a higher price with the retailers.

Locally produced crops and niche cultivars seem to be the main directions for crop selection in the European countries. The Slovenian company Ponika set out to sell fresh-cut herbs in their 400m^2-sized system since providing niche products in the Slovenian market with no other local producer of fresh-cut herbs. The company based their rationale on three main reasons. The first was that the data available, albeit scarce, from US aquaponics farms showed that fresh-cut herbs seemed to be the crops that succeed well in aquaponics and gained a higher price in the market. The second was the years of positive experience within the small-scale DIY aquaponics garden with these crops. In addition, the third was the positive feedback received from fresh-cut herbs distributors in Slovenia regarding the interest in the crops. The company set out to produce fresh-cut herbs and managed to sell them to Slovenian gastro-distributors for two seasons, narrowing the initial number of crops from six to three: fresh-cut chives, basil and mint. Other fresh-cut herbs they tested proved to be either too sensitive, or there was too small and infrequent demand on the market. The plan was to first sell the fresh-cut herbs to the gastro-distributors and then gradually proceed to selling these to large-scale retail chains. The reality, however, showed significantly large risks in the production (e.g. powdery mildew with basil and yellow tips with chives) and too small a system to be able to secure a steady uninterrupted production as requested by the large-scale distribution chains. Although the margins would be higher, Ponika never started to sell to the retail chains since the contracts with large retail chains included financial penalties in the case where the farm could not deliver the orders. Additionally, the retail chains made weekly orders thus not allowing for appropriate planning, and in some cases, the excess that was not sold had to be collected by the farmer, and this discarded produce was expected to be deducted from the overall order even though the over-ordering was on the side of the retailer.

The main reasons for the Slovenian-based company Ponika to stop their operation were the combination of high risks accruing from the labour needed to cut, screen and package the produce combined with the average gained price of 8 EUR / kg of fresh-cut herbs (packaged in 100 g bundles) not providing enough of an economic return to cover the extra workload. Since the company was the only Slovenian company in the Slovenian market, the gastro-distributers were willing to take their produce over the imported produce – yet only if the prices were equal to what the prices of international competitors were on the market. With a high percentage of fresh-cut herbs sold in the European market being delivered from North Africa, the high extent of labour costs for aquaponically produced fresh-cut herbs meant that a small-scale system could not compete with the prices as set by the extensive fresh-cut herbs farms in warmer regions with lower labour costs, even when including transportation costs. This shows that even when there is a niche in a local market, there are often specific reasons for local producers not filling the niche. In the case of fresh-cut herbs in Slovenia, this was the high cost of labour for too small a niche market.

18.6 Aquaculture Side of Commercial Aquaponics in Europe

Starting a business in temperate climate regions of Europe or Northern America requires a larger investment since the systems have to be kept frost-free requiring more electrical energy for plant lighting when operated throughout the year. In Europe, there are two strong horticultural production powerhouses, one in Westland/NL and the other in Almeria, southern Spain. The market concentration is high and contribution margins are slim. As a result, some aquaponic producers presumed that in aquaponics the contribution margin from aquaculture is more interesting than that of horticulture, which is probably why some of the few commercial operators chose to oversize the aquaculture part of the setup. This can lead to technical issues because a larger quantity of nutrients than required by the plant side is being produced in the aquaculture side. The excess process water has to be discarded

(Excursion Graber 2016 and Interview Echternacht 2018), putting the sustainability claims of aquaponics in question. Christian Echternacht from ECF reports that the contribution margin of the aquaculture has been overestimated in early calculations, rendering the oversizing of the aquaculture part of the farm counterproductive for the overall profitability of the farm.

Numerous different fish species have been reported to be produced in commercial aquaponics in Europe. Popular species for aquaponics production are tilapia, African catfish, largemouth bass, jade perch, carp and trout. There is no known commercial aquaponics farm currently rearing European catfish, but researchers at the University of Applied Sciences of South Westphalia (Morgenstern et al. 2017) found this species to be suitable for aquaponics production. The selection of fish species is influenced by a large number of different project-specific parameters. Most importantly of course are the market needs, price and distribution options. Within Europe, coastal regions have a traditionally strong market for marine fish with a diverse set of species and products. This creates a marketing challenge for freshwater aquaculture production. Ivo Haenen from *Uit je Eigen Stad*, Rotterdam, and Ragnheidur Thorarinsdottir from *Samraekt Laugarmyri*, Iceland, talked about this effect in their interviews. Rotterdam customers are used to a rich and diverse supply of marine fish products, making it difficult to market freshwater tilapia and African catfish. The marine wild catch tradition is so ingrained into Icelandic culture that the aquaculture aspect of aquaponics is probably not going to be actively promoted in future aquaponics projects.

Tilapia, one of the most commonly used fish species in aquaponics in the USA (Love et al. 2015), is a fish species that is not commonly known in Europe. As the experiences from NerBreen in Spain show, European tilapia aquaponics producers face a double marketing challenge: Their marketing attention needs to be put not only towards building customer awareness on the benefits of aquaponics production but also on the benefits of this relatively unknown fish species.

The suitability of the selected species for elevated water temperatures is another important factor. Fish are poikilothermic; thus their growth and consequently their production yields speed up with higher water temperatures. But elevated water temperatures require more energy, which, depending on the selected energy source for heating the process water, is connected with higher operational costs. Therefore, the positive effect of higher yields has to be balanced with the elevated costs for heating the water. From this perspective, it is desirable to tap into the potential of residual heat usage from adjacently located power plants or industries. These locations, however attractive and sensible they may be from an economical and ecological point of view, might pose a challenge for the overall marketing of the farm and its products. Industry sites are usually not idyllic and emotionally attractive, and worse still in case of anaerobic sewage plants or similar industries, they might even appear to be repulsive. Consequently, the available locations, and the context the farm can reasonably be placed in, is one factor for species selection.

The influence the different fish species have on plant yield and quality has not yet been completely researched. Knaus and Palm (2017) conducted experiments comparing plant yield in two identical aquaponics systems with identical operating parameters rearing tilapia and carp and found that plant performance with tilapia was better than with carp. These results show that there is indeed a difference in fish-plant interaction, but these have not been researched for a wider range of different species. In addition, the potential for fish polyculture, where two or more different fish species are reared in the same aquaculture cycle, has not yet been systematically researched.

One of the important operational factors for the fish selection is juvenile availability. Most of the commercial aquaponics producers buy juveniles from hatcheries. One notable exception is the company Aqua4C in Belgium that produces jade perch juveniles and uses these fish in their aquaponics system. A common recommendation is to select a species with at least two known suppliers with significantly larger capacity than the projected demand for the aquaponics farm. The rationale behind this recommendation is risk mitigation. If the supplier of juveniles experiences production issues and cannot deliver, the whole aquaponics production is in jeopardy.

As with the horticulture side of aquaponics, similarly the aquaculture part of aquaponics faces high technical risks, such as the death of the fish due to electricity outages, as reported by both Ponika from Slovenia and NerBreen in Spain. Ivo Haenen, former operator of the aquaponics system of the Urban Farm "Uit je Eigen Stad" in Rotterdam, reports that the heating system of the initial system setup was not dimensioned appropriately. An unexpected period of cold weather led to lower than tolerable process water temperatures resulting in losses in the aquaculture part of the operation. These kinds of instances have to be attributed to the pioneering character of early commercial aquaponics operations in Europe. The presented cases illustrate why Lohrberg and his team classified aquaponics in the "experimental" category of the seven identified business models of urban agriculture (Lohrberg et al. 2016).

18.7 Public Acceptance and Market Acceptance

The future of aquaponics production depends on public perception and the associated social acceptance in important stakeholder groups (Pakseresht et al. 2017). In addition to potential aquaponics plant operators, players at the wholesale and retail level as well as gastro-distributors and collective catering are important actors in supply chains. Moreover, consumers are key actors as they bring in the money into the supply chain at its end. Even though they have no direct economic stakes in aquaponics production, the general public as well as political and administrative bodies are important aspects to consider. The necessity of involving the aforementioned stakeholders is based on studies such as Vogt et al. (2016), who show that suitable framework conditions are an important basis for the establishment of innovative processes in food value chains. Technical developments without involving stakeholders run the risk of non-acceptance at the end of the research and development pipeline. In general, they build on a comprehensive understanding of a marketing philosophy with a multi-stakeholder approach.

For aquaponics, there is still no knowledge about the conditions that promote the dissemination of this technology. Although the technology used in aquaponics installations for freshwater fish farming in tanks is also used in aquaculture, until now this is unknown to a large part of society (Miličić et al. 2017). With regard to consuming plants from aquaponics, there is scepticism regarding their contact with fish water (Miličić et al. 2017). Preliminary studies based on a small sample regarding the acceptance of aquaponics products by potential consumers indicate that the requirements for products from aquaponics facilities go far beyond what the previous purchasing behaviour of fish products suggests (Schröter et al. 2017c). Based on the results of Schröter et al. (2017a), first hints on the effect of information on the acceptance process are available. These need to be further explored by means of perceptual and impact analyses of various information and presentation variants (e.g. textual facts, images, word-image content) and validation on the basis of representative samples. In addition, previous research has focused on citizens in general and on potential consumers. Studies on the acceptance of other important stakeholders such as potential plant operators, food retailers and public catering as well as political and regulatory actors and the general public are lacking completely.

First analyses of the consumers' response on aquaponics indicate that consumers showed a positive attitude towards aquaponics, with food safety issues being the major consumer concern in Canada (Savidov and Brooks 2004). Initial preliminary work on the willingness to pay for fish products from aquaponics was carried out by Mergenthaler and Lorleberg (2016) in Germany and Schröter et al. (2017a, b) on the basis of non-representative samples in Germany. Part of these studies show a relatively high willingness to pay for fish products from aquaponics. However, these results are based on small samples and cannot be generalised because the willingness to pay has been compiled from a specialist target group (see Mergenthaler and Lorleberg 2016) or in connection with the visit to a greenhouse for tropical and subtropical plants grown using aquaponics (Schröter et al. 2017a, b).

According to Tamin et al. (2015), aquaponics products are green products. A product is defined as *green* when it includes significant improvements in relation to the environment compared to a conventional product in terms of the production process, consumption and disposal (Peattie 1992). Based on the "theory of planned behaviour (TPB)", consumer acceptance of aquaponics products as innovative green products has been examined by Tamin et al. (2015) with closed-ended questionnaires in Malaysia. From a set of different behaviour-influencing factors (relative advantage, compatibility, subjective norm, perceived knowledge, self-efficacy and trust), two factors have been identified as having a significant impact: Relative advantage and perceived knowledge. The *relative advantage* describes how far buying behaviour is influenced by superior product qualities compared to conventional products. The aquaponics products were perceived fresh and healthy, and this perception led to a buying advantage. The *perceived knowledge* relates to how much the customer knows about the production method. The more the customers were familiar with the method, the more likely they were willing to buy aquaponics products. There was no correlation in the category subjective norm, which relates to how much the buying decision is influenced by the opinion of friends and family. Interestingly there was no correlation for the factor of *compatibility*. This factor relates to how much the product buying experience is compatible with the customer lifestyle. It seems as if the product to market process in Malaysia is not very different for aquaponics products and conventional products. So while it is questionable whether the results of this study can be safely transferred to European markets, a base message is that education about the production method and communicating the beneficial effects regarding the freshness of the food and the benefits for the environment are important marketing activities (Tamin et al. 2015).

Zugravu et al. (2016) surveyed the purchase of aquaponics products in Romania. Customers were influenced by friends and family. This dimension, subjective norm, showed no correlation in the Malaysian survey. The survey finds that consumers have a general good overall image of aquaponics. They think that the products are good for their health and that they are fresh. The paper describes a discrepancy between the perception of fish from aquaculture and wild catch and the perception of aquaculture. Retailers think that farmed fish can have a negative image, but actually aquaculture itself does not really have a pronounced image. The lack thereof is perceived by retailers as a marketing risk, yet it is described in the paper as giving the potential for positive branding through targeted communication. As a recommendation, the paper concludes that the retailers should build on the trust that the consumers showed when purchasing these fish products and should label the aquaculture fish as "healthy and fresh" (Zugravu et al. 2016).

Interestingly both Tamin and Zugravu had a significant higher questionnaire return count from women (Tamin et al. 2015; Zugravu et al. 2016). This raises the question about gender differences in aquaponics marketing. Although quantitative studies include gender as an independent variable in their analyses, no systematic and consistent patterns have been found yet. This asks for more research explicitly addressing gender aspects.

According to Echternacht from ECF in Germany, whose main business model is to set up aquaponics systems, marketing is the component that is usually most underestimated by their potential clients. ECF Farmsystems builds on this experience and surveys their potential customers for their intended marketing and distribution goals. If they have an existing business with actual production and established marketing channels, then the customer is very interesting. Idealistic customers who think that the products are going to market themselves are treated with caution.

Depending on the intended target group, different scales of production units might be favourable. Whilst some consumer segments prefer small-scale production possibly linked to short transportation distances and local production, there might be other consumer segments more interested in resource efficiency and low-cost production which can be realised in rather large-scale production units linked to waste energy and waste heat sources. Results from Rostock show (Palm et al. 2018) that small-scale systems with simple technology can make sense. Medium-scale systems require all the maintenance and the operational expenses of larger-scale systems, but do not have the benefit and output of large-scale systems. Conclusions from their experience show that you should either go small and achieve high prices in local markets or go to larger-scale systems with respective exploitation of economies of scale allowing for price reductions. Bioaqua from the UK is one of the rare European aquaponics companies that decided to follow the path of small-scale production with simpler and cheaper systems and providing the added value via catering and finding niche products for direct distribution to restaurants.

There might be other consumer segments displaying high preferences for fish welfare who therefore have to be targeted with fish from production units that conform to these ideals. As Miličić et al. (2017) show, consumers can express unexpected aversions, such as vegans expressing highly negative attitudes towards aquaponics. As pointed out in the literature, some facets of aquaponics may arouse high emotional involvement, such as the aesthetics of the aquaponics system (Pollard et al. 2017), level of mechanisation (Specht et al. 2016), soilless crop production (Specht und Sanyé-Mengual 2017), fish welfare (Korn et al. 2014), concerns about health risks due to the water recirculation system (Specht und Sanyé-Mengual 2017), or negative emotions bordering on disgust, because fish excrement is used as fertiliser for vegetables (Miličić et al. 2017). In this context, the perception and evaluation of aquaponics and its products may be based on unconscious processes rather than on careful consideration of logical arguments.

For some consumer segments, plants from aquaponics are innovative and interesting, and for others the link between fish and plant production might not be acceptable. This is also shown by ECF in Berlin: ECF decided to modify their initial production and marketing strategy. In the beginning, they attempted to produce a wide range of crops and market them directly on location. Nevertheless, according to Christian Echternacht (Interview Feb 2018) the marketing effort is simply too large. From their experience, the customers do not want to visit too many locations with only a few products at each location. Therefore, ECF decided to produce only one crop, basil, that is being marketed through a supermarket chain. Their experience as well as more comprehensive literature reviews shows that depending on the degree of meeting customer expectations, different levels of willingness to pay can be achieved and therefore achievable market prices are highly context specific.

Similarly, Slovenian-based company Ponika first attempted direct distribution of their fresh-cut herbs to restaurants in Ljubljana. But, just as with individual customers, restaurants were also averse to direct ordering even if the price was lower. For the restaurant managers, the time and effort needed to order individual products was much too high a price to pay, and they were not willing to order directly. They preferred to stay within their own gastro-distributors, whereby they could make their overall purchase in just one order.

The experiences of ECF from Berlin and Ponika from Slovenia described above are in line with previous experiences in the marketing of organic food products. To sell these products locally through direct distribution will only be possible for a small part of the products. Even though many consumers want to buy local and/or organic products, they often want to make their purchases as conveniently as possible. This means that shopping has to be efficient in order to fit into their daily schedule. As shown by Hjelmar (2011) for organic products, the availability of these food products is important for consumers because most of the consumers are pragmatic. They do not want to go to several stores in order to get what they want. They want to buy their products conveniently in a nearby supermarket and if the supermarket does not have a wide selection of organic products, many consumers end up by buying conventional products (Chryssohoidis and Krystallis 2005). Similar experiences can be described for consumers buying regional products in Germany (Schuetz et al. 2018). The same will presumably apply to aquaponics-grown products. If these products will not be available in supermarkets, aquaponics will probably remain niche production.

Organic food shoppers constitute a special potential target group of aquaponics. Indoor production of vegetables might require less or no pesticide applications, but soilless cultivation of plants is not an option in today's legislation on organic agriculture (cf. Chap. 19 of this book). Therefore, aquaponics in its strictest sense will not provide the necessary characteristics to be eligible for certification as organic production and no organic labels would be allowed on aquaponics products. Therefore, either policymakers have to be lobbied to induce changes in organic legislation, or organic shoppers have to be educated in this rather complicated issue. This aspect is also important in the regard that organic classified products usually achieve higher market prices than conventional products, and such certification would make the aquaponics systems more economically viable. If aquaponics-grown products can be sold at the same prices as organic products, under certain conditions, the payback period of aquaponics systems can be reduced by less than half (Quagrainie et al. 2018).

Besides product marketing, services surrounding aquaponics production can generate additional income streams. The high level of innovativeness of aquaponics generates high levels of interest, which can be exploited in different service offers which included paid-for aquaponics visits, workshops and consultancy services around the establishment of new aquaponics systems. There are several examples of aquaponics facilities venturing in this direction:

- ECF provides business consultancy for the establishment of new aquaponics systems.
- UrbanFarmers, Den Hague, offered paid visits to the facility as well as an event location. (Note: The project has now ceased).

Besides adjusting production systems to customer expectations in a comprehensive marketing concept, communication strategies also play a role. Up to now, knowledge about aquaponics in the society is weak (Miličić et al. 2017; Pollard et al. 2017). When acquiring information, different variants of information and of information representation will significantly influence the public perception of this innovative technology. To satisfy the stakeholders' information demands, different channels of communication and different information materials can be used.

Diversification strategies are required that include workshops, visitor guides and other services. There are opportunities for new and alternative business ideas. Examples of innovative communication approaches by some commercial aquaponics operators show the specific challenges associated with aquaponics:

1. ECF, Berlin: Choosing red variety of tilapia. Branding as "Rosébarsch" at the beginning of sales. Inspired by a customer's branding in a restaurant, ECF rebranded to "Hauptstadtbarsch" (capital city perch) in the meantime (Interview Echternacht 2018). Thereby regional branding is put in the foreground of communication rather than the inherent product quality oriented at the colour of the fish meat.
2. Aqua4C, Belgium: They introduced jade perch from Oceania into the European market. Aqua4C developed a branding as "Omega Baars" thus taking a novel food approach. They implicitly market the regionally unknown fish species as healthy while carefully avoiding to make any health claims.
3. Ponika, Slovenia: Marketing of produce from aquaponics has been difficult. The situation was complicated as the aquaponics farm was located far away from the market which was too far for a quick visit and also no other attraction was nearby. They concluded that without the possibility of visits, it would be difficult to secure the farm additional revenue sources or marketing directly to consumers. In their marketing approach, they thus first targeted gastro-distributors with a focus on quality and local production for a competitive price. They did not dare to target individual consumers via supermarket chains due to having too small of a system and subsequent inability to secure a steady, large enough volumes of production. Thus they sold fresh-cut herbs directly to gastro-distributors, whereby the price and local production played the most important role. Their experience showed that gastro-distributors liked the story of innovative food production and they liked helping young people in their start-up business. So they were supportive in the sense that they adopted their purchasing process by taking up the produce when it was available and ordering from foreign sellers when it was not. In general, however, they were not very interested in the sustainability character of aquaponics – in other words, they did not care how the fresh-cut herbs were produced but rather that they were locally produced and had appealing packaging (1 kg and 1/2 kg) where the local character of the production was emphasised. Thus in their experience with retailers, a company story of young innovators worked the best. Customers at the retail level furthermore did not like the connection with hydroponics as they mixed aquaponics and hydroponics. In Slovenia, the customers are wary of hydroponics, and the Ponika company needed to tackle the challenge of changing the consumer's perception from hydroponics, which has a negative image as being "unnatural", into aquaponics and create a positive image of aquaponics. Additionally, the selection of fresh-cut herbs for the individual consumers proved to be problematic, since the health benefits were not important enough in fresh-cut herbs as people just do not eat that much of those to care enough about, for example, pesticide-free production.

4. NerBreen, Spain: NerBreen is focused more on the aquaculture element of their business, since 70% of their business model represents the revenue from selling the fish. Yet they provide extensive marketing of both the fish and the vegetables. In both cases, they try to target the individual consumers via retail chains, preferably those retailers that target consumers who are willing to pay the premium price for higher local quality, aiming for approximately 20% higher prices than average. They face challenges in the marketing of both vegetables and tilapia. With vegetables they focused on fresh garlic and cherry tomatoes because they could reach higher prices due to smaller competition in those areas. Their marketing efforts include well-designed packaging with leaflets, whereby they explain the sustainability benefits of aquaponics. Here, they focus both on securing premium quality and adding additional story to their branding. When selling tilapia, they face a bigger challenge. The Spanish consumers currently have a negative perception of tilapia since they either mistake it with pangasius which is considered as a cheap and low-quality fish or they think it is imported from intensive aquaculture from the Far East and similarly ancd supposedly to be lower in quality. Within their marketing efforts, NerBreen thus needed to change this negative image and is focusing on providing information on the fact that this tilapia is locally produced, whereby both the water quality and the fish feed quality are of highest consideration, resulting in a high-quality fish product.
5. Urban Farmers, Netherlands: In an effort to generate additional income streams, they established visits of aquaponics production facilities. It is, however, currently questionable if the visitor business is economically sustainable. Questions arise whether the visitor stream will wane when the hype around aquaponics settles or once "everyone" has already seen it. Apart from visitors, other income streams are already tapped: Rooftop Farms offer gardening workshops. *(It should be noted that Urban Farmers in the Hague ceased to trade.)*

Print media and social media are suitable for public education, as well as thematic workshops, guided farm visits and tastings of aquaponics products (Miličić et al. 2017). However, information provision will be successful only if it meets the information needs of the target audience. Stakeholders, such as representatives of national governments, different associations (e.g. organic farming associations), plant operators or plant manufacturers are probably more interested in comprehensive factual information. For citizens' and consumers' information, focusing on emotion and entertainment could be more attractive. With regard to this target audience, pictures combined with concise text messages are particularly suitable for information transfer. For these stakeholders beyond conscious information perception and information processing, also unconscious effects play an important role. Different frames, that mean different presentation formats of the same information, can influence the recipient's behaviour in different ways (Levin et al. 1998). For a better understanding of the unconscious processes that may influence the stakeholders' behaviour, neuroeconomic research methods in association with traditional methods of market research are useful tools. Eye tracking makes it possible to answer questions regarding visual perception in an objective way. Combined with other empirical methods of communication research, especially qualitative and quantitative surveys, it is possible to conduct complex perception and impact analyses. As a pilot study by Schröter and Mergenthaler (2018a, b) shows, attitudes towards different aquaponics systems are related to the gaze behaviour of the study participants whilst viewing information material about aquaponics.

This underlines the importance of a careful and target group-oriented design of information material about aquaponics. The possible solutions are that either production planning has to accommodate and to add the additional revenue sources or direct marketing by growing a large variety of different crops, thereby further complicating the production process. Yet as, for example, ECF in Germany shows, they started with a variety of vegetables but decided to focus only on basil and sell it through one large retail chain. Another possibility is to build strategic alliances with other regional producers in order to achieve innovative marketing and distribution strategies. In general, however, we can conclude that the marketing aspect of commercial aquaponics is one of its most important challenges and one in which European aquaponics farms had to undergo a number of changes in attempts to try and find the right product-market fit. It remains to be seen, however, whether this product-market fit has been found and how stable it will remain.

18.8 Conclusion and Outlook

As discussed in this chapter, economic evaluations of aquaponic systems are still a very complex and difficult task at present. Although aquaponics is sometimes presented as an economically superior method of food production, there is no evidence for such generalised statements. Up to now, there is hardly any reliable data available for a comprehensive economic evaluation of aquaponics. That is partly because there is not "one aquaponics system", but there exist a variety of different systems operating in different locations under different conditions. For example, factors such as climatic conditions, which mainly affect the energy consumption of the systems, wage levels, workload required for operating the systems, and legal conditions have to be considered on the cost side. On the revenue side, factors such as the chosen fish-plant combination with its specific product prices, the option to manage the systems as organic production as well as the long-term public acceptance of the aquaponics systems and their products have an impact on the economic assessment. Not least, the economic evaluation of aquaponics in its strictest sense should be done in comparison to recirculating aquaculture systems and hydroponics systems as stand-alone systems.

Aquaponics constitutes a major communication challenge as a rather unknown food production system with high innovation levels and in most cases with high technological inputs. As food consumption in advanced societies is increasingly linked to some form of naturalness, major challenges in the communication of aquaponics systems and products can be expected. The limited evidence available suggests that this challenge can be managed under certain framework conditions but this needs high time, as well as financial and creative inputs. It has to be acknowledged that the reported high prices for aquaponics products only come at considerable costs of brand establishment. As any economic viability of aquaponic systems will critically depend on achievable prices, more research is needed to understand the different determinants of the customers' willingness to pay for aquaponics products.

Location decisions for aquaponics farming are a key determinant of economic viability as many production factors related to aquaponics production are not flexible in terms of space. This relates particularly to land. Aquaponics as a land-efficient production system can only count on this advantage in land-scarce regions. Comparatively, rural areas with relatively low land prices therefore cannot generate sufficient incentives unless there are other site-specific advantages, for example, waste energy supply from biogas plants. Though being land-efficient in general, aquaponics in urban contexts still competes for highly limited land resources. In functional markets, land would be allocated to those activities with the highest profits per unit of land and it is highly questionable if aquaponics will be able to compete with very efficient industrial or service-oriented activities in urban contexts. Therefore, aquaponics seems to fit only in urban areas that provide aquaponics with a competitive advantage over competing potential activities.

Extending the definition of aquaponics and including aquaponics farming as introduced by Palm et al. (2018) might align aquaponics much closer to traditional economic farming analyses. This wider definition of aquaponics refers to process water that is used for fertilisation combined with irrigation on fields. With this wider interpretation of aquaponics, it becomes possible to produce staple foods in aquaponic production systems. Since the nutrient absorption capacity of the agricultural area might be limited in some regions, this definition implicitly positions aquaponics as a competitor to pig, beef and poultry production. Since aquaculture uses less resources than pigs, beef and poultry with respect to the final output, this could become a viable option.

In traditional economic farm analyses, there is a strong technological and conceptual separation of animal and plant production. With less technological interaction in aquaponics farming as compared to aquaponics in its stricter senses, there will also be less complex economic evaluation as the two systems of fish and plant production can be modelled separately. To link the systems economically, system internal prices would have to be determined, e.g. prices for nutrients brought from fish production to fields of plant production.

Another issue is the prices obtained for final products from aquaponics farming. Evidence on achievable prices from such kind of production systems are completely lacking thus restricting reliable estimations on the economic viability as yet. With a stronger separation of fish and plant production, it might be feasible to use prices for conventional aquaculture products and conventional prices from plant production. This would assume that there is no price premium for aquaponics farming. To test whether this really is the case, price experiments combined with different communication tools have to be implemented.

From a communications perspective, there is the question of the perception of aquaponics farming as being superior to traditional farming approaches. At first glance, aquaponics farming might look like conventional livestock keeping only using a different type of animal. Communication efforts will have to focus on the higher efficiency levels of aquaculture as compared to other types of livestock production. Advertising plant products from aquaponics farming as being superior to products from conventional plant production might be a challenge and requires further in-depth analyses. However, one advantage with regard to communication might be the fact that a stronger separation of fish and plant production in aquaponics farming could facilitate organic certification. Organic labelling is expected to be a further advantage in communication efforts related to aquaponics. (It should be note that in the UK, at least, organic certification is tied to growing produce in soil and thus a different and special type of certification may need to be identified. For more on the issue of aquaponics and organic certification, see Chap. 19).

Finally, it is important to note that the interviewed European aquaponic companies, even those who abandoned their commercial aquaponics farming, remain hopeful for the future of aquaponics. They opted for aquaponics because of its sustainability potential and they still see that potential. They, acknowledge, however, that adoption of aquaponics is a gradual and long-term process, which cannot be just simply repeated in different locations but should be adapted to the local environment. As such, aquaponics remains one of the potentially sustainable technologies of the future, one which cannot (yet) be said to be able to properly compete on the market with its competitors, but one that will continue to need more public support and one whose adoption is determined not only by its commercial advantages, but much more on the public determination and goodwill. As stated in Chap. 16 of this publication, asking the question "under what circumstances can aquaponics outcompete traditional large scale food production methods?" is not the same as asking "to what extent can aquaponics meet the sustainability and food security demands of our age".

References

ACAC Association of Commercial Aquaponics Companies (2018) 10 Dec 2018. https://euaquaponicshub.wordpress.com/association-of-commercial-aquaponics-companies-acac/

Adler PR, Harper JK, Wade EM (2000) Economic analysis of an aquaponic system for the integrated production of rainbow trout and plants. Int J Recirculating Aquac 1(15):15–34

Baker AAC (2010) Preliminary development and evaluation of an aquaponic system for the American Insular Pacific. [Honolulu]:[University of Hawaii at Manoa], [December 2010]

Bailey DS, Ferrarezi RS (2017) Valuation of vegetable crops produced in the UVI Commercial Aquaponic System. Aquacult Rep 7:77–82

Bailey DS, Rakocy JE, Cole WM, Shultz KA, St Croix U (1997) Economic analysis of a commercial-scale aquaponic system for the production of tilapia and lettuce. In: Tilapia aquaculture: proceedings of the fourth international symposium on Tilapia in aquaculture, Orlando, Florida, pp 603–612

BORISplus (2018) Amtliche Informationen zum Immobilienmarkt in NRW. 16 Mar 2018. https://www.boris.nrw.de/borisplus/?lang=en

Bosma R (2016) The economic feasibility of aquaponics: a post-hoc cost-benefit analysis of investing in a fish vegetable farm near Dumaguete, Philippines. 11th Asian Fisheries and Aquaculture Forum, Bangkok, 2016-8-6/2016-8-6, 10 Dec 2018. http://library.wur.nl/WebQuery/wurpubs/512805

Bosma RH, Lacambra L, Landstra Y, Perini C, Poulie J, Schwaner MJ, Yin Y (2017) The financial feasibility of producing fish and vegetables through aquaponics. Aquacult Eng 78:146–154. https://doi.org/10.1016/j.aquaeng.2017.07.002

Bunyaviroch C (2013) Aquaponic systems in Puerto Rico: assessing their economic viability. PhD dissertation, Worcester Polytechnic Institute

Chryssohoidis GM, Krystallis A (2005) Organic consumers' personal values research: testing and validating the list of values (LOV) scale and implementing a value-based segmentation task. Food Qual Pref 16:585–599

Destatis (2015) Statistisches Bundesamt. https://www.destatis.de/. Accessed 10 Nov 2015

Engle CR (2015) Economics of aquaponics. SRAC Publication – Southern Regional Aquaculture Center. No.5006. 15 Feb 2018 https://srac.tamu.edu/serveFactSheet/282

Goddek S, Delaide B, Mankasingh U, Ragnarsdottir K, Jijakli H, Thorarinsdottir R (2015) Challenges of sustainable and commercial aquaponics. Sustainability 7:4199–4224. https://doi.org/10.3390/su7044199

Goddek S, Espinal C, Delaide B, Jijakli M, Schmautz Z, Wuertz S, Keesman K (2016) Navigating towards decoupled aquaponic systems: a system dynamics design approach. Water 8:303. https://doi.org/10.3390/w8070303

Graber A, Junge R (2009) Aquaponic Systems: Nutrient recycling from fish wastewater by vegetable production. Desalination 246:147–156. https://doi.org/10.1016/j.desal.2008.03.048

Hjelmar U (2011) Consumers' purchase of organic food products. A matter of convenience and reflexive practices. Appetite 56:336–334

IndustryARC (2012) Aquaponics market by equipment (grow lights, pumps, valves, greenhouse, water heaters, aeration systems); Components (rearing tanks, settling basins, sump, biofilter, hydroponics); End user (Commercial, home, community) Forecast (2013–2018), 24 Mar 2018. https://www.slideshare.net/rajesh193k/aquaponics-market-size-and-forecast-2018

IndustryARC (2017) Aquaponics market by equipment (Grow lights, pumps, valves, greenhouse, water heaters, aeration systems); Components (Rearing tanks, settling basins, sump, biofilter, hydroponics); End user (Commercial, home, community) – Forecast (2015–2021), 1 Feb 2018. http://industryarc.com/Report/22/global-commercial-aquaponics-market.html

Karimanzira D, Keesman KJ, Kloas W, Baganz D, Rauschenbach T (2016) Dynamic modeling of the INAPRO aquaponic system. Aquacult Eng 75:29–45. https://doi.org/10.1016/j.aquaeng.2016.10.004

Knaus U, Palm HW (2017) Effects of the fish species choice on vegetables in aquaponics under spring-summer conditions in northern Germany (Mecklenburg Western Pomerania). Aquaculture 473:62–73. https://doi.org/10.1016/j.aquaculture.2017.01.020

Korn A, Feucht Y, Zander K, Janssen M, Hamm U (2014): Entwicklung einer Kommunikationsstrategie für nachhaltige Aquakulturprodukte. 29 Oct 2017. http://orgprints.org/28279/1/28279-11NA040-066-uni-kassel-ti-2014-fischlabelling.pdf

Körner O, Holst N (2017) An open-source greenhouse modelling platform. Acta Horticult 241–248. https://doi.org/10.17660/ActaHortic.2017.1154.32

Körner O, Gutzmann E, Kledal PR (2017) A dynamic model simulating the symbiotic effects in aquaponic systems. Acta Horticult:309–316. https://doi.org/10.17660/ActaHortic.2017.1170.37

KTBL (2009) Gartenbau: Produktionsverfahren planen und kalkulieren. Kuratorium für Technik und Bauwesen in der Landwirtschaft. KTBL

Laidlaw J (2013) From Innovation to sustainability: Surviving five years as an urban aquaponics social enterprise. 13 Feb 2018. http://citiesprogramme.org/wp-content/uploads/2016/08/laidlaw_from_innovation_to-sustainability-web.pdf

Levin IP, Schneider SL, Gaeth GJ (1998) All frames are not created equal: a typology and critical analysis of framing effects. Organ Behav Hum Decis Process 76(2):149–188

Lewis WM, Yopp JH, Schramm HL, Brandenburg AM (1978) Use of hydroponics to maintain quality of recirculated water in a fish culture system. Trans Am Fish Soc 107:92–99. https://doi.org/10.1577/1548-8659(1978)107

Leyer M, Hüttel S (2017) Performance analysis with DEA, process mining and business process simulation on a livestock process. In: Proceedings of the international multiconference of engineers and computer scientists, 13 Dec 2018. https://pdfs.semanticscholar.org/2d2d/ef068539537127feed0080655cdba4266ce6.pdf

Lohrberg F, Lička L, Scazzosi L, Timpe A (eds) (2016) Urban agriculture Europe. Jovis, Berlin

Love DC, Fry JP et al (2014) An international survey of aquaponics practitioners. PLoS One 9(7). https://doi.org/10.1371/journal.pone.0102662

Love DC, Fry JP, Li X, Hill ES, Genello L, Semmens K, Thompson RE (2015) Commercial aquaponics production and profitability: findings from an international survey. Aquaculture 435:67–74. https://doi.org/10.1016/j.aquaculture.2014.09.023

Mackowiak C, Owens L, Hinkle C, Prince R (1989) Continuous hydroponic wheat production using a recirculating system. NASA Technical Memorandum TM 102784. 3 Mar 2018. https://ntrs.nasa.gov/archive/nasa/casi.ntrs.nasa.gov/19900009537.pdf

Medina M, Jayachandran K, Bhat MG, Deoraj A (2015) Assessing plant growth, water quality and economic effects from application of a plant-based aquafeed in a recirculating aquaponic system. Aquacult Intl 24(1):415–427. https://doi.org/10.1007/s10499-015-9934-3

Mergenthaler M, Lorleberg W (2016) Voruntersuchung zur Zahlungsbereitschaft für frischen und geräucherten europäischen Wels aus einer Aquaponik-Pilotanlage. In: Notizen aus der Forschung, Fachbereich Agrarwirtschaft, Soest,14/2016

Miličić V, Thorarinsdottir R, Santos M, Hančič M (2017) Commercial aquaponics approaching the European market: to consumers' perceptions of aquaponics products in Europe. Water 9:80. https://doi.org/10.3390/w9020080

Morgenstern R, Lorleberg W, Biernatzki R, Boelhauve M, Braun, J., Haberlah-Korr, V., 2017. Pilotstudie "Nachhaltige Aquaponik-Erzeugung für Nordrhein-Westfalen." Fachhochschule Südwestfalen, Soest. ISBN:978-3-940956-66-8

Naegel LCA (1977) Combined production of fish and plants in recirculating water. Aquaculture 10:17–24. https://doi.org/10.1016/0044-8486(77)90029-1

Orsini F, Dubbeling M, de Zeeuw H, Gianquinto G (2017) Rooftop urban agriculture. Springer

Pakseresht A, McFadden BR, Lagerkvist CJ (2017) Consumer acceptance of food biotechnology based on policy context and upstream acceptance: evidence from an artefactual field experiment. Eur Rev Agricult Econom 44:757–780. https://doi.org/10.1093/erae/jbx016

Palm HW, Knaus U, Appelbaum S, Goddek S, Strauch SM, Vermeulen T, Haïssam Jijakli M, Kotzen B (2018) Towards commercial aquaponics: a review of systems, designs, scales and nomenclature. Aquacult Intl. https://doi.org/10.1007/s10499-018-0249-z

Peattie K (1992) Green marketing: The M and E handbook series. Pitman Longman Group, London

Petrea SM, Coadă MT, Cristea V, Dediu L, Cristea D, Rahoveanu AT, Zugravu AG, Rahoveanu MMT, Mocuta DN (2016) A comparative cost – effectiveness analysis in different tested aquaponic systems. Agricult Agricult Sci Proc 10:555–565. https://doi.org/10.1016/j.aaspro.2016.09.034

Pollard G, Ward J, Koth B (2017) Aquaponics in urban agriculture: social acceptance and urban food planning. Horticulturae 3:39. https://doi.org/10.3390/horticulturae3020039

Quagrainie KK, Flores RMV, Kim H-J, McClain V (2018) Economic analysis of aquaponics and hydroponics production in the U.S. Midwest. J Appl Aquacult 30:1–14

Savidov N, Brooks A (2004) Evaluation and development of aquaponics production and product market capabilities in Alberta. Crop Diversification Centre South, Alberta Agriculture, Food and Rural Development

Schröter I, Mergenthaler M (2018a) Eye-tracking als neuroökonomische Methode zur Erfassung von Wahrnehmungsprozessen von Informationen über Aquaponik. Notizen aus der Forschung Nr. 2/2018. Fachbereich Agrarwirtschaft, Soest

Schröter I, Mergenthaler M (2018b) Technik oder Natur: Wahrnehmung und Bewertung von Aquaponik-Varianten. Notizen aus der Forschung Nr. 1/2018. Fachbereich Agrarwirtschaft, Soest

Schröter I, Hüppe J-H, Lorleberg W, Mergenthaler M (2017a) Einflussfaktoren auf die Zahlungsbereitschaft für Fisch nach dem Besuch einer Aquaponikanlage. In: Notizen aus der Forschung, Fachbereich Agrarwirtschaft, Soest.71

Schröter I, Hüppe J-H, Lorleberg W, Mergenthaler M (2017b) Kenntnisse über Aquaponik und Zahlungsbereitschaft für Fisch nach dem Besuch einer Polykulturanlage mit Aquaponik. In: Notizen aus der Forschung, Fachbereich Agrarwirtschaft, Soest. 70

Schröter I, Lorleberg W, Mergenthaler M (2017c) Erwartungen potentieller Konsumenten an Fischprodukte aus Aquaponik in einer Sondierungsstichprobe. In: Notizen aus der Forschung, Fachbereich Agrarwirtschaft, Soest,69

Schuetz K, Voigt L, Mergenthaler M (2018) Regionale Lebensmittel zwischen Anspruch und Wirklichkeit: Verbrauchererwartungen, Begriffs- und Qualitätsverständnisse regionaler Vermarktungsinitiativen und verbraucherpolitische Implikationen. KVF working paper

Smeets PJ (2010) Expedition Agroparks; Research by Design Into Sustainable Development and Agriculture in the Network Society. Wageningen Academic Publications

Smil V (2008) Energy in nature and society: general energetics of complex systems. MIT Press, Cambridge, MA

Specht K, Sanyé-Mengual E (2017) Risks in urban rooftop agriculture: Assessing stakeholders' perceptions to ensure efficient policymaking. Environ Sci Policy 69:13–21. https://doi.org/10.1016/j.envsci.2016.12.001

Specht K, Weith T, Swoboda K, Siebert R (2016) Socially acceptable urban agriculture businesses. Agron Sustain Dev 36(1):17. https://doi.org/10.1007/s13593-016-0355-0

Tamin M, Harun A, Estim A, Saufie S, Obong S (2015) Consumer Acceptance towards Aquaponic Products. IOSR J Bus Manage (IOSR-JBM) 17:49–64. https://doi.org/10.9790/487X-17824964

Thorarinsdottir RI (2015) Aquaponics guidelines, University of Iceland: Iceland. Available online: http://skemman.is/en/stream/get/1946/23343/52997/1/Guidelines_Aquaponics_20151112.pdf

Tokunaga K, Tamaru C, Ako H, Leung P (2015) Economics of small-scale commercial aquaponics in Hawaii. J World Aquacult Soc 46:20–32

Villarroel (2017) Map. Aquaponics Map (COST FA1305). https://www.google.com/maps/d/u/0/viewer?ll=35.35294037658608%2C0.45745135072172616&z=4&mid=1bjUUbCtUfE_BCgaAf7AbmxyCpT0. Accessed 8 Sep 2018

Villarroel M, Junge R, Komives T, König B, Plaza I, Bittsánszky A, Joly A (2016) Survey of Aquaponics in Europe. Water 8:468. https://doi.org/10.3390/w8100468

Vogt L, Schütz K, Mergenthaler M, (2016) Regionalisierung von Lieferketten in der Ernährungswirtschaft–Herausforderungen und Ansatzpunkte

Zugravu AG, Rahoveanu MMT, Rahoveanu AT, Khalel MS, Ibrahim MAR (2016) The perception of aquaponics products in Romania. Risk Contempo Econom 3:525–530

Chapter 19
Aquaponics: The Ugly Duckling in Organic Regulation

Paul Rye Kledal, Bettina König, and Daniel Matulić

Abstract Due to the cyclic or systemic nature of both aquaponics and organic production, organic certification appears to be a natural step for a researcher, system designer or commercial-oriented aquaponics producer to engage in. However, the underlying principles and justifications of aquaponics and organic production differ considerably between respectively a technological- and a soil-based understanding of nutrient cycles and long-term sustainability in food production. These principles are confirmed in both the organic regulation regime of the EU and USA, and presently leave the question ambiguously open as to whether aquaponics as a food production system can be recognized and certified as organic. Despite an openness in the organic regulation for new knowledge, adaptations and innovations, the organic sector itself has shown a reluctance to recognize more knowledge-based intensive speciality crops and technologies. This is particularly difficult with respect to small organic sub-sectors such as horticulture and aquaculture production. Both are very specific subsystems of the agricultural sector, where aquaponics potentially would belong at the intersection between organic greenhouse horticulture and organic aquaculture. Organically certified aquaponics would therefore need to establish a niche within the organic sector. So in order to move forward, there is a great need for a more serious but open-minded exchange and discussion among the aquaponics and organic sub-sectors themselves to explore the potential but also limitations of their respective production models. However, between the two food production systems, there should be room for debate with a view to finding new and feasible roles for aquaponics in the organic community.

P. R. Kledal (✉)
Institute of Global Food & Farming, Hellerup, Denmark
e-mail: paul@igff.dk

B. König
Faculty of Life Sciences, Thaer-Institute, Horticultural Economics and IRI THESys, Humboldt Universität zu Berlin, Berlin, Germany

D. Matulić
Department of Fisheries, Beekeeping, Game management and Special Zoology, Faculty of Agriculture, University of Zagreb, Zagreb, Croatia
e-mail: dmatulic@agr.hr

S. Goddek et al. (eds.), *Aquaponics Food Production Systems*,
https://doi.org/10.1007/978-3-030-15943-6_19

Keywords Aquaponics · Organic certification · EU organic regulation · US organic regulation · Recirculating aquaculture · Hydroponics

19.1 Introduction

Aquaponics is an integrated closed-loop multi-trophic food production system that combines elements of a recirculating aquaculture system (RAS) and hydroponics (Endut et al. 2011; Goddek et al. 2015; Graber and Junge 2009). Aquaponics is therefore discussed as a sustainable eco-friendly food production system, where nutrient-enriched water from fish tanks is recirculated and used to fertilize vegetable production beds, thus making good use of the valuable nutrients that in conventional aquaculture systems are discarded (Shafahi and Woolston 2014) and presents a potential solution to an environmental problem usually referred to as eutrophication of aquatic ecosystems.

Organic agriculture is also based on natural principles of recirculation and resource minimization as defined by the International Federation of Organic Agriculture Movements (IFOAM 2005):

> *Organic Agriculture is a production system that sustains the health of* ***soils, ecosystems*** *and* ***people****. It* ***relies on ecological processes, biodiversity and cycles adapted to local conditions****, rather than the use of inputs with adverse effects. Organic Agriculture combines* ***tradition, innovation*** *and* ***science*** *to benefit the shared environment and promote* ***fair relationships*** *and a good* ***quality of life*** *for all involved.*

Due to the cyclic or systemic nature of both production systems, obtaining organic certification could appear to be a natural step for a researcher, system designer or commercially oriented aquaponics producer to engage in. On the other hand, the underlying principles of aquaponics and organic production differ considerably. From a research perspective, one could argue that the discussion on aquaponics being organic or not demonstrates an interesting case between two poles of agro-food-systems-thinking, namely, agro-industrial (conventional) and agro-ecological (organic). Somehow aquaponics needs to find its place in this continuum.

The aim of this chapter is to shed light on the current *status quo* regarding the barriers of certifying aquaponics as organic food production, and to discuss the underlying principles, contradictions and views upon its sustainability. It will also discuss possible future scenarios emerging out of the links between the rationales and implementations of the two production systems. We can consider both organic farming and aquaponics as food production systems, because both farmers and aquaponics producers are faced with complex decision-making situations involving the balancing of interconnected and interrelated inputs, external factors (environment, markets, value chains, etc.) and management procedures in order to produce food.

19.1.1 Aquaponic Production Systems and the Technology Applied

Present-day aquaponic production systems are generally categorized according to the type of technology applied to the plant production part, and whether the integration is coupled into one single loop between the plants and the fish or decoupled into separate loops. The most common technologies applied in the plant production are: (1) The Deep Water Culture (DWC) or in the literature often called UVI because originally developed at the University of Virgin Island, (2) NFT (Nutrient Fluid Technology) and (3) 'Flood and Ebb'. The DWC and NFT are the most common or 'classic' technologies applied to the plant production, and often the fish and plant production are connected into one dependent loop of water and nutrient flow. This singular connection and interdependency of the whole system is a factor that increases risks tremendously, and is a prime barrier to establishing large-scale commercial production.

The main differences between the first two systems are related to how the plants are grown.

In the DWC system, the plant bed is the floating system where plants are grown on RAFTs (usually polystyrene) floating in long tanks of variable width, acting both as an extensive biofilter and a water buffer, regulating temperature and pH fluctuations.

In the NFT system, plants are grown in hydroponic plastic pipes, well known from modern horticulture. A thin layer of nutrient water supplied to the pipes feeds the plants. In both cases of the NFT and DWC, the holes for the plants are fixed, which constrains the producer as to what kind of plants can be produced. In some aquaponics systems, both plant-growing technologies from the DWC and NFT are applied at the same time giving more flexibility and security while running a singular aquaponics loop (Kledal and Thorarinssdottir 2018).

In the third category, 'Flood and Ebb', plants are grown in pots placed on (often moveable) plant tables, and then fed two or three times a day by flooding the tables for 5–10 min. The plant tables provide the option of producer with flexibility in the choice of plants grown and the size of the pots, as well as the prospect of using soil and this opening up the possibility of organic certification.

In recent years, decoupled aquaponics is starting to emerge as the production system applicable to large-scale commercial aquaponics production (Fig. 19.1). In decoupled aquaponics the fish and plant production, each has their own loop of water supply, but is also connected to each other via a fertilizer tank supplying nutrient deficiencies to the plants. In this way, the dependency between the fish and plant production has been removed, but the symbiotic benefits are kept allowing for investment in large-scale commercial production. There is currently some debate about the advantages of circulating, or coupled vs. decoupled aquaponics (Goddek et al. 2016). However, there is not yet a consensus about the status of decoupled systems since they could be considered as just another plant nutrient supply method, as long as the water does not circulate back to the fish (Junge et al. 2017).

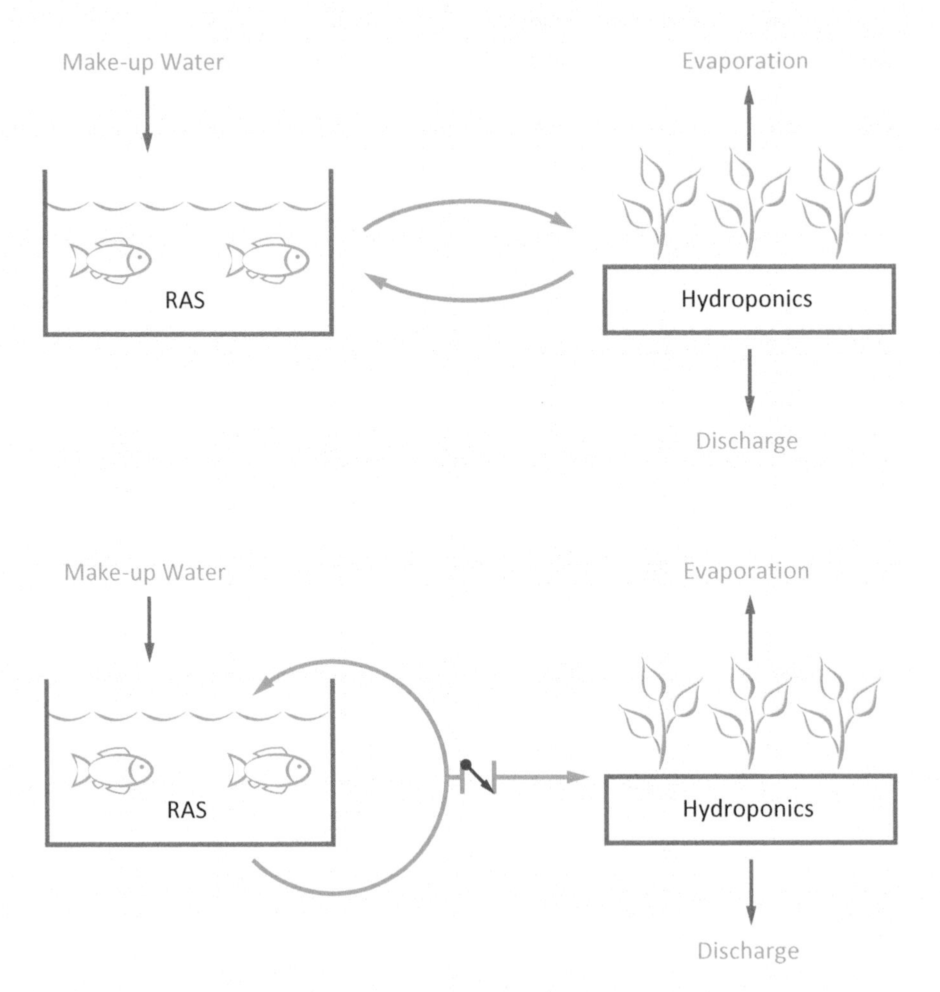

Fig. 19.1 (**a**) coupled (**b**) decoupled aquaponics system. (Adapted from Peterhans 2015)

Aquaponics is in general receiving increased interest globally as a sustainable food production method, and with the prospects of promoting aquaponics on a commercial scale, there is an equivalent interest in receiving organic certification with its related price premiums on the produce. According to Kledal and Thorarinssdottir (2018) and König et al. (2018), the commercially available systems of aquaponics are mainly based on the research carried out by Rakocy and his co-workers (Rakocy 1999a, b, 2002, 2009; Rakocy et al. 2001, 2004, 2006, 2009), as previously described (the classic production technologies 1 and 2 above) where plants and fish are connected in a singular dependent loop. Likewise, most production systems, whether they are based on DWC, NFT or moveable plant tables with 'Flood and Ebb', are not using an organic growth media, hence already excluding themselves from obtaining an organic certification for the horticultural part. Regarding aquaponics production, access to organic fish feed is in general only available for a few commercial freshwater fish species. So, aside from the various input factors, there are constraints to starting an organic aquaponics production on a commercial scale.

So we can see that despite the growing interest in marketing aquaponics as an environmental-friendly food production system the present organic legislation regime in the EU and USA prohibits aquaponics from being recognized as such (NOSB 2017). However, the discussion is not finalized, and while it is ongoing, some private certification agencies in the US allow for the certification of the vegetables as organic (Friendly Aquaponics 2018).

19.1.2 Aquaponics and the EU Organic Regulation Regime

In the EU, the present regulatory framework for organic fish and horticultural production is regulated by the Council Regulation (EC) No. 834/2007, whereas more detailed rules are regulated by the Commission Regulations (EC) No. 889/2008 and (EC) no. 710/2009. However, the EU organic regulatory regime does *not* have any standards or regulations for certifying aquaponics as organic. The organic regulation is based on the aims and principles of recognizing organic farming as a natural resource-based food production (Lockeretz 2007; Aeberhardt and Rist 2008). This is backed up in the implementing regulation annex by the exclusion of inputs not allowed for organic farming. Consequently, aquaponics farming systems using the RAS technology and soilless vegetable production (hydroponics) cannot be certified as organic under the present EU organic regulation.

However, among aquaponics practitioners, there is continuous discussion about aquaponics and organic certification. First of all, the rapid industrial developments within fish farming and the market diversification and demand for organic products make it economically desirable to qualify for the organic price premium as one way to reimburse the high capital investments required for commercial aquaponics. Second, it appears only natural to link an environmental-friendly food production such as aquaponics, to already well-established certification labels and consumer perceptions of a sustainable food production rather than engaging in the high transaction costs of creating a whole new food label.

In view of the current discussions on limited resources for food production, animal welfare, the increasing pressure on the sustainability of the aquatic environment, paralleled with the ongoing technological progress within aquaponics, this article asks why aquaponics cannot be certified organic.

In the following paragraph, the organic rules and regulations under the EU regime creating barriers to aquaponics will be examined.

19.2 Organic Regulations

19.2.1 Organic Rules in Horticulture

The hydroponic production technology in the absence of an organic growth media cannot be certified as organic, which has proven to be a long-time effective barrier for the conversion of existing greenhouse vegetable producers to organic farming schemes (König 2004). For horticultural products, the specific EU regulation preventing products produced under 'classical' aquaponics systems to obtain an organic certification are the following:

834/2007 Regulation (12): ....Plants should preferably be fed through the soil eco-system and not through soluble fertilizers added to the soil

889/2008 Article. (4): Organic farming is based on nourishing the plants primarily through the soil ecosystem. Therefore hydroponic cultivation, where plants grow with their roots in an inert medium feed with soluble minerals and nutrients, should not be allowed.

Since aquaponics is based on using fish sludge as a source of fertilizing the plants, the absence of mineral fertilizers would at first seem like a step towards organic production. However, the 'classical' aquaponics production systems started using components from the soilless hydroponic technology, and therefore the plants produced under such a system *cannot* be certified as organic. In order to understand this prohibition in organic regulation, it is helpful to remember that hydroponics was developed and adopted by growers as a response to the challenges greenhouse growers met in intensive soil-based vegetable cultivation systems, e.g. enrichment of the soil with soil-borne pathogens. In contrast, the organic horticulture approach departs from the question of how greenhouse farming has to look like in order to avoid these challenges. Their starting point is instead to change the management of the soil rather than inventing a production technology without soil.

In addition to this general principle of soil-based production, organic horticulture can be considered a specialized niche within organic farming offering a considerable variety of crops. The legislation for fruit vegetables, such as tomato, cucumber, pepper, eggplant, etc. prescribes cultivation in natural soil. Plants sold with the soil, such as seedlings or potted herbs, can be certified as organic. The prerequisite is that the plant could continue growing at the customer's greenhouse or kitchen window. This means, that herb bunches, salads cut off from the roots need to be grown in soil in order to qualify for organic certification. Inputs allowed for organic production are regulated in the implementing regulation. For Germany, Switzerland and the Netherlands, the testing and approval of inputs for organic production is maintained by FiBL (Research Institute of Organic Agriculture), who are currently aiming at developing a European List on inputs certified as suitable for organic status.

The nutrient supply in organic greenhouse production is a challenge. Not only are mineral fertilizers common in hydroponic production systems not allowed but, in the special case of the German organic farmer associations (going beyond EU legislation), also hydroxylates that are of animal origin (interview with organic extension service). Greenhouse growers who have invested in infrastructure sealing the natural soil with permanent greenhouse flooring, have faced a long-time effective barrier for the conversion of existing greenhouse infrastructure to organic farming schemes, except for potted herbs (König 2004). New investments in greenhouse infrastructure have contributed to the increase of organic fruit and vegetable production in the last years, e.g. in Germany. However, for these modern organic growers, aquaponics does not yet provide any solution as they are looking for answers into the areas of suitable soils, improved crop rotation, effective microorganisms, compost and the like.

Horticulture is facing the general challenge that the organic EU regulation is not very detailed in this area. Theoretically, this leaves room for new production approaches such as aquaponics. However, at this stage of development within both commercial organic horticulture and aquaponics, the start-up costs for the producers are immensely high let alone the search for information on production management, prohibitions, potential yields, etc. In the end, the suitability of innovative production systems is left to the decision of the local certification authority on a project-by-project basis (König et al. 2018).

However, since the starting point in organic agriculture is about soil-based production and the fact that horticulture, aquaculture and aquaponics are small sub-sectors; the EU regulation regime on organic production may not be something that is expected to change in the near future.

19.2.2 Organic Rules in Aquaculture

For organic aquaculture, the production is regulated by Commission Regulations of 889/2008 and 710/2009. In parr. 11. Commission Regulation (2009), recirculating technologies are clearly prohibited in organic aquaculture, except for the specific production in hatcheries and nurseries making and selling fingerlings for further growth in open-air pond systems.

> *Parr.* 11.
>
> *Recent technical development has led to increasing use of closed recirculation systems for aquaculture production, such systems depend on external input and high energy but permit reduction of waste discharges and prevention of escapes. Due to the principle that organic production should be as close as possible to Nature, the use of such systems should not be allowed for organic production until further knowledge is available. Exceptional use should be possible only for the specific production situation of hatcheries and nurseries.*

Since recirculating technology is at the core of the aquaponics production system, it is at present not possible to get a complete organic certification of an aquaponics system, if all of the finishing produces is to be sold for the consumer market.

Likewise, the organic regulation on fish density in open ponds and marine cages are mainly made to secure a minimal discharge of fish manure to the aquatic environment. Questions on fish welfare are therefore an indirect matter related to their well-being based upon the level of freshwater exchange in the ponds. The stocking density in organic aquaculture systems is often 1/4 to 1/3 of that in modern RAS systems, and therefore from an economical aspect, not very cost-effective for this technology. At the same time, we need research and development on animal welfare indicators as well as on feasible and meaningful animal welfare monitoring tools as a prerequisite to discussing specific stocking densities. Only then will we be able to assess the potential economic viability of the aquaculture part of an organic aquaponics system (Ashley 2007; Martins et al. 2012).

19.2.3 Aquaponics and the USA Organic Regulation Regime

As in Europe, there is an ongoing discussion in the USA about how to handle soilless or soil replacement approaches for nutrient provision to plants as a mean of a resource-efficient food production and their inclusion or exclusion from the organic certification scheme. Despite these discussions, the state of the art is somewhat similar undecided to that in Europe, but practices differ: recently, the Crops Subcommittee of the National Organic Standards Board provided a suggestion to make aeroponics, aquaponics and hydroponics prohibited practices under Sect. 205.105 of the USDA Organic Regulations (NOSB 2017). This decision was rejected with 8:7 votes, yet not achieving 10 votes to make the decision a recommendation of the NOSB to USDA. Only the rejection of aeroponics found sufficient votes (14 out of 15, NOSB 2017). Therefore, the USDA Agricultural Marketing Service is only reviewing the recommendation to exclude aeroponics from organic certification (AMS 2018, p.2). This NOSB decision had been forced due to non-harmonized practices in the past among accredited organic certification agencies, whereby some of them certified hydroponics as organic under the National Organic Program (NOP) while others did not. These differing practices can be seen as a result of a long discussion process with no clear conclusions, ending with eight certifiers certifying hydroponic operations as organic in 2010 and a 33% increase of organic certified hydroponic producers (NOSB 2016: Hydroponic and aquaponics subcommittee report). Already back in 2010, the NOSB had received recommendation for a federal rule concerning greenhouse production systems, indicating basically that '*Growing media shall contain sufficient organic matter capable of supporting natural and diverse soil ecology. For this reason, hydroponic and aeroponic systems are prohibited.*', yet this explicit prohibition did not enter current law (NOSB 2010, 2016:122). Instead, the more open definition of organic production from 2002 was in place, where organic production is '*[a] production system that is managed in accordance with the Act and regulations in this part to respond to site-specific conditions by integrating cultural, biological, and mechanical practices that foster cycling of resources, promote ecological balance, and conserve biodiversity*' (NOSB 2016, p. 7). The hydroponic and aquaponics subcommittee concludes that '*Under current law and clarification from NOP/USDA, hydroponic and aquaponic production methods are legally allowed for certification as USDA Organic as long as the producer can demonstrate compliance with the USDA organic regulations.*'

(NOSB 2016, p. 10–11). However, the difficulty is that organic production is about management of the soil, whereas hydroponics is a system managing fertilizers. By not addressing this difference could lead to some ambiguity and potential negative consequences for the support of organic certification by farmers and consumers (AMS 2016). In the (NOSB 2016, Alternative Labeling Subcommittee Report), other experts presented a range of ideas as to how labels within the USDA organic scheme or outside could appear. Because of a lack of standards and norms, which is a necessary basis for labels, the group did not arrive at a consensus. The opinion was that, if aquaponics were included, or an extra label added, among the already existing great diversity between the different organic productions systems, it would both challenge the certification process as well as being a source of confusion for consumers. Interestingly, the suggestions of label alternatives under the USDA organic umbrella, or in addition to it, highlight the anecdotal evidence that the principle of aquaponics farms seems to be appealing to consumers, and that they do not need to be certified organic to be viable (NOSB 2016, Alternative Labeling Subcommittee Report, p.5).

In summary, the NOSB (2016) provides a detailed process description from the early 1990s until today, which reflects also different opinions of stakeholders involved in this discussion. The Organic Foods Production Act of 1990 (OFPA) builds on this basis for the development of federal US organic certification for the NOSB, and since then the discussion about allowing greenhouse production systems for organic certification or not has been in place (NOSB 2016). By now, there is agreement in the discussion which recognizes that the roots of organic farming lie in the concern about soil fertility and soil quality. All organic farming practices and standards developed are based upon this premise, and any discussions about its further development have to start from this point of view.

In the discussion, there are more open questions involved as to whether hydroponics could be called organic or not. The comparison of conventional and organic farming, in the case of horticultural greenhouse crops, hinges on some poorly researched or still controversially discussed issues (NOSB 2017):

The type of farming practice may also explain the differences found in organic and conventional products, e.g. lower content of secondary plant metabolites of conventionally grown greenhouse vegetables compared to organic vegetables from field farming. Allowing hydroponics to be certified as organic, this currently communicated added value of organic products could not be communicated to consumers anymore as added value unambiguously.

An important source of nutrients in hydroponic systems is hydrolysed soybean meal, which US growers import from Europe in order to ensure GMO-free sourcing compatible with organic standards. This impinges negatively upon the overall sustainability.

One principle of organic farming is dealing with resilience, which is doubted for hydroponic and aquaponics farming systems as they are highly dependent on an external energy supply (anecdotal observations). Opponents state that organic farms are likewise not ‘resilient’ against severe natural disasters, yet both groups remain somewhat unclear about their resilience concept when applied to these production systems.

A comparison of processes at the root surface, i.e. the microbial environment in soil versus water and the nutrient uptake, is an open question and opponents argue that literature on this topic is perceived as not sufficient.

In contrast, all the arguments that can be found in Europe as to why hydroponics or aquaponics should be certified as organic are also brought into the discussion in the US. The most remarkable point is, however, the lack of data on the direct comparison of systems in order to be able to evaluate the mentioned impacts and advantages systematically. In summary, the NOSB rejected to label hydroponic or aquaponics systems as organic in general because (NOSB 2017, p. 70–71):

> *§ 6513 Organic Plan: "An organic plan shall contain provisions designed to foster soil fertility, primarily through the management of the organic content of the soil through proper tillage, crop rotation, and manuring...An organic plan shall not include any production or handling practices that are inconsistent with this chapter."*
>
> • *§ 205.200 General: " Production practices implemented in accordance with this subpart must maintain or improve the natural resources of the operation, including soil and water quality."*
>
> • *§ 205.203 Soil fertility and crop nutrient management practice standard: (a) "The producer must select and implement tillage and cultivation practices that maintain or improve the physical, chemical, and biological condition of soil and minimize soil erosion." (b) "The producer must manage crop nutrients and soil fertility through rotations, cover crops, and the application of plant and animal materials." (c) "The producer must manage plant and animal materials to maintain or improve soil organic matter content..."*

Later, in the year 2016, definitions were given for hydroponics, aquaponics and aeroponics, stating for aquaponics that (NOSB 2017, p. 82):

> *Aquaponic production is a form of hydroponics in which plants get some or all of their nutrients delivered in liquid form from fish waste. Aquaponics is defined here as a recirculating hydroponic system in which plants are grown in nutrients originating from aquatic animal waste water, which may include the use of bacteria to improve availability of these nutrients to the plants. The plants improve the water quality by using the nutrients, and the water is then recirculated back to the aquatic animals.*

The NOP has strict standards for handling animal manure in terrestrial organic production, but no such standards exist to ensure the safety of plant foods produced in the faecal waste of aquatic vertebrates. Also, the NOP has not yet issued standards for organic aquaculture production, upon which aquaponics plant production would be dependent. '*The Crops Subcommittee is opposed to allowing aquaponic production systems to be certified organic at this time. If aquaculture standards are issued in the future, and concerns about food safety are resolved, aquaponics could be reconsidered.*' (NOSB 2017, p. 82).

In there is the 'Naturally Grown' certification, a peer-review, grassroot certification, which explicitly includes aquaponics (https://www.cngfarming.org/aquaponics). This certification involves a catalogue with criteria from January 2016. Only the vegetable produce is certified, not the fish part because at the moment it (e.g. fish feed) does not meet the general criteria for livestock certification. The criteria regulate the following aspects: System Design and Components, Materials for Main System Components and Growing Media/Root Support, Water Sources, Monitoring, Inputs for pH Adjustment, Waste Use & Disposal, Crop Production and Management, Fish Management, Location and Buffers, Energy and Record Keeping. The scheme is based on peer inspection schemes and does not allow the usage of synthetic pesticides and fungicides, copper-based pesticides, petrochemical-based pesticides or fungicides. It does not regulate the components on the plant part, but assesses its functions: water regulation, aeration, degassing, biofiltrations and removal of fish waste solids.

In summary, there are individual organic certification bodies in the US certifying aquaponics (parts) as organic production, but there are also cases reported of farmers claiming organic production without being certified (Friendly Aquaponics 2018). After this chapter goes to press there may be new developments affecting the organic certification topic. In addition, the issue of urban farming being declared as farming, and hence being eligible for agricultural funds might get a clearer status with the pending new US Farm Bill.

19.3 Discussion and Conclusions

This chapter has attempted to clarify the regulatory aspects relevant to understanding why aquaponics presently is not eligible for organic certification in the EU and the USA. As in the EU, the main paradigm behind organic farming in the USA is briefly, to manage soils in a natural way. In the EU, organic certification decisions for organic aquaponics are not carried out by local authorities, whereas the USA has seen a growth in this type of action in the last few years as well as an increase in private peer-review certifications and decisions of individual organic certification agencies.

In principle, all EU organic regulations are open to adaptation as soon as there is new scientific evidence, as stated in paragraph 24.

Parr. 24

Organic aquaculture is a relatively new field of organic production compared to organic agriculture, where long experience exists at the farm level. Given consumers' growing interest in organic aquaculture products further growth in the conversion of aquaculture units to organic production is likely. This will soon lead to increased experience and technical knowledge. Moreover, planned research is expected to result in new knowledge in particular on containment systems, the need of non- organic feed ingredients, or stocking densities for certain species. New knowledge and technical development, which would lead to an improvement in organic aquaculture, should be reflected in the production rules. Therefore provision should be made to review the present legislation with a view to modifying it where appropriate.

So, the horticultural, aquacultural and organic sectors would need to organize themselves, integrating knowledge from different domains. Yet, there are difficulties in convening such knowledge-intensive discussions as the experts and the communities of practice are fragmented and dispersed. Moreover, it is a knowledge-intensive endeavour: the NOSB subcommittee on Hydroponic and aquaponics stated that an in-depth justification would need more time (NOSB 2017; Hydroponic & Aquaponic Subcommittee Report, p. 2). In the EU, the most important Organic Research Institute (FiBL) involved in regulation and input testing is based in Switzerland. However, with its new organizational umbrella now located in Brussels, circumstances are expected to improve in the coming years. Still, to review all the details of every component of hydroponic and aquaculture will raise new questions that only a very few organic greenhouse and aquaculture experts in Europe will be able to answer. Aquaponics actors might then be in a situation to discuss (a) the circular economy approach, e.g. from the point of view of Life Cycle Assessment and in terms of the construction and production components used to compare it with a soil-based production and (b) to consider the question of how does aquaponics improve the on-site (sustainability) situation of fish and plant production. Asking such questions may stimulate ideas for new system-designed aquaponics that might be perceived as interesting by the organic community. However, it may also raise new (or in fact old) barriers to the implementation of large-scale organic production, e.g. thinking of organic pots that do not pose a challenge to the potting machine for aquaponics salads and fish welfare indicators in order to develop knowledge-based species-specific stocking densities. In this way, system design changes may raise new research questions.

For the time being, more collaborative research and development into developing aquaponics systems for the organic sector might be an interesting pathway that would best be discussed and developed among open-minded experts and growers from aquaponics, organic greenhouse production and organic aquaculture. To conclude, there is a great need for a knowledge exchange and discussion among the aquaponics and organic niches to explore the potentials and limitations of their respective production models, and reach some kind of consensus as to whether there is a future role for recirculating aquaponics systems in the organic community and what organic aquaponics could actually look like. But with the diverse visions of aquaponics systems by entrepreneurs, farmers, researchers and communities already in place, what would certify them all as organic mean for communication to the consumer, for marketing and for achieving sustainability goals? In the USA as in Europe, the question is who would benefit from organic certification of aquaponics? At present, aquaponics seems somewhat to be an ugly duckling both within the conventional and the organic agriculture regimes, but in the future it may turn into a beautiful sustainable swan—and may be certified organic?!

References

Aeberhardt A, Rist S (2008) Koproduktion von Wissen in der Entwicklung des Biolandbaus – Einflüsse von Marginalisierung, Anerkennung und Markt. In: Mayer J, Alföldi T, Leiber F, Dubois D, Fried P, Heckendorn F, Hillmann E, Klocke P, Lüscher A, Riedel S, Stolze M,

Strasser F, van der Heijden M, und Willer H (Hrsg.) 2009. Werte – Wege – Wirkungen: Biolandbau im Spannungsfeld zwischen Ernährungssicherung, Markt und Klimawandel. Beiträge zur 10. Wissenschaftstagung Ökologischer Landbau, ETH Zürich, 11.-13. Februar 2009. http://orgprints.org/14377/1/Aeberhard_14377.pdf

AMS (2016) Memorandum to the national organic standards board. Hydroponic and Aquaponic Task Force Report. Agricultural Marketing Service. https://www.ams.usda.gov/sites/default/files/media/2016%20Hydroponic%20Task%20Force%20Report.PDF. Accessed 2 Feb 2019

AMS (2018) Memorandum to the National Organic standards board. Response to National Organic Standards Board Recommendations (Fall 2017 Meeting). https://www.ams.usda.gov/sites/default/files/media/NOPResponsetoNOSBFall2017.pdf. Accessed 2 Feb 2019

Ashley PJ (2007) Fish welfare: current issues in aquaculture. Appl Anim Behav Sci 104(3–4):S. 199–S. 235

Commission Regulation (2008) (EC) no 889/2008 of 5 September 2008 laying down detailed rules for the implementation of council regulation (EC) no 834/2007 on organic production and labelling of organic products with regard to organic production, labelling and control. Off J Eur Union 2008(1):84

Commission Regulation (2009) (EC) No 710/2009 of 5 August 2009 amending Regulation (EC) No 889/2008 laying down detailed rules for the implementation of Council Regulation (EC) No 834/2007, as regards laying down detailed rules on organic aquaculture animal and seaweed production. Off J Eur Union L 204:15–34

Endut A, Jusoh A, Ali N, Nik WBW (2011) Nutrient removal from aquaculture wastewater by vegetable production in aquaponics recirculation system. Desalin Water Treat 32:422–430. https://doi.org/10.5004/dwt.2011.2761

Friendly Aquaponics (2018) USDA organic certification of aquaponic systems. https://www.friendlyaquaponics.com/organic-certification/. Accessed 21 Dec 2017

Goddek S, Delaide B, Mankasingh U, Ragnarsdottir K, Jijakli H, Thorarinsdottir R (2015) Challenges of sustainable and commercial aquaponics. Sustainability 7:4199–4224. https://doi.org/10.3390/su7044199

Goddek S, Espinal CA, Delaide B, Jijakli MH, Schmautz Z, Wuertz S, Keesman KJ (2016) Navigating towards decoupled aquaponic systems: a system dynamics design approach. Water 8:303

Graber A, Junge R (2009) Aquaponic systems: nutrient recycling from fish wastewater by vegetable production. Desalination 246:147–156

IFOAM (2005) Definition of organic farming. https://www.ifoam.bio/en/organic-landmarks/definition-organic-agriculture. Accessed 2 Feb 2019

Junge R, König B, Villarroel M, Komives T, Haïssam Jijakl M (2017) Strategic points in aquaponics. Water 9:182

Kledal PR, Thorarinssdottir R (2018) Aquaponics: a new niche for sustainable aquaculture, 173-190. In: Hai FI, Visvanathan C, Boopathy R (eds) Sustainable aquaculture. Springer, Cham. https://doi.org/10.1007/978-3-319-73257-2

König B (2004) Adoption of sustainable production techniques: structural and social determinants of the individual decision making process. In: Proceedings of the 15th international symposium on horticultural economics and management, Berlin, Aug 29–Sept 3 2004

König B, Janker J, Reinhardt T, Villarroel M, Junge R (2018) Analysis of aquaponics as an emerging technological innovation system. J Clean Prod 180:232–243. https://doi.org/10.1016/j.jclepro.2018.01.037

Lockeretz W (2007) Organic farming. An international history. CAB International, Wallingford, p 275

Martins CIM, Gakhardo L, Noble C, Damsgard B, Spedicato MT, Zupa W, Beauchaud M, Kulczykowska E, Massabuau J-C, Carter T, Planellas SR, Kristiansen T (2012) Behavioural indicators of welfare in farmed fish. Fish Physiol Biochem 38

NOSB (2010) Formal recommendation by the national Organic Standards Board (NOSB) to the National Organic Program (NOP). https://www.ams.usda.gov/sites/default/files/media/NOP%20Final%20Rec%20Production%20Standards%20for%20Terrestrial%20Plants.pdf. Accessed 2 Feb 2019

NOSB (2016) Hydroponic and aquaponic subcommittee report 2016. Hydroponic & Aquaponic Subcommittee Report. Preserving a philosophy while embracing a changing world, In: NOSB 2016: National Organic Standards Board (NOSB) Hydroponic and Aquaponic Task Force Report. https://www.ams.usda.gov/sites/default/files/media/2016%20Hydroponic%20Task%20Force%20Report.PDF. Accessed 21 Dec 2017

NOSB (2017) National organic standards board crops subcommittee proposal hydroponics and container – growing recommendations. August 29, 2017. https://www.ams.usda.gov/sites/default/files/media/CSHydroponicsContainersNOPFall2017.pdf. Accessed 21 Dec 2017

Peterhans H (2015) Aquaponic nutrient model. Thesis project. MS Thesis. Wageningen University. p 37

Rakocy J (1999a) Aquaculture engineering – the status of aquaponics part 1. Aquaculture Magazine, pp 83–88

Rakocy J (1999b) Aquaculture engineering – the status of aquaponics part 2. Aquaculture Magazine, pp 64–70

Rakocy J (2002) Hydroponic lettuce production in a recirculating fish culture system. Virgin Islands

Rakocy J (2009) Ten guidelines for aquaponic systems. US Virgin Islands

Rakocy J, Bailey D, Shultz K, Cole W (2001) Evaluation of a commercial-scale aquaponic unit for the production of Tilapia and lettuce. University of the Virgin Islands, St. Croix

Rakocy J, Bailey D, Shultz C, Thoman E (2004) Tilapia and vegetable production in the UVI aquaponic system. University of the Virgin Islands

Rakocy J, Masser M, Losordo T (2006) Recirculating aquaculture tank production systems: aquaponics–integrating fish and plant culture. Southern Regional Aquaculture Center

Rakocy J, Bailey D, Shultz C, Danaher J (2009) Fish and vegetable production in a commercial aquaponic system: 25 years of research at the University of the Virgin Islands. Kingshill, Virgin Islands, USA

Shafahi M, Woolston D (2014) Aquaponics: a sustainable food production system. ASME International Mechanical Engineering Congress and Exposition. Volume 3: Biomedical and Biotechnology Engineering. Montreal, Quebec, Canada, November 14–20, 2014

Chapter 20
Regulatory Frameworks for Aquaponics in the European Union

Tilman Reinhardt, Kyra Hoevenaars, and Alyssa Joyce

Abstract This chapter provides an overview of the regulatory framework for aquaponics and the perspectives for European Union (EU) policy. Using Germany as an example, we analyze the specific regulations concerning construction and operation of aquaponic facilities and the commercialization of aquaponic products. We then show how aquaponics fits in with different EU policies and how it might contribute to EU sustainability goals. In the end, we provide some recommendations on how institutional conditions could be improved for aquaponics as an emerging technological innovation system.

Keywords Aquaponics · Law · Regulatory framework · Organic production · Animal welfare · Food labeling · Food safety policies

20.1 Introduction

Regulatory frameworks can have a decisive influence on the implementation of sustainable technologies. However, there are currently no specific regulations or policies for aquaponics in the European Union (EU) or most of its member states. One of the reasons perhaps is that it falls at the intersection of various larger fields (industrial aquaculture, wastewater recycling, hydroponics, urban aquaculture), wherein producers are subject to a variety of potentially disparate and conflicting regulations. The following chapter provides an overview of the regulatory

T. Reinhardt (✉)
GFA Consulting Group, Berlin, Germany
e-mail: tilman.reinhardt@gfa-group.de

K. Hoevenaars
AquaBioTech Group, Mosta, Malta
e-mail: kyh@aquabt.com

A. Joyce
Department of Marine Science, University of Gothenburg, Gothenburg, Sweden
e-mail: alyssa.joyce@gu.se

S. Goddek et al. (eds.), *Aquaponics Food Production Systems*,
https://doi.org/10.1007/978-3-030-15943-6_20

framework for aquaponics and gives some perspectives on how the development of aquaponics could be supported through EU policy. It builds on the work by Koenig et al. (2018) who have analyzed aquaponics through the theoretical framework for emerging technological innovation systems (see Bergek et al. 2008) and have shown how development pathways for this aquaponics might be influenced by institutional conditions.

The first section provides an overview of specific regulations that govern each step in development of aquaponic enterprises, i.e., construction, operation, and commercialization. It analyzes how this regulatory framework provides incentives or disincentives for individual entrepreneurs and market actors to invest in aquaponics. The second section analyzes how aquaponics fits in with different EU policies and how aquaponics can contribute to achieving EU sustainability goals. It then shows how the policies and strategies need to be redefined in order to provide better opportunities in this sector. In the third section, we draw conclusions from lessons learned in the first sections in order to provide policy recommendations.

Note: The first section summarizes the findings of a legal guideline on the feasibility of aquaponics projects in Germany. Detailed references to German regulations and case law have been left out for better readability. A German version with references to specific provisions and relevant case law is available on request from the authors. Parts of the second sub-chapter have been published in Eco cycles. Reference: Hoevenaars, K., Junge, R., Bardocz, T., and Leskovec, M. 2018. EU policies: New opportunities for aquaponics. Eco cycles 4(1): 10–15. DOI: 10.109040/ecocycles.v4il.87.

20.2 Legal Framework for Aquaponics

In this first section, our goal is to provide an overview of relevant regulations for the construction and operation of aquaponics facilities and the marketing of aquaponically produced products. We focus specifically on Germany, as it is impossible to extrapolate across the EU given that several important regulations, especially regarding zoning and construction, have not been harmonized across the EU. Although we focus on the German context, similar findings regarding planning law have also been reported in other countries (Joly et al. 2015).

20.2.1 Regulations on Construction

Aquaponic facilities must comply with various planning, building and water regulations, many of which do not fall under EU competence. In Germany, the general framework for planning and water law is harmonized at the national level, while building and local water-use regulations are determined at the state level, with urban and regional planning covered at the municipal level.

20.2.1.1 Planning Law

Planning law regulates the use of the soil and the area-related requirements for construction projects. There is a major distinction between projects in outlying and inner urban areas.

According to Section 35 of the German Building Code, outlying areas should be kept free of buildings and are reserved for certain uses, such as agriculture or renewable energy production. Whether or not aquaponics constitutes agriculture in that sense remains an unanswered question: while courts have ruled that soilless cultivation of vegetables such as hydroponics can be considered *agriculture*, the case is less clear for aquaculture in indoor facilities with no connection to the natural water cycle. The definition of agriculture in Section 201 of the Building Code only recognizes fisheries. Most courts therefore view recirculatory aquaculture systems as commercial rather than agricultural enterprises. Recently, however, the administrative court of Hamburg has ruled that a plant for fish and crustacean production can be considered agricultural, if the majority of the required feed could theoretically be produced on the agricultural land, belonging to the farm regardless of the type of fish produced, or whether feed is actually produced on the farm. This exception might not be viable however in cases, where agriculturally sourced feed is not used at all. In practice, aquaculture operations were often set up in connection with biogas plants. As farmers received an additional bonus on the feed-in-tariff for *cogenerating* plants (i.e. plants that also produce heat), there was an incentive to install heat-absorbing aquaculture next to the biogas plant.

Additional restrictions may apply in protected areas. The construction of aquaculture facilities is seen as problematic especially adjacent to natural water bodies. Exceptions for agriculture are only available for existing facilities. This has created a number of problems in traditional fishery areas, such as Mecklenburg, where many professional fishermen have an interest and necessary skills to operate ancillary businesses such as aquaculture or aquaponics (Paetsch 2013). Given that aquaponic systems do not rely on the natural water cycle, they could provide a creative possibility for new enterprise if their benefits were assessed and recognized by relevant authorities.

However, regardless of their size, aquaponics facilities do not require an environmental impact assessment, which is only a requirement for fish farms that discharge waste into surface waters.

20.2.1.2 Urban Areas

Many proponents envision aquaponics as a possibility for urban agriculture, given that commercial facilities can be built on rooftops or unused warehouses to allow for direct delivery of products to supermarkets in urban centers. Semicommercial systems can also be located in residential areas (*backyard aquaponics*). Under German planning law, the permitting of a facility depends on its classification and the area where it is located. Commercial aquaponics farms can be classified as either commercial or horticultural businesses. As such, they are generally not allowed in residential areas. In villages and mixed-use areas, both commercial and horticultural enterprises are allowed. In commercial and industrial areas, only commercial, but not horticultural, businesses are possible.

As aquaponic facilities have comparatively few noise and odor issues, they may be allowed on an exceptional basis even in areas where they are not currently admissible under planning laws. However, obtaining an exception creates additional administrative burdens and uncertainty, which could present an obstacle to scaling-up of the technology. Project-specific planning allows for cooperation with planning authorities but, in practice, is only relevant for large-scale projects due to the costs involved.

Backyard aquaponics plants might be allowed in all zones under the exception for *ancillary facilities for keeping of small animals.* However, ancillary facilities must be noncommercial and are interpreted differently by different district authorities. Some municipalities take a rather restrictive approach and only allow traditional forms of small animals such as dogs, chicken, pigeons, etc.

20.2.1.3 Building Law

The structural-technical requirements for buildings and administrative procedures for obtaining building permits are regulated by state-level building regulations, and while building codes follow, a so-called model building code, there can be substantial differences between states.

All construction products must comply with EU Regulation 305/2011, which requires a Declaration of Conformity to technical standards. For small wastewater treatment plants, the technical standard EN 12566 CEN applies. Rooftop systems may require special facilities for fire protection and affect minimum clearances. The static stability of the building must not be affected.

Although some of the components of an aquaponics facility, especially green houses or water tanks, do not need an individual building permit, the installation of a commercial food production system will usually require a building permit, particularly if the building has served a different purpose before. The procedure to obtain this permit might constitute a significant administrative and financial hurdle. Once obtained, however, it can also be seen as providing increased stability for outside investors given regulations will be viewed as having been met.

20.2.1.4 Water Law

Aquaponic systems do not necessarily depend on the use of surface water. Ideally water leaves an aquaponic system only via evapotranspiration or as water retained in the vegetables produced. We would argue that such facilities therefore should not require a permit under the water act or wastewater regulations. This could provide a major regulatory advantage compared, for instance, to traditional aquaculture or aquaculture, for which increasingly restrictive water and wastewater regulations constitute a significant obstacle to new enterprises. Savings on wastewater discharge fees present an incentive for implementing such systems.

However, it is not totally certain, if courts would follow this line of argument. Water might be considered wastewater, the moment it has been used for aquaculture. It would then be subject to the rules on wastewater disposal, which generally require disposal through centralized facilities. For example, the Higher Administrative Court of Berlin recently disallowed the use of a reed bed to clear domestic grey and blackwater, where reed was later used for energy use. The court explicitly stated that no right to multiple use of the water exists under German Water Law. In this case, a special permit for decentralized wastewater disposal would be required, and the entire aquaponics installation would have to comply with rules on wastewater disposal facilities.

The only real waste produced in aquaponic systems is filter sludge (which can be avoided, if an additional cycle for remineralization of this sludge is integrated or sludge is degraded on-site, e.g., through vermicomposting: if the filter sludge can be used on-site, no registration under the fertilizer law is required (see below)). If used outside the premises, regulations related to organic waste or sewage sludge disposal (more restrictive) apply. Filter sludge would be considered sewage sludge if the aquaponic system as a whole is considered a *wastewater treatment plant:* whether this applies in practice remains to be determined.

20.2.1.5 Conclusions on Construction

Fish farms present few problems in regard to noise and odor. One might therefore assume that aquaponic systems could be permitted with more ease than other animal production facilities. However, aquaponics does not fit well into the German legal framework.

As aquaponic production does not depend on the use of soil, installations may not be “agricultural enough” for outlying areas, i.e., agricultural lands. On the other hand, aquaponics might be “too agricultural” for urban areas, as urban agriculture is not considered a relevant category under German planning law. In particular, aquaponics may be generally inadmissible in commercial, industrial, and residential areas.

Commercial aquaponics facilities always require a building permit even if they are installed in pre-existing buildings which themselves do not require new building permits.

Aquaponics pioneers with very visible urban projects such as ECF or Urban Farmers seem to have coped well with the existing regulatory framework. However, planning law issues could present a relevant problem for scaling up the technology, in which case projects need to be developed in close consultation with the authorities in order to avoid future conflicts and to provide certainty for investors.

A major regulatory advantage of aquaponics may lie in the fact that little or no wastewater is produced, thus reducing the need for wastewater removal. Wastewater permits and fees have been reported to be significant obstacles for conventional fish farmers. As wastewater fees will probably be calculated according to pollution loading in the future, they may form an even stronger incentive to think about alternative types of wastewater disposal in the future (Schendel 2016). However as the water law generally does not provide for multiple uses, a legal clarification would be very important to create certainty for producers.

Apart from this, regulatory conditions in the German water sector do not particularly favor innovation. German water law strictly adheres to the paradigm of centralized sewage and generally does not allow for decentralized recycling of material flows and other forms of "creative ecology." Unlike in the waste sector, where the regulatory framework has given strong incentives to the private sector to consider waste as a resource, regulation of wastewater sector does not create incentives for the private sector to create and implement innovative recycling technologies.

20.2.2 Regulations on Aquaponic Production

Aquaponic production is subject to regulations for plant and for animal production at all stages of production and processing. Under the regulatory approach "from farm to fork," many relevant regulations have been harmonized at the European level (especially through the so-called EU hygiene package). However, a few exemptions exist for small producers selling directly to customers.

20.2.2.1 Hydroponic Production

Hydroponic production is subject to comparatively few regulations: growing media need an EU approval. Using fish wastes as fertilizer requires no authorization under German fertilizer laws if those fish wastes are derived from aquaculture.

The most significant restrictions concern the use of pesticides (note: in single-cycle aquaponic systems, pesticide use is inherently limited due to the pesticides' toxicity for fish; however, pesticide use is possible in multi-cycle (decoupled) aquaponic systems where water does not return from the plants to the fish components). The German Plant Protection Act generally imposes *integrated pest management* which means that preventive measures and the promotion of natural response mechanisms (e.g., suitable locations, substrates, varieties, seeds and fertilizers, as well as physical and biological control measures) have to be given priority before pesticide use. The use of invasive species for biological pest control is prohibited.

Pesticides may only be used by qualified personnel. Only pesticides that are approved under European regulation (EC) 1107/2009 may be used. Regulation (EC) 1107/2009 also includes rules on the use, storage, and disposal of pesticides.

Before harvesting, certain waiting periods must be observed. Residues in vegetables may not exceed certain *maximum residue levels* (MRLs). A free online database of MRLs is provided by the Directorate General for Health and Consumers (DG SANCO).

20.2.2.2 Aquaculture

In contrast to horticulture, aquaculture is carefully regulated through many different regulations. However a distinct *aquaculture law* neither exists at the national nor European level, and state-level fishery laws only regulate fishing in natural water bodies.

20.2.2.3 Cultivation of Non-native Species

The most commonly cultivated fish in aquaponics systems are tropical species such as tilapia or African catfish. However, the complex rules of Regulation (EC) 708/2007 concerning the use of alien species in aquaculture generally do not apply to closed recirculating aquaculture facilities (registered in a directory of recirculating aquaculture facilities). Some countries (e.g., Spain and Portugal, but not Germany or France), however, have decided to ban some types of exotic fish outright, which also affects the possibilities of cultivating them in closed facilities.

20.2.2.4 Regulation on Fish Diseases in Aquaculture

All aquaculture producers are subject to German regulation on fish diseases, which implements European Directive 2006/88/EC on animal health requirements for aquaculture animals and products thereof and on the prevention and control of certain diseases in aquatic animals (Ministry of Agriculture of Bavaria 2010). Under this regulation, aquaculture operations generally require permits by the local veterinary authorities. However, producers who only sell small quantities of fish directly to consumers or to local retailers only need to register certain information such as name and address, location and size of the operation, source of water supply, amount of fish held, and fish species.

Most importantly, the regulation on fish diseases imposes a duty on operators of fish farms to inform the local veterinary authorities in the event of a suspected disease outbreak. Veterinary authorities may then implement necessary control measures, which in some cases may involve destroying the entire stock should there be concern of disease spread.

Note: European animal health law, which was previously regulated relatively confusing in some 400 individual acts, is unified under Regulation (EU) 2016/429. However, the regulation only enters into force only on 21 April 2021. The content concerning fish diseases will not change (Art. 173 et seq. Regulation (EU) 2016/429).

20.2.2.5 Regulations on Animal Feed

The ability to source sustainable animal feeds is a key prerequisite for sustainable food production. In comparison to land animals, fish have a much better feed conversion rate; however, many higher trophic-level fish species require a certain portion of their feed to be derived from protein and fat from animal origin (e.g., fish meal). Feeding insects or insect larvae to fish is often seen as a possible way to increase the sustainability of aquaculture. Insects can be cultivated using waste organic nutrients, in some cases derived from animal wastes, including offal.

Animals for human consumption however must not be fed proteins of animal origin (with the exception of fish protein) according to Regulation (EC) 999/2001, which was implemented in reaction to BSE crisis in the 1990s. While it is sometimes argued that the ban of feed sources from animal protein should not apply to insects, which were not considered as potential feed sources in 2001, the use of insect feed is disallowed in practice by the German veterinary authorities.

Currently, some pet food is already produced using insect proteins (e.g., dog food from Brandenburg-based start-up Tenetrio). Given the increased interest in using insect protein for animal feed, several legislative changes have been made at the European level to allow for the feeding of insect protein based on adaptations to the existing regulatory frameworks (Smith and Pryor 2015). Since 2017, a so-called risk profile of the European Food Safety Authority (EFSA) is available (EFSA Journal 2015; 13 (10): 4257). Insects may be allowed as feed in aquaculture from 2018. However, some restrictions remain in place: in particular insects to be used as animal feed must not have fed on human or ruminant waste products. Insect production also poses some addition open regulatory questions, e.g., welfare issues regarding standard procedures for killing.

20.2.2.6 Regulation on Animal Welfare

Compared to other livestock, there are far few animal welfare regulations (Chap. 17) for handling and killing of fish. Although it is generally accepted that fish can feel pain, there is a lack of scientific evidence to justify restrictions for animal welfare (Studer/Kalkınç 2001). On the European level, there are only a few non-binding recommendations initiated in 2006 by the European Commission. According to Article 22 of these recommendations, a revised version based on new scientific evidence was anticipated by 2011, but to date EFSA has only published species-specific recommendations for certain types of fish, as well as special provisions on the transport of fish. Article 25 lit f – h. and Annex XIII of Regulation (EC) 889/2008 on organic production and labeling of organic products also contains species-specific rules on stocking density. As organic labeling is not available for fish from recirculatory aquaculture (see below), these rules are not relevant for aquaponics. Most private certification standards also do not consider animal welfare aspects (Stamer 2009).

Under Section 11 of the German law on animal welfare, keeping animals for commercial purposes generally requires a permit. To obtain this permit, one must demonstrate appropriate training or previous professional experience in animal husbandry and show that the production system provides adequate nutritional and housing facilities (Windstoßer 2011). Operations are considered commercial when expected sales exceed € 2000 per year.

According to Section 11 para 1 no. 8 TierSchG, no permit is needed for the commercial keeping of "farm animals." Whether fish can be considered *farm animals* in this sense is unclear. Exceptions to the animal welfare law are generally interpreted strictly: species are only considered *farm animals* if the necessary skills for keeping them can be acquired anytime, anywhere and there exists sufficient experience regarding the keeping of a species (Windstoßer 2011). This may not be the case for some types of tropical fish that differ fundamentally from native species (e.g., *Arapaima*, whose use in aquaculture is currently being explored at IGB Berlin).

The Administrative Court of Cologne has recently examined animal welfare aspects when ruling on the admissibility of a so-called fish spa, where Kangal fish were kept with the goal of using them to clean human feet. Operators of this fish spa were able to prove through veterinary reports that animal welfare was not being compromised and as such, a permit was granted.

20.2.2.7 Regulation on Fish Slaughter

Animal slaughter is regulated by European Regulation (EU) 1099/2009 as well as the German Decree of 20.12. 2012 (Federal Law Gazette I, p. 2982).

According to Recital 11 of Regulation (EU) 1099/2009, fish are physiologically different from terrestrial animals, and thus farmed fish can be slaughtered and killed with fewer animal welfare restrictions, in this case with specific implications for the inspection process. Furthermore, research on the stunning of fish is far less developed than for other farmed species. Separate standards should be established on the protection of fish at killing. Therefore, provisions applicable to fish should, at present, be limited to the key principle.

Under the general rule of Art. 3 Section 1 of Regulation (EU) 1099/2009, animals shall be spared any avoidable pain, distress, or suffering during their killing and related operations. However, there is no explicit obligation to stun fish before slaughter. That said, EU member states may maintain or adopt national rules aimed at ensuring more extensive protection of animals at the time of killing than those contained in this Regulation (Art. 26).

For instance, in Germany, slaughtering fish is subject to stricter conditions than those imposed by Regulation (EU) 1099/2009: as such, all types of fish except flatfish and eel have to be stunned before killing. Those doing the killing need a certificate of competence. Appropriate stunning methods may differ from species to species, with implications for producers: in Switzerland, aquaculture producer reportedly had to close his operation as the local veterinary authorities did not allow stunning using the ice water method, which he was employing. Defining appropriate killing methods for different fish species is the subject of an ongoing research project funded by BLE at the University of Veterinary Medicine Hannover and may lead to more restrictive regulations in the future.

Animal welfare aspects may also restrict certain forms of marketing and selling of fish. For example, the Higher Administrative Court of Bremen has forbidden placing farmed fish into ponds, from which were to be fished by recreational anglers as this was considered "unnecessarily harmful."

20.2.2.8 Regulation on Vocational Training of Fish Farmers

The German federal regulation on the vocational training of fish farmers does not mention aquaponics. Some private companies offer courses on aquaponics on the German market. However, it is not clear whether such courses are considered sufficient to obtain necessary permits (e.g., for pesticide use, commercial keeping of animals, slaughtering, etc.).

20.2.2.9 Hygiene Law

Hygiene law is harmonized on the European level through Regulations (EC) 852/2004, 853/2004, and 854/2004.

As a general rule, all food business operators, regardless of product, have to comply with EU hygiene law. As such, they must comply with the general standards of hygiene and management in Annex I and II of Regulation 852/2004, including basic requirements on production processes and personal hygiene, as well as appropriate waste treatment. They have to keep a register of the origin of animal feeds, as well as the use of pesticides and veterinary drugs. Measures to avoid risks must be documented in an appropriate manner.

According to Annex II Chapter IX no. 3 of Regulation (EC) 852/2004, food has to be protected against any contamination at all stages of production, processing, and distribution. Under Regulation (EC) No 178/2002, contamination may refer to any biological, chemical, or physical agent in a food or a condition of a foodstuff, which can cause adverse health effects. Food business operators must implement and maintain a HACCP (hazard analysis and critical control points) system, which has to be certified by accredited certification bodies. Details are agreed upon with local authorities.

EU hygiene legislation does not apply to the direct supply of small quantities of primary products to the final consumer or to local retail establishments. Small amounts are defined as household amounts for direct delivery to consumers or for local retail as for usual daily consumption. Primary production in the case of fish includes catching, slaughtering, bleeding, heading, gutting, removing fins, refrigeration, and wrapping. Activities such as flash-freezing, filleting, vacuum packing, or smoking will result in the fish no longer being considered primary local production.

Under German law certain restrictions on food hygiene exist however even for direct local food suppliers (Annex 1 LMHV).

Registration or authorization requirements in Regulation (EC) No 853/2004 depend on volume and type of processing. No registration or authorization is required for the supply of primary products in household quantities directly at the place of production, processing, or storage (including nearby markets). It is also possible to supply retail establishments (supermarkets, restaurants), consumers, or restaurants within a radius of 100 km. If primary products are delivered to end consumers, or to restaurants in larger quantities, the company must register and demonstrate their ability to meet food hygiene requirements. If more than one-third of the animal-derived products are sold to retail outlets outside the region (radius of 100 km), a public health license is also required.

20.2.2.10 Conclusions on Production

The legal requirements for aquaponic production are not higher than for the production of fish or vegetables. However, the large number of applicable laws reflects the complexity of aquaponics.

Compared to livestock farming, aquaculture may appear less regulated, especially in the area of animal welfare law. However, legal "gray areas" and the corresponding uncertainty are not always to the advantage of producers. Without established administrative practices, there is a considerable risk of conflicts (c.f. for the cited case in Switzerland, where an aquaculture operation was shut down, because the producer was not allowed to kill the fish in specific ways). In addition, the large number of applicable regulations can be burdensome, especially where European and national regulations coexist (e.g., on animal welfare or hygiene). As there is no harmonized law on aquaculture in Germany, producers need various permits from different authorities. Authorities often have little experience with nontraditional aquaculture, and as such, uncertain administrative requirements can be discouraging for entrepreneurs. Given the relative newness of commercial aquaponics, potential producers are strongly advised to contact local authorities at an early stage. In the case of larger, commercial plants, operators should probably contact veterinary and hygiene authorities prior to beginning construction.

The increasingly strict requirements of European hygiene laws can also constitute a significant burden, especially for small enterprises who wish to market directly to consumers or local restaurants (Schulz et al. 2013). However, whether exemptions for direct sellers are of practical help to aquaponic operators remains to be determined. The few existing aquaponics facilities in Germany currently demonstrate the necessity for producers to depend on a variety of sales channels and the need to create various forms of ancillary revenue (guided tours, secondary processing, cooking classes, etc.). Exemptions for direct selling therefore may become irrelevant if hygiene standards have to be met for other reasons.

20.2.3 Commercialization

Commercialization of aquaponic products is affected by different regulatory regimes. Hygiene regulations not only concern food production but also food retailing (see above). Business and tax laws, labeling regulations, or special certifications, such as the EU organic labeling regulations, may also be relevant.

20.2.3.1 Business and Tax Law

Agriculture is privileged in a number of ways under German business law: the marketing of self-produced agricultural products through farm stores, from the field or from a market stall, is not considered a business under German law and therefore requires no registration. This exception extends to the first stage of processing, i.e., cleaning and gutting, filleting and smoking in the case of fish, or in the case of fruits and vegetables, peeling, chopping, cooking, as well as the production of juices and wine (Chamber of Agriculture Rhineland-Palatinate 2015). Direct selling of agricultural products is also exempt from legal restrictions on opening times and the Sunday sales ban. Considering the low costs and low administrative requirements of a business registration, however, these privileges may not constitute a relevant advantage.

Tax privileges for agricultural production may be of greater practical importance. Regardless of size, aquaculture operations are subject to so-called average-rate VAT taxation, which offers a greatly reduced effective rate of value-added-tax that enables local producers to sell at more competitive prices relative to international imports.

In the income tax code, there are considerable privileges for "small farms" (i.e., turnover <500,000 €, farm size <20 hectares without special uses). If certain area limits are respected (600 m^2 of under-glass vegetables, 1600m^2 of ponds) income from aquaculture and vegetable growing is not taxed at all; even if the cultivated areas exceed these limits, effective tax rates are extremely low. As a result, the operation of aquaponics for small farmers might essentially be considered tax-free.

20.2.3.2 Regulations on Food Labeling

Rules on food labeling have largely been harmonized at the European level through European Regulation 1169/2011 on food packaging. Besides the formal rules, however, voluntary labeling standards play an even bigger role in the marketplace (see Sodano et al. 2008). In the case of the EU organic label, the voluntary standard is also regulated by law. In other cases, the rules of private certification schemes have to be followed.

20.2.3.3 General Labeling Rules

The general rules for selling packaged products are laid down in European Regulation 1169/2011 (e.g., the duty to include a list of ingredients, etc.). As a general rule, Art. 7 para. 1 Regulation (EU) 1169/2011 bans misleading claims on food packaging.

In addition to these general rules, Regulation (EU) 1379/2013 contains special rules on consumer information regarding aquaculture products. For example, under Art. 38 of Regulation (EU) 1379/2013, the member state or the third country, in which the aquaculture product has acquired more than half of its final weight, must be correctly named on the label.

20.2.3.4 EU Organic Regulation

Products from aquaponics are not eligible for labeling as organic under the current EU regulations for organic products. Article 4, Regulation (EC) 889/2008, explicitly forbids the use of hydroponics in organic farming. Recital 4 states that organic/biological crop production is based on the principle that plants obtain their food primarily from the soil ecosystem. For aquaculture products, Art. 25 g Regulation (EC) 710/2009 forbids the use of closed-circuit systems, and according to Recital 11 of Regulation (EC) 710/2009, this follows from the principle that organic production should be as close to nature as possible. These rules will not change in the new EU organic labeling regulation adopted in 2018, which will enter into force in 2021.

Laws preventing organic certification of hydroponic products are not shared by countries such as the USA and Australia, where hydroponic/aquaponic products can be certified organic.

20.2.3.5 Private Labels

Although there are currently no specific certification schemes for aquaponics, a number of certifications are available for aquaculture. Private certifications are usually "awarded" for a certain period of time, if private certain certification bodies can verify that productions comply with the criteria defined by their labeling standard. Private labeling schemes are usually purely contractual, with standards set by private institutions, which are subject only to general legal obligations (e.g., antitrust law). While the design of certification systems is increasingly in the purview of European legislation, including certification for aquaculture (cf. Commission report on options with regard to the allocation of EU eco-label for fishery and aquaculture products from 05.18.2016, COM see. (2016) 263 final), no concrete obligations for labeling schemes for aquaculture exist to date under European law.

Labeling schemes are essential, especially in business-to-business (B2B) relationships between producers/processors and retailers. Increasingly, however, certification also plays a role in marketing to final consumers (B2C). Beside qualitative aspects, B2C often certify compliance with certain environmental and social standards. Certification schemes can vary greatly in standards, verification modalities, and cost.

Among existing certification schemes, the Aquaculture Stewardship Council (ASC) may be most relevant for aquaponics producers. This certification has been awarded by the Aquaculture Stewardship Council (ASC) since 2010, as a complement to the better-known Marine Stewardship Council (MSC) program. The ASC is a formally independent, private, nonprofit organization initiated by the WWF and responsible for developing quality, ethical, and sustainability standards with scientific input. Private certification companies (e.g., in Germany TÜV NORD) are accredited by the ASC to confirm compliance with these standards. ASC standards currently exist for the following species: abalone, trout, shrimp, salmon, clams, catfish, and tilapia. Standards, audit manuals, and audit preparation checklists are freely available on the ASC website, and standardized procedures have also been made public. The ASC standard has different priorities, for instance, for the EU organic label (e.g., GMO-based animal feed is not prohibited under ASC).

Aquaculture certifications are also offered by the Cologne-based quality assurance and certification system GLOBALG.A.P. While GLOBALG.A.P. usually focuses on B2B certification for quality assurance in the food retail sector, a consumer label called GGN is also awarded for farmed fish and is often used, for instance, for fish that are not eligible for organic certification because they are wild caught or produced in aquaponics. Consumers can access information via www.my-fish.info.

20.2.3.6 Market Organizations

Regulation (EU) 1379/2013 on the common organization of markets in fishery and aquaculture products contains detailed rules on the establishment, recognition, objectives, and actions of professional organizations, i.e., producer organizations (Art. 6 ff.) and inter-branch organizations (Art. 11 ff.).

According to Article 8 (3) of Regulation (EU) 1379/2013, producer organizations in the field of aquaculture may, inter alia, make use of the following measures: promoting sustainable aquaculture, notably in terms of environmental protection, animal health, and animal welfare; collecting information on marketed products, including economic information on sales and production forecasts; collecting environmental information; planning the management of aquaculture activities of their members; and supporting programs for professionals to promote sustainable aquaculture products. According to Article 15 of Regulation 1379/2013, producers' organizations may also receive financial support from the EU's maritime and fisheries policies.

Measures taken by inter-branch organizations include, for example, the promotion of aquaculture products of the European Union in a nondiscriminatory manner by using, for example, certification and designations of origin, quality seals, geographical designations, traditional specialties guaranteed, and sustainability merits (Art. 13 lit a Regulation 1379/2013). Inter-branch organizations may adopt rules for the production and marketing of aquaculture products which are stricter than the provisions of European Union or national law (Art. 13 lit c Regulation 1379/2013).

Recognition as a producer organization can have far-reaching consequences; member states may, under certain conditions, make the rules agreed upon within a producer organization binding on all producers in the area (Art. 22 Regulation 1379/2013). Also, agreements, decisions, or concerted practices agreed upon within an inter-branch organization can be made binding on other operators (Article 23 of Regulation 1379/2013). According to Art. 41 VO 1379/2013, producer organizations are largely exempt from antitrust law.

So far, there are no professional associations in the field of aquaponics in the European Association of Fish Producers Organizations (EAPO). However, in 2018 an association for aquaponics has been founded in Germany (http://bundesverband-aquaponik.de/), and an EU Aquaponics Association (EUAA) based in Vienna has been founded at the initiative of several stakeholders assembled in the EU COST Action.

20.2.3.7 Conclusions on Commercialization

In terms of business and tax law, there are various privileges that could theoretically be exploited by aquaponics operators. Tax benefits could be especially interesting for outside investors. However, it remains to be seen whether certain conditions regarding the legal form of an enterprise and required investment volumes prevent operators from claiming these benefits. So far, the best-known urban aquaponics projects in Germany have not been profitable, so the issue of paying taxes has not arisen.

Under business and tax laws, the thresholds for privileges of small-scale installations and direct marketers are not congruent with the thresholds under hygiene law. A detailed review of operating and marketing concepts is therefore required in each individual case.

The EU organic label is currently out of the question for aquaponic products. However, there are increasingly private opportunities for certification.

20.3 Aquaponics and EU Policies

National policies can only be analyzed for each individual country. We therefore concentrate on relevant EU policies.

20.3.1 Overview of Relevant EU Policies

The Common Fisheries Policy (CFP) and the Common Agricultural Policy (CAP) apply to the aquaculture and hydroponics components of aquaponics, respectively (European Commission 2012, European Commission 2013). Policies on food safety, animal health and welfare, plant health, and the environment (waste and water) also apply.

20.3.1.1 Common Agricultural Policy

The Rural Development Policy, also referred to as the second pillar of CAP, focuses on increasing competitiveness and promoting innovation (Ragonnaud 2017). Each member state has at least one rural development program. Most countries have set goals to provide training, restructure and modernize existing farms, set up new farms, and reduce emissions. Measures against excessive use of inorganic fertilizers were introduced in the CAP as well as environmental policies and are regulated through the EU's Nitrates Directive (Directive 91/676/EEC 1991) and the Water Framework Directive (WFD).

20.3.1.2 Common Fisheries Policy

The CFP reform and strategic guidelines for the sustainable development of EU aquaculture were issued by the Commission to assist EU countries and stakeholders with challenges that the sector is facing. The emphasis is on facilitating implementation of the Water Framework Directive as it relates to aquaculture (European Commission 2013).

The CFP requires the development of a multiannual national strategic plan in each member state with strategies to promote and develop the aquaculture sector (European Commission 2016). Taking into account their different histories and cultivated species, each member state can support their existing aquaculture technologies but also develop new ones, such as aquaponics. This strategy should lead to an increase in production and reductions in dependence on imports. The main actions planned by member states are simplification of administrative procedures, coordinated spatial planning, enhancement of competitiveness, and promotion of research and development.

In the framework of the CFP, an Aquaculture Advisory Council (AAC) has been established. The main objective of the AAC is to provide advice and recommendations to European institutions and member states on issues related to sustainable development of the aquaculture sector (Sheil 2013).

A goal of both CFP and CAP is to increase competitiveness and sustainability of aquaculture and agriculture, respectively (Massot 2017). One of the objectives in the CFP is to exploit competitive advantage by obtaining high-quality, health, and environmental production standards.

20.3.1.3 EU Food Safety Policy

The goal of the food safety policy of the EU is to ensure safe and nutritious food from healthy animals and plants while supporting the food industry (European Commission 2014). The integrated food safety policy also includes animal welfare and plant health. In the strategy for animal welfare, there is an action on the welfare of farmed fish; however, there are no specific rules in place (European Commission 2012).

20.3.1.4 Environmental Policies

Environmental impacts of aquaculture are regulated under a range of EU legal requirements including water quality, biodiversity, and pollution. Environmental policies relevant for aquaponic operators are the strategy on the prevention and recycling of waste (European Commission 2011) and the seventh Environment Action Program (EAP) under the EU Environmental Policy (European Union 2014).

20.3.2 How Aquaponics Can Contribute to Goals in EU Policies and Strategies

Aquaponics can contribute to the development goals mentioned in these policies, with the main factors being reductions of water use and waste from fish production through nutrient recycling. Discharged water is converted into a resource and solid wastes can be upgraded as plant fertilizers. Because modern aquaponics is based on recirculating aquaculture systems, these operations are relatively independent of their location and can contribute to regional food production and value chains even in urban areas. Open aquaculture systems have constraints: water resource use, pollution, localized reduction in benthic biodiversity, significant dredging of water bodies, physical modification of land, changes in water flow, and introduction of alien species (European Union 2016). However, mitigation of most constraints is possible in aquaponic systems. Compared to hydroponic systems, aquaponics reduces the use of mineral, often unsustainably mined, fertilizers.

One of the priorities in the strategic guidelines on aquaculture is to improve access to space and water (European Commission 2013). Competition among different stakeholders and often strict environmental rules limit the further development of open aquaculture systems inside the EU. However, aquaponic systems can be located almost anywhere, including deserts and degraded soil and salty, sandy islands, since a closed-loop uses a minimum of water. Therefore, it can utilize space that is not suitable for other food production systems, like rooftops, abandoned industrial sites, and generally nonarable or contaminated land. Since aquaponics reuses 90–95% of the water, it relies much less on water availability compared to other systems like open aquaculture, hydroponics, and irrigation agriculture.

Just like in recirculating aquaculture systems, a benefit of larger commercial aquaponics systems is the possibility to obtain a high level of biosecurity, in which environmental conditions can be fully controlled ensuring a healthy environment for the fish (Badiola et al. 2012), thus minimizing the risk for diseases and parasite outbreaks (Yanong and Erlacher-Reid 2012). Because of the higher control on production, risk of losses is lower (Yanong and Erlacher-Reid 2012), which can provide aquaponic farmers with a competitive advantage over traditional farmers. On the other hand, using one nitrogen source to culture two products (Somerville et al. 2014) increases the investment risk as both fish and plant production must be maximized in order to make a profit. However, if this is done successfully, combined with the positive perceptions in Western markets of more environmentally friendly products, high revenues can be achieved (Somerville et al. 2014).

An objective in the strategy on the prevention and recycling of waste (European Commission 2011) includes introducing life-cycle thinking that considers a range of environmental impacts. It mentions that the prevention of waste is the priority, followed by reuse, recycling, recovery, and last disposal. Also, one of the priority areas in the seventh EAP targets transformation of the EU into a resource-efficient, low-carbon economy with a special focus on using waste as a resource (European Union 2014). Aquaponics systems minimize waste output (Goddek et al. 2015). The water in aquaponics systems is recirculated, thus wastewater is minimized. By using the fish process water for plant nutrition, organic waste from aquaculture is reused in the hydroponics component of the aquaponics system. The solid waste produced in an aquaponics system can be mineralized and returned to the system or utilized as compost for soil-based agriculture. Aquaponics also promotes local food production, thereby minimizing transport costs. Lastly, placing aquaponics farms in urban settings, it can provide ecological value in cities and play a role in adaptation to climate change.

20.3.3 Financial Support by the EU

The Seventh Framework Program (under the Multiannual Financial Framework of the European Commission) funded a couple of projects related to aquaponics. The EU Framework Program Horizon 2020 (challenge 2 "Food security, sustainable agriculture and forestry, marine and maritime and inland water research, and the bioeconomy" and challenge 5 "Climate action, environment, resource efficiency and raw materials") provides funding to several aquaponics initiatives including COST (European Cooperation in Science and Technology) Action FA1305 "The EU Aquaponics Hub: Realising Sustainable Integrated Fish and Vegetable Production for the EU" to promote innovation and capacity building by a network of researchers and commercial aquaponics companies.

Other possible funding opportunities for aquaponic development projects under the Multiannual Financial Framework of the European Commission include the European Innovation Partnership *Agricultural Productivity and Sustainability*

(EIP-AGRI), *a long term EU-Africa research and innovation partnership on food and nutrition security and sustainable agriculture* (LEAP-AGRI), the European Innovation Council pilot *Small and Medium-Sized Enterprises Instrument* (SME Instrument), the ERANET MED initiative *Partnership on Research and Innovation in the Mediterranean Area* (PRIMA), and the *European Maritime and Fisheries Fund* (EMFF). The EMFF can support research institutions and universities as well as companies; however, it requires different rates of co-funding.

20.3.4 Conclusions on the Overall EU Policy Landscape

None of the EU policies and guidelines so far explicitly mentions aquaponics. According to DG MARE, regulations on aquaponics need to be resolved within the individual member states (COST Action FA1305 2017), e.g., involving action resulting from the respective national strategic plans. Even though there is no explicit EU framework for aquaponics, it is an innovative agricultural system that can contribute to many priorities set through EU policies and strategies. EU support through financial measures is assisting further development of the technology. However, this mostly targets research projects, while the sector also needs assistance for commercial development through support of proof-of-concept projects. As a matter of fact, there are so far very few successful commercial aquaponics systems operating in the EU, so currently, there might not be a necessity for an aquaponics policy. However, ultimately recognizing and covering the technology in existing policies will be beneficial for the development of the sector.

20.4 Overall Conclusions and Policy Recommendations

Aquaponics is not only at the nexus of different technologies but also at the nexus of different regulatory and policy fields. While it may provide solutions to various sustainability goals, it seems to fall in the cracks between established legal and political categories. To add to the complexity, the development of aquaponics is affected by regulation from different levels of government. For example, facilitation of urban agriculture has to come from the national or even subnational level, as the EU has no competence in planning law. Major regulatory incentives for the implementation of aquaponic technology could probably be set in water law, which falls under national and EU competence. Implementation of aquaponics could gain significant traction, if aquaculture operations had the obligation or at least financial incentives to deal with wastewater themselves. However this would require a major change in the current regulatory approach.

In the theory on technological innovation systems (TIS), an "institutional alignment" in the formative phase of a TIS is seen as critical. Only if institutions are sufficiently aligned will markets form and provide space for entrepreneurial experimentation to determine commercially viable paths for implementing the technology (Bergek et al. 2008). For institutional alignment to take place, the proponents of the new technology need to be sufficiently organized to contribute to a process of "legitimation" of their technology (Koenig et al. 2018).

As a first step, we therefore recommend that proponents reach out, create, and strengthen links to various stakeholders and among themselves in the relevant professional communities in order to build a case for making aquaponics a legitimate activity. The newly founded EU Aquaponics Association could play a large role in this process. A further step could be the development of certifiable standards in cooperation with established certification systems. With the current lack of a coherent legal framework, such standards would give producers, consumers, administrators, and other relevant parties (e.g., outside investors or insurers, who doubt the safety of aquaponic products) a framework for understanding quality and risks. Such standards could be adapted flexibly to the practical demands of producers. Eventually formal regulations could build on such standards, as they do in other regulatory domains.

On the European level, stakeholders should push for greater recognition of the potential benefits of aquaponics in different policy areas. The EU must provide critical financial support, since commercial implementation of aquaponics is still in its infancy. The EU should also provide a forum for the exchange of best practices for regulatory issues such as construction and wastewater that fall under the competence of member states.

On the national level, stakeholders have to push for a coherent and accessible regulatory framework that is adapted to the realities of modern aquaculture and sets incentives for "creative ecology." Significant advances might even be realized on a subnational level, where political resistance may be overcome more easily. Further research should concentrate on the regulatory strategies of different countries to identify best practices.

References

Badiola M, Mendiola D, Bostock J (2012) Recirculating Aquaculture Systems (RAS) analysis: main issues on management and future challenges. Aquac Eng 51:26–35. https://doi.org/10.1016/j.aquaeng.2012.07.004

Bergek A, Jacobsson S, Carlsson B, Lindmark S, Rickne A (2008) Analyzing the functional dynamics of technological innovation systems: a scheme of analysis. Res Policy 3(37):407–429

COST Action FA1305 (2017) Report on a Workshop with representatives from DG MARE, DG AGRI and DG RTD. 5 p. Available online: http://euaquaponicshub.com/hub/wp-content/uploads/2017/07/Workshop-notes.pdf

Directive 91/676/EEC (1991) Council Directive of 12 December 1991 concerning the protection of waters against pollution caused by nitrates from agricultural sources. Off J Eur Communities 8 p. Available online: http://eur-lex.europa.eu/legal-content/EN/TXT/?uri=celex:31991L0676

European Commission (2011) Report from the commission to the European parliament, the council, the European economic and social committee and the committee of the regions on the Thematic Strategy on the Prevention and Recycling of Waste. 11 p

European Commission (2012) Communication from the commission to the European Parliament, the council, the European Economic and social committee and the committee on the European Union Strategy for the Protection and Welfare of Animals 2012–2015. 12 p. Available online: https://ec.europa.eu/food/sites/food/files/animals/docs/aw_eu_strategy_19012012_en.pdf

European Commission (2013) Communication from the commission to the European Parliament, the council, the European Economic and social committee and the committee of the regions; Strategic guidelines for the sustainable development of EU aquaculture. 12 p. Available online: http://eur-lex.europa.eu/legal-content/EN/TXT/PDF/?uri=CELEX:52013DC0229&from=EN

European Commission (2014) The European Union explained -food safety. 16 p. Available online: https://europa.eu/european-union/topics/food-safety_en

European Commission (2016) Summary of the 27 Multiannual national aquaculture plans. 12 p. Available online: https://ec.europa.eu/fisheries/sites/fisheries/files/docs/body/27-multiannual-national-aquaculture-plans-summary_en.pdf

European Union (2012) The common agricultural policy-a story to be continued. Publicaetions Office of the European Union. 23 p. Available online: http://ec.europa.eu/agriculture/50-years-of-cap/files/history/history_book_lr_en.pdf

European Union (2014) Living well, within the limits of our planet, 7th EAP-The new general Union Environment Action Programme to 2020. 4 p. Available online: http://ec.europa.eu/environment/pubs/pdf/factsheets/7eap/en.pdf

European Union (2016) Commission staff working document on the application of the Water Framework Directive (WFD) and the Marine Strategy Framework Directive (MSFD) in relation to aquaculture. 36 p. Available online: http://ec.europa.eu/environment/marine/pdf/SWD_2016_178.pdf

Goddek S, Delaide B, Mankasingh U, Ragnarsdottir KV, Jijakli H, Thorarinsdottir R (2015) Challenges of sustainable and commercial aquaponics. Sustainability 7:4199–4224. https://doi.org/10.3390/su7044199

Joly A, Junge R, Bardocz T (2015) Aquaponics business in Europe: some legal obstacles and solutions. Ecocycles 1(2):3–5. https://doi.org/10.19040/ecocycles.v1i2.30

König B, Janker J, Reinhardt T, Villarroel M, Junge R (2018) Analysis of aquaponics as an emerging technological innovation system. J Clean Prod 180:232–243

Massot A (2017) The Common Agricultural Policy (CAP) and the treaty. Fact sheets on the European Union. European Parliament. Available online: http://www.europarl.europa.eu/atyourservice/en/displayFtu.html?ftuId=FTU_3.2.1.html

Paetsch U (2013) Jahresfischereitag und Jahreshauptversammlung des Landesverbandes der Binnenfischer MV e.V. – Güstrow, 25. Februar 2013: Bericht des Präsidiums für das Jahr 2012, Fischerei & Fischmarkt in MV 2/2013, 6–9

Ragonnaud G (2017) Second pillar of the CAP: rural development policy. Fact sheets on the European Union. European parliament. 4 p. Available online: http://www.europarl.europa.eu/atyourservice/en/displayFtu.html?ftuId=FTU_3.2.6.html

Schulz K, Weith T, Bokelmann W, Petzke N (2013) Urbane Landwirtschaft und "Green Production" als Teil eines nachhaltigen Landmanagements. Discussion paper 6. Müncheberg

Schendel F (2016) Abwasserabgabe – wann kommt eine Reform? Natur und Recht:166–171

Sheil S (2013) Strategic guidelines for EU aquaculture. Library briefing. 6 p. Available online: http://www.europarl.europa.eu/eplibrary/Strategic-guidelines-for-aquaculture-in-the-EU.pdf. Accessed on 05 Oct 2017

Somerville C, Cohen M, Puntanella E, Stankus A, Lovatelli A (2014) Small-scale aquaponic food production. FAO Fisheries and Aquaculture Technical paper 589. 288 p. Available online: http://www.fao.org/3/a-i4021e.pdf

Smith R, Pryor R (2015) Deliverable 5.1 – Mapping Exercise Report with regard to current Legislation & Regulation: Europe and Africa & China, 2015

Sodano V, Hingley M, Lindgreen A (2008) The usefulness of social capital in assessing the welfare effects of private and third-party certification standards: trust and networks. Br Food J 110 (4/5):493–513

Stamer A (2009) Betäubungs- & Schlachtmethoden für Speisefische, FiBL, Available online: http://orgprints.org/16511/1/stamer-2009-literaturstudie_fischschlachtung-FiBL_Bericht.pdf
Windstoßer C (2011) Rechtliche Voraussetzungen für die Errichtung von Kreislaufanlagen, Präsentation im Rahmen des Seminars "Fischproduktion in Kreislaufanlagen – Prinzipien, Wirtschaftlichkeit, Zukunft" FFS Baden-Württemberg Available online: http://www.landwirtschaft-bw.info/pb/site/lel/get/documents/MLR.LEL/PB5Documents/lazbw_ffs/Windsto%C3%9Fer_Rechtliche%20Voraussetzung.pdf
Yanong RPE, Erlacher-Reid C (2012) Biosecurity in aquaculture, Part 1: An overview. SRAC Publication No. 4707

Chapter 21
Aquaponics in the Built Environment

Gundula Proksch, Alex Ianchenko, and Benz Kotzen

Abstract Aquaponics' potential to transform urban food production has been documented in a rapid increase of academic research and public interest in the field. To translate this publicity into real-world impact, the creation of commercial farms and their relationship to the urban environment have to be further examined. This research has to bridge the gap between existing literature on growing system performance and urban metabolic flows by considering the built form of aquaponic farms. To assess the potential for urban integration of aquaponics, existing case studies are classified by the typology of their building enclosure, with the two main categories being greenhouses and indoor environments. This classification allows for some assumptions about the farms' performance in their context, but a more in-depth life cycle assessment (LCA) is necessary to evaluate different configurations. The LCA approach is presented as a way to inventory design criteria and respective strategies which can influence the environmental impact of aquaponic systems in the context of urban built environments.

Keywords Aquaponics classification · Urban aquaponics · Enclosure typologies · Greenhouses · Indoor growing · Controlled environment agriculture · Life cycle assessment

G. Proksch (✉) · A. Ianchenko
Department of Architecture, College of Built Environments, University of Washington, Seattle, WA, USA
e-mail: prokschg@uw.edu

B. Kotzen
School of Design, University of Greenwich, London, UK
e-mail: b.kotzen@greenwich.ac.uk

S. Goddek et al. (eds.), *Aquaponics Food Production Systems*,
https://doi.org/10.1007/978-3-030-15943-6_21

21.1 Introduction

Aquaponics has been recognized as one of "ten technologies which could change our lives" by merit of its potential to revolutionize how we feed growing urban populations (Van Woensel et al. 2015). This soilless recirculating growing system has stimulated increasing academic research over the last few years and inspired interest in members of the public as documented by a high ratio of Google to Google Scholar search results in 2016 (Junge et al. 2017). For a long time, aquaponics has been primarily practiced as a backyard hobby. It is now increasingly used commercially due to strong consumer interest in organic, sustainable farming methods. A survey conducted by the CITYFOOD team at the University of Washington in July 2018 shows that the number of commercial aquaponic operations has rapidly increased over the last 6 years. This focused search for aquaponic operations identified 142 active for-profit aquaponic operations in North America. Based on online information, 94% of the farms have started their commercial-scale operation since 2012; only nine commercial aquaponic farms have been in operation for more than 6 years (Fig. 21.1).

Most of the surveyed aquaponic operations are located in rural areas and are often connected to existing farms to take advantage of low land prices, available

Fig. 21.1 Existing aquaponic practitioners in North America, 142 commercial companies (red) and 17 research centers (blue), (CITYFOOD, July 2018)

Fig. 21.2 Aquaponics across Europe: 50 research centers (blue) and 45 commercial companies (red). (EU Aquaponics Hub 2017)

infrastructure, and conducive building codes for agricultural structures. Regardless, a growing number of aquaponic operations are also located in cities. Due to their relatively small physical footprint and high productivity, aquaponic operations are well suited to practice in urban environments (Junge et al. 2017). Surveys undertaken under the auspices of the European Union (EU) Aquaponics Hub in 2017 identified 50 research centers and 45 commercial companies operating in the European Union (Fig. 21.2). These companies range in size from small to medium sized.

21.1.1 Aquaponics in Urban Environments

Space is a valuable commodity in cities. Urban farms have to be resourceful to find available sites such as vacant lots, existing rooftops, and underutilized warehouses that are affordable for an agricultural business (de Graaf 2012; De La Salle and Holland 2010). Urban aquaponic farms need to balance higher production costs with competitive marketing and distribution advantages that urban locations offer. The largest benefit for locating aquaponic operations in cities is a growing consumer market with an interest in fresh, high-quality and locally grown produce. When complying with local regulations for organic produce, urban farms can achieve premium prices for their aquaponically grown leafy greens, herbs, and tomatoes (Quagrainie et al. 2018). Unlike hydroponics, aquaponics also has the capacity to produce fish, further enhancing economic viability in an urban setting which often has diverse dietary needs (König et al. 2016). Urban aquaponic farms can also save some operational costs by reducing transportation distance to the consumer and reducing the need for crop storage (dos Santos 2016).

Urban environmental conditions can also be advantageous for aquaponic farms. Average temperatures in cities are higher than in rural surroundings (Stewart and Oke 2010). In colder regions particularly, farms can benefit from a warmer urban climate, which can help reduce heating demand and operational costs (Proksch 2017). Aquaponic farms that are integrated with the building systems of a host building can further utilize urban resources such as waste heat and CO_2 in exhaust air to benefit the growth of plants as an alternative to conventional CO_2 fertilization. Urban farms can also help mitigate the negative aspects of the urban heat island effect during the summer months. The additional vegetation, even if grown in greenhouses, helps to reduce the ambient temperature through increased evapotranspiration (Pearson et al. 2010). In aquaponics, the use of recirculating water infrastructure reduces overall water consumption for the production of both fish and lettuce and can, therefore, have a positive effect on the urban water cycle. Aquaponically-grown produce strives to close the nutrient cycle, thereby avoiding the production of agricultural run-off. Through smart resource management within major environmental systems, aquaponics helps to reduce excessive water consumption and eutrophication usually created by industrial agriculture.

21.1.2 Aquaponics as Controlled Environment Agriculture (CEA)

Traditional agricultural techniques to extend the natural crop-growing season range from minimal environmental modifications, such as temporary hoop houses used on soil-based fields, to full environmental control in permanent facilities that allow for year-round production regardless of the local climate (Controlled Environment Agriculture 1973). The latter strategy is also known as controlled environment agriculture (CEA) and includes both greenhouses and indoor growing facilities. In addition to controlling the indoor climate, CEA also significantly reduces the risk of crop loss to natural calamities and the need for herbicides and pesticides (Benke and Tomkins 2017). Most aquaponic operations are conceived as CEA since they combine two complex growing systems (aquaculture and hydroponics), which both require controlled growing conditions to guarantee optimal productivity. Additionally, CEA enables year-round production to amortize high investment in aquaponic infrastructure and achieve premium crop prices at the market outside of the natural growing season. The performance of aquaponic farm enclosures is highly dependent on local climate and seasonal swings (Graamans et al. 2018).

As aquaponics is a relatively young discipline, most of the existing research is focused at the system level – for example, studies evaluating the technical integration of aquaculture with hydroponics in different configurations (Fang et al. 2017; Lastiri et al. 2018; Monsees et al. 2017). Whilst individual aquaponic system components and their interactions can still be further optimized for productivity, their performance within a controlled environment envelope has not been comprehensively addressed. Recent research in CEA has begun to assess hydroponic system performance in tandem with built environment performance, although there is only one study to date that models aquaponic system performance in a controlled envelope (Benis et al. 2017a; Körner et al. 2017; Molin and Martin 2018a; Sanjuan-Delmás et al. 2018).

21.1.3 Aquaponics Research Collaborations

The current expansion in interest in aquaponics led to the creation of several interdisciplinary aquaponics related research collaborations funded by the European Union (EU). The COST FA1305 project, which created the EU Aquaponics Hub (2014–2018) brought together aquaponics research and commercial producers to better understand the state of the art in aquaponics and to generate coordinated research and education efforts across the EU and around the world. Innovative Aquaponics for Professional Application (INAPRO) (2014–2017), a consortium of 17 international partners, aimed to advance current approaches to rural and urban aquaponics through the development of models and construction of prototypical greenhouses. The project CITYFOOD (2018–2021) within the Sustainable Urban Growth Initiative (SUGI), co-funded by the EU, Belmont Forum, and respective science foundations, investigates the integration of aquaponics in the urban context and its potential impact on global challenges of the food-water-energy nexus.

21.2 Classification of Controlled Environment Aquaponics

The term aquaponics is used to describe a wide range of different systems and operations, greatly varying in size, technology level, enclosure type, main purpose, and geographic context (Junge et al. 2017). The first version of the classification criteria for aquaponic farms included stakeholder objectives, tank volume, and parameters describing aquaculture and hydroponic system components (Maucieri

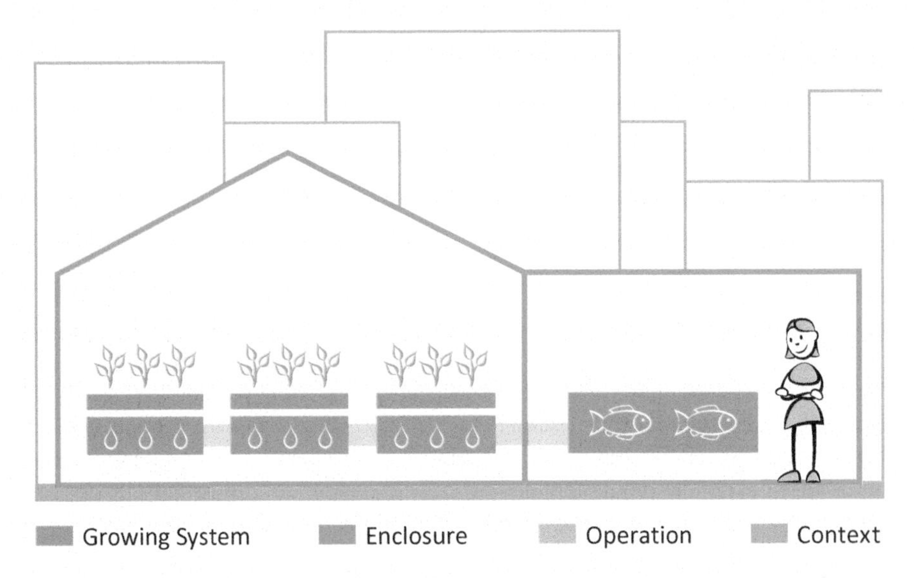

Fig. 21.3 Classification criteria for identifying aquaponic farm types

et al. 2018). Additional work was undertaken by a large group of researchers to further define aquaponics and to present a nomenclature based on international consensus (Palm et al. 2018). This led to a comprehensive discussion on system types and scales and most importantly a definition of aquaponics which is: "*the majority (> 50%) of nutrients sustaining the optimal plant growth must derive from waste originating from feeding the aquatic organisms.*" However, both definitions focus on the growing systems and do not consider other essential aspects of a functioning commercial aquaponics farm. As aquaponic operations become part of local economies, classification criteria identified by interdisciplinary research in fields like architecture, economics, and sociology will also become essential.

This classification proposal focuses on the emerging field of commercial aquaponic operations through the lens of the built environment. The key characteristics that describe an aquaponic operation fall into four different categories: growing system, enclosure, operation, and context (Fig. 21.3). These categories for classification criteria impact one other across scales, where growing system configurations can affect the contextual performance of the farm as a business, or local market demands can determine the type of crop grown in the aquaponic system. Some farm classification criteria are relevant on all scales, such as "size," measured in tank volume, growing area, number of employees, and annual revenue (Table 21.1).

- *Growing system* classification criteria describe the configuration of the interconnected aquaculture and hydroponic system. This includes specifications for the physical components that enable water and nutrient recirculation (such as water tanks, filters, pumps, and piping), living organisms that transform available nutrients at different stages (including fish, plant, and microorganism species) and

Table 21.1 Possible classification criteria for aquaponic farm types

Growing system	Enclosure	Operation	Context
Aquaculture system type	Enclosure typology	Purpose	Geographical location
Fish species	Structural system	Stakeholders	Physical context
Water temperature	Envelope assembly cover material	Business model	Environmental impact
Filtration system	Heating/cooling systems	Labor distribution	Socioeconomic context
Feed type	Light source	Funding type	Social impact
Hydroponic system type	Ventilation system	Marketing scheme	
Crop species	Host building integration	Distribution model	
Water distribution system			

values describing the physical performance of the system, such as temperature, pH levels, oxygen/carbon content, and electrical conductivity (Alsanius et al. 2017).

- *Enclosure* classification criteria define characteristics of the buildings that house the growing systems, at the next scale. Most aquaponic farms use CEA enclosures that vary by identifying typology (such as a greenhouse or warehouse), structural system, heating and cooling systems, lighting, ventilation, and humidity control systems (Benke and Tomkins 2017).
- *Operations* classification criteria describe how each aquaponic farm operates as a business and farm, which includes human expertise and labor input necessary for growing and selling produce. Criteria in this section include funding type, business structure and management, labor requirements and division, marketing scheme, produce distribution model, and overall purpose of the aquaponics facility.
- *Context* classification criteria, at the largest scale, describe the geographic location, physical context, urban integration, and overall social impact of aquaponic farms. Context criteria describe how an aquaponic farm is part of the urban food chain and built environment, capable of influencing economic growth, social involvement and large-scale environmental impacts on a city-wide scale (dos Santos 2016).

21.3 Enclosure Typologies and Case Studies of Commercial Farms

This further investigation focuses on defining aquaponic classification criteria at the enclosure level to complement existing system-level definitions. The enclosure types discussed here work with different construction systems, levels of technological control, passive climate control strategies, and energy sources to achieve an appropriate indoor climate. The best application of each enclosure typology depends primarily on the size of operation, geographic location, local climate, targeted fish and crop species, required parameters of the systems it houses, and the budget. This study identifies five different enclosure typologies and defines the characteristics of indoor spaces that house aquaculture infrastructure.

21.3.1 Greenhouse Typologies

This classification includes four categories of greenhouses – medium-tech greenhouses, passive solar greenhouses, high-tech greenhouses, and rooftop greenhouses – that are applicable to commercial-level aquaponic operations (Table 21.2). Existing greenhouses may not exactly fit a single typology, but fall within a spectrum from medium-tech to high-tech by selectively incorporating active and passive environmental control techniques.

Medium-Tech Greenhouses Greenhouses with intermediate levels of technology to control the indoor climate include freestanding or gutter-connected Quonset (Nissen hut type), hoop house (polytunnel) and even-span greenhouses. They are usually covered with double polyethylene film (PE) or rigid plastic panels, such as acrylic panels (PMMA) and polycarbonate panels (PC). These greenhouses are less expensive to install, though film cladding needs to be replaced frequently due to rapid deterioration caused by constant exposure to UV radiation (Proksch 2017). These greenhouses protect crops from extreme weather events and to some extent pathogens, but they offer only a limited level of active climate controls. Instead, they rely on solar radiation, simple shading systems, and natural ventilation. With their limited ability to modify growing conditions within a certain range, medium-tech greenhouses are rarely used for housing aquaponic farms in cold climates. This is because the high initial investment into the hydroponic and aquaculture components requires a stable environment and reliable year-round production to be commercially viable.

Aquaponic operations in warmer climates have successfully demonstrated the use of medium-tech greenhouses that employ evaporative cooling and simple heating systems. For example, Sustainable Harvesters in Hockley, Texas, USA uses a simple Quonset greenhouse (12,000 sf/1110 m^2) for year-round lettuce production without relying on extensive supplemental heating or lighting. Ouroboros Farms in Half Moon Bay, California, USA uses an existing greenhouse (20,000 sf/1860 m^2) to produce lettuce, leafy greens, and herbs (Fig. 21.4). Due to the mild climate, the farm uses primarily static shading and little supplemental heating and cooling. Both farms, as many smaller medium-tech operations, place their fish tanks in the same greenhouse space as the hydroponic crop growing system. The farms grow fish species that tolerate a wide temperature range (tilapia) and shade aquaculture tanks to prevent overheating and algae growth.

Table 21.2 Comparison of case studies by enclosure typologies

CEA type	Case studies	Construction system	Controls	Growing season[a] and latitude	Hardiness zone[b]
Medium-tech greenhouses	Ouroboros Farms, Half Moon Bay, CA, USA (20,000 sf/1860 m²)	Existing, gutter-connected GH with two even spans, clad with single-pane glass, fish tanks in GH	Static shading, shading curtains	319 days/ 10.6 months	10a
					30 to 35 °F
				37.5° N	−1.1 to 1 .7 °C
	Sustainable Harvesters, Hockley, TX, USA (12,000 sf/1110 m²)	Quonset frame, multi-tunnel (3) GH, clad with PE- film and rigid plastic panels, fish tanks in GH	Evaporative cooling, forced ventilation	272 days/ nine months	8b
					15 to 20 °F
				30.0° N	−9.4 to 6 .7 °C
Passive solar greenhouses	Aquaponic solar greenhouse, Neuenburg am Rhein, Germany (2000 sf/180 m²)	(Chinese) solar greenhouse, with adobe wall as additional thermal mass, clad with ETFE film, fish tanks in GH	Custom-built photovoltaic modules for shading and energy production	202 months/ 6.6 months	8a
					10–15 °F
				47.8° N	−12.2 to −9 .4°C
	Eco-ark greenhouse at Finn & Roots, Bakersfield, VT, USA (6000sf/ 560 m²)	Solar greenhouse, earth sheltered, steep angle of south facing roof (ca. 60°), thick insulation, special solar collecting glazing, fish tanks in northern, subterranean side	Wood-fuelled radiant heat, energy curtain, ventilation with stack-effect, supplemental LED lighting	108 days/ 3.6 months	4a
					−30 to −25 °F
				44.8° N	−34.4 to −31.7 °C
High-tech greenhouses	Superior Fresh Farms, Hixton, WI, USA (123,000 sf/11,430 m²)	Venlo-style, gutter-connected, (20 × 41 bays), clad with glass, fish tanks in separate building	Computer-controlled CEA environment, supplemental LED lighting,	122 days/ 4.1 months	4b
					−25 to −20 °F
				44.4° N	−31.7 to −28.9 °C
	Blue Smart Farms, Cobbitty, NSW, Australia (53,800 sf/5000 m²)	Venlo-style, gutter-connected, (14 × 18 bays), clad with glass, two-story construction, fish tanks on the lower level	Computer-controlled CEA environment biological pest control	300 days/ 10 months	9b
					25 to 30 °F
				34.0°S	−3.9 to −1.1 °C

(continued)

Table 21.2 (continued)

CEA type	Case studies	Construction system	Controls	Growing season[a] and latitude	Hardiness zone[b]
Rooftop greenhouses	Ecco-jäger Aquaponik Dachfarm, Bad Ragaz, Switzerland (12,900 sf/1200 m^2)	Venlo-style, gutter-connected, (7 × 13 bays), clad with glass, fish tanks on the lower level	CEA environment, supplemental LED lighting, use of exhaust heat from cooling facility	199 days/ 6.6 months	7b
					5 to 10 °F
				47.0° N	−15.0 to −12.2 °C
	BIGH's Ferme abattoir, Brussels, Belgium (21,600 sf/2000 m^2)	Venlo-style, gutter-connected, (15 × 10 bays), clad with glass, fish tanks on the lower level	CEA environment, supplemental LED lighting	224 days/ 7.3 months	8b
					15 to 20 °F
				50.8° N	−9.4 to −6 .7 °C
Indoor growing spaces	Urban Organics, Schmidt's Brewery, St. Paul, MN, USA (87,000 sf/8080 m^2)	Steel-frame warehouse, highly insulated, stacked growing, fish tanks in separate space	Fluorescent UV lighting, computer-controlled CEA environment	140 days/ 4.7 months	4b
					−25 to −20 °F
				45.0° N	−31.7 to 28.9 °C
	Nutraponics, Sherwood Park, AB, Canada (10,800 sf/1000 m^2)	Steel-frame warehouse, highly insulated, stacked growing, fish tanks in separate space	LED lighting, computer-controlled CEA environment	121 days/ 4 months	4a
					−30 to −25 °F
				53.5° N	−34.4 to −31.7 °C

[a]Frost-free growing season, National Gardening Association, Tools and Apps, https://garden.org/apps/calendar/
[b]Based on the USDA Hardiness Zone Map, which identifies the average annual minimum winter temperature (1976–2005), divided into 10° F zones. Plant Maps, https://www.plantmaps.com/index.php

Passive Solar Greenhouses This greenhouse type is designed to be solely heated by solar energy. Substantial thermal mass elements, such as a solid north-facing wall, store solar energy in form of heat that is then re-radiated during colder periods at night. This approach buffers air temperature swings and can reduce or eliminate the need for fossil fuels. Solar greenhouses have a transparent south-facing side and an opaque, massive, highly insulated north-facing side. The integration of large volumes of water in form of fish tanks is an asset for the thermal performance of this greenhouse type. Furthermore, the tanks can be located in areas of the greenhouse that are less suited for plant cultivation or partly submerged into the ground for added thermal stability.

Fig. 21.4 Ouroboros Farms (Half Moon Bay, California, USA)

The Aquaponic solar greenhouse (2000 sf/180 m^2), developed and tested by Franz Schreier, has proven as a suitable environment for housing a small aquaponic system in southern Germany. The greenhouse collects solar energy through its south-facing arched roof and wall clad with ethylene tetrafluoroethylene (ETFE) film. Heat is stored in partially submerged fish tanks, floor, and adobe-clad northern wall to be dissipated at night. The greenhouse's custom-built photovoltaic (PV) panels transform solar radiation into power. Located in the colder climate of Vermont, USA, the Eco-Ark Greenhouse at the Finn & Roots farm (6000 sf/560 m^2) houses an aquaponic system that works with a similar passive solar approach. The greenhouse has a steep (approx. 60°) south-facing transparent roof with special solar-collecting glazing (Fig. 21.5). Its highly insulated, opaque northern side is submerged into a hillside and houses the fish tanks. In addition to these passive controls, the Eco-Ark has a radiant floor heating that supplements heating during the coldest seasons.

High-Tech Greenhouses Venlo-style, high-tech greenhouses that feature a high level of technology to control the indoor climate are the standard for commercial-scale hydroponic CEA. High-tech greenhouses are characterized by computerized controls and automated infrastructure, such as automatic thermal curtains, automatic lighting arrays, and forced-air ventilation systems. These technologies enable a high level of environmental control, though they come at the cost of high energy consumption.

Fig. 21.5 Eco-Ark Greenhouse at Finn & Roots Farm (Bakersfield, Vermont, USA)

Some large-scale commercial aquaponic farms use this greenhouse typology for their plant production, such as Superior Fresh farms, located in Hixton, Wisconsin, USA (123,000 sf/11,430 m^2), with the aquaculture systems housed in a separate opaque enclosure. Automated supplemental LED lighting and heating enables Superior Fresh farms to cultivate leafy greens year-round despite lack of daylight in the winter, where the natural, frost-free growing season lasts only 4 months. Automated systems for internal climate control allow high-tech greenhouses to be operated anywhere in the world – Blue Smart Farms greenhouse uses an array of sensors to optimize shading during hot Australian summers.

Thanet Earth, the largest greenhouse complex in the UK, is located in the southeast of England. Its five greenhouses cover more than 17 acres (7 hectares) each, growing tomatoes, peppers, and cucumbers using hydroponics (Fig. 21.6). This enterprise is powered by a combined heat and power system (CHP) that provides power, heat, and CO_2 for the greenhouses. The CHP system operates very efficiently and channels excess energy to the local district by feeding it into the local power supply grid. In addition, computer-controlled technologies such as energy curtains, high-intensity discharge supplemental lighting, and ventilation regulate the indoor growing conditions.

Rooftop Greenhouses This most recent type includes greenhouses built on top of host buildings, either as retrofits of existing structures or as part of new construction. Due to high land costs, saving space is increasingly important to aquaponic farms in urban contexts. Connecting a greenhouse to an existing building is one strategy for urban farmers looking to revitalize underused space and find a central location in the city. Rooftop greenhouses are already used by commercial-scale hydroponic growers but are a relatively rare enclosure type for aquaponic farms due to the

Fig. 21.6 Thanet Earth, state of the art greenhouses with combined heat and power provision, (Isle of Thanet in Kent, England, UK)

additional weight of water which can strain existing structures beyond their loading capacity. The few rooftop aquaponic farms that currently exist prioritize lightweight water distribution systems (nutrient film technique or media-based growing rather than deep water culture) and locate their fish tanks on the level below the crop growing space due to relatively decreased demand for natural light.

Two rooftop farms with high-tech aquaponic systems have recently opened in Europe. Both consulted with Efficient City Farming (ECF) farm systems consultants in Berlin. Ecco-jäger Aquaponik Dachfarm in Bad Ragaz, Switzerland sits on top of a distribution center of a family-owned produce company. The Venlo-style rooftop greenhouse (12,900 sf/1200 m^2) is located on a two-story depot building; the fish tanks are installed on the floor below the greenhouse. By growing leafy greens and herbs on their rooftop, Ecco-jäger reduces the need for transportation and can offer produce immediately after harvest. In addition, the farm takes advantage of waste heat generated by its cold storage to heat the greenhouse. BIGH's Ferme Abattoir (21,600 sf/2000 m^2) is a larger version of a similar Venlo-style rooftop greenhouse (Fig. 21.7), which occupies the roof of the Foodmet market hall in Brussels, Belgium. These early examples point to further potential to optimize both aquaponic and envelope performance through connecting water, energy, and air flows between farm and host building, known as building-integrated agriculture (BIA). Currently, research is being done on the flagship hydroponic integrated rooftop greenhouse located on the building shared by the Institute of Environmental Science and Technology (ICTA) and the Catalan Institute of Paleontology (ICP) at the Autonomous University of Barcelona (UAB) to dermine

Fig. 21.7 BIGH Ferme Abattoir with the high-tech greenhouse in the background (Brussels, Belgium)

the benefits of full building integration, although no such example exists in the field of aquaponics to determine the benefits of full building integration, although no such example exists in the field of aquaponics.

21.3.2 Indoor Growing Type

Indoor growing spaces rely exclusively on artificial light for plant production. Often, these growing spaces are highly insulated and clad in an opaque material, originally intended as storage or industrial manufacture rooms. Indoor growing spaces typically have better insulation than greenhouses due to the envelope material, though cannot rely on daylighting or natural heating. The assumption is that this typology is better suited to extreme climates, where temperature swings are of larger concern than lighting (Graamans et al. 2018), though more conclusive research is needed.

Urban Organics operates two commercial-scale indoor growing aquaponic farms within two refurbished breweries in the industrial core of St. Paul, Minnesota, USA. The two farms cultivate leafy greens and herbs in stacked growing beds illuminated by fluorescent grow lights (Fig. 21.8). Their second site allows Urban Organics to tap into the brewery infrastructure around an existing aquifer; the aquifer water needs minimal treatment and is supplied at 10 °C to arctic char and rainbow trout

Fig. 21.8 Urban Organics (St. Paul, Minnesota, USA)

tanks. Using existing structures lowered construction costs for Urban Organics and offered the opportunity to revitalize a struggling area of the city. In an even colder climate, Nutraponics grows leafy greens in a warehouse on a rural parcel 40 km outside Edmonton, Alberta, Canada. Since local produce is highly dependent on seasonal temperature swings, Nutraponics gains a competitive edge in the market by employing LED lighting to accelerate crop growth year-round (Fig. 21.9).

21.3.3 Enclosures for Aquaculture

The enclosures for the aquaculture component of aquaponic operations are technically not as demanding as the enclosure design for the hydroponic components since fish do not require sunlight to thrive. Nevertheless, control over indoor growing conditions enables farmers to optimize growth, reduce stress, and draw up precise schedules for fish production which gives their stock a competitive edge in the market (Bregnballe 2015). Aquaculture space enclosures are mainly required to keep water temperatures stable. Fish tanks should be able to support comfortable water temperature ranges for specific fish species, warm-water fish 75–86°F (24–30°C) and cold-water fish 54–74°F (12–23°C) (Alsanius et al. 2017). Water and room temperature can be controlled most efficiently if fish tanks are housed in well-

Fig. 21.9 Nutraponics (Sherwood Park, Alberta, Canada)

insulated space with few windows to minimize solar gains during the summer months and temperature losses when the outside temperature drops (Pattillo 2017) as demonstrated in the set-up of the INAPRO enclosure. The large volume of water required for fish cultivation needs to be considered from an architectural perspective, as it carries consequences for structural and conditioning systems within a building.

21.4 Assessing Enclosure Typologies and Possible Applications

The actual performance of aquaponic farms depends on many case-specific factors. Some preliminary conclusions about enclosure typologies' advantages, challenges, and possible applications can be drawn from the comparison of a relatively small set of case studies. An empirical study of a more significant number of existing case studies will be needed to establish a correlation between enclosure type, geographic location, and commercial success.

Medium-tech greenhouses offer a commercially-feasible option for aquaponic operations only in temperate climates with mild winters and moderate summers, due to their limited environmental control capability. In locations that do not require much heating and cooling, farms using this greenhouse typology can operate in a resource-efficient manner with lower upfront investment for their enclosure. These farms usually operate on a lower budget and include the fish tanks in the same greenhouse, which limits their selection of fish species to those with a large temperature tolerance and draws their commercial focus to the production of lettuce, leafy greens, and herbs.

Passive solar greenhouses rely on passive systems, specifically the use of thermal mass, to control the indoor climate. The use of this typology for aquaponic systems is advantageous since the large volume of water in the fish tanks provides additional thermal mass. Due to their energy efficiency, they are often used in northern latitudes where conventional greenhouses would require a high level of supplemental heating. However, operating any greenhouse in those regions relies on the use of supplemental lighting due to low light levels and short daylight hours during the winter season. Although passive solar greenhouses in Europe and North America are currently used on a small experimental scale, the more general successful application of these single-slope, energy-efficient greenhouses on 1.83 million acres (0.74 million hectares) of farmland in China shows that this typology can be successfully implemented on a large scale (Gao et al. 2010).

High-tech greenhouses, especially large Venlo-style, gutter-connected systems, are the industry standard for commercial hydroponic production. The largest well-funded commercial aquaponic farms use this typology for their hydroponic growing systems in conjunction with a separate enclosure for their aquaculture infrastructure. This setup guarantees the highest level of environmental control as well as crop and fish productivity. Technically, this type of greenhouse can be operated anywhere, as long as the revenue produced pays for the high energy and operation costs in extreme climates. However, this type of operation may not be environmentally sensitive in some northern latitudes due to the extensive need for heating and supplemental lighting. The exact environmental footprint of a high-tech greenhouse can only be assessed on a per-project basis and depends mostly on the quality of energy sources used for supplemental heat and light.

Most *rooftop greenhouses* are Venlo-style high-tech greenhouses constructed on rooftops. Whilst similar benefits and challenges apply, the construction of rooftop greenhouses is even more expensive than that of regular high-tech greenhouses, primarily due to building codes and architectural requirements. The structural system of rooftop greenhouses is often over-dimensioned to comply with building codes for commercial office buildings, which are stricter than building code requirements for agricultural structures. Furthermore, aquaponic operations on rooftops need additional infrastructure to access the roof and comply with fire and egress regulations, which has generated a sprinkler equipped-greenhouse in a recent example (Proksch 2017). The most promising application of rooftop greenhouses is on top of host buildings in urban centers. Urban roofs often offer ample access to sunlight, which greenhouses require to function effectively – a resource that is usually lacking, or at least is not consistent due to shadowing, at ground level in dense urban areas (Ackerman 2012). If purposefully designed, host buildings can offer other resources such as exhaust heat and CO_2 that can make the operation of a rooftop aquaponic farm more feasible. This type of integration with the host building can generate energy and environmental synergies that improve the performance of both greenhouse and host building.

Fig. 21.10 INAPRO aquaponics enclosure with two sections, opaque for fish and greenhouse for plants (Murcia, Spain)

Indoor growing spaces depend entirely on artificial lighting and active control systems for heating, cooling, and ventilation, which results in a high level of energy consumption, environmental footprint, and operation cost. This typology is most applicable in areas with cold winters and short growing seasons, where the natural exposure to sunlight and heat gain is low and extensive supplementation is needed to operate a commercial aquaponics greenhouse. The use of an opaque enclosure allows high levels of insulation, which reduces heat loss during winter months and provides autonomy from external temperature swings. Besides its dependence on electrical lighting, indoor growing exceeds the productivity of greenhouses as measured in other resources, such as water, CO_2, and land area (Graamans et al. 2018). Additionally, the production per unit of land area can be much higher through the use of stacked growing systems. Regarding the urban integration of aquaponics in cities, indoor grow spaces allow for the adaptive reuse of industrial buildings and warehouses, which can reduce the up-front cost for the construction of the enclosure and support the integration of aquaponic farms in underserved neighborhoods.

The Innovative Aquaponics for Professional Applications (INAPRO, 2018) project set-up included the comparison of the same state of the art aquaponic system and greenhouse technology, across a number of sites in Germany, Belgium, and Spain. The aquaponics system located in China was housed in a passive solar greenhouse. The INAPRO aquaponics facilities in Europe utilized a glass-clad greenhouse type for plant production and an industrial type shed component for fish tanks and filtration units (Fig. 21.10). The INAPRO project demonstrates that greenhouse technologies need to be adapted and chosen to suit local climate conditions. The Spanish INAPRO team found, that the selected enclosure was well suited for the cooler northern Europe regions, but not the warmer, Mediterranean regions in southern Europe. This observation highlights the importance of more research on the performance of greenhouses typologies to advance the field of commercial aquaponics operations.

While the comparison of the different typologies reveals certain performance patterns between typology, location, and investment (Table 21.3), for a comprehensive understanding of farm performance and environmental impact, a more robust system for the analysis and design of farm enclosures is needed.

21.5 Impact Assessment as a Design Framework

The growth of aquaponics and generalized claims that aquaponics is more sustainable than other forms of food production has stimulated discussion and research into how sustainable these systems actually are. Life cycle assessment (LCA) is one key quantification method that can be used to analyze sustainability in both agriculture and the built environments by evaluating environmental impacts of products throughout their lifespan. For a building, an LCA can be divided into two types of impact – *embodied* impact which includes material extraction, manufacture, construction, demolition and disposal/reuse of said materials, and *operational* impact which refers to building systems maintenance (Simonen 2014). Similarly, conducting an assessment of an agricultural product can be also divided into the *structural* impact of the building envelope and system infrastructure, *production* impact associated with continuous cultivation and *post-harvest* impact of packaging, storage, and distribution (Payen et al. 2015). Conducting an LCA of an aquaponic farm requires the simultaneous understanding of both building and agricultural impacts since there is an overlap in the envelope's *operational* phase with a crop's *production* phase. The way a building operates its heating, cooling, and lighting systems directly influences the cultivation of the crop; conversely, different types of crops require different environmental conditions. Numerous studies exist comparing LCA results for different building types situated in different contexts (Zabalza Bribián et al. 2009). Similarly, LCA has been used by the agricultural sector to compare efficiencies for different crops and cultivation systems (He et al. 2016; Payen et al. 2015). Evaluating the performance of controlled environment agriculture and aquaponics in particular requires a skillful integration of the two methodologies into one assessment (Sanyé-Mengual 2015).

The proposed aquaponic farm LCA framework (Fig. 21.11) is intentionally broad to capture a wide range of farm typologies found in the field. In order to apply the results of LCA to existing farms, factors such as climate and economic data must be included to validate environmental assessment (Goldstein et al. 2016; Rothwell et al. 2016)

The following section discusses a collection of aquaponic farm enclosure design strategies based on the LCA inventory of aquaponic farms that synthesizes existing

Table 21.3 Comparison of controlled environment agriculture typologies

CEA type	Benefits	Challenges	Cost and revenue[a]
Medium-tech greenhouses	Relies almost entirely on solar energy, low additional energy requirement	Limited environmental control options, susceptible to environmental fluctuations	Lower up-front/construction cost, (approx. 30–100 $/m^2$)
	Less reliance on non-renewable materials and energy sources	Only applicable to fish species with a large temperature tolerance, (if tanks are in the greenhouse)	
Passive solar greenhouses	Relies on passive systems, uses thermal mass, (including the fish tanks) to buffer temperature swings	Control with passive systems needs more experience and deliberate design	Lower up-front/construction cost, (approx. 30–100 $/m^2$)
	Low energy consumption, potentially without the need for any fossil fuel	Require supplemental lighting, if located in northern latitudes due to low light levels	
High-tech greenhouses	Highest levels of controls	Relies on active systems for heat, cooling, ventilation and supplemental lighting	High up-front/construction cost, (approx. 100–200 $/m^2$ and more)
	High productivity with the potential to scale up	High energy consumption and operation cost	
Rooftop greenhouses	Highest levels of controls	Relies on active systems for heat, cooling, ventilation and supplemental lighting	Very high up-front/construction cost (approx. 300–500 $/m^2$)
	High productivity		
	Potential for energetic and environmental synergies, if integrated with host building	High energy consumption and operation cost	
		Requires code compliance at the level of commercial office buildings	
		Transport of supplies to rooftop is an infrastructural challenge	
Indoor growing spaces	Adaptive reuse of industrial buildings possible	Depends entirely on electrical lighting and active control systems for heating, cooling, and ventilation	Up-front/construction cost can be lower if existing building can be used
	High productivity per unit of footprint though stacked growing systems	High energy consumption and operation cost	Cost depends also on the growing system, stacking multiple levels
	High level of insulation possible		
	Reduced heat loss during winter months		

[a]Based on Proksch (2017)

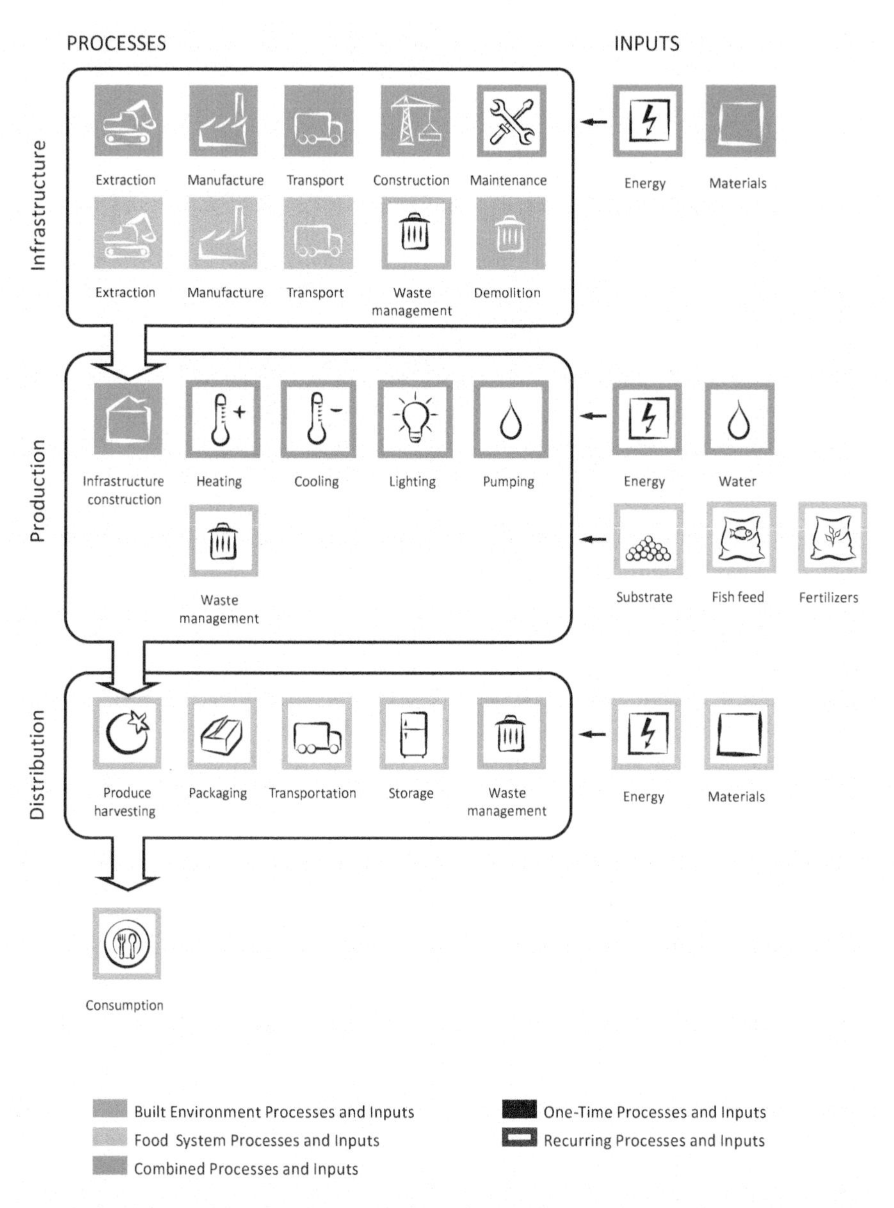

Fig. 21.11 Example of an integrated LCA process including building and aquaponic system performance. (Based on Sanyé-Mengual et al. 2015).

literature with case studies and suggests directions for future work. The unique integration of aquaponic and building-related impacts is of particular interest.

21.5.1 Embodied Impacts: Embodied Energy and Embodied Carbon

Structure Materials and Construction Embodied energy is the calculation of the sum of energy used to extract, refine, process, transport, produce, and assemble a material or product. Embodied carbon is the amount of CO_2 emitted to produce the same material or product. Compared to conventional open-field agricultural operations, the embodied impact of a controlled environment growing system is greater due to increased material extraction and manufacture at the construction stage (Ceron-Palma et al. 2012). For example, in the ICTA-ICP rooftop greenhouse, the structure of the envelope generates 75% more Global Warming Potential (GWP) than a soil-based multi-tunnel greenhouse structure due to the quantity of polycarbonate used in construction (Sanyé-Mengual et al. 2015). Similarly, a building-integrated greenhouse simulation situated in Boston resulted in increased environmental impacts at the construction stage, due to the extraction of iron ores for the manufacture of structural steel (Goldstein 2017). Embodied impacts associated with controlled environment envelopes can be mitigated through smart material use (given that building code adjustments are made to avoid over-sizing structural members) but would nevertheless surpass those of traditional agriculture. Growing food in a constructed envelope will always be more resource-intensive at the beginning compared to simply planting vegetables in an open field, though will also dramatically increase the amount of food that can be produced per area footprint in the same timeframe.

To avoid structure-related environmental impacts, some aquaponic operations make use of existing buildings instead of constructing a new envelope. Urban Organics in St. Paul, Minnesota, USA refurbished two brewery buildings as their indoor growing spaces. In another example of adaptive reuse, The Plant in Chicago, Illinois, USA operates its food incubator and urban farm collective in a 1925 factory building previously used by Peer Foods as a meat-packaging facility (Fig. 21.12). Existing insulation and refrigeration equipment were repurposed to control temperature fluctuations in the experimental aquaponic facility.

Aquaponic Equipment and Substrate When integrated into buildings, the material choice for aquaponic tanks becomes an important design consideration, since it may limit assembly and transport into the building. For example, polyethylene parts can be assembled on-site using plastic welding, but this is not possible with fiberglass parts (Alsanius et al. 2017). Furthermore, the manufacture of aquaponic system equipment can be a significant contributor to overall environmental impact – for

Fig. 21.12 The Plant (Chicago, Illinois, USA)

example, glass fiber-reinforced polyester used for the 100 m^3 water tank at the ICTA-ICP rooftop greenhouse is responsible for 10–25% of environmental impact at the manufacturing stage (Fig. 21.13). The choice of substrate for plants in an aquaponic system has a weight ramification for the structure of the host building, but also contributes to environmental impact. In a recent study done on aquaponics integrated with living walls, mineral wool, and coconut fiber performed comparably, despite one being compostable and the other being single-use (Khandaker and Kotzen 2018).

Structure and Equipment Maintenance Initial material selection for aquaponic equipment and envelope components determines the long-term upkeep of aquaponic farms. Manufacturing more durable materials such as glass or rigid plastics requires a greater initial investment of environmental resources than plastic films; however, films require replacement more frequently – for example, glass is expected to remain functional for 30+ years, whilst more conventional coated polyethylene film can only last 3–5 years before becoming too opaque (Proksch 2017). Depending on the intended lifespan of an aquaponic system envelope, it may be more advantageous to choose a material with a shorter lifespan, and a lesser manufacturing impact. ETFE film used in the Aquaponic solar greenhouse is a promising compromise between longevity and sustainability, although further research is needed. Standard aquaponic equipment consists of water tanks and piping. Piping for aquaponic systems is often manufactured from PVC, which produces a significant environmental impact in its manufacturing process but does not require replacement for up to 75 years. Some aquaponic suppliers offer bamboo as an organic alternative.

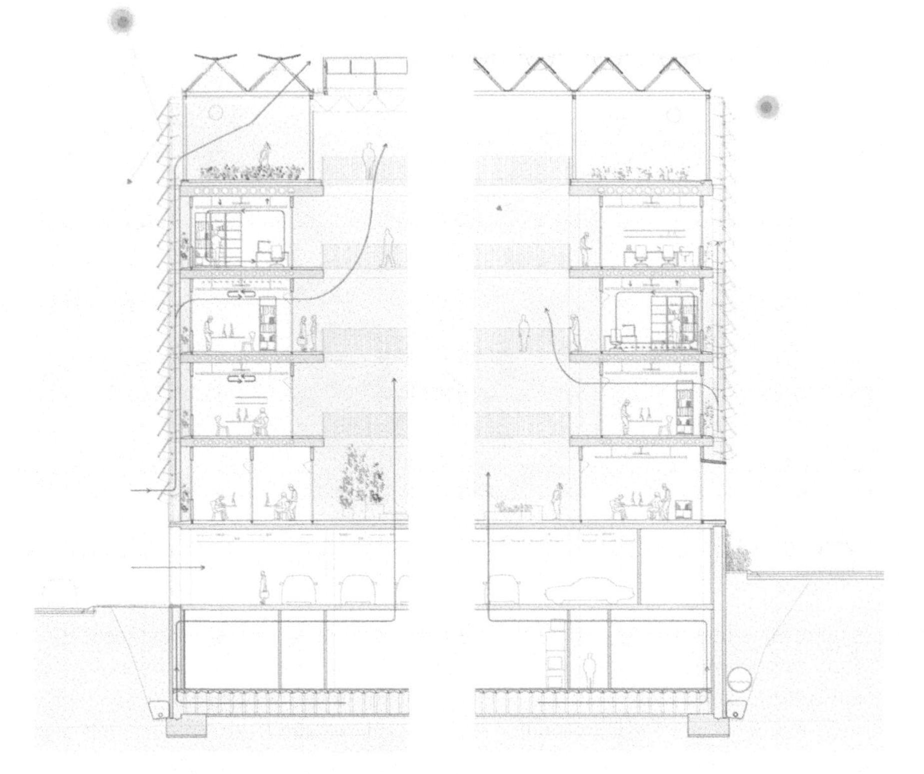

Fig. 21.13 Building section with rooftop greenhouses by Harquitectes, ICTA-ICP building (Bellaterra, Spain)

21.5.2 Operational Impacts

Energy In 2017, 39% of total energy consumption within the United States corresponded to the building sector (EIA). The agricultural sector accounted for approximately 1.74% of total U.S. primary energy consumption in 2014, relying heavily on indirect expenditures in the form of fertilizers and pesticides (Hitaj and Suttles 2016). Energy efficiency is a well-established field of research within both the built environment and agriculture, often defining the operational impacts of a product, building, or farm in the overall LCA (Mohareb et al. 2017). Integrating building and agricultural energy use can optimize the performance of both (Sanjuan-Delmás et al. 2018).

Heating Energy requirements for heating growing spaces are of particular interest in the northern climates, where extending a naturally short growing season gives building-integrated aquaponic farms a competitive edge in the market (Benis and Ferrão 2018). However, in colder climates, energy consumption by active heating systems is a significant contributor to overall environmental impact – in an assessment of conditioned growing spaces in Boston, Massachusetts, heating costs neutralized the benefits of eliminating food miles in the urban food chain (Benis et al. 2017b; Goldstein 2017). This does not hold true in Mediterranean climates, where climatic conditions are conducive to agriculture and where nearly year-round and conventional greenhouse structures can rely on passive solar heating (Nadal et al. 2017; Rothwell et al. 2016).

In both cold and warm climates, integrating controlled environment growing systems on existing rooftops can provide insulation to the host building – a farm in Montreal, Quebec reports to capture 50% of the greenhouse heating needs from the existing host structure, thereby reducing heating load (Goldstein 2017). Lighting systems can also be partially responsible for satisfying heating demand in interior vertical growing applications such as plant factories or shipping containers (Benis et al. 2017b).

Residual heat capture is another promising design strategy that can optimize the performance of both the host structure and the growing system. Post-occupancy studies of the experimental rooftop greenhouse at the ICTA-ICP in Bellaterra, Spain indicate that the integration of the building with the greenhouse delivered an equivalent carbon savings of 113.8 kg/m^2/year compared to a conventional free-standing greenhouse heated with oil (Nadal et al. 2017). Without intervention from active heating, ventilation and air conditioning (HVAC) systems, the thermal mass of the host laboratory/office building raised the greenhouse temperature by 4.1°C during the coldest months, enabling the cultivation of the tomato crop year-round.

Cooling In Mediterranean and tropical climates, artificial cooling is often a requirement to grow produce year-round. In a rooftop greenhouse simulation, cooling loads represented up to 55% of total farm energy demands in Singapore and in the more temperate climate of Paris, 30% (Benis et al. 2017b). Cooling energy demands are especially high in arid climates, which can benefit the most from cutting conventional transportation costs for perishable produce (Graamans et al. 2018; Ishii et al. 2016). Evaporative cooling, fog cooling, and shading are some strategies for lowering temperatures in aquaponic farms and improving farm performance in terms of yield.

Building-integrated aquaponic systems have the advantage of storing thermal mass in fish tanks to alleviate cooling as well as heating loads. In cases where this mode of passive cooling does not satisfy the cooling demand, evaporative cooling is most commonly used. The Sustainable Harvesters greenhouse produces lettuce for the Houston, Texas, USA area year-round by using a fan and pad cooling system, a subset of evaporative cooling technology. Hot air from outside the envelope first passes through a wet cellulose medium before entering the growing space. As a result, the interior air is cooler and more humid. Evaporative cooling is most effective in dry climates but requires high water use, which may be a limitation to farms in arid areas of the world.

Fog cooling is an alternative strategy. In a fog-cooled greenhouse, plants are periodically misted with water from overhead sprinklers/misters until the space reaches the desired temperature for cultivation. Fog cooling uses less water than evaporative cooling but increases the relative humidity of a growing space. If paired with the right ventilation strategy, fog cooling can be a water-saving technology particularly suited to arid regions (Ishii et al. 2016). Additionally, fog cooling decreases the rate of evapotranspiration in plants, which is critical to optimizing plant metabolism in aquaponic systems (Goddek 2017). The flagship greenhouse of Superior Fresh farms uses a computerized fog-cooling system to maintain cultivation temperatures during the hot season.

Shading devices can also contribute to lowering greenhouse temperatures. Traditionally, the seasonal lime whitewashing of greenhouses was used to reduce solar radiation levels during the hottest months (*Controlled Environment Agriculture* 1973). However, shading can be integrated with other building functions. A promising shading strategy is using semi-transparent photovoltaic modules to simultaneously cool the space and produce energy (Hassanien and Ming 2017). The Aquaponic solar greenhouse combines its photovoltaic array with shading functionality; it uses rotating aluminium panels as shading devices that operate as solar collectors with the help of mounted photovoltaic cells. The integrated photovoltaic system then transforms excess solar radiation into electrical energy.

Lighting The main advantage of greenhouses over indoor growing spaces is their ability to capitalize on daylight to facilitate photosynthesis. However, farms in extreme climates may find that satisfying heating or cooling loads for a transparent envelope is not financially feasible; in this case, farmers may choose to cultivate crop in indoor growing spaces with an insulated envelope (Graamans et al. 2018). Aquaponic farms that operate in indoor growing spaces rely on efficient electrical lighting to produce crops.

Many advances in contemporary farm lighting originated in Japanese plant factories, used to optimize plant yields in dense hydroponic systems by replacing sunlight with engineered light wavelengths (Kozai et al. 2015). Currently, LED lighting is the most popular choice for electrical horticultural lighting systems. They are 80% more efficient than high-intensity discharge lamps and 30% more efficient than their fluorescent counterparts (Proksch 2017). LED lighting continues to be investigated to optimize energy efficiency and crop yield (Zhang et al. 2017). Large-scale greenhouses like Superior Fresh, Wisconsin, USA rely on computerized, supplemental lighting regimes to extend the photosynthesis period of its crop in northern latitudes.

Energy Generation Constrained by the same factors as all CEA, the energy management of an aquaponic farm depends on exterior climate, crop selection, the production system, and structure design (Graamans et al. 2018). Growing produce through aquaponics is not inherently sustainable if not managed properly – all of the factors above can affect energy efficiency for the better or worse (Buehler and Junge 2016). In many cases, CEA is more energy-intensive than conventional open-field agriculture; however, higher energy expenditures may be justified if the way we source energy shifts toward renewable sources and efficient strategies for heating, cooling, and lighting are incorporated into the design of the farm.

Photovoltaic (PV) power generation can play an important part in offsetting operational impacts for controlled environment aquaponics, reducing environmental strain. In an example of a high-tech greenhouse in Australia, using energy from a PV array caused a 50% reduction in lifecycle greenhouse gas emissions compared to the conventional grid scenario (Rothwell et al. 2016). Renewable energy generation can be combined with aquaponic farms, space permitting – for example, the Lucky Clays Fresh aquaponic greenhouse on a rural farm in North Carolina runs on energy generated by wind turbines and photovoltaic panels that are situated elsewhere on the owner's land parcel.

Water Water use efficiency has been often cited as a major benefit of CEA and hydroponic systems (Despommier 2013; Specht et al. 2014). Aquaponic systems are even better suited to increase water efficiency – where 1 kg of fish produced in a conventional aquaculture system requires between 2500 and 375,000 L, the same amount of fish raised in an aquaponic system requires less than 100 L (Goddek et al. 2015). Rainwater capture and greywater reuse have been proposed as two strategies to offset the watershed impacts of operating a hydroponic or aquaponic farm even further. At the existing ICTA-ICP greenhouse, 80–90% of the water needs for the production of tomatoes in an aggregate hydroponic system were covered by rainwater capture within a year of operation (Sanjuan-Delmás et al. 2018). However, the ability of rainwater capture to meet crop demand depends on the climatic context. In a study evaluating the viability of rooftop greenhouse production on existing retail parks in eight cities around the world, seven met crop self-sufficiency through rainwater capture – only Berlin did not (Sanyé-Mengual et al. 2018).

Some existing CEA facilities already reuse greywater to improve efficiency (Benke and Tomkins 2017). However, greywater reuse in an urban context is currently limited due to lacking regulatory support and currently-lacking research on the health risks of using greywater in agriculture. A pilot of greywater reuse, the Maison Productive in Montréal collects greywater from household uses to supplement its rainwater collection to irrigate gardens and a communal greenhouse for food production that nine residential units share (Thomaier et al. 2015). With further advances in policy on the treatment of greywater, building-integrated aquaponics can tap into the existing water cycle instead of relying on municipal sources.

From an architectural standpoint, water distribution in an aquaponic system is likely to present a structural challenge. Aquaponic fish tanks weigh more than hydroponic grow beds and may limit what types of structures are feasible for retrofitting an aquaponic farm. The growing medium also requires consideration – deep water culture (DWC) systems require a large and heavy volume of water, whilst nutrient film technique (NFT) systems are lightweight but expensive to manufacture (Goddek et al. 2015).

Nutrients Compared to conventional open-field farming, CEA reduces the need for fertilizers and pesticides, as the farmer can physically separate the crop from harsh external conditions (Benke and Tomkins 2017). However, due to the density of an aquaponic system, plant or fish diseases can spread quickly if a pathogen infiltrates the space. Preventative options such as the use of predator insects or tight environmental control measures such as a "buffer" entryways can avert this risk (Goddek et al. 2015).

The integration of different fish and crop nutrient needs is a challenge in single-recirculating aquaponic systems (Alsanius et al. 2017). Generally, plants require higher nitrogen concentrations than fish can withstand and careful crop and fish selection can match nutrient requirements to optimize yields, but is still difficult to achieve. Decoupled systems (DRAPS) have been proposed to separate the aquaculture water cycle from the hydroponic one to achieve desired nutrient concentrations, but is not yet commonly applied in commercial farms (Suhl et al. 2016). Urban Organics based in St. Paul, Minnesota, USA chose to develop a DRAPS system for their second farm to optimize both crop and fish yields and avoid crop loss in case of nutrient imbalances within fish tanks. ECF Farm in Berlin, Germany, and Superior Fresh farms in Wisconsin, USA also operate decoupled systems to optimize fish and plant growth.

Alternatively, aquaponic nutrient cycles can be optimized through the introduction of an anaerobic reactor to transform solid fish waste into plant-digestible phosphorus (Goddek et al. 2016). Currently, The Plant in Chicago, USA is planning to operate an anaerobic digester which may play a part in optimizing nutrient cycles for crop growth. The mechanical system requirements for DRAPS and anaerobic digestion will influence the performance as well as the spatial layout of an aquaponic farm.

21.5.3 End-of-Life Impacts

Materials Waste Management A theoretical advantage of CEA over open-field farming is the ability to control materials waste runoff, preventing leaching (Despommier 2013; Gould and Caplow 2012). A tight envelope can play a role in efficient materials waste management. One pathway of recycling organic waste matter to improve building performance is the use of plant stalks for the production of insulating biochar, although this research is in early stages (Llorach-Massana et al. 2017). Additionally, considering the incorporation of waste management components such as a filtration bed, an anaerobic digester or a heat recovery ventilator into the enclosure design at an early stage can close energy, nutrient, and water loops for the farm.

Distribution Chains Packaging has been a hotspot in various farm LCAs assessing the impact of production. It is responsible for as much as 45% of the total impact for a tomato in Bologna, Italy, and is the largest contributor to the environmental impacts of indoor hydroponic systems in Stockholm, Sweden (Molin and Martin 2018b; Orsini et al. 2017; Rothwell et al. 2016). Siting aquaponic farms close to consumers can reduce the need for packaging, storage, and transport as with other forms of urban agriculture, if local retailers and distributors collaborate with farmers (Specht et al. 2014). Unfortunately, due to consumer acceptance, most large-scale retailers currently require standard plastic packaging for aquaponic produce to be sold alongside conventional brands - – therefore, selecting a site close to a consumer market for controlled environment aquaponics does not guarantee significant changes in the overall performance of the farm.

Reduced transportation, or food-miles, is often cited in the literature as a major advantage of urban agriculture (Benke and Tomkins 2017; Despommier 2013; Sanjuan-Delmás et al. 2018). However, it is important to note that the relative contribution of shortened transportation chains varies on a case-by-case basis. In Singapore, where nearly all food has to be imported from neighboring countries, cutting transportation chains makes sense financially and in terms of environmental impact (Astee and Kishnani 2010). The same cannot be said for Spain, where the conventional supply chain of tomatoes from farm to city is already short (Sanjuan-Delmás et al. 2018). Cities with the longest supply chains can benefit from localized food production, but the benefits of cutting transportation must be weighed against operational and embodied impacts. In the case of Boston, the benefits of reduced transportation were entirely negated by the impact of heating and operating a greenhouse inside the city (Goldstein 2017). Despite long conventional food supply chains, transportation impacts were similarly insignificant in the bigger picture of CEA performance in Stockholm (Molin and Martin 2018a).

Consumption and Diet Aquaponic farms in cities can alter urban diets, which play a significant role in the environmental impact of food consumption (Benis and Ferrão 2017). Meat consumption via the conventional chain produces the largest share of the current environmental footprint and seeking protein alternatives has the potential for a larger impact than the widespread implementation of urban agriculture (Goldstein 2017). Since aquaponics produces fish as well as vegetables, this potential to change protein diets on a large scale should not be ignored in larger assessments of environmental performance.

21.6 Integrated Urban Aquaponics

When deliberately designed with respect to environmental impact, aquaponic farms can become part of a resource-efficient urban food system. No aquaponic farm operates in isolation since when crops are harvested and reach the farm gate, they enter a larger socioeconomic food network as fish and produce is distributed to customers. At this stage, the performance of aquaponic farms is no longer confined to the growing system and envelope – economics, marketing, education, and social outreach are also involved. Urban aquaponic farms will need to operate as competitive businesses and good neighbors to be successfully integrated into city life.

21.6.1 Economic Viability

The economic viability of aquaponic farms depends on many contextual factors where both local conventional fish production chains and open-field farming must be matched (Stadler et al. 2017). While aquaponics requires a relatively costly initial investment, it may outperform conventional farming during the production and distribution phase where the design of the recirculating water system reduces water costs, and greatly reduces the need for fertilizers, which usually comprise between 5% and 10% of overall farm costs (Hochmuth and Hanlon 2010). However, estimating the economic viability of aquaponic farms is particularly challenging due to the range of dynamic factors affecting performance including the local price for labor and energy being two examples (Goddek et al. 2015). In an economic analysis of aquaponic farms in the Midwestern United States, labor constituted 49% of all operational costs despite the assumption that only minimum wages would be paid. In reality, the wide range of expertise required to operate an aquaponic system will likely warrant higher wages in an urban farm scenario (Quagrainie et al. 2018).

Site selection and envelope design have a direct relationship to the profitability of an aquaponic farm by affecting operation efficiency and how broad the potential market can be. Aquaponic farms located in urban environments can tap into multiple markets outside agricultural production, where many aquaponic farms offer tours, workshops, design consulting services, and supply backyard aquaponic systems for hobbyists. Integrating agriculture with other types of spaces within urban environments can contribute to the financial health of aquaponic farms. The ECF aquaponic farm is located on the work yard of the industrial landmark building Malzfabrik, Berlin, Germany, which operates a cultural center and houses work spaces for artists and designers.

21.6.2 Accessibility and Food Security

Urban agriculture is often cited as a strategy to provide fresh food for underserved communities located in food deserts, yet few commercial urban farms target this demographic, proving that commercial-scale urban agriculture can be just as exclusionary as conventional supply chains (Gould and Caplow 2012; Sanyé-Mengual et al. 2018; Thomaier et al. 2015). Aquaponic farms that use high-tech infrastructure try to redeem their high investments by achieving premium prices in urban markets, though aquaponics can also stem from grassroots and hobbyist applications. Aquaponics may also have the potential to increase food security for urban residents. This is evidenced in the lasting legacy of Growing Power, a non-profit organization that until recently, ran an urban farm in Milwaukee, Wisconsin, USA started by Will Allen in 1993. Many current aquaponic farmers attended Growing Power's workshops, in which Allen championed an aquaponic model that gives back to the surrounding community by means of community-supported agriculture boxes and classes. Initiated by Growing Power's educational programs, other aquaponic non-profit organizations have taken up to the torch such as Dre Taylor with Nile Valley Aquaponics in Kansas City, Kansas, USA. This farm aims to provide 100,000 pounds (45,400 kg) of local produce to the surrounding community in an award-winning new campus for the expanding farm (Fig. 21.14).

Fig. 21.14 Proposed Nile Valley Aquaponics campus (Kansas City, Kansas, USA) by HOK Architects

21.6.3 Education and Job Training

Aquaponics can be used as an educational tool to promote systems thinking and environmental mindfulness (Junge et al. 2014; Specht et al. 2014). In urban applications, aquaponic systems could be used to raise awareness of ecological cycles much like existing soil-based farms (Kulak et al. 2013). The Greenhouse Project in New York City translates this into a new approach to science education in public schools. The Greenhouse Project aims to build 100 rooftop greenhouses on public schools as science classrooms. These greenhouses, customized for their dual mission of growing and learning, all include an aquaponic system. However, aquaponic systems also require greater collaboration between existing academic disciplines in order to move forward in this new multidisciplinary academic field (Goddek et al. 2015). The collaboration of aquaculture and horticulture specialists, engineers, business strategists, and built environment professionals amongst many others is necessary to turn aquaponics into an important contributor to sustainable urban development.

21.7 Conclusions

There is an array of criteria that contribute to the performance of each farm and their number grows with the number of disciplines involved in this the interdisciplinary field of aquaponics. Of note is an earlier study that has provided a definition of aquaponics and a classification of the types of aquaponics based on size and system

(Palm et al. 2018). Many criteria for the analysis of the enclosure type identified in this study stem from immediate farm context – local climate, the quality of the built environment context, energy sourcing practices, costs, market, and local regulatory frameworks. An aquaponic greenhouse in a rural context performs differently than one in a city, just as farms in arid climates do not share the same requirements as their counterparts in colder areas. In general, greenhouses classified as medium-tech and passive solar offer a lower cost, environmentally sustainable enclosure option, currently only used by smaller aquaponic operations. However, due to their intentionally limited level of technical environmental controls, they only perform well in specific climate zones. In comparison, high-tech and rooftop greenhouses can be technically implemented anywhere, though in extreme climate conditions they generate high operational costs and larger environmental footprints. Recent case studies show that indoor growing facilities can be financially feasible, but due to their exclusive reliance on electrical lighting, their resource use efficiency and environmental footprint are of concern. Further research is needed to establish the relationship of specific aquaponic farms and their enclosures to existing resource networks. This work can help connect aquaponics to research done on urban metabolism.

Other criteria determining farm typology and performance are internal. These include environmental control levels, crop and fish selection, aquaponic system type and scale and enclosure type and scale. Taking on an integrated LCA approach, the relationship between all factors have to be assessed throughout the lifespan of the farm, from cradle to grave. Life cycle assessment of aquaponic farms must include both building impacts and growing system impacts since there is overlap in the farm operation phase. A series of promising strategies in heating, cooling, lighting, and material design can improve overall farm efficiency throughout the entire lifespan of the farm. Beyond accounting for environmental impact, LCA can become a design framework for horticulture experts, aquaculture specialists, architects, and investors.

Continuing to survey existing commercial aquaponic farms is important to validate LCA models, identify strategies, and cataloguing aquaponic operations emerging on a larger scale. Combining modeling with case study research on controlled environment aquaponics has the potential to connect aquaponics to the larger scope of urban sustainability.

Acknowledgments The authors of this study acknowledge the financial support of the National Science Foundation (NSF) under the umbrella of the Sustainable Urbanization Global Initiative (SUGI) Food Water Energy Nexus and the support of all CITYFOOD project partners for providing ideas and inspiration.

References

Ackerman K (2012) The potential for urban agriculture in New York City: growing capacity, food security and green infrastructure report. Columbia University Urban Design Lab, New York

Alsanius BW, Khalil S, Morgenstern R (2017) Rooftop aquaponics. In: Rooftop urban agriculture, urban agriculture. Springer, Cham, pp 103–112. https://doi.org/10.1007/978-3-319-57720-3_7

Astee LY, Kishnani NT (2010) Building integrated agriculture: utilising rooftops for sustainable food crop cultivation in Singapore. J Green Build 5:105–113. https://doi.org/10.3992/jgb.5.2.105

Benis K, Ferrão P (2017) Potential mitigation of the environmental impacts of food systems through urban and peri-urban agriculture (UPA) – a life cycle assessment approach. J Clean Prod 140:784–795. https://doi.org/10.1016/j.jclepro.2016.05.176

Benis K, Ferrão P (2018) Commercial farming within the urban built environment – taking stock of an evolving field in northern countries. Glob Food Sec 17:30–37. https://doi.org/10.1016/j.gfs.2018.03.005

Benis K, Reinhart C, Ferrão P (2017a) Development of a simulation-based decision support workflow for the implementation of building-integrated agriculture (BIA) in urban contexts. J Clean Prod 147:589–602. https://doi.org/10.1016/j.jclepro.2017.01.130

Benis K, Reinhart C, Ferrão P (2017b) Building-integrated agriculture (BIA) in urban contexts: testing a simulation-based decision support workflow. Presented at the Building Simulation 2017, San Francisco, USA, p 10. https://doi.org/10.26868/25222708.2017.479

Benke K, Tomkins B (2017) Future food-production systems: vertical farming and controlled-environment agriculture. Sustain Sci Pract Policy 13:13–26. https://doi.org/10.1080/15487733.2017.1394054

Bregnballe J (2015) A guide to recirculation aquaculture: an introduction to the new environmentally friendly and highly productive closed fish farming systems. Food and Agriculture Organization of the United Nations: Eurofish, Copenhagen

Buehler D, Junge R (2016) Global trends and current status of commercial urban rooftop farming. Sustainability 8:1108. https://doi.org/10.3390/su8111108

Ceron-Palma I, Sanyé-Mengual E, Oliver-Solà J, Rieradevall J (2012) Barriers and opportunities regarding the implementation of rooftop eco.greenhouses (RTEG) in mediterranean cities of Europe. J Urban Technol 19:87–103. https://doi.org/10.1080/10630732.2012.717685

Controlled Environment Agriculture (1973) A global review of greenhouse food production (No. 89), Economic Research Service. U.S. Department of Agriculture

de Graaf PA (2012) Room for urban agriculture in Rotterdam: defining the spatial opportunities for urban agriculture within the industrialised city. In: Sustainable food planning: evolving theory and practice. Wageningen Academic Publishers, Wageningen, pp 533–546. https://doi.org/10.3920/978-90-8686-187-3_42

De La Salle JM, Holland M (2010) Agricultural urbanism. Green Frigate Books

Despommier D (2013) Farming up the city: the rise of urban vertical farms. Trends Biotechnol 31:388–389. https://doi.org/10.1016/j.tibtech.2013.03.008

dos Santos MJPL (2016) Smart cities and urban areas–aquaponics as innovative urban agriculture. Urban For Urban Green 20:402–406. https://doi.org/10.1016/j.ufug.2016.10.004

EU Aquaponics Hub (2017) COST Action FA1305, Aquaponics map (Cost FA1305), https://www.google.com/maps/d/u/0/viewer?ll=50.77598474809961%2C12.62131196967971&z=4&mid=1bjUUbCtUfE_BCgaAf7AbmxyCpT0

Fang Y, Hu Z, Zou Y, Zhang J, Zhu Z, Zhang J, Nie L (2017) Improving nitrogen utilization efficiency of aquaponics by introducing algal-bacterial consortia. Bioresour Technol 245:358–364. https://doi.org/10.1016/j.biortech.2017.08.116

Gao L-H, Qu M, Ren H-Z, Sui X-L, Chen Q-Y, Zhang Z-X (2010) Structure, function, application, and ecological benefit of a single-slope, energy-efficient solar greenhouse in China. HortTechnology 20:626–631

Goddek S (2017) Opportunities and challenges of multi-loop aquaponic systems. Wageningen University, Wageningen

Goddek S, Delaide B, Mankasingh U, Ragnarsdottir KV, Jijakli H, Thorarinsdottir R (2015) Challenges of sustainable and commercial Aquaponics. Sustainability 7:4199–4224. https://doi.org/10.3390/su7044199

Goddek S, Schmautz Z, Scott B, Delaide B, Keesman KJ, Wuertz S, Junge R (2016) The effect of anaerobic and aerobic fish sludge supernatant on hydroponic lettuce. Agronomy 6:37. https://doi.org/10.3390/agronomy6020037

Goldstein BP (2017) Assessing the edible city: environmental implications of urban agriculture in the Northeast United States. Technical University of Denmark, Lyngby

Goldstein B, Hauschild M, Fernández J, Birkved M (2016) Testing the environmental performance of urban agriculture as a food supply in northern climates. J Clean Prod 135:984–994. https://doi.org/10.1016/j.jclepro.2016.07.004

Gould D, Caplow T (2012) Building-integrated agriculture: a new approach to food production. In: Metropolitan sustainability: understanding and improving the urban environment. Woodhead Publishing Limited, Cambridge, pp 147–170

Graamans L, Baeza E, van den Dobbelsteen A, Tsafaras I, Stanghellini C (2018) Plant factories versus greenhouses: comparison of resource use efficiency. Agric Syst 160:31–43. https://doi.org/10.1016/j.agsy.2017.11.003

Hassanien RHE, Ming L (2017) Influences of greenhouse-integrated semi-transparent photovoltaics on microclimate and lettuce growth. Int J Agric Biol Eng 10:11–22. https://doi.org/10.25165/ijabe.v10i6.3407

He X, Qiao Y, Liu Y, Dendler L, Yin C, Martin F (2016) Environmental impact assessment of organic and conventional tomato production in urban greenhouses of Beijing city, China. J Clean Prod 134:251–258. https://doi.org/10.1016/j.jclepro.2015.12.004

Hitaj C, Suttles S (2016) Trends in U.S. agriculture's consumption and production of energy: renewable power, shale energy, and cellulosic biomass, Economic information bulletin, no. 159. USDA/Economic Research Service, Washington, DC

Hochmuth GJ, Hanlon EA (2010) Commercial vegetable fertilization principles 17

INAPRO - Innovative Aquaponics for Professional Applications (2018). http://inapro-project.edu

Ishii M, Sase S, Moriyama H, Okushima L, Ikeguchi A, Hayashi M, Kurata K, Kubota C, Kacira M, Giacomelli GA (2016) Controlled environment agriculture for effective plant production systems in a semiarid greenhouse. JARQ 50:101–113. https://doi.org/10.6090/jarq.50.101

Junge R, Wilhelm S, Hofstetter U (2014) Aquaponic in classrooms as a tool to promote system thinking. In: Transmission of innovations, knowledge and practical experience into everyday practice. Presented at the Conference VIVUS – on agriculture, environmentalism, horticulture and floristics, food production and processing and nutrition, Naklo, Slovenia, p 11

Junge R, König B, Villarroel M, Komives T, Jijakli MH (2017) Strategic points in aquaponics. Water 9:182. https://doi.org/10.3390/w9030182

Khandaker M, Kotzen B (2018) The potential for combining living wall and vertical farming systems with aquaponics with special emphasis on substrates. Aquac Res 49:1454–1468. https://doi.org/10.1111/are.13601

König B, Junge R, Bittsanszky A, Villarroel M, Komives T (2016) On the sustainability of aquaponics. Ecocycles 2(1):26–32. https://doi.org/10.19040/ecocycles.v2i1.50

Körner O, Gutzmann E, Kledal PR (2017) A dynamic model simulating the symbiotic effects in aquaponic systems. Acta Hortic 1170:309–316. https://doi.org/10.17660/ActaHortic.2017.1170.37

Kozai T, Niu G, Takagaki M (2015) Plant factory: an indoor vertical farming system for efficient quality food production. Academic

Kulak M, Graves A, Chatterton J (2013) Reducing greenhouse gas emissions with urban agriculture: a life cycle assessment perspective. Landsc Urban Plan 111:68–78. https://doi.org/10.1016/j.landurbplan.2012.11.007

Lastiri DR, Geelen C, Cappon HJ, Rijnaarts HHM, Baganz D, Kloas W, Karimanzira D, Keesman KJ (2018) Model-based management strategy for resource efficient design and operation of an aquaponic system. Aquac Eng 83:27. https://doi.org/10.1016/j.aquaeng.2018.07.001

Llorach-Massana P, Lopez-Capel E, Peña J, Rieradevall J, Montero JI, Puy N (2017) Technical feasibility and carbon footprint of biochar co-production with tomato plant residue. Waste Manag 67:121–130. https://doi.org/10.1016/j.wasman.2017.05.021

Maucieri C, Forchino AA, Nicoletto C, Junge R, Pastres R, Sambo P, Borin M (2018) Life cycle assessment of a micro aquaponic system for educational purposes built using recovered material. J Clean Prod 172:3119–3127. https://doi.org/10.1016/j.jclepro.2017.11.097

Mohareb E, Heller M, Novak P, Goldstein B, Fonoll X, Raskin L (2017) Considerations for reducing food system energy demand while scaling up urban agriculture. Environ Res Lett 12:125004. https://doi.org/10.1088/1748-9326/aa889b

Molin E, Martin M (2018a) Assessing the energy and environmental performance of vertical hydroponic farming (No. C 299). ICL Swedish Environmental Research Institute, ICL Swedish Environmental Research Institute

Molin E, Martin M (2018b) Reviewing the energy and environmental performance of vertical farming systems in urban environments (No. C 298). ICL Swedish Environmental Research Institute, ICL Swedish Environmental Research Institute

Monsees H, Kloas W, Wuertz S (2017) Decoupled systems on trial: eliminating bottlenecks to improve aquaponic processes. PLoS One 12:e0183056. https://doi.org/10.1371/journal.pone.0183056

Nadal A, Llorach-Massana P, Cuerva E, López-Capel E, Montero JI, Josa A, Rieradevall J, Royapoor M (2017) Building-integrated rooftop greenhouses: an energy and environmental assessment in the mediterranean context. Appl Energy 187:338–351. https://doi.org/10.1016/j.apenergy.2016.11.051

Orsini F, Dubbeling M, de Zeeuw H, Prosdocimi Gianquinto GG (2017) Rooftop urban agriculture, Urban agriculture (springer (firm)). Springer, Cham

Palm HW, Knaus U, Appelbaum S, Goddek S, Strauch SM, Vermeulen T, Jijakli M, Kotzen B (2018) Towards commercial aquaponics : a review of systems, designs, scales and nomenclature. Aquac Int 26(3):813–842. ISSN 0967-6120. https://doi.org/10.1007/s10499-018-0249-z

Pattillo DA (2017) An overview of aquaponic systems: aquaculture components (No. 20), NCRAC Technical Bulletins. North Central Regional Aquaculture Center

Payen S, Basset-Mens C, Perret S (2015) LCA of local and imported tomato: an energy and water trade-off. J Clean Prod 87:139–148. https://doi.org/10.1016/j.jclepro.2014.10.007

Pearson LJ, Pearson L, Pearson CJ (2010) Sustainable urban agriculture: stocktake and opportunities. Int J Agric Sustain 8:7–19. https://doi.org/10.3763/ijas.2009.0468

Proksch G (2017) Creating urban agriculture systems: an integrated approach to design. Routledge, New York

Quagrainie KK, Flores RMV, Kim H-J, McClain V (2018) Economic analysis of aquaponics and hydroponics production in the U.S. Midwest. J Appl Aquac 30:1–14. https://doi.org/10.1080/10454438.2017.1414009

Rothwell A, Ridoutt B, Page G, Bellotti W (2016) Environmental performance of local food: trade-offs and implications for climate resilience in a developed city. J Clean Prod 114:420–430. https://doi.org/10.1016/j.jclepro.2015.04.096

Sanjuan-Delmás D, Llorach-Massana P, Nadal A, Ercilla-Montserrat M, Muñoz P, Montero JI, Josa A, Gabarrell X, Rieradevall J (2018) Environmental assessment of an integrated rooftop greenhouse for food production in cities. J Clean Prod 177:326–337. https://doi.org/10.1016/j.jclepro.2017.12.147

Sanyé-Mengual E (2015) Sustainability assessment of urban rooftop farming using an interdisciplinary approach. Universitat Autònoma de Barcelona, Bellaterra

Sanyé-Mengual E, Oliver-Solà J, Montero JI, Rieradevall J (2015) An environmental and economic life cycle assessment of rooftop greenhouse (RTG) implementation in Barcelona, Spain. Assessing new forms of urban agriculture from the greenhouse structure to the final product level. Int J Life Cycle Assess 20:350–366. https://doi.org/10.1007/s11367-014-0836-9

Sanyé-Mengual E, Martinez-Blanco J, Finkbeiner M, Cerdà M, Camargo M, Ometto AR, Velásquez LS, Villada G, Niza S, Pina A, Ferreira G, Oliver-Solà J, Montero JI, Rieradevall J (2018) Urban horticulture in retail parks: environmental assessment of the potential implementation of rooftop greenhouses in European and south American cities. J Clean Prod 172:3081–3091. https://doi.org/10.1016/j.jclepro.2017.11.103

Simonen K (2014) Life cycle assessment. Routledge, London

Specht K, Siebert R, Hartmann I, Freisinger UB, Sawicka M, Werner A, Thomaier S, Henckel D, Walk H, Dierich A (2014) Urban agriculture of the future: an overview of sustainability aspects of food production in and on buildings. Agric Hum Values 31:33–51. https://doi.org/10.1007/s10460-013-9448-4

Stadler MM, Baganz D, Vermeulen T, Keesman KJ (2017) Circular economy and economic viability of aquaponic systems: comparing urban, rural and peri-urban scenarios under Dutch conditions. Acta Hortic 1176:101–114. https://doi.org/10.17660/ActaHortic.2017.1176.14

Stewart ID, Oke TR (2010) Thermal differentiation of local climate zones using temperature observations from urban and rural field sites. Presented at the Ninth symposium on urban environment, Keystone, CO, p 8

Suhl J, Dannehl D, Kloas W, Baganz D, Jobs S, Scheibe G, Schmidt U (2016) Advanced aquaponics: evaluation of intensive tomato production in aquaponics vs. conventional hydroponics. Agric Water Manag 178:335–344. https://doi.org/10.1016/j.agwat.2016.10.013

Thomaier S, Specht K, Henckel D, Dierich A, Siebert R, Freisinger UB, Sawicka M (2015) Farming in and on urban buildings: present practice and specific novelties of zero-acreage farming (ZFarming). Renewable Agric Food Syst 30:43–54. https://doi.org/10.1017/S1742170514000143

Van Woensel L, Archer G, Panades-Estruch L, Vrscaj D, European Parliament, Directorate-General for Parliamentary Research Services (2015) Ten technologies which could change our lives: potential impacts and policy implications: in depth analysis. European Commission/EPRS European Parliamentary Research Service, Brussels

Zabalza Bribián I, Aranda Usón A, Scarpellini S (2009) Life cycle assessment in buildings: state-of-the-art and simplified LCA methodology as a complement for building certification. Build Environ 44:2510–2520. https://doi.org/10.1016/j.buildenv.2009.05.001

Zhang H, Burr J, Zhao F (2017) A comparative life cycle assessment (LCA) of lighting technologies for greenhouse crop production. Journal of Cleaner Production, Towards eco-efficient agriculture and food systems: selected papers addressing the global challenges for food systems, including those presented at the Conference "LCA for Feeding the planet and energy for life" (6–8 October 2015, Stresa & Milan Expo, Italy) 140:705–713. https://doi.org/10.1016/j.jclepro.2016.01.014

Part V
Aquaponics and Education

Chapter 22
Aquaponics as an Educational Tool

**Ranka Junge, Tjasa Griessler Bulc, Dieter Anseeuw,
Hijran Yavuzcan Yildiz, and Sarah Milliken**

Abstract This chapter provides an overview of possible strategies for implementing aquaponics in curricula at different levels of education, illustrated by case studies from different countries. Aquaponics can promote scientific literacy and provide a useful tool for teaching the natural sciences at all levels, from primary through to tertiary education. An aquaponics classroom model system can provide multiple ways of enriching classes in Science, Technology, Engineering and Mathematics (STEM), and the day-to-day maintenance of an aquaponics can also enable experiential learning. Aquaponics can thus become an enjoyable and effective way for learners to study STEM content, and can also be used for teaching subjects such as business and economics, and for addressing issues like sustainable development, environmental science, agriculture, food systems, and health. Using learner and teacher evaluations of the use of aquaponics at different educational levels, we attempt to answer the question of whether aquaponics fulfils its promise as an educational tool.

R. Junge (✉)
Institute of Natural Resource Sciences Grüenta, Zurich University of Applied Sciences, Wädenswil, Switzerland
e-mail: ranka.junge@zhaw.ch

T. G. Bulc
Faculty of Health Sciences, University of Ljubljana, Ljubljana, Slovenia
e-mail: tjasa.bulc@zf.uni-lj.si

D. Anseeuw
Inagro, Roeselare, Belgium
e-mail: info@inagro.be

H. Yavuzcan Yildiz
Department of Fisheries and Aquaculture, Ankara University, Ankara, Turkey
e-mail: yavuzcan@ankara.edu.tr

S. Milliken
School of Design, University of Greenwich, London, UK
e-mail: S.Milliken@greenwich.ac.uk

S. Goddek et al. (eds.), *Aquaponics Food Production Systems*,
https://doi.org/10.1007/978-3-030-15943-6_22

Keywords Aquaponics · Education · Aquaponics course · Vocational training · Higher education · Survey

22.1 Introduction

Aquaponics is not only a forward-looking food production technology; it also promotes scientific literacy and provides a very good tool for teaching the natural sciences (life and physical sciences) at all levels of education, from primary school (Hofstetter 2007, 2008; Bamert and Albin 2005; Bollmann-Zuberbuehler et al. 2010; Junge et al. 2014) to vocational education (Baumann 2014; Peroci 2016) and at university level (Graber et al. 2014).

An aquaponic classroom model system provides multiple ways of enriching classes in Science, Technology, Engineering, and Mathematics (STEM). The "hands-on" approach also enables experiential learning, which is the process of learning through physical experience, and more precisely the "meaning-making" process of an individual's direct experience (Kolb 1984). Aquaponics can thus become an enjoyable and effective way for learners to study STEM content. It can also be used for teaching subjects such as business and economics, addressing issues such as sustainable development, environmental science, agriculture, food systems, and health (Hart et al. 2013).

A basic aquaponics can be built easily and inexpensively. The World Wide Web is a repository of many examples of videos and instructions on how to build aquaponics from a variety of components, resulting in a range of different sizes and set-ups. Recent investigations of one such prototype micro-aquaponics showed that despite being small, it can mimic a full-scale unit and it is an effective teaching tool with a relatively low environmental impact (Maucieri et al. 2018). However, implementing aquaponics in classrooms is not without its challenges. Hart et al. (2013) report that technical difficulties, lack of experience and knowledge, and maintenance over holiday periods can all pose significant barriers to teachers using aquaponics in education, and that disinterest on the teacher's part may also be a crucial factor (Graham et al. 2005; Hart et al. 2014). Clayborn et al. (2017), on the other hand, showed that many educators are willing to incorporate aquaponics in the classroom, particularly when an additional incentive, such as hands-on experience, is provided.

Wardlow et al. (2002) investigated teachers' perceptions of the aquaponic unit as a classroom system and also illustrated a prototype unit that can easily be constructed. All teachers strongly agreed that bringing an aquaponics unit into the classroom is inspiring for the students and led to greater interaction between students and teachers, thereby contributing to a dialogue about science. On the other hand, it is unclear exactly how the teachers and students made use of the aquaponics and the instructional materials offered. Hence, the information needed to evaluate the impact of aquaponics classes on meeting the objectives of the students' curricula is still missing. In a survey on the use of aquaponics in education in the USA (Genello et al.

2015), respondents indicated that aquaponics were often used to teach subjects, which are more exclusively focused on STEM topics. Aquaponics education in primary and secondary schools is science-focused, project-oriented, and geared primarily toward older students, while college and university aquaponics were generally larger and less integrated into the curriculum. Interdisciplinary subjects such as food systems and environmental science were taught using an aquaponics more frequently at colleges and universities than they were at schools, where the focus was more often on single discipline subjects such as chemistry or biology. Interestingly, only a few vocational and technical schools used aquaponics to teach subjects other than aquaponics. This indicates that for these educators, aquaponics is a stand-alone subject and not a vehicle to address STEM or food system topics (Genello et al. 2015).

While the studies mentioned above reported aquaponics as having the potential to encourage the use of experimentation and hands-on learning, they did not evaluate the impact of aquaponics on learning outcomes. Junge et al. (2014) evaluated aquaponics as a tool to promote systems thinking in the classroom. The authors reported that 13–14 year old students (seventh grade in Switzerland) displayed a statistically significant increase from pre- to post-test for all the indices measured to assess their systems thinking capacities. However, since the pupils did not have any prior knowledge of systems thinking, and since there was no control group, the authors concluded that supplementary tests are needed to evaluate whether aquaponics has additional benefits compared to other teaching tools. This issue was addressed in the study by Schneller et al. (2015) who found significant advances in environmental knowledge scores in 10–11 year old students compared to a control group of 17 year olds. Moreover, when asked for their teaching preferences, the majority of students indicated that they preferred hands-on experiential pedagogy such as aquaponics or hydroponics. The majority of the students also discussed the curriculum with their families, explaining how hydroponic and aquaponics work. This observation extends the belief that hands-on learning using aquaponics (and hydroponics) not only has a stimulating impact on teachers and students, but also leads to intergenerational learning.

The objective of this chapter is to provide an overview of possible strategies for implementing aquaponics in curricula at different levels of education, illustrated by case studies from different countries. Based on evaluations conducted with some of these case studies, we attempt to answer the question of whether aquaponics fulfils its promise as an educational tool.

22.2 General Scenarios for Implementing Aquaponics in Curricula

The introduction of aquaponics into schools may be an aspiration, but in many countries, primary and secondary schools have rigid curricula with learning objectives that must be met by the end of each school year. Commonly, these objectives,

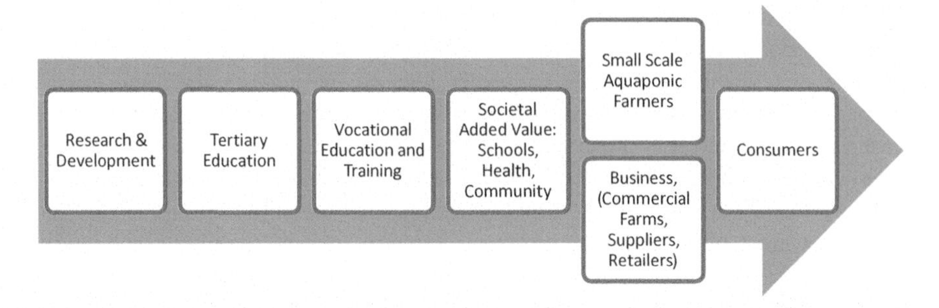

Fig. 22.1 An aquaponics can address various goals or stakeholders by offering to develop key competences in appropriate educational and training processes. (Modified after Graber et al. 2014)

called attainment terms or outcome competencies, are course-specific and defined by the education authorities. Thus, this calls for a well-thought-out strategy to successfully introduce an aquaponics in school classes. In comparison, colleges and universities have more freedom to map out their own curricula.

22.2.1 Which Types of Aquaponics Are Suitable for Education?

There are, as stated above, many aquaponics described and illustrated on the web. It is also possible to purchase a kit, or have a complete system delivered and installed. However, building an aquaponics is in itself a valuable educational experience, and the fact that it is not delivered to the classroom ready-made adds to its instructional value.

An aquaponics can address various goals or stakeholders (Fig. 22.1). To attain all of these, the components of a system have to fulfill various requirements (Table 22.1). The choice of what kind of aquaponics is suitable for a particular institution should result from a realistic assessment of its facilities and the educational objectives.

Maucieri et al. (2018) proposed a general classification of aquaponics according to different design principles. While a system can simultaneously fulfill several objectives, including greening and decoration, social interaction, and food production, here we assume that the main objective is education. If we follow the classification of Maucieri et al. (2018), which categorizes the aquaponics according to several categories (stakeholder, size), several distinct options for choosing a suitable aquaponics emerge (Table 22.2). Any decision has to be made within the limits of the available budget, though it is possible to construct a system at very low cost.

Table 22.1 General requirements for three types of educational aquaponics

Aspect	Research (basic and applied)	Tertiary education (BSc, MSc, PhD)	Societal added value: Education (primary and secondary), vocational training, communication, health benefits
Access	Good access for daily work and monitoring,	Good access for daily work and monitoring; Good access for groups	Good access for groups
Size	Reasonable size for scaling-up for potential commercial farms (depending on the crop)	Reasonable size for a good overview of different cultivation options	Reasonable size for a good overview
Construction	Easy remodeling[a]	Easy remodeling	Mostly commercial off-the-shelf elements
Climate control	Advanced	Basic	Basic
Diversity of production methods	Variable according to current research projects[b]	Variable to high: from basic (demonstration of system) to cutting edge (research)	High: from basic (demonstration of system) to cutting edge (demonstration of potential)
Recycling, closed-loop systems	Quantitative importance: improving the ecological footprint and thus reducing costs	Quantitative and qualitative importance	Qualitative importance: demonstration of ecological principles
Provision of energy from renewable sources	Quantitative importance: improving the ecological footprint and thus reducing costs	Quantitative and qualitative importance	Qualitative importance: demonstration of ecological principles
Rainwater harvesting, treatment, and use	Quantitative importance: improving the ecological footprint and thus reducing costs	Quantitative and qualitative importance	Qualitative importance: demonstration of ecological principles

Modified after Graber et al. (2014)
[a]allows testing of different set-ups
[b]from state of the art (aligned with current practices of professional vegetable growers and fish farmers) to cutting edge (testing innovative production methods)

Additional questions to be asked before installing an aquaponics are

- What size of system to choose? The size of the system will most probably increase in relation to the age of the students: smaller systems in kindergarten and larger systems in high school.
- Where is the system to be placed? Micro-systems (Table 22.2) can be placed in a classroom. However, very small and small systems (Table 22.2) require more space and perhaps a greenhouse will need to be constructed to house these.
- Is the system going to be a temporary or a permanent feature? If it is going to be a permanent feature, who will take care of the system during the holidays? If it is going to be a temporary feature, the institution might consider borrowing an

Table 22.2 Suitability of different design options for an educational aquaponics. The green color denotes the most suitable options, yellow options are less suitable, while red options are not suitable for the majority of cases

Size	Operational mode of the aquaculture compartment	Water cycle management	Water type	Type of implemented hydroponic system	Use of space
XXS micro systems (<5 m^2)	Extensive[a]	Closed loop systems[c]	Freshwater (Tap or Well)	Grow beds with different media	Horizontal
XS very small (5-50 m^2)	Intensive[b]	Open loop systems[d]	Rainwater	Ebb-and-flow system	Vertical
S small (51-200 m^2)			Salt / Brackish water	Grow bags	
M medium (201-1000 m^2)				Drip irrigation	
L large (>1000 m^2)				Deep water cultivation (rafts)	
				Nutrient film technique (NFT)	

[a]Extensive (fish density is mostly under 10 kg/m^3 and allows for integrated sludge usage in grow beds).
[b]Intensive (fish density requires additional sludge separation; however, the sludge has to be treated separately).
[c]Closed loop ("coupled" systems): after the hydroponic component, the water is recycled to the aquaculture component.
[d]Open loop or end-of pipe ("decoupled" systems): after the hydroponic component, the water is either not or only partially recycled to the aquaculture component.

aquarium from an aquarist among the staff or the students, who would also be able to give advice on fish care.

- Are the fish going to be harvested? Animal welfare should always be observed and killing the fish should be done according to animal protection laws (Council of the European Union 1998). Children might have problems in killing and eating a living animal, which resembles Dory (from the movie finding Nemo). If the fish are not going to be harvested, then goldfish or Koi are a good option.
- Are the plants going to be harvested and eaten? If yes, then suggestions for using the produce need to be prepared. If not, then consider using ornamental plants instead.

22.2.2 How to Embed Aquaponics as a Didactic Tool?

An aquaponics with living fish and plants obviously provides the potential for long-term engagement compared to conventional single discipline scientific experiments. While this is a manifest asset for progressive and continuous experiential learning, it has been indicated that safeguarding the teacher's interest in the long run and the provision of learning material are key challenges to successfully incorporating aquaponics in school classes (Hart et al. 2013; Clayborn et al. 2017).

Ideally, the model aquaponics should be embedded in different classes in a way that it facilitates attaining course-specific educational goals. Subjects, which promote an understanding of natural cycles, waste recycling, and environmental protection, are the most obvious. However, aquaponics can also be used in other subjects, such as art, social sciences, and economics. The examples discussed in Examples 22.1, 22.2, 22.3, 22.4, 22.5, 22.6 and 22.7 below provide an insight into the versatility of aquaponics in education.

Active aquaponics can be used for teaching over different time periods, and accordingly there are distinct scenarios:

(a) Over one term, 1–2 classes per week (8–12 weeks) (see Examples 22.1 and 22.3)
(b) As a half- to one-day educational activity (see Example 22.4)
(c) As a Science Week or Project Week on 2–5 consecutive days (see Example 22.2)
(d) As an extracurricular activity, during one term of 10–15 weeks
(e) As a permanent feature for the whole school, thus providing a focal "conversation piece" and study/research facility for several classes (see Examples 22.5 and 22.6, Graber et al. 2014)

22.3 Aquaponics in Primary Schools

According to the International Standard Classification of Education (UNESCO-UIS 2012), primary education (or elementary education in American English) at ISCED level 1 (first 6 years) is typically the first stage of formal education. It provides children from the age of about 5–12 with a basic understanding of various subjects, such as maths, science, biology, literacy, history, geography, arts, and music. It is therefore designed to provide a solid foundation for learning and understanding core areas of knowledge, as well as personal and social development. It focuses on learning at a basic level of complexity with little, if any, specialization. Educational activities are often organized with an integrated approach rather than providing instruction in specific subjects.

The educational aim at ISCED level 2 (further 3 years) is to lay the foundation for lifelong learning and human development upon which education systems may then expand further educational opportunities. Programs at this level are usually organized around a more subject-oriented curriculum.

According to the United Nations Children's Fund (UNICEF 2018), providing children with primary education has many positive effects, including increasing environmental awareness.

At primary school age, children's rich but naïve understandings of the natural world can be built on to develop their understanding of scientific concepts. At the same time, children need carefully structured experiences, taking into account their prior knowledge, instructional support from teachers, and opportunities for sustained engagement with the same set of ideas over longer periods (Duschl et al. 2007). One

Fig. 22.2 (**a**) Opening ceremony in the school of Älandsbro, (**b**) The simple aquaponics at Älandsbro, (**c**) Older students making observations for the "Recirculation Book," (**d**) Model built by the younger students during an arts class

way of providing sustained and continuous engagement can be through the building, management, and maintenance of an aquaponics.

Key advice for introducing aquaponics to primary school students is as follows:

- Low-tech and robust classroom systems favor the engagement of both the teacher and the students and are most effective for this stage of education (Example 22.1, Fig. 22.2b).
- Productivity is not a central issue but demonstrating the laws of nature (cycling of nutrients, energy flow, population dynamics, and interactions within the ecosystem) is. Therefore, sufficient effort needs to be put into developing learning materials to meet the goals of the curriculum.
- From an educational point of view, understanding the chemical, physical, and natural processes in an aquaponics, albeit through trial and error, is more important than achieving a perfectly run system.

- Include a wide range of activities: drawing the plants and animals, keeping a class journal, measuring the water quality, monitoring the fish (size, weight, and well-being), feeding the fish, cooking the produce, role playing, writing, prose, poetry, and song.

Example 22.1 Aquaponics at Älandsbro skola, Primary School (Sweden)
The 10-month project, which was a part of the FP6 project "Play with Water," started with an opening ceremony in September (Fig. 22.2a) and ended before the summer holidays. Several teachers and about 90–100 students aged between 9 and 12 years were involved and were very enthusiastic about the project. The school used readily available materials to create two simple table-top systems with an aquarium, fish and plants (Fig. 22.2b). Before the start of the learning activities, the students filled in a questionnaire (see the section on Assessment), which was then repeated at the end of term. After an introduction to aquaponics, the students planted the system and populated it with fish.

A diary, called a "Recirculation Book," was kept for each aquarium. Students made daily notes about the systems (Fig. 22.2c). They recorded the pH, temperature, nitrate and nitrite concentrations, the length of the plants, the activity of the fish, and when they added food and water into the system. They also made drawings and described any significant events, which occurred.

Different classes in the school then had different weeks where they took over daily responsibility for the systems. The younger students took care of one system, while the second system was used by the older children.

The aquaponic units were used for teaching different subjects. The younger students worked with the concept of recirculation by building cardboard models of an aquaponics, with tubes, pumps, fish, and plants (Fig. 22.2d). They also worked with paintings, drama, and music in order to increase their understanding of the relationship between the plants and the fish.

The older students collected information about pH, temperature, nitrate, nitrite, and other changes in the "Recirculation Book." They were pursuing different themes, for example: (i) the water cycle in a global perspective; (ii) the everyday use of water in a house; (iii) the different appearances, smells, and tastes of water; (iv) fish biology and ecology; (v) other ecological recirculation systems; and (vi) the importance of water on a global scale. They also taught the younger pupils, for example, by explaining recirculation systems, or demonstrating small experiments.

For the evaluation, see Sect. 22.8.1.2.

Example 22.2 Two-day Aquaponic-Centered Course in Waedenswil, Switzerland

Over 2 days, 16 students (aged 13–14 years) and their teacher from the Gerberacher School visited the Zurich University of Applied Sciences (ZHAW) campus in Waedenswil, where an undergraduate student had prepared a two-day program about the importance of water using aquaponics as a focus (Fig. 22.6). Learning progression was assessed by means of a questionnaire (pre-activity and post-activity).

Day 1:

- Welcome address, explanation of the course schedule.
- Knowledge test (what do students know about aquaculture, recycling, plant nutrition, ecosystems, etc.)
- The concept of "systems" explained through simple analogy with hammer as an example of a system (The hammer is made of two parts: the handle and the head. If the parts are separated, the hammer cannot function. So, the hammer is more than just the sum of its parts, it is a system.)
- Assessment of the understanding of systems before the teaching unit: What is a system? Fill in the gap text.
- Introduction to aquaponics and ecosystems. The students learn what an ecosystem is and understand that individual systems are integrated into it.
- Visit to the demonstration aquaponics (Fig. 22.3a).
- Expanding the knowledge: The importance of proteins in food. Discuss and fill in the gap text.
- Expanding the knowledge: The benefits of a closed water cycle (Discuss and fill in the Worksheet).
- Practical work: Construction of two simple aquaponics, adding plants (basil), measuring nitrite and pH.
- Basics of tilapia (*Oreochromis niloticus*) and basil, *Ocimum basilicum* biology.
- Global importance of water (Role-playing game, Worksheets).
- Time for questions.

Day 2:

- Why is water conservation necessary? How many people perish due to lack of drinking water? (Mathematics Task).
- Measure the pH and nitrite content in the aquaria. Students learn how to carry out an Aqua-Test and what the values indicate.
- Answer repetition questions (Card game with rewards, Worksheets).
- Practical work: Transfer tilapia from the aquarium to the aquaponics. Feed the fish. Fill in the fish observation sheet.
- Draw a poster of Aquaponics, explaining the important terms (Fig. 22.3b).
- Final knowledge test and evaluation of the learning unit (see also Sect. 22.8.1.3).

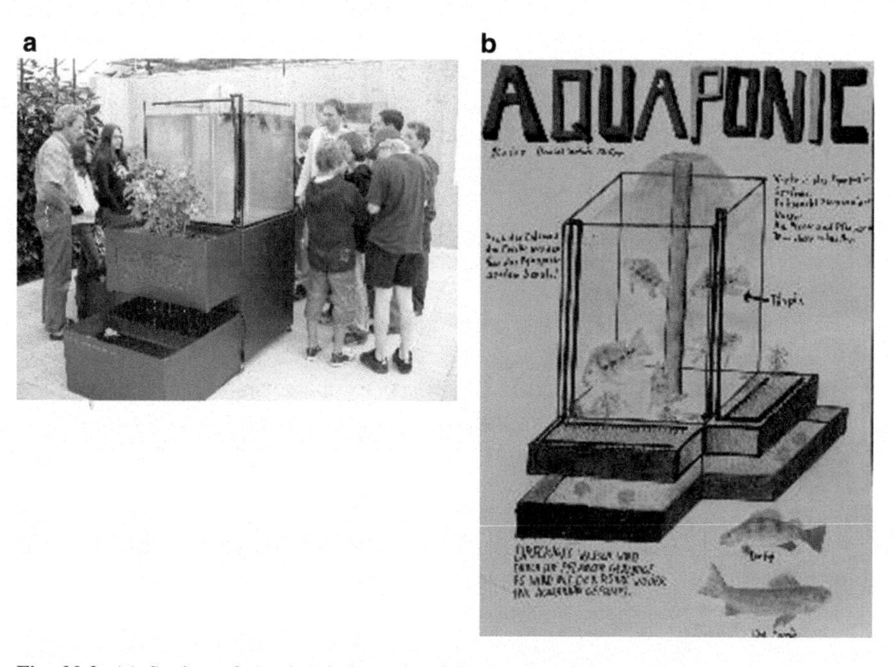

Fig. 22.3 (a) Students from the sixth grade of Gerberacher School visiting the demonstration aquaponics at Zurich University of Applied Sciences (Waedenswil, Switzerland). (b) A poster designed by the same students, explaining the basics of aquaponics

22.4 Aquaponics in Secondary Schools

According to the ISCED classification (UNESCO-UIS 2012), secondary education provides learning and educational activities building on primary education and preparing for both first labor market entry as well as post-secondary non-tertiary and tertiary education. Broadly speaking, secondary education aims to deliver learning at an intermediate level of complexity.

While at primary education level, students are mainly directed toward observational and descriptive exercises on organisms and processes in an aquaponics, students from secondary schools can be educated in understanding dynamic processes. Aquaponics enables this increased complexity and fosters system thinking (Junge et al. 2014).

Example 22.3 One Semester Course in a Grammar School in Switzerland

Hofstetter (Hofstetter 2008) implemented aquaponic teaching units in a Grammar School (German: Gymnasium) in Zurich and tested the hypothesis that incorporating aquaponics into teaching has a positive influence on systems thinking (Ossimitz 2000) among the students. Gymnasium students in

(continued)

Example 22.3 (continued)
Switzerland belong to an above-average section of the student population: they have very good grades, are used to autonomous work, and show consistent ability and general interests in different issues. Three seventh grade classes were involved, with a total of 68 students (32 female, 36 male), aged between 12 and 14 years old.

Six simple, small aquaponics were constructed according to the general description in Bamert and Albin (2005) (Fig. 22.4). The students were responsible for the construction, operation, and monitoring of the systems. They were provided with the necessary materials and built the aquaculture and hydroponic units. Tomato (*Solanum lycopersicum*) and basil (*Ocimum basilicum*) seedlings were planted in expanded clay beds. Each aquarium was stocked with two common rudd (*Scardinius erythrophthalmus*) caught in a nearby pond and returned there after the experiment.

Each system was monitored daily, and the following operations were carried out: measuring plant height, observing plant health, measuring fish feed and feeding the fish, monitoring fish behavior, measuring water temperature, and topping up the aquarium with water. All the measurements and observations were documented in a diary, which also served to transfer information between the three groups that worked on the same system.

The teaching sequence (Table 22.3) took place between October 2007 and January 2008. Several themes were introduced in the lessons basic system concepts (relationship between system components, concepts of feedback, and self-regulation), and basic knowledge about aquaponics. All the teaching units are described in detail in Bollmann-Zuberbuehler et al. (2010). The effect of the teaching sequence on systems thinking competences was assessed at the beginning and at the end of the sequence (See Sect. 22.8.1.4) and was described in detail in Junge et al. (2014).

Example 22.4 EXPLORLabor: Science Day at Zurich University of Applied Sciences for the Students of Secondary Schools, Switzerland
Twenty students aged 18–19 (11th school year, majoring in Biology & Chemistry) from the Cantonal School at Menzingen visit the Zurich University of Applied Sciences (ZHAW) every year for an Aquaponics Workshop. The program varies slightly from year to year, depending on the current experiments in the Aquaponics Lab.

(continued)

Example 22.4 (continued)
Program:

- Greetings: Introduction to the workshop.
- E-learning video: Introduction to aquaponics.
- Tour of the aquaponic demo facility; discussion of appropriate behavior in the experimental facility.
- Learning measurement methods. Division into 4 teams.
- Tour of the Aquaponics Lab, consisting of 4 systems (three aquaponic and one hydroponic). Each team gathered data from one system.

Assignment:

1. Measuring water quality in different parts of the aquaponic and hydroponic systems (fish tank, biofilter, and sump) using the handheld multi-electrode meter (Hach Lange GmbH, Rheineck, CH) to measure the temperature (T), pH, oxygen content, and electric conductivity (EC).
2. Using the Dualex-Clip to measure Nitrogen Balance Index (NBI), Chlorophyll Content (CHL), Flavonoid Content (FLV), and Anthocyanin Content (ANTH) of leaves of three lettuces.
3. Filling in the data in the pre-prepared Excel spreadsheet.
4. Back to the classroom: Calculating if there are differences between the lettuce plants that grow in aquaponic and hydroponic systems, comparison of the data, and discussions.

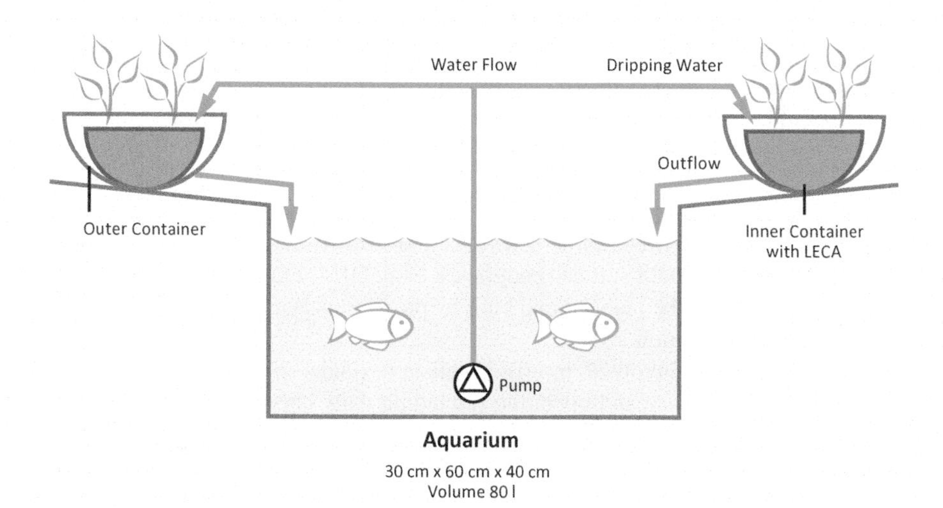

Fig. 22.4 Simple classroom aquaponics. (Adapted after Bamert and Albin 2005). The plants grow in the containers filled with light expanded clay aggregate (LECA) that is usually used in hydrocultures

Table 22.3 Sequence of teaching units in three classes of seventh grade students during one semester course in a Grammar School in Switzerland

Teaching unit	Number of lessons	Methods	Content
TU1	1	Survey of existing knowledge	Pre-activity Test
TU2	4	Lecture by teacher, research, & presentations by students	System basics
TU3	2	Lecture by teacher, student assignment	"Connection circle" tool allows the students to draw a diagram of a system (adopted from Quaden and Ticotsky 2004)
TU4	2	Discovery learning Presentations by students	Planning an aquaponics: sub-units, connections
TU5	2	Problem-based learning (PBL)	Defining the main indicators of the system: Fish and plants and their interactions
TU6	3	Discovery learning	Monitoring the aquaponics
TU7	3	Presentations by students	Drawing a diagram of the interconnections in the aquaponics
TU8	1	Survey of knowledge	Post-activity test
TU9	2	Aquaponic party	Harvest, preparation of salad, eating

Modified after Junge et al. (2014)

22.5 Aquaponics in Vocational Education and Training

UNESCO-UIS/OECD/EUROSTAT (2017) defines vocational education programs as "*designed for learners to acquire the knowledge, skills and competencies specific to a particular occupation, trade, or class of occupations or trades. Successful completion of such programs leads to labour market relevant, vocational qualifications acknowledged as occupationally-oriented by the relevant national authorities and/or the labour market*" (UNESCO, 2017).

In order to educate future aquaponic farmers and aquaponic technicians, the training has to include the professional operation of aquaponics. Therefore, the training environment needs to be state-of-art. However, the setting does not have to be large: 30 m^2 should suffice (Podgrajsek et al. 2014, Examples 22.5 and 22.6). Such systems should be planned and built by professionals as they require complex monitoring and operation.

Students can be involved in: (i) installation (under professional guidance); (ii) general maintenance and operation (including daily checks and cleaning); (iii) operation of the hydroponic subsystem (planting, harvesting, integrated pest management, climate control, adjustment of pH and nutrient levels, etc.); (iv) operation of the aquaculture sub-system (fish feeding, fish weight determinations, adjustment of pH levels, etc.); (v) monitoring of parameters (water quality, fish growth and health, plant growth, and quality); and (vi) harvesting and post-harvest operations.

The European Union invested in the development of vocational education via the Leonardo Program, and more recently ERASMUS+. These programs have funded several projects that included the implementation of aquaponics, including the Leonardo da Vinci Transfer of Innovation Project (Lifelong Learning Programme) "Introducing Aquaponics in VET: Tools, Teaching Units and Teacher Training" (AQUA-VET)'. The project prepared a curriculum for vocational education in aquaponics and the results are available at www.zhaw.ch/iunr/aquavet. The teaching units were tested at three vocational schools in Italy, Switzerland (Baumann 2014), and Slovenia (Peroci 2016). As part of this project an aquaponics unit was constructed at the Biotechnical Centre Naklo vocational school in Slovenia (Example 22.5). Another example is the aquaponic unit built at the Provinciaal Technisch Insituut, a horticulture school in Belgium (Example 22.6).

Example 22.5 Aquaponics at Biotechnical Centre Naklo in Slovenia
The aquaponics (Fig. 22.5a) was constructed in 2013 within the framework of the Leonardo da Vinci project "AQUA-VET" (Krivograd Klemenčič et al. 2013; Podgrajšek et al. 2014). An aquaponics course module was developed and taught to a class of 30 students in the second year of their Environmental Technician program (Peroci 2016). The aim was to investigate the possibility of including aquaponics in the regular program of secondary vocational education in biotechnical sciences, and in the standard curriculum of professional skills needed for the national vocational qualification of an "Aquaponic farmer." The course consisted of six lessons (45 min each), of which four were dedicated to the theory of aquaponics and two to practical training (Fig. 22.5b).

Lesson 1 Aquaponics: definition, introduction to aquaculture and hydroponics, operation of an aquaponics.

Lesson 2 Microorganisms: (i) the role of microorganisms: useful microorganisms, the nitrogen cycle, and the importance of biofilters in aquaponics; (ii) monitoring of selected parameters, monitoring plan, protocols, and evaluation of the results.

Lesson 3 Fish: structure and functioning of the fish body, selection of fish species suitable for growing in aquaponics, feeding methods, fish diseases and injuries, and fish breeding.

Lesson 4 Plant anatomy, selection of plant species suitable for growing in aquaponics, the role of plants in aquaponics, identification of plant diseases, and appropriate plant protection strategies.

Practical work (2 h) Students worked in two groups (observation, monitoring, discussion, presentation) at the aquaponic unit at the Biotechnical Centre Naklo.

The learning activities covered various skill levels, providing well-designed lessons for both the theoretical and practical parts. Learning progression was assessed by means of several questionnaires (see Sect. 22.7.2).

Example 22.6 Aquaponics at Provinciaal Technisch Insituut, a Horticulture School in Belgium

The Provinciaal Technisch Instituut (PTI) in Kortrijk pioneered classroom aquaponics in Belgium at the vocational education level. The project started in 2008, when fish tanks were introduced into the greenhouse that is used for teaching practical courses to students in agronomy and biotechnology (Fig. 22.6).

Initially, the aquaponics was used on an ad hoc basis in a number of classes, for example, to teach plant and fish biology, water chemistry, etc., but it was only after a couple of years that aquaponics became structurally embedded in a number of course modules. The main challenge was to find the time to translate the governmental attainment goals into the aquaponics course modules. This challenge illustrates the importance of providing sufficient support both in terms of meeting the needs of the curriculum, and in tackling operational and organizational obstacles for new aquaponic initiatives in schools.

From the beginning, the aquaponics at PTI also acted as a pilot in applied research projects with universities (Ghent University, KU Leuven, ULG Gembloux), university colleges (HoWest, HoGent, Odisee), research institutes (Inagro, PCG), and private companies (aqua4C, Agriton, Lambers-Seeghers, Vanraes automation). In effect, the school has become a valuable partner in multiple national and international projects. Pupils at PTI are involved not only in the day-to-day operational work but also in data collection for experiments coordinated by the academic researchers. This collaboration creates a unique opportunity for the students to familiarize themselves with the activities of university colleges and universities, and may stimulate the students to progress to higher education.

Fig. 22.5 (**a**) Aquaponics at the Biotechnical Centre Naklo in Slovenia. (Photo: Jarni 2014). (**b**) Practical work at the aquaponic unit of the Biotechnical Centre Naklo. (Photo Peroci 2016)

Fig. 22.6 A view into the greenhouse of Provinciaal Technisch Insituut (PTI). Fish tanks (containing *Scortum barcoo*) are located below the drip-irrigated tomato gullies. In the middle of the greenhouse, Australian crayfish (*Cherax quadricarinatus*) are grown in a series of aquaria

22.6 Aquaponics in Higher Education

Higher education programs need to be adapted to meet the expectations of the new millennium, such as long-term food security and sovereignty, sustainable agriculture/food production, rural development, zero hunger, and urban agriculture. These important drivers mean that higher education institutions involved in the areas of food production can play a key role in the teaching of aquaponics through both capacity development and knowledge creation and sharing. Additionally, it is clear that the interest in teaching and learning aquaponics is increasing (Junge et al. 2017).

At universities and colleges, aquaponics is usually taught as part of agriculture, horticulture, or aquaculture courses and the context for course content development in higher education is specific to each institution's internal and external dynamics. The main challenge in designing courses at higher education level is the interdisciplinary nature of aquaponics, as prior knowledge of both aquaculture and horticulture is essential. While some studies investigated the use of aquaponics in education (Hart et al. 2013; Hart et al. 2014; Junge et al. 2014; Genello et al. 2015) and a number of on-line courses are available, a course outline for aquaponics at the tertiary level at a main-stream does not yet exist, or at least hasn't been published. For tertiary level aquaponics courses to be implemented in the EU, the Bologna Process, which underlines the need for meaningful implementation of learning outcomes in order to consolidate the European Higher Education Area (EHEA), needs to be followed. Learning outcomes are (i) statements that specify what a learner will know or be able to do as a result of a learning activity; (ii) statements of what a learner is expected to know, understand, and/or be able to demonstrate after completing a process of learning; and (iii) are usually expressed as knowledge, skills, or attitudes (Kennedy 2008).

Table 22.4 and Example 22.7 introduce two conceptual frameworks for teaching aquaponics. Both courses are considered to be worth 5 ECTS credits (European Credit Transfer System), which correspond to a study load of approximately 150 h.

Example 22.7 Project Aqu@teach, an Erasmus+ Strategic Partnership in Higher Education (2017–2020)

The core task of the project is to devise an aquaponics curriculum (150 h of student's workload corresponding to 5 ECTS credits and a supplementary entrepreneurial skills module (60 h), which will be taught by means of blended learning. Blended learning (combining digital media and the Internet with classroom formats that require the physical co-presence of teacher and students) offers alternative pathways to gain knowledge and involve students in creating content. This also improves the preparation of students for their lessons, and fosters their motivation, so that interactions with the teacher can be devoted to in-depth learning and the development of practical skills. Information and Communication Technologies (ICTs) are particularly valuable for teaching aquaponics as they enable effective presentation of systems and processes, such as simulation modeling (graphic, numerical) of the parameters (weight of fish/feeds input/surface area of aquaponic beds, etc.). Students taking the Aqu@teach course will use e-portfolios (Mahara programme) to document their progress in learning. The curriculum will include the following modules:

	Module	No. of hours
1	Aquaponic technology	8
2	Aquaculture	12
3	Fish anatomy, health and welfare	8
4	Fish feeding and growth	10
5	Nutrient water balance	5
6	Hydroponics	13
7	Plant varieties	10
10	Integrated pest management	8
9	Monitoring of parameters	8
10	Food safety	12
11	Scientific research methods	10
12	Design and build	16
13	Urban agriculture	10
14	Vertical aquaponics	8
15	Social aspects of aquaponics	12

(continued)

Example 22.7 (continued)

The use of blended learning to teach a unique multidisciplinary curriculum will enable HE students from a variety of different academic disciplines to join together physically and virtually to gain professional and transversal skills desired by employers. They will gain advanced skills in the circular economy, environmental and ecological engineering, and closed-loop production systems (energy, water, waste), using aquaponics as an example of good practice. At the end of the project in 2020, the module guide and curriculum will be made available on the project website (https://aquateach.wordpress.com/), along with a toolbox of innovative didactic techniques appropriate for teaching aquaponics, a textbook and teaching materials, and a best practice guide for teaching aquaponics. The aquaponics curriculum and supplementary entrepreneurial skills curriculum will be freely accessible as an interactive online aquaponics course.

Table 22.4 Proposed aquaponics course outline at university level (5 ECTS). The flexible framework contains two key topics (hydroponics and aquaculture) and is clustered into six learning areas

Course title:	**Aquaponics**	
Course Description:	Aquaponics is a food production method that combines hydroponics and aquaculture to form a system that re-circulates the water and nutrients and grows terrestrial and aquatic plants including algae and aquatic organisms while minimizing waste discharge. This course allows students to use the technical skills acquired to set up an integrated system. It equips them with the knowledge needed to be able to undertake and be aware of critical aspects of aquaponics.	
Entry Level:	**BSc or MSc**	
Unit Name	**1. Aquaponics**	**2. Aquaponics Operations**
Unit Purpose	To understand system design and management, components, and construction techniques.	To understand water characteristics, the microbiological and biochemical cycles (e.g., the nitrogen cycle) within an aquaponics, and interactions between water and plants.
Recommended prior knowledge and skills:	Basic knowledge of biology and agriculture (horticulture and aquaculture).	Basic knowledge of water quality, aquatic organisms, aquatic microbiology.
Learning outcomes	Students should be able to	Students should be able to
	explain the characteristics of an aquaponics;	explain water quality parameters;
	explain the types of aquaponics;	explain biochemical cycles and microbial transformations;
	explain the construction techniques; and	identify criteria for fish and plant production;
	describe the operational components.	calculate all relevant fish growth parameters; and
		explain harvesting and processing.

(continued)

Table 22.4 (continued)

Knowledge and/or skills:	On completion of the unit, students should be able to design an aquaponics.	On completion of the unit, students should be able to
		describe water quality criteria;
		analyze the main water quality; parameters (dissolved oxygen, pH, ammonia, nitrite, nitrate);
		explain the effects of water temperature on fish and plants;
		explain the optimum water conditions for fish;
		explain the interaction between water and plants;
		explain feeding strategies; and
		compute the required feed rations
Evidence requirements:	Students will be required to evaluate the different types of aquaponics in terms of their relative advantages and disadvantages.	Students will be required to
		Apply a calculation for stocking density and feeding regime using fish size, water volume, and water quality parameters.
Unit Name	**3. Plant Production**	**4. Aquatic Organisms Production**
Unit Purpose	To demonstrate the ability to care for plants in order to maintain optimum growth and health while considering pruning, planting, and irrigation.	To understand the growth of aquatic organisms such as fish and crustaceans and their requirements in an aquaponic system.
	To describe the optimal conditions for plant growth.	
Recommended prior knowledge and skills:	Basic knowledge of horticulture.	Basic physiology of aquatic organisms, nutritional concepts, and reproduction of fish and crustaceans in aquaculture.
	Computer/technical literacy.	
Learning outcomes	Students should be able to	Students should be able to
	describe how to plant a plant;	explain the growth of aquatic organisms, nutritional principles in aquaculture, fingerling production, broodstock management, breeding/fry sex reversal; and
	describe plant growth requirements;	explain the principles of aquaculture.

(continued)

Table 22.4 (continued)

	identify the most suitable plants for the aquaponics;	
	describe seed production techniques;	
	describe transplantation techniques;	
	allocate suitable plants for different seasons;	
	define water pressure; and	
	flow rate and how to calculate these;	
	explain how to control pests; and	
	explain harvesting techniques.	
Knowledge and/or skills:	On completion of the unit, students should be able to explain	On completion of the unit, students should be able to:
	seedling production	describe the functioning of RAS
	plant growth	identify the fish/crustacean species suitable for aquaponics
	disease and insect control	identify the growth requirements of different aquatic organisms
	harvesting	
Evidence requirements:	Students will be required to	Students will be required to
	Compare different plants regarding their suitability for aquaponics	Describe the role of aquaculture within an aquaponics
Unit Name	**5. Economics of Aquaponics**	**6. Risk Management Strategy**
Unit Purpose	To understand the processes required to set up an economically viable system	To understand the risk management elements including risk identification, analysis, responses, and control
Recommended prior knowledge and skills:	Basic mathematical and statistical knowledge	Basic mathematical knowledge
Learning outcomes	Students should be able to	Students should be able to
	explain profitability and sustainability;	identify the potential risks;
	estimate the depreciation, capital expenditure, operating expenses, sales, profit and loss statement;	develop a risk management plan;
	explain the calculations used to determine cash flows;	analyze quantitative and qualitative risks; and
	discuss the balance of the budget; and	monitor and control the risks.
	rate the financial indicators.	

(continued)

Table 22.4 (continued)

Knowledge and/or Skills:	On completion of the unit, students should be able to create an economic feasibility model.	On completion of the unit, students should know how to create a probability impact matrix relating to risks.
Evidence requirements:	Students will be required to	Students will be required to
	Create a feasibility model using different financial indicators based on a case study	Create a probability impact matrix of risk based on a case study
Outcome Assessment Activities:	Oral/computer presentation, written report, classroom quizzes.	

22.7 Does Aquaponics Fulfill Its Promise in Teaching? Assessments of Teaching Units by Teachers

22.7.1 Teacher Interviews in Play-With-Water

Aquaponic teaching units were assessed in the FP6 project "Play-With-Water" on seven separate occasions in three countries (Sweden, Norway, Switzerland). This involved six schools (1 school in Norway, 1 in Sweden, and 4 in Switzerland) where the age of students ranged between 7 and 14 years. Six teachers were asked to keep a diary, which they then used to answer an online questionnaire complemented with phone interviews, which are summarized in Table 22.5.

Feedback from teachers on their experience with the aquaponics indicated that some issues were too complex for primary schools. The "Play-With-Water" experiments such as those available on the project website (www.zhaw.ch/iunr/play-with-water/) may be more appropriate for use in secondary education. The learning

Table 22.5 Summarized answers of the six interviewed teachers regarding the advantages and disadvantages of using aquaponics as a teaching tool

What are the main advantages?	Number of mentions	What are the main disadvantages?	Number of mentions
Suitable to learn system thinking	3	None.	2
Facilitates teamwork	2	High time requirements.	2
mobilization of students	2	High knowledge requirements.	2
Provides diversity in teaching	2	Difficult concepts & language.	1
Motivating for students	1	Sensitive for pests.	1
Motivating for teachers	1	Students were not always paying attention.	1
Transfer between different subjects possible	1		
Versatile: several possible educational objectives	1		

materials contain descriptions of complex processes and ecological interactions that require a deeper knowledge of natural sciences such as chemistry or biology than can be expected at primary school. If the material is to be used by teachers, it needs to provide the information in a classroom-ready format. Explanations of chemical and biological processes such as nitrification ought to be greatly simplified.

22.7.2 Comprehensive Study of the Potential for Including Aquaponics in Secondary Vocational Education in Slovenia

Peroci (2016) investigated a series of aspects related to the potential for including aquaponics in the educational process of secondary vocational education in Slovenia (Fig. 22.7). This included

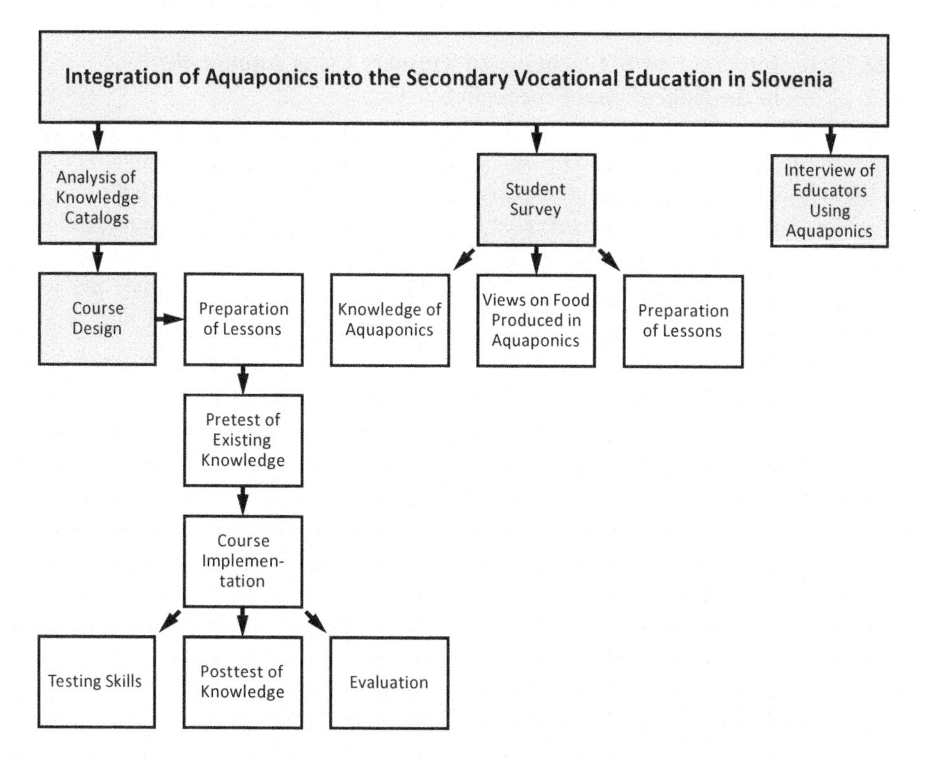

Fig. 22.7 The general structure of the study of Peroci (2016) about the potential for including aquaponics in the educational process of secondary vocational education in Slovenia

(i) Analysis of catalogs of vocational secondary education in biotechnical fields in order to assess the compatibility of these educational programs with learning objectives related to aquaponics.
(ii) Design of a short Aquaponic Educational Course including the definition of learning outcomes (knowledge and skills). The didactic material for experiential learning was tested and evaluated by a class of students at the Biotechnical Centre Naklo (Precedent 5, Sect. 22.8.2).
(iii) Survey of the knowledge of, and attitudes toward, aquaponics in biotechnical schools in Slovenia by students attending the programs for land managers, horticultural technicians, technicians in agriculture and management, and environmental technicians, in order to evaluate students' attitudes toward this type of food production (see Sect. 22.8.2.) The list of potential candidates for participation in the survey was prepared based on a review of secondary schools by the Ministry of Education, Science and Sport of the Republic of Slovenia.
(iv) Semi-structured interviews with teachers at relevant schools, examining the implementation of aquaponics as a learning tool in Slovenia (Sect. 22.7.2.1).

22.7.2.1 Interview with Teachers and Trainers Using Aquaponics in the Educational Process in Slovenia

In order to investigate the use of aquaponics as a learning tool in Slovenia, Peroci (2016) conducted semi-structured interviews (45–60 min) with five teachers.

The analysis of interviews revealed the following reasons for using aquaponics: (i) possibility for experiential learning, (ii) flexible installation that can be adapted to the education goal, (iii) a good way of teaching about food production, and for teaching STEM subjects. These were very similar to reasons revealed by interviews in North America conducted by Hart et al. (2013). However, in contrast, in the interviews conducted in Slovenia, two reasons for using aquaponics were absent: fun, and developing responsibility and compassion for living organisms.

Based on the analysis of the interviews related to the three aquaponics units used for education in Slovenia, the future implementation of aquaponics as a learning tool needs to focus on the following steps:

1. Developing a set of learning outcomes that can be achieved using an aquaponic unit
2. Designing the aquaponic teaching unit, which facilitates learning outcomes and competencies that students must gain in order to become an "Aquaponic Farmer"
3. Establishing a link between teachers and trainers (kindergartens, primary schools, secondary schools, universities), local communities, companies, and individuals involved in aquaponics
4. Developing guidelines for integrating aquaponics in the learning process
5. Performing workshops for the design, construction, operation, and maintenance of an aquaponics

22.8 Does Aquaponics Fulfill Its Promise in Teaching? Evaluation of Students' Responses to Aquaponics

22.8.1 EU FP6 Project "WasteWaterResource"

The aim of the Waste Water Resource project was to assemble, develop, and assess teaching and demonstration material on ecotechnological research and methods for pupils aged between 10 and 13 years (http://www.scientix.eu/web/guest/projects/project-detail?articleId=95738). The teaching units were assessed in order to improve the methods and content and maximize learning outcomes. Based on discussions with educational professionals, the assessment was based on a simple approach using questionnaires and semi-structured interviews. Teachers assessed the units by answering the online questionnaire (see Sect. 22.7.1). The aquaponic units were evaluated in Sweden (in the Technichus Science Center, and in Älandsbro skola in Härnösand), and in Switzerland.

22.8.1.1 Technichus Science Center, Sweden

Between 2006 and 2008, an aquaponic unit was installed at Technichus, a science center in Härnösand, Sweden (www.technichus.se). The questionnaire was placed beside the system so that the visiting students could answer the questions at any time. It consisted of 8 questions (Fig. 22.8).

The answers showed that the students understood how the water in the system was re-circulated. They understood less well how nutrients were transported within the system and the contents of the nutrients and, interestingly, one in four students did not know that the plants growing in the aquaponic unit were edible.

22.8.1.2 Älandsbro skola, Sweden

The questionnaire used in Älandsbro skola was first explained by the teacher in order to ensure that the students would understand the questions. The questions were answered before the project started and at the end of the project.

	I agree (%)	I mostly agree (%)	I mostly disagree (%)	I disagree (%)
I understand that the plants in this aquarium get the nutrients from the fish.	33	21	21	25
I know the content of the nutrient, which comes from the fish.	29	14	24	33
I know that the water recirculates from the aquarium to the plants and back to the aquarium.	43	14	33	10
I know that the plants need light in order to grow.	59	9	5	27
I understand what a circulating system is.	61	11	6	22
I think it is good that you can cultivate plants together with fish.	48	19	19	14
I know that you can eat the plants and fruits which grow above the fish in the aquarium.	50	20	5	25
I know that this aquarium with fish and plants are a recirculating system.	43	10	33	14

Fig. 22.8 Questionnaire and the frequency of answers of the 24 students (aged from 8 to 17 years) visiting the exhibition in Technichus, Sweden

On average, there were 28% more correct answers to the general questions about nutrient requirements of plants and fishes after the teaching unit. As expected, and similar to the findings of Bamert and Albin (2005), the increase in knowledge was evident.

The conclusions of the investigation were that (i) working with aquaponics has a great potential to help pupils attain relevant learning goals in the Swedish curriculum for biology and natural sciences; (ii) the teachers thought that the work gave natural opportunities to talk about cycling of matter and that it attracted the pupils' interest; (iii) the questionnaires showed that a large number of pupils had changed their opinion about the needs of fish and plants before and after they worked with the system; and (iv) the interviews with the older pupils showed that they had acquired good knowledge about the system.

Even more important, all the people involved (teachers and students) found that aquaponics provided the means to expand the horizon of the discipline, in a refreshing and effective way.

22.8.1.3 Comparison of the Success of Aquaponics in Classes from Urban and Rural Environments in Switzerland

Bamert (2007) compared the effects of teaching with classroom aquaponics to students aged 11–13 years in two different environments in Switzerland. The School in Donat, Grisons Canton, is situated in the rural alpine region, where the students mostly live on nearby farms. Many of these farms are organic, so these students knew certain concepts about cycles in nature from their everyday life. There were 16 students, aged 11–13 years, in the joint class of fifth and sixth grade. Their mother tongue is Rhaeto-Romanic, but the aquaponics classes were given in German.

The School in Waedenswil, on the other hand, is situated in the greater Zürich area. The students mostly grew up in an urban environment and had less experience of nature compared with the students from Donat. Because the students from Donat stated that the theoretical part was rather difficult, nitrification was not explained in Waedenswil (Example 22.2). Also, one must consider that the teaching unit was spread over 11 weeks in Donat, while it was performed as a 2-day workshop in Wädenswil.

Answers to questions about what they liked/disliked most about the aquaponics lessons are presented in Fig. 22.9. While the rural students were most fascinated by the system itself, the urban students were mostly fascinated by the fish. Generally, fish were the biggest motivator in both classes. Netting the fish, transporting, feeding, and just observing them were all very popular activities. The thirst for knowledge about fish mainly involved questions about reproduction, growth, etc.

22.8.1.4 Promoting Systems Thinking with Aquaponics in Switzerland

The effect of the teaching sequence described in Example 22.3 on systems thinking competencies was assessed at the beginning and at the end of the sequence. The

Donat (N=16)	everything	nothing	the system	fish	survey	lecture notes	developing countries	puzzles	theory	final party/game
exciting	31	0	38	13	6	6	13	19	6	13
boring	0	50	0	6	13	19	0	0	0	0
difficult	0	13	0	0	19	0	0	0	50	0
fascinating	0	0	94	6	0	0	0	0	0	0
interesting	0	0	13	50	0	0	0	0	6	0

Wädenswil (N=20)	everything	nothing	the system	fish	survey	lecture notes	developing countries	puzzles	theory (simplified)	final party/game
exciting	30	0	5	55	0	0	0	5	0	5
boring	0	40	0	0	15	10	0	0	5	0
difficult	0	40	0	0	15	0	0	0	20	0
fascinating	0	15	30	20	0	0	10	0	0	0
interesting	0	0	15	40	0	0	10	0	10	0

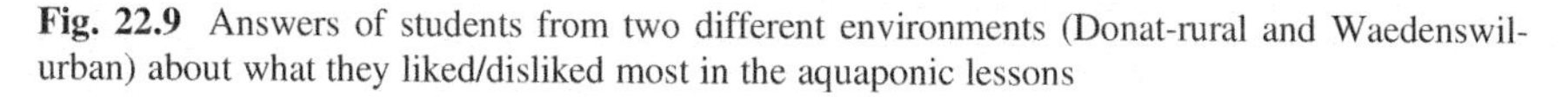

Fig. 22.9 Answers of students from two different environments (Donat-rural and Waedenswil-urban) about what they liked/disliked most in the aquaponic lessons

ability of students to think in a systemic way instead of linear succession improved significantly in the post-test compared to the pre-test.

Systems thinking is one of the key competencies in the complex world (Nagel and Wilhelm-Hamiti 2008), and is necessary in order to gain an overview of the underlying systems of the real world, because most problems are complex and require a systemic approach to develop a viable solution.

Systems thinking includes four central dimensions (Ossimitz 1996; Ossimitz 2000): (i) thinking in models; (ii) interconnected thinking; (iii) dynamic thinking (thinking about dynamic processes, such as delays, feedback loops, oscillations); and (iv) manipulation of systems, which implies the ability for practical system management and system control. Classroom aquaponics mostly concern interconnected thinking and thinking in models. Interconnected thinking involves identification and appraisal of direct and indirect effects, particularly with regard to identifying feedback loops, construction, and the understanding of networks and of cause and effect.

The main goal of the teaching sequence "Classroom aquaponics" described in Example 22.3 was to enable students to adopt tools, which can help them to examine complex problems. The hypothesis tested was that incorporating aquaponics into teaching units would have a positive influence on the systems thinking abilities of the pupils.

All the 68 students performed a test at the beginning and at the end of the teaching sequence. The pre- and post-test were identical and contained a short text about life as a farmer, which animated the students to think about farmers and their behavior. It ended with the question: "Why did the farmer put manure on his fields?" The pupils answered with a drawing and/or a description of the reasons. The answers of the students were evaluated according to the method outlined by Bollmann-Zuberbuehler et al. (2010), which allows a qualitative method to be used with quantitative results (for more details on this, see also Junge et al. 2014).

Generally, the delineation of systems shifted from a qualitative description to a more schematic description and became more complex in the post-test. When numerical scores were assigned to each level of drawing (Table 22.6), an interesting

Table 22.6 Identification of the delineation of the system representations

Delineation	Description	Score
No drawing	No representation at all	1
Schematic representation	Schemes without logical connection	2
Figure with stages	Logical sequence with a minimum of 3 stages	3
Other representation types	All other representations, which could not be clearly allocated	4
Linear Graph	Contains at least 1 chain of events	5
Effect diagram	Contains in addition at least 1 junction	6
Network diagram	Contains in addition at least 1 feedback loop and/or cycle	7

Table 22.7 Comparison of the median delineation scores between the pre- and post-test

	Pre-activity test (median)	Post-activity test (median)	Change
Girls	2.5	7	4.5
Boys	2	7	5

pattern emerged (Table 22.7). While both genders reached the median level of 7, meaning that the majority of drawings contained at least one loop and/or cycle, at the end of the teaching sequence, the change was more marked among the boys, who started at a lower level. This indicated that boys profited more from hands-on experience than girls.

In the next step, the Complexity index, Interconnection index, and the Structure index were calculated (for details, see Junge et al. 2014).

The complexity index (German: Komplexitätsindex, KI) shows how many system concepts the student implemented:

$$\mathrm{KI} = \mathrm{variables} + \mathrm{arrows} + \mathrm{chain\ of\ events} + \mathrm{junction} + \mathrm{feedback\ loops} \quad (22.1)$$

The interconnection index (Vernetzungsindex, VI) shows the frequency of the connections between the variables:

$$\mathrm{VI} = 2\ \mathrm{x\ arrows}/\mathrm{variables} \quad (22.2)$$

The structure index (Strukturindex, SI) shows how many complex system concepts the student implemented in the representation:

$$\mathrm{SI} = (\mathrm{chains\ of\ events} + \mathrm{junctions} + \mathrm{feedback\ loops})/\mathrm{variables} \quad (22.3)$$

The students found more system concepts and knew more about system variables at the post-test than in the pre-test, a fact reflected by all indices applied (Fig. 22.10).

These results appear to support the hypothesis that incorporating aquaponics into teaching has a positive influence on the systems thinking capabilities of students, and that the devised "Classroom Aquaponic Sequence" was successful in training students in systems thinking.

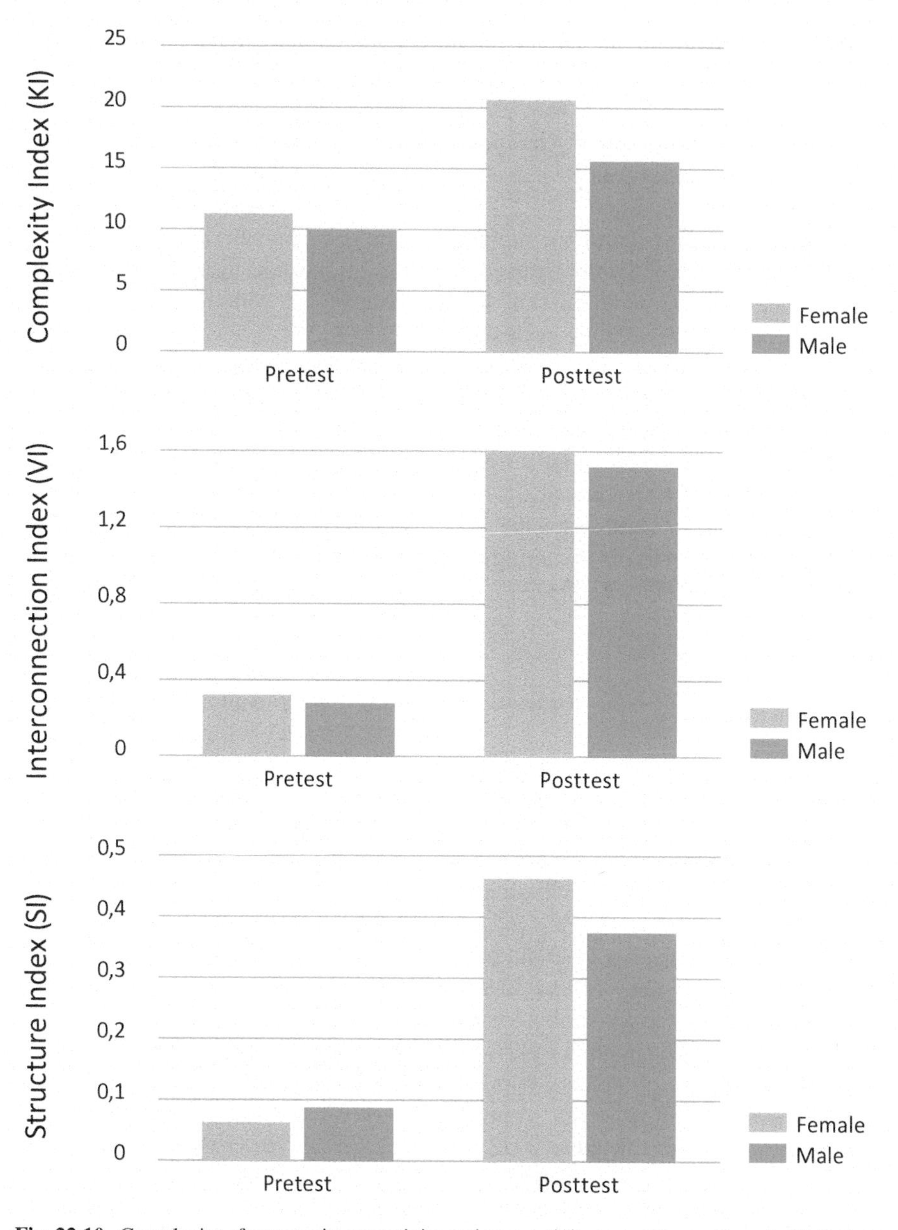

Fig. 22.10 Complexity of answers in pre-activity and post-activity tests. Above: Complexity Index (KI), centre: Interconnection Index (VI), below: Structure Index (SI)

22.8.2 *Evaluation of the Aquaponics Teaching Unit in Vocational Education in Slovenia*

22.8.2.1 Evaluation of the Aquaponics Course, Biotechnical Centre Naklo, Slovenia

The learning progression of the short aquaponic course within the study of Peroci (2016) (see Precedent 5) was assessed by means of questionnaires: (i) pre-test/post-test; (ii) test of the acquired skill level in connection with food production in aquaponics; and (iii) teaching evaluation.

The influence of various factors on the popularity of the lessons and the practical work was evaluated. Students named several factors as being crucial for their interest in the aquaponics course. The most relevant factors were: more relaxed teachers (80%); entertainment (76%); attractive location of the practical work (72%); contact with nature (68%); active practical work (64%); and use of interesting new methods (56%). Generally, students rated the more interesting lessons as those that were less difficult (e.g., the lesson "Monitoring water quality and bacteria" was less interesting and most difficult) (Fig. 22.11).

22.8.2.2 Survey of Knowledge and Attitudes Toward Aquaponics

Peroci (2016) investigated knowledge, attitudes toward food produced, and interest in the use of aquaponics among students at 8 secondary vocational schools in biotechnical fields within the educational programs for land manager (1st–third year), horticultural technician (1st–fourth year), technician in agriculture and management (1st–fourth year), and environmental technician (1st–fourth year) during 2015 and 2016.

The survey involved a 15-minute questionnaire, with closed-ended answers (yes or no). The survey showed that 42.9% of 1049 students had already heard about aquaponics. They had learnt about it at school (379 students), from the media (79), from peers and acquaintances (42), from advertisements (18), when visiting the aquaponics (12), at agricultural fairs (2), and in aquaristic (1). Most of the positive answers were from students from the Biotechnical Center Naklo where the aquaponics was constructed in 2012 (Podgrajšek et al. 2014) and aquaponics was already integrated in the learning process; 28% of respondents lacked any knowledge about aquaponics and 19.8% of respondents said they would choose the aquaponics course over other modules, mostly because of its interdisciplinary nature and due to its sustainable and creative approach. The students also expected that after attending such a course, they would have better chances of finding a job. Most students liked the practical work, and 10.7% of respondents said they would like to volunteer by maintaining the aquaponics and that they would like to set up their own aquaponics. The analysis regarding the interest of students in producing food using aquaponics showed that they liked this idea. However, they were not sure if they

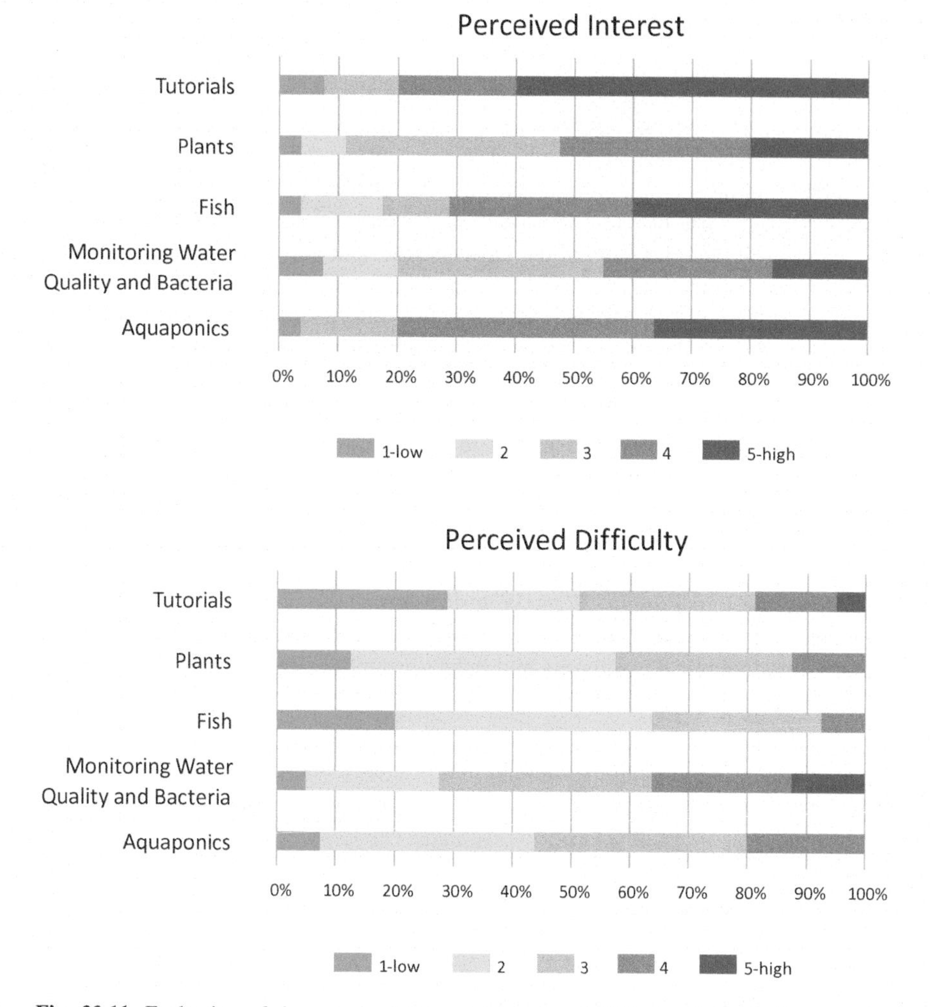

Fig. 22.11 Evaluation of the perceived interest (above) and difficulty (below) of aquaponics lessons at the vocational school in Naklo, Slovenia. (Modified after Peroci 2016)

would eat the fish and vegetables produced in this way, mostly because they had no previous experience of eating food produced in an aquaponics. Based on these results, we can assume that the production of food in aquaponics will be well accepted by the students of secondary vocational schools in biotechnical fields. This is important as these students are the next generation of entrepreneurs, farmers, and technicians who will not only generate, make, and evolve aquaponics in the future but also help generate the confidence in aquaponics among stakeholders so that it becomes part of food production in Slovenia in the future.

22.9 Discussion and Conclusions

An aquaponics is a perfect example of a system that can bring nature closer to a classroom and can be used as the starting point for a host of educational activities at both primary and secondary school levels. A model system, together with corresponding didactic methods, serves to make natural processes more tangible to pupils. This, in turn, helps to develop the necessary competencies for dealing with the complexity and problems of the environment, and promotes a sense of responsibility toward humanity. Creating the opportunity for hands-on experience with nature and natural elements such as water, fish, and plants also develops environmental consciousness and a greater understanding of the potential for practical solutions and a willingness to act on this knowledge.

In this chapter, we have presented various case studies of the use of aquaponics at different educational levels, and also a number of examples that assess the benefits of introducing aquaponics into schools.

While for each separate study the assessment methods were in themselves logical, and provided interesting insights, clear-cut comparisons across the studies are not practical, because the methods were not, or were only partly, comparable.

During the FP6 Project "WasteWaterResource" the pedagogical specialists in the team voiced some critical comments about using questionnaires to measure impacts on ecological awareness and behavior among students (Scheidegger and Wilhelm 2006):

- In multiple choice questionnaires, students tend to provide the answers that they think the teacher would like to hear.
- Children often have difficulties ranking their answers to questions such as "*how was my motivation in the Aquaponic lectures?*" (1: very low to 5: extremely high).
- The answers are highly influenced by the teacher and the current objectives of education.

Therefore, it is questionable whether quantitative survey methods are appropriate for revealing the potential effects, and whether they provide realistic data on the perceptions of the students.

It seems to be more appropriate to focus on qualitative assessment methods such as semi-structured interviews with the teachers, or the process of self-observation according to the action research method outlined by Altrichter and Posch (2007). Teachers are practitioners who have long-term experience of dealing with students and can therefore provide better and deeper information on a potential impact than a survey can reveal. A deeper interview or dialogue with the teachers will also provide information on critical issues of the learning systems and ideas on its further development. The research question "*how did the teachers perceive the material?*" seems therefore to provide much more useful information than the question "*what was the impact on the students?*"

A key issue for the successful dissemination of new teaching units appears to be a robust integration of the units into the national school frameworks. The feedback from the schools strongly indicates that teachers have very limited time to find and initiate new ideas and teaching materials. They usually use already established information portals that provide the material in a form that corresponds to the national education plan and is ready-made for a particular school level. There is therefore a need to establish cooperation with the key players in the national pedagogical frameworks. In order to better evaluate the impacts of aquaponics on STEM subjects, environmental and other learning outcomes, a comparative study between educational institutions where they used aquaponics as a teaching tool based on the same and well-designed research methods and addressing various teaching goals would be needed.

Acknowledgments This work was partly supported by funding received from the COST Action FA1305 "The EU Aquaponics Hub—Realising Sustainable Integrated Fish and Vegetable Production for the EU."

We acknowledge the contribution of the EU (FP6-2004-Science-and-society-11, Contract Number 021028) to the project "WasterWater Resource," and thank the entire team, especially Nils Ekelund, Snorre Nordal, and Daniel Todt.

We acknowledge the contribution of the EU (Leonardo da Vinci transfer of innovation project, Agreement Number - 2012-1-CH1-LEO05-00392) to the project Aqua-Vet, and thank the entire team, especially Nadine Antenen, Urška Kleč, Aleksandra Krivograd Klemenčič, Petra Peroci, and Uroš Strniša.

References

Altrichter H, Posch P (2007) Lehrerinnen und Lehrer erforschen ihren Unterricht. Verlag Julius Klinkhardt, Bad Heilbrunn

Bamert R (2007) Kinderleichte Wasseranalysen und Bonitierungsmethoden als Schlüssel zum Verständnis der Prozesse in Aquaponic. Diploma Thesis. Hochschule Wädenswil HSW. 71 pp. Available online: https://www.zhaw.ch/storage/lsfm/dienstleistung/schulen/play-with-water/kinderleichte-wasseranalysen.pdf. Accessed on 28 Feb 2018

Bamert R, Albin V (2005) Aquaponic als Unterrichtsmodell. Term Thesis. Hochschule Wädenswil HSW. 90 pp. Available online: https://www.zhaw.ch/storage/lsfm/dienstleistung/schulen/play-with-water/unterrichtsmodell-aquapoinic.pdf. Accessed on 18 Feb 2018

Baumann K (2014) Adaptation der Unterrichtsmaterialen der FBA (Fachbezogene Berufsunabhängige Ausbildung) Aquakultur für die Berufsschulen. Term Thesis. Zurich University of applied Sciences (ZHAW), 48 pp., Waedenswil

Bollmann-Zuberbuehler B, Frischknecht-Tobler U, Kunz P, Nagel U, Wilhelm Hamiti S (2010) Systemdenken foerdern: Systemtraining und Unterrichtsreihen zum vernetzten Denken: 1.-9. Schuljahr. Published by Schulverlag plus, Bern. 94 pp. ISBN: 978-3-292-00628-8

Clayborn J, Medina M, O'Brien G (2017) School gardening with a twist using fish: encouraging educators to adopt aquaponics in the classroom. Appl Environ Edu Commun 16(2):93–104. https://doi.org/10.1080/1533015X.2017.1304837

Council of the European Union (1998) Council Directive 98/58/EC of 20 July 1998 concerning the protection of animals kept for farming purposes. Off J Eur Commun:23–27. Available online: http://eur-lex.europa.eu/eli/dir/1998/58/oj. Accessed 22 Mar 2018

Duschl RA, Schweingruber HA, Shouse AW (eds) (2007) Taking science to school: learning and teaching sciences in Grades K-8. committee on science learning, Kindergarten through Eighth Grade, 404p. ISBN 0-309-66069-6

Genello L, Fry JP, Frederick JA, Li X, Love DC (2015) Fish in the classroom: a survey of the use of Aquaponics in education. Eur J Health Biol Edu 4(2):9–20. https://doi.org/10.12973/ejhbe.2015.213p

Graber A, Antenen N, Junge R (2014) The multifunctional aquaponic system at ZHAW used as research and training lab. In: Maček Jerala M, Maček MA (eds) Conference VIVUS: transmission of innovations, knowledge and practical experience into everyday practice, Collection of Papers, Strahinj, 14–15. Biotehniški center Naklo, Strahinj, pp 245–255. ISBN 978-961-93564-4-9

Graham H, Beall DL, Lussier M, McLaughlin P, Zidenberg-Cherr S (2005) Use of school gardens in academic instruction. J Nutr Edu Behav 37(3):147–151

Hart ER, Webb JB, Danylchuk AJ (2013) Implementation of aquaponics in education: an assessment of challenges and solutions. Sci Edu Int 24(4):460–480. Available online: https://files.eric.ed.gov/fulltext/EJ1022306.pdf

Hart ER, Webb JB, Hollingsworth C, Danylchuk AJ (2014) Managing expectations for aquaponics in the classroom: enhancing academic learning and teaching an appreciation for aquatic resources. Fisheries 39(11):525–530. Available online: https://fisheries.org/docs/wp/AFS-Fisheries-November-2014.pdf#page=48. 25 Dec 2017

Hofstetter U (2007) Aquaponic – ein Unterrichtsmodul über den geschlossenen Kreislauf von Wasser und Nährstoffen. Term Thesis, Zürcher Hochschule für Angewandte Wissenschaften ZHAW, Wädenswil. 61 pp. Available online: https://www.zhaw.ch/storage/lsfm/dienstleistung/schulen/play-with-water/aquaponic-unterrichtsmodul-geschlossener-kreislauf.pdf. Accessed 20 Feb 2018

Hofstetter U (2008) Aquaponic im Unterricht, Bachelor Thesis, Zürcher Hochschule für Angewandte Wissenschaften ZHAW, Wädenswil. 105 pp. Available online: https://www.zhaw.ch/storage/lsfm/dienstleistung/schulen/play-with-water/aquaponic-im-unterricht.pdf. Accessed on 18 Feb 2018

Junge R, König B, Villarroel M, Komives T, Jijakli MH (2017) Strategic Points in Aquaponics. Water 9(3):182. https://doi.org/10.3390/w9030182

Junge R, Wilhelm S, Hofstetter U (2014) Aquaponic in classrooms as a tool to promote system thinking. In: Maček Jerala M, Maček MA (eds) Conference VIVUS: transmission of innovations, knowledge and practical experience into everyday practice, Collection of Papers, Strahinj, 14–15. Biotehniški center Naklo, Strahinj, pp 234–244. ISBN 978-961-93564-4-9

Kennedy G (2008) Writing and using learning outcomes, EUA Bologna Handbook. RAABE Publishing, Berlin

Kolb D (1984) Experiential learning as the science of Learning and development. Prentice Hall, Englewood Cliffs

Krivograd Klemenčič A, Jarni K, Griessler Bulc T (2013) Akvaponika kot izobraževalno orodje v poklicnem in strokovnem izobraževanju. (Aquaponics as an educational tool in vocational and professional education). Didakta, Vol. XXII, Nr. 167, pp 42–45

Maucieri C, Forchino AA, Nicoletto C, Junge R, Pastres R, Sambo P, Borin M (2018) Life cycle assessment of an aquaponic system built using recovered material for learning purposes. J Clean Prod 172:3119–3127. https://doi.org/10.1016/j.jclepro.2017.11.097

Nagel U, Wilhelm-Hamiti S (2008) Komplexität erproben und erlernen – Elemente einer Didaktik des systemischen Denkens. Available online: http://www.umweltbildung.at/cms/download/359.pdf. 08 Mar 2018

Ossimitz G (1996) The development of systems thinking skills using system dynamics modeling tools. http://webdoc.sub.gwdg.de/ebook/e/gdm/1996/ossimitz.pdf. 07 Mar 2018

Ossimitz G (2000) Entwicklung systemischen Denkens. Theoretische Konzepte und empirische Untersuchungen, Profil Verlag, München Wien, pp 256. ISBN-13: 978-3890194943

Peroci P (2016) Vključevanje akvaponike v učni proces srednješolskega poklicnega izobraževanja v Sloveniji. (Inclusion of aquaponics in the educational process of secondary vocational education in Slovenia.). Master Thesis, Univerza v Mariboru, Fakulteta za naravoslovje in matematiko, Oddelek za biologijo. Maribor, Slovenia. 186 pp. Available at https://www.researchgate.net/publication/307476342

Podgrajšek B, Schmautz Z, Krivograd Klemenčič A, Jarni K, Junge R, Griessler Bulc T (2014) Preliminary monitoring of an aquaponic system in Biotechnical Centre Naklo. Moje podeželje 5 (9):10–11. ISSN 1855-9204

Quaden R, Ticotsky A (2004) The shape of change. Creative Learning Exchange, Acton. Available http://www.clexchange.org/cleproducts/shapeofchange.asp

Scheidegger B, Wilhelm S (2006) Bildungsarbeit mit Aquaponic-Systemen. Unpublished. Available online: https://www.zhaw.ch/storage/lsfm/dienstleistung/schulen/play-with-water/Scheidegger___Wilhelm_2006__Umweltbildung_mit_Aquaponic.pdf. Accessed on 07 Mar 2018

Schneller AJ (2015) A case study of indoor garden-based learning with hydroponics and aquaponics: evaluating pro-environmental knowledge, perception, and behavior change. Appl Environ Edu Commun 14:256–265. https://doi.org/10.1080/1533015X.2015.1109487

UNESCO-UIS (2012) International standard classification of education ISCED 2011. UNESCO Institute for Statistics, Montreal, Quebec, Canada. ISBN 978–92–9189-123-8. 88 pp. Available online: http://uis.unesco.org/sites/default/files/documents/international-standard-classification-of-education-isced-2011-en.pdf. Accessed on 05 Jan 2018

UNESCO-UIS/OECD/EUROSTAT (2017) Data collection on formal education. Manual on concepts, definitions and classifications. Available online: http://uis.unesco.org/sites/default/files/documents/uoe2016manual_11072016_0.pdf. Accessed on 20 Feb 2018

UNICEF (2018) Goal: achieve universal primary education. Available online: https://www.unicef.org/mdg/education.html. Accessed 11 Mar 2018

Wardlow GW, Johnson DM, Mueller CL, Hilgenberg CE (2002) Enhancing student interest in the agricultural sciences through aquaponics. J Nat Resour Life Sci Edu 31:55

Chapter 23
Food, Sustainability, and Science Literacy in One Package? Opportunities and Challenges in Using Aquaponics Among Young People at School, a Danish Perspective

Bent Egberg Mikkelsen and Collins Momanyi Bosire

Abstract The call for sustainable food production and consumption has led to an increased interest and new policy measures to support the circular economy and climate-smart farming practices. The merits of aquaponics and closed-loop nutrient cycling systems are increasingly being examined in terms of sustainable productivity in various settings including urban environments. Aquaponics also has the potential to be applied as a learning tool for people of all ages but especially for young people at school. This chapter studies the potential of aquaponics to teach food and science literacy and the use of the technology as an educational tool in primary school. The chapter draws on data from the Growing Blue & Green (GBG) program carried out in cooperation among Aalborg University, Copenhagen, municipal schools and their teachers and a private aquaponic enterprise. The chapter draws on three empirical studies including an exploratory study on the educational opportunities at school, a feasibility study carried out among teachers, as well as the educational Growing Blue & Green (eGBG) study, in which a digital-based regulation component was added. The conclusion is that low-cost versions of aquaponics have considerable potential for supportive learning in elementary school. Preliminary findings furthermore suggest that fitting the setup with easy-to-install intelligent sensors and devices offers the opportunity to provide learning about food, sustainability, and a basic understanding of the control and management of biological systems in one package.

Keywords Aquaponics · Sustainability education · Science literacy · Food literacy · Education · Primary school

B. E. Mikkelsen (✉)
Integrated Food Studies, Aalborg University Copenhagen, Copenhagen, Denmark
e-mail: bemi@learning.aau.dk

C. M. Bosire
Department of Learning & Philosophy, Aalborg University, Copenhagen, Denmark

S. Goddek et al. (eds.), *Aquaponics Food Production Systems*,
https://doi.org/10.1007/978-3-030-15943-6_23

23.1 Introduction

Sustainable food production and consumption are important societal challenges. Growing populations, scarcity of arable land, and urbanization are all important factors in this area, leading to a growing interest in new sustainable food production technologies, which are not necessarily confined to maritime or rural settings. Aquaponics is one of these technologies that has received increased attention, in particular, since it can easily be applied also in urban environments. These new technologies, including aquaponics, also offer new opportunities as learning tools for people of all ages, but it is particularly appealing for young people at school. This chapter reports on the findings from the educational Growing Blue & Green (GBG) program that has been developed and tested in educational settings in the Greater Copenhagen area. Additional studies also suggest that there appears to be potential for using aquaponics as a key way of learning about sustainable food production in a wide spectrum of academic disciplines at school since it can readily be integrated into the existing educational curriculum. A few studies have examined the application of aquaponics in an educational context. Graber et al. (2014) studied the potential of aquaponics as a food production method for urban areas teaching seventh-grade pupils sustainability issues in science classes. The idea behind the concept was to introduce and train students on "systems thinking" by combining fish and plant growing. Junge et al. (2014) showed that the students' ability to think in a systematic way improved significantly as a result. The study also suggested that building on social learning in groups the students developed greater teamwork skills. However, apart from these examples, aquaponics literature is relatively limited and most of the available articles have focused on the technological aspects of the systems. This chapter attempts to fill this knowledge gap by exploring the opportunities for integrating aquaponic technology into school learning and to uncover some of the constraints as well as the opportunities.

The chapter draws on three empirical cases, where aquaponics has been applied in primary school settings in the Greater Copenhagen area. This includes an exploratory study on the educational opportunities at school (Bosire et al. 2016), a feasibility study carried out among teachers (Bosire and Sikora 2017), as well as the insights from the first part of an educational Growing Blue & Green study (Toth and Mikkelsen 2018).

The aim of this chapter is thus to introduce and discuss educational aquaponic interventions in which aquaponics is used in an elementary school in Denmark and to discuss how theory can be taken into practice. The chapter discusses the potential of aquaponics to be able to contribute to the fostering of a deeper learning about urban, sustainable food production among young school children in school settings, as well as the potentials for creating digital literacy through the addition of a self-regulating digital sensing and maintenance tool, i.e., the eGBG tool.

23.2 Conceptual Foundation

Supporting Sustainable Development (SD) of the food system through educational efforts can be expected to be a good investment, as school children are the future policy makers and producers.

According to Shephard (2008), educators and particularly higher educators have traditionally focused on the cognitive domain of learning with no much emphasis being put on primary education. We hold the view that using appropriate learning tools at primary school level can be an essential pillar to bringing about long-term positive change in societies. These can be realized though alternative learning and teaching approaches, different from traditional deductive approaches such as "learning by doing" and "experiential learning" pioneered by Dewey (1997) in his work experience and education. In our research work, we present a type of extracurricular dimensional perspective, where we add to pupils' learning outcomes by tapping into the affective domain, which focuses on interests, attitudes, appreciations, values, changing behaviors, and emotional sets or biases (Shephard et al. 2015). Practical aquaponics promises to deliver a hands-on problem-based inductive learning tool for education.

Study cases all build on the idea of Service Learning (SL), where students use academic knowledge to address community needs, and the knowledge triangle (education, research, and innovation), which are part of the teaching at the Integrated Food Studies (IFS) program at Aalborg University (Mikkelsen and Justesen 2015). The IFS also uses Problem-Based Learning (PBL), where learning is approached with open-ended problems with no absolute right answer, as well SL approaches. SL is a pedagogical approach that is rooted in PBL as well as in the experiential learning approach (McKay-Nesbitt et al. 2012). Using the SL approach, students are expected to become involved in projects based on the needs, wishes, and demands of local communities. The recent interest in reforming educational practices and strategies makes the use of aquaponics an important component in the educational context timely and relevant. Additionally, the use of inductive methods such as PBL and discipline-based learning (Wood 2003: Armstrong 2008) as well as experiential learning (Beard 2010; McKay-Nesbitt et al. 2012), where everyday life problems and questions are used to inform the learning process, is spreading. These concepts are all favorable for aquaponic teaching. Furthermore, the idea of SL is compatible with the aquaponic teaching concept and the recent Danish School Reform (Danish Ministry of Education 2014) that present guidelines on how to integrate the practical and theoretical aspects of the curriculum.

While there are several aquaponic systems that can be supplied by manufacturers and/or bespoke systems designed by consultants, aquaponic technology in principle is rather simple. The basic principles can thus be well understood by students, and the systems can be designed, built, and monitored by students using a range of materials and methods, ranging from the basic to the sophisticated. Taking this

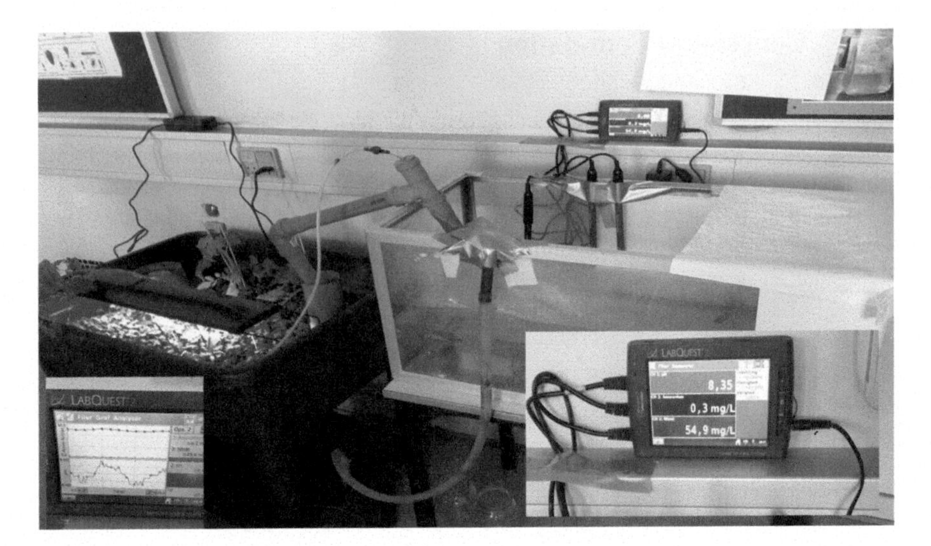

Fig. 23.1 The aquaponic learning and experimental mock-up. The illustration shows the setup including aquarium fish tank and the monitoring devices that are used to measure the equilibrium of the whole system. The last part is the core of the learning goal for students. (Pictures: courtesy of Lija Gunnarsdottir)

premise, aquaponics is thus a technology that is highly suited to the knowledge triangle approach. Education can be enhanced through the creation of links between the three sides of the knowledge triangle, i.e., education, research, and innovation. Innovative thinking on how education for sustainability could be implemented using practical educational tools leads the educator toward aquaponics: a food production method which is essentially a symbiotic integration of two mature disciplines – recirculating aquaculture and hydroponics in one production system, where the live fish generate nutrients for plant production. A simple aquaponic system unit, such as the one shown in Fig. 23.1, was set up at an elementary school in Copenhagen. The figure illustrates some of the basic components used with a brief detail on its working principle: a simple aquarium where water in the fish tank is kept at a constant height through appropriate design for the comfort of the fish. Through some pumping action from a sump tank situated below the grow bed, excess water containing fish waste is cycled through plant grow beds, where bacteria and other microbes are hosted.

The sump and grow bed act together as mechanical and biofilters, respectively, by removing solids and dissolved waste.

The setup in Fig. 23.1 illustrates a practical educational example, focusing on sustainability since it provides a practical example of how the goals set out under Sustainable Development Goals (SDGs) in UN Agenda 2030 for SD can be addressed (UN 2015b). Goal number 2, which targets ending hunger, achieving food security and improved nutrition, and promoting sustainable agriculture, and goal number 4, which focuses on ensuring inclusive and equitable quality education and the promotion of lifelong learning opportunities for all (UN 2015b). These crucial issues can be included in the Problem-Based Learning (PBL) approach that has been developed in the GBG case. Based on a shared firm belief in having technological solutions for the problems of contemporary food systems, the GBG approach contributes to a demonstration of "ecological modernization" in food production processes. Through the development of the didactic for the eGBG themes of sustainability and food literacy, it became clear that for such a system to bring about change, there is a need for the right platform through which knowledge and skills can be exchanged among young people and their teachers in the school setting.

Other studies have shown that the lack of food and nutrition literacy among young people is of growing concern (Vidgen and Gallagos 2014; Dyg and Mikkelsen 2016). This is particularly concerning, as the conventional ways of food production and the current persistent drivers of science and technology have fueled unsustainable global exploitation of earth's resources leading to numerous challenges within the food system (FAO 2010; UNDP 2016). In addition, the increase in world population and the rapid urbanization have overloaded the food system. The United Nations predicts that world population will increase by more than 1 billion people within the next 15 years, reaching 8.5 billion in 2030. Of these, the majority (66%) is forecasted to be living in cities by the year 2050 (UN 2015a). These trends in combination with the growth in unhealthy eating habits and nutrition-related disorders have made a new approach to food nutrition and agri-literacy at school imperative.

The insights from the GBG project and the results from numerous interviews with both teachers and students showed that the successful application of aquaponic technology is dependent on the careful planning and maintenance of the system. The digital version of the GBG – the eGBG – was developed to address these challenges and to use the related opportunities in promoting digital literacy in school. The idea of the eGBG takes inspiration from the idea of self-regulation in biological systems. It is conceptually based on the idea of autopoiesis: referring to a system capable of reproducing and maintaining itself. The term first introduced in 1972 by biologists Maturana and Varela (1980) describes the self-maintaining chemistry of living cells, and ever since then, the concept has been applied in a wide array of fields such as cognition, systems theory, and sociology. In the eGBG study, illustrated by the setup and components in Fig. 23.2, water quality, temperature, dissolved oxygen, CO2, pH, ammonia, and nitrite content are measured with sensors using an electronic and digitalized setup, followed by appropriate automated regulation and adjustments to the required or set levels. This system used alongside a basic maintenance regime better enables children to learn Information and Communications Technology (ICT), together with Science, Technology, Engineering, and Mathematics (STEM) subjects in addition to a wider understanding of sustainable urban farming and animal welfare practices. The eGBG minimizes human error and reduces the amount of critical resources such as the physical labor and hours that would otherwise be required for the care and maintenance of a balanced aquaponic system.

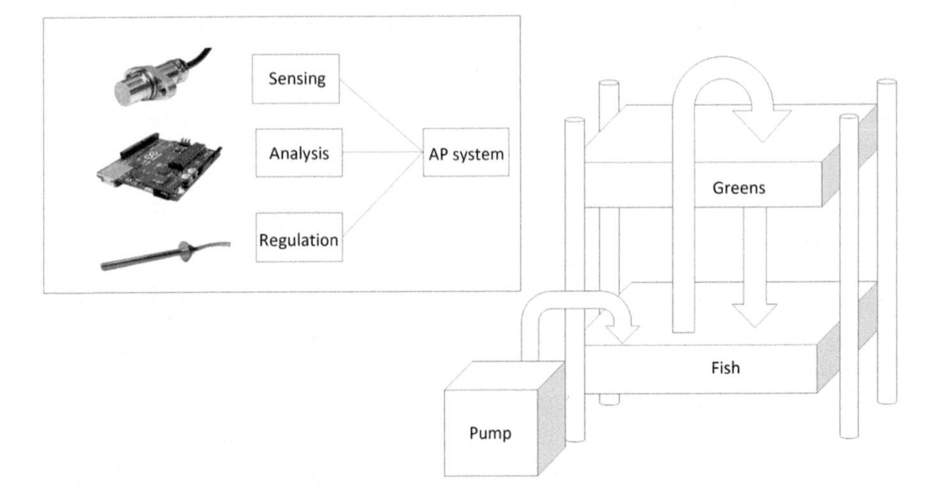

Fig. 23.2 The experimental eGBG setup. The illustration shows the two parts of the system. The aquaponic system itself and the measuring devices and the minicomputer used to follow the biological condition of the eGBG system

23.3 Methods

In the context of this chapter, three data sources were used including (a) an exploratory study on the educational opportunities at school (Bosire et al. 2016), (b) a feasibility study carried out among teachers (Bosire and Sikora 2017), and (c) the eGBG study (Toth and Mikkelsen 2018).

The first study (a) was carried out as an exploration of the opportunities and challenges of using aquaponics as an educational tool. The study aimed at investigating to what extent it makes sense to use aquaponics in school teaching. Data from three (N = 3) independent qualitative interviews were collected. The informants were (1) a biology teacher engaged in natural science teaching at primary school; (2) a consultant entrepreneur, which is an aquaponic expert, too; and (3) one local aquaponic bio-farmer. The data analysis procedure was inspired by the future workshop approach (Jungk and Müllert 1987), leading to a categorization and evaluation according to the three categories of critique, fantasy, and strategy.

In the second study (b), a feasibility study was carried out at the Blågaard Public School located within the Copenhagen Municipality in cooperation with two biology teachers and a physics teacher and with approval from the school administration. The local aquaponic bio-farmer and expert was also involved. A low-cost aquaponic facility for teaching was developed using a simple do-it-yourself (DIY) food production system and off-the-shelf components. The idea for this design and construction was to illustrate that this kind of technology can readily be undertaken, and it is not only for advanced urban growing but that it also has potential to be used as a science-teaching tool in humbler settings such as a local school. Since the budget of the school is limited, the overall goal was to complete the project at low cost and to carefully fit the system within the requirements of the existing curricula.

In the third study (c), a digital component was added to an improved version of the aquaponic system and the eGBG was introduced. The eGBG is a learning program based on simple aquaponics, and it is designed to create learning insights among adolescents. The program's special focus is on teaching principles of sustainable food production in cities and at the same time to facilitate ICT learning. The program's didactics is aimed at showing how a biological system like aquaponics can be controlled, maneuvered, and self-regulated using sensors and feedback mechanisms. This is done by connecting sensors that measure temperature, pH, and nutrient balance via a digital interface such as Arduino. The eGBG developed a simple urban farming tool based on a learning package for schools where students can learn about this technology in biology classes. By studying how the sensors work, they have the ability to learn how ICT can be integrated to monitor and control a living biological system.

The eGBG educational program can be used both in an interdisciplinary course with ICT as a theme or in the subjects of biology, physics, and chemistry. The components of the eGBG are for a low-cost aquaponic system that has been developed for the school context as described previously. Some of the key elements were supplied by BioTeket, which is a company with a social and cultural remit with an emphasis on environmental technology. BioTeket offers a series of workshops and events, giving the citizens of Copenhagen an opportunity to gain experience with sustainable urban life. The assembly was done under the company's technical supervision. According to the national curricula, ICT in elementary school is taught not as a stand-alone subject, but in a transversal manner spanning several subjects. Combining smart and sensor-based control and biological system therefore seems straightforward for this requirement. Urban farming technologies require a monitoring system with a multitude of sensors since maintaining a system in balance requires continuous measurement of temperature, pH, etc. To meet this requirement, the eGBG was developed in cooperation between Aalborg University, a municipal school in Albertslund, and the enterprise BioTeket. The development process was configured as an action research study where data was collected along with the development process.

23.4 Findings and Discussion

1. From the first study, the findings showed that visions of a new way of teaching with inclusion of the modern technology could be perceived as an advantage in influencing transformational processes at school. Nevertheless, this process requires some critical, practical, and theoretical considerations for implementation of the system to make it successful and sustainable in the long term. Some of the positive issues from the users' perspectives included a wide range of application in the subjects of biology, mathematics, science, and more. Reduction

pollution and efficient resource usage; flexibility of the system setup, e.g., on rooftops; and the production of (organic-like*) twin products (fish and plant foods). Potential limitations included time constraints, lack of financial resources, as well as the need for frequent care and maintenance. (* In the EU, current legislation provides that only vegetal produce grown in soil may be considered "organic." This is not the case, e.g., in the USA, where aquaponic produce can be grown organically and legally sold as being organic.)

2. From the second study (b), the feasibility study, the experiences of the study indicated that the learning concept, the overall idea, and the didactics fit well into the educational curricula and also with the projects that the school had already planned to undertake in the field of sustainability. The experience showed that such teaching needs to be carefully planned well ahead of time. Furthermore, the idea of a knowledge triangle approach, bringing service learning, university research, a small enterprise, and the learning staff into an informal project and innovation network, is a fruitful way of organizing the undertaking. In addition, the initiative enjoys the support of the municipality that sees entrepreneurship and innovative learning approaches as important objectives.
3. The third study (c), the eGBG study, showed that the school was supportive and already had newly purchased sensors to measure pH, temperature, CO_2, and dissolved oxygen (DO). Therefore, the data could be conducted with minimal new effort for training, as the teaching staff were already well prepared to collect data digitally. The school, at the point of project startup, was already planning to measure nitrate and ammonia using the sensors, since the basic concept of the teaching was to increase knowledge, skill, and competency in relation to the nitrogen cycle. The idea of creating aquaponic technology and applying it in the teaching was readily accepted by the school since the neighboring school already had that kind of an AP system up and running.

Acknowledgments Thanks to biology teachers Mette and Else at Blågård School in Copenhagen Municipality, to Lilja Gunnarsdottir and the teachers at Herstedlund school, and to Inge Christensen from the Nature Centre in Albertslund municipality. Thanks also to Viktor Toth, a student at Integrated Food Studies, Aalborg University, for providing data from the eGBG study. Thanks also to Tomasz Sikora and Kathrine Breidahl from the Integrated Food Studies that participated in the fieldwork. Thanks also to the owner and CEO Lasse Antoni Carlsen of Bioteket, Copenhagen, for providing components and guidance in developing the GBG program.

References

Armstrong EG (2008) A hybrid model of problem-based learning. In: Boud D, Feletti G (eds) The challenge of problem-based learning. Routledge, London

Beard C (2010) The experiential learning toolkit: blending practice with concepts. Kogan Page, London

Bosire CM, Sikora TA (2017) Blue-green farms of the future: using aquaponics at Primary School to Foster sustainable development (Master thesis). Retrieved from http://projekter.aau.dk/projekter/files/259876456/Collins_Bosire_Tomasz_Sikora__Blue_Green_Farms_of_the_Future_EMT.pdf

Bosire CM, Breidahl KS, Sikora TA, Mikkelsen BE (2016) Education for sustainability and food literacy – assessing opportunities and challenges for using aquaponics among young people at Schools. In: Mikkelsen BE, Ofei KT, Tvedebrink TDO, Romani AQ, Sudzina F (eds) Proceedings from 10th international conference on culinary arts and sciences, July 5–7th 2017 Aalborg University Copenhagen – Exploring Future Foodscapes, Published by Captive Food Studies. AAU Aalborg, p 250–267

Danish Ministry of Education (2014) Improving the public school– overview of reform of standards in the Danish Public School (primary and lower secondary education) The Danish Ministry of Education. Available at http: www.uvm.dk/publicschool. ISBN (Electronic version): 978-87-603. Accessed 24 May 2016

Dewey J (1997) Experience and education. Kappa Delta Pi, Touchstone. (Original work published 1938)

Dyg PM, Mikkelsen BE (2016) Cooperation models, motivation and objectives behind farm–school collaboration: case insights from Denmark. Int J Soc Agric Food 23(1):41–62

Food and Agriculture Organization (2010) Sustainable diets and biodiversity: directions and solutions for policy, research and action. International scientific symposium, biodiversity and sustainable diets united against hunger. Rome, Italy, 3–5 November 2010

Graber A, Antenen N, Junge R (2014) The multifunctional aquaponic system at ZHAW used as research and training lab. Available at http://pd.zhaw.ch/publikation/upload/207534.pdf. Accessed 22 May 2016

Junge R, Wilhelm S, Hofstetter U (2014) Aquaponic in classrooms as a tool to promote system thinking. Available at http://www.adam-europe.eu/prj/10804/prj/Paper_VIVUSConf_Junge_et_al.pdf. Accessed 23 May 2016

Jungk R, Müllert N (1987) Future workshops: how to create desirable futures. Institute for Social Inventions, London

Maturana H, Varela F (1980) Realization of the living. In: Boston Studies in the Philosophy of Science, vol 42. Reidel, Dordrecht

McKay-Nesbitt J, DeMoranville CW, McNally D (2012) A strategy for advancing social marketing: social marketing projects in introductory marketing courses. J Soc Mark 2(1):52–69. https://doi.org/10.1108/20426761211203256

Mikkelsen BE, Justesen J (2015) Multi disciplinarily in higher educations for the captive foodscape profession: case of Denmark. In: Proceedings of the international conference on culinary arts and eating in society. Montclair State University, Montclair, pp 262–263

Shephard K (2008) Higher education for sustainability: seeking affective learning outcomes. Int J Sustain High Educ 9(1):87–98

Shephard K, John H, Brent L, Miranda M, Sheila S, Liz S, Mick S, Mary F, Tim J, Lynley D (2015) Seeking learning outcomes appropriate for ‘education for sustainable development’ and for higher education. Assess Eval High Educ 40(6):855–866

Toth V, Mikkelsen BE (2018) Growing blue & green – exploring the potentials of teaching sustainable urban food the smart way. Protocol for extended master thesis at Integrated Food Studies, Aalborg University

United Nations (2015a) World population prospects, the 2015 revision vol. 2: Demographic Profiles. Available at http://esa.un.org/unpd/wpp/Publications/Files/WPP2015_Volume-II-Demographic-Profiles.pdf. Accessed 8 Mar 2016

United Nations (2015b) Transforming our world: the 2030 agenda for sustainable development. Available at https://sustainabledevelopment.un.org/post2015/transformingourworld. Accessed 3 Mar 2017

United Nations Development Programme (2016) Overview: human development report 2016 human development for everyone (online). Available at http://hdr.undp.org/sites/default/files/HDR2016_EN_Overview_Web.pdf. Accessed 24 Mar 2017

Vidgen HA, Gallegos D (2014) Defining food literacy and its components. Appetite 76:50–59

Wood DF (2003) Problem based learning. BMJ 326(7384):328–330. https://doi.org/10.1136/bmj.326.7384.328

Chapter 24
Aquaponics and Social Enterprise

Sarah Milliken and Henk Stander

Abstract This chapter presents some examples of recent initiatives by social enterprises using aquaponics. Aquaponics offers an innovative form of therapeutic horticulture, which can provide employment and promote well-being for people with disabilities. If implemented as a program to be managed by local communities, aquaponic systems also have the potential to address issues such as food security and food sovereignty, especially in urban areas. Increasing public familiarity with aquaponics has seen a number of social ventures being set up around the world. However, the viability of these depends not only on stakeholder commitment, thorough market analysis, clear governance structures, and a robust business plan but also on external factors, such as the local political context and regulations.

Keywords Health · Well-being · Skills · Food security · Food sovereignty

24.1 Introduction

Social enterprises, as distinct from traditional private or corporate enterprise, aim to deliver products and services that cater to basic human needs. For a social enterprise, the primary motivation is not maximizing profit but building social capital; economic growth is therefore only part of a much broader mandate that includes social services such as rehabilitation, education and training, as well as environmental protection. There is growing interest in aquaponics among social enterprises, because it represents an effective tool to help them deliver their mandate. For example, aquaponics can integrate livelihood strategies to secure food and small

S. Milliken (✉)
School of Design, University of Greenwich, London, UK
e-mail: S.Milliken@greenwich.ac.uk

H. Stander
Department of Animal Sciences (Division of Aquaculture), University of Stellenbosch, Stellenbosch, South Africa
e-mail: hbs@sun.ac.za

S. Goddek et al. (eds.), *Aquaponics Food Production Systems*,
https://doi.org/10.1007/978-3-030-15943-6_24

incomes for landless and poor households. Domestic production of food, access to markets, and the acquisition of skills are invaluable tools for securing the empowerment and emancipation of women in developing countries, and aquaponics can provide the foundation for fair and sustainable socioeconomic growth (Somerville et al. 2014). This chapter presents some examples of recent initiatives by social enterprises using aquaponics.

24.2 Health, Well-being, and Skills

Aquaponics offers an innovative form of therapeutic horticulture, a nature-based approach that can promote well-being for people with mental health problems through using a range of green activities such as gardening and contact with animals. Over the past decade, a number of social enterprises have emerged that provide therapeutic horticulture programs for improving the well-being of local communities. The social enterprise approach builds on "social firms" by facilitating people with mental health problems to develop new skills and re-engage with the workplace. A social firm is a specific type of social enterprise where the social mission is to create employment, work experience, training, and volunteering opportunities, within a supportive and inclusive environment, for people who face significant barriers to employment and in particular for people with a disability (including mental ill health, and learning disability), abuse issues, a prison record, or homeless issues (Howarth et al. 2016).

There are particular qualities of the plant–person relationship that promote people's interaction with their environment and hence their health, functional level, and subjective well-being. Plants are seen to bestow non-discriminatory rewards on their carer without imposing the burden of an interpersonal relationship and, by responding to care or neglect, can immediately reinforce a sense of personal agency. The efficacy of practicing horticulture in a group context has also been demonstrated. Many people with mental and physical health problems face social exclusion because they do not have equal access to opportunities in society, including paid employment, housing, education, and leisure. Social networks such as those provided by community horticulture initiatives can act as buffers to stressors, provide a structure for acquiring skills, and validate and enhance an individual's self-worth (Diamant and Waterhouse 2010; Fieldhouse 2003).

To date there are few examples of social enterprises using aquaponics for therapeutic horticulture. In the United States, a small farming business called Green Bridge Growers in Indiana (www.greenbridgegrowers.org) is growing produce all year-round, primarily using aquaponics. The company now employs a number of individuals with autism spectrum disorder (ASD) and finds that the scheduling, precision, and monitoring required in aquaponics perfectly match with their skills (Fasciglione 2015). A core value of the business is to engage their workforce through leadership training, active participation, and team building, and provide them with the opportunity to gain new skills and competencies. Similarly, the ACRES Project (Adults Creating Residential and Employment Solutions; https://acresproject.org/aquaponics) in Pennsylvania uses aquaponics to provide horticultural therapy, employment, and community integration for adults with autism and intellectual disabilities. They are involved in all facets of the aquaponic system, from care and maintenance to harvest and sales, and the scheduled procedures and daily routines that aquaponics requires provide them with the stability and structure that they find reassuring. By fostering social, vocational, and self-advocacy skills, ACRES therefore uses aquaponics to help autistic individuals optimize their potential, develop practical life skills, increase social capacity, and transition to work and independence.

The FabLab Nerve Centre in Northern Ireland has set up a social enterprise aquaponic farm to teach people with learning difficulties entrepreneurial and digital skills. Using state-of-the-art digital equipment, such as 3D printers, CNC routers, and laser cutters, students will receive hands-on training and experience in a range of digital design and making techniques that will allow them to design, build, and operate an aquaponic farm. As part of the project, a newly created social enterprise will be developed by the young people, allowing them to sell the produce from the farm to local businesses, thereby developing their skills in social entrepreneurship, business, and marketing (www.nervecentre.org/news/fablab-nerve-centre-launches-aquaponic-digital-farm).

Solutions for Change, a social enterprise which is dedicated to solving family homelessness, runs Solutions Farms in California (www.solutionsfarm.org). The aquaponic farm provides training for homeless families in growing tilapia and seasonal leafy greens and herbs, which are then sold to local restaurants, markets, and schools. It functions as a laboratory for teaching important work values and preparing people for re-entry into the workplace, thereby raising hope, as well as produce.

Asociacíon Huerto Lazo (www.huertolazo.eu) is a social enterprise in the province of Malaga, Spain, which offers internships to young people from troubled backgrounds. The interns are given practical training in aquaponics in a safe environment. The catfish, tilapia, and tench are sold to El Sollo restaurant in Fuengirola (Fig. 24.1).

Fig. 24.1 Aquaponic facilities at Asociacíon Huerto Lazo – anticlockwise from top left: catfish tanks in the aquaponic greenhouse; tilapia tanks with *Gynostemma pentaphyllum*, which is sold for medicinal purposes; the water filtration tanks at Huerto Lazo; Ulrich Eich demonstrating his aquaponic system (Photographs: Sarah Milliken)

24.3 Food Security and Food Sovereignty

Food security exists when all people, at all times, have physical and economic access to sufficient, safe, and nutritious food that meets their dietary needs and food preferences for an active and healthy life (Allison 2011). There are four food security pillars, which define, defend, and measure food security status locally, nationally, and internationally. These are food availability, food accessibility, food utilization, and food stability. Food availability is achieved when nutritious food is available at all times for people to access, while food accessibility is achieved when people at all times have the economic ability to obtain nutritious food available according to their dietary preferences. Food utilization is achieved when all food consumed is absorbed and utilized by the body to make a healthy active life possible, and food stability is achieved when all the other pillars are achieved (Faber et al. 2011).

Urban and peri-urban agriculture are increasingly recognized as a means by which cities can move away from current inequitable and resource-dependent food systems, reduce their ecological footprint, and increase their liveability (van Gorcum et al. 2019; Dubbeling et al. 2010). On account of being almost completely dependent on produce imported from other regions, urban consumers are particularly vulnerable to food insecurity. For those of low socioeconomic status, this dependence means that any fluctuation in food prices translates into limited purchasing power, increased food insecurity, and compromised dietary options. Community-based aquaponics enterprises offer a new model for blending local agency with scientific innovation to deliver food sovereignty and food security, by re-engaging and giving communities more control over their food production and distribution (Laidlaw and Magee 2016). If implemented as a program to be managed by local people, aquaponic systems have the potential to address food sovereignty. In turn, food security is boosted by consuming the fish, which is a significant source of protein, essential amino acids, and vitamins. Even when consumed in small quantities, fish can improve dietary quality by contributing essential amino acids, which are often missing or underrepresented in vegetable-based diets.

British social enterprise Byspokes Community Interest Company (CIC) set up a pilot aquaponic system and training program at the Al-Basma Centre in Beit Sahour, Occupied Palestinian Territories (OPT), a region where availability of space for food production is a serious problem, particularly in the urban areas and refugee camps. Even in agricultural areas, land access is being lost through Israeli controls and through effective annexation by the Israeli "Security Fence." Aquaponics therefore offers a water- and space-efficient solution to growing fresh, local produce, including a high-quality protein source (fish), thereby helping to combat malnutrition and food insecurity, while at the same time providing new opportunities for income generation. 40% of the population in the OPT (25% in the West Bank) are classed as "chronically food insecure", and unemployment stands at around 25%, with highs of 80% in some refugee camps. From an economic viewpoint, the project demonstrated that an aquaponic system could contribute significantly to household incomes and so help lift families out of poverty, while also providing a range of fresh vegetables and fish to families least able to afford such high-quality food (Viladomat and Jones 2011).

Since 2010, the Food and Agriculture Organization (FAO) of the United Nations has been implementing an *Emergency Food Production Support Project* for poor families in the Gaza Strip, where 11 years of Israeli sea, land, and air blockade, combined with low rainfall resulting in drought, have severely compromised the possibilities for domestic food production in one of the most densely populated areas of the world. With so many restrictions, fresh vegetables are expensive and hard to find. 97% of the Gaza Strip population are urban or camp dwellers and therefore do not have access to land. Poverty affects 53% of the population, and 39% of families headed by women are food insecure. Enabling families to produce their own affordable fresh food is therefore a highly appropriate and effective response to the current situation. Food-insecure female-headed households living in urban areas were given rooftop aquaponic units, and other units were installed in educational and community establishments. Having an aquaponic unit on their roof means that the women can simultaneously improve their household food security and income while still taking care of their children and homes. All of the beneficiaries have increased their household food consumption as a result (FAO 2016).

Through its Adaptive Agriculture Program, INMED Partnerships for Children is dedicated to establishing sustainable food programs that improve food security, conserve natural resources, promote strategies for adapting to climate change, and provide opportunities for income generation in developing countries. INMED has developed a simple and affordable aquaponic system for small-scale farmers, schools, government institutions, and home gardeners using easily accessible off-the-shelf local materials. Over the past decade, INMED has established a highly successful *Adaptive Aquaculture and Aquaponics Program* in South Africa, Jamaica, and Peru. In South Africa, INMED focuses on achieving food security and sustainable income generation by strengthening local capacity to understand and address climate change, while resolving interrelated issues of environmental degradation, increasing water scarcity, and poverty. It offers business-planning links to markets and assistance with applications for development grants and loans to expand enterprises. At the core of this far-reaching vision, in addition to intensive traditional cultivation, is aquaponics. Several projects have been successfully implemented in different provinces in the country. An aquaponic system was installed at the Thabelo Christian Association for the Disabled in a remote area of the Venda region in Limpopo province. Because INMED's system requires no heavy labour or complex mechanical systems, it is ideal for individuals with disabilities and those unable to perform traditional farming activities. Since the installation, the co-op has increased its revenue by more than 400%. Co-op members receive stable monthly salaries and have invested in breeding animals for additional revenue. Communities that have embraced this new way of farming have strengthened their ability to ensure food security and to provide new and adaptive opportunities for income generation (https://inmed.org) (Fig. 24.2).

Another good example of community upliftment in South Africa is Eden Aquaponics (www.edenaquaponics.co.za). Eden Aquaponics (Pty) Ltd. is the brainchild of Jack Probart who, with the realization that food security is fast becoming just as vital as a healthy economy, had the vision of developing a commercial business with a community focus. Using aquaponics to produce fish and vegetables in the Eden area of the Garden Route in the Western Cape, Eden Aquaponics supplies fish for consumption, as well as fingerlings for fish farming, and grows a variety of organic vegetables for distribution to the local farmers' markets, restaurants, and retailers. The Community Upliftment division manufactures and installs customized commercial systems of various sizes including DIY backyard aquaponic equipment, and supplies seedlings and fingerlings. They also teach less fortunate communities to become self-sufficient in growing, marketing, and selling their produce, thereby enabling previously unemployed people to develop skills, self-confidence, self-esteem, and the ability to provide for themselves.

Food Basket for Africa (www.foodbasketforafrica.org.za) is a similar initiative. They focus on men's development work, which is done through Men's Fraternity, by providing social and emotional counselling to men in rural areas. Men's Fraternity works in association with Effective Living Centres, a reputable counselling organization. Food Basket for Africa runs a number of agricultural development projects in rural areas around Southern Africa. Food tunnels (not hydroponic but mainly

Fig. 24.2 Some illustrations of INMED's Community Projects in South Africa. (Photographs supplied by Janet Ogilvie from INMED)

wicking beds) are placed in communities, which adopt them, and initial training for tunnel care watering and fertilizing techniques is given. An aquaponic system was commissioned at one of the projects, in Kommetjie, Cape Town.

Issues of food security and food sovereignty are not only pertinent to the developing world. In Seville, Spain, social enterprise Asociacíon Verdes del Sur has set up an aquaponic greenhouse in the grounds of a school in Polígono Sur, the most socially deprived part of the city which is characterized by long-term unemployment and a high incidence of drug-related crime. The aquaponic unit is used as part of an environmental education program for local residents, including teaching the benefits of eating locally grown fresh food and developing skills for the unemployed (http://huertosverdesdelsur.blogspot.com). A prototype domestic unit has also been set up in the house of one of the local residents, Soledad (Fig. 24.3).

The Well Community Allotment Group (Crookes Community Farm) is a social enterprise run by volunteers in Sheffield, UK, that is on a mission to connect the local community with their food by actively involving them in its production, and by educating them about the benefits of local food. In 2018, the association was awarded an Aviva Community Fund Award in order to build an aquaponic unit which will be used to educate individuals, schools, youth groups, and other organizations (https://www.avivacommunityfund.co.uk/voting/project/pastwinnerprojectview/17-6291).

Fig. 24.3 Aquaponic facilities in Polígono Sur – anticlockwise from top left: the aquaponic greenhouse at the school; Soledad with a frozen tilapia raised in her domestic unit; tomatoes and an aubergine saved for their seeds; the domestic aquaponic unit. (Photographs: Sarah Milliken)

In the United States, a number of social enterprises using aquaponic systems have been set up across the country as part of a growing social movement focusing on using urban agriculture to increase food security and community cohesion. One of the first was Growing Power, which was founded by Will Allen in 1995, with the objective of using urban agriculture as a vehicle for improving food security in central Milwaukee and for the long-term strengthening of its neighborhoods, and to give inner-city youngsters an opportunity to gain life skills by cultivating and marketing organic produce. Growing Power provided facilities or land, guidance in food growing, and overall project maintenance, and the produce was either donated to meal programs and emergency food providers or sold by the youngsters at local farm shops and farmers' markets, with the stipulation that one-quarter of the proceeds be returned to the local community (Kaufman and Bailkey 2000). By all accounts, Growing Power was doing exactly what they had set out to do: they were feeding, training, and exposing thousands of people to a more autonomous relationship with their food. But while their mission was being fulfilled, it carried significant costs. More money was exiting than entering Growing Power's doors, and by 2014, the social enterprise had a debt of more than $2 million (Satterfield 2018). Faced with insurmountable debt and legal pressure, Growing Power was eventually dissolved in 2017. However, the legacy of the enterprise lives on in the form of other social ventures that were inspired to start similar initiatives. One such venture which acknowledges Will Allen's influence is the Rid-All Green Partnership in Cleveland, Ohio, whose mission is to educate the next generation to not only learn to grow and eat fresh foods but also to operate and grow their own businesses in the food industry, ranging from selling fresh produce and fish to food distributors to processing and packaging fresh food products (https://www.greennghetto.org).

The urban agriculture movement in the United States has been fuelled by the US Department of Agriculture (USDA) Community Food Project (CFP) competitive grant program, which was established in 1996 with the aim of fighting food insecurity through the development of community food projects that promote the self-sufficiency of low-income communities. Since 1996, this program has awarded approximately $90 million in grants. One social enterprise which has benefited from this scheme is Planting Justice (www.plantingjustice.org) which built an aquaponic system on a vacant lot in East Oakland, California, which is run by former prison inmates. Twelve living wage jobs have been created, 5000 pounds (2268 kilos) of free produce has been given to the community, and the project has put $500,000 in wages and $200,000 in benefits back into the neighborhood (New Entry Sustainable Farming Project 2018).

The GrowHaus (https://www.thegrowhaus.org) was founded in 2009 as a social enterprise, which focuses on healthy, equitable, and resident-driven community food production; 97% of the food consumed in Colorado is produced out of state, and the neighborhood where The GrowHaus is located has been designated a "food desert" based on characteristics of low income, race/ethnicity, long distance to a grocery store, lack of access to fresh affordable food, and dependence on public transportation. The residents have come to rely on fast food, convenience stores, petrol stations, and food banks for the majority of their food staples. Due to these factors, many people face significant challenges in terms of food security and access, resulting in dramatic increases in related health issues. Initially in partnership with Colorado Aquaponics (www.coloradoaquaponics.com), and since 2016 independently, the GrowHaus operates a 3200 square foot (297 square meter) aquaponic farm, and the produce is sold through a weekly farm fresh food basket program at a price point comparable to Walmart, as well as to restaurants, with a portion donated to the local community. To help the transition to healthier eating, the GrowHaus also organizes free training and community events focused around food. In the 2016–2017 fiscal year, the GrowHaus generated an income of $1,204,070, of which $333,534 was earned income, and $870,536 was raised through government grants, charitable foundations, corporate contributions, and individual donations. With operating costs of $934,231, the net annual income was $269,839 (https://www.thegrowhaus.org/annual-report).

Trifecta Ecosystems (formerly Fresh Farm Aquaponics; http://trifectaecosystems.com) was founded in 2012 in Meriden, Connecticut. Their mission is to address urban food security by creating incentives for communities to grow their own food while also raising awareness about sustainable farming through education, workshops, and city projects. The enterprise employs six staff who provide aquaponic systems to organizations for educational purposes, workforce development, therapeutic gardening, and high-quality food production. The aquaponic systems range from commercial scale production facilities to small educational units for use in classrooms. In 2018, the South Central regional Water Authority (RWA) awarded a $500,000 grant to facilitate the creation of a series of custom-controlled environment agriculture aquaponic systems, an urban farming technology platform, and workforce training programs aimed at improving food security.

The SchoolGrown social enterprise (www.schoolgrown.org) was set up in 2014 by aquaponics enthusiasts who felt that children weren't getting enough hands-on experiences growing food and learning about their connection to the world about them. Situated next to the commercial aquaponics operation at Ouroboros Farms, California, the aquaponics 'classroom' is run by volunteers and is used to provide training. Their main focus, however, is on spreading aquaponic systems to schools and communities around the United States in order to teach sustainable agricultural practices, environmental stewardship, and resource conservation, and at the same time produce fresh and local food, thereby building a deeper connection between communities and the food they eat. The LEAF (Living Ecosystem Aquaponic Facility) is a 1800 square foot (167 square meter) greenhouse with a solar-powered aquaponic system that was specifically designed for this purpose. Costing $75,000, which includes salaries for two part-time staff responsible for maintaining the system and harvesting, the greenhouses are funded by a combination of a Community Supported Agriculture (CSA) vegetable box scheme, local community or business sponsorship, and crowdfunding. Each LEAF is intended to be financially self-sustaining through the generation of revenue from the produce.

24.4 The Viability of Aquaponics Social Enterprises

The examples above illustrate some of the different business models adopted by aquaponics social enterprises. Whether they will continue to thrive and grow or, like Growing Power, ultimately fail, remains to be seen. In the case of Growing Power, potential reasons for its collapse include Will Allen's inability to empower and retain an operational management team, and a lack of oversight by board members, which compromised the organization's financial health (Satterfield 2018). An in-depth analysis of two aquaponics social enterprises conducted in 2012–2013 revealed four distinct factors what were significant to their survival (Laidlaw and Magee 2016). Sweet Water Organics (SWO) began as an urban aquaponic farm in a large, disused, inner city industrial building in Milwaukee in 2008. It was funded primarily by its founders in order to develop creative capacity, employment opportunities, and chemical-free, fresh, and affordable food for the local community. In 2010, a new organization, Sweet Water Farms (SWF), was split from SWO, with the idea that they would grow as a mutually supportive, cohesive hybrid organization, including both a for-profit commercial urban farm (SWO) and a not-for-profit aquaponics "academy" (SWF). SWF managed volunteer operations and hosted training and education programs at the Sweet Water urban farm, while developing programs on a local (Milwaukee and Chicago), regional, national, and international scale. Sweet

Water had a loyal following among local restaurateurs and fresh food stores for its lettuce and sprouts produce, and sold its fish to a single wholesaler. However, the hybrid not-for-profit/for-profit enterprise model proved to be challenging, as both sides of the organization struggled to identify their role in relation to the other. While each side had a different structure relating to their operational character, and although their operations frequently overlapped, their strategic planning and visions sometimes did not. After 3 years of operation, SWO had still not managed to make a profit, and in 2011 the Milwaukee municipal government awarded a $250,000 loan on condition that 45 jobs would be created by 2014. In October 2012, SWO had 11–13 permanent employees, but was still being sustained through loans financing and equity investment. By June 2013, as loan repayments fell due and the job creation targets were not met, the for-profit arm of Sweet Water went into liquidation, and SWF took over as the primary operator of the Sweet Water urban farm. Currently, SWF operates entirely as an educational and advisory enterprise run by volunteers and a small team of part-time employees, and no longer supplies restaurants with produce (Laidlaw and Magee 2016).

The Centre for Education and Research (CERES) in Melbourne, Australia, opened its aquaponics facility in 2010. The system was designed as a suboptimized commercial system with the production capacity to support a single wage for the farmer who maintains it. This wage varies based on how much he/she produces, with the vegetables being sold through the CERES Fair Food organic box delivery service. The scale of the operation does not generate a return that would permit the setting up of a fish-processing facility. Stakeholders at Sweet Water Farms and CERES identified that the principal factor behind their survival was ongoing commitment, in the form of continued support of personnel with technical and business management skills combined with an enduring leadership, and the willingness of the stakeholders to remain involved and prepared to cooperate without strong financial incentives. The second factor was the local political context. While the city of Milwaukee supported Sweet Water both through policy initiatives and direct financial aid, which allowed it to expand its fixed assets and human resources, build market awareness, and acquire a sizeable regular commercial customer base, the CERES project had little such support, beyond an initial grant, and it had struggled to generate revenue, which would have allowed it to expand. Costs of compliance and licensing also made it difficult to engage with local markets in more than a token way, which dampened its motivation to market and sell the produce, and made it untenable for the operation to develop beyond a small part-time income-generating enterprise. The third factor was the availability of markets for urban aquaponics produce. While urban aquaponics is attractive to a customer base that is increasingly responsive to issues of food security and ethical consumption, such as in Milwaukee, this was not the case in Melbourne. The final factor was diversification. Both CERES and SWO/SWF benefitted from translating social and technical experimentation into a range of training and educational services. SWO/SWF, being a larger concern, obviously had greater capacity for developing these services, and these proved vital in sustaining the social enterprise when commercial plans failed to materialize (Laidlaw and Magee 2016).

24.5 Conclusions

In "Ten technologies which could change our lives" (European Parliamentary Research Service, 2015), aquaponic systems were singled out as a solution for developing innovative and sustainable food sources for Europe which, through shortening of supply chains, could improve food security and food systems resilience. However, the technology is still newly emerging and as yet relatively undeveloped, and as the study by Laidlaw and Magee (2016) highlights, the viability of an aquaponics social enterprise depends not only on stakeholder commitment, thorough market analysis, clear governance structures, and a robust business plan, but also on external factors, such as the local political context and regulations.

References

Allison EH (2011) Aquaculture, fisheries, poverty and food security. Security 61

Diamant E, Waterhouse A (2010) Gardening and belonging: reflections on how social and therapeutic horticulture may facilitate health, wellbeing and inclusion. Br J Occup Ther 73(2):84–88. https://doi.org/10.4276/030802210X12658062793924

Drangert JO (1998) Fighting the urine blindness to provide more sanitation options. Water SA 24:157–164

Dubbeling M, de Zeeuw H, van Veenhuizen R (2010) Cities, poverty and food: multi- stakeholder policy and planning in urban agriculture. Practical Action Publishing Ltd., Colchester

European Parliamentary Research Service (2015) Ten technologies which could change our lives: Potential impacts and policy implications. http://www.europarl.europa.eu/EPRS/EPRS_IDAN_527417_ten_trends_to_change_your_life.pdf

Faber M, Witten C, Drimie S (2011) Community-based agricultural interventions in the context of food and nutrition security in South Africa. S Afr J Clin Nutr 24:21–30

FAO (2016) Urban agriculture in the Gaza Strip through vertical gardens and aquaponics. Food and Agriculture Organization of the Unuted Nations, Rome. http://www.fao.org/fileadmin/templates/FCIT/PDF/Fact_sheet_on_aquaponics_Final.pdf

Fasciglione M (2015) Corporate social responsibility and the right to employment of persons with disabilities. In: Fina VD, Cera R (eds) Protecting the Rights of People with Autism in the Fields of Education and Employment. Springer, Cham, pp 171–193. https://doi.org/10.1007/978-3-319-13791-9_7

Fieldhouse J (2003) The impact of an allotment group on mental health clients' health, wellbeing and social networking. Br J Occup Ther 66(7):286–296

Howarth ML, McQuarrie C, Withnell N, Smith E (2016) The influence of therapeutic horticulture on social integration. J Public Ment Health 15(3):136–140. https://doi.org/10.1108/JPMH-12-2015-0050

Kaufman J, Bailkey M (2000) Farming inside cities: entrepreneurial urban agriculture in the United States. Lincoln Institute of Land Policy working paper. http://www.urbantilth.org/wp-content/uploads/2008/10/farminginsidecities.pdf

Laidlaw J, Magee L (2016) Towards urban food sovereignty: the trials and tribulations of community-based aquaponics enterprises in Milwaukee and Melbourne. Local Environ 21 (5):573–590. https://doi.org/10.1080/13549839.2014.986716

New Entry Sustainable Farming Project (2018) Community food projects: indicators of success FY 2017. https://nesfp.org/sites/default/files/uploads/ios_2017_final_draft_april_13_2018.pdf

Satterfield S (2018) Behind the rise and fall of Growing Power. Public Radio International, March 2018. https://www.pri.org/stories/2018-03-16/behind-rise-and-fall-growing-power

Somerville C, Cohen M, Pantanella E, Stankus A, Lovatelli A (2014) Small-scale aquaponic food production. Integrated fish and plant farming. FAO fisheries and aquaculture technical paper no. 589. FAO, Rome

van Gorcum et al (2019) https://link.springer.com/article/10.1007/s10499-019-00379-1

Viladomat L, Jones P (2011) Development of aquaponic systems for space and water efficient food production in the Occupied Palestinian Territories – Pilot project final report. http://www.byspokes.org/wp-content/uploads/documents/basma_report.pdf

Correction to: Aquaponics Food Production Systems

Simon Goddek, Alyssa Joyce, Benz Kotzen, and Gavin M. Burnell

Correction to:
S. Goddek et al. (eds.), *Aquaponics Food Production Systems*, https://doi.org/10.1007/978-3-030-15943-6

This book was inadvertently published with a typo in Chapter 3 Page 37; NO2 should have been printed as NO_2.

Also in Chapter 7 Page 197, incorrect DOI was printed. Correct DOI is https://doi.org/10.1016/j.aquaculture.2018.03.021

Both corrections have now been incorporated and are correctly appearing in Chapter 3 page 37 and Chapter 7 page 197.

The updated version of these chapters can be found at
https://doi.org/10.1007/978-3-030-15943-6_3
https://doi.org/10.1007/978-3-030-15943-6_7

S. Goddek et al. (eds.), *Aquaponics Food Production Systems*,
https://doi.org/10.1007/978-3-030-15943-6_25